P9-DVT-965

Print and Media Resources

For the Instructor

The goal of the instructor's resource package is to provide you with a valuable source of ideas and resources to enrich your instruction and assessment efforts. The items listed here serve not only as a source of images, questions, and activities but as a springboard for your own ideas.

Instructor's Resource CD-ROM (0-13141021-0) and Instructor's Resource Library (www.prenhall.com/belk)

The Instructor's Resource CD-ROM and the Instructor's Resource Library provide a fully searchable and integrated collection of resources (available in two locations) to help you make efficient and effective use of your lecture preparation time, as well as to enhance your classroom presentations and assessment efforts. In short, you have everything you need at your fingertips. Both resources feature:

- *Presentation Gallery*: Designed to make the preparation of your lecture presentation faster and easier. Presentation resources include over 1000 jpeg files of illustrations, tables, and photos from the text; PowerPoint slides with all labeled and unlabeled images embedded; and animations of major concepts in .swf and .mov format.
- *Assessment Gallery*: Contains a wealth of new ideas to help you determine your students' level of understanding throughout the course. This reservoir of ideas and resources includes projects, group and collaborative activities, discussion topics and questions, demonstrations, worksheets, and a test bank complete with two types of questions for each chapter: Analyzing and Applying the Basics and Connecting the Science. All are in easy-to-use, editable Word documents.
- *Course Management Resources Gallery*: Supplies WebCT- and BlackBoard-ready resources to help you manage your course. These resources are easy to incorporate into any existing course or can be used to construct a new one. In addition, a prepared Prentice Hall CourseCompass management system is available with flexible customization.
- *Research Navigator*™: Another powerful tool designed to make your lecture and assessment preparation easier. Research Navigator includes three extensive databases of credible and reliable source material including EBSCO Academic Journal and Abstract Database, *The New York Times* Search by Subject Archive, and "Best of the Web" Link Library.
- *Student Study Gallery*: Provides all of the questions and activities found on the Student Companion Web site and in the Student Study Guide as editable Word documents. This resource allows you access to the review materials available to students.
- *Search Engine*: A tool that lets you find relevant resources via a number of different parameters, such as key terms, learning objectives, figure number, and resource type (e.g., Media Activities).

Instructor's Resource Guide (0-13141022-9)

The printed Instructor's Resource Guide offers a portable alternative to the resources on the Instructor's Resource CD-ROM and the Instructor's Resource Library. It includes assessment ideas and resources ranging from discussion questions, to demonstrations, to group and collaborative activities. The Instructor's Resource Guide also features a list of the digital resources available with *Biology: Science for Life*.

Transparency Pack (0-13141025-3)

Even in this digital age, transparencies are still an efficient and effective way to visually reinforce your lecture. The pack includes 250 four-color transparencies that have been selected from the text, including all of the illustrations, and have been enlarged for large lecture-hall viewing. These images enhance your classroom presentations with enlarged labels and increased color saturation.

Supporting *Science for Life*

For the Student

The goal of the student resource package is to provide opportunities to exercise scientific reasoning skills and apply biological knowledge to real problems. The items listed here offer students many tools for review that are compatible with a wide variety of learning styles.

Student Companion Web Site (www.prenhall.com/belk)

The Student Companion Web Site to *Biology: Science for Life* provides students with the opportunity to review biological concepts and practice problem-solving skills. The focus of the questions and activities is on application, critical thinking, problem solving, analysis, and synthesis. All of the questions and activities provide students with valuable feedback in the form of hints and coaching that identify their areas of weakness and provide them with guidance on how to improve and where to find additional information about these areas. The Student Companion Web Site features:

- *Chapter Outlines*: Available on the Web site, this feature provides a brief overview of the book's chapter contents and major concepts addressed.
- *Learning Objectives*: These objectives address what the students should be able to understand after reading and completing the activities for each chapter.
- *Media Activities*: These activities offer students a visual view of concepts and test their knowledge using a wide variety of activities integrated into each chapter.

  Animations: These activities demonstrate dynamic concepts and processes to better show changes over time and complex interactions. Each animation is followed by interactive activities designed to help students achieve mastery of chapter-specific learning objectives.

  Explore the Issue and Explore the Science: These activities offer in-depth exploration of each chapter's major issue and scientific concepts through Web sites, articles, and essay questions. Each activity reinforces the student's need to apply biological knowledge and scientific reasoning to real-world issues.
- *Self Test*: The self-test questions allow students to test both their mastery of the major concepts in the chapter as well as their understanding of the connections these biological concepts have to their lives. Divided into two different types—Analyzing and Applying the Basics, and Connecting the Science—these questions contain helpful hints and coaching that provide students with valuable feedback to help focus their time most effectively.
- *Essay Challenge*: These high-level essay questions focus on Analyzing and Applying the Basics, and Connecting the Science. Hints and suggested answers accompany every question.
- *Issues Update*: This tool links to on-line articles and Web sites that address the major issues and science introduced in a chapter. Links are updated each semester to keep current with recent research and writings.
- *Research Navigator*: This tool equips students with the means to start a research assignment or research paper or to access full text articles. It is complete with extensive help on the research process and three exclusive databases of credible and reliable source material, including the EBSCO Academic Journal and Abstract Database, *The New York Times* Search by Subject Archive, and "Best of the Web" Link Library, Research Navigator™, and enables students to efficiently and effectively make the most of their research time and stay up-to-date on the issues.
- *Science Skills*: This resource features Chemistry and Math Review to help students recap basic knowledge in these areas. It also provides an extensive collection of links and articles on evaluating information and avoiding misinformation.
- *Student Study Tips*: This tool offers extensive resources on how to prepare for tests, get the most out of lectures, and make the best use of study time.

Student Study Guide (0-13141505-0)

The Student Study Guide serves as the print version of many of the resources found on the Student Companion Web Site, including chapter outlines, learning objectives, self tests, Essay Challenge, Explore the Science, and Explore the Issues. For the student who is always on the go, this guide offers a portable alternative to our media resources.

Laboratory Program

Biology: Science for Life Laboratory Manual

This inquiry-driven laboratory manual, designed specifically for the non-science major, reinforces and extends the key biological concepts from *Biology: Science for Life*. These fifteen exercises take a process-oriented approach and often lead to open-ended results. As with the text, the laboratory exercises are connected to compelling stories and provide instructors with an excellent springboard for discussion on the role of biological research in contemporary society.

Symbiosis: The Prentice Hall Custom Laboratory Program for Biology (www.pearsoncustom.com/database/symbiosis/ph.com)

With *Symbiosis: The Prentice Hall Custom Laboratory Program for Biology*, instructors can select from a wide variety of biology, microbiology, or anatomy and physiology labs to build a custom lab manual that exactly matches their content needs and course organization. By visiting the Web site, instructors can select from an extensive list of Prentice Hall laboratory publications or from Pearson Custom Publishing's established library of biology labs. Using the tools provided in our *Lab Ordering and Authoring Kit*, instructors have the ability to develop the best possible lab manual for their courses.

BIOLOGY

BIOLOGY
Science for Life

Colleen Belk
University of Minnesota–Duluth

Virginia Borden
University of Minnesota–Duluth

Pearson Education, Inc.
Upper Saddle River, NJ 07458

Library of Congress Cataloging-in-Publication Data

Belk, Colleen M.
Biology : science for life / Colleen Belk, Virginia Borden.
p. cm.
ISBN 0-13-089241-6
1. Biology. I. Borden, Virginia. II. Title.

QH307.2.B43 2004
570--dc21

2003052834

Executive Editor: Teresa Ryu Chung
Editor in Chief, Life and Geosciences: Sheri L. Snavely
Development Editor: Rebecca Strehlow
Illustration Designer: Kim Quillin
Production Editor: Tim Flem/PublishWare
Media Editor: Travis Moses-Westphal
Assistant Editor: Colleen Lee
Art Director: Jonathan Boylan
Art Editor: Adam Velthaus
Managing Editor, Audio/Video Assets: Patricia Burns
Senior Marketing Manager: Shari Meffert
Director of Marketing for Science: Linda Taft MacKinnon
Marketing Director: John Tweeddale
Vice President of Production & Manufacturing: David W. Riccardi
Executive Managing Editor: Kathleen Schiaparelli
Director of Creative Services: Paul Belfanti
Director of Design: Carole Anson
Page Composition: PublishWare
Manager of Formatting: Jim Sullivan
Manufacturing Manager: Trudy Pisciotti
Buyer: Alan Fischer
Editor in Chief of Development: Carol Trueheart
Assistant Managing Editor, Science Media: Nicole Bush
Media Production Editor: Rich Barnes
Supplements Production Editor: Becca Richter
Editorial Assistants: Nancy Bauer, Mary Kuhl
Marketing Assistant: Juliana Tarris
Cover Designer: Joseph Sengotta
Interior Designers: Joseph Sengotta, Dina Curro
Illustrators: Stephen Graepel, Quade Paul, Emiko Paul
Photo Research: Yvonne Gerin
Photo Research Administrator: Beth Brenzel
Cover Photographs: Top left: Volker Steger/Photo Researchers, Inc.; Top middle: David McGlynn/Getty Images, Inc.; Top right: AP/Wide World Photos; Bottom left: Prof, P. Motta/Photo Researchers, Inc.; Bottom middle: D. Cavgnaro/DRK Photo; Bottom right: Alfred Pasieka/Photo Researchers, Inc..

PEARSON Prentice Hall

Pearson Education, Inc.
Upper Saddle River, New Jersey 07458

Printed in the United States of America

10 9 8 7 6 5 4 3 2 1

ISBN 0-13-089241-6

Pearson Education Ltd., *London*
Pearson Education Australia Pty., Limited, *Sydney*
Pearson Education Singapore, Pte. Ltd
Pearson Education North Asia Ltd., *Hong Kong*
Pearson Education Canada, Ltd., *Toronto*
Pearson Educación de Mexico, S.A. de C.V.
Pearson Education—Japan, *Tokyo*
Pearson Education Malaysia, Pte. Ltd.

"Because science,

told as a story, can intrigue and inform the non-scientific minds among us, it has the potential to bridge the two cultures into which civilization is split—the sciences and the humanities. For educators, stories are an exciting way to draw young minds into the scientific culture."

E. O. Wilson

Brief Contents

Preface

To the Student

As you worked your way through high school, or otherwise worked to prepare yourself for college, you were probably unaware that an information explosion was taking place in the field of biology. This explosion, brought on by advances in biotechnology and communicated by faster, more powerful computers, has allowed scientists to gather data more quickly and disseminate data to colleagues in the global scientific community with the click of a mouse. Every discipline of biology has benefited from these advances, and today's scientists collectively know more than any individual could ever hope to understand.

Paradoxically, as it becomes more and more difficult to synthesize huge amounts of information from disparate disciplines within the broad field of biology, it becomes more vital that we do so. The very same technologies that led to the information boom, coupled with expanding human populations, present us with complex ethical questions. These questions include whether or not it is acceptable to clone humans, when human life begins and ends, who owns living organisms, what our responsibilities toward endangered species are, and many more. No amount of conceptual understanding alone will provide satisfactory answers to these questions. Addressing these kinds of questions requires the development of a scientific literacy that surpasses the rote memorization of facts. To make decisions that are individually, socially, and ecologically responsible, you must not only understand some fundamental principles of biology but also be able to use this knowledge as a tool to help you analyze ethical and moral issues involving biology.

To help you understand biology and apply your knowledge to an ever-expanding suite of issues, we have structured each chapter of *Biology: Science for Life* around a compelling story in which biology plays an integral role. Through the story you will not only learn the relevant biological principles but you will also see how science can be used to help answer complex questions. As you learn to apply the strategies modeled by the text, you will begin developing your critical thinking skills.

By the time you have read the last chapter, you should have a clear understanding of many important biological principles. You will also be able to think like a scientist and critically evaluate which information is most reliable instead of simply accepting all the information you read in the paper or hear on the radio or television. Even though you may not be planning to be a practicing biologist, well-developed critical thinking skills will enable you to make decisions that affect your own life, such as whether or not to take nutritional supplements, and decisions that affect the lives of others, such as whether or not to believe the DNA evidence presented to you as a juror in a criminal case.

It is our sincere hope that understanding how biology applies to important personal, social, and ecological issues will convince you to stay informed about such issues. On the job, in your community, at the doctor's office, in the voting booth, and at home reading the paper, your knowledge of the basic biology underlying so many of the challenges that we as individuals and as a society face will enable you to make well-informed decisions for your home, your nation, and your world.

To the Instructor

Colleen Belk and Virginia Borden have collaborated on teaching the nonmajors biology course at the University of Minnesota–Duluth for over a decade. This collaboration has been enhanced by their differing but complementary areas of expertise. In addition to the nonmajors course, Colleen Belk teaches General Biology for majors, Genetics, Cell Biology, and Molecular Biology courses. Virginia Borden teaches General Biology for majors, Evolutionary Biology, Plant Biology, Ecology, and Conservation Biology courses.

After several somewhat painful attempts at teaching all of biology in a single semester, the two authors came to the conclusion that this strategy was not effective. They realized that their students were more engaged when they understood how biology directly affected their lives. Colleen and Virginia began to structure their lectures around stories they knew would interest students. When they began letting the story drive the science, they immediately noticed a difference in student interest, energy, and willingness to work harder at learning biology. Not only has this approach increased student understanding, it has increased the authors' enjoyment in teaching the course—presenting students with fascinating stories infused with biological concepts is simply a lot more fun. This approach served to invigorate their teaching. Knowing that their students are learning the biology that they will need now and in the future gives the authors a deep and abiding satisfaction.

By now you are probably all too aware that teaching nonmajor students is very different from teaching biology majors. You know that most of these students will never take another formal biology course, therefore your course may be the last chance for these students to see the relavance of science in their everyday lives and the last chance to appreciate how biology is woven throughout the fabric of their lives. You recognize the importance of engaging these students because you know that these students will one day be voting on issues of scientific importance, holding positions of power in the community, serving on juries, and making healthcare decisions for themselves and their families. You know that your students' lives will be enhanced if they have a thorough grounding in basic biological principles and scientific literacy.

Themes in *Science for Life*

Helping nonmajors to appreciate the importance of learning biology is a difficult job. We have experienced the struggle to actively engage students in lectures and to raise their scientific literacy and critical thinking skills, and it seems that we were not alone. When we asked instructors from around the country what challenges they faced while teaching the nonmajors introductory biology course, they echoed our concerns. This book was written to help you meet these challenges.

The Story Drives the Science. We have found that students are much more likely to be engaged in the learning process when the textbook and lectures capitalize on their natural curiosity. This text accomplishes this by using a story to drive the science in every chapter. Students get caught up in the story and become interested in learning the biology so they can see how the story is resolved. This approach allows us to cover the key areas of biology, including the unity and diversity of life, cell structure and function, classical and molecular genetics, evolution, and ecology, in a manner that makes students want to learn. Not only do students want to learn, this approach allows students to both connect the science to their everyday lives and integrate the principles and concepts for later application to other situations. This approach will give you flexibility in teaching and will support you in developing students' critical thinking skills.

The Process of Science. This book also uses another novel approach in the way that the process of science is modeled. The first chapter is dedicated to the scientific method and hypothesis testing, and each subsequent chapter weaves the scientific method and hypothesis testing throughout the story. The development of students' critical thinking skills is thus reinforced for the duration of the course. Students will see that the application of the scientific method is often the best way to answer questions raised in the story. This practice not only allows students to develop their critical thinking skills but, as they begin to think like scientists, helps them understand why and how scientists do what they do.

Integration of Evolution. Another aspect of *Biology: Science for Life* that sets it apart from many other texts is the manner in which evolutionary principles are integrated throughout the text. The role of evolutionary processes is highlighted in every chapter, even when the chapter is not specifically focussed on an evolutionary question. For example, when discussing infectious diseases, the evolution of antibiotic-resistant strains of bacteria is addressed. With evolution serving as an overarching theme, students are better able to see that all of life is connected through this process.

Pedagogical Elements

Open the book and flip through a few pages and you will see some of the most inviting, lively, and informative illustrations you have ever seen in a biology text. The illustrations are inviting because they have a warm, hand-drawn quality that is clean and uncluttered. The liveliness of the illustrations is accomplished with vivid colors, three-dimensionality, and playful compositions. Most importantly, the illustrations are informative, not only because they were carefully crafted to enhance concepts in the text but also because they employ techniques like the "pointer" that help draw the students' attention to the important part of the figure (see page 3). Likewise, tables are more than just tools for organizing information; they are illustrated to provide attractive, easy references for the student. We hope that the welcoming nature of the art and tables in this text will encourage nonmajors to explore instead of being overwhelmed before they even get started.

In addition to lively illustrations, this text also strives to engage the nonmajor student through the use of analogies. For example, the process of translation is likened to baking a cake, and the heterozygote advantage is likened to the advantage conferred by having more than one pair of shoes (see pages 166 and 381). These clever illustrations are peppered throughout the text.

Students can reinforce and assess what they are learning in the classroom by reading the chapter, studying the figures, reviewing the key terms, and answering the end-of-chapter questions. We have written these questions in every format likely to be used by an instructor during an exam so that students have practice answering many different types of questions. We have also included "Connecting the Science" questions that would be appropriate for essay exams, class discussions, or use as topics for term papers.

Supplements

Development of the supplements package that accompanies *Biology: Science for Life* began several years ago. A group of talented and dedicated biology educators teamed up with us to build a set of resources that equip nonmajors with the tools to achieve scientific literacy that will allow them to make informed decisions about the biological issues that affect them daily. In each chapter, a variety of resources are tightly integrated with the text through specific chapter learning objectives. The student resources offer opportunities to exercise scientific reasoning skills and to apply biological knowledge to real problems and issues within the framework of these learning objectives. The instructor resources provide a valuable source of ideas for educators to enrich their instruction and assessment efforts. Available in print and media formats, the *Biology: Science for Life* resources are easy to navigate and support a variety of learning and teaching styles.

We believe you will find that the design and format of this text and its supplements will help you meet the challenge of helping students both succeed in your course and develop science skills—for life.

Acknowledgments

The Reviewers

Each chapter of this book was thoroughly reviewed several times as it moved through the development process. Reviewers were chosen on the basis of their demonstrated talent and dedication in the classroom. Many of these reviewers were already trying various approaches to actively engage students in lectures, and to raise the scientific literacy and critical thinking skills among their students. Their passion for teaching and commitment to their students was evident throughout this process. These devoted individuals scrupulously checked each chapter for scientific accuracy, readability, and coverage level. In addition to general reviewers, we also had a team of expert reviewers evaluate individual chapters to ensure that the content was accurate and that all the necessary concepts were included.

All of these reviewers provided thoughtful, insightful feedback, which improved the text significantly. Their efforts reflect their deep commitment to teaching nonmajors and improving the scientific literacy of all students. We are very thankful for their contributions to *Biology: Science for Life*.

Karen Aguirre	Clarkson University
Susan Aronica	Canisius College
Mary Ashley	University of Chicago
Thomas Balgooyen	San Jose State University
Donna Becker	Northern Michigan University
Lesley Blair	Oregon State University
Susan Bornstein-Forst	Marian College
James Botsford	New Mexico State University
Bryan Brendley	Gannon University
Peggy Brickman	University of Georgia
Carole Browne	Wake Forest University
Neil Buckley	State University of New York, Plattsburgh
Suzanne Butler	Miami-Dade Community College
David Byres	Florida Community College
Peter Chabora	Queens College
Mary Colavito	Santa Monica College
Walter Conley	State University of New York, Potsdam
Melanie Cook	Tyler Junior College
George Cornwall	University of Colorado
Angela Cunningham	Baylor University
Garry Davies	University of Alaska, Anchorage
Miriam del Campo	Miami-Dade Community College
Veronique Delesalle	Gettysburg College
Beth De Stasio	Lawrence University
Donald Deters	Bowling Green State University
Douglas Eder	Southern Illinois University, Edwardsville
Deborah Fahey	Wheaton College
Richard Firenze	Broome Community College
David Froelich	Austin Community College
Anne Galbraith	University of Wisconsin, La Crosse
Wendy Garrison	University of Mississippi
Robert George	University of North Carolina, Wilmington

Sharon Gilman	Coastal Carolina University
John Green	Nicholls State University
Robert Greene	Niagara University
Bruce Goldman	University of Connecticut, Storrs
Eugene Goodman	University of Wisconsin, Parkside
Tamar Goulet	University of Mississippi
Mark Grobner	California State University, Stanislaus
Stan Guffey	University of Tennessee, Knoxville
Mark Hammer	Wayne State University
Blanche Haning	North Carolina State University
Patricia Hauslein	St. Cloud State University
Stephen Hedman	University of Minnesota–Duluth
Julie Hens	Yale University
Leland Holland	Pasco-Hernando Community College
Jane Horlings	Saddleback Community College
Michael Hudecki	State University of New York, Buffalo
Laura Huenneke	New Mexico State University
Carol Hurney	James Madison University
Jann Joseph	Grand Valley State University
Michael Keas	Oklahoma Baptist University
Karen Kendall-Fite	Columbia State Community College
David Kirby	American University
Dennis Kitz	Southern Illinois University, Edwardsville
Jennifer Knapp	Nashville State Technical Community College
Loren Knapp	University of South Carolina
Phyllis Laine	Xavier University
Tom Langen	Clarkson University
Lynn Larsen	Portland Community College
Mark Lavery	Oregon State University
Mary Lehman	Longwood College
Doug Levey	University of Florida
Jayson Lloyd	College of Southern Idaho
Paul Lurquin	Washington State University
Douglas Lyng	Indiana University/Purdue University
Michelle Mabry	Davis and Elkins College
Ken Marr	Green River Community College
Kathleen Marrs	Indiana University/Purdue University
Steve McCommas	Southern Illinois University, Edwardsville
Colleen McNamara	Albuquerque TVI
John McWilliams	Oklahoma Baptist University
Diane Melroy	University of North Carolina, Wilmington
Joseph Mendelson	Utah State University
Hugh Miller	East Tennessee State University
Stephen Molnar	Washington University
Bertram Murray	Rutgers University
Ken Nadler	Michigan State University
Joseph Newhouse	California University of Pennsylvania
Jeffrey Newman	Lycoming College

Kevin Padian	University of California–Berkeley
Javier Penalosa	Buffalo State College
Rhoda Perozzi	Virginia Commonwealth University
John Peters	College of Charleston
Patricia Phelps	Austin Community College
Calvin Porter	Xavier University
Linda Potts	University of North Carolina, Wilmington
Gregory Pryor	University of Florida
Laura Rhoads	State University of New York, Potsdam
Laurel Roberts	University of Pittsburgh
Deborah Ross	Indiana University/Purdue University
Michael Rutledge	Middle Tennessee State University
Wendy Ryan	Kutztown University
Christopher Sacchi	Kutztown University
Jasmine Saros	University of Wisconsin, La Crosse
Ken Saville	Albion College
Robert Schoch	Boston University
Robert Shetlar	Georgia Southern University
Thomas Sluss	Fort Lewis College
Douglas Smith	Clarion University of Pennsylvania
Sally Sommers Smith	Boston University
Amanda Starnes	Emory University
Timothy Stewart	Longwood College
Shawn Stover	Davis and Elkins College
Bradley Swanson	Central Michigan University
Martha Taylor	Cornell University
Alice Templet	Nicholls State University
Nina Thumser	California University of Pennsylvania
Alana Tibbets	Southern Illinois University, Edwardsville
Jeffrey Travis	State University of New York, Albany
Robert Turgeon	Cornell University
James Urban	Kansas State University
John Vaughan	St. Petersburg Junior College
Martin Vaughan	Indiana State University
Paul Verrell	Washington State University
Tanya Vickers	University of Utah
Janet Vigna	Grand Valley State University
Don Waller	University of Wisconsin, Madison
Jennifer Warner	University of North Carolina, Charlotte
Lisa Weasel	Portland State University
Carol Weaver	Union University
Frances Weaver	Widener University
Elizabeth Welnhofer	Canisius College
Wayne Whaley	Utah Valley State College
Vernon Wiersema	Houston Community College
Michelle Withers	Louisiana State University
Art Woods	University of Texas, Austin
Elton Woodward	Daytona Beach Community College

Supplement Authors

Print and media supplements were prepared by a very creative, energetic, and fun team of nonmajors biology instructors from colleges and universities across the country. Early in the development process we attended a workshop with them in Cambridge, Massachusetts, to discuss the goals of the supplements. We had a great time working with this good-natured group. It was a joy spending time with people who care so much about their students. This very productive workshop led to a truly collaborative effort to address the needs of the instructors and students—their contributions energized the project tremendously. As a result, students will see dynamic animations of many complex processes and will have the opportunity to practice newly learned skills. The work of these instructors helped ensure that the supplements were reinforcing the chapter learning objectives. We cannot thank them enough.

Supplement Contributors

Scott Cooper	University of Wisconsin, La Crosse
Anne Galbraith	University of Wisconsin, LaCrosse
David Howard	University of Wisconsin, La Crosse
Tom Langen	Clarkson University
John McWilliams	Oklahoma Baptist University
Diane Melroy	University of North Carolina, Wilmington
Jennifer Miskowski	University of Wisconsin, La Crosse
Laura Rhoads	State University of New York, Potsdam
Janet Vigna	Grand Valley State University
Jennifer Warner	University of North Carolina, Charlotte

Media Reviewers

Steve Berg	Winona State University
Carole Browne	Wake Forest University
Gregory Pryor	University of Florida
Nina Thumser	California University of Pennsylvania
Frances Weaver	Widener University

Supplement Reviewers

Deborah Fahey	Wheaton College
Stan Guffey	University of Tennessee, Knoxville
Karen Kendall-Fite	Columbia State Community College
Mary Lehman	Longwood University
Michelle Mabry	Davis and Elkins College
Calvin Porter	Xavier University
Michael Rutledge	Middle Tennessee State University

The Book Team

When we set out to write this book, we would not have predicted that we would so thoroughly enjoy the experience. Our enjoyment stems directly from the enthusiasm and talent of the Prentice Hall team. It has been an honor to work with all of these talented, dedicated people.

The book team came together due to the efforts of our editor Teresa R. Chung. Teresa is a woman of tremendous vision, insight, integrity, humor, energy, and style. She has guided every aspect of this project from its inception to its delivery. It was heartening to be in such capable hands and to be able to thoroughly trust your editor's judgment. It was also a pleasure to work with someone who is so cheerful and upbeat. For keeping us on track and inspiring us to do our best work, we sincerely thank her.

Another important book team member was Becky Strehlow, who served as our Development Editor. She has been with us from the very beginning—reading every word from a student's perspective and helping us effectively address issues raised by the reviewers. Her keen insights and hard work are very much appreciated.

What a gift it was to work with our illustration designer, Dr. Kim Quillin. Her artistic sensibilities and understanding of biology provided a synergy between art and science rarely seen in textbooks. Kim's pioneering, ingenious, and tireless work will help innumerable undergraduates understand science. We are extremely thankful to have had the opportunity to work with her.

Media Editor Travis Moses-Westphal was the wizard behind our media and has brought so much creativity to the entire package. Both he and Assistant Editor Colleen Lee managed to beautifully address the challenges facing instructors teaching this course through the supplements and to build a team of talented and creative supplement contributors. We were very lucky to have them aboard.

At the very early stages of production, this text and its illustrations and images were in the hands of four very capable people. Art Director Jonathan Boylan guided the book design with much talent and creativity. Art Editor Adam Velthaus skillfully managed the production of the illustration program. Copyeditor Jocelyn Phillips did an excellent job of working the text into its final form, making sure no mistakes crept in. Yvonne Gerin, Photo Researcher, has located most of the striking images in the text. She did an excellent job of translating our photo wishes into beautiful images.

We cannot emphasize enough the contributions of illustrators Steve Graepel, Quade Paul, and Emi Paul. The illustrations in this text are both beautiful and informative. We appreciate not only the artistic talent that this team of illustrators brought to this project, but also the hard work and flexibility necessary to make the illustrations as accurate as possible.

Tim Flem was the Production Editor for this text. He managed to seamlessly coordinate the work of the copyeditor, photo researcher, illustrators, and authors under a tight schedule. Tim stands out from the crowd because he has turned this juggling act into a craft, and makes the job look so easy.

Shari Meffert, Senior Marketing Manager, has been a very enthusiastic promoter of this text. She strategically planned every step to ensure that every nonmajors biology professor got an opportunity to evaluate this text. We appreciate her savvy, enthusiasm, and dedication.

This book is dedicated to our families, friends, and colleagues who have endured our inability to get our minds around anything but *Biology: Science for Life* for the past three years. Having loving families, great friends, and a supportive work environment enabled us to make this heartfelt contribution to nonmajors biology education.

Colleen Belk and Virginia Borden
University of Minnesota-Duluth

Print and Media Resources Supporting *Science for Life*

For the Instructor

The goal of the instructor's resource package is to provide you with a valuable source of ideas and resources to enrich your instruction and assessment efforts. The items listed here serve not only as a source of images, questions, and activities but as a springboard for your own ideas.

Instructor's Resource CD-ROM (0-13141021-0) and Instructor's Resource Library (www.prenhall.com/belk)

The Instructor's Resource CD-ROM and the Instructor's Resource Library provide a fully searchable and integrated collection of resources (available in two locations) to help you make efficient and effective use of your lecture preparation time, as well as to enhance your classroom presentations and assessment efforts. In short, you have everything you need at your fingertips. Both resources feature:

- *Presentation Gallery*: Designed to make the preparation of your lecture presentation faster and easier. Presentation resources include over 1000 jpeg files of illustrations, tables, and photos from the text; PowerPoint slides with all labeled and unlabeled images embedded; and animations of major concepts in .swf and .mov format.

- *Assessment Gallery*: Contains a wealth of new ideas to help you determine your students' level of understanding throughout the course. This reservoir of ideas and resources includes projects, group and collaborative activities, discussion topics and questions, demonstrations, worksheets, and a test bank complete with two types of questions for each chapter: Analyzing and Applying the Basics and Connecting the Science. All are in easy-to-use, editable Word documents.

- *Course Management Resources Gallery*: Supplies WebCT- and BlackBoard-ready resources to help you manage your course. These resources are easy to incorporate into any existing course or can be used to construct a new one. In addition, a pre-prepared Prentice Hall CourseCompass management system is available with flexible customization.

- *Research Navigator*™: Another powerful tool designed to make your lecture and assessment preparation easier. Research Navigator includes three extensive databases of credible and reliable source material including EBSCO Academic Journal and Abstract Database, *The New York Times* Search by Subject Archive, and "Best of the Web" Link Library.

- *Student Study Gallery*: Provides all of the questions and activities found on the Student Companion Web site and in the Student Study Guide as editable Word documents. This resource allows you access to the review materials available to students.

- *Search Engine*: A tool that lets you find relevant resources via a number of different parameters, such as key terms, learning objectives, figure number, and resource type (e.g., Media Activities).

Instructor's Resource Guide (0-13141022-9)

The printed Instructor's Resource Guide offers a portable alternative to the resources on the Instructor's Resource CD-ROM and the Instructor's Resource Library. It includes assessment ideas and resources ranging from discussion questions, to demonstrations, to group and collaborative activities. The Instructor's Resource Guide also features a list of the digital resources available with *Biology: Science for Life*.

Transparency Pack (0-13141025-3)

Even in this digital age, transparencies are still an efficient and effective way to visually reinforce your lecture. The pack includes 250 four-color transparencies that have been selected from the text, including all of the illustrations, and have been enlarged for large lecture-hall viewing. These images enhance your classroom presentations with enlarged labels and increased color saturation.

For the Student

The goal of the student resource package is to provide opportunities to exercise scientific reasoning skills and apply biological knowledge to real problems. The items listed here offer students many tools for review that are compatible with a wide variety of learning styles.

Student Companion Web Site (www.prenhall.com/belk)

The Student Companion Web Site to *Biology: Science for Life* provides students with the opportunity to review biological concepts and practice problem-solving skills. The focus of the questions and activities is on application, critical thinking, problem solving, analysis, and synthesis. All of the questions and activities provide students with valuable feedback in the form of hints and coaching that identify their areas of weakness and provide them with guidance on how to improve and where to find additional information about these areas. The Student Companion Web Site features:

- *Chapter Outlines*: Available on the Web site, this feature provides a brief overview of the book's chapter contents and major concepts addressed.

- *Learning Objectives*: These objectives address what the students should be able to understand after reading and completing the activities for each chapter.

- *Media Activities*: These activities offer students a visual view of concepts and test their knowledge using a wide variety of activities integrated into each chapter.

  Animations: These activities demonstrate dynamic concepts and processes to better show changes over time and complex interactions. Each animation is followed by interactive activities designed to help students achieve mastery of chapter-specific learning objectives.

Explore the Issue and Explore the Science: These activities offer in-depth exploration of each chapter's major issue and scientific concepts through Web sites, articles, and essay questions. Each activity reinforces the student's need to apply biological knowledge and scientific reasoning to real-world issues.

- *Self Test*: The self-test questions allow students to test both their mastery of the major concepts in the chapter as well as their understanding of the connections these biological concepts have to their lives. Divided into two different types—Analyzing and Applying the Basics, and Connecting the Science—these questions contain helpful hints and coaching that provide students with valuable feedback to help focus their time most effectively.

- *Essay Challenge*: These high-level essay questions focus on Analyzing and Applying the Basics, and Connecting the Science. Hints and suggested answers accompany every question.

- *Issues Update*: This tool links to on-line articles and Web sites that address the major issues and science introduced in a chapter. Links are updated each semester to keep current with recent research and writings.

- *Research Navigator*: This tool equips students with the means to start a research assignment or research paper or to access full text articles. It is complete with extensive help on the research process and three exclusive databases of credible and reliable source material, including the EBSCO Academic Journal and Abstract Database, *The New York Times* Search by Subject Archive, and "Best of the Web" Link Library, Research Navigator™, and enables students to efficiently and effectively make the most of their research time and stay up-to-date on the issues.

- *Science Skills*: This resource features Chemistry and Math Review to help students recap basic knowledge in these areas. It also provides an extensive collection of links and articles on evaluating information and avoiding misinformation.

- *Student Study Tips*: This tool offers extensive resources on how to prepare for tests, get the most out of lectures, and make the best use of study time.

Student Study Guide (0-13141505-0)

The Student Study Guide serves as the print version of many of the resources found on the Student Companion Web Site, including chapter outlines, learning objectives, self tests, Essay Challenge, Explore the Science, and Explore the Issues. For the student who is always on the go, this guide offers a portable alternative to our media resources.

Laboratory Program

Biology: Science for Life Laboratory Manual

This inquiry-driven laboratory manual, designed specifically for the non-science major, reinforces and extends the key biological concepts from *Biology: Science for Life*. These fifteen exercises take a process-oriented approach and often lead to open-ended results. As with the text, the laboratory exercises are connected to compelling stories and provide instructors with an excellent springboard for discussion on the role of biological research in contemporary society.

Symbiosis: The Prentice Hall Custom Laboratory Program for Biology (www.pearsoncustom.com/database/symbiosis/ph.com)

With *Symbiosis: The Prentice Hall Custom Laboratory Program for Biology*, instructors can select from a wide variety of biology, microbiology, or anatomy and physiology labs to build a custom lab manual that exactly matches their content needs and course organization. By visiting the Web site, instructors can select from an extensive list of Prentice Hall laboratory publications or from Pearson Custom Publishing's established library of biology labs. Using the tools provided in our *Lab Ordering and Authoring Kit*, instructors have the ability to develop the best possible lab manual for their courses.

Media Activities

We have created a media support package for *Biology: Science for Life* that will help instructors teach the course and help students achieve biological and scientific literacy through the exploration of contemporary issues and relevant biological concepts. Each activity specifically addresses selected chapter learning objectives through animation and exploration of the Internet. The learning objectives for each activity are listed in the activity and in the Instructor's Resource Manual, making it easy to identify and assign useful activities. Activities are identified in the book by the tabs in the margins and by a brief description at the end of each chapter. Instructors have access to the complete activities on the Instructor's Resource Library and CD-ROM that they can integrate into their presentations.

WWW

Media Activity 1.1: Hypothesis Testing

Essays

Contents

Unit Four Health and Disease

Chapter 12

Gender and Athleticism Developmental Biology, Reproductive Anatomy, and Endocrinology 302

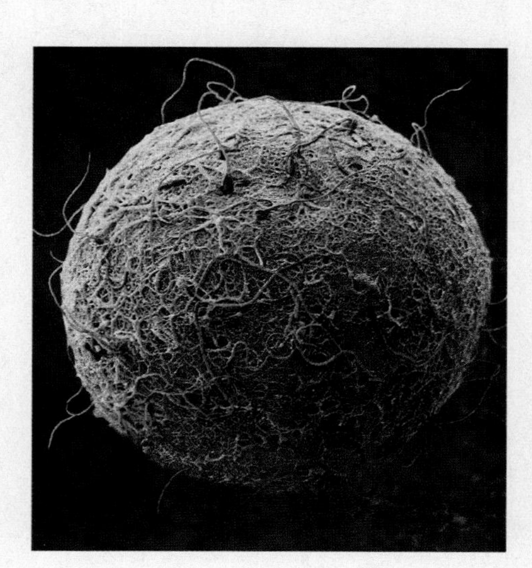

Chapter 13

Attention Deficit Disorder Brain Structure and Function 330

Unit Five Ecology and Environment

CHAPTER 1

Can Science Cure the Common Cold?

Introduction to the Scientific Method

Jake has *another* cold!
What should he do?

"Jake, take massive doses of Vitamin C."

"Jake, drink echinacea tea!"

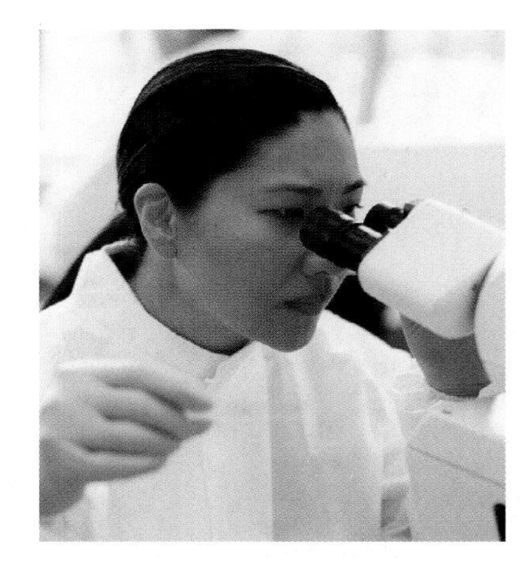

How would a scientist determine which advice is best?

Jake is in bad shape. He has a big exam coming up in his Abnormal Psychology class, a paper due in his Nineteenth-century American Writers course, and he needs to put in extra hours at his job at the pizzeria to make this month's rent payment. On top of everything, Jake has a nasty head cold—his third one this semester. "I'm not going to make it to my junior year if I keep getting sick like this!" he moans to all who will sympathize.

Jake's complaints have brought him endless advice. "Take massive doses of vitamin C—it works for me. I haven't been sick all year," gloats his Biology lab partner. "My sister goes to a chiropractor, and he does some body adjustments that improve her immune system," says one of his basketball teammates. "Take zinc lozenges." "Stop eating so much fried food." "Meditate for a half-hour every day and visualize your strong immune-system warriors." "Drink echinacea tea," says his sister. "Exercise more." "Drop a class." "Have your Ayurvedic balance evaluated." And from his mom, "Wear a hat and gloves when you go outside in the cold—and call me more often!"

What is Jake to do? All the advice he has been getting is from well-meaning, intelligent people; but it is impossible to follow all of these prescriptions—some are even contradictory. If Jake is like most of us, he will follow the advice that makes the most sense to him, and if that doesn't work, he'll try another remedy. Jake might increase his intake of vitamin C and decrease the amount of fried food in his diet. If he gets another cold anyway, he could toss the vitamin C tablets and return to his favorite fast-food place, and then try drinking echinacea tea to minimize its effects.

Jake's testing of different cold preventatives and treatments is the kind of science we all do daily. We see a problem, think of a number of possible causes, and try to solve the problem by addressing what we feel is the most likely cause. If our solution fails to work, we move to another possible solution that addresses other possible causes.

Jake's brand of science may eventually give him an answer to his question about how to prevent colds. But he won't know if it is the best answer unless he tries out all the potential treatments. We already know that Jake does not have time for that. Luckily for him, and for all of us, legions of professional scientists spend their time trying to answer questions like Jake's. Scientists use the same basic process of testing ideas about how the world works and discarding (or modifying) ideas that are inadequate.

There are, however, some key differences between the ways scientists approach questions and the daily scientific investigations illustrated by Jake's quest for relief. This chapter will introduce you to the process of science as it is practiced in the research setting, and will help you understand how to evaluate scientific claims by following Jake's quest for relief from the common cold.

1.1 The Process of Science

The statements made by Jake's friends and family about what actions will help him remain healthy (for example, his mother's advice to wear a hat) are in some part based on the advice-giver's understanding of how our bodies resist colds. Ideas about "how things work" are called **hypotheses**. Or, more formally, a hypothesis is a proposed explanation for one or more observations. All of us generate hypotheses about the causes of some phenomenon based on our understanding of the world (Figure 1.1). When Jake's mom tells him to dress warmly in order to avoid colds, she is basing her advice on her belief in the following hypothesis: Becoming chilled makes an individual more susceptible to becoming ill.

The hallmark of science is that hypotheses are subject to rigorous testing. Therefore, scientific hypotheses must be **testable**—it must be possible to evaluate the hypothesis through observations of the measurable universe. Not all hypotheses are testable. For instance, the statement that "colds are generated by disturbances in psychic energy" is not a scientific hypothesis, since psychic energy cannot be seen or measured—it does not have a material nature. In addition, hypotheses that require the intervention of a *supernatural* force cannot be tested scientifically. If something is supernatural, it is not constrained by the laws of nature, and its behavior cannot be predicted using our current understanding of the natural world.

Scientific hypotheses must also be **falsifiable**, that is, able to be proved false. The hypothesis that exposure to cold temperatures increases your susceptibility to colds is falsifiable, because we can imagine an observation would cause us to reject this hypothesis (for instance, the observation that people exposed to cold temperatures do *not* catch more colds than people protected from

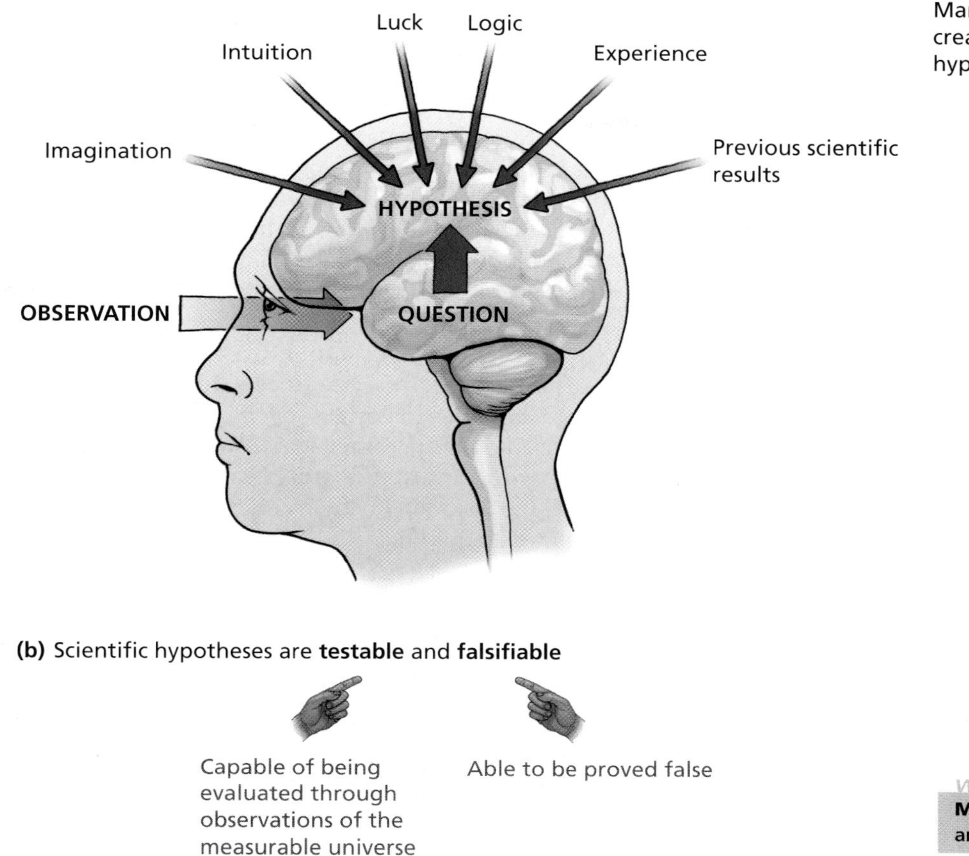

Figure 1.1 Hypothesis generation. Many different factors, both logical and creative, influence the development of a hypothesis.

WWW **Media Activity 1.1A** **Hypothesis Formation and Testing**

chills). However, hypotheses that are judgments, such as "It is wrong to cheat on an exam," are not scientific, since different people have different ideas about right and wrong. It is impossible to falsify these types of statements.

The Logic of Hypothesis Testing

Of all the advice Jake has heard, he is inclined toward that given by his lab partner. She insisted that taking vitamin C supplements was keeping her healthy. Jake also recalls learning about vitamin C in his Human Nutrition class last year. In particular, he remembers that:

1. Fruits and vegetables contain lots of vitamin C.
2. People with diets rich in fruits and vegetables are generally healthier than people who skimp on these food items.
3. Vitamin C is known to be an anti-inflammatory agent, reducing throat and nose irritation.

Given his lab partner's experience and what he learned in class, Jake makes the following hypothesis:

Consuming vitamin C decreases the risk of catching a cold.

This hypothesis makes sense. After all, Jake's lab partner is healthy and Jake has made a logical case for why vitamin C is good cold prevention. This certainly seems like enough information on which to base his decision about how to proceed—he should start taking vitamin C supplements if he wants to avoid future colds. However, a word of caution: Just because a hypothesis seems logical does not mean that it is true.

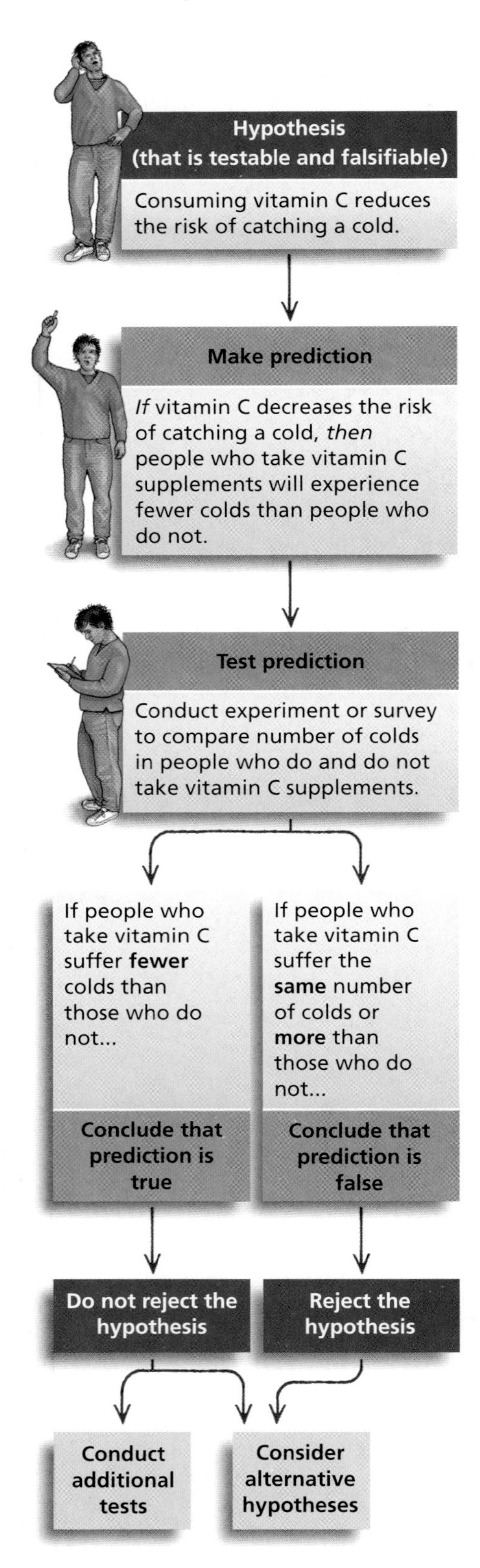

Figure 1.2 Hypothesis testing. Tests of hypotheses follow a logical path. This flow chart illustrates the process.

WWW
Media Activity 1.1B Spontaneous Generation and Pasteur's Experiments

Consider the ancient hypothesis that the sun revolves around Earth, asserted by Aristotle in approximately 350 B.C. This hypothesis was logical, based on the observation that the sun appeared on the eastern horizon every day at sunrise and disappeared behind the western horizon at sunset. For two thousand years, this hypothesis was considered to be "a fact" by nearly all of Western society. To most people, the hypothesis made perfect sense, especially since the common religious belief in Western Europe was that Earth had been created and then surrounded by the vault of heaven. It was not until the early seventeenth century that this hypothesis was falsified as the result of observations made by Galileo Galilei of the movements of Venus. Galileo's work helped to confirm Nicolai Copernicus' more modern hypothesis that Earth revolves around the sun.

So even though Jake's hypothesis about vitamin C is perfectly logical, it needs to be tested. Hypothesis testing is based on a process called **deductive reasoning** or *deduction*. Deduction involves making a specific *prediction* about the outcome of an action or test based on observable facts. The prediction is the result we would expect from a particular test of the hypothesis.

Deductive reasoning takes the form of "if/then" statements. A prediction based on the vitamin C hypothesis could be:

> *If* vitamin C decreases the risk of catching a cold, *then* people who take vitamin C supplements with their regular diets will experience fewer colds than people who do not take supplements.

Deductive reasoning, with its resulting predictions, is a powerful method for testing hypotheses. However, the structure of such a statement means that hypotheses can be clearly rejected if untrue, but impossible to prove if they are true (Figure 1.2). This shortcoming is illustrated using the "if/then" statement above.

Consider the possible outcomes of a comparison between people who supplement with vitamin C and those who do not: People who take vitamin C supplements may suffer through more colds than people who do not, they may have the same number of colds as people who do not supplement, or supplementers may in fact experience fewer colds. What do these results tell Jake about his hypothesis?

If people who take vitamin C have more colds, or the same number of colds as those who do not supplement, the hypothesis that vitamin C alone provides protection against colds can be rejected. But what if people who supplement with vitamin C *do* experience fewer colds? If this is the case, should Jake be out proclaiming the news, "Vitamin C—A Wonder Drug that Prevents the Common Cold"? No, he should not. Jake needs to be much more cautious than that; he can only say that he has supported and not disproven the hypothesis.

Why is it impossible to say that the hypothesis that vitamin C prevents colds is true? Primarily because there could be other factors (that is, there are *alternative hypotheses*) that explain why people with different vitamin-taking habits are different in their cold susceptibility. In other words, demonstrating the truth of the *then* portion of a deductive statement does not guarantee that the *if* portion is true.

Consider the alternative hypothesis that frequent exercise reduces susceptibility to catching a cold. Perhaps people who take vitamin C supplements are more likely to engage in regular exercise than those who do not supplement. What if the alternative hypothesis were true? If so, the prediction that people who take vitamin C supplements experience fewer colds than people who do not supplement would be true, but not because the original hypothesis (vitamin C reduces the risk of cold) is true. Instead, people who take vitamin C supplements experience fewer colds than people who do not supplement because they are more likely to exercise, and it is exercise that reduces cold susceptibility.

A hypothesis that seems to be true because it has not been rejected by an initial test may be rejected later based on the results of a different test. As a matter of fact, this is the case for the hypothesis that vitamin C consumption reduces susceptibility to colds. The argument for the power of vitamin C was popularized in 1970 by the Nobel Prize-winning chemist Linus Pauling in his book *Vitamin C and the Common Cold*. Pauling based his assertion that large doses of vitamin C reduce the incidence of colds by as much as 45% on the results of a few studies that had been published since the 1930s. However, repeated careful tests of this hypothesis have since failed to support it. In many of the studies Pauling cited, it appears that one or more alternative hypotheses may explain the difference in cold frequency between vitamin C supplementers and nonsupplementers. Today, most researchers studying the common cold agree that the hypothesis that vitamin C prevents colds has been convincingly falsified.

The Experimental Method

Is Jake out of luck even before he starts his evaluation of research on the prevention of the common cold? Even if one of the hypotheses about cold prevention is supported, does the difficulty of eliminating alternative hypotheses mean that he will never know which approach is truly best? The answer is "yes and no." Hypotheses cannot be proven absolutely true; it is always possible that the true cause of a particular phenomenon may be found in a hypothesis that has not yet been evaluated. However, in a practical sense, a hypothesis can be proven beyond a reasonable doubt. One of the most effective ways to test many hypotheses is through rigorous scientific experiments.

Experiments are contrived situations designed to test specific hypotheses. Generally, an experiment allows a scientist to control the conditions under which a given phenomenon occurs. Having the ability to manipulate the environment enables a scientist to minimize the number of alternative hypotheses that may explain the result. The information collected by scientists during hypothesis testing is known as **data**. Data collected from experiments should allow researchers to either reject or support a hypothesis.

Not all scientific hypotheses can be tested through experimentation. For instance, hypotheses about the origin of life or the extinction of the dinosaurs are usually not testable in this way. These hypotheses must instead be tested via careful observation of the natural world. Not all testable hypotheses are subjected to experimentation either—the science that *is* performed is a reflection of the priorities of the decision-makers in our society (Essay 1.1). Hypotheses about the origin and prevention of colds can and are tested experimentally, however.

Experimentation has enabled scientists to prove beyond a reasonable doubt that the common cold is caused by a virus. A virus has a very simple structure—it typically contains a short strand of genetic material and a few chemicals called proteins encased in a relatively tough outer shell composed of more proteins and sometimes a fatty membrane. Biologists disagree over whether viruses should be considered living organisms. Since a virus must enter, or infect, a cell in order to reproduce, some biologists refer to them as "subcellular infectious particles." Of the over 200 types of viruses that are known to cause varieties of the common cold, most infect the cells in our noses and throats. The sneezing, coughing, congestion, and sore throat characteristic of infection by most cold viruses appear to be the result of the body's immune response to a viral invasion (Figure 1.3).

The role of viruses in colds is generally accepted as a fact for two reasons. First, all reasonable alternative hypotheses about the causes of colds (for instance, exposure to cold air) have been rejected in numerous experimental tests, and second, the hypothesis has *not* been rejected after carefully designed experiments measuring cold incidence in people exposed to purified virus samples. "Truth" in science can therefore be defined as *what we know and understand based on all available information*. If a hypothesis appears to explain all instances

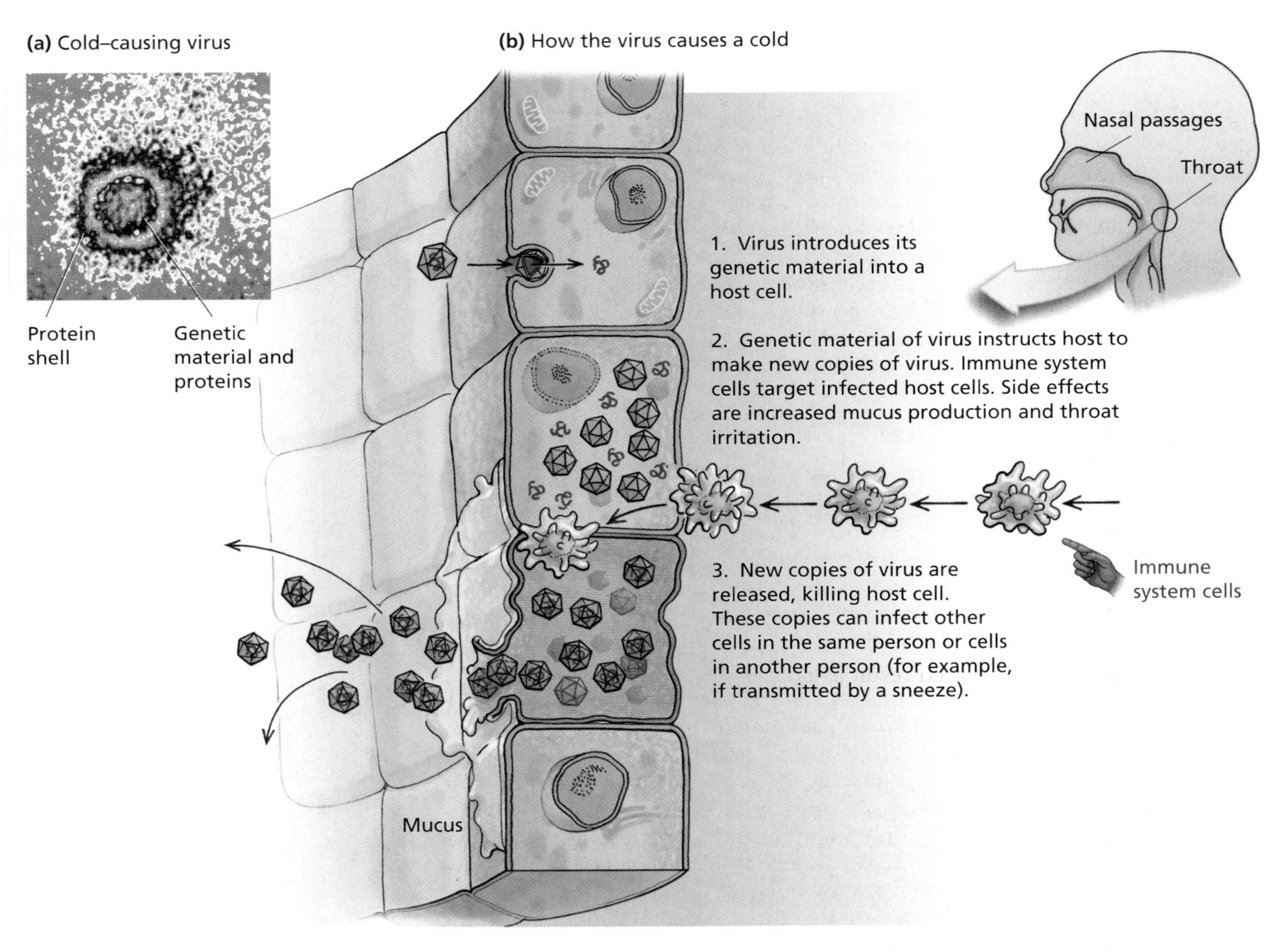

(a) Cold–causing virus
(b) How the virus causes a cold
Protein shell
Genetic material and proteins
Nasal passages
Throat
1. Virus introduces its genetic material into a host cell.
2. Genetic material of virus instructs host to make new copies of virus. Immune system cells target infected host cells. Side effects are increased mucus production and throat irritation.
3. New copies of virus are released, killing host cell. These copies can infect other cells in the same person or cells in another person (for example, if transmitted by a sneeze).
Immune system cells
Mucus

Figure 1.3 A cold-causing virus. (a) An image from an electron microscope of a typical rhinovirus, one of the many viruses that cause the common cold. (b) A rhinovirus causes illness by invading cells in the lining of the nose and throat, and using those cells as "factories" to make virus copies. Cold symptoms result when our immune systems attempt to control and eliminate this invader.

of a particular phenomenon, and has been repeatedly tested and supported, it may eventually be accepted as accurate. However, even the strongest scientific hypotheses may potentially be replaced by better explanations.

Controlled Experiments **Control** has a very specific meaning in science. A control subject for an experiment is an individual who is similar to an experimental subject, except that the control is not exposed to the experimental treatment. Measurements of the control group are used as baseline values for comparison to measurements of the experimental group.

One of the suggestions Jake received to reduce his suffering was to drink echinacea tea. *Echinacea purpurea*, a common North American prairie plant, has been touted as a treatment to reduce the likelihood as well as the severity and duration of colds (Figure 1.4). Jake's sister's suggestion of echinacea tea was based on the results of a scientific study showing that people who drank echinacea tea felt that it was 33% more effective at reducing symptoms. The "33% more effective" is in comparison to the opinions of people about the effectiveness of a tea that did not contain *Echinacea* extract; that is, the results from the control group (Figure 1.5). Jake is intrigued by this result—perhaps if he cannot avoid catching a cold, he can reduce its effects once it has started.

Figure 1.4 ***Echinacea purpurea*, an American coneflower.** Extracts from the leaves and roots of this plant are among the most popular herbal remedies sold in the United States.

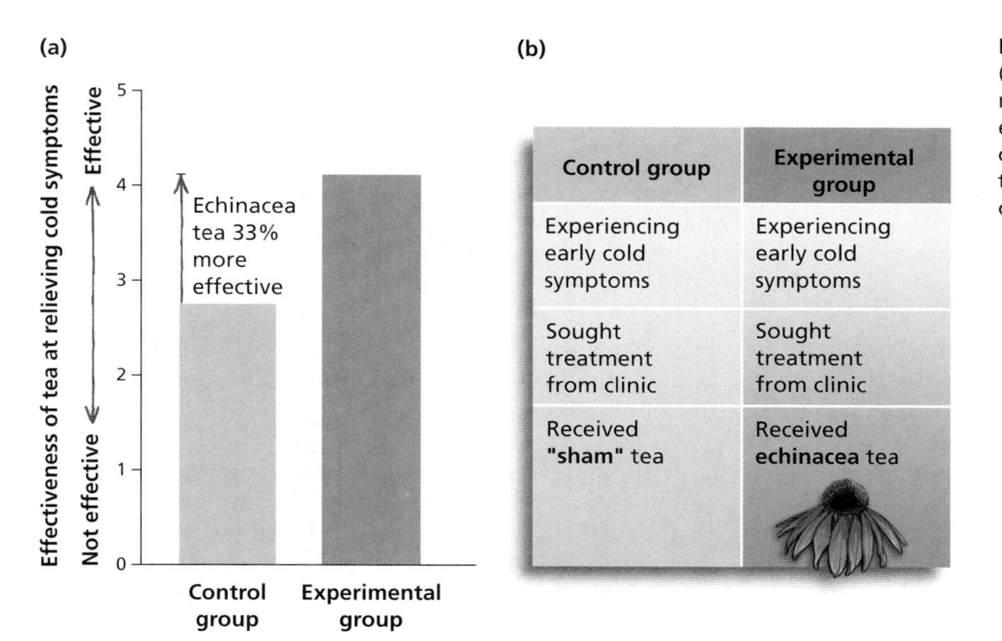

Figure 1.5 A controlled experiment. (a) A graph of the results of an experiment on the effectiveness of drinking echinacea tea. (b) Experimental and control groups were similar and were treated identically except for the type of tea they consumed.

A good controlled experiment eliminates as many alternative hypotheses that could explain the observed result as possible. The first step is to select a pool of subjects in such a way as to eliminate differences in participants' ages, diets, stress levels, and likelihood of visiting a health care provider. The most effective way of doing this is the **random assignment** of individuals to these categories. For example, a researcher might put all the volunteers' names in a hat, draw out half, and designate these people as the experimental group and the remainder as the control group. Random assignment helps reduce the likelihood that there is a systematic difference between the experimental and control groups. In the echinacea tea study that Jake's sister had told him about, members of both the experimental and control group were female employees of a nursing home who sought relief from their colds at their employer's clinic. Imagine what would happen if the colds experienced in the nursing home changed over the course of the experiment—that is, one cold virus affected a number of individuals for a few weeks, and then a different cold virus affected other individuals in the next few weeks. If the researchers had simply assigned the first 25 visitors to the clinic to the control group and the next 25 to the experimental group, they would run the risk of the two groups actually experiencing different colds as well as drinking different teas. To avoid this kind of problem, the volunteers were randomly assigned into either the experimental or control group.

The second step in designing a good control is to attempt to treat control subjects and experimental subjects identically during the course of the experiment. In this study, all participants received the same information about the purported benefits of echinacea tea, and during the course of the experiment, all participants were given tea with instructions to consume five to six cups daily until their symptoms subsided. However, individuals in the control group received "sham tea" that did not contain *Echinacea* extract. This sham tea would be equivalent to "sugar pills," or *placebos*, that are given to control subjects when testing a particular drug. Employing a placebo generates only one consistent difference between individuals in the two groups—in this case the type of tea they consumed.

Good controls are the basis of *strong inference*. In the echinacea tea study, the data indicated that cold severity was lower in the experimental group compared to those who received the sham tea. Because their study utilized controls, the researchers can have high confidence that the reason the two groups would differ is if *Echinacea* extract relieved cold symptoms. Because their control had greatly reduced the likelihood that alternative hypotheses could explain their results,

Essay 1.1 The Social Context of Science

How might society influence the general direction of scientific research? The opinions and worldviews of researchers interact with the views of the directors of government funding agencies, legislators, and business organizations that make grants for research. Through these channels, both the questions scientists may test and the ways in which they may be tested are heavily influenced by the society that surrounds them.

Consider the following example. Depression is a disorder that affects nearly 19 million Americans, and billions of dollars have been spent on research. Much of this funding has helped researchers understand changes in brain chemistry and to design effective drug therapies to treat depression. However, we know that major risk factors for depression in the United States include gender (depression is twice as common among women as among men), societal status (risk of depression is greater among ethnic minorities), and geographic location (city dwellers are more likely to become depressed than rural residents). These risk factors suggest that, in addition to biology, environmental conditions probably play some role in the origin of depression. Despite these observations, until recently there has been relatively little research on techniques of preventing depression, even among these high-risk groups. A review of the medical literature reveals six times as many research papers on using drug therapy to treat depression as on the prevention of depression.

Because depression has long been thought of as a disease of the individual, research has focused on what makes depressed individuals "different" and how we can treat these differences. If depression had been seen as a disease stemming from a reaction to poor local conditions, the research focus might then have been on what makes an environment likely to lead to depression, and how the environment could be modified to reduce the risk of depression.

At least part of the reason for approaching depression as a "brain disease" is that much of the funding for research comes from pharmaceutical companies. These companies will only realize a profit if they can develop drug treatments. They will naturally be less interested in research on prevention if it involves nonpharmaceutical interventions. The result is many different drug therapies to treat depression, but very little specific advice on how to reduce the risk of experiencing depressive disorders.

However, the influence of economics and politics also means that citizens of the United States can have a profound effect on the direction of science by working with their elected officials to increase the federal funding for certain areas of research. Activists in the 1980s and 1990s, for instance, were successful in obtaining major increases in funds for breast cancer and AIDS research. These successes remind us that all citizens—scientist and nonscientist alike—have the power to affect the progress of science. It is our responsibility to use that power wisely and well.

the researchers could strongly infer that they were measuring a real, positive effect of echinacea tea on colds.

The study described above supports the hypothesis that echinacea tea reduces the severity of colds. However, it is extremely rare that a single experiment will cause the scientific community to accept a hypothesis beyond a reasonable doubt. Dozens of studies, each using different experimental designs, have investigated the effect of *Echinacea* extract on common colds and other infections. Some of these studies have shown a positive effect, but others have shown none. In the medical community as a whole, the jury is still out regarding the effectiveness and appropriate use of this popular herb.

Minimizing Bias in Experimental Design Scientists and human research subjects may have strong opinions about the veracity of a particular hypothesis even before it is tested. These opinions may cause participants to influence, or *bias*, the results of an experiment—often unwittingly.

One potential source of bias is *subject expectation*, which is sometimes called the "onstage effect." Individual experimental subjects may consciously or unconsciously model the behavior they feel the researcher expects from them. For example, an individual who knew she was receiving echinacea tea may have felt confident that she would recover more quickly. This might cause her to underreport her cold symptoms. This potential problem is avoided by designing a *blind experiment*, where individual subjects are not aware of exactly what they are predicted to experience. In experiments on drug treatments, this means not telling participants whether they are receiving the drug or a placebo.

Another source of bias arises when a researcher makes consistent errors in the measurement and evaluation of results. This phenomenon is called *observer bias*. In the echinacea tea experiment, observer bias could take various forms. Expecting a particular outcome might lead a scientist to give slightly different instructions about what symptoms constituted a cold to subjects who received echinacea tea. Or, if the researcher expected people who drank echinacea tea to experience fewer colds, she might make small errors in the measurement of cold severity that influenced the final result. To avoid the problem of experimenter bias, the data collectors themselves should be "blind." Ideally, the scientist, doctor, or technician applying the treatment does not know which group (experimental or control) any given subject is part of until after all data have been collected (Figure 1.6). Blinding the data collector ensures that the data are *objective*, in other words, without bias.

We call experiments **double blind** when *both* the research subjects and the technicians performing the measurements are unaware of either the hypothesis or whether a subject is in the control or experimental group. Double-blind experiments nearly eliminate the effects of human bias on results. When both researcher and subject have few expectations about the hypothesized outcome of a particular experimental treatment, the results obtained from the experiment should be considered more credible.

Using Correlation to Test Hypotheses

Well-controlled experiments can be difficult to perform when humans are the experimental subjects. As you can see from the echinacea tea study, the requirement that both experimental and control groups be treated nearly identically means that some people receive no treatment. In the case of cold sufferers, who have limited means of reducing cold duration and severity, the placebo treatment does not substantially hurt those who receive it. However, placebo treatments are impossible or unethical in many cases. For instance, imagine testing the effectiveness of a birth control drug by giving one group of women the drug and comparing their rate of pregnancies to another group of women who thought they were getting the drug but who were actually getting a placebo!

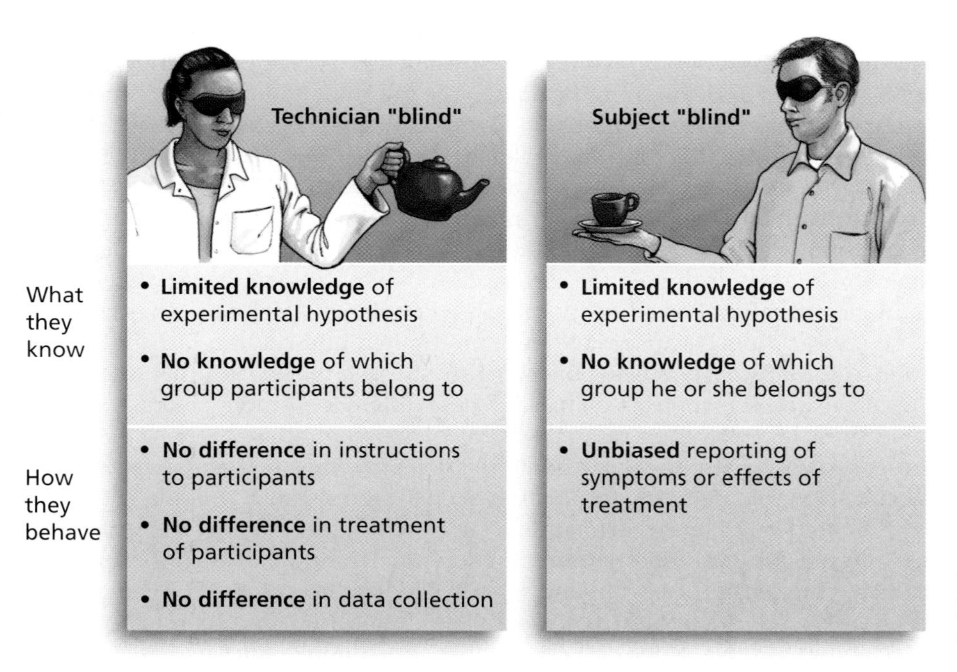

Figure 1.6 Double-blind experiments. Double-blind experiments result in more objective data.

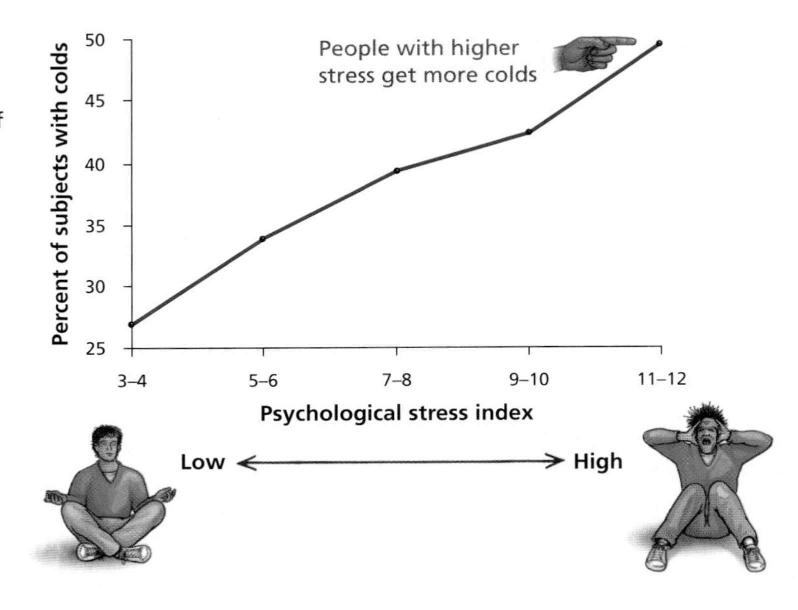

Figure 1.7 Correlation between stress level and illness. This graph summarizes the results of an experiment that compared rates of virus infection in groups of individuals with different self-reported stress levels. The graph indicates that people experiencing higher levels of stress become infected by a virus more often than people experiencing low levels of stress.

When controlled experiments are difficult or impossible to perform, scientists will test hypotheses using **correlations**. A correlation is a relationship between two variables. Suggestions that Jake reduce his workload, exercise more, or spend more time with his mom to reduce his susceptibility to colds are based on a correlation between high levels of psychological stress and increased susceptibility to cold-virus infection (Figure 1.7). This correlation was generated by researchers who collected data on a number of individuals' psychological stress levels before giving them nasal drops containing a cold virus. Doctors later reported on the incidence and severity of colds among participants in the study.

Let's examine the data presented in Figure 1.7. The horizontal axis of the graph, or *x* axis, contains a scale of stress level—from a low stress level on the left edge of the scale to a high stress level on the right. The vertical axis of the graph, the *y* axis, indicates the percentage of study participants who developed "clinical colds"; that is, colds reported by their doctors. Each point on the graph represents a group of individuals and tells us what percentage of people in each stress category had clinical colds. The line connecting the five points on the graph illustrates a correlation—the relationship between stress level and susceptibility to cold virus infection. Because the line rises to the right, these data tell us that people who have higher stress levels typically experience more colds. In fact, it appears from the data in the graph that individuals experiencing high levels of stress are more than twice as likely to become ill. But does this relationship mean that high stress causes increased cold susceptibility?

In order to conclude that stress causes illness, we need the same assurances that are given by a controlled experiment. In other words, we must assume that the individuals measured for the correlation are similar in every way, except for their stress levels. Is this a good assumption? Not necessarily. Most correlations cannot control for alternative hypotheses. People who feel more stressed may have poorer diets because they feel time-limited and rely on fast food more often. Alternatively, people who feel highly stressed may be in situations where they are exposed to more cold viruses. These differences among people who differ in stress level may also influence their cold susceptibility (Figure 1.8). Therefore, even with a strong correlational relationship between the two factors, we cannot strongly infer that stress *causes* decreased resistance to colds.

Researchers who use correlational studies do their best to ensure that their subjects are similar in many characteristics. For example, this study on stress and cold susceptibility evaluated whether individuals in the different stress categories were different in age, weight, sex, education, and their exposure to infected individuals. None of these other factors differed among low-stress and

(a) Does high stress cause high cold frequency?

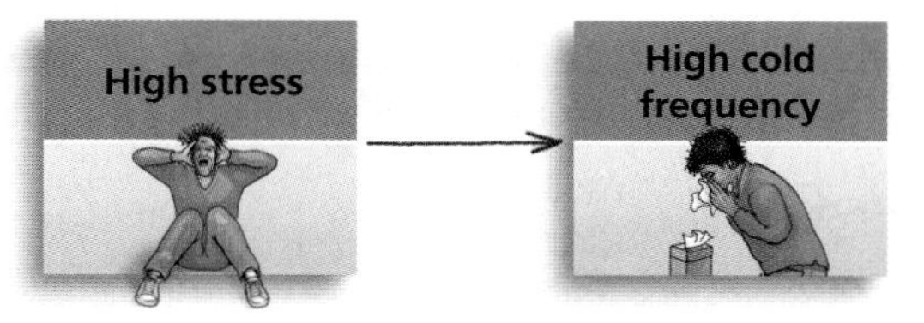

(b) Or does one of the causes of high stress cause high cold frequency?

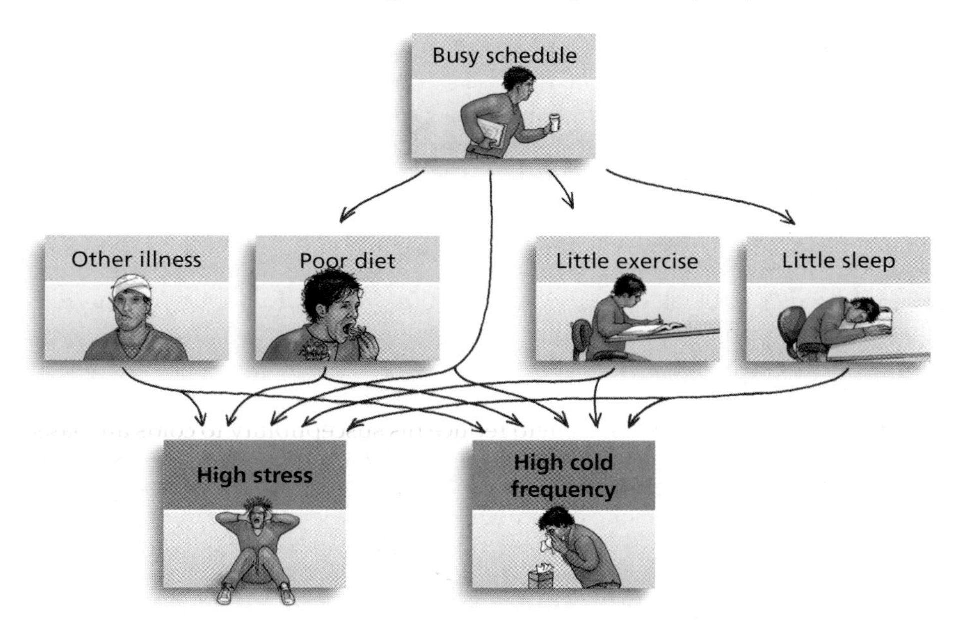

Figure 1.8 Correlation does not signify causation. Does high stress cause high cold frequency? Or does one of the *causes* of high stress cause high cold frequency? A correlation typically cannot eliminate all alternative hypotheses.

high-stress groups. Eliminating some of the alternative hypotheses that could explain this correlation increases the strength of the inference that high stress levels truly do increase susceptibility to colds. However, people with high-stress lifestyles still may be fundamentally different from those with low-stress lifestyles, and it is possible that one of those important differences is the real cause of disparities in cold frequency.

You may see from the above discussion that it is difficult to demonstrate a cause-and-effect relationship between two factors simply by showing a correlation between them. In other words, *correlation does not equal causation*. For example, a commonly understood correlation exists between exposure to cold air and epidemics of the common cold. It is true that as outdoor temperatures drop, the incidence of colds increases. But numerous controlled experiments indicate that chilling does not increase susceptibility to colds. Instead, cold outdoor temperatures mean increased close contact with other people (and their viruses). Despite the correlation, cold air does not cause colds—exposure to viruses does.

Understanding Statistics

As Jake reviews scientific literature on cold prevention and treatment, he might come across statements about the "significance" of the effects of different cold-reducing measures. For instance, one report may state that factor A reduced cold severity, but that the results of the study were "not significant." Another study may state that factor B caused a "significant reduction" in illness. Jake might then assume that this means factor B will help him feel better, while factor A will have little effect. He finally has an answer! Well, no—unfortunately for Jake, in scientific studies "significance" is defined a bit differently from its daily usage. To evaluate the scientific use of the term *significance*, Jake needs a basic understanding of statistics.

Statistics is a specialized branch of mathematics used in the evaluation of experimental data. An experimental test utilizes a small subgroup, or **sample**, of a population. *Descriptive statistics* helps researchers summarize data from the sample—for instance, we can describe the average, or *mean*, length of colds experienced by experimental and control groups. *Inferential statistics* allows scientists to extend the results they summarize from their sample to the entire population. Inferential statistics takes the form of **statistical tests**. When scientists conduct an experiment, they hypothesize that there is a true, underlying effect of their experimental treatment on the entire population. An experiment on a sample of a population can only estimate this true effect, but statistical tests help scientists evaluate whether the results of a single experiment demonstrate the true effect of a treatment. In the experiment with the echinacea tea, statistical tests tell us if the experimental result of a 33% reduction in cold severity is an indication of how well echinacea tea works or if it might be due to chance differences between the experimental and control group.

We can explore the role statistical tests played in a study on another proposed treatment to reduce the severity of colds—lozenges containing zinc. Some forms of zinc can block certain common cold viruses from entering the cells that line the nose. This observation led scientists to hypothesize that consuming zinc at the beginning of a cold decreases the number of cells that become infected, which in turn decreases the length and severity of cold symptoms. To test this hypothesis, a group of researchers at the Cleveland Clinic performed a study using a sample of 100 of their employees who enrolled in the study within 24 hours of developing cold symptoms. The researchers randomly assigned subjects to control or experimental groups. Members of the experimental group received lozenges containing zinc, while members of the control group received placebo lozenges. Members of both groups received the same instructions about use of the lozenges and were asked to rate their symptoms until they had recovered. The experiment was double-blind.

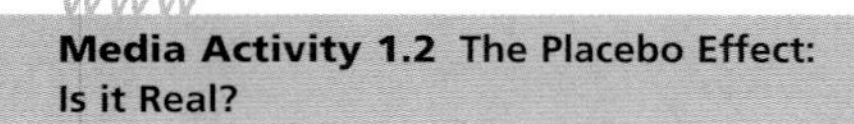

WWW
Media Activity 1.2 The Placebo Effect: Is it Real?

When the data from the experiment were summarized, the researchers observed that the mean length of time to recovery was more than three days shorter in the zinc group than in the placebo group (Figure 1.9). Superficially, this result appears to support the hypothesis. However, a statistical test is necessary because, even with well-designed experiments, chance will always result in some difference between the control and experimental groups. The effect of chance on experimental results is known as **sampling error**. Even if there is *no* true effect of an experimental treatment, the results observed in the experimental and control groups will never be exactly the same.

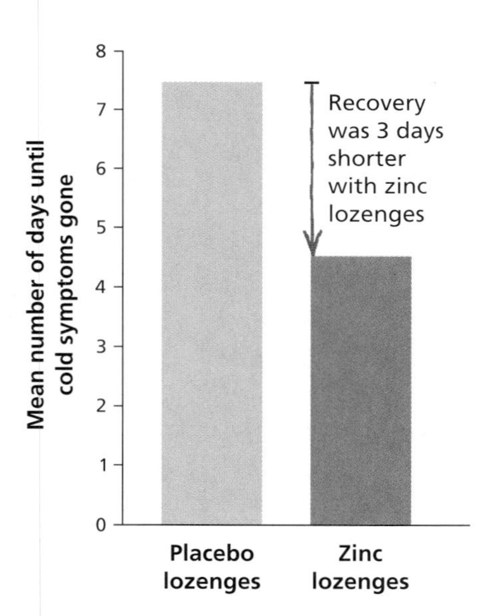

Figure 1.9 Zinc lozenges reduce the duration of colds. This graph illustrates the results of an experiment on the effectiveness of zinc lozenges on decreasing cold duration. Individuals in the experimental group had colds lasting about 4½ days as opposed to approximately 7½ days for the placebo group.

We know that people differ in their ability to recover from a cold infection. If we give zinc lozenges to one volunteer and placebo lozenges to another, it is likely that they will have colds of different lengths. But even if the zinc-taker had a shorter cold than the placebo-taker, you would probably say that the test did not tell us much about our hypothesis—the zinc-taker might just have had a less severe cold for other reasons. Now imagine that we had five volunteers in each group and saw a difference. Or that the difference was only one day instead of three days. Statistical tests allow researchers to look at their data and determine how likely it is that the result is due to sampling error.

Statistical tests actually evaluate the *null hypothesis*. "Null" means zero, and the null hypothesis is that there is zero difference between the experimental and control populations. In other words, the experimental treatment has no effect. In this case, the null hypothesis is that there is no difference in the length of colds experienced by people who take zinc lozenges and those who take placebo lozenges. A statistical test allows the researchers to evaluate whether the observed data are consistent with this null hypothesis. The logic behind this approach is as follows: As the data from the control and experimental groups diverge from each other, the null hypothesis becomes less and less credible. If the difference between results in the experimental group and results in the control group becomes large enough, the investigator must reject the null hypothesis. In the case of the experiment with zinc lozenges, the statistical test indicated that there was a low probability, less than one in 10,000 (0.01%), that the experimental and control groups were so different simply by chance. In other words, the null hypothesis above is very unlikely to be true, and the result is **statistically significant**.

One characteristic of experiments influencing the power of statistical tests is *sample size*—the number of individuals in the experimental and control groups. A larger sample size minimizes the chance of sampling error. In addition, the more participants there are in a study, the more likely it is that researchers will see a true effect of an experimental treatment, if one exists. If the sample size is large, any difference between an experimental and control group is more likely to be statistically significant.

Since both sample size and the strength of an experimental treatment affect statistical significance, it is not equivalent to *practical significance*. If the effect of a treatment is real but minor, an experiment with a very *large* sample size may return a statistically significant result, but that result means little in practice. Conversely, if the effect of a treatment is real, but the sample size of the experiment is *small*, a single experiment may not allow researchers to reject the null hypothesis. The relationship between hypotheses, experimental tests, sample size, and statistical significance is summarized in Figure 1.10.

Statistical significance by itself is not a sufficient measure of the accuracy of an experiment, and all statistical tests operate with the assumption that the experiment was *designed and carried out correctly*. In other words, a statistical test evaluates the chance of sampling error, not observer error, and a statistically significant result should never be taken as the last word on an experimentally tested hypothesis. An examination of the experiment itself is required. In the test of the effectiveness of zinc lozenges, the experimental design minimized the likelihood

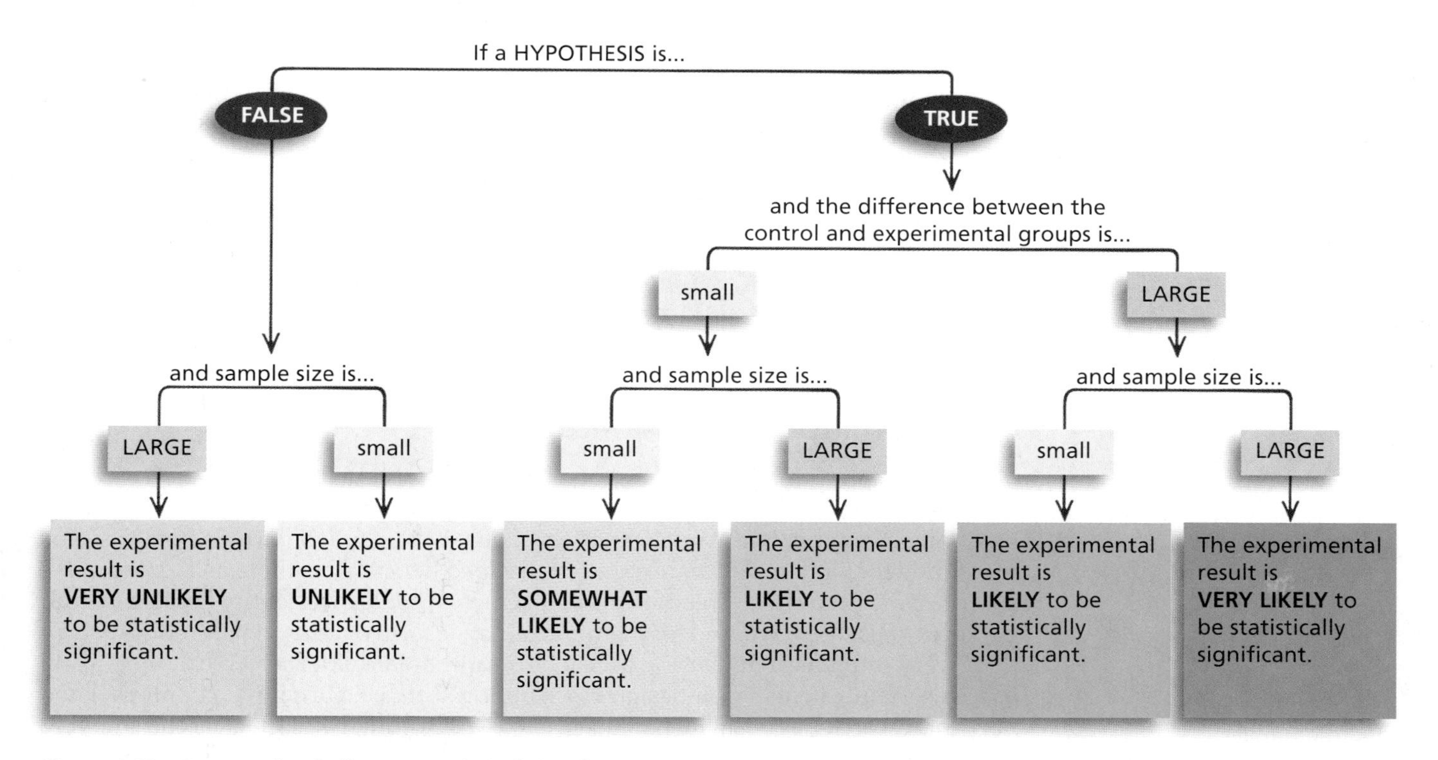

Figure 1.10 Factors that influence statistical significance. This flowchart summarizes the relationship between the true effect of a treatment and the sample size of an experiment on the likelihood of obtaining statistical significance. A large sample size can detect a statistically significant effect, even if the difference is small and of little practical significance. A small sample size might fail to detect a true effect of a treatment.

that alternative hypotheses could explain the results by randomly assigning subjects to treatment groups, using an effective placebo, and blinding both the data collectors and the subjects. Given such a well-designed experiment, this statistically significant result allows researchers to strongly infer that consuming zinc lozenges reduces the duration of colds.

There is one final caveat however. A statistically significant result is defined as one that has a 5% probability or less of being due to chance alone. If all scientific research uses this same standard, as many as one in every 20 statistically significant results (that is, 5% of the total) is actually reporting an effect that is *not real*. An experiment with a statistically significant result will still be considered to support the hypothesis. However, the small but important probability that the results are due to chance explains why one supportive experiment is usually not enough to convince all scientists that a hypothesis is accurate. Even with a statistical test indicating that the result had a likelihood of less than 0.01% of occurring by chance, Jake should begin to feel assured that taking zinc lozenges will reduce the duration of his colds only after locating additional tests of this hypothesis that give similar results. In fact, scientists continue to test this hypothesis, and there is still no consensus among them about the effectiveness of zinc as a cold treatment.

WWW
Media Activity 1.3 Evaluating Health Information from the Internet

1.2 Evaluating Scientific Information

Given the challenges inherent in establishing scientific "truth"—the rigorous requirements for using controls to eliminate alternative hypotheses, and the problem of sampling error—we can see why definitive scientific answers to our questions are slow in coming. A well-designed experiment can certainly allow us to approach the truth. Looking at reports of experiments critically can help us make well-informed decisions about actions to take. However, Jake's busy schedule hampers a thorough evaluation of all of the current scientific research on cold prevention from **primary sources** written by the researchers themselves and reviewed within the scientific community (Figure 1.11). The process of *peer review* helps increase confidence in scientific information because other scientists critique the results and conclusions of an experiment before it is published in a professional journal. These journals, such as *Science, Nature*, the *Journal of the American Medical Association*, and hundreds of others, represent the first and most reliable source of current scientific knowledge.

If he's like most of us, Jake will get his scientific information from **secondary sources**, such as books, news reports, and advertisements. How can he evaluate information in this context?

Information from Anecdotes

Information about dietary supplements such as echinacea tea and zinc lozenges is often in the form of **anecdotal evidence**—meaning that the advice is based on one individual's personal experience. Jake's biology lab partner's enthusiastic plug for vitamin C, because she felt it helped her, is an example of a *testimonial*—a common form of anecdote. Advertisements that use a celebrity to pitch a product "because it worked for them" are a classic form of testimonial. You should be very cautious about basing decisions on anecdotal evidence, which is not in any way equivalent to well-designed scientific research. For example, countless hours of research have established that there is a clear link between cigarette smoking and lung cancer. Although everyone has heard anecdotes of someone's grandpa who was a pack-a-day smoker and lived to the age of 94, the risk of premature death due to smoking is very well established. While anecdotes may indicate that a product or treatment has merit, only well-designed tests of the hypothesis can help determine if it is likely to be safe and effective for most people.

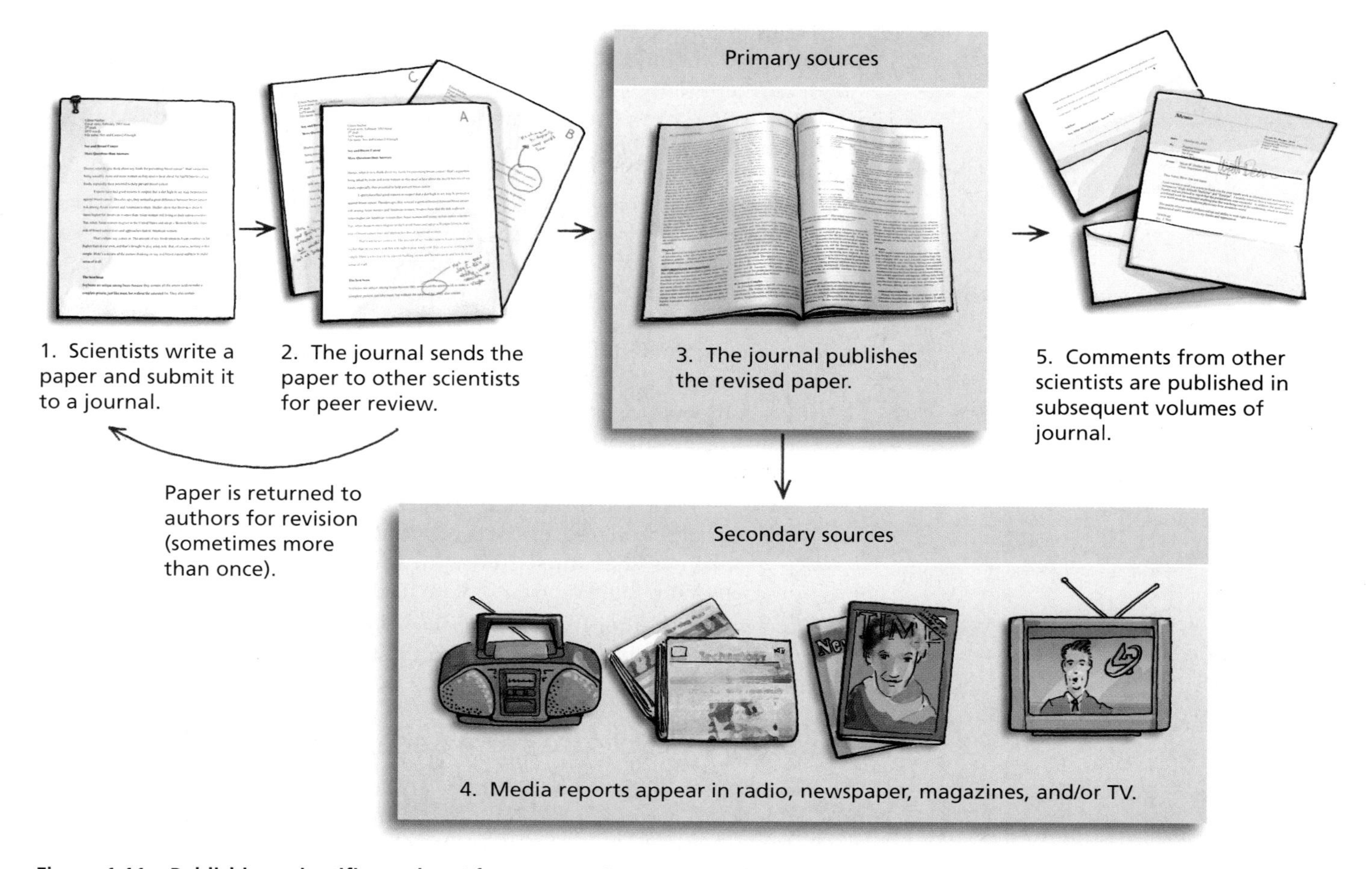
Primary sources
1. Scientists write a paper and submit it to a journal.
2. The journal sends the paper to other scientists for peer review.
3. The journal publishes the revised paper.
5. Comments from other scientists are published in subsequent volumes of journal.
Paper is returned to authors for revision (sometimes more than once).
Secondary sources
4. Media reports appear in radio, newspaper, magazines, and/or TV.

Figure 1.11 Publishing scientific results. After an experiment is complete, the researchers write a scientific paper for publication in a journal. Both before and after publication, the paper is reviewed by other scientists who evaluate the research presented in the paper and the researchers' conclusions. Peer review provides checks and balances that help maintain the integrity of the science presented.

Science in the News

Popular news sources provide a steady stream of health information. However, stories about research results in the general media rarely contain information about the adequacy of controls, the number of subjects, or the experimental design. How can anyone evaluate the quality of research that supports statements like these? "Supplement Helps Melt Fat and Build Muscle," or "Curry Spice Might Prevent Bowel Cancer".

First, you must consider the source of media reports. Certainly news organizations will be more reliable reporters of fact than entertainment tabloids, and news organizations with science writers should be considered better reporters of the substance of a study than those without. Television talk shows, which need to fill airtime, regularly have guests who promote a particular health claim. Too often, these guests may be presenting information that is based on anecdotes or an incomplete survey of the primary literature, as well as work that has not been subjected to peer review.

Paid advertisements are a legitimate means of disseminating information. However, claims in advertising should be very carefully evaluated. Our pursuit of health fuels a multibillion-dollar industry—companies that succeed need to be very effective at getting the attention of consumers. While advertisements of over-the-counter and prescription drugs must conform to rigorous government standards regarding the truth of their claims, advertisements for herbal supplements, many health food products, and diet plans have lower standards.

Be sure to examine the fine print—advertisers often are required to clarify the statements made in bold type in their ads.

Another commonly used source for health information is the Internet. As you know, anyone can post information on the Internet. Typing in "common cold prevention" on a standard Web search engine will return thousands of Web pages—from highly respected academic and government sources to small companies trying to sell their products, or individuals who have strong, sometimes completely unsupported, ideas about cures. Often it can be difficult to determine the reliability of a well-designed Web site. Here are some things to consider when using the Web as a resource for health information:

1. Choose sites maintained by prestigious medical establishments, such as the National Institutes of Health (NIH), or the Mayo Clinic.
2. It costs money to maintain a Web site. Consider whether the Web site seems to be promoting a product or agenda. Advertisements for a specific product should alert you to a Web site's bias.
3. Check the date when the Web site was last updated, and whether the page has been updated since its original posting. Science and medicine are disciplines that must frequently incorporate new data into hypotheses. A reliable Web site will be updated often.
4. Determine whether unsubstantiated claims are being made. Look for references, and be suspicious of any studies that are not from peer-reviewed journals.

Understanding Science from Secondary Sources

Once you are satisfied that a media source is relatively reliable, examine the scientific claim that it is presenting. Begin by using your understanding of experimental design to evaluate what is being presented. Does the story about the claim present the results of a scientific study, or is it built around an untested hypothesis? Is the story confusing correlation with causation? Does it seem that the information is applicable to non-laboratory situations, or is it based on results from test-tube or animal studies? Look for clues about how well the reporters did their homework. Scientists usually discuss the limitations of their research in their papers; are these cautions noted in an article or television piece? If not, the reporter may be overemphasizing the applicability of the results.

Then, note if the scientific discovery itself is controversial. That is, does it reject a hypothesis that has long been supported? Does it concern a subject that is controversial in human society (like racial differences or homosexuality)? Might it lead to a change in social policy? In these cases, be extremely cautious. New and unexpected research results must be evaluated in light of other scientific evidence and understanding. Reports that lack comments from other experts in related fields may omit important problems with a study, or fail to place the study in context with other research.

Finally, realize that even among the most credible organizations, the news media generally highlights only stories about experiments that editors and producers find newsworthy (see Essay 1.1). As we have seen, scientific understanding accumulates relatively slowly, with many tests of the same hypothesis finally leading to the "truth." News organizations are also more likely to report a study that supports a hypothesis rather than one that gives less supportive results, even if both types of studies exist. And even the most respected media sources may not be as thorough as readers would like. For example, a recent review published in the *New England Journal of Medicine* evaluated the news media's coverage of new medications. Of 207 randomly selected news stories, only 40% that cited experts who had financial ties to a drug disclosed this relationship. This potential conflict of interest may influence how credible the expert is. Another 40% of the news stories did not give a numerical analysis of the drugs' benefits. The majority of news reports also failed to distinguish between the absolute benefits (how many people were helped by the drug), and relative benefits (how many people were helped by the drug relative to other therapies for the condition). The *Journal*'s review reminds us that we need to be cautious when reading or viewing news reports on scientific topics.

Even after you have followed all of these guidelines, you still may find situations where reports on several scientific studies seem to give conflicting and confusing results. This could mean one of two things: Either the reporter is not giving you enough information, in which case you may want to read the researchers' papers yourself, or the researchers themselves are just as confused as you are. This is part of the nature of the scientific process—early in our search for understanding of a phenomenon, many hypotheses are proposed and discussed, some are tested and rejected immediately, and some are supported by one experiment but later rejected by more thorough experiments. It is only by clearly understanding the process and pitfalls of scientific research that you can distinguish "what we know" from "what we don't know."

1.3 Is There a Cure for the Common Cold?

So where does our discussion leave Jake? Will he ever find the best way to prevent a cold or reduce its effects? In the United States over one billion cases of the common cold are reported per year, costing billions of dollars in medical visits, treatment, and lost work days. Consequently, there is an enormous effort to find effective protection from the different viruses that cause colds. Despite all of the research and the emergence of some promising possibilities, the best prevention method is still the old standby—keep your hands clean. Numerous studies have indicated that rates of common-cold infection are 20–30% lower in populations who employ effective hand-washing procedures. Cold viruses can survive on surfaces for many hours; if you pick them up from a surface on your hands and transfer them to your mouth, eyes, or nose, you may inoculate yourself with a seven-day sniffle.

Of course, not everyone gets sick when exposed to a cold virus. The reason Jake has more colds than his lab partner might not be because of a difference in personal hygiene. The correlation that showed a relationship between stress and cold susceptibility appears to have some merit. Research indicates that among people exposed to viruses, the likelihood of ending up with an infection increases with high levels of psychological stress—something that Jake is clearly experiencing. Research also indicates that vitamin C intake, diet quality, exposure to cold temperatures, and exercise frequency appear to have *no effect* on cold susceptibility, although, along with echinacea tea and zinc lozenges, there is some evidence that vitamin C may reduce cold symptoms after infection. Table 1.1 summarizes our current understanding of the factors that may prevent and minimize the effects of infection with a common cold virus. Surprisingly, scientists are still a long way from "curing" the common cold.

Factors that **do** affect cold susceptibility	Factors that **do not** affect cold susceptibility	Factors that shorten cold duration
• Exposure to cold virus	• Vitamin C	• Zinc lozenges (?)
• Psychological stress	• Diet quality	• Vitamin C (?)
• Hand washing	• Exercise	• Echinacea tea (?)
	• Exposure to cold temperatures	

Table 1.1 Has science cured the common cold? A summary of the current state of knowledge about factors that may increase cold susceptibility and decrease cold duration. Question marks denote that not all scientists agree. As you can see, the scientific effort to cure, or at least minimize the effects of, the common cold is far from over.

So, as Jake reviews scientists' careful research on the prevention of colds, he will find that he can forgo the vitamin C supplements, remain fashionably mittenless, and continue eating fries with his chicken sandwiches without affecting his chances of getting another cold. But Jake will also learn that he should keep his hands clean and maybe drop an activity from his schedule if he wants to stay healthy. He feels better already.

CHAPTER REVIEW

Summary

- Science is a process of testing statements about how the natural world works—called hypotheses. Scientific hypotheses must be testable and falsifiable. Hypotheses are tested via the process of deductive reasoning, which allows researchers to make specific predictions about expected observations. Absolutely proving hypotheses is impossible. However, well-designed scientific experiments allow researchers to strongly infer that their hypothesis is correct.
- Controlled experiments test hypotheses about the effect of experimental treatments by comparing a randomly assigned experimental group with a control group. Controls are individuals who are treated identically to the experimental group except for application of the treatment. Bias in scientific results can be minimized with double-blind experiments that keep subjects and data collectors unaware of which individuals belong in the control or experimental group.
- Some hypotheses about human health are difficult to test with experiments. These hypotheses may be tested using a correlational approach, which looks for associations between two factors. A correlation can show a relationship between two factors, but it does not eliminate all alternative hypotheses.
- Statistics help scientists evaluate the results of their experiments, by determining if results appear to reflect the true effect of an experimental treatment on a sample of a population. A statistically significant result is one that is very unlikely to be due to chance differences between the experimental and control group. A statistical test indicates the role chance plays in the experimental results; this is called sampling error. Even when an experimental result is highly significant, hypotheses are tested multiple times before scientists come to consensus on the true effect of a treatment.
- Primary sources of information are experimental results published in professional journals and reviewed by other scientists before publication. Most people get their scientific information from secondary sources, such as the news media. Being able to evaluate science from these sources is an important skill. Anecdotal evidence is an unreliable means of evaluating information, and media sources are of variable quality—distinguishing between news stories and advertisements is important when evaluating the reliability of information. The Internet is a rich source of information, but users should look for clues to a particular Web site's credibility.
- Stories about science should be carefully evaluated for information on the actual study performed, the universality of the claims made by the researchers, and other studies on the same subject. Sometimes confusing stories about scientific information are a reflection of controversy within the scientific field itself.

Key Terms

anecdotal evidence p. 14
control p. 6
correlation p. 10
data p. 5
deductive reasoning p. 4
double blind p. 9
experiment p. 5
falsifiable p. 2
hypotheses p. 2
primary sources p. 14
random assignment p. 7
sample p. 12
sampling error p. 12
secondary sources p. 14
statistically significant p. 13
statistical test p. 12
statistics p. 12
testable p. 2

Learning the Basics

1. What characteristics distinguish a hypothesis that is testable by science?

2. What is a controlled experiment?

3. How does double-blinding decrease the amount of bias introduced into experimental results?

4. What does statistical significance mean?

5. What are the advantages and disadvantages of using correlations to test hypotheses?

6. A scientific hypothesis is _____.
   - **a.** an opinion
   - **b.** a proposed explanation for an observation
   - **c.** a fact
   - **d.** easily proved true
   - **e.** an idea proposed by a scientist

7. Which of the following is a prediction of the hypothesis: Eating chicken noodle soup is an effective treatment for colds?
   - **a.** People who eat chicken noodle soup have shorter colds than people who do not eat chicken noodle soup.
   - **b.** People who do not eat chicken noodle soup experience unusually long and severe colds.
   - **c.** Cold viruses cannot live in chicken noodle soup.
   - **d.** People who eat chicken noodle soup feel healthier than people who do not eat chicken noodle soup.
   - **e.** Consuming chicken noodle soup causes people to sneeze.

8. When both the subjects in an experiment and the technicians who are measuring and recording data do not know which individuals are in the experimental group and which are in the control group, we call the experiment _____.
   - **a.** controlled
   - **b.** a placebo
   - **c.** biased
   - **d.** double-blind
   - **e.** falsifiable

9. Control subjects in an experiment _____.
   - **a.** should be similar in most ways to the experimental subjects
   - **b.** should not know whether they are in the control or experimental group
   - **c.** should have essentially the same interactions with the researchers as the experimental subjects
   - **d.** help eliminate alternative hypotheses that could explain experimental results.
   - **e.** all of the above

10. A relationship between two factors, for instance between outside temperature and number of people with active colds in a population, is known as a(n) _____.
    - **a.** significant result
    - **b.** correlation
    - **c.** hypothesis
    - **d.** alternative hypothesis
    - **e.** experimental test

11. If the results of an experiment are exactly what was predicted by the hypothesis _____.
    - **a.** the hypothesis is proved
    - **b.** the alternative hypotheses are falsified
    - **c.** the hypothesis is supported
    - **d.** the hypothesis was scientific
    - **e.** none of the above

12. Statistical tests tell us _____.
    - **a.** if an experimental treatment showed more of an effect than would be predicted by chance
    - **b.** if a hypothesis is true
    - **c.** whether an experiment was well designed
    - **d.** if the experiment suffered from any bias
    - **e.** how similar the sample was to the population it was drawn from

13. A primary source of scientific results is _____.
    - **a.** the news media
    - **b.** anecdotes from others
    - **c.** articles in peer-reviewed journals
    - **d.** the Internet
    - **e.** all of the above

14. A celebrity promoting a product, saying "It worked for me," is an example of a(n) _____.
    - **a.** anecdote
    - **b.** primary source
    - **c.** unbiased source of information
    - **d.** peer-reviewed claim
    - **e.** scientific test

Analyzing and Applying the Basics

1. Which of the following statements are written as scientific hypotheses? (If they are not, can you revise them to be testable and falsifiable statements?)

   People from Minnesota are better than people from North Dakota.

   People from Minnesota are more favored by God than people from Iowa.

   People from Minnesota have larger diameter heads than people from Michigan.

   People from Minnesota like snow more than do people from Wisconsin.

2. There is a strong correlation between obesity and the occurrence of a disease known as Type II diabetes—that is, obese individuals have a higher instance of diabetes than non-obese individuals. Does this mean that obesity causes diabetes? Explain.

3. To test the hypothesis that changes occurring in boys' brains before birth make them better at math than girls, researchers gave a large sample of eighth-grade boys and girls a math test. Boys did significantly better than girls on the test. Can the researchers strongly infer the truth of their hypothesis? Explain.

4. In an experiment on the effect of vitamin C on reducing the severity of cold symptoms, college students visiting their campus health service with early cold symptoms either received vitamin C or treatment with over-the-counter drugs. Students then reported upon the length and severity of their colds. The timing of dosages and the type of pill were very different, thus both the students and the clinic health providers knew which treatment they were receiving. This study reported that vitamin C significantly reduced the length and severity of colds experienced in this population. Why might this result be questionable, given the experimental design?

5. Samuel George Morton published data in the 1840s reporting differences in brain size among human races. His research indicated the Europeans had larger brains than Native Americans and Africans. His measures of brain size were based on skull volume calculated by packing individual skulls with mustard seed and then measuring the volume of the seeds they contained. When the biologist Stephen Jay Gould reexamined Morton's data in the 1970s, he found that Morton systematically erred in his measurement—consistently underestimating the size of the African and Native American skulls. According to Gould, Morton appeared not to realize that he was affecting his own results to support his hypothesis that Europeans had larger brains than the other groups. How could Morton have designed his experiment to minimize the effect of this bias on his results?

Connecting the Science

1. Do you think that reporters should be required to give more complete information in stories about research in health and science, or do you think it is up to the public to be able to critically analyze media reports on these subjects?

2. Much of the research on common cold prevention and treatment is performed by scientists employed or funded by drug companies. Often these companies do not allow scientists to publish the results of their research for fear that competitors at other drug companies will use this research to develop a new drug before they do. Should our society allow scientific research to be owned and controlled by private companies?

3. Should society put restrictions on what kinds of research are performed by government-funded scientists? For example, many people believe that there should be restrictions on research performed on tissues from human fetuses, because they believe that this research would justify abortion. If a majority of Americans feel this way, should government avoid funding this research? Are there any risks associated with *not* funding research with public money?

Media Activities

Media Activity 1.1 Hypothesis Testing
Estimated Time: 5 minutes
The animation illustrates the process of hypothesis testing and experimental design using historical experiments as examples.

Media Activity 1.2 The Placebo Effect: Is It Real?
Estimated Time: 10 minutes
Explore the issue of the Placebo Effect and the controversy behind it.

Media Activity 1.3 Evaluating Health Information from the World Wide Web
Estimated Time: 10 minutes
Explore the science of vitamin C, the evidence of its effectiveness in fighting colds, and the validity of information found on the Internet.

CHAPTER 2

The Only Diet You Will Ever Need

Cells and Metabolism

There is a high premium placed on thinness in our culture.

2.1 Nourishing Your Body

2.2 Converting Food into Energy

2.3 Body Fat and Health

Most college students are, for the first time, making all their own choices about food.

Making unhealthy choices now can lead to future struggles with obesity...

...or anorexia.

The average college student spends a lot of time thinking about his or her body and ways to make it more attractive. While most people realize that there are a wide range of body types and sizes, attractiveness tends to be more narrowly defined by the images of men and women we see in the popular media; and nearly all media images equate attractiveness and desirability with this limited range of body types.

For men, the image of a tall, broad-shouldered, muscular man with so little body fat that every muscle is visible is portrayed as most desirable.

The standards for female beauty are unforgiving also. Images of female beauty are almost exclusively women with small hips; long, thin limbs; large breasts; and no body fat. This is virtually the only image of female beauty seen in fashion magazines, on billboards, on television, and in the movies.

Into this milieu steps the average college student—worried about appearance, trying to find time to study, exercise, and socialize and now making all his or her own decisions about food, often on a limited budget.

Making choices that are good for long-term health is not easy. The typical dining-center meal-plan choices, often greasy and fat-laden, are available in unlimited portions. The difficulty of making healthful choices is compounded by the presence of campus snack shops, vending machines, and conveniently located fast-food restaurants that offer time-pressed students easily accessible, inexpensive foods containing little nutrition.

Coupling tremendous pressure to be thin with a glut of readily available unhealthful foods can lead to the establishment of unhealthful eating habits that persist far beyond college life. In many cases, these conflicting pressures can lead students to develop eating habits that result in a lifelong battle with obesity or starvation, along with their associated health risks.

Learning about the kinds and amounts of foods you should be eating, and understanding how much body fat is right for you, will help you make good decisions about eating that will set you up for a lifetime of good health.

WWW
Media Activity 2.2 Personal Diet Analysis

2.1 Nourishing Your Body

To achieve a healthful diet you must eat the proper *balance* of different kinds of foods so that your body has the raw materials it requires to build, maintain, and repair its basic units, the **cells**. You also need to eat the right *amount* of food so that you have enough energy to power your daily activities without having too much or too little stored as fat.

Energy is measured in units called *calories*. A calorie is the amount of energy required to raise the temperature of one gram of water 1° Celsius. In scientific literature, energy is usually reported in kilocalories, and one kilocalorie equals 1000 calories of energy. However in physiology, the prefix *kilo* is dropped, and a kilocalorie is referred to as a **Calorie** (with a capital C).

Balancing Nutrients

Nutrients are substances found in food that provide the energy or structural materials required for normal growth, maintenance, and repair. These include carbohydrates, proteins, fats, vitamins, minerals, and fiber.

Nutritionists have devised a set of general guidelines to help determine whether you are achieving the right balance of nutrients. These guidelines are based on six different food groups: (1) milk and dairy products; (2) meats; (3) vegetables; (4) fruits; (5) breads, cereals, and grains; and (6) fats, oils, and sweets. As Figure 2.1 shows, these food groups are arranged as a pyramid. The amount of space allocated for each group represents the relative amount of your diet they should comprise. The largest is taken up by breads, cereals, and grains, which are rich in carbohydrates.

Carbohydrates Foods such as bread, cereal, rice, and pasta, as well as fruits and vegetables, are rich in sugars called **carbohydrates**. Carbohydrates are the major source of energy for cells. A healthful diet obtains up to 60% of its Calories from carbohydrate sources.

Carbohydrates are composed of carbon, hydrogen, and oxygen in the ratio CH_2O. Glucose is $6(CH_2O)$ or $C_6H_{12}O_6$. Glucose is a simple sugar, or **monosaccharide**, which consists of a single ring-shaped structure. **Disaccharides** are two rings joined together—lactose, found in milk, is a disaccharide composed of glucose and galactose. The sugar you bake with or sprinkle on cereal is sucrose, a disaccharide composed of glucose and fructose (Figure 2.2).

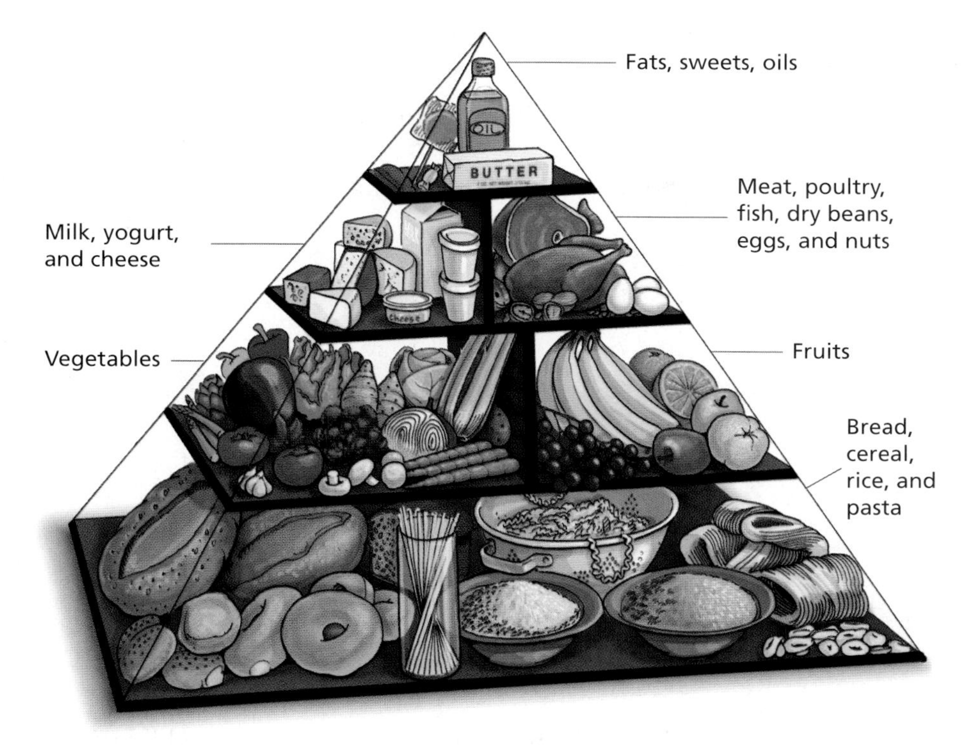

Figure 2.1 Food-guide pyramid. This pyramid represents the most recent recommendations by U.S. government nutritionists for healthful eating.

Joining many individual subunits, or *monomers*, together produces *polymers* (*poly* means "many"). Polymers of sugar monomers are called **polysaccharides**—these multi-subunit sugars can be composed of many different branching chains of sugar monomers and are also called **complex carbohydrates**.

Complex carbohydrates are often involved in storing energy for later use. Plants, such as potatoes, store their excess carbohydrates as polymers of *starch*. Animals store their excess carbohydrates as *glycogen* in muscles and the liver. Both starch and glycogen are polymers of glucose (Figure 2.3).

Nutritionists recommend that carbohydrates in a healthful diet be mostly in the form of complex carbohydrates, and consumption of refined and processed sugars should be minimized. The complex carbohydrates you find in fruits, vegetables, and grains also contain many vitamins and minerals, as well as fiber—the refined sugars that you find in processed foods and sweets are a source of Calories, but do not supply vitamins, minerals, or fiber.

Dietary fiber, also called roughage, is composed mainly of complex carbohydrates that humans cannot digest into component monosaccharides. For this

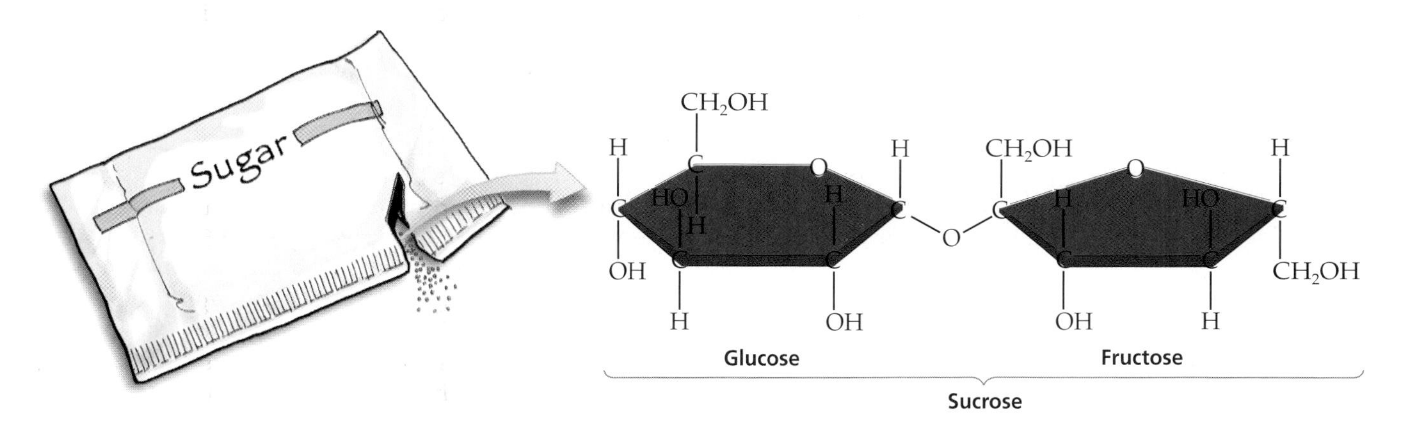

Figure 2.2 Sucrose. Sucrose is a disaccharide formed when two monosaccharides, glucose and fructose, are joined.

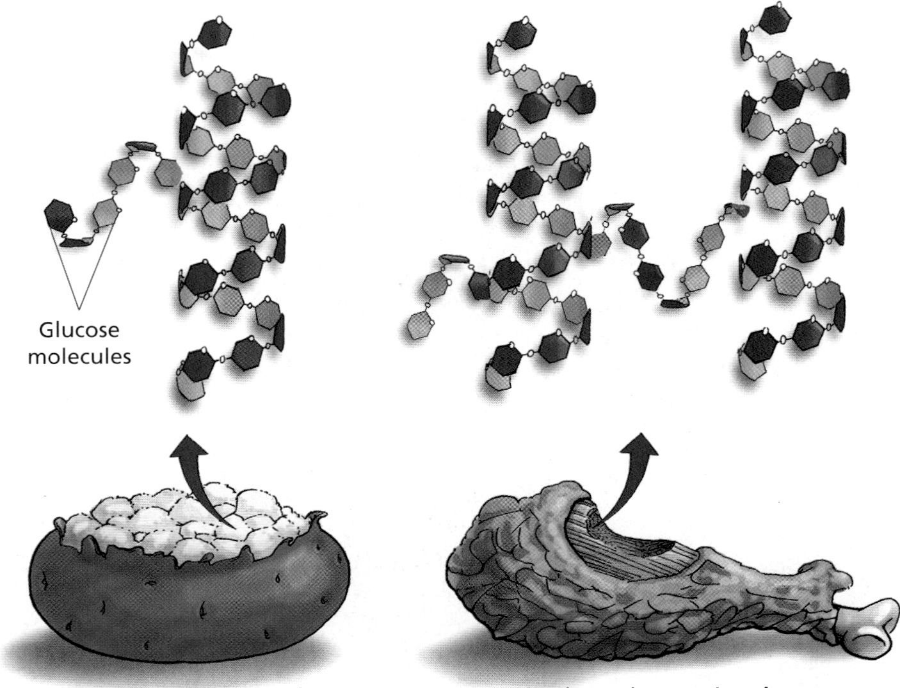

Figure 2.3 Starch and glycogen. Plants store their excess carbohydrates as starch, and animals store excess carbohydrates as glycogen. The glucose molecules are abbreviated as hexagons here.

reason, dietary fiber is passed into the large intestine; some fiber is digested by bacteria living there, and the remainder gives bulk to the feces.

Although fiber is not digested, it is still an important part of a healthful diet. It lowers total cholesterol without changing the level of HDL, or "good" cholesterol while lowering LDL, the "bad" form of cholesterol-carrying molecules. (Cholesterol is discussed more fully in Section 2.3.) Fiber may also decrease your risk of various cancers. Fruits and vegetables tend to be rich in dietary fiber.

Proteins While carbohydrates are located at the base of the food pyramid, *proteins* occupy the smaller space in the middle of the pyramid. Protein-rich foods include beef, poultry, fish, beans, eggs, nuts, and dairy products such as milk, yogurt, and cheese. Your body requires proteins for a wide variety of processes. **Proteins** are important structural components of cells. In fact, they make up half the dry weight of a cell—muscles are largely composed of proteins. Proteins called *enzymes* help regulate all the chemical reactions that build up and break down molecules inside your cells. Proteins also function as hormones that send chemical messages throughout your body. A healthful diet will obtain about 15% of its Calories from protein.

Proteins are large molecules made of monomer subunits called **amino acids**. There are 20 commonly occurring amino acids. Like carbohydrates, amino acids are made of carbons, hydrogens, and oxygens, but they have an additional amino (NH_3^+) group and various *side groups*. Side groups are chemical groups that give amino acids different chemical properties (Figure 2.4a).

Polymers of amino acids are sometimes called *polypeptides* because the name for the chemical bond joining adjacent amino acids is a *peptide bond* (Figure 2.4b). Different amino acids are joined together to produce different proteins in much the same manner that children can use differently shaped beads to produce different structures (Figure 2.4c). Each amino acid has unique chemical properties. Since different proteins are composed of different amino acids, each protein has unique chemical properties. Your body synthesizes most of the amino acids it needs. Those your body cannot synthesize are called *essential*

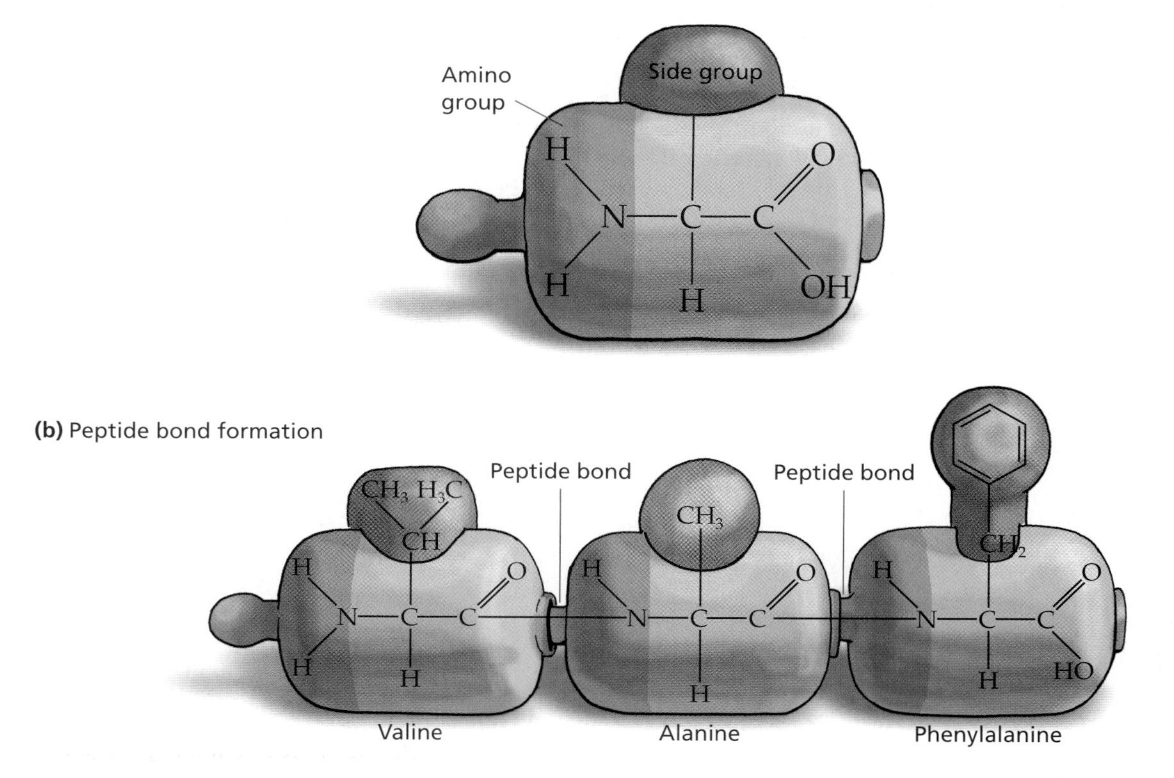

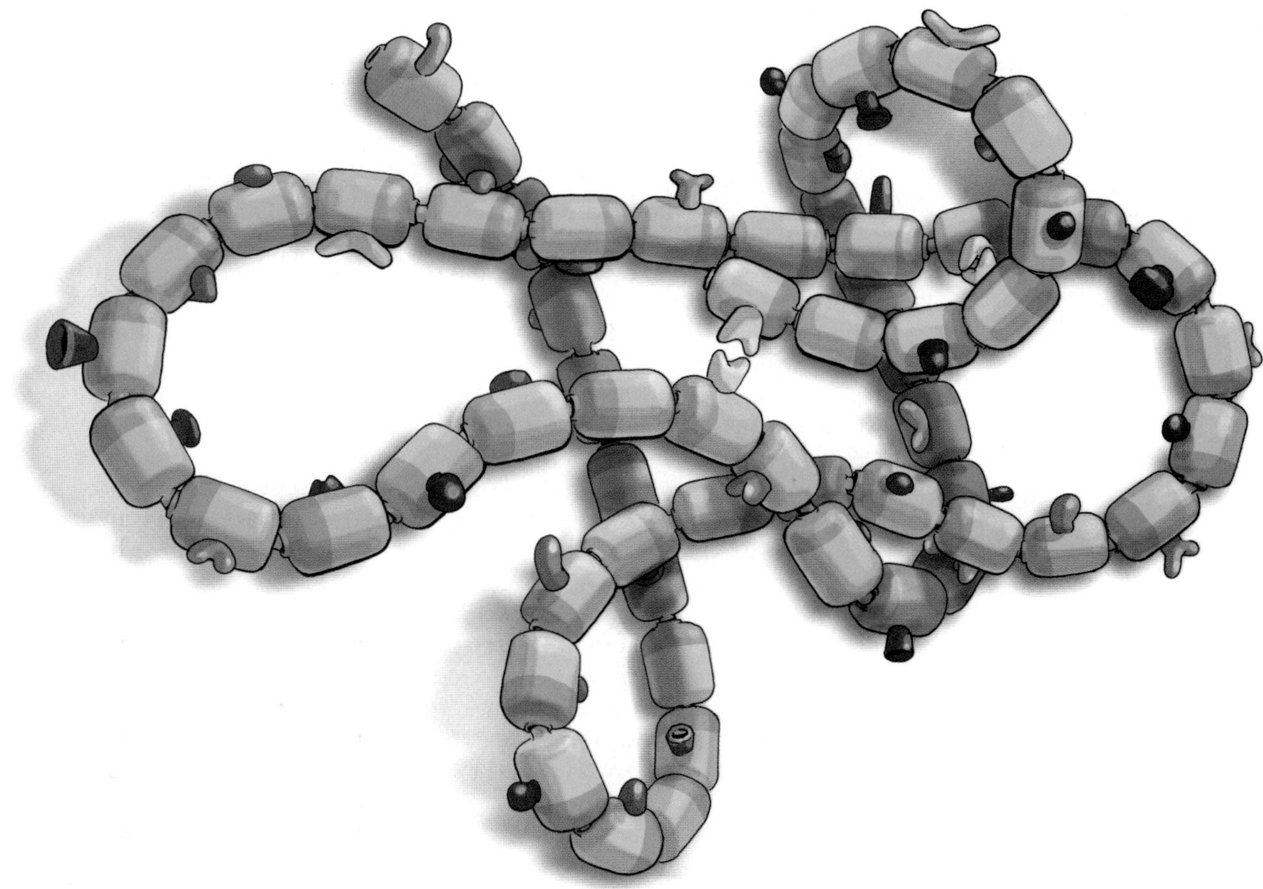

Figure 2.4 Amino acids. (a) All amino acids have the same general formula, but different side groups. (b) Amino acids are joined together by chemical bonds called peptide bonds. Long chains of these are called polypeptides. (c) Polypeptide chains fold upon themselves to produce proteins, and different combinations of amino acids produce distinct proteins.

(a) Fat within muscle

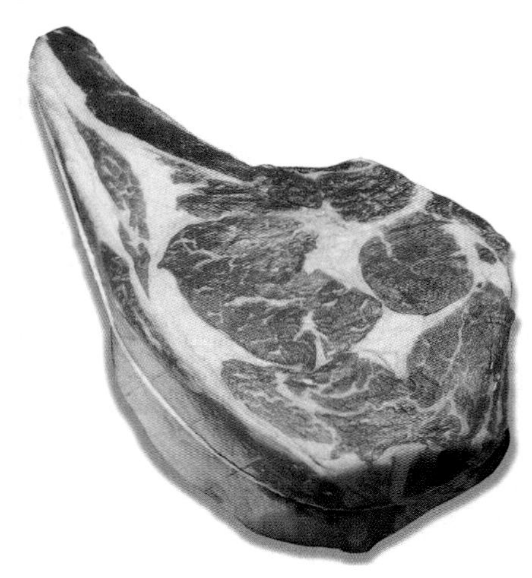

(b) Fat on surface of muscle

Figure 2.5 Fat storage. Fat can be intertwined with muscle tissue, as seen in this marbled piece of beef (a), or on the surface, as seen on this chicken breast (b).

amino acids and must be supplied by the foods you eat. *Complete proteins* contain all the essential amino acids your body needs—proteins obtained by eating meat are more likely to be complete than those obtained by eating plants; plant proteins can often be missing one or more essential amino acids.

In the past, some nutritionists believed that vegetarians might be at risk for deficiencies in certain amino acids. However, scientific studies have shown that there is little cause for concern. If a vegetarian's diet is rich in a wide variety of plant-based foods, the body will have little trouble obtaining all the amino acids it needs to build proteins.

A more contemporary issue is whether vegetarians are deficient in vitamin B_{12}. This vitamin is required for the production of red blood cells and helps maintain proper nerve function. Vegetarians can increase their B_{12} intake by drinking soy milk or taking vitamin supplements.

Even though it is possible to obtain all the proteins you need by eating a variety of plants, Americans tend to eat a lot of meat as well. In addition to being rich in proteins, meat tends to be rich in *fat*. A quick glance at Figure 2.1 shows that the dietary requirements for fat are minimal.

Fats The body uses **fat** as a source of energy. Gram for gram, fat contains a little more than twice as much energy as carbohydrate and protein. Foods that are rich in fat include meat, milk, cheese, vegetable oils, and nuts. Muscle is often surrounded by stored fat—some animals store fat throughout muscle, leading to the marbled appearance of some red meat. Others, chickens for example, store it on the surface of the muscle where it can be more easily removed for cooking (Figure 2.5). Humans store fat just below the skin to help insulate the body from cold weather, and to store energy in case of famine. Some scientists believe that prehistoric humans often faced times of famine and may have evolved to crave fat. Another benefit of fat is its ability to help cushion and protect vital organs.

Fat, like carbohydrate and protein, contains carbon, hydrogen, and oxygen, but in different proportions. The structure of a fat is that of a three-carbon glycerol molecule with up to three long chains of hydrogens and carbons, called *hydrocarbons*, attached to it. The long hydrocarbon chains are called *fatty acid tails* of the fat molecule (Figure 2.6a). The fatty acid tails are unable to dissolve in water because the hydrogens and carbons are too busy interacting with each other to form bonds with water molecules. This leads to one of the characteristic properties of fat, its inability to dissolve in water.

The carbon atoms in fats—and in all molecules—are able to make chemical bonds with up to four other atoms. Each carbon in the fatty acid tail is bound to two other carbons, one in front of it and one behind it in the hydrocarbon chain. When a particular carbon is double-bonded to one of these carbons, it can only make one other bond to a hydrogen. When the carbon is single-bonded to the carbon on either side of it, it can make bonds with two hydrogen atoms.

When carbons are bound to as many hydrogens as possible, it is said to be a **saturated fat** (saturated in hydrogens). When there are carbon–carbon double bonds, the fat is not saturated in hydrogens, and it is therefore an **unsaturated fat** (Figure 2.6b). When there are many unsaturated carbons, the fat is a polyunsaturated fat.

The double bonds in unsaturated fats make the structures kink instead of lying flat. This prevents the adjacent fat molecules from packing tightly together, so unsaturated fat tends to be liquid at room temperature. Cooking oil is an example of an unsaturated fat. Unsaturated fats are more likely to come from plant sources. Saturated fats, with their absence of carbon–carbon double bonds, do pack tightly together to make a solid structure. This is why saturated fats, such as butter, are solid at room temperature (Figure 2.6c).

Commercial food manufacturers sometimes add hydrogen bonds to unsaturated fats by adding hydrogen gas to vegetable oils under pressure. This process retards spoilage and solidifies liquid oils, thereby making food seem less greasy. Margarine is vegetable oil that has undergone this *hydrogenation* process.

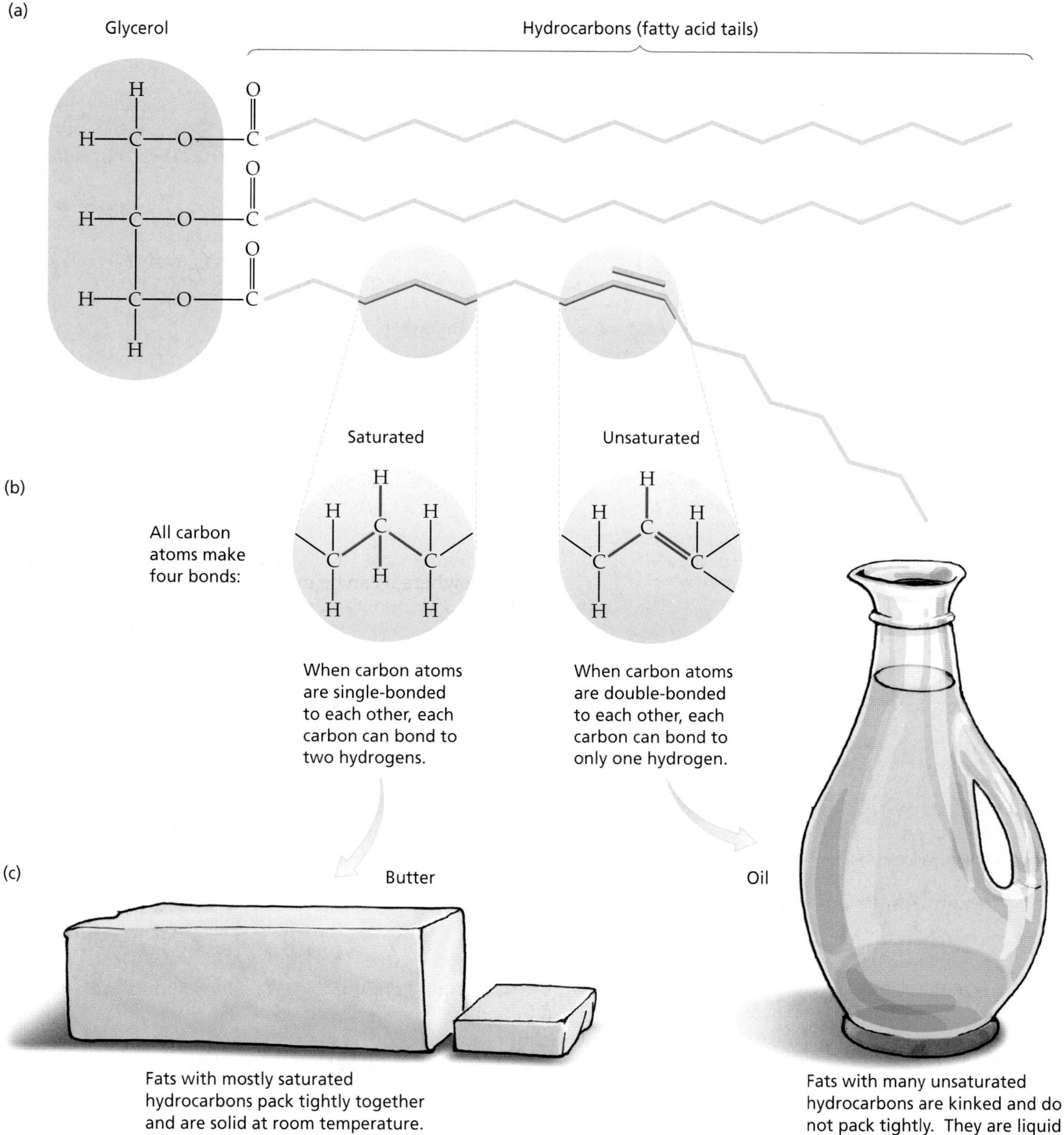

Figure 2.6 Fat structure. (a) Fats are long chains of hydrocarbons (fatty acids) attached to a 3-carbon skeleton called glycerol. (b) Carbon can make chemical bonds with up to four other atoms. The carbon atoms in a saturated fat are bonded to four other atoms. The carbon atoms in an unsaturated fat are double-bonded to other carbon atoms. (c) Saturated fats are solid at room temperature. Vegetable oil, an unsaturated fat, is liquid at room temperature.

When hydrogen atoms are added to the same side of the carbon–carbon double bond, they are said to be in the *cis* configuration—naturally occurring unsaturated fats have their hydrogen atoms in *cis* configuration. When hydrogen atoms are added on opposite sides of the double bond, they are said to be in the *trans* configuration. During hydrogenation, there is no way to control whether hydrogen

Water-soluble vitamins

- Small organic molecules (containing carbon)
- Will dissolve in water
- Cannot be synthesized by body
- Supplements packaged as pressed tablets
- Excesses usually not a problem since water-soluble vitamins are excreted in urine, not stored.

Vitamin	Sources	Functions	Effects of Deficiency
Thiamin (B_1)	Pork, whole grains, leafy green vegetables	Required component of many enzymes	Water retention and heart failure
Riboflavin (B_2)	Milk, whole grains, leafy green vegetables	Required component of many enzymes	Skin lesions
Folic Acid	Dark green vegetables, nuts, legumes (dried beans, peas, and lentils), whole grains	Required component of many enzymes	Neural-tube defects, anemia, and gastrointestinal problems
B_{12}	Chicken, fish, red meat, dairy	Required component of many enzymes	Anemia and impaired nerve function
B_6	Red meat, poultry, fish, spinach, potatoes, and tomatoes	Required component of many enzymes	Anemia, nerve disorders, and muscular disorders
Pantothenic acid	Meat, vegetables, grains	Required component of many enzymes	Fatigue, numbness, headaches, and nausea
Biotin	Legumes, egg yolk	Required component of many enzymes	Dermatitis, sore tongue, and anemia
C	Citrus fruits, strawberries, tomatoes, broccoli, cabbage, green pepper	Collagen synthesis, improves iron absorption	Scurvy and poor wound healing
Niacin (B_3)	Nuts, leafy green vegetables, potatoes	Required component of many enzymes	Skin and nervous system damage

Table 2.1 Vitamins. A variety of water- and fat-soluble vitamins perform many functions.

atoms are added in the *cis* or *trans* configuration. The long-term health risks of consuming foods rich in *trans*-fatty acids, common in fast foods, are unknown.

Since fat contains more Calories per gram than carbohydrate and protein, and because excess fat intake is associated with several diseases, nutritionists recommend that you limit the number of Calories obtained from fat to less than 25% of your daily intake.

Along with fat, sweets (highly processed, sugar-rich foods) occupy a very small portion of the food pyramid since both foods provide no real nutrition, just Calories. Sweets are often referred to as "empty" Calories because consuming them provides Calories but no vitamins and minerals.

Vitamins and Minerals **Vitamins** are organic substances (*organic* means "carbon containing") that are only required in small amounts. Vitamins are not destroyed during use, nor are they burned for energy. Most function as **coenzymes**, molecules that help enzymes, thus speeding up the body's chemical reactions. When a vitamin is not present in sufficient quantities, deficiencies can result. These deficiencies can affect every cell in your body because many different enzymes, all requiring the same vitamin, are involved in many different bodily functions. Vitamins also help with the absorption of other nutrients; for example,

Fat-soluble vitamins

- Small organic molecules (containing carbon)
- Will not dissolve in water
- Cannot be synthesized by body (except Vitamin D)
- Supplements packaged as oily gel caps
- Excesses can cause problems since fat-soluble vitamins are not excreted readily

Vitamin	Sources	Functions	Effects of Deficiency	Effects of Excess
A	Leafy green and yellow vegetables, liver, egg yolk	Component of eye pigment	Night-blindness, scaly skin, skin sores, and blindness	Drowsiness, headache, hair loss, abdominal pain, and bone pain
D	Milk, egg yolk	Helps calcium be absorbed, and increases bone growth	Bone deformities	Kidney damage, diarrhea, and vomiting
E	Dark green vegetables, nuts, legumes, whole grains	Required component of many enzymes	Neural-tube defects, anemia, and gastrointestinal problems	Fatigue, weakness, nausea, headache, blurred vision, and diarrhea
K	Leafy green vegetables, cabbage, cauliflower	Helps blood clot	Bruising, abnormal clotting, and severe bleeding	Liver damage, and anemia

Table 2.1 ***(continued)***

vitamin C increases the absorption of iron from the intestine. Some vitamins may even help protect the body against cancer and heart disease, and may slow the aging process.

Vitamin D is the only vitamin your cells can synthesize. (Sunlight is required for synthesis, therefore people living in cold climates can develop deficiencies in vitamin D.) All other vitamins must be supplied by the foods you eat. Many vitamins are water-soluble, so boiling causes them to leach out into the water—this is why it is a good idea to use fresh vegetables or include the vitamin-rich broth of canned vegetables when making soup. Steaming vegetables is a better way of cooking them because it preserves their vitamin content. The water-soluble vitamins are more likely than fat-soluble vitamins to be the source of dietary deficiencies since the body does not store them. Vitamins A, D, E, and K are fat soluble and build up in stored fat—allowing an excess of these vitamins to accumulate in the body can be toxic. Table 2.1 lists some vitamins and their roles in the body.

Minerals are substances that do not contain carbon, but are essential for many cell functions—because they lack carbon, they are said to be *inorganic*. Minerals are important for proper fluid balance, muscle contraction, conduction of nerve impulses, and building bones and teeth. Calcium, chloride, magnesium, phosphorus, potassium, sodium, and sulfur are all minerals. Like some vitamins, minerals are water soluble and are lost during boiling. Also like vitamins, minerals aren't synthesized in the body and must be supplied through your diet. Table 2.2 lists the various functions of minerals your body requires and what happens when there is a deficiency or an excess in various minerals.

Processed vs. Whole Foods It is best to limit your consumption of *processed foods* in general. Food that has undergone extensive processing has been stripped of much of its nutritive value. For example, refined flour has had

Minerals

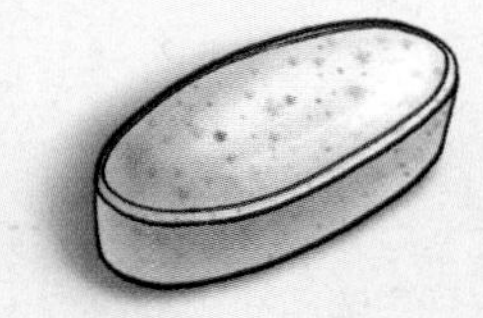

- Will dissolve in water
- Inorganic elements (do not contain carbon)
- Cannot be synthesized by body
- Supplements packaged as pressed tablets

Mineral	Sources	Functions	Effects of Deficiency	Effects of Excess
Calcium	Milk, cheese, dark green vegetables, legumes	Bone strength, blood clotting	Stunted growth, osteoporosis	Kidney stones
Chloride	Table salt, processed foods	Formation of HCl in stomach	Muscle cramps, reduced appetite, poor growth	High blood pressure
Magnesium	Whole grains, leafy green vegetables, legumes, dairy, nuts	Required component of many enzymes	Muscle cramps	Neurologic disturbances
Phosphorus	Dairy, red meat, poultry, grains	Bone and tooth formation	Weakness, bone damage	Impaired ability to absorb nutrients
Potassium	Meats, fruits, vegetables, whole grains	Water balance, muscle function	Muscle weakness	Muscle weakness, paralysis, and heart failure
Sodium	Table salt, processed foods	Water balance, nerve function	Muscle cramps, reduced appetite	High blood pressure
Sulfur	Meat, legumes, milk, eggs	Components of many proteins	None known	None known

Table 2.2 Minerals. The minerals we require and their roles in the body.

the bran and germ removed during processing, resulting in the loss of many vitamins and minerals, and much of the fiber.

Foods that have not been stripped of their nutrition by processing are called *whole foods*. Eating a wide variety of whole foods such as fruits, vegetables, and grains will provide you with a much better chance of achieving a healthful diet than eating highly refined, fatty foods, low in complex carbohydrates and vitamins—also called "junk food."

In addition to eating a well-balanced diet rich in unprocessed foods, it is also important to eat the right amount of food. All food, whether carbohydrate, protein, or fat, can be turned into fat when too much is consumed. On the other hand, not consuming enough Calories is the source of many serious health problems.

Balancing Energy

Balancing energy means eating the correct amount of food to maintain health. When foods are eaten, they are broken down into their component subunits and can be used to make a form of energy the cell can use (Figure 2.7). Cells power their activities by using a chemical called *ATP* as their energy currency. (ATP is discussed in detail in Section 2.2.) When the supply of Calories is greater than the demand, the excess Calories are stored by the body as fat.

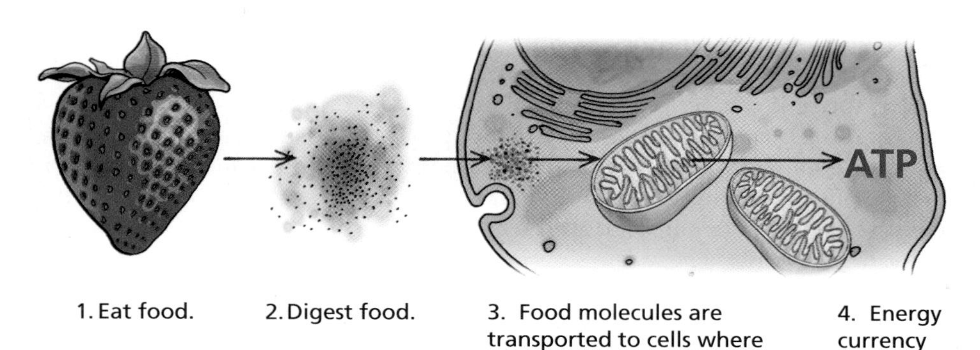

Figure 2.7 Digestion. The food you eat is broken down by the digestive system and transported into cells, where nutrients are used to make energy (ATP).

WWW
Media Activity 2.1A Overview of Energy Acquisition

The amount of fat a given individual will store is partly dependent upon the rate of their metabolism. **Metabolism** is a general term used to describe all of the chemical reactions occurring in the body, but here we are more concerned with *metabolic rate* and its effect on a healthy diet.

Metabolic Rate The **metabolic rate** of an individual is a measure of a person's energy use. This rate changes according to the activity level of the individual. For example, a person requires less energy when asleep than he or she does when exercising. The **basal metabolic rate** represents the resting energy use of an awake, alert person. The average basal metabolic rate is 70 Calories per hour or 1680 Calories per day. However, this is only an estimate, because many factors influence a given individual's basal metabolic rate. These include exercise habits, body weight, gender, age, and genetics.

Exercise requires energy, which allows you to consume more Calories without having to store them. As for body weight, a heavy person utilizes more Calories during exercise than a thin person does. Figure 2.8 shows the number of Calories used per hour for various activities based on body weight.

Males require more Calories per day than females because testosterone, a hormone produced in larger quantities by males, increases the rate at which fat breaks down. Men also have more muscle than women, and muscle is costly to maintain in terms of energy.

Age and genetics also play a role in metabolic rate. Two people of the same size and sex, who consume the same number of Calories and exercise the same amount, will not necessarily store the same amount of fat. The rate at which the foods you eat are metabolized slows as you age, and some people are simply born with lower basal metabolic rates. This happens if the *enzymes* involved in regulating all of the body's chemical reactions are slower.

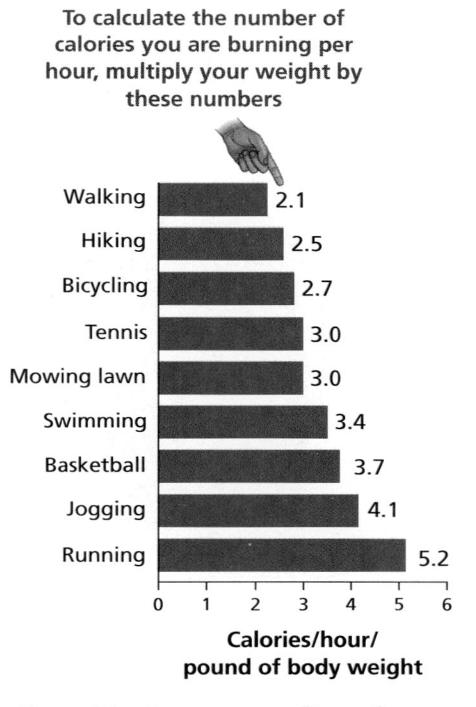

Figure 2.8 Energy expenditures for various activities. This bar graph can assist you in determining how many calories you burn.

Enzymes All metabolic reactions are regulated by proteins called **enzymes** that speed up, or **catalyze**, the rate of reactions. Different enzymes catalyze different reactions, a property called **specificity**. Enzymes are usually named for the reaction they catalyze, and end in the suffix *-ase*. For example, sucrase is the enzyme that breaks down table sugar (sucrose).

The specificity of an enzyme is the result of its shape—different enzymes have different shapes because they are composed of different amino acids. The 20 amino acids, with their side groups, are arranged in distinct orders for each enzyme, producing enzymes of all shapes and sizes.

The chemicals that are metabolized—either built up or broken down—by an enzyme-catalyzed reaction are called the enzyme's **substrate**. The specificity of an enzyme for its substrate occurs because the enzyme can only bind to a substrate whose shape conforms to the enzyme's shape. The region on the enzyme where the substrate binds is called the enzyme's **active site**. The enzyme binds its substrate, helps convert it to a reaction product, and then resumes its original shape so it can perform the reaction again (Figure 2.9).

Lactose intolerance is a common dietary problem caused by an enzyme deficiency. People with lactose intolerance are unable to digest large amounts of lactose (the most common sugar in milk) which is the result of a shortage of the enzyme *lactase*, typically produced by the cells of the small intestine. This enzyme breaks down lactose into simpler forms that can then be absorbed into the bloodstream. When there is not enough lactase available to digest lactose, nausea, cramps, bloating, gas, and diarrhea can occur. For reasons scientists do not yet understand, the lactase enzyme becomes less efficient as we age.

The speed and efficiency of the many different enzymes will lead to an overall increase or decrease in metabolic rate. Thus, when you say that your metabolism is slow or fast, you are actually referring to the speed at which enzymes catalyze chemical reactions in your body.

The properties of metabolic enzymes, like those of all proteins, are determined by the genes that encode them. Genes are passed from parents to children; they are large molecules located inside your cells that contain protein-building instructions. This is why the genes you inherit influence your rate of fat storage and utilization. In addition, if an enzyme requires a vitamin to act as a coenzyme and that vitamin is not available, metabolism may be slowed.

All of these variables help explain why some people seem to eat and eat and never gain an ounce, while others struggle with their weight for their entire lives. To obtain a rough measure of how many Calories you should consume per day, multiply the weight you wish to maintain by 15, and add the number of Calories you burn during exercise. If you are trying to lose weight, decrease your caloric intake or increase your exercise level. Losing one pound of fat requires you to burn 3500 Calories.

Regardless of your metabolic rate, all food must be converted into a form of energy your body can use.

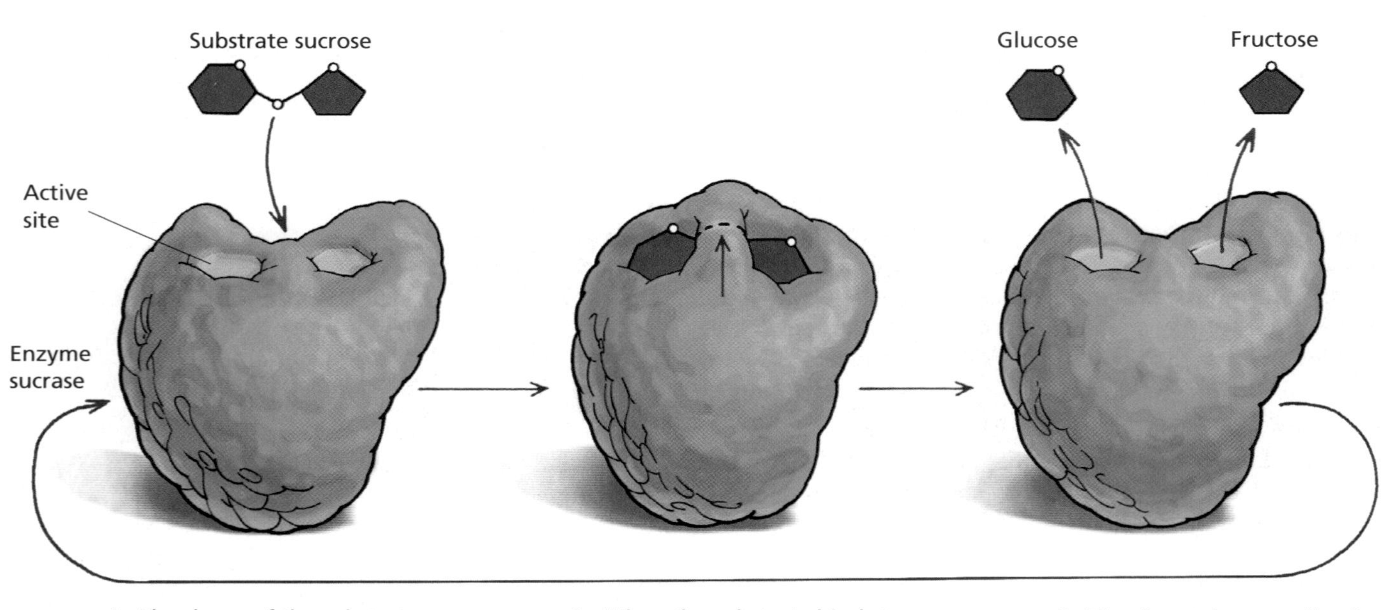
Substrate sucrose
Glucose
Fructose
Active site
Enzyme sucrase
1. The shape of the substrate matches the shape of the enzyme's active site.
2. When the substrate binds to the active site, the enzyme changes shape and the bond between the sugars is stressed.
3. The shape change splits the substrate and releases the two subunits. The enzyme is able to perform the reaction again.

Figure 2.9 Enzymes. The enzyme sucrase is cleaving (splitting) the disaccharide sucrose into its monosaccharide subunits, fructose and glucose. The enzyme can then be recycled to perform the same reaction over and over again.

2.2 Converting Food into Energy

Your body cannot use the foods you eat directly. First, the energy in food is converted into energy present in the chemical bonds of **adenosine triphosphate (ATP)**. This process requires that food first be digested by the digestive system.

The Digestive System

The digestive system is a group of organs that work together to break down food (Figure 2.10). Any group of organs that works together to perform a body function is called an *organ system*. (Table 2.3 lists the functions of various organ systems.)

Media Activity 2.1B **Structure and Function of the Digestive System**

(a) Digestive system

1. **Mouth**
   - Teeth reduce the size of food, increasing surface area available for digestion by enzymes.
   - Enzymes in saliva then start breaking down carbohydrates.
2. **Pharynx (throat)**
   - A flap called the epiglottis blocks the opening to the windpipe when we swallow so that our food goes into our esophagus rather than our lungs.
3. **Esophagus**
   - Transports food to stomach by rhythmic waves of muscle contraction called peristalsis.
4. **Stomach**
   - Acidic gastric juices start breaking down foods.
   - The enzyme pepsin breaks down proteins.
   - Mucous prevents gastric juices from digesting stomach.
   - Pyloric sphincter regulates movement of food from stomach to small intestine.
5. **Small intestine**
   - Where most digestion of carbohydrates, proteins, fats, and nucleic acids occurs.
   - Nutrients are absorbed into the bloodstream.
6. **Large intestine (colon)**
   - Water is reabsorbed.
   - Undigested materials (such as fiber) produce feces.

Accesory organs to small intestine:

Liver
- Produces bile which aids absorption of fats.

Gall bladder
- Stores bile and empties into small intestine.

Pancreas
- Produces digestive enzymes.
- Produces a buffer that neutralizes acidity of stomach acid.

(b) Absorption of nutrients into bloodstream

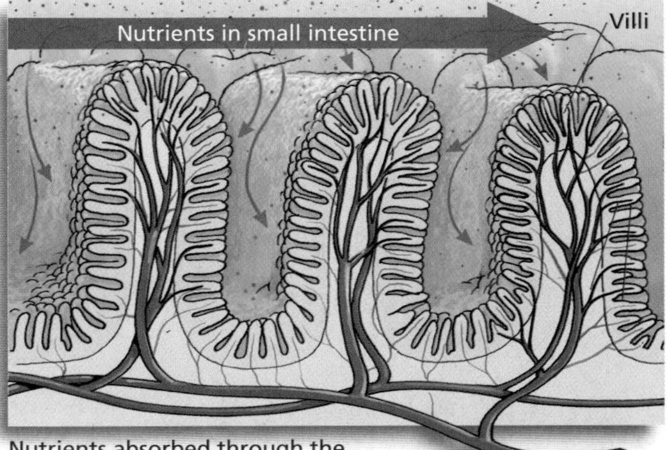

Nutrients absorbed through the intestinal lining into the bloodstream for transport to cells throughout the body.

Figure 2.10 **The digestive system.** The digestive system facilitates the breakdown of food (a) and its absorption into the bloodstream (b).

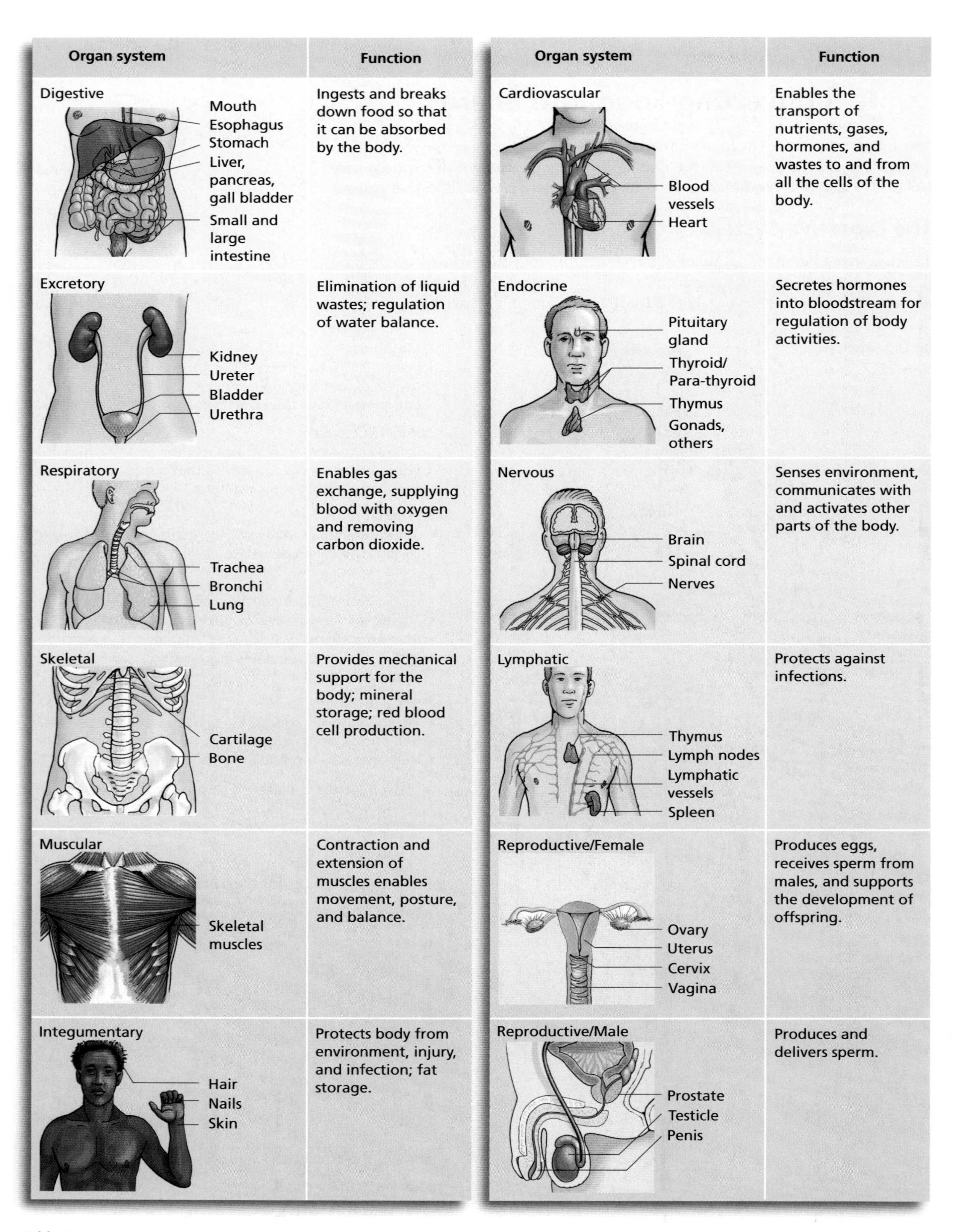

Organ system	Function	Organ system	Function
Digestive: Mouth, Esophagus, Stomach, Liver, pancreas, gall bladder, Small and large intestine	Ingests and breaks down food so that it can be absorbed by the body.	Cardiovascular: Blood vessels, Heart	Enables the transport of nutrients, gases, hormones, and wastes to and from all the cells of the body.
Excretory: Kidney, Ureter, Bladder, Urethra	Elimination of liquid wastes; regulation of water balance.	Endocrine: Pituitary gland, Thyroid/Para-thyroid, Thymus, Gonads, others	Secretes hormones into bloodstream for regulation of body activities.
Respiratory: Trachea, Bronchi, Lung	Enables gas exchange, supplying blood with oxygen and removing carbon dioxide.	Nervous: Brain, Spinal cord, Nerves	Senses environment, communicates with and activates other parts of the body.
Skeletal: Cartilage, Bone	Provides mechanical support for the body; mineral storage; red blood cell production.	Lymphatic: Thymus, Lymph nodes, Lymphatic vessels, Spleen	Protects against infections.
Muscular: Skeletal muscles	Contraction and extension of muscles enables movement, posture, and balance.	Reproductive/Female: Ovary, Uterus, Cervix, Vagina	Produces eggs, receives sperm from males, and supports the development of offspring.
Integumentary: Hair, Nails, Skin	Protects body from environment, injury, and infection; fat storage.	Reproductive/Male: Prostate, Testicle, Penis	Produces and delivers sperm.

Table 2.3 Organ systems. Illustrations and descriptions of various systems.

When you swallow chewed food, it moves from your mouth to your esophagus. The esophagus brings food to the stomach, where the breakdown of food into its subunits begins in earnest. Food then travels through the intestines; substances that cannot be further broken down pass through the large intestine, and exit the body, and substances that can be digested move from the small intestine through the bloodstream to individual cells (Figure 2.10b).

Cells

All living things are made of *cells*, and all living things must be able to obtain energy—in the form of nutrients—from the environment in order to survive. Nutrients absorbed from the small intestine are transported to cells via the bloodstream (Figure 2.11).

All cells are enclosed by a structure called a **plasma membrane** (Table 2.4). The plasma membrane defines the outer boundary of each cell, isolates the cell's contents from the environment, and serves as a selectively permeable barrier that determines which nutrients are allowed into and out of the cell.

The plasma membrane is made of molecules called **phospholipids**, each of which has a *hydrophilic* (water-loving) head and two *hydrophobic* (water-hating) tails. Phospholipids get their name because the group of phosphate-bearing heads is combined with a fat, or *lipid*, molecule. When phospholipid molecules are placed in a watery solution, such as a cell, they orient themselves so that their hydrophilic heads are exposed to the water and their hydrophobic tails are

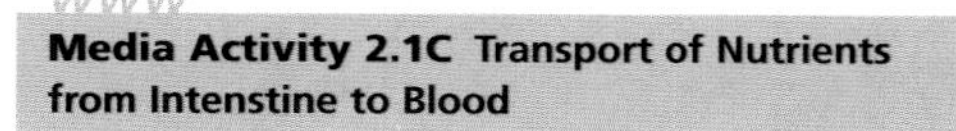

www
Media Activity 2.1C Transport of Nutrients from Intenstine to Blood

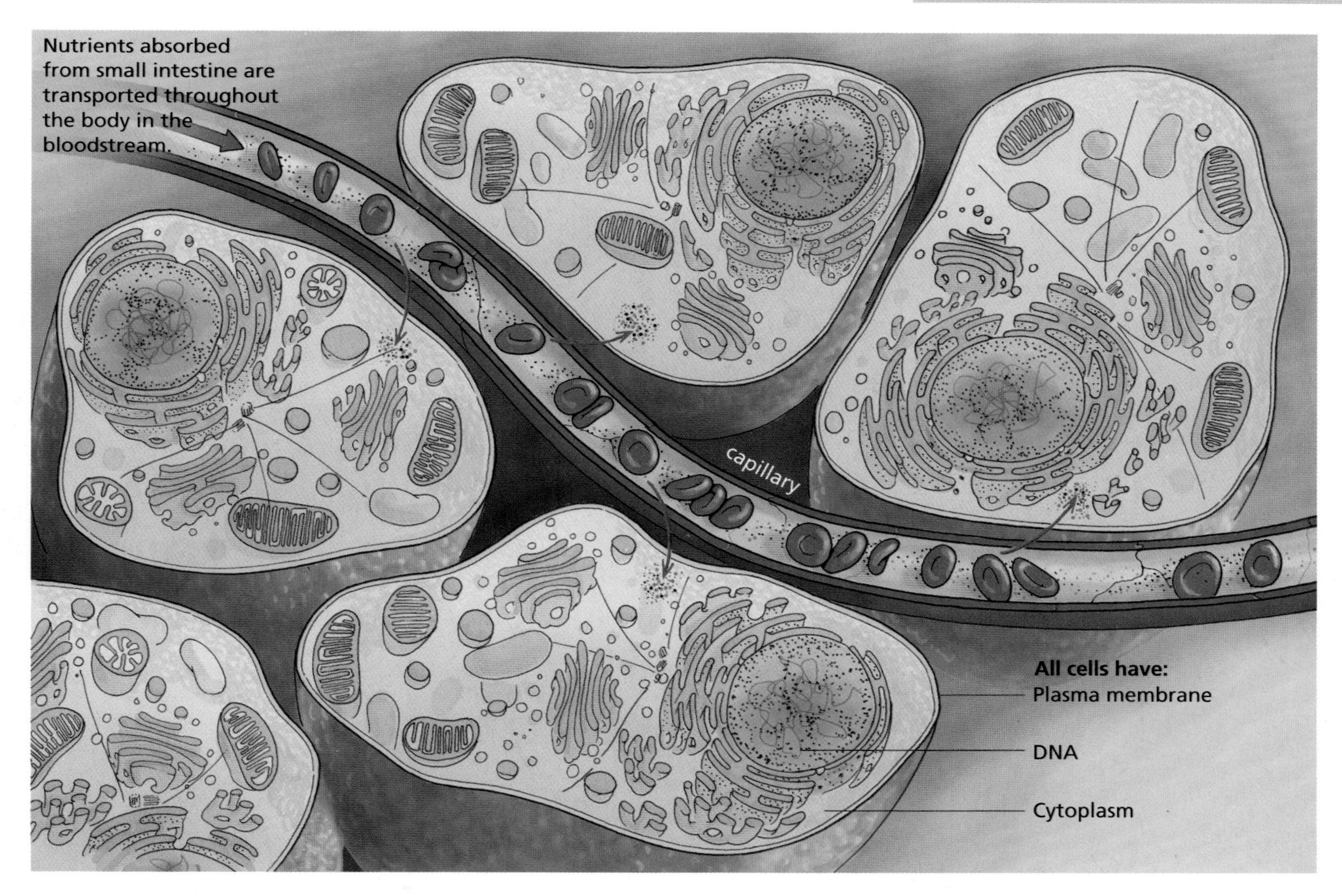

Figure 2.11 Nutrients move from the bloodstream to cells. Food absorbed from the small intestine into the bloodstream makes its way into individual cells. To do so, food must first cross the plasma membrane; once inside a cell, the food can be further broken down to release the energy stored in its chemical bonds.

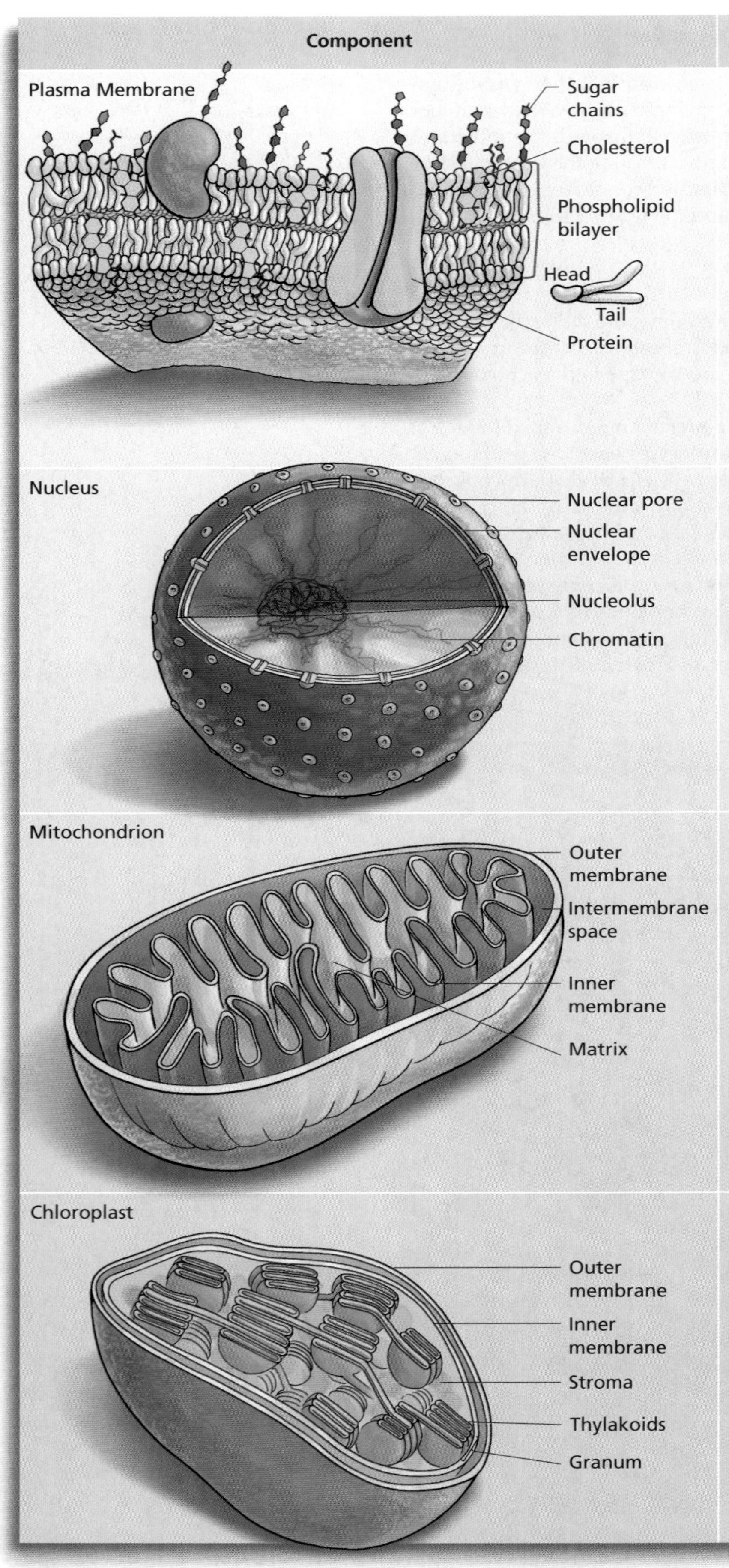

Component	Function
Plasma Membrane	Each cell is surrounded by a plasma membrane. It is composed of a bilayer of phospholipids (tails toward the center), perforated by proteins. Proteins in the bilayer help transport substances across the hydrophobic core of the membrane. Cholesterol helps maintain the fluidity of the membrane by preventing the phospholipids from packing too tightly. The sugar chains function as identification tags, marking cells as a particular cell type (liver cell, heart cell, etc.).
Nucleus	The nucleus is a spherical structure surrounded by two membranes, together called the nuclear envelope. The nuclear envelope is studded with nuclear pores that regulate traffic into and out of the nucleus. Inside the nucleus is chromatin, composed of DNA and proteins. The nucleolus is where ribosomes are produced.
Mitochondrion	Mitochondria are energy-producing organelles composed of two membranes. The inner and outer mitochondrial membranes are separated by the intermembrane space. The highly convoluted inner membrane carries many of the proteins involved in producing ATP. The outer membrane regulates traffic into and out of the mitochondrion. The matrix of the mitochondrion is the location of many of the reactions of cellular respiration.
Chloroplast	An important organelle present only in plant cells, the chloroplast uses the sun's energy to convert carbon dioxide and water into sugars. Each chloroplast has an outer membrane, an inner membrane, a liquid material called the stroma, and a network of flattened membranes called thylakoids that stack on one another to form structures called grana (singular granum).

Table 2.4 Cell components. Illustrations and descriptions of cell components and their function.

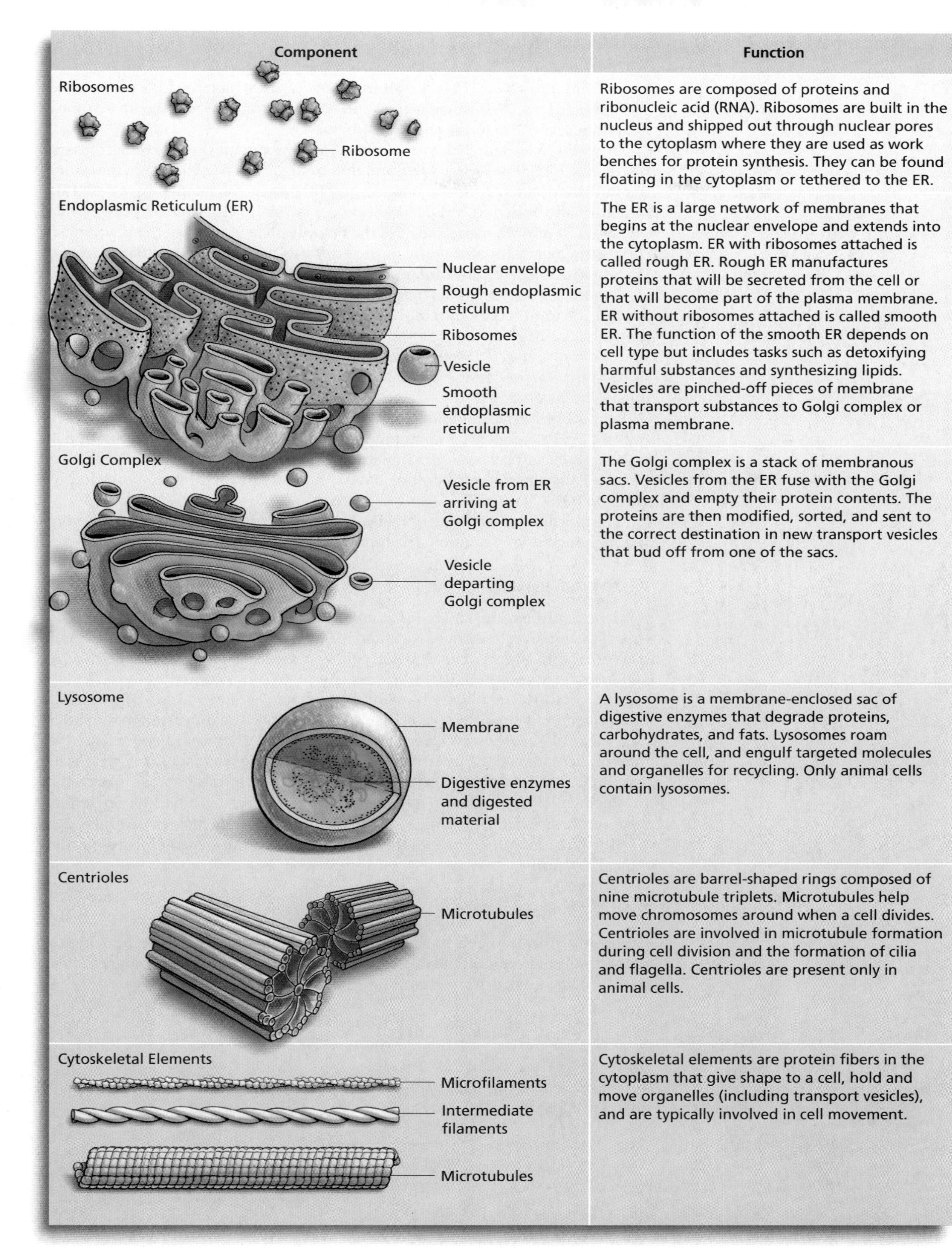

Component	Function
Ribosomes	Ribosomes are composed of proteins and ribonucleic acid (RNA). Ribosomes are built in the nucleus and shipped out through nuclear pores to the cytoplasm where they are used as work benches for protein synthesis. They can be found floating in the cytoplasm or tethered to the ER.
Endoplasmic Reticulum (ER)	The ER is a large network of membranes that begins at the nuclear envelope and extends into the cytoplasm. ER with ribosomes attached is called rough ER. Rough ER manufactures proteins that will be secreted from the cell or that will become part of the plasma membrane. ER without ribosomes attached is called smooth ER. The function of the smooth ER depends on cell type but includes tasks such as detoxifying harmful substances and synthesizing lipids. Vesicles are pinched-off pieces of membrane that transport substances to Golgi complex or plasma membrane.
Golgi Complex	The Golgi complex is a stack of membranous sacs. Vesicles from the ER fuse with the Golgi complex and empty their protein contents. The proteins are then modified, sorted, and sent to the correct destination in new transport vesicles that bud off from one of the sacs.
Lysosome	A lysosome is a membrane-enclosed sac of digestive enzymes that degrade proteins, carbohydrates, and fats. Lysosomes roam around the cell, and engulf targeted molecules and organelles for recycling. Only animal cells contain lysosomes.
Centrioles	Centrioles are barrel-shaped rings composed of nine microtubule triplets. Microtubules help move chromosomes around when a cell divides. Centrioles are involved in microtubule formation during cell division and the formation of cilia and flagella. Centrioles are present only in animal cells.
Cytoskeletal Elements	Cytoskeletal elements are protein fibers in the cytoplasm that give shape to a cell, hold and move organelles (including transport vesicles), and are typically involved in cell movement.

Table 2.4 ***(continued)***

away from the water. They cluster into a *bilayer* in which the tails of the phospholipids interact with themselves and exclude water, while the heads maximize their exposure to the surrounding water.

In addition to being surrounded by a plasma membrane, all cells contain a **nucleus**, which houses **deoxyribonucleic acid**, or **DNA**, a molecule that stores the information required for making all of the proteins needed by a cell. (The structure and function of DNA will be explained in Chapter 6.) Between the nucleus and plasma membrane lies the **cytoplasm**, a watery matrix. Cytoplasm contains water, salts, and many of the enzymes required for cellular reactions. It also houses many subcellular structures called **organelles**.

Organelles are to cells as organs are to the body. Each organelle performs a specific job required by the cell, and all organelles work together to keep an individual cell healthy and to produce the raw materials the cell needs to survive. In many ways, these subcellular efforts mimic the efforts required for more complex processes. Table 2.4 describes the structures and functions of most cellular organelles.

Not all cell types contain every organelle. *Eukaryotes* are organisms with cells that have a nucleus and membrane-bounded organelles—humans and all organisms except bacteria and archaea are made of eukaryotic cells. Other types of cells called *prokaryotes* do not have a nucleus or membrane-bounded organelles (see Chapter 3).

There is one eukaryotic organelle in particular whose function in digestion requires further explanation, the mitochondrion.

Mitochondria

It is in the **mitochondria** (singular, *mitochondrion*) that the nutrients in food released by the digestive system and carried by the bloodstream to cells are finally converted into a form of energy that cells can use—*adenosine triphosphate (ATP)*, an energy-rich molecule that transfers energy to other compounds.

Mitochondria are kidney-bean-shaped organelles, surrounded by two membranes, the inner and outer *mitochondrial membranes*. The inner membrane houses some of the *proteins* involved in producing ATP. The space between the two membranes is called the **intermembrane space**. Inside the inner membrane is the semifluid **matrix** of the mitochondrion where some of the *enzymes* involved in producing ATP are located. The outer membrane serves as a barrier, preventing some substances from gaining access to the mitochondrial insides (see Table 2.4).

Once inside mitochondria, the nutrients in food undergo a process called *cellular respiration*.

Cellular Respiration

Cellular respiration gets its name from the fact that oxygen is used during the process and then carbon dioxide is released, similar to the gas exchange that occurs in your lungs during breathing.

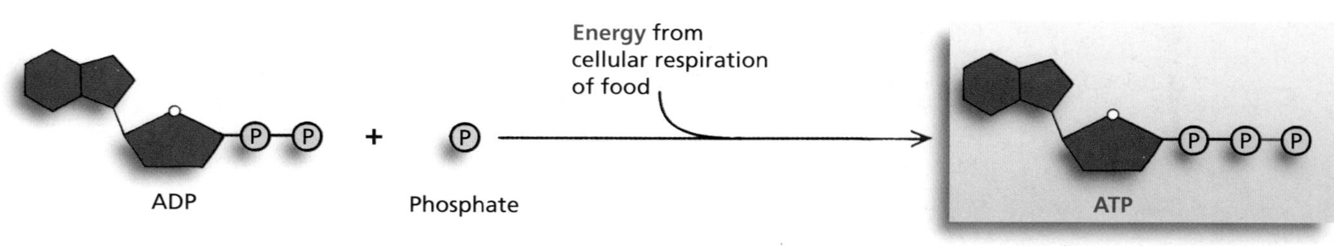

Figure 2.12 Regenerating ATP. ATP is regenerated from ADP and phosphate during the process of cellular respiration.

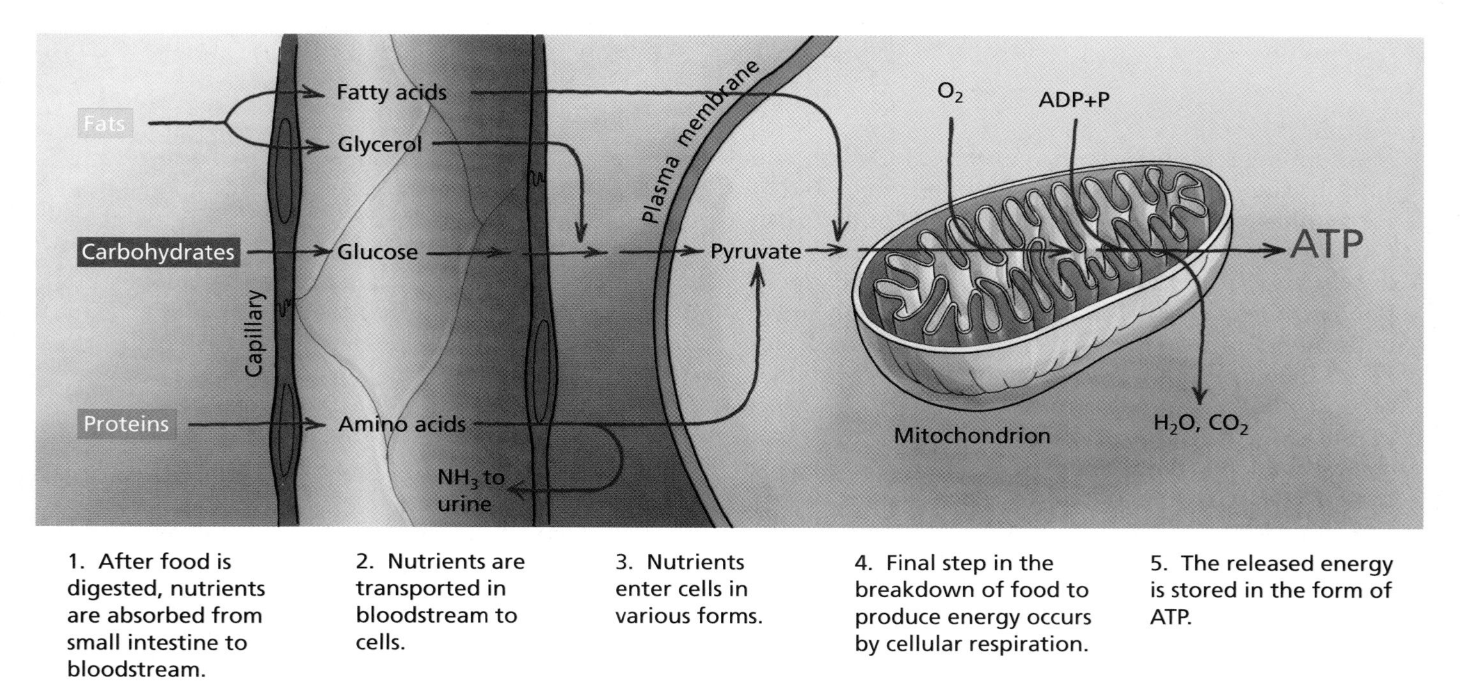
Fats
Fatty acids
Glycerol
Carbohydrates
Glucose
Capillary
Proteins
Amino acids
NH3 to urine
Plasma membrane
Pyruvate
O2
ADP+P
ATP
Mitochondrion
H2O, CO2
1. After food is digested, nutrients are absorbed from small intestine to bloodstream.
2. Nutrients are transported in bloodstream to cells.
3. Nutrients enter cells in various forms.
4. Final step in the breakdown of food to produce energy occurs by cellular respiration.
5. The released energy is stored in the form of ATP.

Figure 2.13 Cellular respiration. Carbohydrates, proteins, and fats are broken down inside a mitochondrion. The energy released by their digestion is used to synthesize ATP.

Media Activity 2.1D ATP Production in the Cell

Cellular respiration occurs in the mitochondria of both plant and animal cells and results in the production of ATP. ATP energizes other compounds through **phosphorylation**, which means it adds a phosphate to them. (You can think of the donated phosphate as a "little bag of energy.") When a molecule, say an enzyme, needs energy, the phosphate group is transferred from ATP to the enzyme, and the enzyme now has the energy it needs to perform its job. When ATP gives away a phosphate group, it becomes *adenosine diphosphate* or *ADP*. ATP is then regenerated from ADP and phosphate by the process of cellular respiration (Figure 2.12).

All of the foods you eat undergo cellular respiration. In this process, cells utilize oxygen and produce carbon dioxide and water (this is why you breathe oxygen in and breathe carbon dioxide out), and the energy stored in the chemical bonds of the nutrients is used to produce ATP (Figure 2.13). The ATP that is produced can then be used to power cellular activities, such as helping an enzyme perform its job (Figure 2.14).

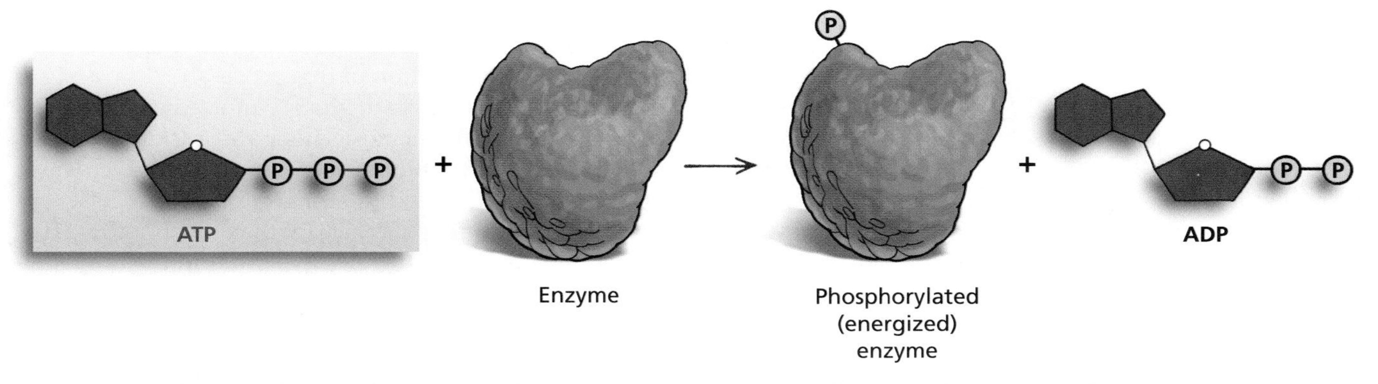
P
P P P
+
ATP
Enzyme
Phosphorylated (energized) enzyme
P P
ADP

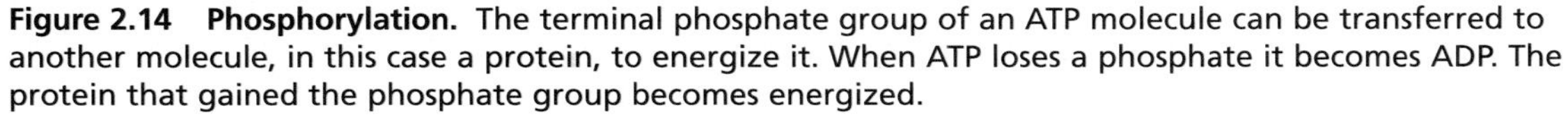

Figure 2.14 Phosphorylation. The terminal phosphate group of an ATP molecule can be transferred to another molecule, in this case a protein, to energize it. When ATP loses a phosphate it becomes ADP. The protein that gained the phosphate group becomes energized.

Essay 2.1 Photosynthesis: How Plants Make Food

The sun is the ultimate source of the calories we consume because sunlight provides the energy required for the synthesis of food made by the plants other organisms consume. Plants transform light energy into chemical energy through the process of *photosynthesis*. Photosynthesis converts *carbon dioxide* (an atmospheric molecule) and water into energy-rich carbohydrates. Plants then use carbohydrates to supply energy to their cells. In other words, plants make their own food. In fact, plants are so effective at converting light energy into chemical energy that they produce enough energy to store it as starch. Other organisms, such as humans, harvest this excess energy.

Photosynthesis takes place inside plant cells in structures called *chloroplasts* (see Table 2.4). Chloroplasts are the organelles that give leaves and other plant structures their green color—the result of the pigment *chlorophyll*. Like all pigments, chlorophyll absorbs light. Light is made up of rays with different colors, or levels of energy, and each energy level has a different wavelength—to the human eye, shorter and middle wavelengths appear violet to green, and longer wavelengths appear yellow to red. Chlorophyll looks green to our eyes because it absorbs the shorter and longer wavelengths of visible light, and reflects the middle (green) range of wavelengths.

When a pigment such as chlorophyll absorbs sunlight, a part of the chlorophyll molecule becomes excited. That is, the light energy is transferred to the molecule and becomes chemical energy. For most pigments, the molecule remains excited for a very brief amount of time, it then "relaxes" and the chemical energy is lost as heat. This is why a surface that looks black (that is, one composed of a pigment that absorbs all visible-light wavelengths) heats up quickly in comparison to a surface that looks white (one that absorbs no visible-light wavelengths). Inside a chloroplast however, the chemical energy of the excited chlorophyll molecules is captured and added to carbon dioxide and water to produce carbohydrate. The equation summarizing photosynthesis is:

$$6(CO_2) + 12(H_2O) \xrightarrow{\text{sun}} C_6H_{12}O_6 + 6(O_2) + 6(H_2O)$$

$$\text{carbon dioxide} + \text{water} \xrightarrow{\text{sun}} \text{glucose} + \text{oxygen} + \text{water}$$

The products of photosynthesis are used as reactants in respiration and vice versa.

Without this reaction in the cholorplasts—where solar energy is transformed into the chemical energy in glucose—cellular respiration could not occur and glucose would not be converted into ATP. For this reason, *all* living things are dependent upon photosynthesis for food—even meat-eating organisms, since they consume organisms that eat plants.

Carbohydrate Metabolism Carbohydrates are broken down in the mitochondria. The equation for this is:

$$C_6H_{12}O_6 + 6O_2 \rightarrow 6CO_2 + 6H_2O$$

$$\text{glucose} + \text{oxygen yields carbon dioxide} + \text{water}$$

Glucose is an energy-rich compound, but the products of its digestion—carbon dioxide and water—are energy poor. The energy released during the conversion of glucose to carbon dioxide and water is used to synthesize ATP. Glucose is produced when energy from the sun is used to drive its synthesis from carbon dioxide and water. Ultimately, all food comes from this process of **photosynthesis**. (See Essay 2.1 for a description of this process.)

To generate energy from glucose, the 6-carbon glucose molecule is first broken down into two, 3-carbon *pyruvate* molecules. This part of the process of cellular respiration actually occurs in the cytoplasm, and is called **glycolysis**. Glycolysis does not require oxygen and is said to be *anaerobic*.

The pyruvate molecules then move into the matrix of the mitochondria. The remaining reactions involved in converting the energy stored in the bonds of pyruvate into the energy stored in ATP are *aerobic*, or oxygen-requiring. During these reactions, the two pyruvates produced by the breakdown of glucose during glycolysis are converted into water and carbon dioxide. Carbon dioxide is removed from the pyruvate molecules, and water (H_2O) is formed when oxygen (O_2) combines with the hydrogen (H) that was removed from the original glucose molecule.

It takes longer to digest complex carbohydrates than simpler sugars because complex carbohydrates have more chemical bonds to break. Endurance athletes will load up on complex carbohydrates for several days prior to a race; this increases the amount of easily accessible energy the athletes can draw on during competition.

Protein Metabolism Most protein is broken down into component amino acids, which are then used to synthesize new proteins. Proteins can also be broken down to supply energy. However, this is only done when fat or carbohydrate is unavailable. In cases of extreme starvation, including that brought on by eating disorders such as *anorexia* and *bulimia,* the body will break down muscle to produce ATP.

The first step in producing energy from the amino acids of a protein is removing the nitrogen-containing amino group of the amino acid. Amino groups are then converted to a compound called urea, which is excreted in the urine. The carbon, oxygen, and hydrogen remaining after the amino group is removed undergo further breakdown and eventually enter the mitochondria, where they too can be used to produce ATP.

Fat Metabolism The subunits of fats, glycerol and fatty acids also enter the mitochondria to produce ATP. Most cells will only break down fat when carbohydrate supplies are depleted. When the number of Calories consumed outstrips those used, the excess units of energy are converted into fat and stored on the body for later use. As we discussed in Section 2.1, too much or too little body fat can cause health problems.

2.3 Body Fat and Health

A clear understanding of how much body fat is healthful is hard to come by because cultural and biological definitions of overweight markedly differ. Cultural definitions of overweight have changed over the years. Men and women who were considered to be attractive in the past might not be seen as meeting today's standards. In the United States, the evolution of this trend has been paralleled by changes in children's action figures and dolls, and by Miss America pageant winners over the last several decades. You need only compare the physiques of action figures from the 1960s and 1970s to the physiques seen on today's action figures to see how the standards have changed for males (Figure 2.15a,b). Standards for

WWW

Media Activity 2.3 **Body Image and Society**

(a) GI Joe has become more muscular over time.

(b) Miss America has become thinner.

Figure 2.15 The perception of beauty. (a) GI Joe in 1964; and (b) in 2002. (c) Miss America, 1964, Donna Axum; and (d) Miss America, 2002, Katie Harman.

women have also changed—the average weight of Miss America contestants has decreased over the last few decades (Figure 2.15c,d). The unrealistic nature of these standards is exemplified by the fact that the average woman in the United States weighs 140 pounds and wears a size 12. The average model weighs 103 pounds and wears a size 4.

The next time you read the newspaper, take note of advertisements for diets featuring men and women of healthful weights promoting diet products. It is often the case that "before" pictures show individuals of healthful weights, and "after" pictures show people that are too thin. With these messages about body fat, it is difficult for the average person to know how much body fat is right for them.

Evaluating How Much Body Fat Is Healthful

A person's gender, along with other factors, determines their ideal amount of body fat. Women need more body fat than men, to maintain their fertility. On average, healthy women have 22% body fat, and healthy men have 14%. To maintain essential body functions women need at least 12% body fat, but not more than 32%; for men the range is between 3% and 29%. This difference between females and males, a so-called sex difference, is due to the fact that women store more fat on their breasts, hips, and thighs than do men. This difference in muscle mass leads to increased energy use by males, since muscles use more energy than fat does.

www
Media Activity 2.4 Measuring Health and Fitness

Women also have an 8% thicker layer of skin under the outer epidermal layer than men. This means that, in a woman and a man of similar strength and body fat, the woman's muscles would look smoother and less defined than the man's would.

A person's frame size also influences body fat—larger-boned people carry more fat. In addition, body fat tends to increase with age.

Measuring Body Fat The simplest way to measure body fat uses a caliper to measure the folds of skin on the triceps (the back of the upper arm), the abdomen (one inch to the right of the belly button), and the front of the thigh. The three measurements are added together and plugged into a formula that produces an estimate of the body fat percentage, which is generalized to the whole body. This technique relies on the assumption that the thickness of subcutaneous (below the skin) fat is proportional to the fat stored deeper within the body. It also requires that a person be trained to find the correct locations on the body to measure body fat, or the results would vary depending upon who performed the test.

A more sophisticated test that does not rely on the assumption that subcutaneous fat is representative of overall body fat or the skill of the person performing the measurements is *bioelectrical impedance*. Bioelectrical impedance requires the use of an apparatus that measures the resistance of body tissues, groups of cells that perform the same function, to the flow of a small, harmless electrical signal (Figure 2.16). Water, muscle, and bone conduct electricity better than fat, so the greater the resistance, the higher the percentage of body fat.

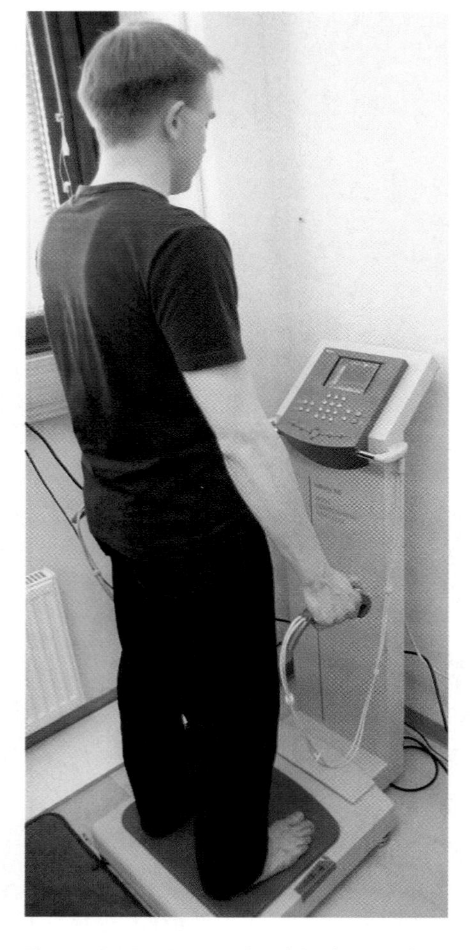

Figure 2.16 Measuring bioelectrical impedance. This instrument accurately measures body fat with a small electrical signal.

Most people don't have easy access to either of these techniques, so we often turn to widely available charts and tables to help us determine if our weight is what it should be.

Determining Ideal Weight Unfortunately, it is a bit tricky to determine what any individual's ideal body weight should be. In the past, you simply weighed yourself and compared your weight to a chart showing a range of acceptable weights for a given height. The weight ranges on these tables were associated with the weights of a group of people who bought life insurance in the 1950s, and whose health was monitored until they died. The problem with using these tables is that the subjects may not have been representative of the whole population. As you learned in Chapter 1, this can lead to erroneous conclusions. People who had the money to buy life insurance tended to have the other benefits of money as well, including easier access to health care, better nutrition, *and* lower body weight. Their longer lives may have had more to do with better health care and nutrition than with their weight.

To deal with some of the ambiguities associated with the insurance company's weight tables, a new measure of weight and health risk, the **body mass index (BMI),** has been developed. BMI is a calculation that uses both height and weight to determine a value that correlates an estimate of body fat with the risk of illness and death. (Table 2.5).

Although the BMI measurement is a better approximation of ideal weight than the insurance charts of old, they are not perfect; BMI still does not account for differences in frame size, gender, or muscle mass. In fact, studies show that as many as one in four people may be misclassified by BMI tables because this measurement provides no means to distinguish between lean muscle mass and body fat. For example, an athlete with a lot of muscle will weigh more than a similarly sized person with a lot of fat, since muscle is heavier than fat.

If your BMI falls within the healthy range (BMI of 19–25) you probably have no reason to worry about health risks from excess weight. If your BMI is high, you may be at increased risk for diseases associated with obesity.

Height	Weight											
4'10"	91	96	100	105	110	115	119	124	129	134	138	143
4'11"	94	99	104	109	114	119	124	128	133	138	143	148
5'0"	97	102	107	112	118	123	128	133	138	143	148	153
5'1"	100	106	111	116	122	127	132	137	143	148	153	158
5'2"	103	109	115	120	126	131	136	142	148	153	158	164
5'3"	107	113	118	124	130	135	141	146	152	158	163	169
5'4"	110	116	122	128	134	140	145	151	157	163	169	174
5'5"	114	120	126	132	138	144	150	156	162	168	174	180
5'6"	117	124	130	136	142	148	155	161	167	173	179	186
5'7"	121	127	134	140	146	153	159	166	172	178	185	191
5'8"	125	131	138	144	151	158	164	171	177	184	190	197
5'9"	129	135	142	149	155	162	169	176	183	189	196	203
5'10"	132	139	146	153	160	167	174	181	188	195	202	207
5'11"	136	143	150	157	165	172	179	186	193	200	208	215
6'0"	140	147	154	162	169	177	184	191	198	206	213	221
6'1"	144	151	159	166	174	182	189	197	205	212	219	227
6'2"	148	155	163	171	179	186	194	202	210	218	225	233
6'3"	151	160	168	176	184	192	200	208	216	224	232	240
6'4"	156	164	172	180	189	197	205	213	221	230	238	246
BMI	19	20	21	22	23	24	25	26	27	28	29	30

Table 2.5 Body Mass Index. A chart based on height and weight correlations.

Obesity

One in four Americans has a BMI of 30 or greater and is therefore considered to be *obese*. This crisis in **obesity** is the result you would expect when constant access to cheap, high-fat, energy-dense, unhealthful food is combined with lack of exercise. Nowhere is this relationship more clearly illustrated than it is in the case of the Pima Indians.

Several hundred years ago, the ancestral population of Pima Indians split into two tribes. One branch moved to Arizona and adopted the American diet and lifestyle; the typical Pima of Arizona gets as much exercise as the average American and also eats a high-fat, low-fiber diet, similar to most Americans. Unfortunately, the health consequences for them are more severe than they are for most Americans—close to 60% of the Arizona Pima are obese and diabetic. In contrast, the Pima of New Mexico maintained their ancestral farming life and their diet is rich in fruits, vegetables, and fiber. The Pima of New Mexico also engage in physical labor close to 22 hours per week and are 60 pounds lighter than their Arizona cousins. Consequently, diabetes is virtually unheard of.

This example illustrates the impact of lifestyle over genetics since the Pima of Arizona share many genes with their New Mexican cousins but have far less healthful lives due to their diet and lack of exercise. The example of the Pima Indians also shows that genes influence body weight, because the Pima of Arizona have higher rates of obesity and diabetes than other Americans whose lifestyle they share. (Genetics is discussed in Chapter 4.)

Whether obesity is the result of genetics, diet, or lack of exercise, the health risks associated with obesity are the same. As your weight increases, so do your risks of diabetes, hypertension, heart disease, stroke, and joint problems. (Excess weight puts extra pressure on the knees, hips, and back that wears away the cartilage protecting them, resulting in joint pain and stiffness.)

Diabetes **Diabetes** is a disorder of carbohydrate metabolism characterized by the impaired ability of the body to produce or respond to *insulin*. Insulin is a hormone secreted by beta cells, which are located within clusters of cells in the pancreas. Insulin's role in the body is to trigger cells to take up glucose so that they can convert the energy-yielding sugar. People with diabetes are unable to metabolize glucose, and as a result, the level of glucose in the blood rises (Figure 2.17).

There are two forms of the disease. Type I, *insulin-dependent diabetes mellitus* (IDDM, formerly referred to as juvenile-onset diabetes), usually arises in childhood. People with IDDM cannot produce insulin because their immune systems mistakenly destroy their own beta cells. When the body is no longer able to produce insulin, daily injections of the hormone are required. Type I diabetes is *not* correlated with obesity.

Type II, *non-insulin-dependent diabetes mellitus* (NIDDM, sometimes called adult-onset diabetes), usually occurs after 40 years of age and is more common in the obese. NIDDM arises from either decreased pancreatic secretion of insulin or reduced responsiveness to secreted insulin in target cells. People with NIDDM are able to control blood-glucose levels through diet and exercise and, if necessary, by insulin injections.

Hypertension **Hypertension**, or high blood pressure, places increased stress on the circulatory system and causes the heart to work too hard. A hypertensive person is six times more likely to have a heart attack than a person with normal blood pressure.

Blood pressure is the force, originated by the pumping action of the heart, exerted by the blood against the walls of the blood vessels. Blood vessels expand and contract in response to this force. Blood pressure is reported as two numbers: The higher number, the *systolic blood pressure*, represents the pressure exerted by the blood against the walls of the blood vessels; the lower number is

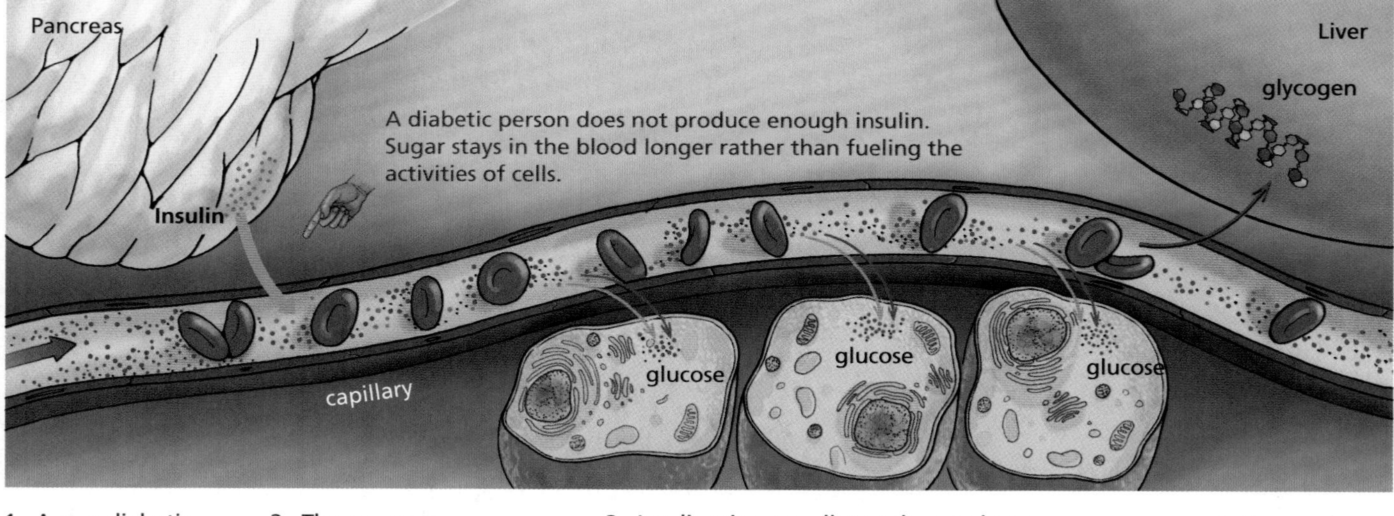
Pancreas
Liver
glycogen
A diabetic person does not produce enough insulin.
Sugar stays in the blood longer rather than fueling the activities of cells.
Insulin
glucose
glucose
glucose
capillary
1. A non-diabetic person has high blood sugar following a meal.
2. The pancreas secretes insulin into the bloodstream.
3. Insulin triggers cells to take up glucose.
4. Excess glucose is stored in liver as glycogen.

Figure 2.17 Diabetes. The cells of the pancreas secrete insulin, which helps glucose move into body cells. Since diabetics produce less insulin, sugar stays in the blood longer.

the *diastolic blood pressure*, and is the pressure that exists between contractions of the heart when the heart is relaxing. Normal blood pressure is around 120/80. Blood pressure is considered to be high when it is persistently above 140/90.

As you put on weight, you gain mostly fatty tissue. Just like tissues in other parts of the body, fat tissue relies on oxygen and nutrients in your blood to produce energy. Excess fat tissue increases the demand for oxygen and nutrients, thus increasing the amount of blood required to nourish your body—more blood traveling through the arteries means added pressure on the artery walls. In addition, excess weight often is associated with an increase in heart rate and a reduction in the capacity of your blood vessels to transport blood—these two factors also increase blood pressure.

Heart Attack and Stroke Heart attack and stroke are more likely in obese people because the elevated blood pressure caused by obesity also damages the lining of blood vessels, increasing the likelihood that cholesterol will be deposited there. Cholesterol-lined vessels are said to be *atherosclerotic.*

Because fats like cholesterol are not soluble in *aqueous* (water-based) solutions, cholesterol is carried throughout the body attached to proteins in structures called **lipoproteins**. **Low-density lipoproteins (LDL)**, have a high proportion of cholesterol (they are low in protein). LDLs distribute both cholesterol synthesized by the liver and cholesterol derived from the diet throughout the body. LDLs are also important for carrying cholesterol to cells, where it is used to help make plasma membranes and hormones. **High-density lipoproteins (HDL)** contain more protein than cholesterol. HDLs scavenge excess cholesterol from the body and return it to the liver, where it is used to make bile. The cholesterol-rich bile is then released into the small intestine and from there much of it exits the body in the feces. The LDL/HDL ratio is an index of the rate at which cholesterol is leaving body cells and returning to the liver.

Your physician can measure your cholesterol level by sampling the amounts of LDL and HDL in your blood. If your *total* cholesterol level is over 200 or your LDL level is above 100, your physician may recommend that you decrease the amount of cholesterol and saturated fat in your diet. This may

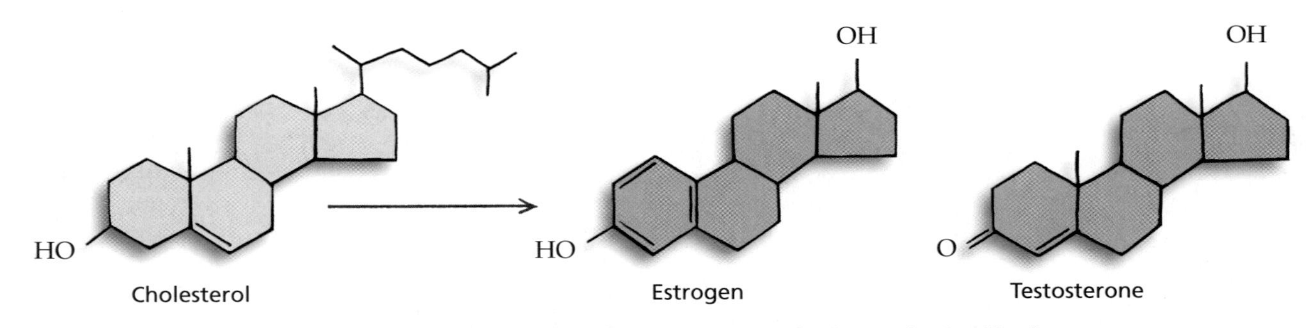

Figure 2.18 Biosynthesis of testosterone and estrogen from cholesterol. The synthesis of both testosterone and estrogen require the use of cholesterol as a starting material.

mean eating more plant-based foods and less meat, since plants do not have cholesterol, and also reducing the amount of saturated fats in your diet. Saturated fat is thought to raise cholesterol levels by stimulating the liver to step up its production of LDLs, and slowing the rate at which LDLs are cleared from the blood.

Before you decide to completely eliminate cholesterol from your diet, however, keep in mind that some cholesterol is necessary—it is present in cell membranes to help maintain their fluidity, and it is the building block for steroid hormones such as estrogen and testosterone (Figure 2.18).

For some people, those with a genetic predisposition to high cholesterol, controlling cholesterol levels through diet is difficult because dietary cholesterol makes up only a fraction of the body's total cholesterol. People with high cholesterol who do not respond to dietary changes may have inherited genes that increase the liver's production of cholesterol. These people may require prescription medications to control their cholesterol levels.

Cholesterol-laden, atherosclerotic vessels increase your risk of heart disease and stroke. Fat deposits narrow your heart's arteries, so less blood can flow to your heart. Diminished blood flow to your heart can cause chest pain, or angina. A complete blockage can lead to a heart attack. Lack of blood flow to the heart during a heart attack can cause the oxygen-starved heart tissue to die, leading to irreversible heart damage.

The same buildup of fatty deposits also occurs in the arteries of the brain. If a blood clot forms in a narrowed artery in the brain, it can completely block blood flow to an area of the brain, resulting in a stroke. If oxygen-starved brain tissue dies, permanent brain damage can result.

Anorexia and Bulimia

Eating disorders that make you underweight cause health problems as severe as those caused by overweight. **Anorexia**, or self-starvation, is rampant on college campuses. Estimates suggest that one in five college women, and one in 20 college men restrict their intake of Calories so severely that they are essentially starving themselves to death. Others allow themselves to eat, sometimes very large amounts of food (called binge eating), but prevent the nutrients from being turned into fat by purging themselves, often by vomiting. Binge eating followed by purging is called **bulimia**.

Anorexia has serious long-term health consequences. Anorexia can starve heart muscles to the point that it develops altered rhythms. Blood flow is reduced and blood pressure drops so much that the little nourishment that is present cannot get to the cells. The lack of fat that accompanies anorexia can also lead to the cessation of menstruation, **amenorrhea**. Amenorrhea occurs when

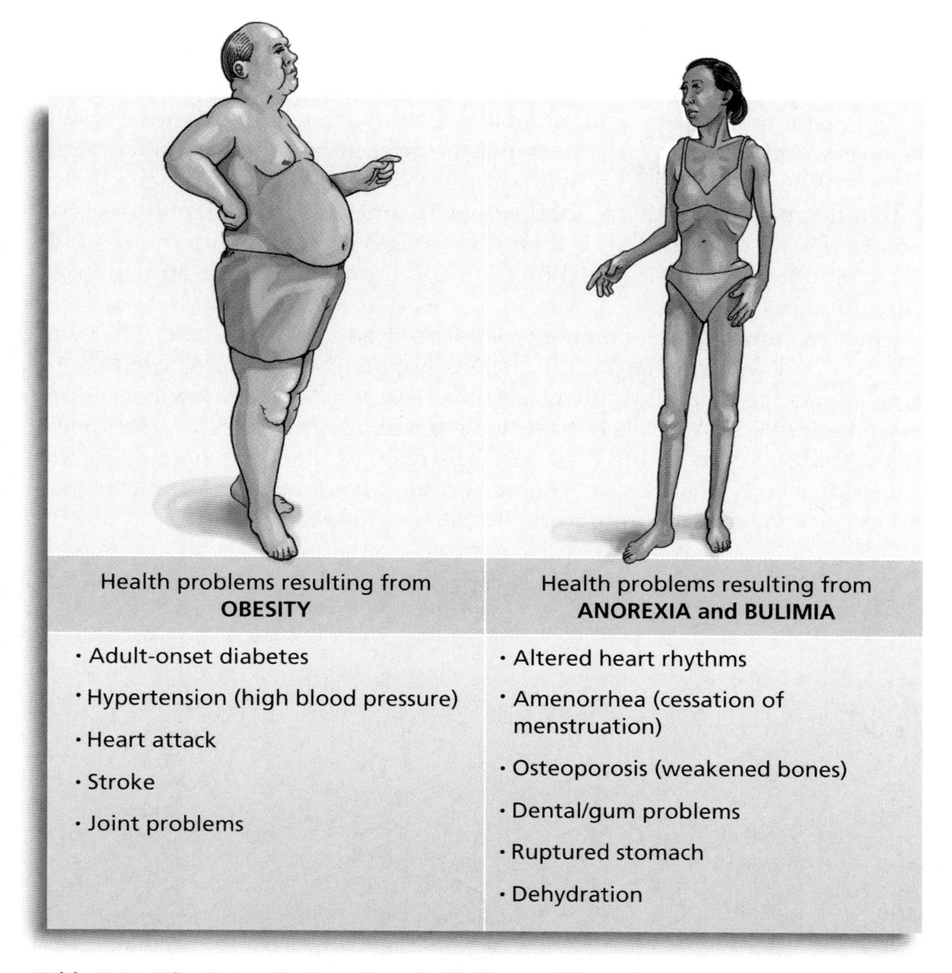

Health problems resulting from **OBESITY**	Health problems resulting from **ANOREXIA and BULIMIA**
• Adult-onset diabetes	• Altered heart rhythms
• Hypertension (high blood pressure)	• Amenorrhea (cessation of menstruation)
• Heart attack	• Osteoporosis (weakened bones)
• Stroke	• Dental/gum problems
• Joint problems	• Ruptured stomach
	• Dehydration

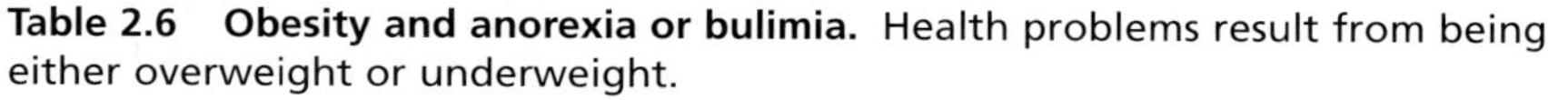

Table 2.6 Obesity and anorexia or bulimia. Health problems result from being either overweight or underweight.

WWW
Media Activity 2.5 Health Problems

a protein called leptin, which is secreted by fat cells, signals the brain that there is not enough body fat to support a pregnancy. Hormones (such as estrogen) that regulate menstruation are blocked, and menstruation ceases. Lack of menstruation can be permanent and causes sterility in a substantial percentage of anorexics.

The damage done by the lack of estrogen is not limited to the reproductive system—bones are affected as well. Estrogen secreted by the ovaries during the menstrual cycle acts on bone cells to help them maintain their strength and size. Anorexics reduce the development of dense bone, and put themselves at much higher risk of breaking their weakened bones, a condition called **osteoporosis**.

In addition to the health problems anorexics face, bulemics can rupture their stomachs from forced vomiting; they often have dental and gum problems caused by stomach acid being forced into their mouths during vomiting, and can become fatally dehydrated.

As you have seen, the health problems associated with obesity, anorexia, and bulimia are quite severe (Table 2.6). You can avoid all of these problems—and improve your overall health—by focusing more on fitness, and less on body weight.

Focus on Fit, not Fat

For most people, slowly working toward being fit rather than trying the latest fad diet is a more realistic and attainable way to achieve the positive health outcomes we all desire. In fact, fitness may be more important than body weight in terms of health. Studies show that fit but overweight people have better health outcomes than unfit slender people. In other words, lack of fitness has a higher risk associated with it than does excess body weight. Therefore, it makes more sense to focus on eating right and exercising than it does to focus on the number on a scale.

When too much time and energy is focused on your body and what you will or will not put in your mouth, there is not much left for seeking out and taking advantage of all that college and life have to offer. Very few people are born to look like the models of beauty we see in advertisements and movies. While it is difficult to ignore the constant stream of messages portraying the media's ideal body as attractive, you must convince yourself that beauty comes from the *presence* of joy, humor, and passion, not the *absence* of fat.

CHAPTER REVIEW

Summary

- Healthful eating means consuming a balanced diet with enough Calories to meet your metabolic needs.
- A healthful diet includes carbohydrates, proteins, and fats that are found in whole foods along with the vitamins and minerals they contain.
- Whole foods are better for your body than processed foods, since processing often removes vitamins and minerals from food.
- Deficiencies in vitamins and minerals can cause a host of health problems, and lead to decreased enzyme function.
- Foods are broken down by the digestive system. From the digestive system, digested nutrients released from foods move out of the intestine to the bloodstream and to cells. To gain access to cells, nutrients move across the cell's plasma membrane. Once inside the cell, nutrients make their way through the cytoplasm to the mitochondria. Mitochondria use the energy stored in the chemical bonds of food to produce ATP, which is used to power all of the cell activities that require energy.
- Plants make food via the process of photosynthesis. During photosynthesis, energy from sunlight is used to power the conversion of carbon dioxide and water into sugar. This occurs in organelles called chloroplasts found in the green parts of plants.
- Cultural and biological definitions of overweight differ.
- Body fat is measured with calipers that determine subcutaneous fat thickness, or an apparatus that uses electrical signals to detect the percentage of body fat.
- Obesity is associated with many health problems including hypertension, heart attack and stroke, diabetes, and joint problems. Anorexia and bulimia are very common on college campuses and result in serious long-term health problems such as osteoporosis.
- Determining your ideal weight using conventional measures is difficult. It makes more sense to focus on becoming fit than on body-fat percentages and body weight.

Key Terms

active site p. 34
adenosine triphosphate (ATP) p. 35
amenorrhea p. 48
amino acid p. 26
anorexia p. 48
basal metabolic rate p. 33
Body Mass Index (BMI) p. 45
bulimia p. 48
Calorie p. 00
carbohydrate p. 24
catalyze p. 33
cell p. 24
cellular respiration p. 40
coenzyme p. 30
complex carbohydrate p. 25
cytoplasm p. 40
deoxyribonucleic acid (DNA) p. 40
diabetes p. 46
disaccharide p. 24
enzyme p. 33
fat p. 28
glycolysis p. 42
high-density lipoprotein (HDL) p. 47
hypertension p. 46
intermembrane space p. 40
lipoprotein p. 47
low-density lipoprotein (LDL) p. 47
matrix p. 40
metabolic rate p. 33
metabolism p. 33
mineral p. 31
mitochondria p. 40
monosaccharide p. 24
nucleus p. 40
nutrients p. 24
obesity p. 46
organelle p. 40
osteoporosis p. 49
phospholipid p. 37
phosphorylation p. 41
photosynthesis p. 42
plasma membrane p. 37
polysaccharide p. 25
protein p. 26
saturated fat p. 28
specificity p. 33
substrate p. 33
unsaturated fat p. 28
vitamin p. 30

Learning the Basics

1. What are the building-block molecules of a carbohydrate, a protein, and a fat?

2. Why is it better to obtain your carbohydrates from an apple, rather than from a candy bar or can of soda?

3. Describe the structure and function of enzymes.

4. Outline the path an avocado travels, from ingestion through the production of ATP. (An avocado contains carbohydrate, protein, and fat.)

5. List the roles of the organelles discussed in this chapter.

6. Describe the structure and function of a cell membrane.

7. Different proteins are composed of different sequences of _____.
a. sugars
b. glycerols
c. fats
d. amino acids

8. Plants store their excess carbohydrates as _____.
a. starch
b. glycogen
c. glucose
d. cellulose
e. fat

9. Proteins may function as _____.
a. parts of a cell
b. hormones
c. energy stores
d. enzymes
e. all of the above

10. Which of the following terms is least like the others in the list below?
a. monosaccharide
b. phospholipid
c. glycogen
d. disaccharide
e. starch

11. A fat molecule consists of _____.
a. carbohydrates and proteins
b. complex carbohydrates only
c. saturated oxygen atoms
d. glycerol and three fatty acids

12. The function of low-density lipoproteins (LDL) is to _____.
a. break down proteins
b. digest starch
c. transport cholesterol from the liver
d. carry carbohydrates into the urine

13. The major process that takes place in the large intestine is _____.
    a. digestion of sugars
    b. breakdown of fats
    c. cellular respiration
    d. absorption of water

14. ATP is produced in the _____.
    a. small intestine only
    b. mitochondria of the small intestine only
    c. mitochondria of all the body's cells
    d. chloroplast

15. ATP provides energy by _____.
    a. metabolizing substances
    b. allowing sugars to cross cell membranes
    c. transferring an energized phosphate group to the substrate
    d. helping to store sugars for later use

16. Which of the following is a *false* statement regarding enzymes?
    a. Enzymes are proteins that speed up metabolic reactions.
    b. Enzymes have specific substrates.
    c. Enzymes supply ATP to their substrates.
    d. An enzyme may be used many times over.

Analyzing and Applying the Basics

1. A friend of yours does not want to eat meat, so she consumes protein powders that she buys at a nutrition store instead. What would be the disadvantages of this practice?

2. Another friend is consuming diet shakes for lunch. Why could this be a bad idea?

3. Two people with very similar diets and similar exercise levels have very different amounts of body fat. What factor is different between these two people?

4. A friend has his cholesterol level checked and tells you that he is really relieved that his cholesterol is normal, that is, under 200. If this is true, would your friend have no health concerns relative to his cholesterol level, or are there more factors he needs to consider?

5. Why might some substances need the help of transport proteins in order to cross the plasma membrane?

Connecting the Science

1. Why do you think that anorexia and bulimia are more common among women than men?

2. What would you say to a friend who believed that he was fat, even though his BMI placed him in the "normal" range? How about a friend who qualifies as obese on a BMI chart but who exercises regularly and eats a well-balanced diet?

3. Some people argue that anorexia and bulimia are less worrisome health problems than obesity. Do you think this is true?

Media Activities

Media Activity 2.1 The Great Harvest: From Food to Energy
Estimated Time: 10 minutes
The animation provides a visual overview of digestion, beginning with the food we eat and proceeding through the breakdown of foods into their chemical components and moving on to how the body's cells use the chemicals to produce energy.

Media Activity 2.2 Personal Diet Analysis
Estimated Time: 20 minutes
Keep a 24-hour food log and examine the nutritional value of the foods you've eaten.

Media Activity 2.3 Body Image and Society
Estimated Time: 10 minutes
Compare current body image ideals with those from previous centuries. The investigation demonstrates that our body fat fixation is as much about culture as it is about science.

Media Activity 2.4 Measuring Health and Fitness
Estimated Time: 10 minutes
Explore methods other than body fat to determine health and fitness and determine what your goals for health should be.

Media Activity 2.5 Health Problems
Estimated Time: 7 minutes
Explore the health problems associated with being over- or underweight and what you should expect of yourself.

CHAPTER 3

Prospecting for Biological Gold

Biodiversity and Classification

This strange Yellowstone hotspring . . .

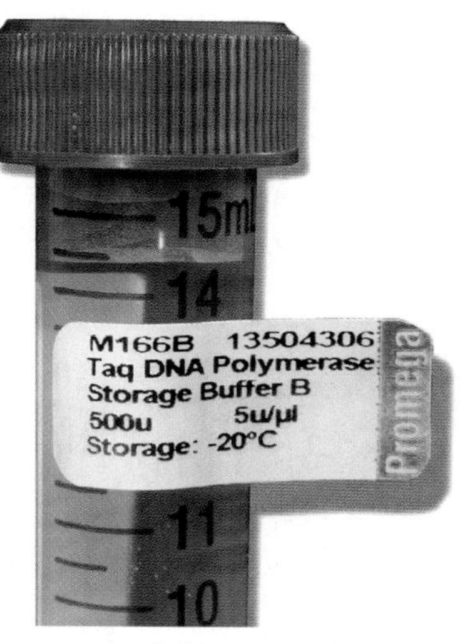

. . . is a source of this extremely valuable chemical, produced by microorganisms.

Scientists are interested in finding other useful products in Yellowstone.

At the end of a narrow foot trail in Yellowstone National Park lies a natural curiosity—Octopus Spring. The boiling hot water of the spring is colored an otherworldly blue. A gooey white crust encircles its main pool, and along the banks of the drainage streams radiating in all directions from it are brightly colored mats and streamers of pink, yellow, green, and orange. Although Octopus Spring is certainly not the most beautiful or dramatic feature of Yellowstone, this relatively small spring looms large in the history of biological discovery and represents a source of continued controversy.

The brilliant colors of Octopus Spring result from large numbers of microscopic organisms living in the water and on nearby surfaces. In the 1960s, Dr. Thomas Brock of the University of Wisconsin was the first to describe this biological community. One of the species he discovered, which he named *Thermus aquaticus* for its affinity for hot water, contained a chemical new to science. This chemical, an enzyme called *Taq* polymerase, can create long chains of DNA at much higher temperatures than other organisms. *Taq* polymerase is now an integral part of the polymerase chain reaction (PCR), a

Are they likely to succeed? Is there much left to discover about life?

high-temperature process used by many laboratories to prepare DNA samples for research or identification. PCR has revolutionized DNA research and made many of the recent advances in genetic technology possible.

Until it was ruled invalid in 1999, the patent for *Taq* polymerase was held by the Swiss pharmaceutical firm Hoffman-LaRoche, and licensing agreements with other companies that used and produced *Taq* polymerase netted the company over $100 million every year. Of this substantial sum, Yellowstone National Park, the National Park Service, and the U.S. Treasury received . . . nothing in royalty payments. Even a small share of the royalties for *Taq* polymerase would have provided funds to improve and manage this heavily used national park.

The managers of Yellowstone Park do not want to miss out on the financial rewards that may come with protecting other valuable species within their borders. In order to capitalize on future discoveries in the park, Yellowstone entered into an agreement in 1997 with Diversa Corporation to identify and describe some of the microscopic species in the park. In return, Diversa agreed to a one-time payment of $100,000 to Yellowstone Park and to provide several thousand dollars for research services. Diversa also agreed to share an undisclosed percentage of royalties from any profitable products that result from their research. The announcement of this deal set off a flurry of criticisms—from environmentalists who fear the disruption of biological communities in the park, to government watchdogs concerned about a few private stockholders profiting from resources taken from a park maintained for the entire public. Diversa has yet to begin exploration in Yellowstone's hot springs, pending the resolution of several legal challenges to their agreement.

Even without the legal challenges, the agreement between Yellowstone's managers and Diversa is a calculated risk by both parties. Diversa is investing nearly a million dollars in this venture, and Yellowstone faces potential damage to the wild and scenic resources the park was designed to protect for the public good. Why are the parties to this agreement willing to take these risks? What is the likelihood of success of Diversa's search for valuable species in Yellowstone? Can there be many organisms humankind has yet to discover? And what do we know about the organisms we have identified?

WWW
Media Activity 3.1A The Tree of Life

3.1 The Organization of Life's Diversity

Diversa Corporation's proposed hunt for new organisms and new uses of known organisms in Yellowstone is called *bioprospecting*. Bioprospectors seek to strike biological "gold" by finding the next penicillin (originally discovered in a fungus), aspirin (produced by willow trees), or *Taq* polymerase in the living world. Yellowstone isn't the only potential source of a mother lode—drug companies are also investing in bioprospecting in the vast Amazonian rain forests, the strange hydrothermal vents of the ocean depths (Figure 3.1), and the bleak expanses of Antarctic ice. Other scientists are also surveying more commonly encountered organisms such as airborne molds and the bacteria that cause tooth decay.

Figure 3.1 A mother lode? The organisms surrounding this deep-sea volcanic site have been known to science for less than thirty years. They represent an intriguing source of unique biological chemicals.

How Diverse Is Life?

The company name "Diversa" reflects the promise and challenge of looking for new chemicals in the natural world. One of the characteristics of life on Earth is that it is full of variety—that is, the living world is diverse. Scientists refer to the variety within and among living species as **biodiversity**. Understanding and appreciating the role of biodiversity in the health of humans and of the planet is an essential aspect of biological science, but Diversa and other bioprospectors are interested in biodiversity for a more utilitarian reason—they are banking on the variety of life to give them chemical "solutions" to human problems.

To bioprospectors, the promise of biodiversity is its great variety, but great variety is also a source of challenge. The number of organisms described by science is between 1.4 to 1.8 million *unique* life forms, called **species**. Even if prospectors spent only a single day screening each species for valuable products, examining them would take more than 5000 years. The variety of species also greatly underestimates the biodiversity within a species (Figure 3.2). Just as the species of tomato can come in a variety of shapes, sizes, and flavors, there may be species where individuals differ greatly in the amount and potency of

Figure 3.2 Biological diversity. These varieties of the common tomato illustrate diversity within a species.

a particular chemical. Bioprospectors could miss a valuable chemical if they only test a small number of individuals from one population of a species.

Biologists disagree about the total number of species that have already been identified and described. Much of this uncertainty stems from the method of storing and cataloguing known species. When biologists identify what appears to be a new species, they collect individual specimens of the organism for storage in specialized museums. Most animal collections are found in natural history museums, while plant repositories are called herbaria (Figure 3.3), and many types of microbes and fungi are kept in specialized facilities called type-collection centers. In order for an organism to be considered a new species by the scientific community, biologists must create a description of the species that clearly distinguishes it from similar species, and they must publish this description in a professional journal. Because there are numerous large natural history museums and herbaria all over the world, along with many different journals, it is often unclear whether a species has already been described. **Systematists** are biologists who specialize in describing and categorizing a particular group of organisms. They evaluate collections to see if there is overlap, but this process is slow and further complicated by the continual discovery and description of new species. The lack of a central resource for species collections and descriptions means that the total number of described species is only an estimate.

The number of known species represents a fraction of the total number of species on Earth. Some scientists estimate that the actual number is closer to 100 million unique species. Most agree that our planet is home to at least 10 million distinct species. Biologists are a long way from knowing all there is to know about the diversity of life.

(a) Natural History Museum **(b)** Herbarium

Figure 3.3 Biological collections. (a) A collection of animals in a natural-history museum. Most collections contain several examples of each species to show the range of variation within the species. (b) Plant specimens are stored in herbaria.

Kingdoms and Domains

Systematists work in the field of **biological classification**, in which they attempt to organize biodiversity into discrete and logical categories. The task of classifying life is much like categorizing books in a library—books can be divided into "fiction" or "nonfiction," and within each of these divisions, more precise categories can be made (for example, nonfiction can be divided into biography, history, science, etc.). The book-cataloguing system used in most public libraries, the Dewey Decimal System, is only one way of shelving books. For instance, academic and research libraries use a different system, developed by the United States Library of Congress. Librarians use the cataloguing system that is appropriate to the collection of books owned by the library and the needs and interests of the library's users; just as there are alternative methods of organizing books, there is more than one way to organize biodiversity to meet differing needs.

Biologists have traditionally subdivided living organisms into great groups that share some basic characteristic. Fifty years ago, most biologists divided life into two categories: plants, for organisms that were immobile and apparently made their own food, and animals, for organisms that could move about and relied on other organisms for food. When it became clear that too many organisms did not fit easily into this neat division of life, some scientists began to argue for a system of five **kingdoms**, in which organisms were categorized according to the type of cell they possessed and their method of obtaining energy. Table 3.1 (page 60) provides an overview of this system.

The five-kingdom system is not perfect either—for instance, the Protista kingdom contains a wide diversity of life forms that have only very superficial similarities. More recently, many biologists have argued that the most appropriate way to classify life is according to historical relationships among organisms. In Chapter 8 we will discuss the theory of *evolution*, which states that all modern organisms represent the descendants of a single set of common ancestors that existed nearly 4 billion years ago. Evidence for shared ancestry includes the universality of many aspects of the cell structures and cellular processes that were described in Chapter 2. The process of divergence from this set of ancestors to the diversity of modern species occurred as millions of branching events—leading biologists to describe it as "the tree of life" (Figure 3.4).

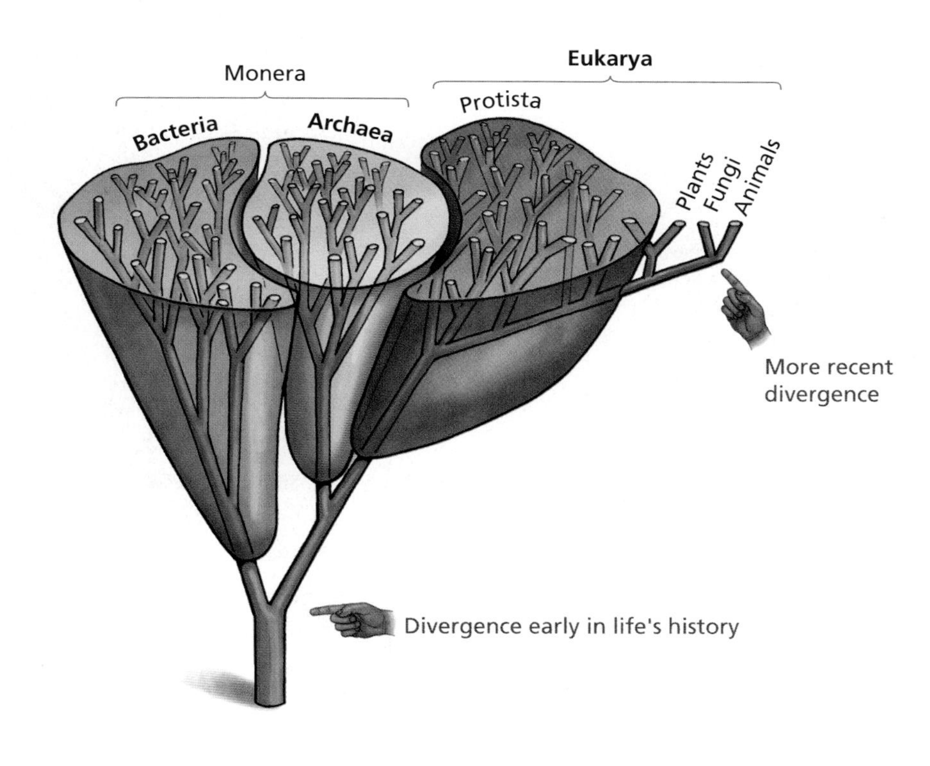

Figure 3.4 The tree of life. This tree exemplifies the current state of knowledge regarding relationships among living organisms. Note that the termination of branch tips are all on the same plane, representing the present time. Living organisms represent a small remnant of all the species that have appeared over Earth's history.

Five-Kingdom System				Three-Domain System
Name	**Characteristics**	**Examples**	**Approximate Number of Known Species**	**Name and Characteristics**
Plantae	Eukaryotic, multicellular, make own food, largely stationary	Pines, wheat, moss, ferns	300,000	**Eukarya** All organisms contain eukaryotic cells.
Animalia	Eukaryotic, multicellular, rely on other organisms for food, mobile for at least part of life cycle	Mammals, birds, fish, insects, spiders, sponges	1,000,000	
Fungi	Eukaryotic, multicellular, rely on other organisms for food, reproduce by spores, body made up of thin filaments called hyphae	Mildew, mushrooms, yeast, *Penicillium*, rusts	100,000	
Protista	Eukaryotic, mostly single–celled forms, wide diversity of lifestyles, including plant–like, fungus–like, and animal–like types	Green algae, *Amoeba*, *Paramecium*, diatoms, chytrids	15,000	The orange boxes indicate the six categories currently used to classify the diversity of life.
Monera	Prokaryotic, mostly single–celled forms, although some form permanent aggregates of cells	*Escherichia coli*, *Salmonella*, *Bacillus anthracis*, *Anabena*, sulfur bacteria	4,000	**Bacteria** Prokaryotes with cell wall containing peptidoglycan. Wide diversity of lifestyles, including many that can make their own food.
		Thermus aquaticus, *Halobacteria halobium*, methanogens	1,000	**Archaea** Prokaryotes without peptidoglycan and with similarities to Eukarya in genome organization and control. Many known species live in extreme environments.

Table 3.1 The classification of life. Until recently, most biologists used the five-kingdom system to organize life's diversity. Now many use the six-category system, which better reflects evolutionary relationship by acknowledging the existence of three major domains as well as four of the kingdoms.

When life is classified according to the relationships among organisms, major groupings correspond to divergences that occurred very early in life's history, and minor groupings correspond to a more recent divergence. Classifying life according to evolutionary relationships may be especially useful to bioprospectors if a close relationship indicates similarity in the compounds produced by living organisms.

Determining the evolutionary relationship among all living organisms requires comparisons of their *DNA*, the basic information molecule found in all living things. As will be discussed in Chapter 7, DNA contains information about how to build an organism. In particular, the sequence of building-block chemicals in a DNA molecule contains information about the sequence of building-block chemicals in *proteins*, and it is proteins that generate the structural and functional parts of cells. Since each species is unique, the set of proteins, and thus the *DNA sequence*, of each species is unique. However, because all species share a common ancestor, all organisms also have basic similarities in their DNA sequences. As a result of the relationship among all living things, the DNA sequences of closely related organisms should be more similar than the DNA sequences of more distantly related organisms (Figure 3.5).

To determine the evolutionary relationship among all modern species, scientists must compare sequences of a segment of DNA that is recognizably similar, and performs a similar function among organisms as diverse as *Homo sapiens* (humans) and *Thermus aquaticus*. The DNA sequence that best fits these criteria is one containing the instructions for making *ribosomal RNA* (rRNA),

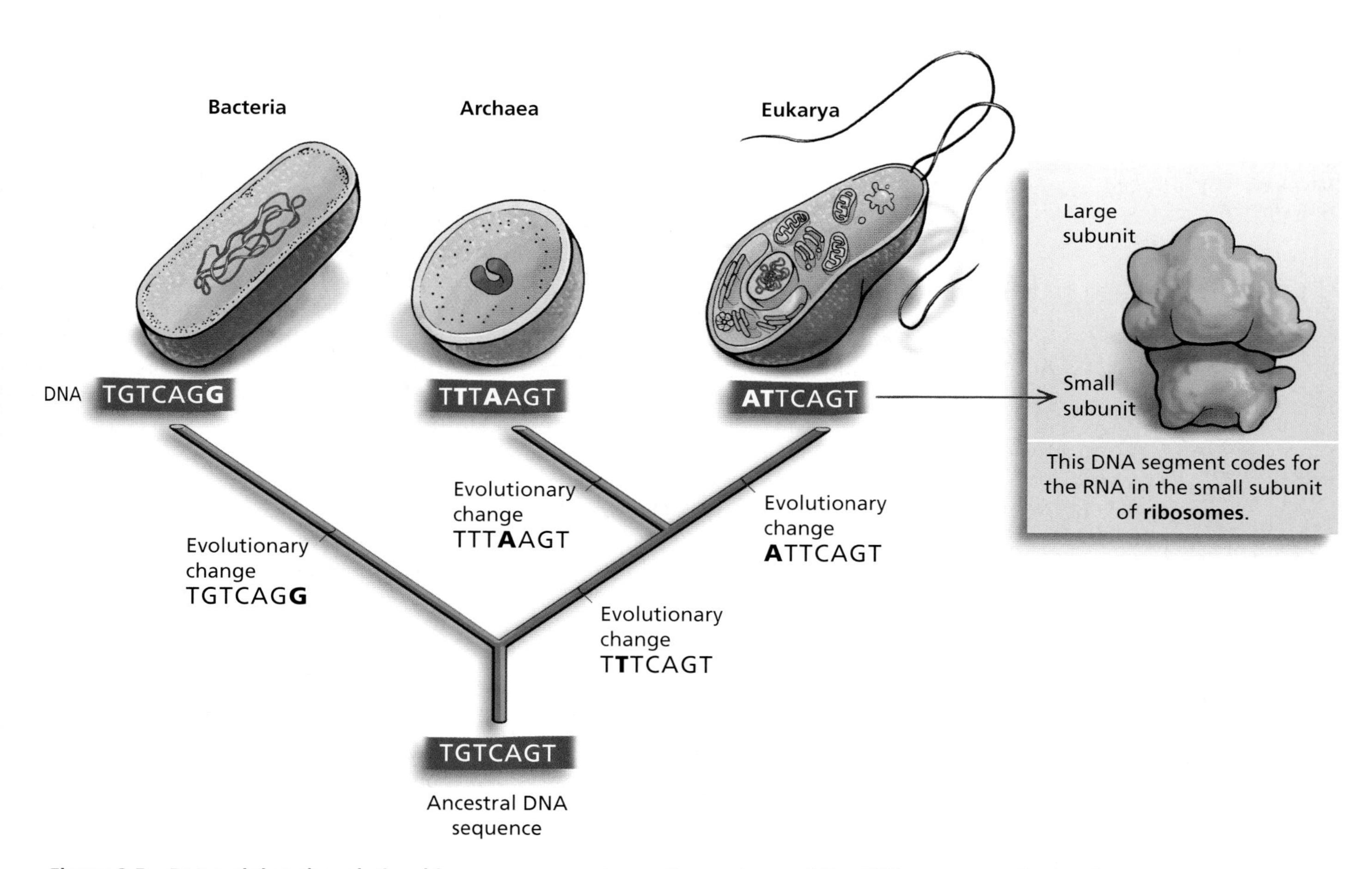

Figure 3.5 Determining the relationships among organisms. Comparisons of the DNA sequences that code for ribosomal RNA in various organisms has helped scientists construct the tree of life.

which is not a protein, but functions as a structural part of a *ribosome*. Ribosomes are cellular structures found both in *prokaryotes* and *eukaryotes*. The function of the ribosome is to translate genetic material into proteins—ribosomes are the fundamental "factories" of all cells. There are several rRNA molecules in each ribosome. The ones primarily used by scientists interested in the relationships among living organisms are found in the *small subunit* of the ribosome. A comparison of the DNA coding for small-subunit rRNAs from myriad organisms yielded the tree diagram in Figure 3.4.

You should note from Figure 3.4 that three of the Kingdoms (Fungi, Animalia, and Plantae) represent relatively recently diverged groups of organisms. What was formerly called the kingdom Monera is actually made up of two groups of organisms that are quite distinct; and the kingdom Protista is a hodgepodge of many very different organisms. To better reflect biological relationship, some biologists have proposed dividing life into three **domains** (represented by the three main branches on the tree—Bacteria, Archaea, and Eukarya), each containing several kingdoms. Until kingdoms within the Bacteria and Archaea are delineated and the relationships among the Protista are determined, most biologists use a six-category system—Protista, Fungi, Plantae, Animalia, Bacteria, and Archaea.

3.2 Locating Valuable Species

In reality, the number of kingdoms and domains into which life is appropriately classified is of minor importance to a bioprospector, but understanding more recent evolutionary relationships may be *very* important, as we will discuss in Section 3.3. However, dividing life into five or six kingdoms simplifies *our* discussion of biodiversity and of where bioprospectors may find valuable resources within the variety of life.

www
Media Activity 3.1B Prokaryotic versus Eukaryotic Cells

Bacteria and Archaea

The most ancient fossilized cells (Figure 3.6), which lived approximately 3.5 billion years ago, are remarkably similar in external appearance to modern **bacteria** and **archaea**. Both bacteria and archaea are **prokaryotes**, meaning that they do not contain a *nucleus*, which provides a membrane-bound, separate compartment for the DNA in *eukaryotes*. Prokaryotes also lack other internal structures bounded by membranes, such as the mitochondria and chloroplasts described in Chapter 2. Although some species may be found in chains or small colonies, as in Figure 3.6, most prokaryotes are **unicellular**, meaning that each cell is an individual organism. These individual cells are hundreds of times smaller than the cells that make up our bodies; for this reason, they are often called *microorganisms* or **microbes**. Their small size, easily accessible DNA, and simple structure make prokaryotes very attractive to bioprospectors, because the process of studying, growing, and manipulating these organisms is less difficult.

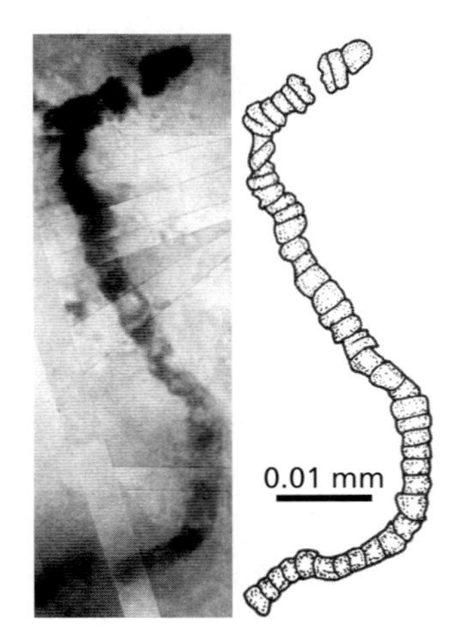

Figure 3.6 The oldest form of life. This photograph of a fossil is accompanied by an interpretive drawing showing the fossil's living form. It was found in rocks dated at 3.465 billion years old.

The relatively simple structure of prokaryotes belies their incredible chemical complexity and diversity. There are prokaryotes that can live on petroleum, others on hydrogen sulfide emitted by volcanoes deep below the surface of the ocean, and some simply on water, sunlight, and air. Prokaryotes are ubiquitous—they are found in and on nearly every square centimeter of Earth's surface, including very hot and very salty places, and even thousands of feet below ground. Prokaryotes are also incredibly numerous—there are more prokaryotes living in your mouth right now than the total number of humans who have ever lived!

Most catalogued prokaryotes are known because they cause disease in humans or crops. While enormous effort is expended to control these organisms, the fact that they can live in and on another living organism is rather amazing.

In order to survive, these organisms must escape eradication by their host's infection-fighting system. The chemicals that allow bacteria to effectively colonize living humans could be useful in treating diseases of the human immune system, and thus represent one source of bacterial biological gold.

Many known bacteria obtain nutrients by decomposing dead organisms. Bacterial species that function as decomposers often have *competitors* for their food sources—other species that also consume the same food source. When many individuals compete for the same resource, those with traits that increase their ability to get a share of the resource are more likely to survive and reproduce. This is the basis of *natural selection*, the process that is the primary cause of the diversification of life. In this process, individuals with effective traits survive and pass these traits on to their offspring, causing the next generation to consist of more individuals with the useful trait. This causes a change in the traits of the population, that is, it has *evolved*. Natural selection is explored more thoroughly in Chapter 9. In the case of bacteria, natural selection has caused the evolution of traits that give them an edge over the competition—often in the form of chemicals such as **antibiotics** that kill or disable other bacteria. Today, more than half of commercial antibiotics are derived from prokaryotes, and bioprospectors are very interested in finding more of these extremely valuable compounds within the kingdom Bacteria.

WWW
Media Activity 3.2 Investigating Antibiotic Resistance

Although superficially similar to Bacteria, domain Archaea differs from it in many fundamental ways, including some of the basic structure of cells and membranes. The known archaea encompass numerous organisms found in extreme environments, including high-salt, high-sulfur, and high-temperature habitats. *Thermus aquaticus*, the source of *Taq* polymerase and a hot-spring dweller, belongs to the Archaean kingdom. *Taq* polymerase is valuable because it operates at high temperature—making it, along with compounds from other hot-spring archaeans, potentially useful in industrial settings. Natural selection of archaea in these and other challenging environments has probably caused the evolution of other unique and useful biological molecules.

Scientists and bioprospectors still have much to learn about bacteria and archaea—beginning with a basic understanding of exactly how diverse they are. For instance, although most archaeans are known from extreme environments, members of this domain are found everywhere they are sought. Some scientists estimate that the number of undescribed prokaryotic species could range up to 100 million. Diversa's focus on the microscopic organisms of Yellowstone reflects the effort drug companies are putting into microbial bioprospecting.

Protista

The kingdom **Protista** is made up of the simplest known **eukaryotes**—organisms made up of cells containing a nucleus and other membrane-bounded internal structures. Most *protists* are single-celled creatures, although there are several that have enormous **multicellular** (many-celled) forms.

The most ancient fossils of eukaryotic cells are approximately 2 billion years old, nearly 1.5 billion years *younger* than the oldest prokaryotic fossils. According to the **endosymbiotic hypothesis** for the origin of protists, eukaryotes were most likely the descendants of a "confederation" of cells. At least some eukaryotic cell structures, including *mitochondria*, the cell's power plants, appear to have descended from bacteria that took up residence inside (*endo*) ancestral eukaryotic cells. The bacteria and primitive eukaryotes both benefited from this arrangement (a *symbiosis*), and over time they became inextricably tied together (Figure 3.7). When biologist Lynn Margulis first popularized this hypothesis in the U.S. in 1981, many of her colleagues were skeptical; but an examination of the membranes, DNA, and ribosomes of mitochondria show clear similarities to the same features in certain bacteria.

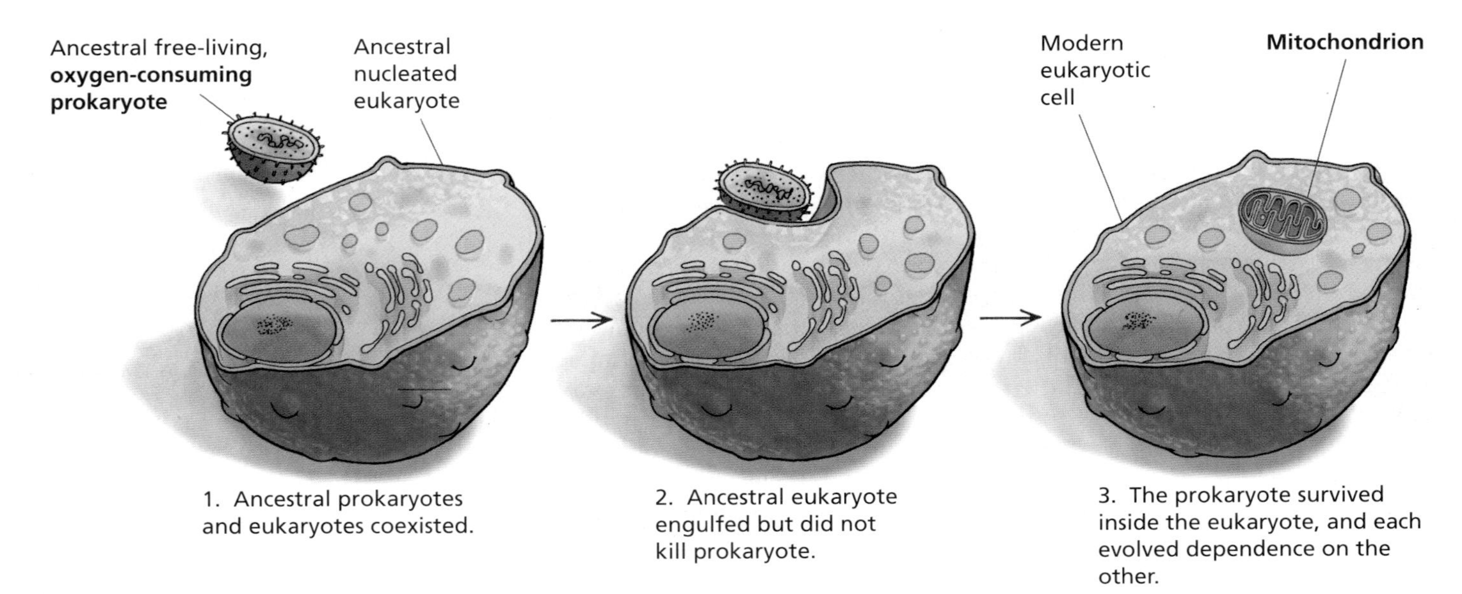
Ancestral free-living, **oxygen-consuming prokaryote**
Ancestral nucleated eukaryote
Modern eukaryotic cell
Mitochondrion
1. Ancestral prokaryotes and eukaryotes coexisted.
2. Ancestral eukaryote engulfed but did not kill prokaryote.
3. The prokaryote survived inside the eukaryote, and each evolved dependence on the other.

Figure 3.7 The evolution of eukaryotes. The leading hypothesis regarding the evolution of eukaryotes is endosymbiosis. Mitochondria (and chloroplasts) appear to be descendants of once free-living bacteria that took up residence within an ancient nucleated cell.

WWW
Media Activity 3.1C Endosymbiotic Theory

Not long after the first eukaryotes appeared, a wide diversity of eukaryotic forms established themselves on Earth. The modern kingdom Protista contains organisms resembling animals, fungi, and plants. Thus it is believed that this diverse kingdom potentially contains a plethora of biologically active compounds associated with these myriad lifestyles (Figure 3.8). As with the prokarya, most of the members of kingdom Protista remain unknown.

The group of protists that has been investigated in most detail by bioprospectors is the **algae**; the only members of this kingdom with the ablity to manufacture food. As with plants, this is accomplished with the aid of sunlight via *photosynthesis* (see Essay 2.1). The photosynthetic production by algae represents a rich and tempting food source to nonphotosynthetic organisms. One strategy that algae and plants have evolved against their *predators* is defensive chemicals. These chemicals make the algae distasteful or even poisonous to the predator—therefore, humans could use them to control *our* predators. For example, extracts of red algae stop the reproduction of a number of human disease organisms in laboratory settings; this effect is presumably related to the algae's defensive chemicals.

The group of organisms commonly referred to as algae is actually made up of several distinct, quite divergent, categories of organisms. Each of these algal divisions have methods of producing and storing food that is quite different from the others; and some of the unique compounds that result from these processes are potentially useful to humans. For example, carageenan, a slimy carbohydrate produced by red algae (see Figure 3.8c), is commercially harvested from ocean algal beds and used as a stabilizer and thickener in foods, medicines, and cosmetics.

WWW
Media Activity 3.1D The Architecture of Animals

Animalia

From the origin of the first prokaryote until approximately 600 million years ago, life on Earth consisted only of single-celled creatures. Then, multicellular organisms first begin to appear in the fossil record. (Note that there is some

(a) Animal-like protist

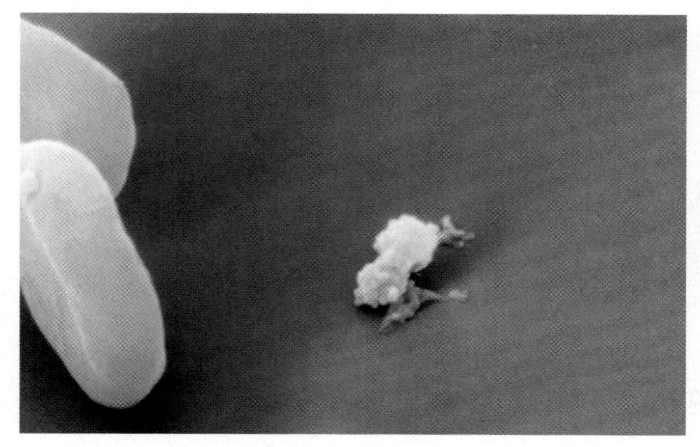

Plasmodium falciparum, the causative agent of malaria. This parasite of humans is of interest because of its ability to outsmart even a healthy immune system.

(b) Fungus-like protist

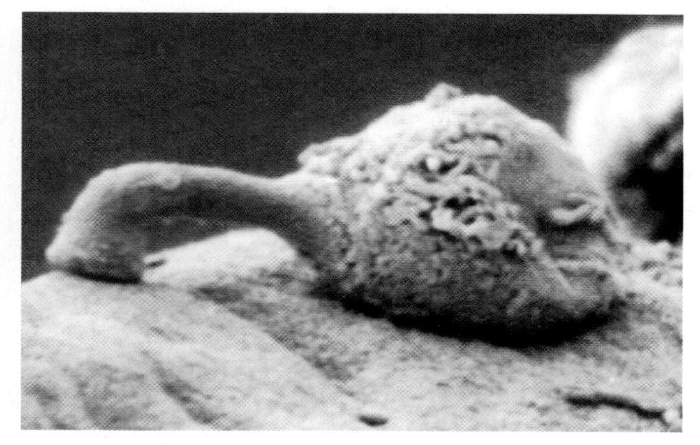

Lagenidium giganteum, used to control mosquito populations.

(c) Plant-like protists

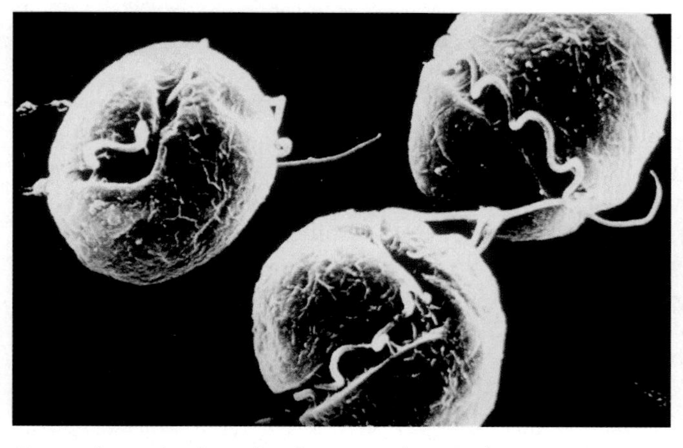

Gonyaulax polyedra, a luminescent alga used to measure levels of toxic materials in ocean sediments.

Chondrus crispus, a red alga that is the source of carageenan.

Figure 3.8 Protista. Protista is the most diverse of life's kingdoms, and includes animal-like forms (a), fungus-like forms (b), and plant-like forms (c). While most plant-like protists are single-celled, there are many large multicellular forms.

disagreement about when multicellular organisms first appeared. In 2002, Australian scientists announced that they had found evidence of multicellular life from at least 1.2 billion years ago. This is nearly twice as old as most other estimates.) The ancient many-celled creatures of 600 million years ago were organisms unlike any modern species, and included giant fronds and ornamented disks (Figure 3.9). Biologists are unsure which of these species is the common ancestor of modern *animals,* multicellular organisms that make their living by ingesting other organisms, and most of whom have the ability to move during at least one stage of their life cycle. Within about 40 million years of the first appearance of the early multicellular organisms pictured in Figure 3.9, *all* modern animal groups had evolved.

The relatively sudden appearance of the modern forms of animals—a period comprising little more than 1% of the history of life on Earth—is referred to as the **Cambrian explosion,** named for the geologic period during which it occurred.

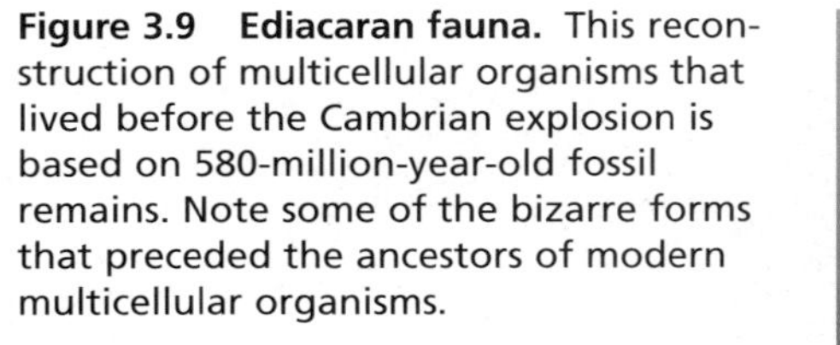

Figure 3.9 Ediacaran fauna. This reconstruction of multicellular organisms that lived before the Cambrian explosion is based on 580-million-year-old fossil remains. Note some of the bizarre forms that preceded the ancestors of modern multicellular organisms.

One of the most compelling questions in biology is the source of this explosion of biodiversity. The sudden evolution of the immense diversity of life from simple, single cells is remarkable. Some scientists hypothesize that the evolution of the animal lifestyle itself—that is, as predators of other organisms—led to the flowering of this diversity.

It can be difficult to conceive that an animal as complex as a human could have evolved from a simple eukaryote ancestor. However, humans are actually not very different from other eukaryotic organisms. By the time the first cell containing a nucleus appeared, all of the complicated processes that take place in modern cells, such as cell division and cellular respiration, had evolved. And by the time the first multicellular animals appeared, many of the complex processes required to maintain these larger organisms, such as communication systems among cells and the formation of organs and organ systems, must have arisen. Although a human and a starfish appear to be very different, the way they develop, and the structures and functions of their cells and common organs are nearly identical. Essay 3.1 should help you appreciate how much time has passed since the initial diversification among animal groups at the Cambrian explosion and the resulting remarkable variety of animals we see today.

While members of the kingdom **Animalia** (Latin *anima*, meaning "breath," or "soul") are relatively similar, there is still significant diversity within the kingdom. We typically picture mammals, birds, and reptiles when we think of animals, but species with backbones (including these and fish and amphibians) represent only 4% of the entire kingdom. A small number of these **vertebrates** have traits interesting to a bioprospector. For instance, poison dart frogs (Figure 3.10) secrete high levels of toxins onto their skin. These toxins are nerve poisons that cause convulsions, paralysis, and even death to their potential predators—in fact, their name derives from the traditional human use of these creatures as a source of toxins to coat the tip of hunting darts. The nerve toxins produced by these frogs are potentially valuable to bioprospectors as sources of potent, nonaddictive painkillers.

Figure 3.10 Poison dart frog. This brightly colored frog contains glands in its skin that release poison when the frog is handled. The frog's bright colors warn potential predators of its toxicity.

Most of the work of bioprospectors in kingdom Animalia focuses on the remaining 96% of known organisms—the **invertebrates** (animals without backbones, Figure 3.11). The vast majority of multicellular organisms on Earth are

(a) Ant (*Pseudomyrmex triplarinus*)

Source of arthritis treatment.

(b) Jellyfish (*Aequorea victoria*)

Source of fluorescent protein, a useful labeling tool in microbiology.

(c) Horseshoe crab (*Limulus polyphemus*)

Source of blood proteins used to test for pathogens in humans.

Figure 3.11 Examples of invertebrates. Most animals are invertebrates, including insects, jellyfish, and horseshoe crabs.

invertebrate animals and most of these animals are insects. Many invertebrates contain chemical compounds not found elsewhere in nature. Invertebrates that produce venoms to repel predators and competitors, such as many species of beetles, ants, bees, wasps, and spiders, are one source of potential drugs. For example, the tropical ant *Pseudomyrmex triplarinus* produces venom that appears to be useful for treating the joint swelling and pain associated with arthritis. Also, ants, bees, wasps, and termites that manage to flourish in crowded colonies may prove to be a source of compounds for reducing the spread of disease in human populations.

The animals that inhabit the oceans' incredibly diverse coral reefs are especially interesting to bioprospectors (Figure 3.12). These biological communities are very crowded with life, and the individuals within them continually interact with predators and competitors. As a result of this challenge, many successful coral-reef organisms contain defensive chemicals that might be useful as drugs. One mantra of reef bioprospectors is, "If it is bright red, slow-moving, and alive, we want it," reflecting the assumption that in order for a marine animal to survive despite being so conspicuous and easy to catch, it must have evolved powerful deterrents to predators. The number of unknown invertebrate species, especially in the oceans, is estimated to be anywhere from 6 to 30 million.

Figure 3.12 A coral reef. This extremely diverse biological community is a rich source of interesting biological chemicals.

Fungi

Early classifications that separated the biological world into two kingdoms, Plantae and Animalia, placed **fungi** in the Plantae kingdom. Like plants, fungi are immobile and many produce organs that function like fruit by dispersing *spores*, cells that are analogous to plant seeds in that they can germinate into new individuals. However, the mushroom you think of when you imagine fungi is a misleading representation of the kingdom. Most of the functional part of fungi

Essay 3.1 Understanding Deep Time

The number and complexity of modern organisms is enormous. How could so much complexity have evolved in such a seemingly short period since the origin of life? Consider the amount of time between the Cambrian explosion, the time all forms of modern animals arose, and today: 530,000,000 years. 530 million does not seem like an unusually large number—in dollars, it is less than 5% of Microsoft-founder Bill Gates's net worth, and it is only a tiny fraction of the $2 trillion annual budget of the United States.

However, if we put 530 million years on a more human scale, that of a single calendar year, the time between the evolution of the first animals and the evolution of humans becomes a bit easier to grasp (Table E3.1). In this table, a single day is equivalent to approximately 1.45 million years. As you can see, the age of a typical college sophomore (20 years) is equivalent to just one *second* in this analogy. All of humankind's recorded history is only approximately nine *minutes*. The diversity of modern species has arisen over a truly enormous time scale.

Table E3.1 530 million years in 365 days. An analogy between the Cambrian explosion and a calendar year. This analogy demonstrates the tiny scale of human history relative to the history of life.

Event	Years Before Present	Corresponds to Calendar Date
College sophomore's birth year	20	December 31, 11:59:59 p.m.
Atomic bomb dropped on Hiroshima	58	December 31, 11:59:57 p.m.
Darwin publishes *Origin of Species*	144	December 31, 11:59:52 p.m.
Columbus arrives in America	511	December 31, 11:59:30 p.m.
Beginnings of Christianity	2000	December 31, 11:58 p.m.
First agriculture	9500	December 31, 11:51 p.m.
First art	28,000	December 31, 11:32 p.m.
First *Homo sapiens*	300,000	December 31, 8 a.m.
First hominids	4.5 million	December 28
First apes	25 million	December 17
Dinosaurs extinct	65 million	November 15
First mammal	200 million	August 18
First animal with backbone	500 million	January 20
Cambrian explosion	530 million	January 1

is made up of very thin, stringy material called *hyphae*, which grows over and within a food source (Figure 3.13). Fungi feed by secreting chemicals that break down the food into small molecules, which they then absorb into the cells of the hyphae. The stringlike form of fungal hyphae maximizes the surface over which feeding takes place—so the vast majority of the "body" of most fungi is microscopic and diffuse. Fungal food sources typically include dead organisms, and the actions of fungi are key to recycling nutrients for plant growth. Fungi are more like animals than plants in that they rely on other organisms for food. In fact, DNA sequence analysis indicates that Fungi and Animalia are more closely related to each other than either kingdom is to the Plantae kingdom (see Figure 3.4).

About one-third of the bacteria-killing antibiotics in widespread use today are derived from fungi. Fungi probably produce antibiotics because their main competitors for food are bacteria. Penicillin, the first commercial antibiotic, is produced by a fungus (Figure 3.14). Its discovery is one of the great examples of good fortune in science.

Before he went on vacation during the summer of 1928, the British bacteriologist Alexander Fleming left a dish containing the bacteria *Staphylococcus aureus* on his lab bench. While he was away, this culture was contaminated by a spore from a *Penicillium* fungus that may have come from a different laboratory in the same building. When Fleming returned to his laboratory, he noticed that the growth of *S. aureus* had been inhibited on the fungus-contaminated culture dish. Fleming had been searching for a method to control bacterial growth, and this chance discovery provided a clue. He inferred that some chemical substance had diffused from the fungus, and he named this antibiotic penicillin, after the fungus itself. The first batches of this bacteria-slaying drug became available during World War II. Many historians believe that penicillin helped the Allies win the war by greatly reducing the number of deaths from infection in wounded soldiers. Since the discovery of penicillin, hundreds of other antibiotics have been isolated from different fungus species.

Some fungi infect living animals, and thus also have potential as sources of drugs. To survive in a living organism, these fungi must be able to escape detection by their host's immune system. Cyclosporin is a drug derived from a fungus that is able to infect live insects by suppressing their host's immune response to ensure their own survival. In an organ-transplant recipient, cyclosporin suppresses the immune response that would be mounted against the foreign organ. Other, related fungi produce substances that interfere with virus reproduction. One of these compounds is used to treat people infected with the virus that causes AIDS.

As with most inconspicuous or hard-to-reach species, the number of unknown fungi is probably much greater than the number that have been identified to date.

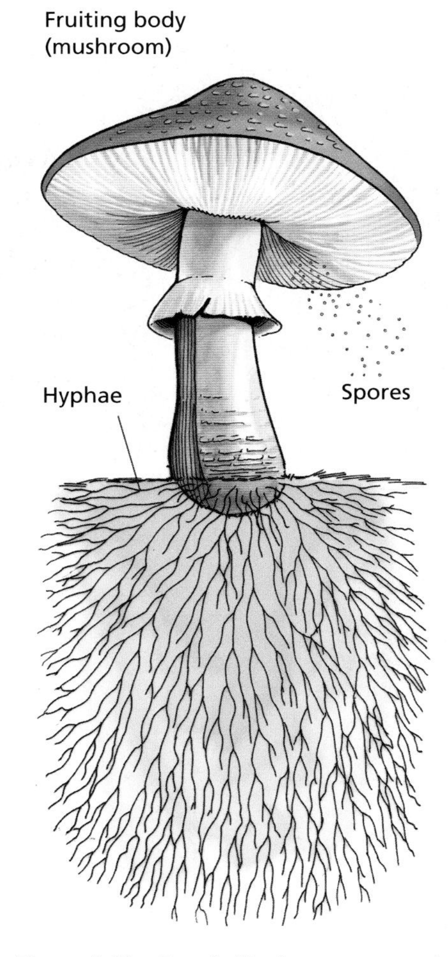

Figure 3.13 Fungi. Hyphae can extend over a large area. The familiar mushroom, as well as the fruiting structures of less-familiar fungi, primarily function only as a method of dispersing spores.

www
Media Activity 3.1E Structure and Reproduction in Fungi

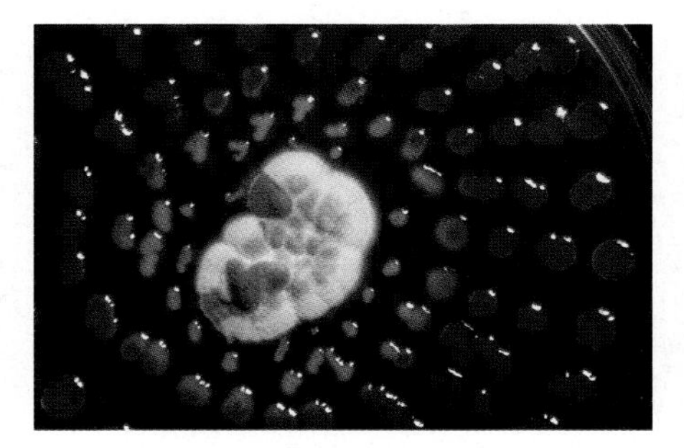

Figure 3.14 Antibiotics from fungi. The pink and grey mold in the center of this petri dish is a colony of *Penicillium*. The red dots on the edges of the dish are colonies of bacteria. The pink dots surrounding the *Penicillium* colony are bacterial colonies that are dying from contact with secretions from the fungi.

Plantae

The kingdom **Plantae** consists of multicellular eukaryotic organisms that make their own food via photosynthesis (see Essay 2.1). Plants have been present on land for over 400 million years, and their evolution is marked by increasingly effective adaptations to the terrestrial environment. The first plants to colonize land were necessarily small and close to the ground, for they had no way to transport water from where it is available in the soil to where it is needed, in the leaves. The evolution of *vascular tissue*, made up of specialized cells that can transport water and other substances, allowed plants to reach tree-sized proportions, and to colonize much drier areas. The evolution of *seeds*, structures that protect and provide a food source for young plants, represented another adaptation to dry conditions on land. However, most modern plants belong to a group that appeared only about 100 million years ago, the **flowering plants**. Like their ancestors, flowering plants possess vascular tissue and produce seeds, but in addition, these plants evolved a specialized reproductive organ, the *flower*. Over 90% of the known plant species are flowering plants (Figure 3.15).

From about 100 million to 80 million years ago, the number of flowering plant families increased from around 20 to over 150. During this time, flowering plants became the most abundant plant group in nearly every habitat. The rapid expansion of flowering plants is called **adaptive radiation**—the diversification of one or a few species into a large and varied group of descendant species. Adaptive radiation typically occurs either after the appearance of an evolutionary breakthrough in a group of organisms or after the extinction of a

(a) Foxglove (*Digitalis purpurea*)

Source of heart drug digitalis.

(b) Aloe (*Aloe barbadensis*)

Source of aloe vera, used to treat burns and dry skin.

(c) Willow (*Salix alba*)

Source of aspirin.

Figure 3.15 Diversity of flowering plants. Some of the plants we see around us every day provide useful medicines.

WWW

Media Activity 3.1F The Evolution of Plant Structure and Reproduction

competing group. The radiation of flowering plants is due to an evolutionary breakthrough—some advantage they had over other plants allowed them to assume roles that were already occupied by other species. Essay 3.2 describes the chronicle of life as a history of the successive adaptive radiations of organisms.

There is still debate among plant biologists, or **botanists**, regarding which of the traits that flowering plants possess gives them an advantage over nonflowering types. Some botanists believe that the unique reproductive cycle of flowering plants—which includes a process of "double fertilization," and oftentimes the assistance of animals—led to their radiation (Figure 3.16). Other botanists think that chemical defenses in these plants, which reduced their susceptibility to predators, provided their edge over other plant groups.

The diversity of chemical defenses in flowering plants makes them particularly interesting to bioprospectors. For instance, curare vines produce toxins that block the ability of nerves to control muscle tissue. Organisms that get this toxin in their bloodstreams become paralyzed, and do little damage to the vine. Doctors use the defensive chemical in curare vines, curarine, as a muscle relaxant during surgery.

The kingdom Plantae is the source of many well-known naturally derived drugs. Aspirin from willow, the heart drug digitalis from foxglove, the anticancer chemical vincristine from the rosy periwinkle, and dozens of other pharmaceutical products are derived directly from plants; hundreds of other drugs based on plant chemicals are produced via manufacturing processes. Many botanists believe that the number of unknown plant species is relatively small—probably a few thousand—but the potential of even the known species as sources of drugs is still mostly unknown.

3.3 Tools of the Bioprospector

Our survey of life's diversity illustrates the essential problem for the bioprospector—there are many more potentially valuable species than resources to find and evaluate them. This challenge has led many drug companies to abandon most of their bioprospecting programs in favor of a strategy called "rational drug design," which allows them to use an understanding of the causes of illness to create synthetic drugs to treat a particular disease. However,

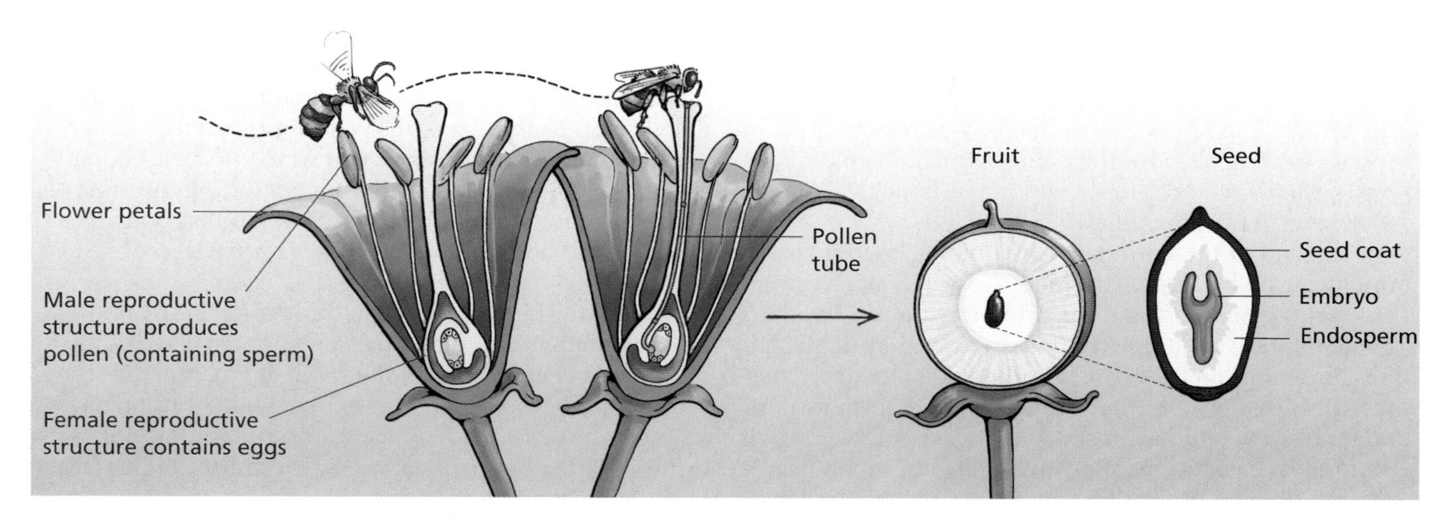
1. **Flower petals** attract insects that move pollen from one flower to another, helping fertilization to occur.
2. **Double fertilization** occurs. The pollen tube carries two sperm. One fertilizes the embryo, and the other fuses with another cell to produce the endosperm, a tissue that nourishes the embryo.
3. **Fruit** consists of seeds packaged in a structure that aids their dispersal, such as tasty flesh or a parachute.
4. **Seeds** contain an embryo and endosperm, and are highly resistant to drying.
Flower petals
Male reproductive structure produces pollen (containing sperm)
Female reproductive structure contains eggs
Pollen tube
Fruit
Seed
Seed coat
Embryo
Endosperm

Figure 3.16 The unique reproduction method of flowering plants. Flowering plants reproduce quite differently from nonflowering plants. These differences, including the production of fruit for dispersal, may have provided their advantage over other plant types, and possibly led to their adaptive radiation.

many scientists have noted that some of the most effective plant-based drugs are so strange in structure that they would never have been designed through this process. Companies like Diversa are banking on finding some of these strange compounds in nature—and they have several techniques to increase their likelihood of success.

Fishing for Useful Products

The National Cancer Institute (NCI) has taken a brute-force approach to screening species for evidence of cancer-suppressing chemicals. NCI scientists receive frozen samples of organisms from around the world, which they then chop up, mix with various chemical solvents, and separate into a number of extracts, each probably containing hundreds of components. These extracts are then tested against up to 60 different types of cancer cells to evaluate their efficacy in stopping or slowing growth of the cancer. Promising extracts are then further analyzed to determine their chemical nature, and chemicals in the extract are tested singly to find the effective compound. This approach is often referred to as the "grind 'em and find 'em" strategy.

To date, this strategy has been effective in identifying one major anticancer chemical—paclitaxel, also known by the trade name Taxol, from the Pacific Yew. Paclitaxel is still produced from the needles of other species of yew trees, and is effective against ovarian cancer, advanced breast cancers, malignant skin cancer, and some lung cancers. Dozens of other less well-known anticancer drugs have been identified by this route as well.

Discovering Relationships Among Species

The grind 'em and find 'em approach works best when researchers are seeking treatments for a specific disease or set of diseases, such as cancer. But most bioprospectors are much more speculative—they are interested in determining

Essay 3.2 Diversity's Rocky Road

Paleontologists study fossils and other evidence of early life, and have been able to piece together the history of life from these data. Early in this reconstruction, they recognized distinct "dynasties" of groups of organisms during different periods. The rise and fall of these dynasties allowed scientists to subdivide life's history into geologic periods. Each period is defined by a particular set of fossils. Table E3.2 gives the names of major geologic periods, their age and length, and major biological events that occurred during them.

The end of the dominance of a biological dynasty is often caused by *mass extinctions*—losses of species that are rapid, global in scale, and affect a wide variety of organisms. For instance, the mass extinction of the dinosaurs (and of 60–80% of *all* organisms) distinguishes the division between the Cretaceous and Tertiary periods. Mass extinctions most probably occur as the result of a global catastrophe. Paleontologists believe that the mass extinction of the dinosaurs was most likely the result of an enormous asteroid strike that occurred off the coast of what is now the Yucatan Peninsula in Mexico. This strike appears to have caused the incineration of large areas of forest in both North and South America and a massive global tidal wave. It also probably threw up an enormous cloud of debris that blocked the sun's light for up to three months, leading to a decade of severe acid rain. The organisms that were fortunate enough to survive through this cataclysm, including our mammal ancestors, formed the basis of modern species. The adaptive radiation of these survivors has led to the current dynasty—The Age of Mammals.

Currently, Earth appears to be experiencing another mass extinction, this time caused by human activity. The current mass extinction is the topic of Chapter 14. The state of biological diversity—and the fate of humans—after this modern mass extinction is in doubt. But if the history of life is any indication, the next great era will be as different as previous ones.

whether an organism contains a chemical that is useful against *any* disease. Doing this effectively requires a more thoughtful approach, taking into account the biology of the species.

One aspect of an organism's biology that can be illuminating to the bioprospector is its **ecology**—that is, its relationship to the environment and other living organisms. Our survey of diversity illustrated some ecological characteristics that increase the likelihood of a species containing valuable chemicals. Some of these characteristics include high levels of competition with bacteria and fungi, susceptibility to predation, and population crowding. An understanding of ecology is useful even within a species. For instance, populations of plants experiencing high levels of insect attack may have more defensive compound present than populations not under attack. Screening organisms whose ecology indicates the probability of defensive or antibiotic compounds is one method bioprospectors use to increase their success.

Another clue to an organism's chemical traits may come from understanding its relationship to other species and the traits found in its closest relatives. This is one reason that some scientists argue that a classification system reflecting evolutionary relationships is more useful than one based on more superficial similarities. The classification of certain birds helps illustrate this point. Vultures (Figures 3.17a and b) are birds that specialize in feeding on dead animals. These birds spend a large amount of time soaring on broad flat wings, have sharp beaks for tearing meat, and regurgitate food to feed their offspring. A nonevolutionary classification places all vultures together. However, research published in the 1970s demonstrated that New World vultures in the Western Hemisphere (Figure 3.17b) appear to be more closely related to storks (Figure 3.17c)—long-legged birds with long beaks that specialize in catching fish in shallow waters—than they are to Old World vultures from the Eastern Hemisphere (Figure 3.17a). Even though a New World vulture *looks* like an Old World vulture, it is much more

Era	Period	Millions of Years Ago	What Life on Earth Was Like
Cenozoic	Quaternary	0	Most modern organisms present.
	Tertiary	1.8	After the extinction of the dinosaurs. Mammals, birds, and flowering plants diversify.
Mesozoic	Cretaceous	65	Massive carnivorous and flying dinosaurs are abundant. Large cone-bearing plants dominate forests. Flowering plants appear.
	Jurassic	144	Huge plant-eating dinosaurs evolve. Forests are dominated by cycads and tree ferns.
	Triassic	206	Early dinosaurs, mammals, and cycads appear on land. Life "restarts" in the oceans.
Paleozoic	Permian	251	Early reptiles appear on land. Seedless plants abundant. Coral and trilobites abundant in oceans. Permian ends with extinction of 95% of living organisms.
	Carboniferous	290	Land is dominated by dense forests of seedless plants. Insects become abundant. Large amphibians appear.
	Devonian	354	Known as the age of fishes. Sharks and bony fish appear. Large trilobites are abundant in the oceans.
	Silurian	408	Life begins to invade land. The first colonists are small seedless plants, primitive insects, and soft-bodied animals.
	Ordovician	439	Life is diverse in the oceans. Cephalopods appear and trilobites are common.
	Cambrian	495	All modern animal groups appear in the oceans. Algae are abundant.
Pre-Cambrian		543	Life is dominated by single-celled organisms in the ocean. Ediacaran fauna appear at the end of the era.
		4500	

Table E3.2 Geological periods. The history of life is divided into four major eras with all but the first era divided into several periods. Periods are marked by major changes in the dominant organisms present on Earth.

(a) Old World vulture

Hooded Vulture (*Necrosyrtes monachus*)

(b) New World vulture

Turkey Vulture (*Cathartes aura*)

(c) Stork

Wood Stork (*Mycteria americana*)

Figure 3.17 The challenge of biological classification. (a) Old World vulture; (b) New World vulture; and (c) a stork. The evolutionary relationship between New World vultures and storks is not evident from their appearance.

similar anatomically, physiologically, and genetically to a stork. An **evolutionary classification** can be quite useful in the study of living organisms; for instance, if scientists wish to know more about the basic biology of New World vultures, they might start by learning what is known about the biology of storks.

Evolutionary classifications are based on the principle that all of the descendant species of a common ancestor will share any biological trait that first appeared in that ancestor. Let us create an imaginary scenario: Imagine a mite species found in college student bedrooms that evolves the ability to use stale corn chips as a food source. We could identify all the descendant species of the corn-chip-eating mite by their ability to eat old nachos. Now imagine that one of the nacho-eating descendant species evolves a coloration that allows it to blend in perfectly against a background of rarely washed socks. All of its descendant species can be identified by this trait. Finally, imagine that one of the sock-colored corn-chip eaters' descendants began reproducing in the bindings of unopened textbooks. Again, any of its descendants will have this trait as well. Because every speciation event involves an evolutionary change, scientists can use these modified traits in modern organisms to reconstruct its **phylogeny**, the evolutionary history of a group of organisms (Figure 3.18).

Of course, in the real world, reconstructing evolutionary relationships is not as simple as our scenario suggests. Descendant species may lose a trait that evolved in their ancestor, or unrelated species may acquire identical traits via a different evolutionary pathway, a process called *convergent evolution*. These occurrences complicate attempts to determine the accurate evolutionary classification of organisms.

Any classification developed by a biologist can be considered to be a hypothesis of the evolutionary relationship among organisms. It is difficult to test this hypothesis directly—scientists have no way of observing the actual speciation events that gave rise to distinct organisms. However, scientists *can* test their hypotheses by using information from both living organisms and fossils. Among living organisms, closely related species should have similar DNA: If the pattern of DNA similarity matches the hypothesized evolutionary relationship, the hypothesis is strongly supported. This is the case with the hypothesized relationship between New World vultures and storks; DNA sequence comparisons indicate that the DNA of New World vultures is more similar to the DNA of storks than to Old World vultures. By examining the fossils of extinct organisms,

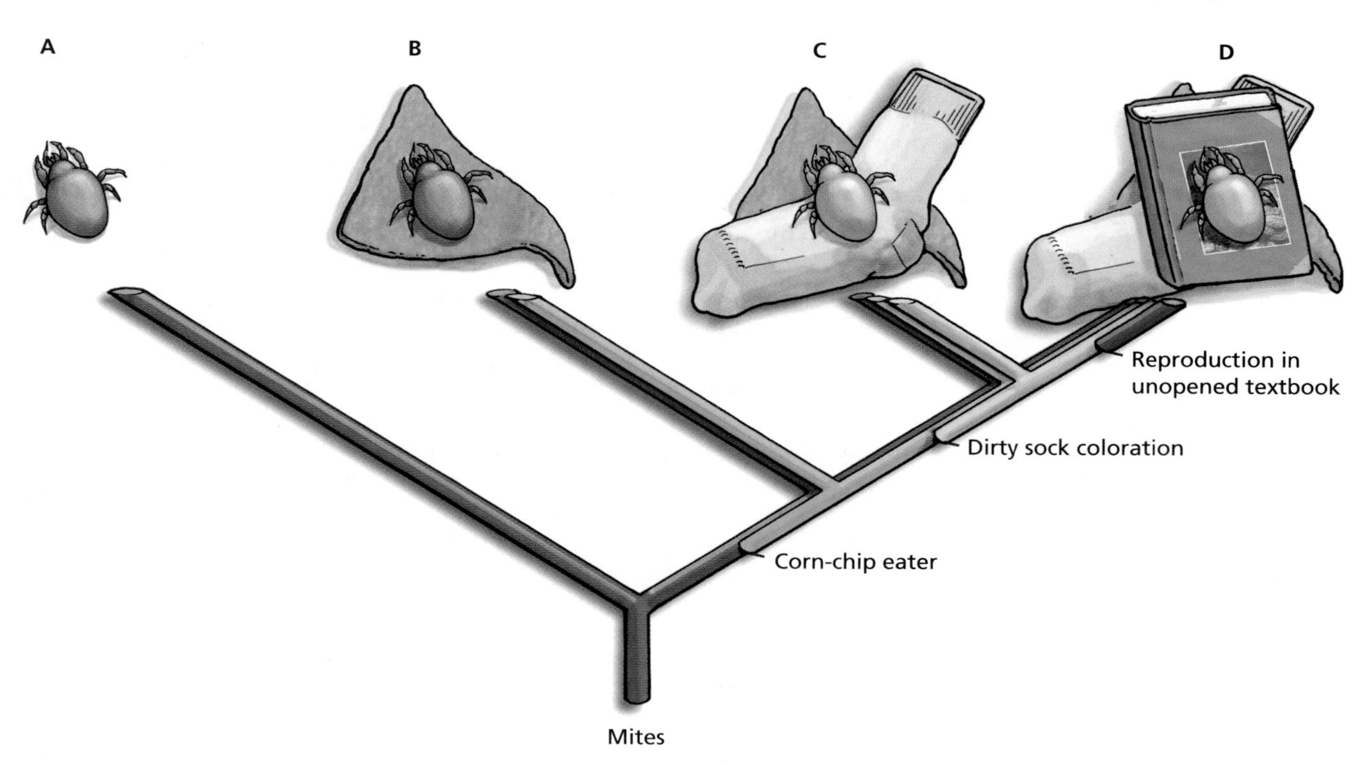

Figure 3.18 Reconstruction of an evolutionary history. This diagram illustrates our imaginary phylogeny. Species B, C, and D all share the corn-chip eating trait, so they share a more recent common ancestor with each other than with species A. Species C and D both have the color of dirty socks, but B does not, indicating that C and D share an even more recent common ancestor.

scientists can deduce the genealogy of related species. For example, fossils of vulture-like birds clearly indicate that this lifestyle evolved independently in both the Old World and the New World. These data allow scientists to strongly infer that the superficial similarities between New World vultures and Old World vultures are a result of convergent evolution.

Once a hypothesis of evolutionary relationship is reasonably well supported by additional data, bioprospectors can use the information gathered about one species in a classification group to predict the characteristics of other species in that group. This helps them determine the likelihood that related species could be additional sources of biological gold.

Learning from the Shaman

This chapter has been describing the search for biologically active compounds in living organisms as a process of scientific exploration—bioprospectors are described as "discovering" new compounds from nature. This is the case in Yellowstone's hot springs; chemicals derived from these organisms were probably truly unknown to humans. In the case of many other species, however, people have known of their usefulness for thousands of years. This knowledge is maintained in the traditions of indigenous people in biologically diverse areas, people who use native organisms as medicines, poisons, and foods. In many cultures, the repository of this traditional knowledge is the medicine man or woman. A *shaman*, as aboriginal healers are often called, can help direct bioprospectors to useful compounds by teaching about their culture's traditional methods of healing. Many of the remedies shamans employ are highly effective against disease. Several bioprospectors have employed a shaman in this manner to increase their chances of finding useful drugs (Figure 3.19).

Figure 3.19 Indigenous knowledge. This shaman of the Matses people of the Amazon rain forest is collecting plants for use in medicines. His intimate knowledge of the natural world is the product of the long history of his people in this diverse environment.

Using the knowledge of native people in developing countries to discover compounds for use in wealthy, developed countries is highly controversial. This process is often referred to as **biopiracy**, because organisms and active compounds discovered by traditional healers can be patented in the United States and Europe, and potentially provide enormous financial rewards to the bioprospector but no return to shamans or their people. The United Nations Convention on Biodiversity has sought to alleviate biopiracy by asserting that each country owns the biodiversity within its borders. However, the United States government has not signed this legally binding document, and thus companies in the United States are not required to abide by its terms. Additionally, even when a country makes a bioprospecting agreement with a pharmaceutical firm, it is unlikely that the indigenous community within the country will benefit in any way from a new drug developed from its store of knowledge. In recent times, indigenous peoples have begun to question the ethics of bioprospecting via a shaman, and several proposed agreements between developing countries and pharmaceutical firms have come under criticism.

The bioprospecting agreement between Diversa and Yellowstone National Park has not escaped charges of biopiracy. While Diversa is not relying on information from indigenous people to help them locate valuable organisms, critics have charged that the managers of Yellowstone are essentially "selling off" organisms and chemical compounds that belong to the American public. In addition, they argue that the action of bioprospecting itself will damage the very resource that provides these remarkable discoveries. The federal courts have dismissed lawsuits against Yellowstone National Park and Diversa that address these points according to current law, but the issue remains an ethical dilemma: What is the responsibility of individuals and corporations profiting from biological diversity to the source and survival of that diversity?

Biological diversity represents an enormous resource for humans, but it also comes with an awesome responsibility. The actions of the United States Congress protected Yellowstone National Park and perhaps ultimately enabled the discovery of *Thermus aquaticus* and *Taq* polymerase. However, thousands of useful compounds are lost every year through the destruction of native habitat, and the loss of indigenous cultures and their shamans. (See Chapter 14 for details on the current biodiversity crisis.) If all of us are to benefit from the organisms and materials provided by the variety of life, we must approach our relationship with these organisms with care and stewardship (Figure 3.20).

WWW
Media Activity 3.3 Bioprospecting in Yellowstone

Figure 3.20 Biodiversity benefits everyone. The abundance of life on our planet provides us with both tangible and intangible benefits.

CHAPTER REVIEW

Summary

- Bioprospectors seek to discover new drugs and other useful chemicals from the diversity of living organisms on Earth. The number of known living species is estimated between 1.4 and 1.8 million, but the total number of species may be as high as 100 million. Organisms are classified in domains according to evolutionary relationship, and kingdoms based on similarities in structure and lifestyle.

- Life on Earth began about 3.5 billion years ago with simple prokaryotes, but it would be 1.5 billion years before eukaryotes evolved. Multicellular organisms did not appear until approximately 600 million years ago, and this advance in form led to the diversity of species on Earth today.

- Bacteria and archaea are prokaryotes, simple single-celled organisms without a nucleus or other membrane-bounded organelles. They are abundant, found in a variety of habitats, and rely on a variety of food sources. Prokaryotes may produce antibiotics or have chemicals that function in extreme conditions.

- The kingdom Protista is a hodge-podge of organisms that are typically unicellular eukaryotes. Eukaryotes probably evolved from symbioses among ancestral eukaryotes and prokaryotes. Algae are especially interesting to bioprospectors because they make defensive chemicals against predators, and produce unique food-storage compounds.

- Animals are multicellular eukaryotes that are motile and rely on other organisms for food. Animal groups evolved in a short period of time known as the Cambrian explosion. Bioprospectors are interested in animals that produce venoms or defensive chemicals.

- Fungi are multicellular eukaryotes that are immobile, rely on other organisms for food, and are made up of thin, thread-like hyphae. Fungi often produce antibiotics that kill their competitors, and some can escape detection by their living host's immune system.

- Plants are multicellular photosynthetic eukaryotes. They have become increasingly adapted to land habitats over time. The diversity of flowering plants may be a result of their production of defensive chemicals.

- Some bioprospectors look for useful products by screening as many compounds as possible against a particular disease. An understanding of the ecological or evolutionary relationships of organisms provides clues to the likelihood and nature of possible chemical compounds in organisms. Studying how indigenous healers use organisms can help bioprospectors identify species that may have useful chemicals. Biopiracy occurs when a small group of people benefit from the knowledge of an indigenous culture. It may also undermine society's efforts to protect biodiversity.

Key Terms

adaptive radiation p. 69
algae p. 64
Animalia p. 66
antibiotic p. 63
archaea p. 62
bacteria p. 62
biodiversity p. 57
biological classification p. 59
biopiracy p. 76
botanist p. 70
Cambrian explosion p. 65
domain p. 62
ecology p. 72
endosymbiotic hypothesis p. 63
eukaryote p. 63
evolutionary classification p. 74
flowering plants p. 69
fungi p. 67
invertebrates p. 66
kingdom p. 59
microbe p. 62
multicellular p. 63
phylogeny p. 74
Plantae p. 69
prokaryotes p. 62
Protista p. 63
species p. 57
systematist p. 58
unicellular p. 62
vertebrates p. 66

Learning the Basics

1. Describe what is meant by the "tree of life." Why is this an effective way to illustrate biodiversity?

2. What characteristics of Bacteria and Archaea make these domains especially interesting to bioprospectors?

3. What characteristics of flowering plants may have driven the diversification of this group of organisms?

4. How is knowledge of the ecology of an organism useful for predicting what types of valuable chemicals it may possess?

5. How are hypotheses about the evolutionary relationships among living organisms tested?

6. Each of the following is considered to be a kingdom except _____.
   a. Animalia
   b. Fungi
   c. Protista
   d. Algae
   e. Plantae

7. According to the theory of evolution, all modern organisms _____.
   a. evolved separately from different ancestors
   b. arose from the same common ancestral species
   c. can be grouped into five or six evolutionary kingdoms
   d. have not changed much over time
   e. were independently created by a supernatural being

8. Comparisons of ribosomal RNA among many different modern species indicate that _____.
   a. there are two very divergent groups of prokaryotes
   b. the kingdom Protista represents a conglomeration of very unrelated forms
   c. fungi are more closely related to animals than plants
   d. a and b are correct
   e. a, b, and c are correct

9. Which of the following characteristics distinguishes prokaryotes from eukaryotes?
   a. Eukaryotes have a nucleus, while prokaryotes do not.
   b. Prokaryotes lack ribosomes, which are found in eukaryotes.
   c. Prokaryotes do not contain DNA, but eukaryotes do.
   d. Eukaryotic organisms are much more widespread than prokaryotes.
   e. Prokaryotes produce antibiotics, eukaryotes do not.

10. The mitochondria in a eukaryotic cell _____.
    a. serves as the cell's power plant
    b. probably evolved from a prokaryotic ancestor
    c. can live independently of the eukaryotic cell
    d. a and b are correct
    e. a, b, and c are correct

11. Most animals _____.
    a. are insects
    b. lack a backbone
    c. are still unidentified
    d. a, b, and c are correct
    e. b and c are correct

12. Fungi feed by _____.
    a. producing their own food with the help of sunlight
    b. chasing and capturing other living organisms
    c. growing upon their food source and secreting chemicals to break it down
    d. filtering bacteria out of their surroundings
    e. producing spores

13. The adaptive radiation of flowering plants _____.
    a. occurred between 100 and 80 million years ago
    b. may have occurred because flowers attract insects which aid their reproduction
    c. may have occurred because flowering plants have unique defensive chemicals
    d. may have occurred because the process of double fertilization allows these plants to produce high-quality seeds
    e. all of the above

14. The relationship of an organism to its environment, including other species, is its _____.
    a. ecology
    b. phylogeny
    c. botany
    d. evolutionary classification
    e. diversity

15. In principle, all of the species that descend from a recent common ancestor _____.
    a. should be identical
    b. should share characteristics that evolved in that ancestor
    c. should be found as fossils
    d. should have the same DNA sequences
    e. should be no more similar than species that are less closely related

Analyzing and Applying the Basics

1. Unless handled properly by living systems, oxygen can be quite damaging to cells. Imagine an ancient nucleated cell that ingests an oxygen-using bacterium. In an environment where oxygen levels are increasing, why might the eukaryotic cell have an advantage if it does not digest the bacterium, but instead provides a "safe haven" for it?

2. One disadvantage of an evolutionary classification system is that the groups do not seem "natural" to us—as in the example of New World vultures seeming to resemble Old World vultures more than storks. Can you think of situations where a nonevolutionary classification system of biodiversity is more useful than an evolutionary one?

3. Imagine that you find an organism that has never been described by science. The organism is made up of several hundred cells, and feeds by anchoring itself to a submerged rock and straining single-celled algae out of pond water. What kingdom would this organism probably belong to, and why do you think so?

4. Imagine two fungi. Both weigh the same, but one consists of a few short, very thick hyphae and the other consists of many long, thin hyphae. Can they both absorb the same amount of food? If not, which fungus is more effective?

5. How might the evolution of toxic antipredator chemicals in a species of plant have led to the plant's adaptive radiation into a number of different species?

6. Humans are grouped into racial categories (White, Black, Asian, Native American, etc.) based on similarities in skin color, eye shape, hair color and texture, and other physical similarities. Is this grouping based on superficial similarity or closer biological relationships of individuals within a race compared to the relationship of individuals in different races? How would you determine if your answer is correct? (Note that Chapter 10 will address this issue.)

Connecting the Science

1. Scientists ridiculed the hypothesis that eukaryotic cells evolved from a "confederation" of cooperating independent cells when it was first proposed. Most biologists still believe that competition for resources among organisms is the primary force for evolution. Do you think biologists' dismissal of the role of cooperation in evolution is a reflection of how life really "works," or do you think that it is a function of scientists' immersion in a culture that values competition over cooperation? Explain your choice.

2. The United States and the European Union allow individuals and companies to patent some types of living organisms. The owners of the patent have exclusive rights to the organism, and may charge users a licensing fee. Should forms of life be "owned"? If there were no ownership of potentially valuable organisms, would drug companies have incentives to engage in bioprospecting?

3. Comment on the agreement between Diversa and the managers of Yellowstone National Park. Is it fair for Yellowstone's managers to grant exclusive bioprospecting rights to a single company? How should the benefits of the biological resources that are protected on public land be distributed?

4. Do we have an obligation to future generations to preserve as much biodiversity as possible, considering that many organisms may contain currently unknown "biological gold"? Would simply preserving the information contained in an organism's genes (in a zoo, or other collection) be good enough, or do we need to preserve organisms in their natural environments?

Media Activities

Media Activity 3.1 The Basis of Life's Diversity
Estimated Time: 10 minutes
Explore the evolutionary origin and the genetic basis of life's diversity.

Media Activity 3.2 Investigating Antibiotic Resistance
Estimated Time: 10 minutes
In this activity, you will read an article from the FDA about antibiotic resistance and answer questions about how to combat antibiotic resistance.

Media Activity 3.3 Bioprospecting in Yellowstone
Estimated Time: 10 minutes
Explore several sides of the bioprospecting debate and express your own opinion.

CHAPTER 4

Are You Only As Smart As Your Genes?

The Science of Inheritance

Could a woman create perfect children if she chooses the right sperm?

4.1 The Inheritance of Traits

4.2 The Role of Genes in Determining Traits

4.3 Genes, Environment, and the Individual

If a woman chooses a donor with the right genes, will her child look like her partner?

If she chooses a donor with the right genes, will her child be a genius?

Or is a child's intelligence more influenced by his environment?

The Fairfax Cryobank is a nondescript brick building located in a quiet, tree-lined suburb of Washington, DC. Stored inside this unremarkable edifice are the hopes and dreams of thousands of women and their partners. The Fairfax Cryobank is a sperm bank; inside its many freezers are vials containing sperm collected from hundreds of men. Women can order these sperm for a procedure called *artificial insemination*, which may allow them to conceive a child despite the lack of a fertile male partner.

Women who purchase sperm from the Fairfax Cryobank have hundreds of potential donors to choose from. The donors are categorized into three classes, and their sperm is priced accordingly. Most women who choose artificial insemination want detailed information about the donor before they purchase a sample; and while all Fairfax Cryobank donors submit to comprehensive physical exams and disease testing, and provide a detailed family health history, not all provide childhood pictures, audio CDs of their voices, or personal essays. Sperm samples from men who did not provide this additional information are sold at a discount because most women seek a donor that matches their partner in interests and aptitude. These women hope that by choosing a donor who is very similar to their partners, they will have children who seem biologically related to them.

However, in addition to the information-rich donors, there is also a set of premium sperm donors referred to by Fairfax Cryobank as its "Doctorate" category. These men are in the process of earning or have completed a doctorate degree. Sperm from this category of donor is 30% more expensive than sperm from the standard donor.

Why would some women be willing to pay significantly more for sperm from a donor who has an advanced degree? Because academic achievement is associated with intelligence, and these women want intelligent children and are willing to pay more to provide them with "extra smart" genes. But are these women putting their money in the right place? Is intelligence about genes, or is it a function of the environmental conditions a baby is raised in? In other words, is who we are a result of our "nature" or our "nurture"? As you read this chapter, you will see that the answer to this question is not a simple one.

4.1 The Inheritance of Traits

Most of us recognize similarities between our birth parents and ourselves. There are also resemblances among members of a family—for instance, all of the children of a single set of parents may have dimples. However, it is usually quite easy to tell siblings apart. Each child of a set of parents is unique, and none of us is simply the "average" of our parents' traits. We are each more of an amalgam—one child similar to her mother in eyes and face shape, another similar to mom in build and hair color.

To understand how your parents' traits were passed to you and your siblings, you need a basic understanding of the human life cycle. A *life cycle* is a description of the growth and reproduction of an individual. Figure 4.1 is a diagram of a very simplified human life cycle. Notice that a human baby is typically produced from

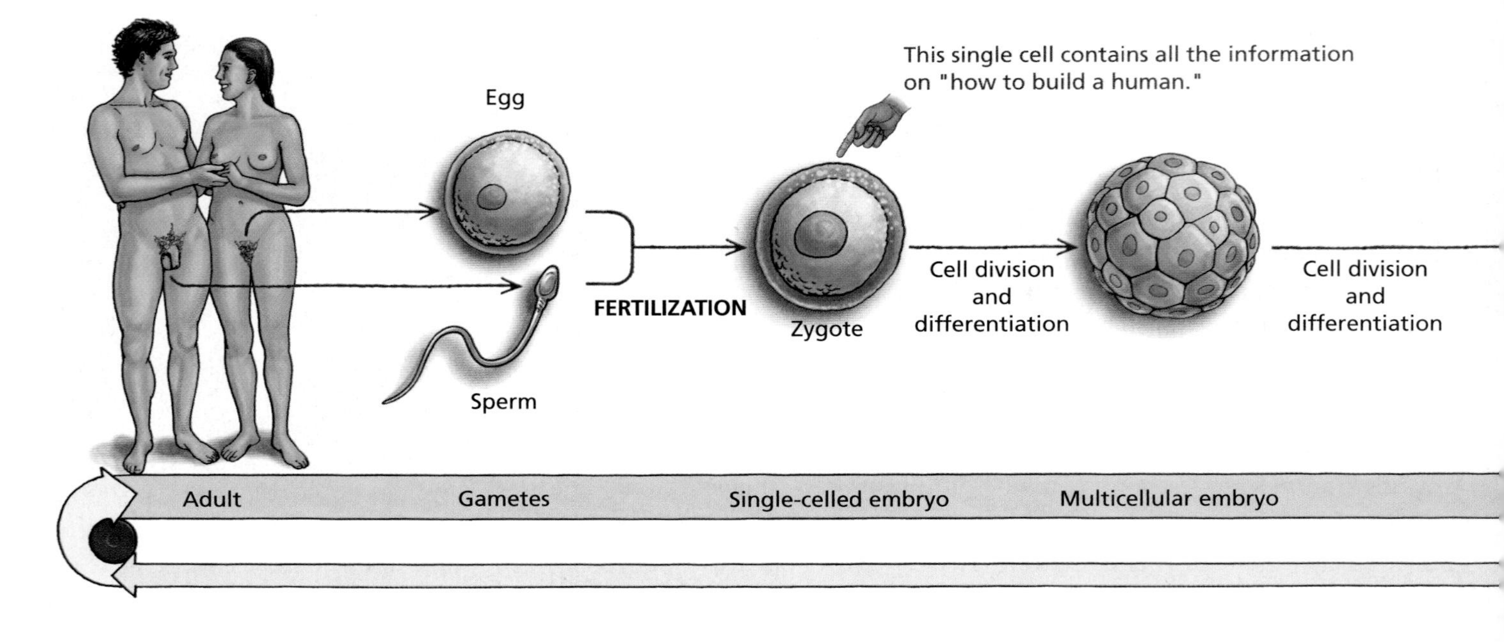

Figure 4.1 The human life cycle. A human baby forms from the fusion of an egg cell from its mother and a sperm cell from its father. The single cell that results from this fusion will grow and divide into billions of cells, each of which carry the same information.

the fusion of a single **sperm** cell produced by the male parent, and a single **egg** cell produced by the female parent. After the egg and sperm, or *gametes*, fuse at **fertilization** the resulting single, fertilized cell (the *zygote*) divides dozens of times to produce *daughter cells*. Each of these daughter cells also divides dozens of times. The cells in this resulting mass then differentiate into specialized cell types, which may continue to divide and organize to produce the various structures of a developing human, called an **embryo**. We are made up of billions of individual cells, all of them the descendants of that first product of fertilization.

The Nature of Genes

Each normal sperm and egg contains information about "how to build a human." A large portion of that information is in the form of **genes**—segments of *DNA* that contain specific pieces of information about the traits of a human. Chapter 2 describes that genes carry instructions about how to make *proteins*. These proteins may be either structural (like the protein that makes up hair), or functional (like the protein lactase, which breaks down milk sugar). Proteins give cells—and by extension, organs and individuals—nearly all of their characteristics. Chapters 5, 6, and 7 explore the physical nature of genes in more detail; here we will use an analogy to describe their function in the inheritance of traits.

Imagine genes as being roughly equivalent to the words used in an instruction manual. Words can have one meaning when they are alone (for instance, saw), another meaning when used in combination with other words (see-saw), or even change meaning in different contexts ("saw the wood" versus "sharpen the saw"). Some words in the manual are repeated frequently, but other words are not. It is the placement of the words and their combination with other words that determines which instruction is given.

Similarly, all cells have the same genes, but it is the combination and use of these genes that determines the activities of a particular cell. For instance, both eye cells and heart cells carry instructions for the protein rhodopsin, which helps detect light, but rhodopsin is utilized only in eye cells, not heart cells. Rhodopsin also requires assistance from another protein to translate the light it senses into the actions of the eye cell. This other protein may also be built in heart cells, but it is combined with a third protein to help coordinate the contractions of the heart muscle. Thus, a protein may serve two different functions, depending

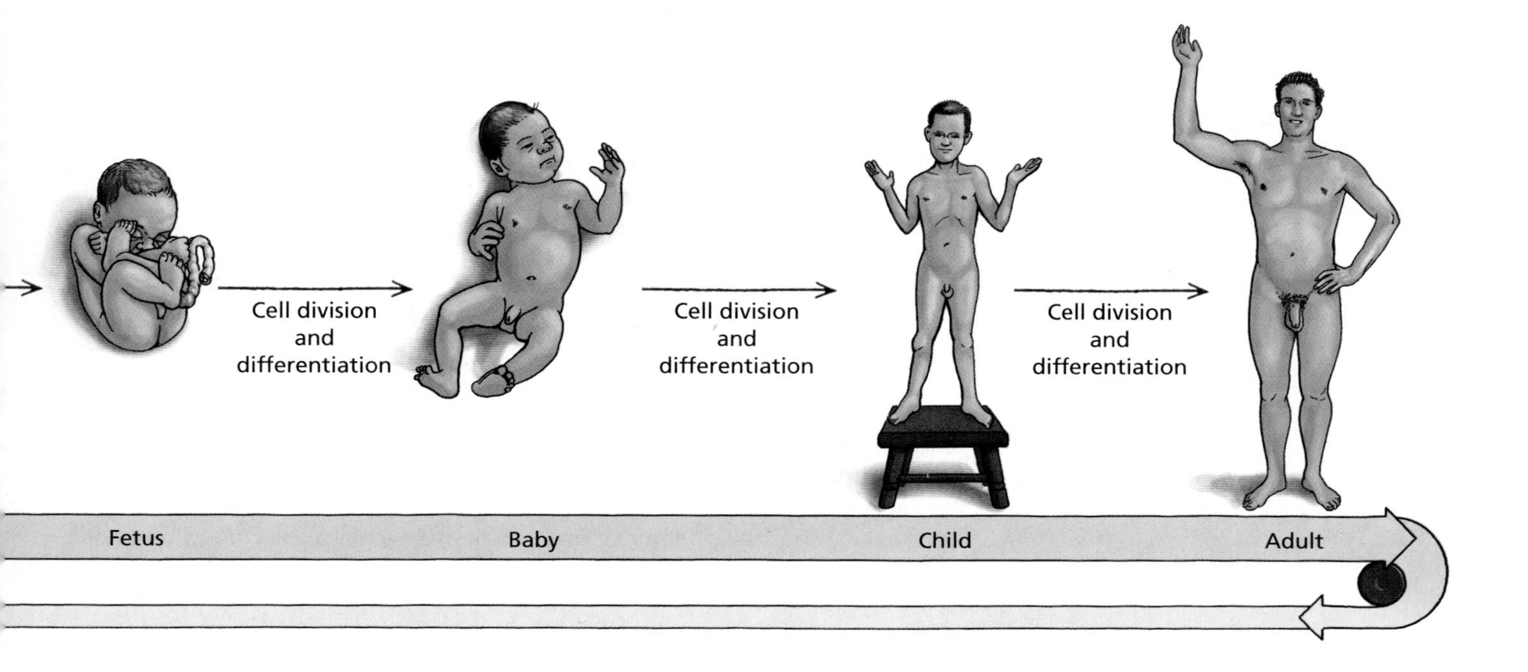

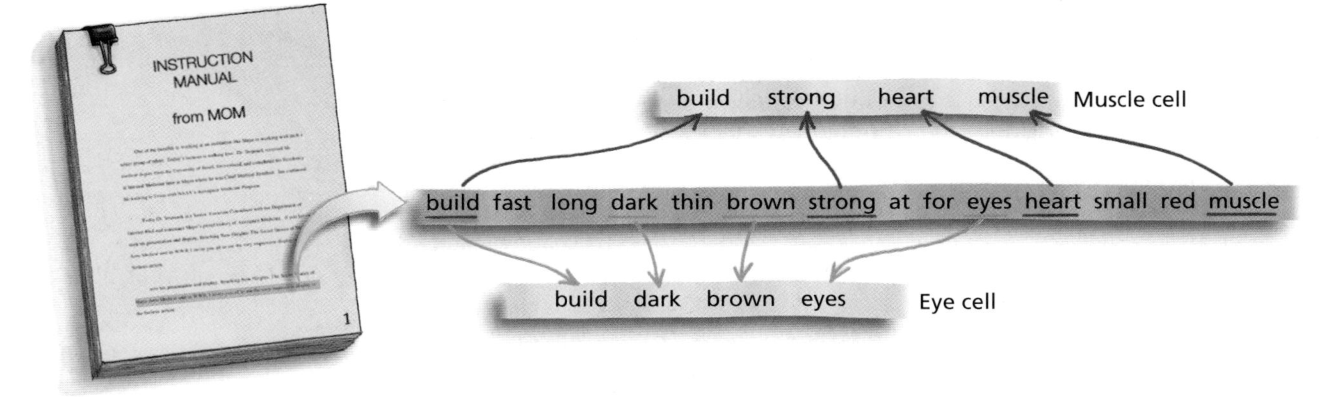

Figure 4.2 Genes as words in an instruction manual. Different words from the manual are used in different parts of the body, and even when the same words are used, they are often used in different combinations. In this way, the manual can provide instructions for making and operating the variety of body parts we possess.

upon its context. Since genes, like words, can be used in many combinations, the instruction manual for building a human is very flexible (Figure 4.2). When parents pass information to their offspring in the form of genes, they each contribute comparable information. Using our analogy above, both parents give a complete instruction manual to their offspring. In humans, all of the words are written on 23 different segments, called *chromosomes*. In our analogy, each chromosome is equivalent to a page in the instruction manual, and two complete sets of chromosomes are required to build a normal human baby (Figure 4.3a).

The information contained on the chromosomes contributed by each parent is comparable, but not identical. You can imagine how a page might change over many generations if each parent had to type a copy for each child. Typographical errors made by a father would be passed on in the manual he gave to his children. If the children made additional typographical errors, the instruction manual they pass on would contain the changes made by their parents, plus those they made themselves. Thus different families would have slightly differing versions of certain words in the manual. The same sorts of "typographical errors" occur when genes are copied and passed on. Changes in genes are called **mutations**, and they result in different versions of a gene, which are called **alleles**. Alleles may be different versions of a gene that produce the same basic effect, or the effect may be very different (Figure 4.3b).

WWW
Media Activity 4.1A The Nature of Inheritance

The Nature of Inheritance

It is the combination of alleles from our mothers and fathers that help determine what traits we have. The environment in which the alleles are used also plays a role. Just as two people reading the same set of instructions may build slightly different dollhouses, embryos developing in slightly different environments may vary in appearance. Part of the reason you do not look exactly like your siblings is because you each developed in dissimilar conditions. These conditions could include your mother's nutrition during pregnancy, the presence of toxic compounds in her environment, or even the number of other siblings in your family at the time of your birth. Two of the important questions in the "nature versus nurture" debate mentioned in the introduction focus upon the relative importance of the environment of development and the aspects of the environment that most strongly affect the result of the instructions.

You also differ in appearance from your siblings because your parents did not give all of their children the exact same set of alleles. Gametes develop from cells that have two copies of each gene; however, each sperm or egg carries only one copy. The process that reduces the number of chromosomes (and thus

(a) Both parents give a complete instruction manual to their offspring.

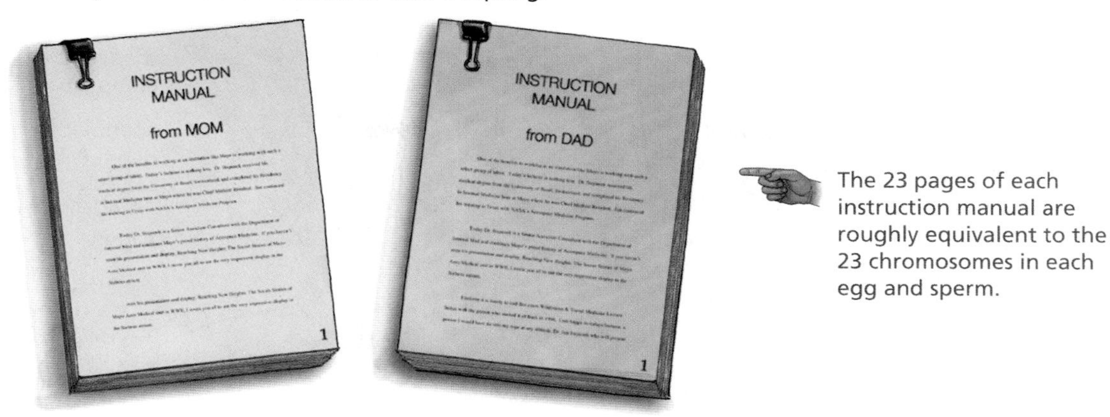

(b) Mutations are misspellings in the instructions.

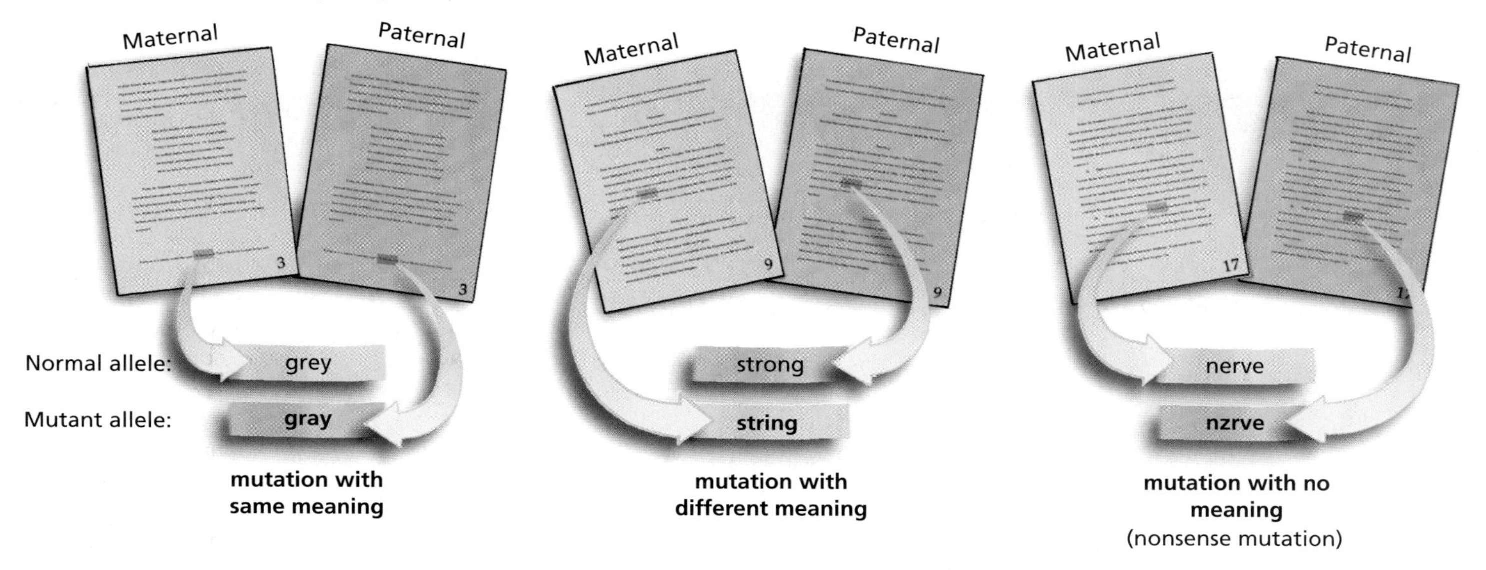

Figure 4.3 The formation of different alleles. (a) Each parent provides a complete set of instructions to each offspring. (b) The instructions are first copied, and different alleles for a gene may form as a result of copying errors. Some of these misspellings do not change the meaning of the word, but some may result in different meanings or have no meaning at all.

genes) by one-half is called *meiosis*, which will be described in detail in Chapter 6. However, our instruction manual analogy should help make it clear why the egg and sperm that produced you were different from the egg and sperm that produced your sibling.

The precursor cell to a human sperm or egg contain two copies of each page of the instruction manual. When a precursor cell develops into a sperm or an egg, it places *one* copy of every page into each of two daughter cells, and each page is placed in a daughter cell independently of all the other pages. In other words, the instruction manual you received from your mother is made up of a combination of pages from the manuals she received from each of her parents. In our analogy, each page is equivalent to a copy of a chromosome. Therefore, each egg or sperm contains a unique subset of chromosomes—and thus a unique subset of the alleles carried by the parent.

The physical process of separating the sets of chromosomes during meiosis results in what is known as **independent assortment**. As a result of independent assortment, an allele for an eye color gene ends up in a sperm or egg *independently* from an allele for the blood-group gene. Since the process of

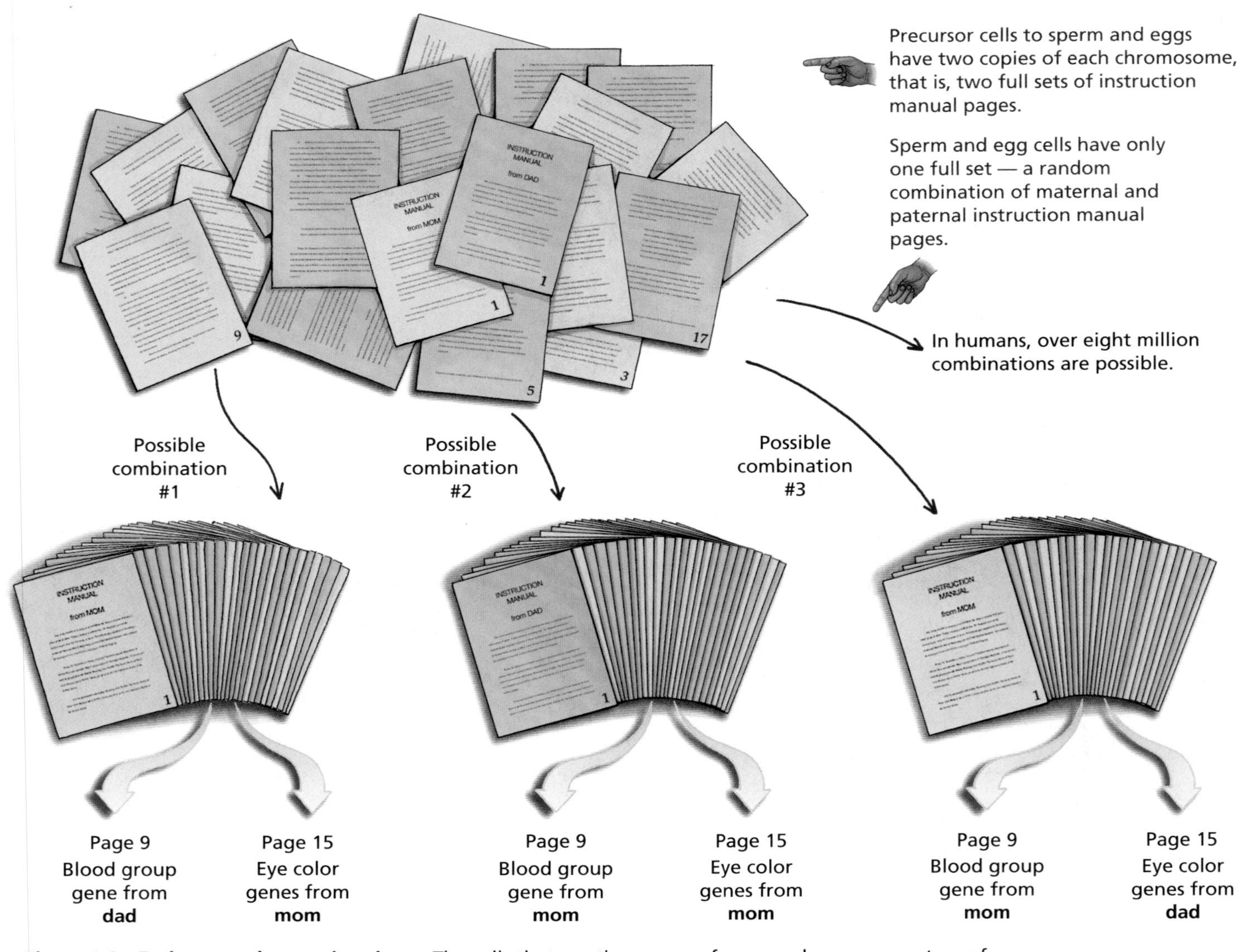

Figure 4.4 Each egg and sperm is unique. The cells that are the source of eggs and sperm carry two of each chromosome—that is, two full copies of the instruction manual. When an egg or sperm is produced, it ends up with only one copy of each page. Since each egg and most sperm are produced independently, the set of pages in each is practically unique.

independently placing chromosomes into daughter cells is repeated every time an egg is produced, the set of alleles each child receives from a mother is different for all of her children. The egg that produced you might have carried the eye color allele from your mom's mom and the blood group from her dad, while the egg that produced your sister might have contained both the allele for eye color and the allele for blood group from your maternal grandmother (Figure 4.4). Due to independent assortment, about 50% of your alleles are identical to the alleles carried by your full sibling—that is, for each gene you have a 50% chance of being like your sister or brother.

In addition to independent assortment, there is another event during meiosis that leads to diversity in egg and sperm. This process is called *crossing over*, and occurs when chromosome pairs "swap" information. In our instruction manual analogy, crossing over is equivalent to tearing a pair of pages in half and reassembling them so that the top part of the page is from one instruction manual and the bottom part is from the other instruction manual. The process of crossing over is discussed in more detail in Chapter 6. For now, it is sufficient to understand that the processes of independent assortment and crossing over create almost limitless variation in eggs or sperm from a single parent.

A Special Case—Identical Twins

Occasionally two children of the same set of parents share 100% of their genes. Unlike most siblings, even nonidentical twins, identical twins are the result of a single fertilization event—the fusion of one egg with one sperm giving rise to two offspring.

Identical twins are often called **monozygotic twins** (*mono* means "one"). Recall that after fertilization, the fertilized egg cell grows and divides, producing an embryo made up of many daughter cells containing the same genetic information. Monozygotic twinning occurs when cells in an embryo separate from each other. If this happens early in development, each cell or clump of cells can develop into a complete individual, yielding twins who carry identical genetic information (Figure 4.5a). In the United States, approximately one person in 150 is an identical twin.

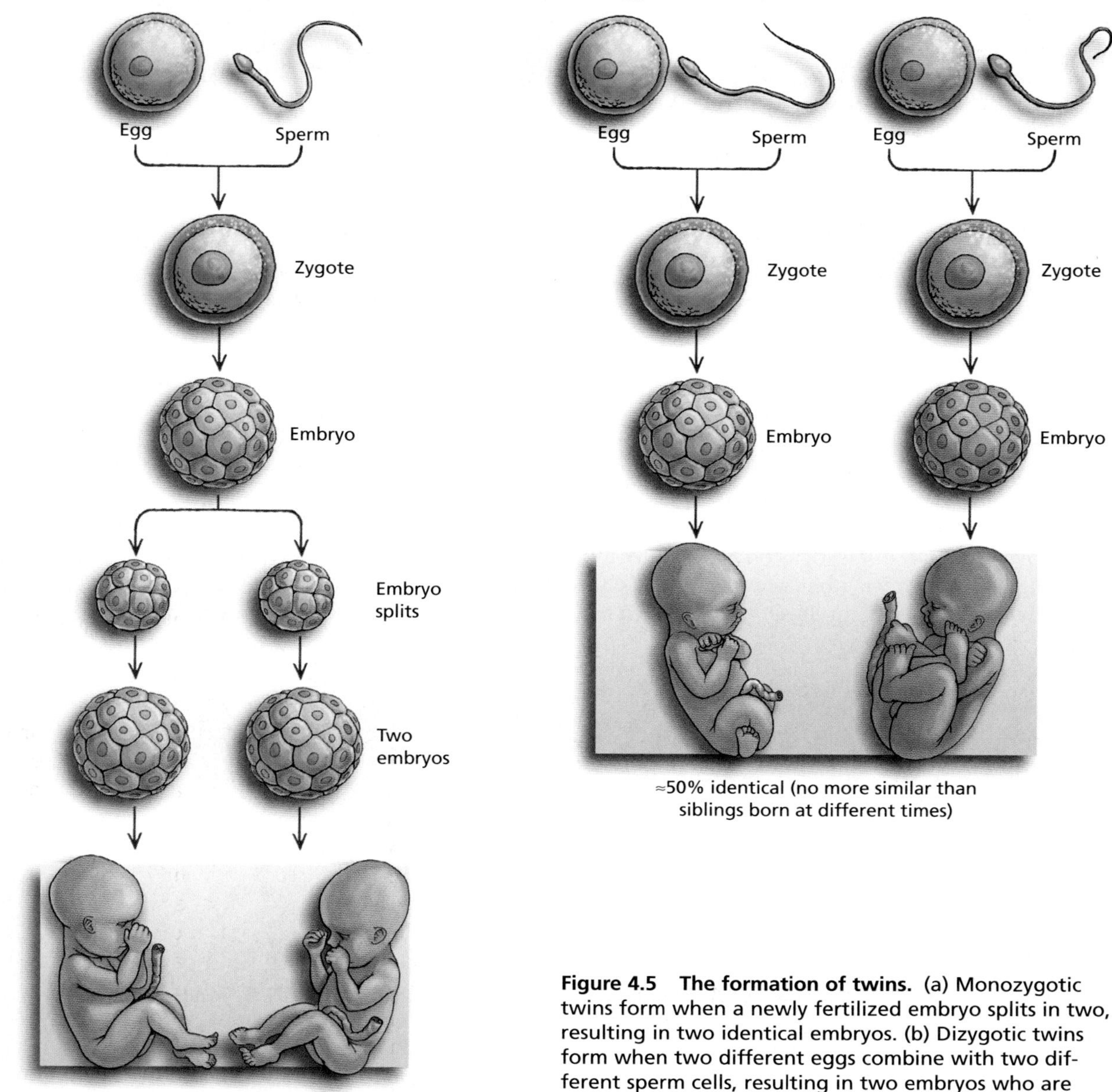

Figure 4.5 The formation of twins. (a) Monozygotic twins form when a newly fertilized embryo splits in two, resulting in two identical embryos. (b) Dizygotic twins form when two different eggs combine with two different sperm cells, resulting in two embryos who are only as similar as siblings.

Essay 4.1 Gregor Mendel

The scientist who first helped explain the manner in which genes are inherited was Johann Gregor Mendel (Figure E4.1A). Due to the impact of his findings in experiments with genetic crosses, traits that are inherited according to the rules elucidated by Mendel's experiments are said to follow a Mendelian pattern of inheritance.

Figure E4.1A Gregor Mendel.

Mendel was born in Austria in 1822. Since his family was poor, he entered a monastery to obtain an education. After completing his monastic studies, Mendel attended the University of Vienna; he studied math and botany there, in addition to other sciences. Mendel attempted to become an accredited teacher, but was unable to pass the examinations. After leaving the university, Mendel returned to the monastery and began his experimental studies of inheritance in garden peas.

Mendel studied close to 30,000 pea plants over a 10-year period. His careful experiments consisted of controlled matings between plants with different traits. Mendel was able to control the types of mating that occurred by hand-pollinating the peas' flowers—for example, by applying pollen, which produce sperm, from a tall pea plant to the pistil, or egg-containing structure, of a short pea plant and then growing the seeds resulting from that cross, he could evaluate the role of each parent in producing the traits of the offspring. Mendel published the results of his studies in 1865, but his contemporaries did not fully appreciate the significance of his work. This was partly because biologists were still wrestling with the implications of Charles Darwin's theory of evolution, published just six years before Mendel's work.

Prior to Mendel, many scientists believed that the traits of a child's parents were blended together, producing a child with characteristics intermediate to those of both parents. In fact, this was Darwin's belief, and it was a major barrier to general acceptance of his ideas on *natural selection* and evolution. Some scientists, the "spermists," believed that sperm contained a completely formed miniature adult, known as a homunculus. Other scientists, the "ovists," believed that it was the egg cell that contained the homunculus. Some even believed that children resemble the parent who initiated the intercourse! In the climate of these seemingly sensational hypotheses, Mendel's patient, scientifically sound experiments showing that both parents contribute equal amounts of genetic information to their offspring went largely unnoticed (Figure E4.1B).

Mendel eventually gave up his genetic studies and focused his attention on running the monastery until his death in 1884. His work was independently rediscovered by three scientists in 1900; only then did its significance to the new science of genetics become apparent.

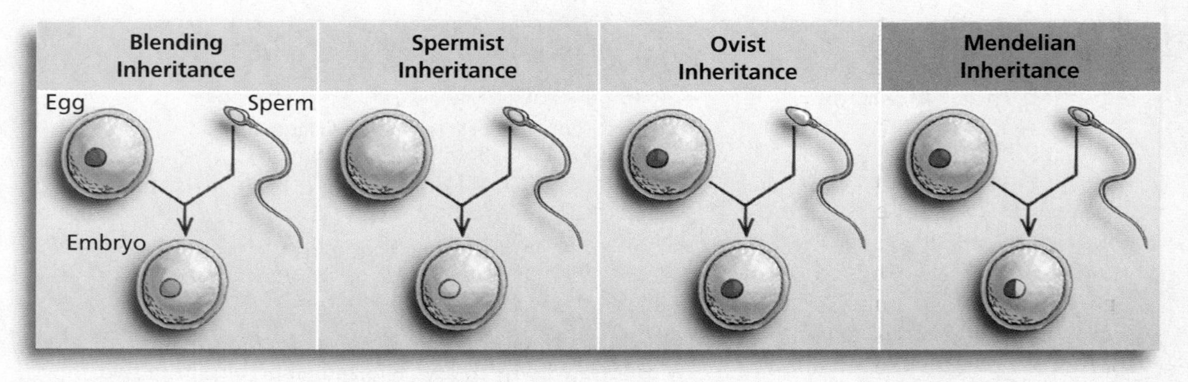

Figure E4.1B Hypotheses about patterns of inheritance.

Nonidentical twins, sometimes called fraternal twins, result from a different process. These twins are called **dizygotic** (*di* means "two"), and most often occur when two separate eggs fuse with different sperm. The resulting embryos, which develop together, are genetically no more similar than siblings born at different times (Figure 4.5b).

Identical twins provide a unique opportunity to study the relative effects of our genes and environment in determining who we are. Since they carry the same genetic information, researchers are able to study how important genes are in determining their health, tastes, intelligence, and personality. We will explore the use of twin studies in the next section.

4.2 The Role of Genes in Determining Traits

www

Media Activity 4.1B Patterns of Inheritance

When the Role of Genes Is Clear

A few human genetic traits have easily identifiable patterns of inheritance. These traits are said to be "Mendelian" because Gregor Mendel (Essay 4.1) was the first person to accurately describe the inheritance of these types of traits. Although Mendel himself did not have an understanding of the chemical nature of genes, he was able to determine how traits were inherited by carefully analyzing the appearance of parents and their offspring. The pattern of inheritance Mendel described occurs primarily in traits that are the result of a single gene with a few distinct alleles. We now know that different alleles correspond to slightly different DNA sequences for the same gene, which result in slightly different proteins.

Not all traits affected by a single gene show straightforward Mendelian inheritance. Most of the Mendelian traits identified in humans are the result of genes with mutant alleles that result in some type of disease or dysfunction. In our instruction manual analogy, an allele that is dysfunctional is equivalent to a misspelled word on a page. Where two copies of each word are available, the correctly spelled word can usually substitute for its misspelled partner without changing the meaning of the instructions. Likewise, since we have two copies of each gene, carrying one dysfunctional allele for a gene is usually not a serious problem—the functional allele acts as a backup for its mutated partner within a cell.

We call the genetic composition of an individual his **genotype**, and his physical traits his **phenotype**. An individual who carries two different alleles for a gene has a **heterozygous** genotype (*hetero* meaning "different," and *zygous* meaning "origin"). For instance, a man who carries one copy of a functional allele and one copy of a dysfunctional allele is heterozygous for that gene. In many cases, this means that his phenotype is normal—he does not show evidence of having the dysfunctional allele. An individual who carries two copies of the same allele has a **homozygous** genotype (*homo* meaning "same"). Imagine a woman who received the same misspelled nonsense word from both of her parents' instruction manuals—that is, she is homozygous for a dysfunctional allele. In this case, it is impossible for her make the correct normal protein, and she will have a disease phenotype (Figure 4.6a). We can also be homozygous with identical *functional* alleles: in this case, there is no disease or dysfunction, and the phenotype is normal.

The effect of a mutant allele on the development of an individual varies. Some mutations are **recessive**, meaning that their effects are invisible as long as a normal allele is present. Often recessive mutations result in nonfunctional alleles. With these types of mutations, a heterozygous individual has a normal phenotype because the normal allele is substituting for its dysfunctional

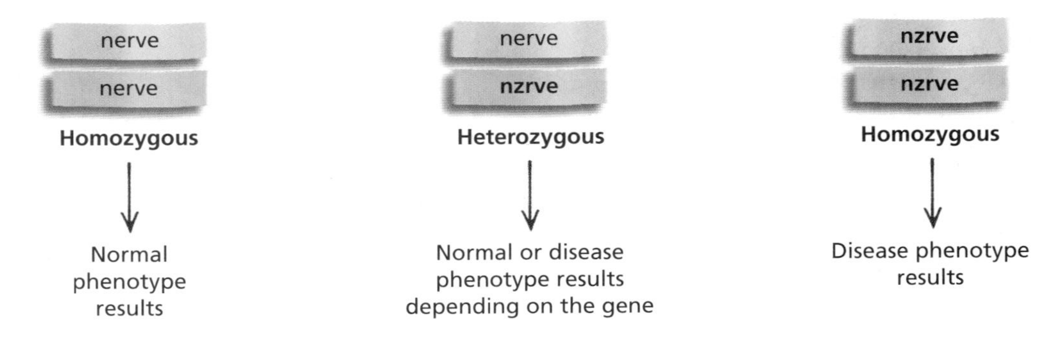

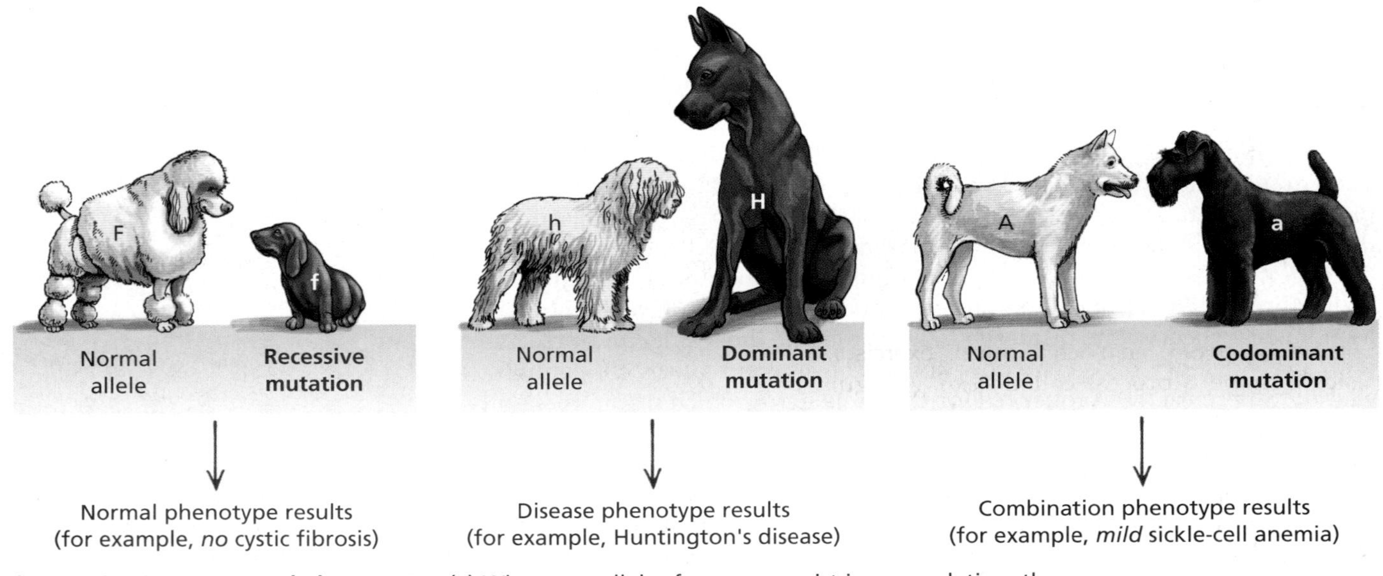

Figure 4.6 Genotypes and phenotypes. (a) When two alleles for a gene exist in a population, there are three possible genotypes and two or three possible phenotypes for the trait. (b) The effect of a mutant allele on a heterozygote individual's phenotype depends on whether the allele is recessive, dominant, or codominant to the normal allele. This figure relates these patterns to particular human genetic conditions.

partner. Other mutations cause a change that essentially overpowers the normal function—these mutations are termed **dominant**, since their effects are seen even when a normal allele is present. A heterozygous individual will have the disease phenotype when he carries a dominant disease allele. Still other mutations result in a new protein with a different, but not dominant, activity compared to the normal protein. This results in **codominant** alleles. An individual who is heterozygous for codominant alleles will express both alleles. See Figure 4.6b to visualize the effects of these three types of mutations.

Scientists have identified genetic conditions in humans that conform to each of these three types of mutations. Cystic fibrosis occurs in individuals with two copies of an allele that instructs for a dysfunctional protein—in particular, a protein that helps transport the element chloride into and out of cells. Affected individuals suffer from progressive deterioration of their lungs and intestines—most children born with cystic fibrosis do not live past their 30s. People who carry one copy of the dysfunctional allele and one copy of the normal allele are not affected. Cystic fibrosis is among the most common genetic diseases in white European populations—nearly one in 2500 individuals is affected with the disease, and one in 25 is heterozygous for the allele. Sperm banks have the ability to test donor sperm for several recessive disorders, including cystic fibrosis. The Fairfax Cryobank tests donor semen for the presence of cystic fibrosis alleles; any men who carry the allele are excluded from their donor program.

Huntington's disease is an example of a genetic condition caused by a dominant allele. Symptoms of Huntington's disease include restlessness, irritability, and difficulty in walking, thinking, and remembering. The disease is progressive and incurable—the nervous, mental, and muscular symptoms gradually become worse, and eventually result in the death of the affected individual. The Huntington's allele causes production of a protein that is toxic when it is at high levels in a cell. Nerve cells in certain areas of the brain are especially susceptible to this toxin, and these cells gradually die off over the course of the disease. Since it produces a toxic protein, an individual needs only one copy of the allele to be affected by the disease—even heterozygotes have the disease. The detailed family medical histories required of sperm bank donors enables Fairfax Cryobank to exclude men with a family history of Huntington's disease from their donor list.

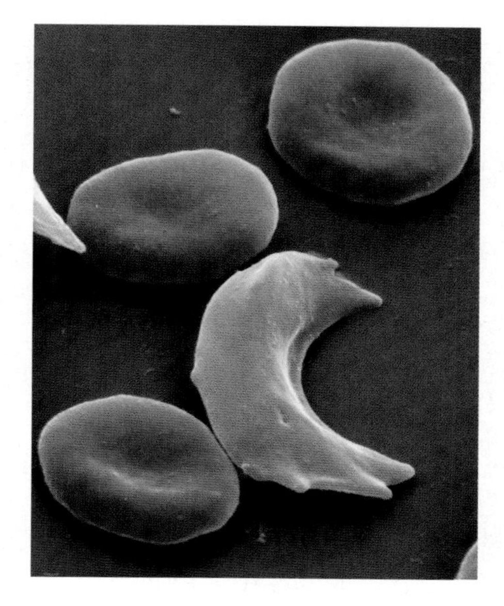

Figure 4.7 The effect of the sickle-cell allele. The sickle-cell anemia allele causes red blood cells to look sickled as seen here, on the right. At left, a normal red blood cell appears round and pillow-like.

www

Media Activity 4.3 Genetics and Sickle Cell Disease

The allele that causes the disease sickle-cell anemia is an example of a codominant mutation. Sickle-cell anemia is a condition that occurs when an individual carries two mutant copies of the gene required to produce normal red blood cells. Red blood cells contain *hemoglobin*, a protein that carries oxygen in the bloodstream. The mutation that causes sickle-cell anemia results in hemoglobin that behaves differently from hemoglobin produced by someone with a normal allele. The abnormal hemoglobin forms rigid structures when red blood cells are low in oxygen (such as during exercise), causing the cells to deform into a sickle shape, a process called *sickling* (Figure 4.7). These cells can become lodged in narrow blood vessels, further blocking the flow of oxygen to tissues. Individuals carrying two copies of the sickle-cell allele are prone to painful and debilitating sickling attacks and often die at an early age. Individuals who have one copy of the sickle-cell allele and one copy of the normal allele make both normal and abnormal hemoglobin. Thus, some of their blood cells will sickle, but not to the degree seen in homozygous individuals.

Heterozygotes with one copy of the sickle-cell allele are generally much less seriously ill than homozygotes for the sickle-cell allele. In some situations, heterozygotes are also healthier than individuals who are homozygous for the normal hemoglobin allele. For example, people with one copy of the sickle-cell allele are resistant to malaria. Among populations in sub-Saharan Africa, Northern India, and around the Mediterranean Sea—where malaria is common—heterozygotes for sickle-cell anemia have higher rates of survival than homozygotes of either type; and because heterozygotes have higher rates of survival, they contribute more children to the next generation than homozygotes do. Since 50% of their children will carry the sickle-cell allele, the allele has stayed common in these populations. Thus, the advantage of the sickle-cell allele results in a high frequency of this allele in human populations that have spent generations in these conditions. The sickle-cell allele is much more common in these populations than we would expect if its only effect was to cause disease. Sperm banks routinely test donor sperm for the presence of the sickle-cell allele and exclude men who carry it from their donor list.

Traits such as cystic fibrosis, Huntington's disease, and sickle-cell anemia are the inevitable result of a mutation in a single gene, and the inheritance of these traits is relatively easy to understand. We can follow the inheritance of the alleles for these types of genes using a tool called a *Punnett square*. A **Punnett square** is a table that lists the different kinds of sperm or eggs parents can produce relative to the gene or genes in question and then predicts the possible outcomes of a **cross**, or mating, between these parents (Figure 4.8).

Imagine a couple in which both members are heterozygotes for cystic fibrosis. If we use the letters *F* and *f* to represent the functional and dysfunctional

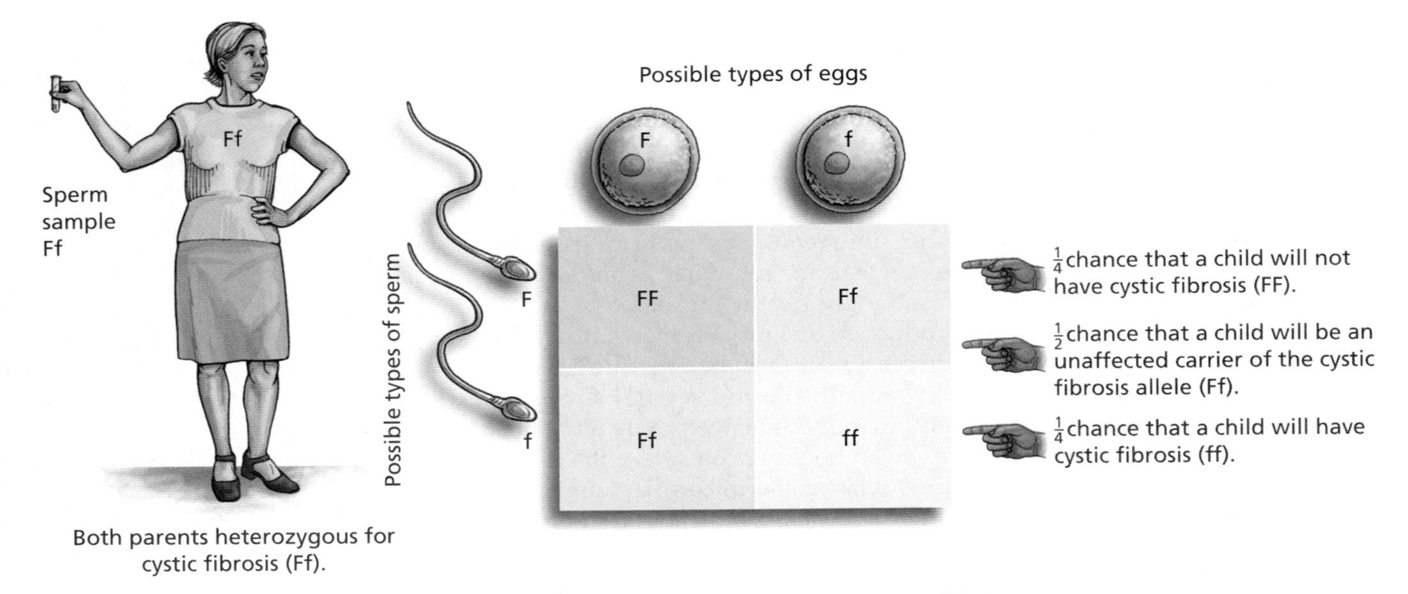

Figure 4.8 What are the risks of accepting sperm from a carrier of cystic fibrosis? This Punnett square helps determine the likelihood that a woman who carries the cystic fibrosis allele would have a child with cystic fibrosis if her sperm donor was also a carrier. With a 25% chance of producing an affected child, most women would consider sperm from a carrier unacceptable.

allele, respectively, a heterozygote would have the genotype *Ff*. The genetic cross could then be symbolized:

$$Ff \times Ff$$

We know that the female in this cross can produce eggs that carry either the *F* or *f* allele, since the process of egg production will separate the two alleles from each other. We place these two egg types across the horizontal axis of what will become a Punnett square. The male in this cross is also heterozygous, so he too can make sperm containing either the *F* or *f* allele. We place the kinds of sperm he can produce along the vertical axis. Thus, the letters on the horizontal and vertical axes represent all the possible types of eggs and sperm that the mother and father can produce by meiosis, if we consider only the gene that codes for the chloride transport protein.

Inside the Punnett square are all the genotypes that can be produced from a cross between these two heterozygous individuals. The content of each box is determined by matching the egg column with the sperm row. Note that for a single gene with two alleles, there are four possible offspring types. The chance of this couple having a child affected with cystic fibrosis is $\frac{1}{4}$, because the *ff* combination of alleles occurs once out of the four possible outcomes. The *FF* genotype is represented once out of four times, and the probability of this couple having a homozygous unaffected child is also $\frac{1}{4}$. The probability of the couple producing a child who is a **carrier** of cystic fibrosis (that is, heterozygous, or *Ff*) is $\frac{1}{2}$, since two of the possible outcomes inside the Punnett square are unaffected heterozygotes—one produced by an *F* sperm and an *f* egg and the other produced by an *f* sperm and an *F* egg.

When parents know which alleles they carry for a single-gene trait, they can easily determine the *probability* that a child they produce will have the disease phenotype. You should note that this probability is generated independently for each child—in other words, *every offspring* of two carriers has a $\frac{1}{4}$ chance of being affected. As more disease alleles are becoming identified, the amount of information about the genes of sperm donors, or any potential parent, may also increase. Unfortunately, this increase in genetic testing is not necessarily parallel to with an increase in our understanding of how most traits develop, as we shall see in the next section.

When the Role of Genes Is Unclear

The single-gene traits discussed in the previous section have a distinct "off or on" character; people either have one phenotype (the disease is present), or the other (the disease is absent). Conditions like this, such as cystic fibrosis, are known as **qualitative traits**. However, many of the traits that interest women who are choosing a sperm donor do not have this off or on character. Traits such as height, weight, eye color, musical ability, susceptibility to cancer, and intelligence, all of which have many possible values, are called **quantitative traits**. Quantitative traits show **continuous variation**—that is, we can see a *range* of phenotypes in a population, for instance from very short people to very tall people. Wide variation in quantitative traits leads to the great diversity we see in the human population (Figure 4.9).

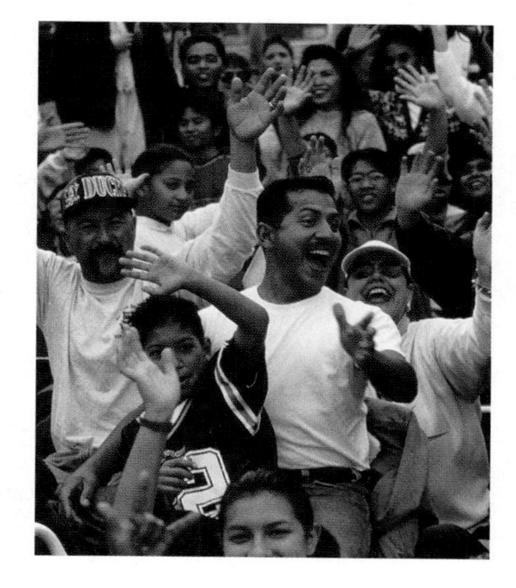

Figure 4.9 Human variation. With a multitude of quantitative physical traits, human beings are quite variable in appearance. Even in a large group of people, it is difficult to find two individuals who are similar in many physical traits.

The distribution of phenotypes of a quantitative trait in a population can be displayed on a graph, and typically takes the form of a bell-shaped curve called a *normal distribution*. Figure 4.10a graphs a normal distribution that represents the height of men in an Amherst College class in 1884. Each column on the graph shows the number of men measured at the height indicated along the bottom of the column. The curved line on the figure is an idealized bell-shaped curve that summarizes this data. A trait that is normally distributed in a population may be described in a number of ways. We are used to thinking about the average, or *mean*, value for data. This is calculated by adding all of the values for a trait in a population, and dividing by the number of individuals in that population. Figure 4.10a shows that the mean height of men in this population is 1.73 meters. However, the average value of a continuously variable trait does not tell you very much about the population. Examine Figure 4.10a closely: Does an average height of 1.73 meters in this particular population imply that most men were this height? Were most men in this population close to the mean, or was there a wide range of heights?

In addition to knowing the mean value for a trait, we also must understand how much *variability* exists in the population for the trait. The amount of variation in a population is described with a mathematical term called *variance*. A low variance for a trait indicates a small amount of variability in the population, while

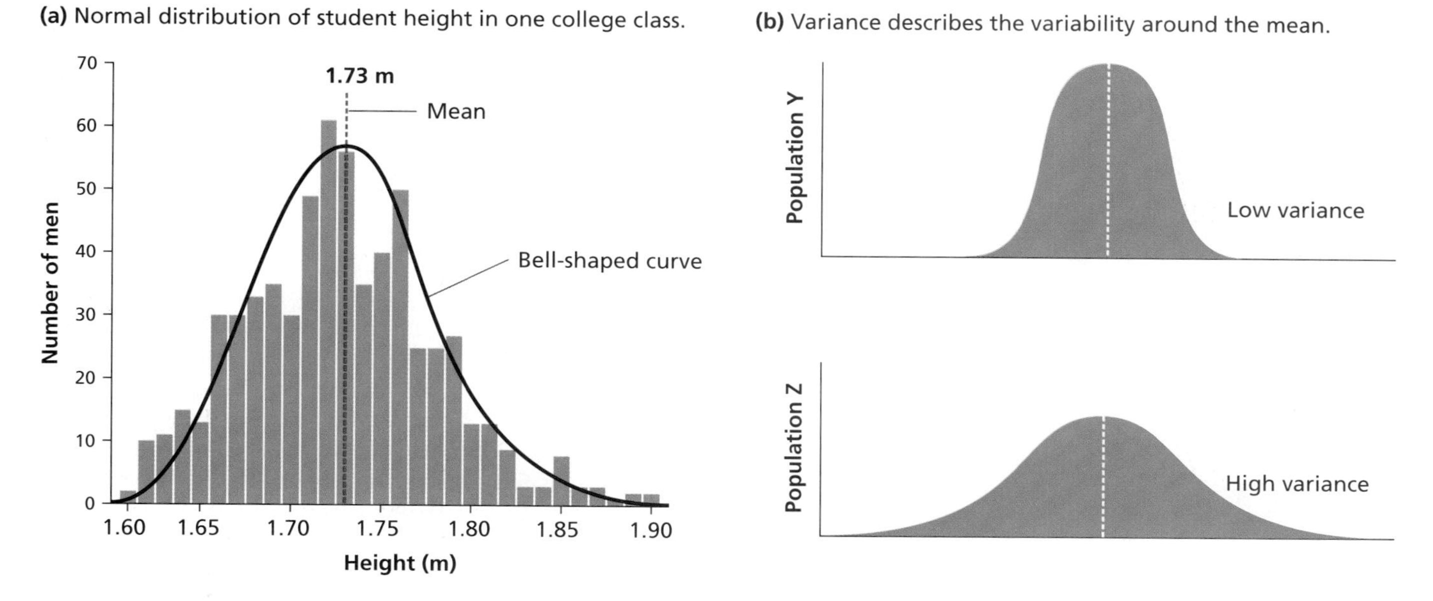

Figure 4.10 A quantitative trait. (a) This graph of the number of men in each category of height is a normal distribution with a center around the mean height of 1.73 m. (b) Though both of these populations have the same mean value, there is much less variation among individuals in population Y than in population Z.

www
Media Activity 4.2 Mouse Genetics

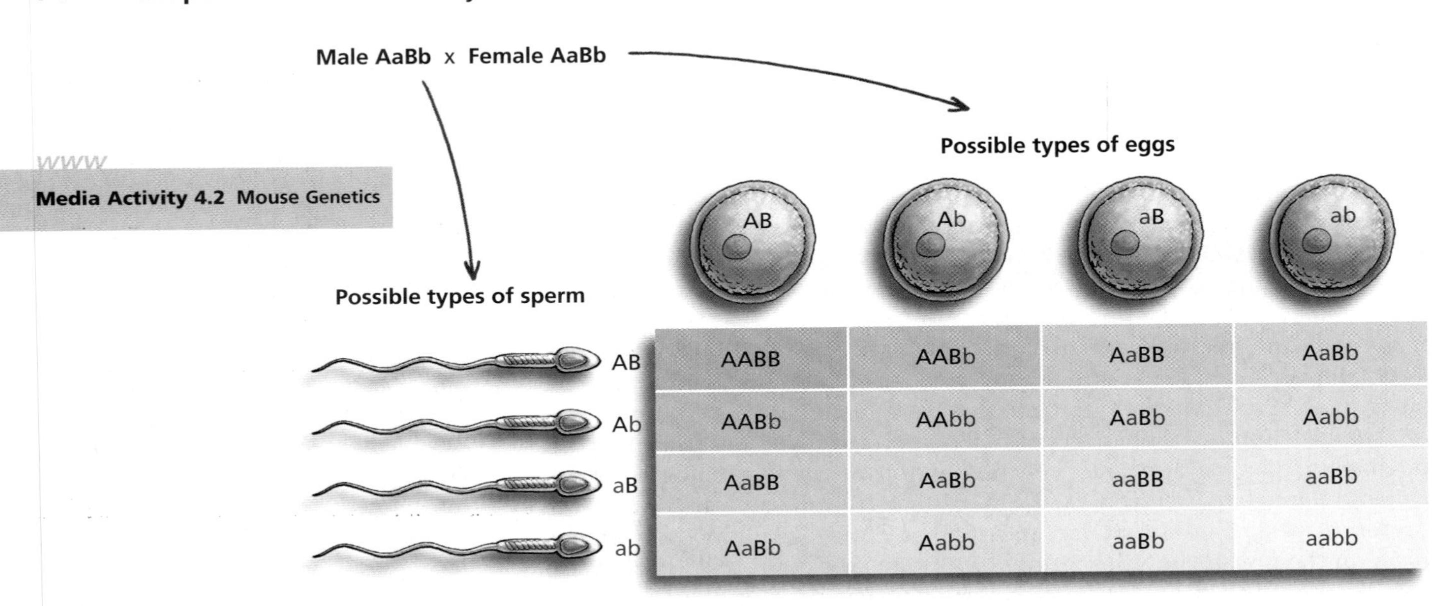

Figure 4.11 A trait determined by two genes can produce a range of genotypes. This Punnett square illustrates the results of a cross between individuals who are heterozygous for two genes. There are four possible gamete genotypes produced; blending these gamete genotypes in the boxes of the square results in nine different genotypes. If each genotype corresponds to a unique phenotype, a range of physical variation can be produced in a polygenic trait.

a high variance indicates a large amount of variability (Figure 4.10b). Scientists who study the inheritance of quantitative traits are interested in determining the genetic basis for the variance in human traits.

Why Traits Are Quantitative One reason we may see a range of phenotypes in a human population is because numerous genotypes exist among the individuals in the population. This happens when a trait is influenced by more than one gene; traits influenced by many genes are called *polygenic traits.* As we saw above, when a single gene with two alleles determines a trait, three possible genotypes are present: AA, Aa, and aa. However, when two genes each with two alleles influence a trait, nine genotypes are possible. This is because each egg and sperm contains one of four possible allele combinations (Figure 4.11). For example, eye color in humans is a polygenic trait influenced by at least three genes. These genes help produce and distribute the pigment melanin to the iris—people with very dark eyes have a lot of melanin in each iris, people with brown eyes have a little less melanin, and blue eyes result when there is very little melanin present. When the genes for eye-pigment production and distribution interact, a range of eye colors, from dark brown to pale blue, is produced in humans. The continuous variation in eye color among people is a result of more than one gene influencing the phenotype.

Another reason that continuous variation may occur in a quantitative trait is due to the influence of environmental factors. In this case, each genotype is capable of producing a range of phenotypes depending upon outside influences. This phenomenon is called *phenotypic plasticity* because the phenotype is plastic, or flexible. Thus, even if all individuals have the same genotype, many different phenotypes can result if they are raised in a variety of environments. A clear example of phenotypic plasticity is shown in Figure 4.12. These identical twins share 100% of their genes, but are quite different in appearance. This is due to variations in their environment—the twin on the bottom smoked and had much greater sun exposure than the twin on top.

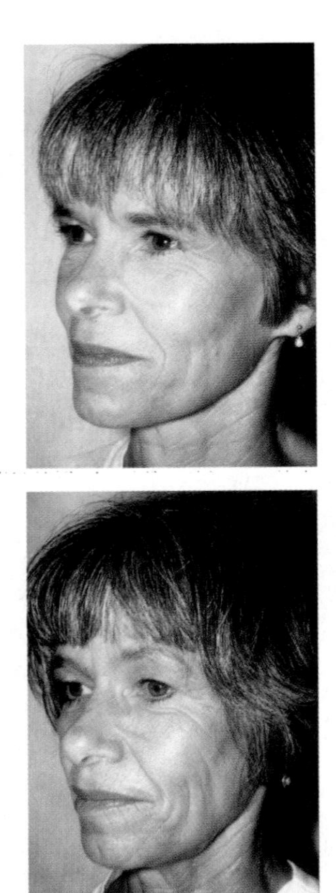

Figure 4.12 An example of phenotypic plasticity. These identical twins have the exact same genotype, but are quite different in appearance due to environmental factors.

Quantitative variation in life expectancy among people of different cultures and economic groups is also largely a result of environmental factors. People born in industrialized countries generally have longer life expectancies than

Botswana 39
Afghanistan 46
Haiti 49
Sudan 57
Iraq 65
Honduras 70
South Korea 74
United States 77
Greece 78
Austria 78
Canada 79
30 40 50 60 70 80
Average life expectancy (years)

Figure 4.13 Life expectancy in selected countries, 2000. This graph shows a 40-year difference in average life expectancy. Differences in the life expectancy of human populations in different countries are entirely due to differences in their environments, particularly income levels.

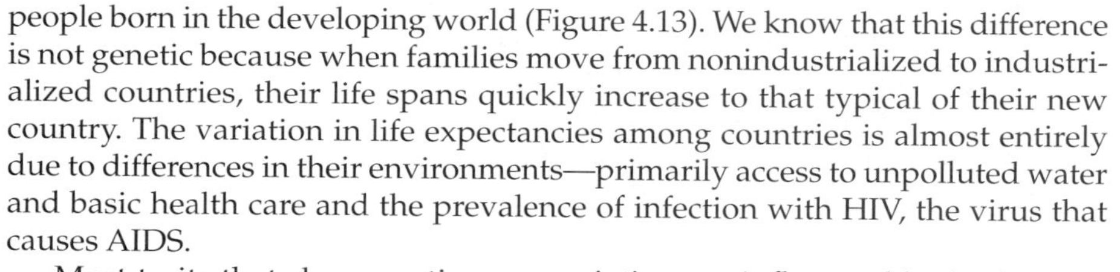

people born in the developing world (Figure 4.13). We know that this difference is not genetic because when families move from nonindustrialized to industrialized countries, their life spans quickly increase to that typical of their new country. The variation in life expectancies among countries is almost entirely due to differences in their environments—primarily access to unpolluted water and basic health care and the prevalence of infection with HIV, the virus that causes AIDS.

Most traits that show continuous variation are influenced by both genes and the effect of differing environmental factors. Skin color in humans is an example of this type of trait. The shade of an individual's skin is dependent upon the amount of melanin present near the skin's surface. A number of genes have an effect on this phenotype—both those that influence melanin production and those that affect the distribution of melanin in the skin. However the environment, particularly the amount of exposure to the sun, also influences the skin color of individuals (Figure 4.14): melanin production increases and any melanin that is present darkens in sun-exposed skin. In climates with warm summers and cool winters, an individual's skin color changes over the course of a year. After many years of intensive sun exposure, skin may become permanently darker. Among people with light skin, the effect of the sun on skin color can be quite dramatic.

Both genetic factors *and* environmental factors influence most traits that are of interest to women choosing sperm donors. Women choosing Doctorate-category sperm donors from Fairfax Cryobank are presumably interested in having smart, successful children, but intelligence has both a genetic and an environmental component. Intelligence partly depends on brain structure and function, and many alleles that interfere with brain structure and function—and thus intelligence—have been identified; but intelligence also depends on environmental factors. For example, if a developing baby is exposed to high levels of cigarette by-products or alcohol before birth, its brain will develop differently, and it may have delayed or diminished intellectual development.

Predicting the Inheritance of Quantitative Traits Unlike qualitative traits, where the relationship between genes and traits is very clear, the inheritance of quantitative traits is difficult to understand. For instance, if the variation among individuals could be a result of many different genes, a variety of environmental effects on phenotype, or (most likely) the interaction of genes and environment, how can we predict if the child of a father with a doctorate will also be capable of earning a doctorate? The most common way in which scientists approach this question is to attempt to determine the relative importance of different alleles in determining variation in phenotype among individuals.

(a) Genes

(b) Environment

Figure 4.14 Skin color is influenced by genes and environment. (a) The difference in skin color between these two women is primarily a result of differences in several alleles that control skin pigment production. (b) The difference in color between the sun-protected and sun-exposed portions of the older man in this picture is entirely the result of environmental effects.

Researchers working with domestic animals and crop plants were the first to develop the scientific model used to measure the importance of genes in determining the value of quantitative traits. These researchers were trying to find the best way to improve production in various agricultural species. For example, farmers who wish to increase their dairy herd's milk production have two basic strategies for doing so: Change the herd's environment by changing the way the cows are reared, housed, and fed; or change the herd genetically by choosing only the offspring of the best milk producers for the next-generation herd. The technique of controlling the reproduction of individual organisms to influence the phenotype of the next generation is known as *artificial selection* (Figure 4.15). Artificial selection is similar to *natural selection*, which is described more thoroughly in Chapter 9.

If milk production in cows is strongly influenced by genes—in other words, it has high **heritability**—then artificial selection is an effective way to boost

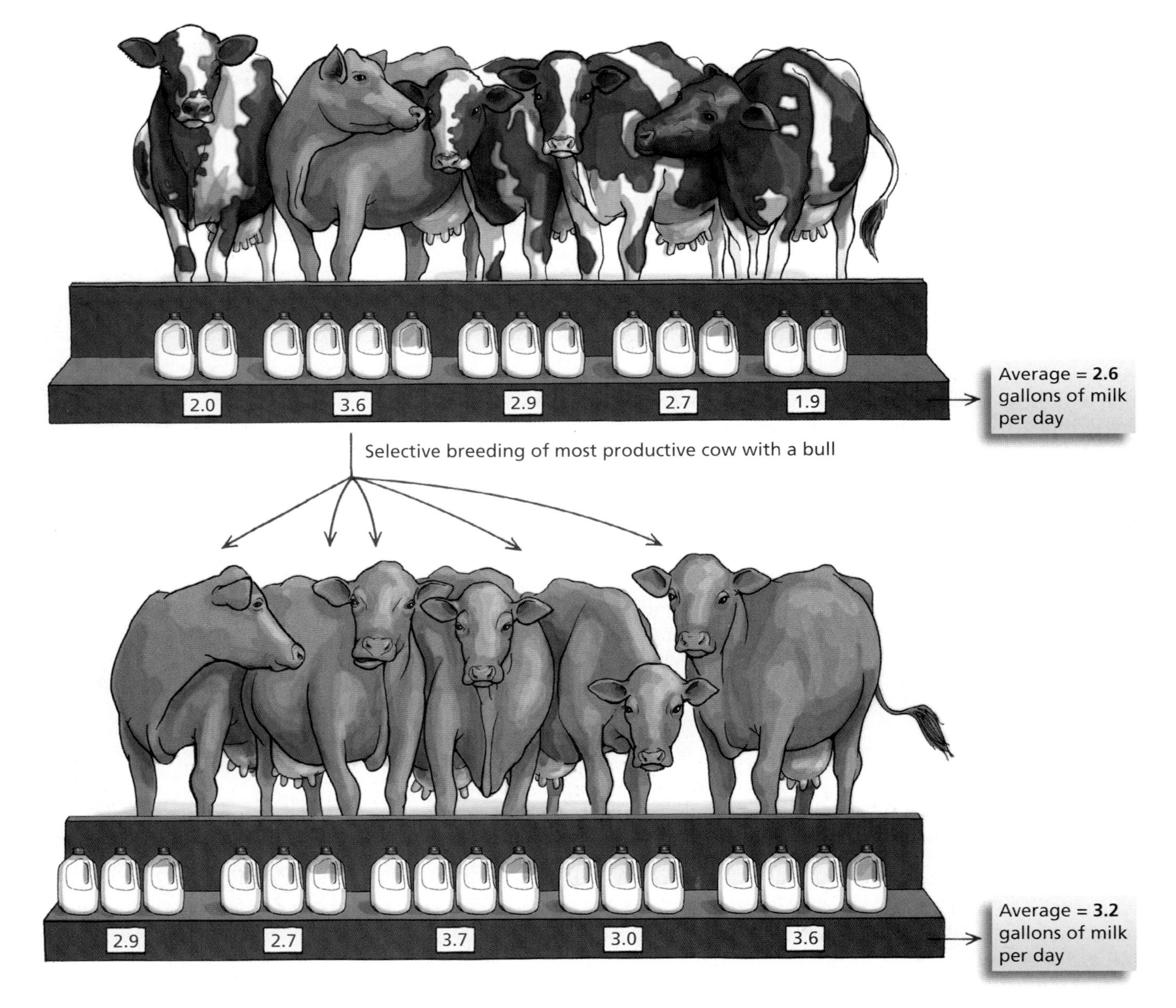

Figure 4.15 Artificial selection increases milk production in cows. Cows that produce exceptional amounts of milk are bred to produce the next generation of dairy cattle. In this example, the female calves of the cow that produces 3.6 gallons of milk daily produce an average of 3.2 gallons of milk per day, 23% more than the previous herd.

milk output. In fact, heritability of milk production can easily be measured by how well a herd responds to artificial selection. If milk production increases in a herd of cows as a result of artificial selection, it is because alleles for proteins that increase milk production in an individual (for instance, alleles for genes that control the size of the udder, or the activity of milk-producing cells) have become more common—that is, the trait is strongly influenced by genes. If milk production does *not* increase as a result of artificial selection, then the alleles the high-production cows possess must not be as important in determining milk output—that is, the trait must be more strongly influenced by the environment than by genes. Scientists have calculated an average heritability of milk production in dairy cattle of 0.30. This means that, in a typical dairy herd, about 30% of the variation among cows in milk production is due to differences in their genes; the remainder of their production variation is due to differences among the cows in their environmental conditions.

Studies of the relative influence of genes and environment that use response to artificial selection cannot be performed in human populations. It is ethically and socially unacceptable to design breeding programs to produce people with various traits, or to select the men and women who will produce the next generation. An alternative strategy is for scientists to compare the value of a trait in children with that of their parents. This comparison takes the form of a *correlation*, where researchers estimate, for instance, how accurately we can predict the height of children if we know the height of their parents.

The correlation between the intelligence of parents and their children helps us determine how important the intelligence of a donor may be to the mental capacity of his children. Intelligence is often measured by performance on an IQ test. Alfred Binet, a French psychologist, developed the *intelligence quotient* (or IQ), in the early 1900s to more efficiently identify Paris schoolchildren who were in need of remedial help in school. The IQ is not an absolute score on a test, it represents a comparison between an individual and his peers in test performance. The average IQ was arbitrarily set at 100 with a standard deviation of 15 points, and IQ tests are designed to produce a normal distribution in scores taken from a sample population.

Binet's IQ test was not based on any theory of intelligence and was not meant to comprehensively measure mental ability, but the tests remain a common way to measure innate, or "natural" intelligence. Even if IQ tests do not really measure general intelligence, IQ scores have been correlated with academic success—meaning that individuals at higher academic levels usually have higher IQs. So, even without knowing their IQ scores, the expectation that donors in the Doctorate category have higher IQs than other available sperm donors is reasonable.

The average correlation between IQs of parents and their children is 0.42—in other words, 42% of IQ variation among individuals is apparently the result of their genes. However, children are typically raised in a similar social and economic environment as their parents. Does the IQ correlation above still clearly tell us the role of genes in determining IQ? Since parents and children are similar in genes *and* environment, comparisons between the two groups do not allow researchers to definitively determine the relative importance of each factor on a given trait. This is the problem found in most arguments about "nature versus nurture"—do children resemble their parents because they are "born that way," or because they are "raised that way"?

The impossibility of using traditional selection studies, and the difficulty of separating genetic and environmental influences in most families compels researchers interested in the heritability of traits in humans to use **natural experiments**. These are situations in which unique circumstances allow a hypothesis test without prior intervention by researchers. Human twins are one source of a natural experiment to test hypotheses about the heritability of quantitative traits in humans.

Estimating the Heritability of Intelligence By comparing monozygotic twins to dizygotic twins, researchers can begin to separate the effects of shared genes from the effects of shared environments. Twins raised in the same family presumably have similar childhood experiences, unlike nontwin siblings, whose social and family environments differ. Any unique social factors that may be associated with being a twin are probably common to both identical and nonidentical twins. Thus, some scientists argue that the only real difference between monozygotic and dizygotic twins is the percentage of genes they share. Recall that monozygotic twins share all of their alleles, while dizygotic twins share, on average, only 50% of their alleles.

Using a formula that includes the correlation of IQ between pairs of monozygotic twins, the correlation of IQ between pairs of dizygotic twins, and relative genetic similarity, the average heritability of IQ calculated from a number of different studies is about 0.52. According to these studies, 52% of the variability in IQ among humans is due to differences in genotypes. It is somewhat surprising that this value is even higher than the 42% calculated from the correlation between parents and children.

However, the heritability value arrived at through twin studies has been criticized by other scientists. One major criticism is that monozygotic twins and dizygotic twins *do* differ in more than just genotype. In particular, identical twins are treated more alike than nonidentical twins. This occurs both because of the greater similarity in appearance of monozygotic twins, and because of the expectation by parents, relatives, friends, and teachers that identical twins are identical in all respects. In fact, a common way scientists determine whether a twin pair is monozygotic is to ask the twins if other people often have trouble telling them apart or comment on their similarities. If monozygotic twins are *expected* to be more alike than dizygotic twins, their IQ scores may be similar because they are continually encouraged to have the same experiences and to achieve at the same level.

Since this difference in treatment of the types of twins cannot be eliminated, researchers studying the heritability of IQ are especially interested in twins that have been raised apart. By comparing identical twins raised in different environments with nonidentical twins that have also been raised apart, the problem of differential treatment of the two *types* of twins is minimized because no one would know that the individual members of a pair have a twin.

The frequency of early twin separation is extremely rare and is becoming even rarer as the policy of keeping adoptive siblings together has become more standard. This infrequency makes the few identical-twin pairs known to be separated especially valuable. Researchers have estimated the heritability of IQ at a remarkable 0.72 in this small sample of twins raised apart. These studies support the hypothesis that differences in our genes explain much of the variation in IQ among people. Table 4.1 summarizes the estimates of IQ heritability and previews the cautions discussed in the next section of this chapter.

4.3 Genes, Environment, and the Individual

Perhaps we can now determine the importance of a sperm-donor father who has earned a doctorate to a child's intellectual development. We know that a sperm donor will definitely influence some of their child's traits—such as eye and skin color, and perhaps even susceptibility to certain diseases. In addition, according to the twin studies discussed above, the donor will probably pass some intellectual traits to the child as well. In fact, the high value for heritability of IQ *appears* to indicate that the environment has relatively little influence in determining an individual's intelligence. It seems that it might be a good idea to pay a premium price for "Doctorate" sperm, after all.

Method of measurement	Result	Warnings when interpreting this result	Warnings that apply to all measurements of heritability
Correlation between parents' IQ and children's IQ in a population.	0.42	• Since parents and children are similar in genes and environment, a correlation cannot be used to indicate the relative importance of genes and environment in determining IQ.	• Heritability values are specific to the populations for which they were measured.
Natural experiment comparing IQ in pairs of identical twins versus nonidentical twins.	0.52	• Identical twins are treated more alike than nonidentical twins. Therefore their environment is different than that of nonidentical twins—the heritability value could be an overestimate.	• High heritability for a trait does not mean that it is not heavily influenced by environmental conditions; we also cannot predict how the trait will respond to a change in the environment.
Natural experiment comparing IQ of individual twins raised apart versus nonidentical twins raised apart.	0.72	• Small sample size may skew results.	• Heritability is a measure of a population, not an individual.

Table 4.1 To what extent is IQ heritable? A summary of various estimates of IQ heritability, their shortcomings, and the problems with using them to understand the role of genes in determining an individual's potential intelligence.

However, we need to be very careful when applying the results of twin studies to questions about the relative value of sperm donors. The results of twin studies actually give us very limited information about how closely a child will match a sperm donor in intelligence and preferences. To understand why, we will take a closer look at the practical significance of heritability.

The Use and Misuse of Heritability

Remember that heritability is a measure of the relative importance of genes in determining variation in qualitative traits among individuals. For example, with a heritability of 0.30, we can say that only 30% of the variation among dairy cow milk production is due to variation in genes among these cows. However, *the calculated heritability value is unique to the population in which it was measured and to the environment of that population*. The specificity of heritability means that we should be very cautious when using heritability to measure the general importance of genes to the development of a trait.

A famous misapplication of heritability comes from the book *The Bell Curve*, by Charles Herrnstein and Richard Murray, published in 1994. In this book, the authors report that IQ scores differ among subpopulations in the United States. Among white Americans, IQ averages around 100; while among African-American populations, IQ averages nearly 15 points lower. Using a conservative estimate of the heritability of intelligence based on the studies described in the previous section (0.60), they argued that the IQ differences between whites and blacks are primarily due to a genetic difference in intelligence between these groups. However, if we evaluate the meaning of heritability carefully, we can see that Herrnstein and Murray's conclusion is flawed. Let us use a thought experiment to clarify the reason for this.

Body weight in laboratory mice has a strong genetic component, with a calculated heritability of about 0.90. In a population of mice where weight is variable, bigger mice have bigger offspring and smaller mice have smaller offspring.

Imagine that we randomly divide a population of variable mice into two groups—one group is fed a rich diet, and the other group is fed a poor diet. The mice are treated identically in all other respects. As anticipated, the mice receiving high levels of calories store some excess as fat, while the mice receiving low levels of calories store very little fat. Keeping the mice in the same conditions, imagine that we allow them to reproduce and then weigh their adult offspring. The mice in the rich-diet environment are twice as heavy, on average, as the mice in the poor-diet environment. If we use Herrnstein and Murray's logic, since the heritability is 0.90 and the average size in the two groups is different, then the groups must be genetically different. However, we know this is not the case—both the heavy and the light mice are offspring of the same original population of parents. These two groups of mice simply live in very different environments (Figure 4.16).

Now extend the same thought experiment to human groups. Say we have two groups of humans, and that we have determined that IQ had high heritability. In this case, one group of people was raised in an enriched environment where most individuals have a high IQ, and the other group was raised

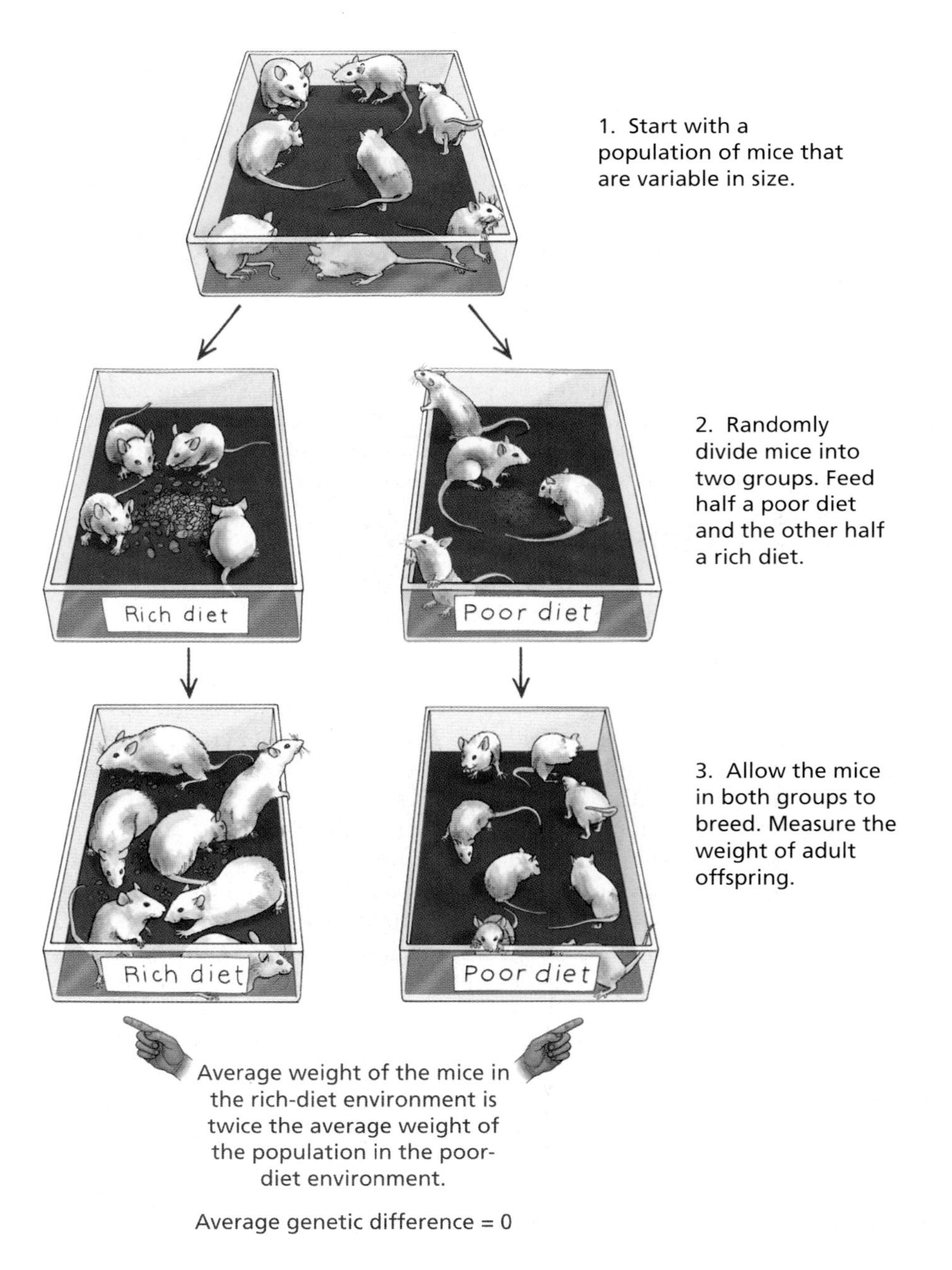

1. Start with a population of mice that are variable in size.
Rich diet
Poor diet
2. Randomly divide mice into two groups. Feed half a poor diet and the other half a rich diet.
Rich diet
Poor diet
3. Allow the mice in both groups to breed. Measure the weight of adult offspring.
Average weight of the mice in the rich-diet environment is twice the average weight of the population in the poor-diet environment.
Average genetic difference = 0

Figure 4.16 The environment can have powerful effects on highly heritable traits. If genetically similar populations of mice are raised in radically different environments, then differences between the populations are entirely due to environment.

in a restricted environment where most individuals have a low IQ. What conclusions could you draw concerning the genetic difference between these two populations? The answer is none—as with the laboratory mice, these differences could be entirely due to environment. Given the history and current social and economic status of African-Americans in the United States, it is certainly possible that their environment is less enriched than the average environment experienced by a white individual. The high heritability of intelligence cannot tell us if two groups with different social environments vary in IQ as a result of differences in genes or differences in environment.

Sometimes heritability is reported as a "percent" of a trait that is due to genetic factors. That is, a heritability of 0.60 is interpreted as meaning that 60% of your IQ is due to genes and the remainder (40%) is due to your environment. This seems to imply that highly heritable traits are not strongly influenced by environmental conditions. In *The Bell Curve*, Herrnstein and Murray state that, given IQ's high heritability, policies that increase financial resources to schools will ultimately fail to increase IQ, since such a predominately genetic trait will not respond well to environmental change. However, intelligence in other animals can be demonstrated to be both highly heritable and strongly influenced by the environment.

The following experiment demonstrates the point that even highly heritable traits are still strongly influenced by environmental conditions: Rats can be bred for maze-running ability. Using the same sort of selection process the dairy farmers above used to produce herds with higher milk production, researchers have produced rats that are "maze-bright" and rats that are "maze-dull." Maze-running ability is highly heritable in the laboratory environment. In a normal laboratory environment, maze-dull rats made an average of 165 mistakes every time they attempted to run the maze, while maze-bright rats made only about 115 errors. However, in different environments, the differences between the maze-dull and maze-bright rats were less extreme (Table 4.2). When both rat populations were raised in environments with very little visual variety (that is, no other rats, no running wheel, and no ability to see activities in the lab), both maze-bright and maze-dull rats made 170 errors per maze. If both types of rats were raised in a high-stimulus, enriched environment (that is, cages with passageways and places to hide), maze-dull rats only made 10 more errors per maze than the maze-bright rats. All rats did better at maze running in enriched environments, regardless of their genotype. In fact, poor maze runners improved to a much greater extent than good maze runners did when the environment was enriched. What this example demonstrates is that we cannot

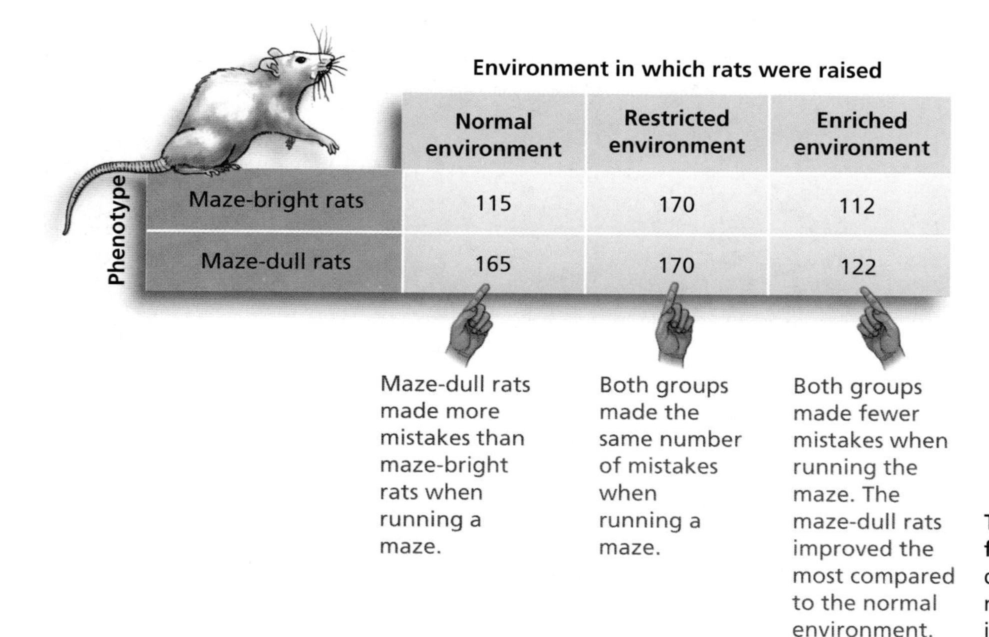

Phenotype	Environment in which rats were raised: Normal environment	Restricted environment	Enriched environment
Maze-bright rats	115	170	112
Maze-dull rats	165	170	122

Maze-dull rats made more mistakes than maze-bright rats when running a maze.
Both groups made the same number of mistakes when running a maze.
Both groups made fewer mistakes when running the maze. The maze-dull rats improved the most compared to the normal environment.

Table 4.2 A highly heritable trait is not fixed in all environments. This table describes the average number of mistakes made by rats of two different genotypes in three different environments.

predict the response of a trait to a change in the environment—even when that trait is highly heritable. Thus, even if IQ has a strong genetic component, environmental factors affecting IQ can have dramatic effects on an individual's intelligence. Enriching the intellectual environment of all children certainly could increase IQ scores across the board, and this might minimize currently observed differences between groups.

Saying that heritability is the percentage of a given trait that is determined by genes also seems to imply that, for traits with high heritability, variability *among individuals* is mostly due to differences in genes. However, heritability is a measure of a population. Thus, even if genes explain 90% of the differences among individuals in a particular environment, the reason one individual differs from another may be entirely a function of environment (see the twins in Figure 4.12). Currently, there is no way to determine if a particular child is a poor student because of his genes, because of a poor environment, or a combination of both factors. There is also no way to predict whether a child produced from the sperm of a man with a doctorate will be an accomplished scholar. All we can say is that given our current understanding of the heritability of IQ and the current social environment, the alleles in Doctorate sperm may *increase the probability* of having a child with a high IQ.

How Do Genes Matter?

We know that genes can have a strong influence on eye color, the risk of genetic diseases such as cystic fibrosis, susceptibility to heart attack, and even on the structure of the brain. But what *really* determines who we are—nature or nurture?

Even with single-gene traits, the outcome of a cross between a woman and a sperm donor is not a certainty—only a probability. Couple this with traits being influenced by more than one gene, and independent assortment greatly increases the offspring types possible from a single mating. Knowing the phenotype of potential parents gives you relatively little information about the phenotype of their children (Figure 4.17). So, even if genes have a strong effect on traits, we cannot "program" the traits of children by selecting the traits of their parents. (Essay 4.2 describes the history of scientific attempts to do this.)

Figure 4.17 The phenotype of children cannot be exactly predicted from the phenotype of parents. This family demonstrates the similarities and differences between children and parents and among siblings.

Essay 4.2 Why Is the "Nature versus Nurture" Debate So Heated?

Discussions of whether personality traits and intelligence are a function of genes or environment are a continual source of often acrimonious debate. At least some of this controversy can be attributed to the concept of *eugenics* in both the United States and Europe.

British statistician Sir Francis Galton coined the term *eugenics* in 1883 to refer to the science of the improvement of the human race by "better breeding." Better breeding implied that the quality of the human species could be improved by using a newfound understanding of the evolution and genetics. Eugenics was the human equivalent of selective animal and plant breeding.

Eugenics was first embraced as a scientific means of halting the stream of impoverished immigrants who came to the United States between 1880 and 1914. These new immigrants arrived principally from eastern and southern Europe, the Balkans, and Russia. They were ethnically and culturally different from earlier waves of foreigners, who had migrated mostly from the countries of western Europe such as Germany, England, Ireland, and Scotland. Many Americans thought these new immigrants were "defective"—less intelligent, more radical, and willing to work for low wages (Figure E4.2). To combat the purported damage these new immigrants would do to the nature of future generations of Americans, eugenicists successfully lobbied for restrictions on immigration from these groups.

Figure E4.2 Immigrants at Ellis Island in 1905.

In the 1920s and 1930s, eugenicists began working on the passage of a number state laws mandating "eugenical sterilization." These laws were intended to eliminate the production of offspring of individuals—typically those in mental institutions or jails—who were considered likely to give birth to defective children. By 1940, more than 30 states had enacted such compulsory sterilization laws. Between 1907 and 1941, more than 60,000 eugenical sterilizations were performed in the United States.

At the same time in Europe, the Nazi government in Germany saw eugenics as a means of maximizing the quality of the Aryan race. American eugenicists collaborated with their German counterparts during this time. Many American eugenicists were enthusiastic about the Nazi eugenics program, which included compulsory sterilization of "defectives" and the elimination of infants with birth defects. Frederick Osborn, the secretary of the American Eugenics Society, declared, "The German sterilization program is apparently an excellent one . . . recent developments in Germany constitute perhaps the most important experiment which has ever been tried."

The Nazi approach to eugenics drove the Holocaust, the systematic extermination of over six million Jews, gypsies, homosexuals, and other supposed defectives—this horrific tragedy immediately discredited eugenics in the United States. The primacy of genes was replaced by the primacy of environment as the key factor in determining personality and intelligence.

The discovery of disease genes in the 1980s and 1990s have led to a new interest in the relationship between genes and behavior. Studies on the genetics of alcoholism, violence, and homosexuality have become front-page news and have raised fears about a resurgence of interest in eugenics. Although it appears unlikely that compulsory sterilization laws will again become widespread, new reproductive technologies (including the use of sperm donors) give parents some ability to choose their offspring's traits. Many people fear that this "silent" eugenics will lead to harsh social penalties for people with easily diagnosed "nonpreferred" physical and mental traits.

In truth, we are really asking ourselves the wrong question when we wonder if nature or nurture has a more powerful influence on who we are. Both our genes and environment have a profound influence on our physical and mental characteristics. Possessing functional genes is imperative to the proper development of a human being—our cells carry instructions for all of the essential characteristics of humanity, but the process of developing from embryo to adult takes place in a physical and social environment that influences how these genes are expressed. Scientists are still a long way from understanding how all of these complex, interacting circumstances result in who we are.

What is the message for women and couples who are searching for a sperm donor from Fairfax Cryobank? The donors in the Doctorate category may indeed have higher IQs than donors in the cryobank's other categories, but there is no real way to predict if a particular child of one of these donors will be smarter than average. According to the current data on the heritability of IQ, sperm from high-IQ donors may "load the dice" and increase the odds of having an offspring with a high IQ, but only if parents provide them with a stimulating, healthy, and challenging environment to mature in—and that would be good for children with any alleles.

CHAPTER REVIEW

Summary

- Children resemble their parents in part because they inherit their parents' genes.
- Most genes are segments of DNA that contain information about how to make proteins.
- Mutations in gene copies can result in slightly different proteins being produced; different gene versions are called alleles.
- Parents contribute a unique subset of alleles to each of their nonidentical twin offspring.
- On average, two offspring of the same set of parents share 50% of their alleles, although identical twins share 100% of their alleles.
- The phenotype of a given individual for a particular gene depends on which alleles it carries (its genotype) and whether the alleles are dominant, recessive, or codominant.
- A Punnett square helps us determine the probability that two parents will produce a child with a particular genotype.
- Many traits show quantitative variation, which results in continuous variation in the population.
- Quantitative variation in a trait may be generated because the trait is influenced by several genes, because the trait can be influenced by environmental factors, or a combination of both factors.
- The role of genes in determining the phenotype for a quantitative trait is estimated by calculating the heritability of the trait.
- Heritability is calculated by examining the correlation between parents and offspring, or by comparing pairs of monozygotic twins to pairs of dizygotic twins.
- Twin studies have revealed that the heritability of IQ is relatively high.
- Calculated heritability values are unique to a particular population in a particular environment. The environment may cause large differences among individuals, even if a trait has high heritability.
- Knowing the heritability of a trait does not tell us why two individuals differ for that trait.
- Our current understanding of the relationship between genes and complex traits does not allow us to predict the phenotype of a particular offspring from the phenotype of its parents.

Key Terms

allele p. 84
carrier p. 92
codominant p. 90
continuous variation p. 93
cross p. 91
dizygotic twins p. 89
dominant p. 90
egg p. 83
embryo p. 83
fertilization p. 83
genes p. 83
genotype p. 89
heritability p. 96
heterozygous p. 89
homozygous p. 89
independent assortment p. 85
monozygotic twins p. 87
mutations p. 84
natural experiments p. 97
phenotype p. 89
Punnett square p. 91
qualitative traits p. 93
quantitative traits p. 93
recessive p. 89
sperm p. 83

Learning the Basics

1. Describe the relationship between a gene, a protein, and a trait.

2. Why are nonidentical-twin children produced by the same set of parents different from each other?

3. What factors cause quantitative variation in a trait within a population?

4. What is heritability?

5. Can we predict the phenotypes of offspring from the phenotype of their parents? If yes, how? If no, why not?

6. A gene _____.
   a. contains information about how to make a protein
   b. may come in different "versions"
   c. is expressed in every cell in the body
   d. a and b are correct
   e. a, b, and c are correct

7. An allele is a _____.
   a. version of a gene
   b. dysfunctional gene
   c. protein
   d. spare copy of a gene
   e. phenotype

8. Sperm or eggs in humans always _____.
   a. each have two copies of every gene
   b. each have one copy of every gene
   c. each contain either all recessive alleles or all dominant alleles
   d. are genetically identical to all other sperm or eggs produced by that person
   e. each contain all of the genetic information from their producer

9. The physical appearance of an individual is _____.
   a. the result of both genotype and environment
   b. its phenotype
   c. completely determined by the genes they carry
   d. a and b are correct
   e. a, b, and c are correct

10. A mistake or "misspelling" that occurs during the copying of gene and results in a change in gene is called a(n) _____.
   a. dominant allele
   b. mutation
   c. mistakes never occur in gene copying
   d. dysfunction
   e. improvement

11. Which of the following genotypes is heterozygous?
   a. *AA*
   b. *Aa*
   c. *a*
   d. *AA BB*
   e. More than one of the above.

12. When the effects of an allele are seen only when an individual carries two copies of the allele, the allele is termed _____.
   a. dominant
   b. incompletely dominant
   c. recessive
   d. codominant
   e. genotypic

13. A quantitative trait _____.
   a. may be one that is strongly influenced by the environment
   b. varies continuously in a population
   c. may be influenced by many genes
   d. has more than a few values in a population
   e. All of the above are correct.

14. All of the following traits are quantitative in human populations *except* _____.
   a. sickle-cell anemia
   b. height
   c. skin color
   d. IQ
   e. susceptibility to cancer

15. When a trait is highly heritable _____.
   a. it is influenced by genes
   b. it is not influenced by the environment
   c. the variance of the trait in a population can be explained primarily by variance in genotypes
   d. a and c are correct
   e. a, b, and c are correct

Genetics Problems

1. A single gene in pea plants has a strong influence on plant height. The gene has two alleles, tall (*T*) which is dominant, and short (*t*) which is recessive. What are the genotypes and phenotypes of the offspring of a cross between a *TT* and a *tt* plant?

2. What are the genotypes and phenotypes of the offspring of *Tt* × *Tt*?

3. Albinism occurs when individuals carry two recessive alleles (*aa*) that interfere with the production of melanin, the pigment that colors hair, skin, and eyes. If an albino child is born to two individuals with normal pigment, what is the genotype of each parent?

4. Huntington's disease is caused by a dominant allele. If a woman with one copy of the Huntington's allele marries a man who is homozygous recessive, what percent of their children are expected to develop Huntington's disease?

5. Flower color in pea plants is controlled by a gene with an incompletely dominant allele. *RR* plants produce red flowers, *rr* plants produce white flowers, and *Rr* plants produce pink flowers.
   - **a.** What percentage of the offspring of a cross between a white-flowered plant and a red-flowered plant are expected to be white?
   - **b.** What percentage of the offspring of a cross between two pink-flowered plants will have white flowers?

6. A cross between a pea plant that produces yellow-colored peas and a pea plant that produces green peas results in 100% yellow-pea offspring.
   - **a.** Which allele is dominant in this situation?
   - **b.** What are the genotypes of the yellow-pea and green-pea plants in the initial cross?

7. A cross between a pea plant that produces yellow peas and a pea plant that produces green peas results in 50% yellow-pea offspring and 50% green-pea offspring. What are the genotypes of the plants in the initial cross?

8. A woman who is a carrier for the sickle-cell allele marries a man who is also a carrier.
   - **a.** What percentage of the woman's eggs will carry the sickle-cell allele?
   - **b.** What percentage of the man's sperm will carry the sickle-cell allele?
   - **c.** The probability they will have a child that carries two copies of the sickle-cell allele is equal to the percent of eggs that carry the allele times the percent of sperm that carry the allele. What is this probability?
   - **d.** Is this the same result you generate when doing a Punnett square of this cross?

9. Blood type in humans is controlled by a single gene with three alleles. Allele I^A produces a protein that results in type A blood, Allele I^B produces a protein that results in type B blood, and allele *i* produces a protein that produces type O blood. Individuals who are homozygous with the *i* allele have type-O blood. Individuals with I^Ai genotype have type-A blood, individuals with the I^Bi genotype have type B blood, and individuals with the I^AI^B genotype have blood type AB. From this information, determine which allele(s) are recessive, which allele(s) are dominant, and the nature of the dominant alleles.

10. The allele *BRCA1* was identified in families with unusually high rates of breast and ovarian cancer. About 85% of women with one copy of the *BRCA1* allele develop one of these cancers in their lifetime.
    - **a.** Is *BRCA1* a dominant or a recessive allele?
    - **b.** How is *BRCA1* different from the typical pattern of Mendelian inheritance?

Analyzing and Applying the Basics

1. Cystic fibrosis occurs when an individual carries two mutated copies of a gene that regulates materials migrating in and out of cells. If two heterozygous individuals already have a child with cystic fibrosis, what is the probability that their next child will be affected? What about a third child? Why is the risk of having another affected child no different after the parents have one affected child?

2. Understanding variance has many practical uses. Figure 4.18 shows the variance in calorie counts in samples of two different food items.

   What is the approximate average number of calories per food item? Which sample has a larger variance? How is this information useful to someone who is counting calories?

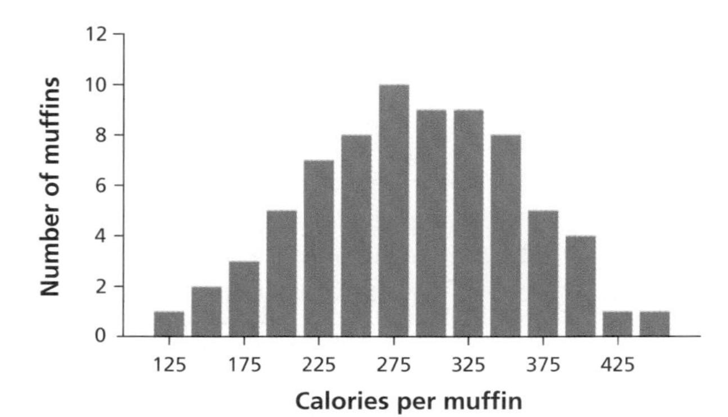

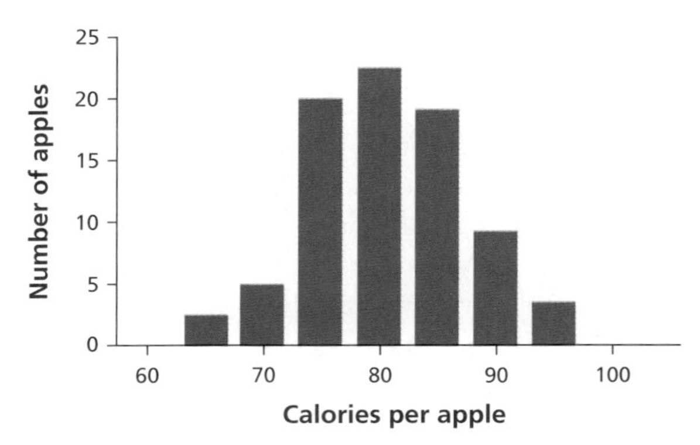

Figure 4.18

3. Does a high value of heritability for a trait indicate that the average value of the trait in a population will not change if the environment changes? Explain your answer.

4. The heritability of IQ has been estimated at about 72%. If John's IQ is 120 and Jerry's IQ is 90, does John have stronger "intelligence" genes than Jerry? Explain your answer.

5. The heritability of IQ was calculated by measuring the correlation of IQ score among pairs of identical twins who were raised apart. The value of this correlation is assumed to be equivalent to the genetic component of IQ. This approach is in contrast to studies of twins raised together, which use a formula to estimate the effect of shared environment by comparing correlations between identical twins and nonidentical twins. What would you need to know about the twin pairs in the raised-apart study to determine if the following assumption is correct: The correlation between the members of the identical twin pairs alone measures the genetic component of IQ?

Connecting the Science

1. If scientists find a gene that is associated with a particular "undesirable" personality trait (for instance, a tendency toward aggressive outbursts) will it mean greater or lesser tolerance toward people with that trait? Will it lead to proposals that those affected by the "disorder" should undergo treatment to be "cured," and that measures should be taken to prevent the birth of other individuals that are also afflicted?

2. *The Bell Curve* was a long and fairly technical book. Despite this, it was a best seller. Why do you think that was the case? Would finding a genetic basis for socioeconomic differences between whites and blacks in the United States mean that social policies in this country should be modified?

3. Does a genetic basis for differences in IQ between people with Down's syndrome and people without this condition mean that we should put fewer resources into education for people with Down's syndrome? How does your answer to this question relate to questions about how we should treat individuals with other genetic conditions?

4. IQ is only one way of measuring intelligence, and having a doctorate is only one measure of success. Given this, do you think that sperm from the Doctorate collection at Fairfax Cryobank should be more costly than non-Doctorate sperm? Why or why not?

Media Activities

Media Activity 4.1 Meiosis and the Human Life Cycle
Estimated Time: 10 minutes
Explore the process of meiosis and follow the fate of two alleles from gamete production to fertilization.

Media Activity 4.2 Mouse Genetics
Estimated Time: 15 minutes
This activity provides an overview of simple Mendelian genetics and the basic laws of probability.

Media Activity 4.3 Genetics and Sickle-Cell Anemia
Estimated Time: 10 minutes
This activity provides the student with an overview of the genetic cause of sickle-cell anemia and explores the function of genes in protein synthesis.

CHAPTER 5

Cancer

The Cell Cycle and Cell Division

Student
Health
Services

Nicole got sick during her junior year of college.

5.1 What Is Cancer?

5.2 Cell Division

5.3 Diagnosis and Treatment

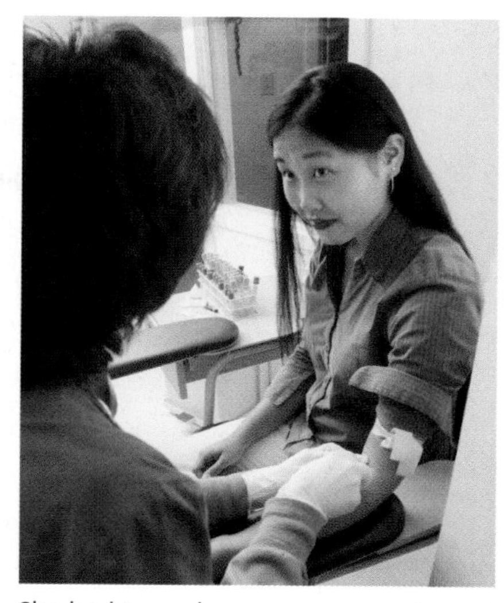

She had to undergo some procedures to see if she had cancer.

Nicole had growths on both of her ovaries—one was cancerous, one was not.

Nicole wants to understand why she got cancer.

Nicole's early college career was similar to that of most students. She enjoyed her independence and the wide variety of courses her majors in biology and psychology required her to take. She worried about her grades and finding ways to balance her course work with her social life. She also tried to find time to lift weights in the school's athletic center and to snowboard at a local ski hill. Some weekends, to take a break from school, she would ride the bus home to see her family.

Managing to get schoolwork done, see friends and family, and still have time left to work out had been difficult, but possible, for Nicole during her first two years at school. That changed drastically in her third school year.

One morning in October of her junior year, Nicole began having episodes of severe abdominal cramping. The first time this happened, she was just beginning an experiment in her cell-biology laboratory course. Hunched over and sweating, she barely managed to make it through the two-hour respiration experiment she and her lab partner were performing. Over the next few days, the cramps intensified so much that she was unable to walk from her apartment to her classes without stopping several times to rest.

Later that week, as she was preparing to leave for class, she had a cramp that was so severe that she had to lie down in the hallway of her apartment.

When her roommate got home a few minutes later, she took Nicole to Student Health Services for an emergency visit. The physician at Health Services first determined that Nicole's appendix had not burst, and then made an appointment for Nicole to see a local gynecologist the next day.

After hearing Nicole's symptoms, Nicole's gynecologist pressed on her abdomen and felt what he thought was a mass on her right ovary. He used a noninvasive procedure called *ultrasound* to try to get an image of her ovary. This procedure requires the use of high-frequency sound waves. These waves, which cannot be heard by humans, were aimed at the ovaries. The pattern of echoes they produced created a picture called a *sonogram*. Healthy tissues, fluid-filled cysts, and tumors all look different on a sonogram.

Nicole's sonogram convinced her gynecologist that she had a large growth on her ovary. He told her that he suspected that this growth was a *cyst*, or fluid-filled sac. Her gynecologist told her that cysts often go away without treatment, but this one seemed to be quite large so it would need to be removed. After her appointment, Nicole went home and called her professors to let them know she would be missing classes for the next week because she would be having surgery.

Even though the idea of having an operation was scary for Nicole, she was relieved to know that the pain would stop; her gynecologist had also assured her that she had nothing to worry about, because cysts are not cancerous. A week after the abdominal cramps began, the cyst and her completely engulfed right ovary were surgically removed through an incision just below her navel. The cystic ovary was then sent to a scientist who specializes in determining whether tissues are cancerous or not. The scientist, called a *pathologist*, determined that Nicole's doctor had been right—she found no sign of cancer.

After the operation, Nicole's gynecologist assured her that the remaining ovary would compensate for the missing ovary by ovulating (producing an egg cell) every month. Her doctor also informed her that he would have to carefully monitor her remaining ovary to make sure that it did not become cystic, or even worse, cancerous. She could not afford to lose another ovary if she wanted to remain fertile and have children some day.

Monitoring her remaining ovary involved monthly visits to her gynecologist's office where Nicole would have her blood drawn and analyzed. The blood would be tested for the level of a protein called CA125, which is produced by ovarian cells. Higher-than-normal CA125 levels usually indicate that the ovarian cells have increased in size or number, and are thus associated with the presence of an ovarian tumor.

Nicole went to her scheduled check-ups for five months after surgery. The day after her March check-up, Nicole received a message from her doctor asking that she come to see him the next day. Because she needed to study for an upcoming exam, Nicole tried to push aside her concerns about the appointment, and by the time she arrived at her gynecologist's office she had convinced herself that nothing serious could be wrong. She thought a mistake had probably been made and that he wanted to perform another blood test.

When her gynecologist entered the exam room, Nicole could tell by his demeanor that something was wrong. He started speaking to her and she began to feel very anxious—he told her that he thought she had a tumor on her remaining ovary, but she could not believe that she had heard him correctly. When her gynecologist said the word *cancer*, she felt as though she was being pulled under water. She could see that he was still talking, but she could not hear or understand him. She was too nauseous to think, so she excused herself from the exam room, took the bus home, and immediately called her mom.

After speaking with her mom, Nicole realized that there were many questions she needed to ask her doctor. She did not understand how it was possible for such a young woman to have lost one ovary to a cyst, and then have a tumor on the other ovary. She wondered how this would be treated, and what her prognosis would be. Despite her background in biology, she did not even really understand what cancer was. Nicole decided to do some research for answers to her questions.

5.1 What Is Cancer?

Cancer is a disease that begins when a single cell escapes from the regulation of its own division. **Cell division** is the process a cell undergoes in order to make copies of itself. This division is normally regulated so that a cell divides only when more cells are required, and when conditions are favorable for division. A cancerous cell is a rebellious cell that divides without instructions from the body.

Unregulated cell division leads to a pile of cells that form a lump or **tumor**. A tumor is a mass of tissue that has no apparent function in the body. Tumors that stay in one place and do not affect surrounding tissues are said to be **benign**. Some benign tumors remain benign, others become cancerous. Tumors that invade surrounding tissues are no longer benign—they are **malignant**, and these invasive tumors are cancers. Cells from the original tumor can break away and start new cancers at distant locations, a process called **metastasis** (Figure 5.1).

Cancer cells can travel virtually anywhere in the body via the *lymphatic* and *circulatory systems*. The lymphatic system collects fluids lost from microscopic blood vessels called *capillaries*. Some blood is lost from the thin walls of

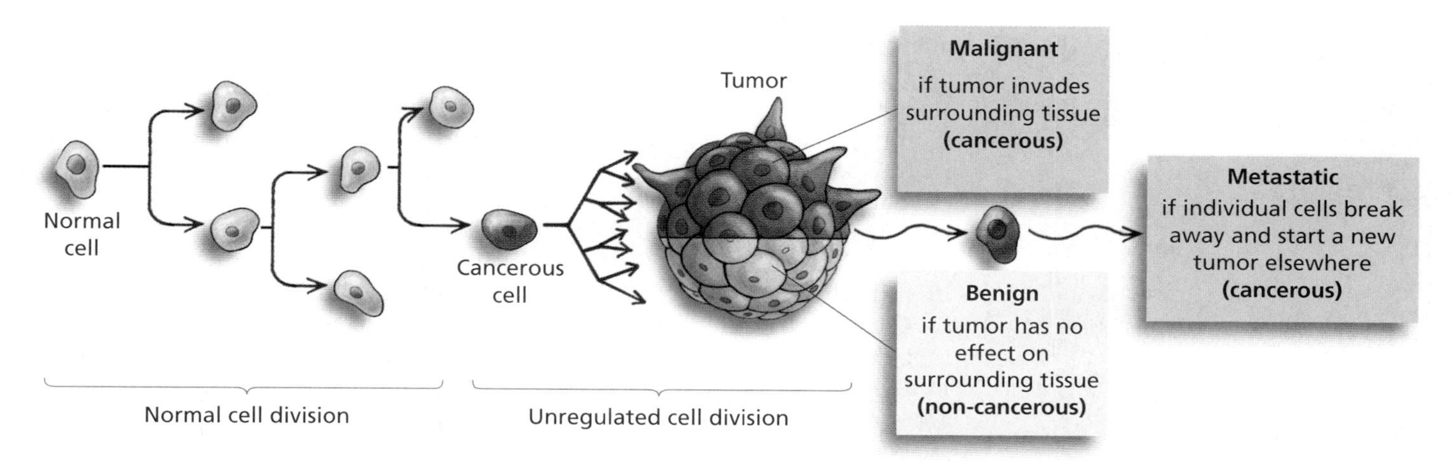

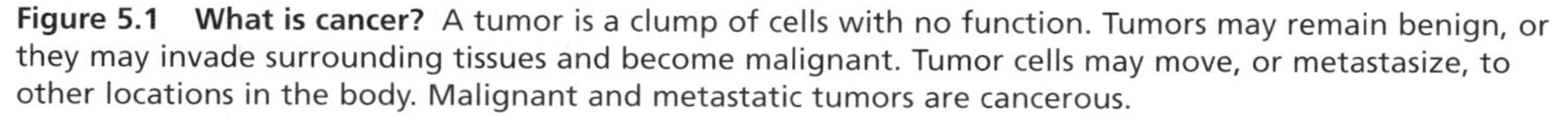

Figure 5.1 What is cancer? A tumor is a clump of cells with no function. Tumors may remain benign, or they may invade surrounding tissues and become malignant. Tumor cells may move, or metastasize, to other locations in the body. Malignant and metastatic tumors are cancerous.

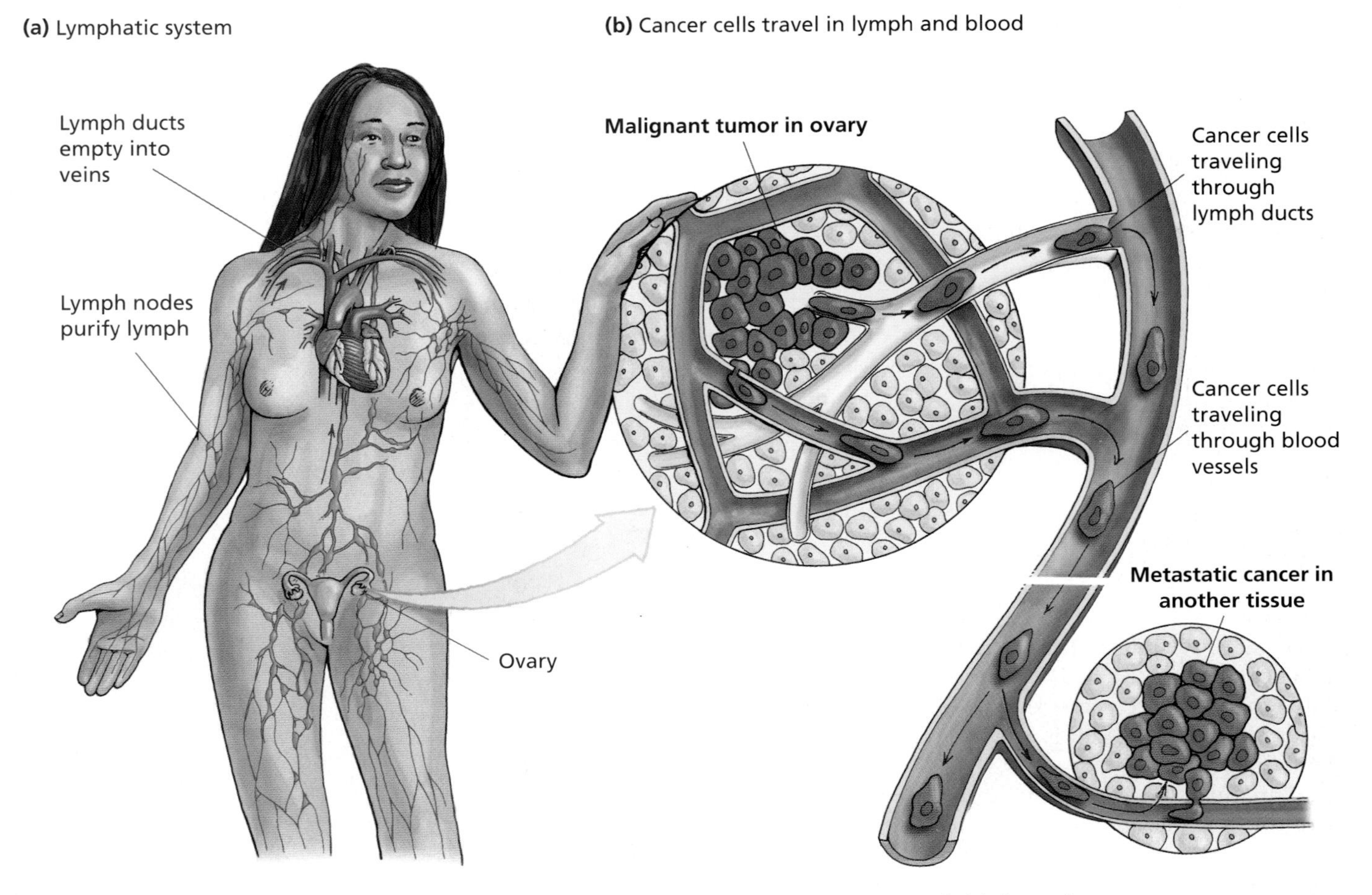

Figure 5.2 Metastasis. (a) The lymphatic system is a series of vessels that remove excess fluids from tissues. Lymph moves throughout the body as a result of pressure applied by muscle contractions near the vessels. (b) The vessels of the circulatory and lymphatic systems provide a pipeline for cancer cells to move to other locations in the body.

the capillaries; the lymphatic system collects the lost fluids (called *lymph*) and returns them to the blood. *Lymph nodes* are small ducts that filter the lymph. When a cancer patient is undergoing surgery, the surgeon will often remove a few lymph nodes to see if any cancer cells are in the lymph. The presence of cancer cells in the lymph nodes indicates that some cells have left the original tumor and are moving through the bloodstream. If this has happened, it is likely that cancerous cells have metastasized to other locations in the body (Figure 5.2).

Cancer cells may also metastasize by gaining access to the circulatory system, which includes blood, vessels to transport the blood, and a heart to pump the blood. Once inside a blood vessel, cancer cells can drift virtually anywhere in the body.

Cancer cells differ from normal cells in three ways: (1) they divide when they should not; (2) they invade surrounding tissues; and (3) they move to other locations in the body. Any tissue that undergoes cell division is susceptible to becoming cancerous.

Nicole's cancer affected ovarian tissue. When an egg cell is released from the ovary during *ovulation*, the tissue of the ovary is perforated (Figure 5.3). Cells near the perforation site undergo cell division to heal the damaged surface of the ovary. For Nicole, these cell divisions may have become uncontrolled, leading to a tumor.

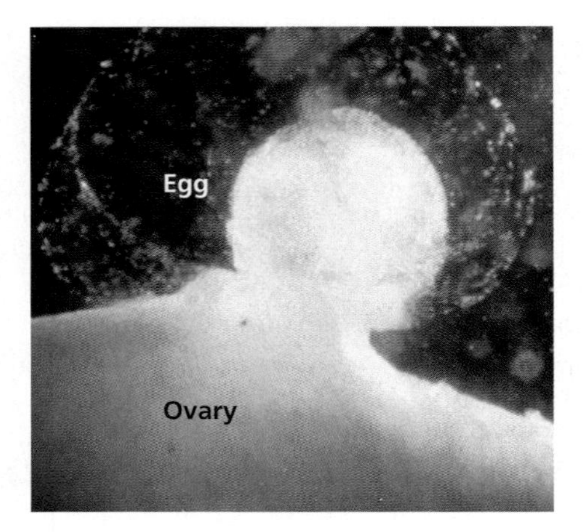

Figure 5.3 Ovulation. When the ovary releases an egg cell, the tissue of the ovary is damaged. Cell division occurs to heal the rupture.

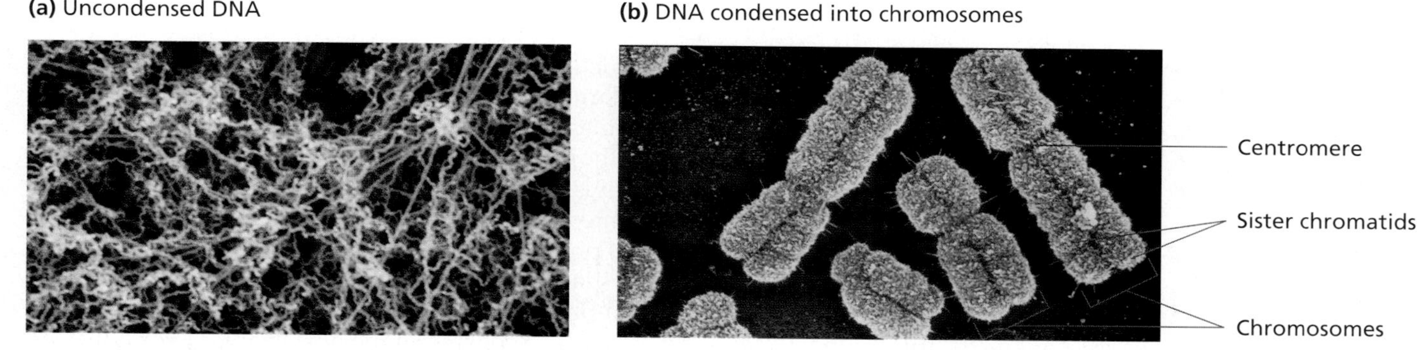

Figure 5.4 DNA condenses at the beginning of mitosis. (a) DNA in its replicated but uncondensed form. (b) During condensation, each copy of DNA wraps itself neatly around many small proteins, forming a condensed structure called a *chromosome*. After DNA synthesis, two identical sister chromatids are produced and joined to each other by a region called the *centromere*.

5.2 Cell Division

The purpose of cell division is to heal wounds, replace damaged cells, and help tissues and organs grow. The **cell cycle** consists of all the events that take place when a cell divides. For one cell to divide itself into two cells, the DNA in the nucleus first must be copied, then each copy must be split into each of two new daughter cells, and finally the original parent cell must be divided in half to form the two separate daughter cells.

DNA is in an uncondensed, stringlike form prior to cell division and must be condensed into more manageable linear structures. These condensed linear DNA structures, called **chromosomes**, are easier to maneuver during cell division and are less likely to become tangled than the uncondensed stringlike chromosomes. A normal human chromosome carries hundreds of genes along its length. When a chromosome is duplicated, a copy is produced that carries the same genes along its length, the copies are called **sister chromatids**. They are attached to each other by a region toward the middle of the chromosome called the **centromere** (Figure 5.4).

The DNA, which makes up the chromosomes, is duplicated during **interphase** of the cell cycle. The DNA is then divided equally between each of two daughter cells during **mitosis**. Separate daughter cells are formed when the cytoplasm splits; this process is called **cytokinesis** (Figure 5.5a).

Media Activity 5.1A Cell Cycle and Division

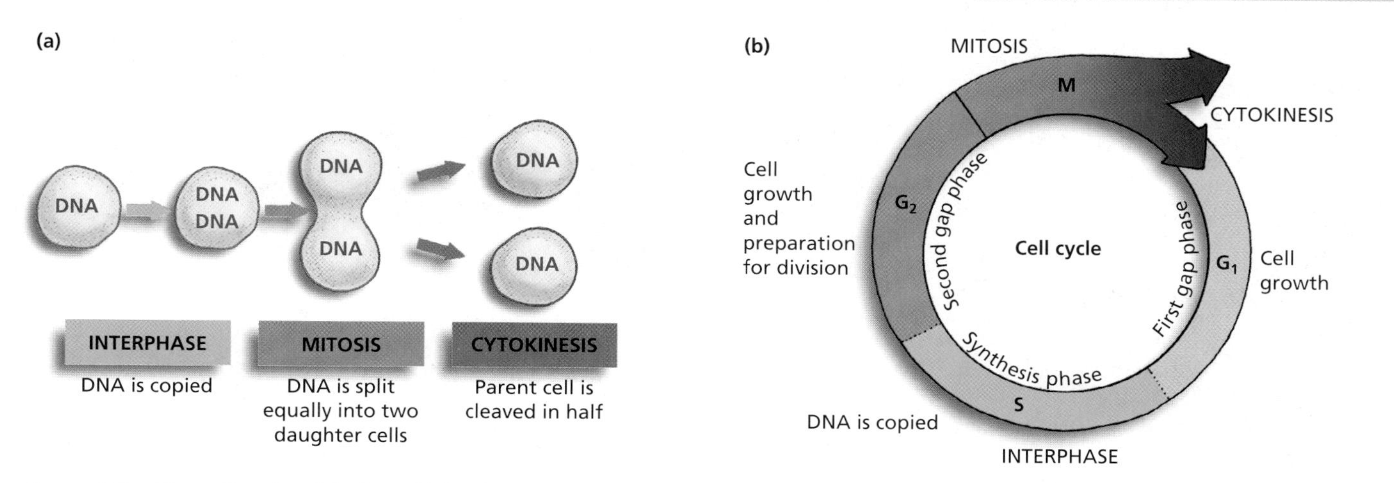

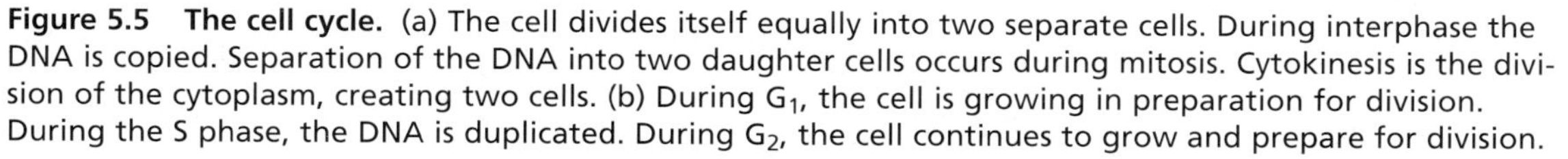

Figure 5.5 The cell cycle. (a) The cell divides itself equally into two separate cells. During interphase the DNA is copied. Separation of the DNA into two daughter cells occurs during mitosis. Cytokinesis is the division of the cytoplasm, creating two cells. (b) During G_1, the cell is growing in preparation for division. During the S phase, the DNA is duplicated. During G_2, the cell continues to grow and prepare for division.

Interphase

A cell spends most of its time in *interphase* (Figure 5.5b). During this phase of the cell cycle, the cell is performing its required functions—different cell types spend varying amounts of time in interphase. Cells that frequently divide spend less time in interphase than those that seldom divide. While in interphase, a cell may also be preparing for division, which includes the replication of DNA. Interphase can be separated into three phases, G_1, S, and G_2.

During the G_1 (first gap) phase, most of the cell's *organelles* are duplicated. Consequently, the cell grows larger during this phase.

During the S (synthesis) phase, the chromosomes are actually copied, or duplicated. (DNA synthesis will be covered in the next chapter.) For now, you can think of DNA synthesis as simply photocopying all of the DNA in the nucleus. Recall that DNA serves as the instructions for building proteins, and that protein-building instructions are called *genes* (Chapter 4).

During the G_2 (second gap) phase of the cell cycle, some of the proteins that will help drive mitosis to completion are produced. The cell continues to grow and prepare for division of the chromosomes that will take place during mitosis.

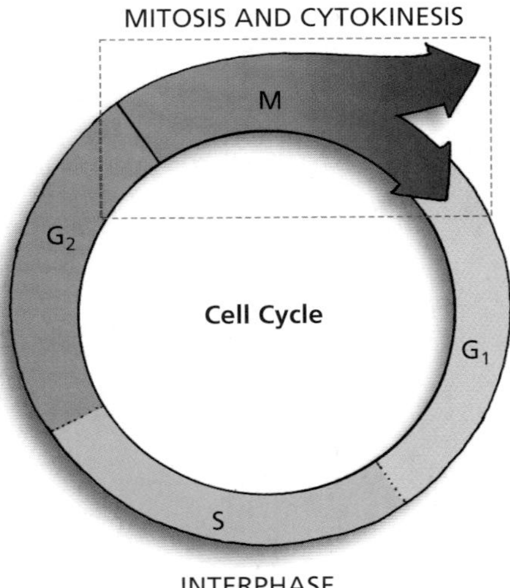
MITOSIS AND CYTOKINESIS
M
G_2
Cell Cycle
G_1
S
INTERPHASE

Mitosis

The movement of the chromosomes into new cells occurs during *mitosis*. The outcome of mitosis and subsequent cytokinesis is the production of genetically identical daughter cells. The daughter cells are exact genetic replicas of the original parent cell. To accomplish this, the sister chromatids of a duplicated chromosome are pulled apart and one copy of each is placed into each newly forming

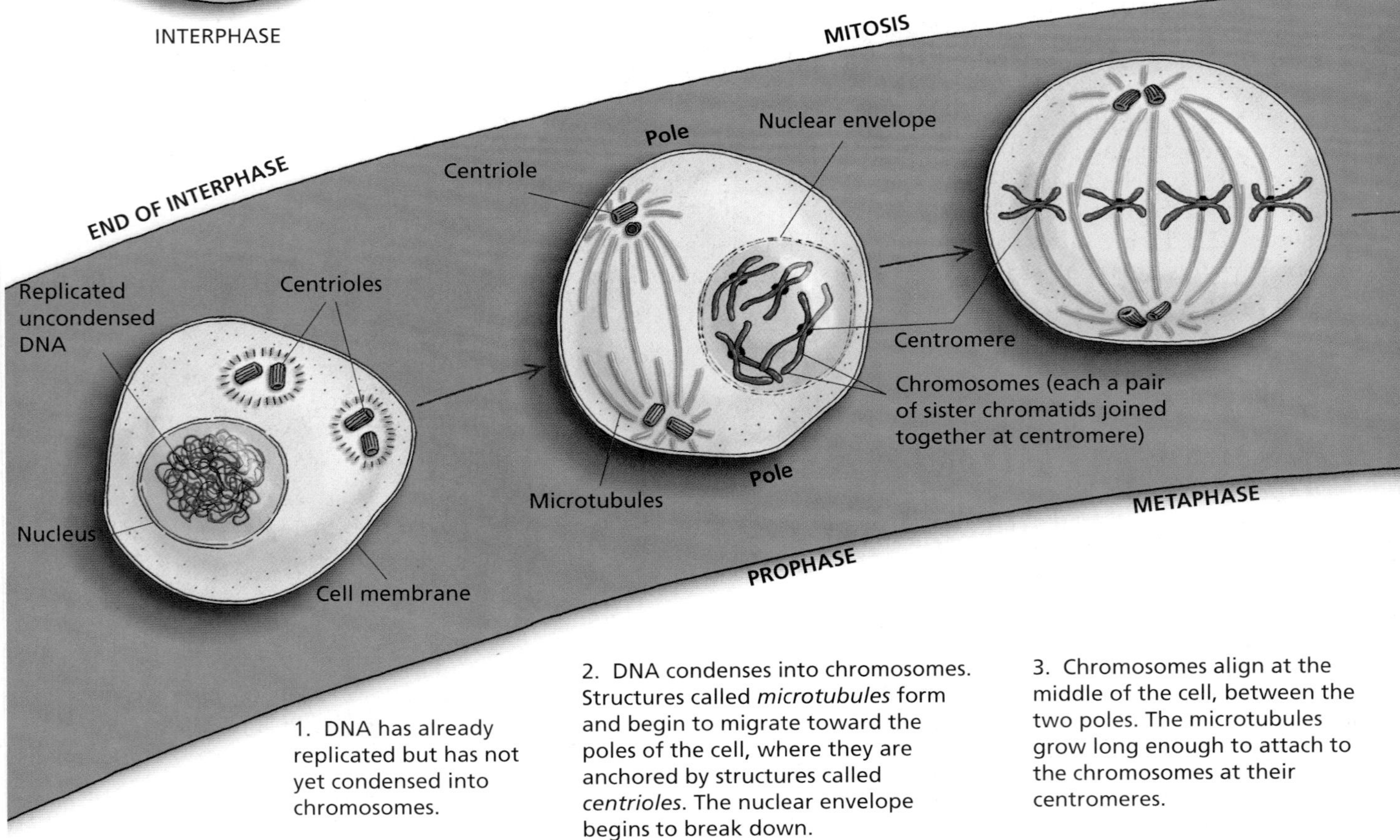
MITOSIS
END OF INTERPHASE
Replicated uncondensed DNA
Centrioles
Nucleus
Cell membrane
Centriole
Pole
Nuclear envelope
Microtubules
Pole
Centromere
Chromosomes (each a pair of sister chromatids joined together at centromere)
PROPHASE
METAPHASE
1. DNA has already replicated but has not yet condensed into chromosomes.
2. DNA condenses into chromosomes. Structures called *microtubules* form and begin to migrate toward the poles of the cell, where they are anchored by structures called *centrioles*. The nuclear envelope begins to break down.
3. Chromosomes align at the middle of the cell, between the two poles. The microtubules grow long enough to attach to the chromosomes at their centromeres.

Figure 5.6 Mitosis and cytokinesis. This diagram illustrates how cell division procedes.

daughter cell. This division of the duplicated chromosomes is accomplished during four stages: *prophase, metaphase, anaphase,* and *telophase* (Figure 5.6).

During **prophase** of mitosis, the duplicated chromosomes condense. This condensing allows them to be moved around in the cell without becoming tangled. Protein structures called **microtubules** also form and begin to grow and radiate out from the poles of the dividing cell. Microtubules are the structures that will move the chromosomes around during cell division by binding to them. The membrane that surrounds the nucleus, the *nuclear envelope*, then begins to break down so that the microtubules can gain access to the duplicated chromosomes. At the top and bottom poles of each dividing cell are structures called **centrioles** that anchor the forming microtubules.

At **metaphase** of mitosis, the duplicated chromosomes are aligned across the middle, or *equator*, of each cell, situated between the two poles. To accomplish this, the microtubules attached to each chromosome at the centromere line the chromosomes up, single file, across the middle of the cell.

During **anaphase** of mitosis, the microtubules shorten and the sister chromatids are separated from each other. The individual chromatids then move to the poles of the dividing cell.

At **telophase**, nuclear envelopes reform around the newly produced daughter nuclei and the chromosomes go back into their uncondensed form. Directly after telophase, cytokinesis occurs.

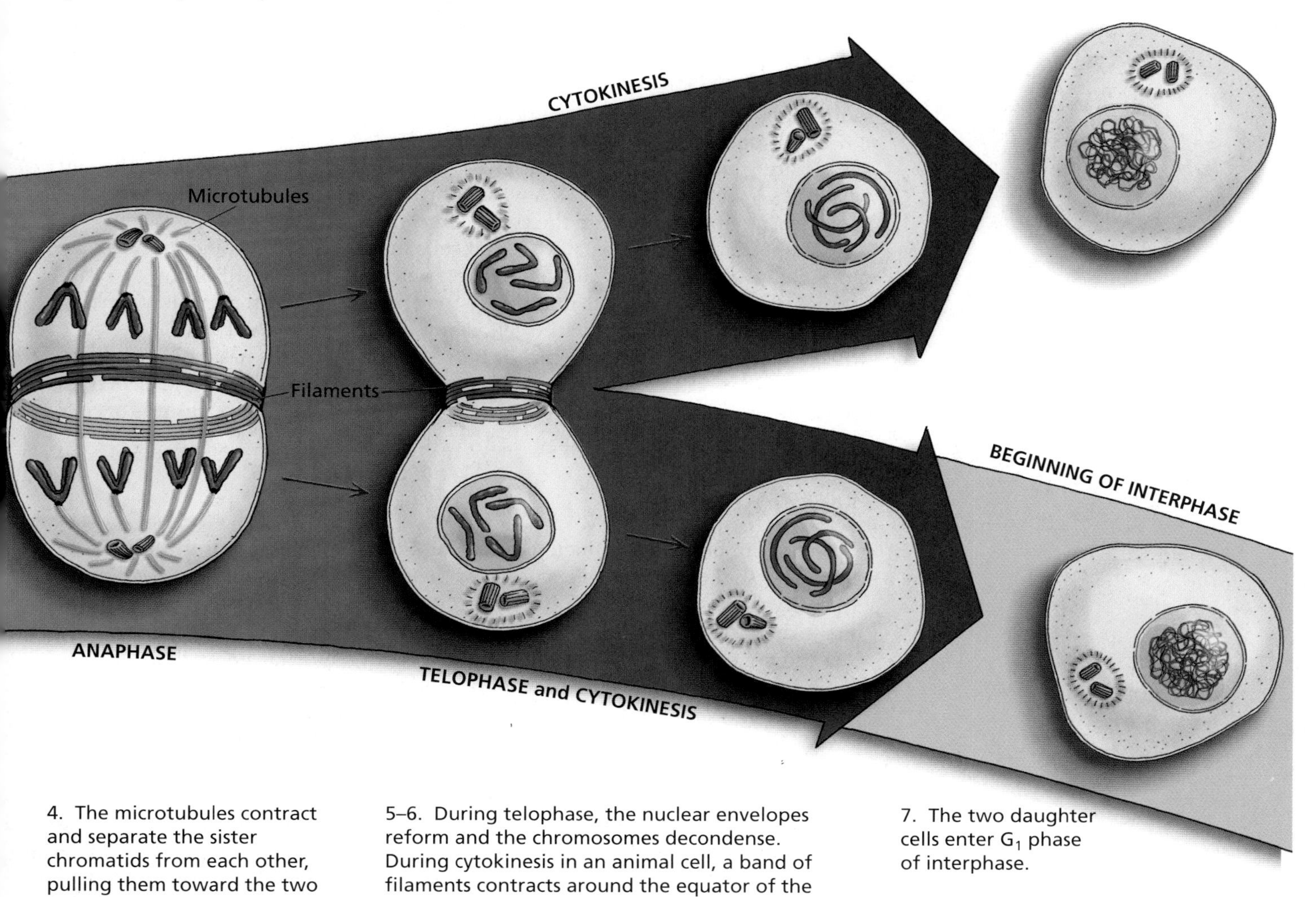

4. The microtubules contract and separate the sister chromatids from each other, pulling them toward the two poles of the cell.

5–6. During telophase, the nuclear envelopes reform and the chromosomes decondense. During cytokinesis in an animal cell, a band of filaments contracts around the equator of the cell, causing two cells to form from the original parent cell.

7. The two daughter cells enter G_1 phase of interphase.

Cytokinesis

Cytokinesis means "cellular movement." During cytokinesis in animal cells, the cytoplasm is divided by a band of proteins that encircle the outside of the cell. This band of proteins contracts to pinch off the two cells that have formed from the original parent cell. Each daughter cell is genetically identical, having its own nucleus with an exact copy of the parent cell's chromosomes, and contains all the necessary organelles and cytoplasm as well. After cytokinesis, the cell reenters interphase and the cell cycle repeats itself.

When cell division is working properly, it is tightly controlled. Cells are given signals for when, and when not, to divide. The normal cells in Nicole's ovary and the rest of her body were responding properly to the signals indicating the rate at which they should divide. However, the cell that started her tumor was not responding properly to these signals.

Control of the Cell Cycle

Instead of proceeding in lockstep fashion through the cell cycle, normal cells halt cell division at what are called **checkpoints**. During this stoppage, proteins survey the cell to ensure that conditions for a favorable cellular division have been met (Figure 5.7a).

Before moving from G_1 of interphase to the S phase, proteins check that the cell is large enough to divide and that all the nutrients the cell will require during division are available. At the G_1 checkpoint, the cell environment is also assessed for the presence of the proteins required to stimulate cells to divide, called **growth factors**—when growth factors are limited, cell division is discontinued.

There is also a checkpoint at G_2. Proteins involved in regulating the cell cycle at G_2 ensure that the DNA was replicated properly and double-check the cell size, again making sure the cell is large enough to divide.

The third and final checkpoint that cells must proceed through occurs during metaphase. Proteins present at metaphase verify that all of the chromosomes are attached to microtubules so that cell division can proceed properly.

If the proteins surveying the cell at any of the three checkpoints determine that conditions are not favorable for division, the process is halted. When these proteins are normal, cell division is properly regulated and all is well. When these cycle-regulating proteins are unable to perform their jobs, clumps of cells can build up and form tumors.

Mutations

Proteins can be rendered ineffective when there are changes, or **mutations**, to the genes that carry the instructions for their synthesis. Recall from Section 4.1 that changes to genes result in changes to protein building instructions, and malformed or nonfunctional proteins can be produced.

The genes that encode, or provide instructions for, proteins that regulate the cell cycle are called **proto-oncogenes** (*proto* means "before," and *onco* means "cancer"). When mutated, these genes are called **oncogenes**. *All* humans carry proto-oncogenes—they are normal genes located on many different human chromosomes. It is only when normal proto-oncogenes undergo mutations and become oncogenes that they become capable of causing cancer.

Many proto-oncogenes encode growth factors, or proteins that respond to the presence of growth factors. A normal growth factor only stimulates cell division when the cellular environment is favorable and all the conditions for division have been met. Mutated proto-oncogenes (oncogenes) can overstimulate cell division and override the G_1 checkpoint (Figure 5.7b).

One of the genes involved in many cases of ovarian cancer is a gene called *Her2*. The *Her2* gene carries the instructions for building a *receptor* for a growth

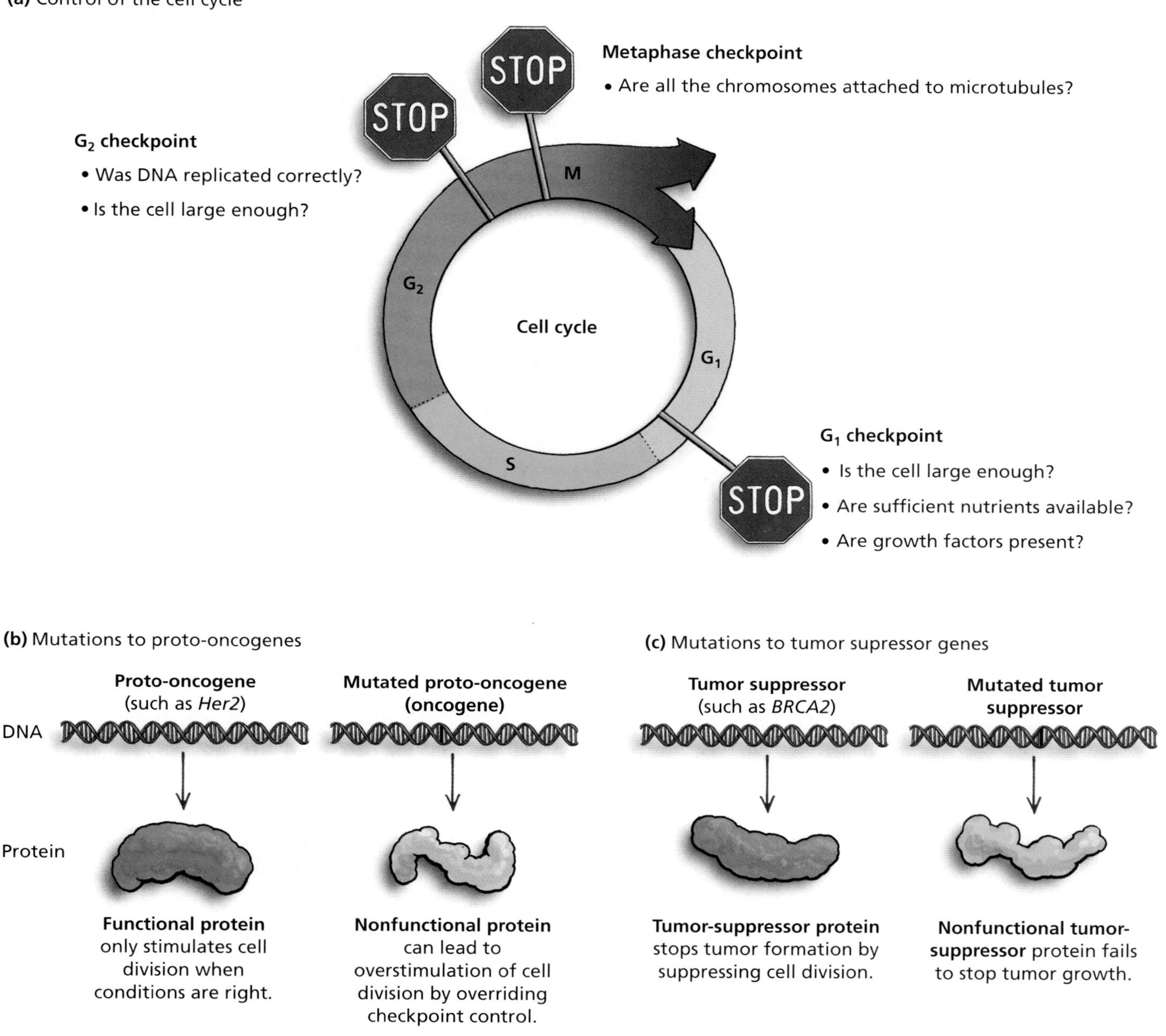

(a) Control of the cell cycle
Metaphase checkpoint
• Are all the chromosomes attached to microtubules?
G_2 checkpoint
• Was DNA replicated correctly?
• Is the cell large enough?
STOP
STOP
M
G_2
Cell cycle
G_1
S
STOP
G_1 checkpoint
• Is the cell large enough?
• Are sufficient nutrients available?
• Are growth factors present?
(b) Mutations to proto-oncogenes
Proto-oncogene (such as *Her2*)
Mutated proto-oncogene (oncogene)
DNA
Protein
Functional protein only stimulates cell division when conditions are right.
Nonfunctional protein can lead to overstimulation of cell division by overriding checkpoint control.
(c) Mutations to tumor supressor genes
Tumor suppressor (such as *BRCA2*)
Mutated tumor suppressor
Tumor-suppressor protein stops tumor formation by suppressing cell division.
Nonfunctional tumor-suppressor protein fails to stop tumor growth.

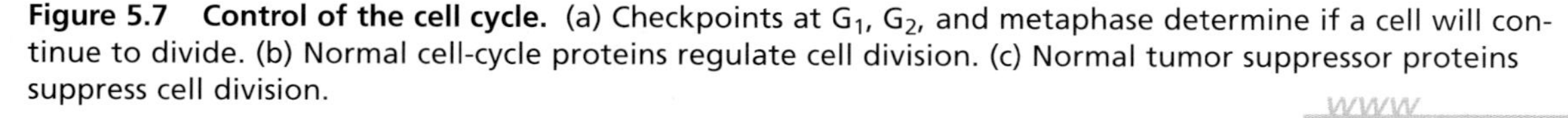

Figure 5.7 Control of the cell cycle. (a) Checkpoints at G_1, G_2, and metaphase determine if a cell will continue to divide. (b) Normal cell-cycle proteins regulate cell division. (c) Normal tumor suppressor proteins suppress cell division.

WWW **Media Activity 5.1B** Mutations and Cancer

factor. When it is normal, the receptor on the surface of the cell signals the inside of the cell to allow division to occur. When mutated, the receptor cell behaves as though there are many growth factors present, even when there are not.

Another class of genes involved in cancer are **tumor suppressors**. These genes, present in all of us, carry the instructions for producing proteins that suppress cell division if conditions are not favorable. These proteins can detect and repair DNA damage. For this reason, normal tumor suppressors serve as backups in case proto-oncogenes undergo mutation. If an oncogene overstimulates cell division, the normal tumor suppressor impedes tumor formation by preventing the mutant cell from moving through a checkpoint (Figure 5.7c).

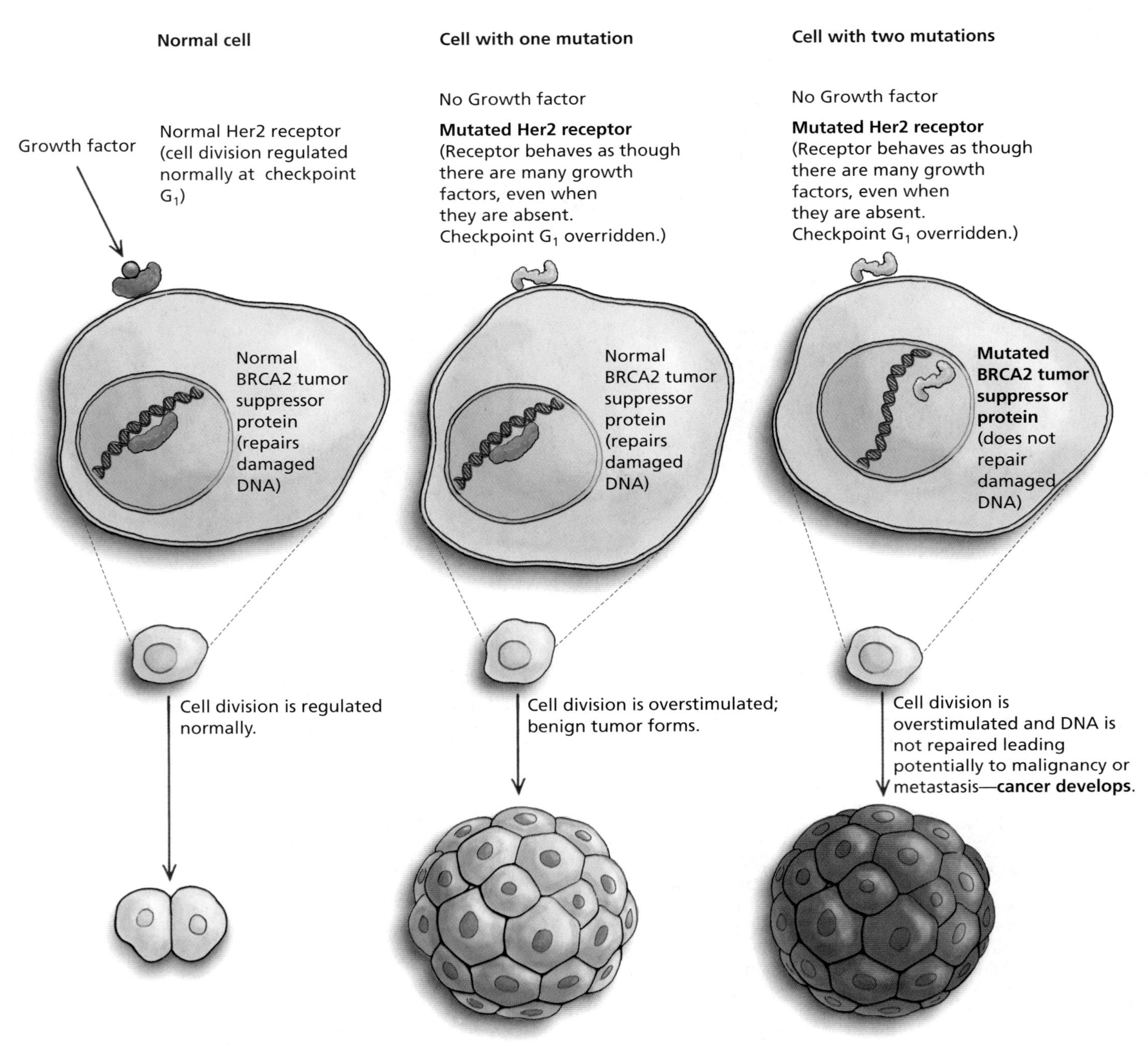
Normal cell
Cell with one mutation
Cell with two mutations
Growth factor
Normal Her2 receptor (cell division regulated normally at checkpoint G_1)
No Growth factor
Mutated Her2 receptor
(Receptor behaves as though there are many growth factors, even when they are absent. Checkpoint G_1 overridden.)
No Growth factor
Mutated Her2 receptor
(Receptor behaves as though there are many growth factors, even when they are absent. Checkpoint G_1 overridden.)
Normal BRCA2 tumor suppressor protein (repairs damaged DNA)
Normal BRCA2 tumor suppressor protein (repairs damaged DNA)
Mutated BRCA2 tumor suppressor protein (does not repair damaged DNA)
Cell division is regulated normally.
Cell division is overstimulated; benign tumor forms.
Cell division is overstimulated and DNA is not repaired leading potentially to malignancy or metastasis—cancer develops.

Figure 5.8 The roles of *Her2* and *BRCA2* in ovarian cancer. When a woman's ovarian cells have normal versions of these genes, cell division is properly regulated. If a cell in the woman's ovary has a *Her2* mutation, cell division is overstimulated, and a benign tumor can form. If a cell in the woman's ovary has both the *Her2* and the *BRCA2* mutations, a cancerous tumor and metastasis is likely.

When a tumor-suppressor protein is not functioning properly, it does not force the cell to stop dividing at the right time. Mutated tumor suppressors also allow cells to override cell-cycle checkpoints. One well-studied tumor suppressor, named *p53*, helps cells to decide whether they should repair damaged DNA at G_2, or throw in the towel and commit cellular suicide if the damage is too severe. Mutations to the gene that encodes *p53* result in damaged DNA being allowed to proceed through mitosis, thereby passing on even more mutations. Over half of all human cancers involve mutations to the gene that encodes *p53*.

A similar phenomenon is often at work in ovarian cancer. Mutations to a tumor suppressor gene are often found in ovarian cells that have become cancerous. Researchers believe that a normal *BRCA2* gene encodes a protein that is involved in helping to repair damaged DNA at G_2. The malformed, mutant version of the protein cannot help to repair damaged DNA, which means that damaged DNA (mutated genes) will be allowed to undergo mitosis, thus passing new mutations to their daughter cells. As more and more mutations occur, the probability that a cell will become cancerous increases. Figure 5.8 summarizes the roles of *Her2* and *BRCA2* in ovarian cancer.

Additional mutations are responsible for the progression of a tumor from a benign state, to a malignant state, to metastasis. For example, some cancer cells are able to stimulate growth of surrounding blood vessels. They do this by secreting a substance that recruits and reroutes blood vessels so that they supply a developing tumor with oxygen (required for cellular respiration) and other nutrients. The formation of new blood vessels is called **angiogenesis**. When a tumor has its own blood supply, it can grow at the expense of other, noncancerous cells. Since the growth of rapidly dividing cancer cells is favored over the growth of normal cells in this process, entire organs can eventually be composed of cancerous cells. When this occurs, an organ can no longer work properly, leading to compromised function or organ failure.

Normal cells also display a property called **contact inhibition,** which prevents them from dividing when doing so would require that they pile up on each other. Conversely, cancer cells continue to divide, forming a tumor (Figure 5.9). Normal cells do need some contact with an underlayer of cells in order to stay in place and are typically held in place by a phenomenon called **anchorage dependence** (Figure 5.10). This requirement for some contact with other cells is overridden in cancer cells, which may leave the original tumor and move to the blood, lymph, or surrounding tissues.

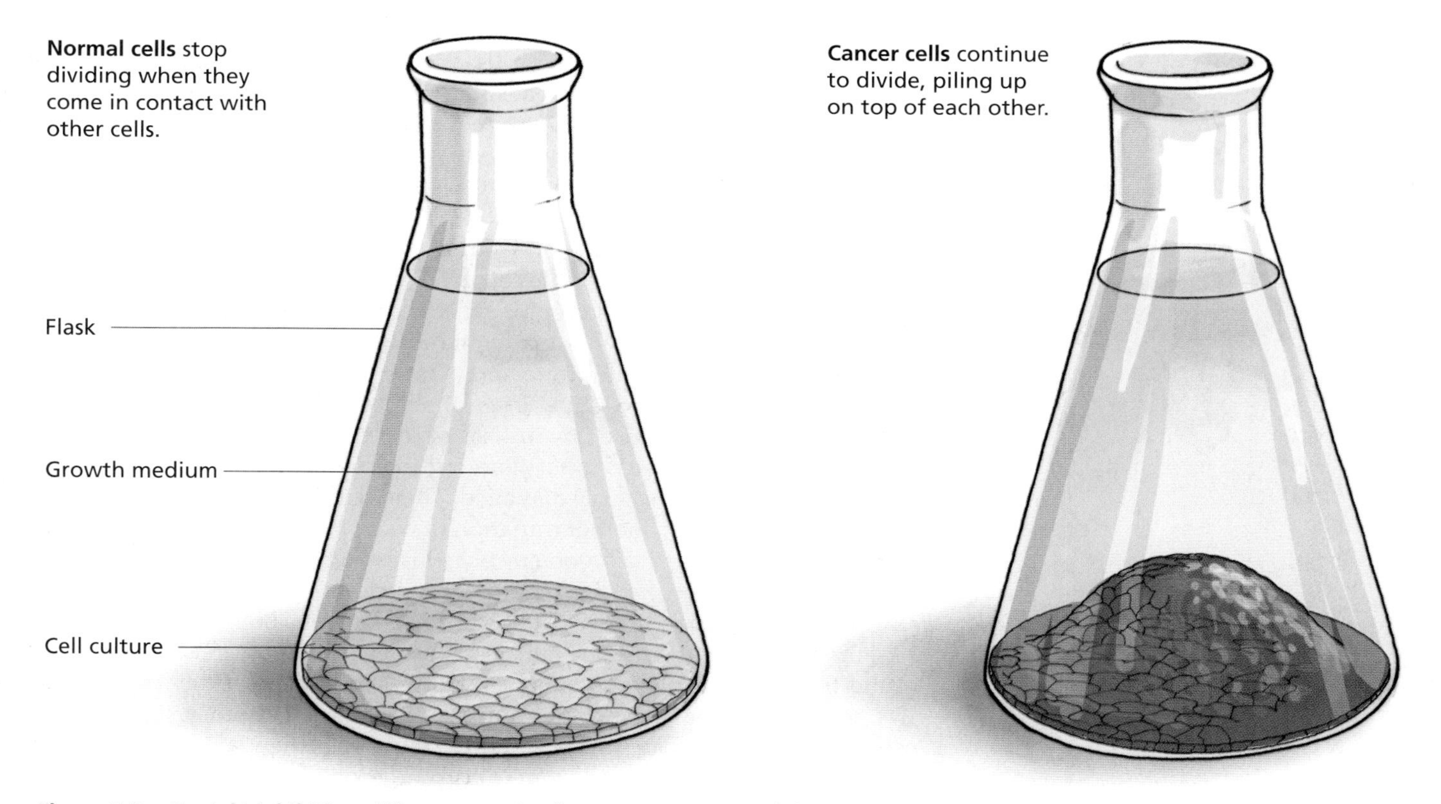
Normal cells stop dividing when they come in contact with other cells.
Flask
Growth medium
Cell culture
Cancer cells continue to divide, piling up on top of each other.

Figure 5.9 Contact inhibition. When normal cells are grown on a solid support such as the bottom of a flask, they grow and divide until they cover the bottom of the dish, and when they come in contact with other cells, they stop dividing. Cancer cells pile up on top of each other.

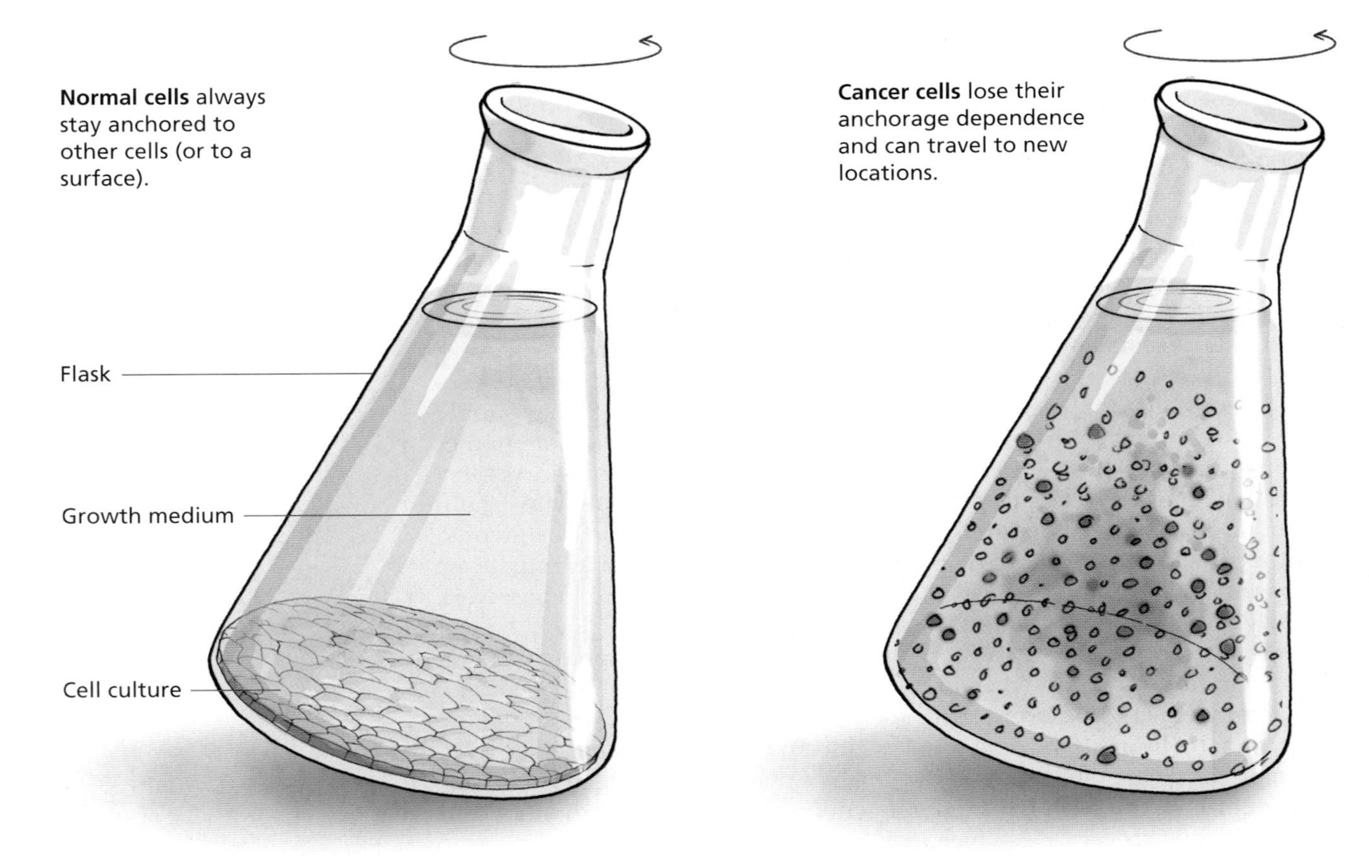
Normal cells always stay anchored to other cells (or to a surface).
Cancer cells lose their anchorage dependence and can travel to new locations.
Flask
Growth medium
Cell culture

Figure 5.10 Anchorage dependence. Cancer cells break away from the site of the original tumor and travel to new locations, where they set up secondary tumors.

Normal cells are programmed to divide a certain number of times—usually 60–70—and then they die. Cancer cells do not obey these life-span limits, instead they are **immortal**. This is because cancer cells can activate a gene that is usually turned off after early development. This gene produces an enzyme called *telomerase*. This enzyme, only active early in development and in cancer cells, allows cells to divide without limit. Cells with active telomerase enzyme are immortal.

In Nicole's case the progression from normal ovarian cell to cancer may have occurred as follows: One single cell in her ovary may have acquired a mutation to its *Her2* growth factor receptor gene. The descendants of this cell would have been able to divide faster than neighboring cells, forming a small benign tumor. Next, a cell within the tumor may have undergone a mutation to its *BRCA2* tumor suppressor, resulting in the inability to fix damaged DNA in the cancerous cells. Cells produced by mitosis of these doubly mutant cells would not only have divided faster, thus enlarging the tumor; they would not have been destroyed if their DNA was damaged. Then, within the population of cells carrying both mutations, another mutation may have occurred, resulting in the tumor's ability to procure a blood supply. Once the tumor had its own blood supply, it had all the oxygen and nutrients it needed to grow at the expense of her surrounding ovarian tissues. At this point, it became likely that even more mutations would occur since the affected cells of her ovary were dividing rapidly, and their DNA-repair proteins were not functioning properly.

If one cell with the former mutations underwent yet another mutation to a gene controlling contact inhibition, that cell would then be able to invade surrounding tissues. If one cell from this population also underwent a mutation to the gene controlling anchorage dependence, it would gain the ability to move away from the original tumor—carrying with it all the mutations the cell has acquired and starting a cancer at the new location. Finally, if any of these multiply mutant cells underwent a mutation that reactivated the telomerase gene, these mutant cells would become immortal.

Some or all of these mutations may have occurred in Nicole's ovary after she was born. It may also be the case that she inherited one or more of these mutations from her parents. Mutations may be induced by exposure to substances that damaged the DNA or chromosomes, called **carcinogens**. For a substance to be considered carcinogenic, exposure to the substance must be correlated with increased risk of cancer. Examples of carcinogens include cigarette smoke, radiation, ultraviolet light, asbestos, and some viruses. If Nicole was not born with a preexisting mutation, she may have been exposed to enough carcinogens in her lifetime to allow enough mutations to accumulate that she got cancer.

Risk Factors

Risk factors are behaviors, exposures, or predispositions that increase the probability of obtaining a given disease. Although she experimented briefly with smoking cigarettes as a freshman, Nicole quit and has not smoked since. She feels lucky that she did not become addicted during that time, but she also realizes that smoking even a very limited number of cigarettes is a risk factor for cancer. People who smoke are at increased risk of developing cancer, since tobacco smoke contains over 20 known carcinogens. Chemicals present in cigarettes have been shown to increase cell division, inhibit DNA repair, and prevent cells from dying when they should. Smoking is such a dangerous habit in part because cigarettes provide so many different opportunities for DNA damage and cell damage that tumor formation and metastasis are quite likely for smokers. In fact, the odds of almost every human cancer are increased when a person smokes cigarettes.

Cancer risk may also be influenced by your diet. Fruits, vegetables, and grains are rich in vitamins, minerals, and fibers. In Chapter 2 you learned that these whole foods are associated with many health benefits. One of these benefits is decreased cancer risk.

The American Cancer Society recommends eating at least five servings of fruits and vegetables every day, and six servings of food from other plant sources such as breads, cereals, grains, rice, pasta, or beans.

Buying fresh fruits and vegetables is expensive on Nicole's limited budget. She also finds it difficult to shop frequently, to have fresh fruits and vegetables available. Thus, Nicole sometimes snacks on highly processed foods that can be purchased in advance and left on the shelf for weeks or months. However, she does try to bring a piece of fruit with her to campus every day, and often stops at the corner store on her way home from classes to buy fresh vegetables for dinner. When she is preparing meals, she washes all fruits and vegetables to remove leftover pesticide residues, because she has heard that pesticides have been shown to be carcinogenic.

Regular exercise is already part of Nicole's life. She skis during the winter and lifts weights year round. Exercise decreases the risk of most cancers, partly because exercise keeps the immune system functioning well. The immune system helps destroy cancer cells when it can recognize them as foreign to the host body. Unfortunately, since cancer cells are actually your own body's cells run amok, the immune system cannot always differentiate between normal cells and cancer cells.

Increasing age is also a risk factor for cancer. As you age, your immune system weakens, and its ability to distinguish between cancer and noncancer cells decreases. This is why many cancers, including ovarian cancer, are far more likely in elderly people. Additional factors that help explain the increasing cancer risk with increasing age include simple probability: If we are all exposed to carcinogens, the longer we are alive, the greater the probability that some of those carcinogens will mutate genes involved in regulating the cell cycle. Also, since multiple mutations are necessary for a cancer to develop, it often takes many, many years to progress from the initial mutation to a tumor, and then to malignant cancer.

Essay 5.1 Cancer Risk and Detection

Over one million people in the United States will be diagnosed with cancer this year. There are some steps you can take to decrease the odds that one of those diagnosed will be you.

Don't use tobacco. The use of tobacco of any type, either cigarettes, cigars, or chewing tobacco, increases your risk of many cancers. While smoking is the cause of 90% of all lung cancers, it is also the cause of about one third of all cancer deaths. Cigar smokers have increased rates of lung, larynx, esophagus, and mouth cancers. Chewing tobacco increases the risk of cancers of the mouth, gum, and cheeks. Also, reduce your exposure to second-hand smoke. People who don't smoke but are exposed to secondhand smoke have increased lung cancer rates. Not using tobacco, or quitting now, and avoiding second-hand smoke, is one of the most important health decisions you can make.

Eat a variety of healthy foods. Your diet should be rich in fruits and vegetables. Try to eat five or more servings of fruits and vegetables each day. In addition, your diet should be rich in other foods from plant sources, such as grains and beans. Plant foods are low in fat and contain lots of minerals and fiber. Therefore, a plant-based diet tends to be low in fat and high in fiber. A high fat (greater than 15% of calories obtained from fat) low fiber (less than 30 grams per day) diet is associated with increased cancer risk.

If you drink alcohol, drink in moderation. Drinking alcohol is associated with increased cancer risk. Men should have no more than two alcohol drinks per day, women one or none. People who both drink and smoke are greatly increasing their risk of cancers of the mouth, esophagus, and larynx.

Exercise and maintain a healthful weight. Obesity is associated with increased risk for many cancers, including breast, uterine, colon, gallbladder, and prostate cancers. Physical activity is an important part of controlling your weight. Try to be physically active for 30 minutes or more on most days of the week.

Don't get sunburned. Skin cancer is one of the most common and preventable kinds of cancer. Most skin cancer occurs on parts of the body that usually aren't covered with clothing when you are outside, so cover exposed areas, use sunscreen, and don't use tanning beds.

Undergo regular screening and self-examination. Cancer treatment is far more effective when the cancer is discovered early. Be aware of changes in your body. The following mnemonic device may help you recall the general warning signs of cancer:

Change in bowel or bladder habit.
A sore that does not heal.
Unusual bleeding or discharge.
Thickening or lump.
Indigestion or difficulty swallowing.
Obvious change in wart or mole.
Nagging cough or hoarseness.

Table E 5.1 on pages 123–124 summarizes risk factors and detection methods for specific cancers.

Each type of cancer has its own particular risk factors and methods of detection (see Essay 5.1). The risk of ovarian cancer is thought to decrease if ovulation is interrupted; the birth-control pill prevents pregnancy by preventing ovulation. Nicole has been on the birth-control pill since she was a teenager, because she had fairly severe acne. Acne occurs when an oily substance called sebum builds up in the sebaceous glands (pores) of the skin. Androgens, produced in larger quantities in males than females, act on sebaceous glands to increase the rate of sebum production. The severity of acne increases at puberty because androgen levels increase in both boys and girls at this time. The estrogen present in the birth control pill counteracts the effects of androgens, and thereby decreases the severity of acne.

When ovulation is interrupted, as it is when a woman is on the birth-control pill or is pregnant, ovarian cancer risk decreases for many years thereafter. This may be because, in the absence of ovulation, there is no damaged ovarian tissue, hence no need for extra cell divisions.

Because a given type of cancer is caused by multiple mutations, there are many combinations of mutations that can cause one type of cancer. Therefore, treatment options are different for each individual. A treatment that works for one woman with ovarian cancer might not work for another woman with ovarian cancer, because a different suite of mutations has lead to the development of the same cancer in each woman.

Cancer Location	Risk Factors	Detection	Comments
Ovary 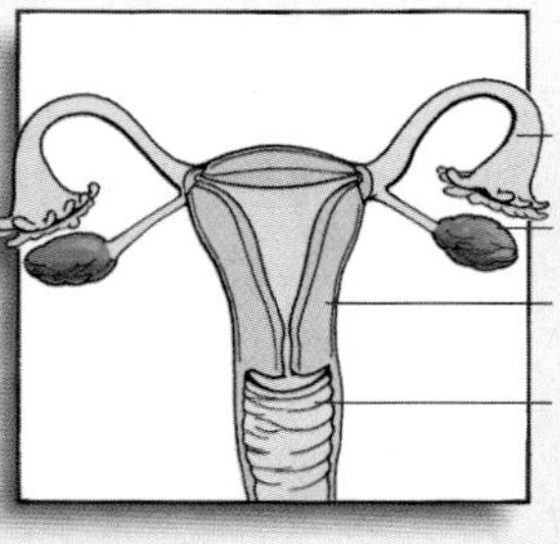 Oviduct Ovary Uterus Vagina	• Smoking • Mutation to *BRCA2* gene • Advanced age • Oral contraceptive use and pregnancy decrease risk	• Blood test for elevated CA125 level • Gynecological exam	• Fifth leading cause of death among women in the United States
Breast 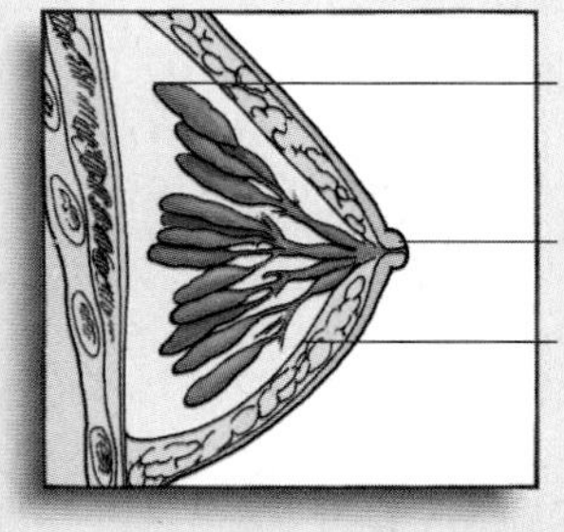 Milk-producing glands Nipple Fatty tissue	• Smoking • Mutation to *BRCA1* gene • High-fat, low-fiber diet • Use of oral contraceptives may slightly increase risk.	• Monthly self exams, look and feel for lumps or changes in contour • Mammogram	• Only 5% of breast cancers are due to *BRCA1* mutations • Second-highest cause of cancer-related deaths • 1% of breast cancer occurs in males
Cervix 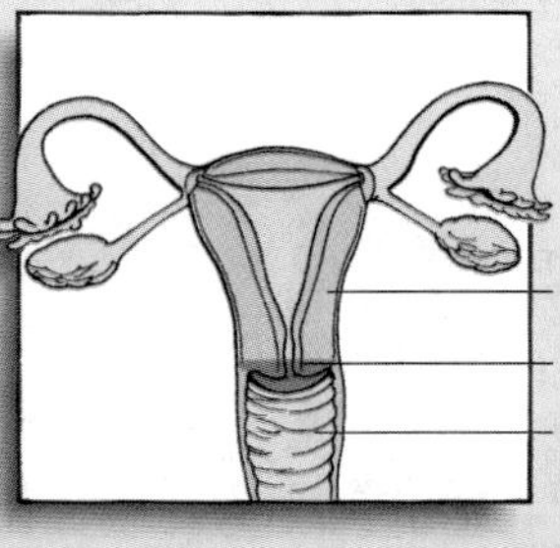 Uterus Cervix Vagina	• Smoking • Exposure to sexually transmistted Human Papilloma Virus (HPV) • Micronutrients such as vitamins C and E may decrease risk.	• Annual Pap-smear tests for the presence of pre-cancerous cells	• Precancerous cells can be removed by laser surgery or cryotherapy (freezing) before they become cancerous.
Skin 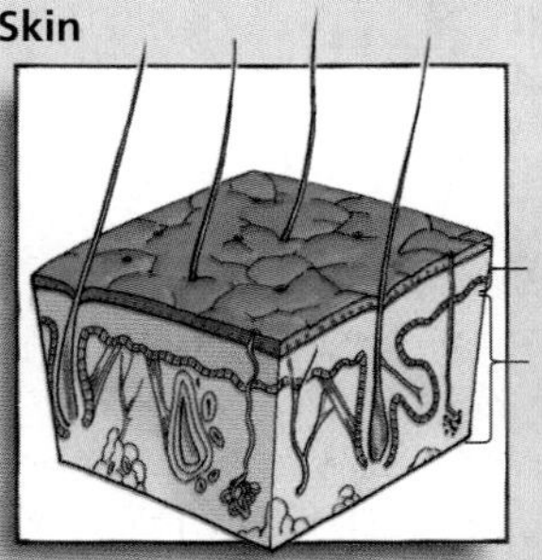 Epidermis Dermis	• Smoking • Fair skin • Exposure to ultraviolet light from the sun or tanning beds	• Monthly self-exams, look for growths that change in size or shape	• Skin cancer is the most common of all cancers and is usually curable if caught early.
Lung 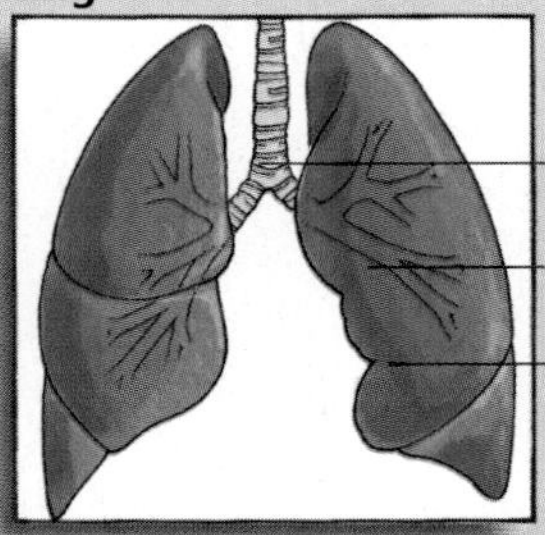 Trachea Bronchi Lungs	• Smoking • Exposure to second-hand smoke • Asbestos inhalation	• X-ray	• Lung cancer is the most common cause of death from cancer, and the best prevention is to quit, or never start, smoking.

Table E5.1 Cancer risk and detection ***(continued on next page).***

Cancer Location	Risk Factors	Detection	Comments
Colon and rectum Small intestine Colon Rectum 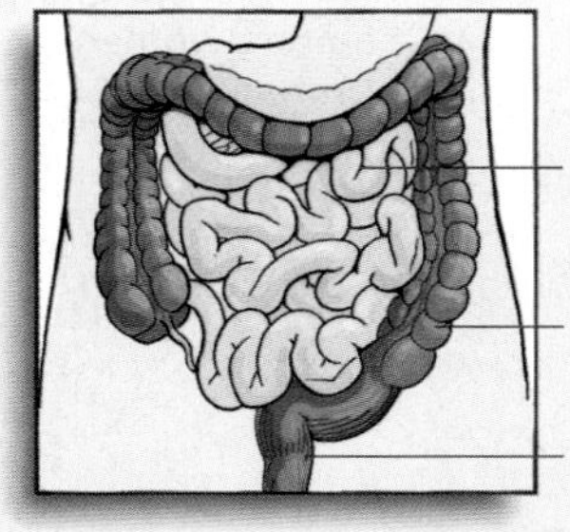	• Smoking • *Polyps* in the colon • Advanced age • High-fat, low-fiber diet	• Change in bowel habit • Colonoscopy is an examination of the rectum and colon using a lighted instrument.	• Benign buds called polyps can grow in the colon; removal prevents them from mutating and becoming cancerous.
Prostate 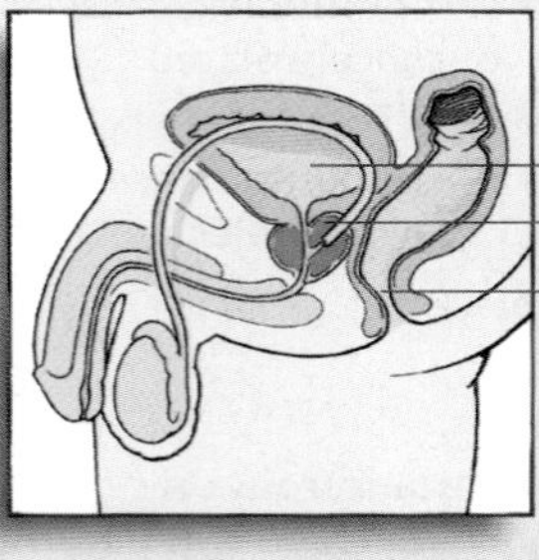 Bladder Prostate Rectum	• Smoking • Advanced age • High-fat, low-fiber diet	• Blood test for elevated level of prostate specific antigen (PSA) • Physical exam by physician, via rectum	• More common in African-American men than Asian, white, or Native American men.
Testicle 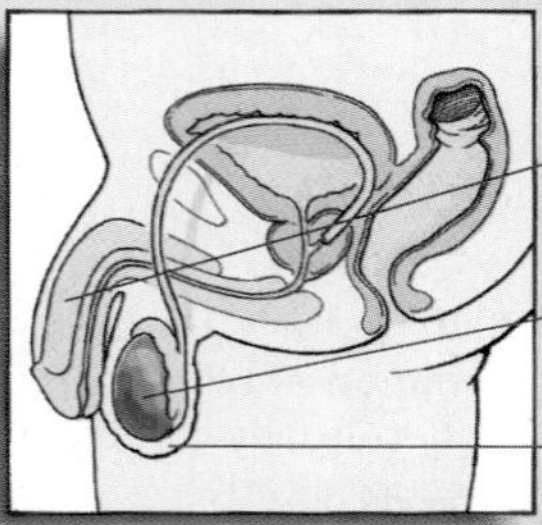 Penis Testicle Scrotum	• Abnormal testicular development	• Monthly self exam, inspect for lumps and changes in contour	• Testicular cancer accounts for only 1% of all cancers in men, but is the most common form of cancer found in males between the ages of 15 and 35.
Blood (Leukemia) 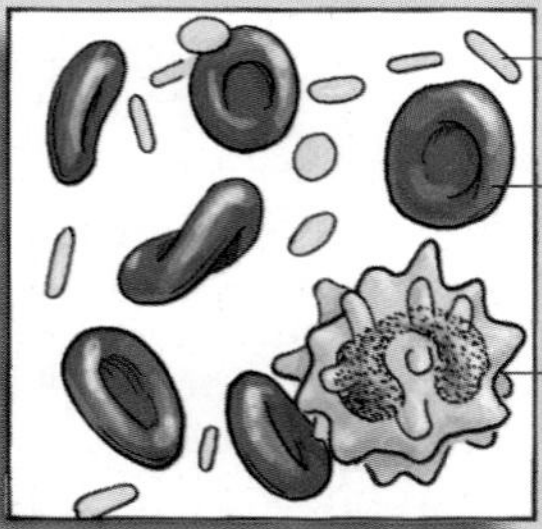 Platelet Red blood cell White blood cell	• Exposure to high-energy radiation such as that produced by atomic bomb explosions in Japan during World War II	• A sample of blood is examined under a microscope.	• Cancerous white blood cells cannot fight infection efficiently: people with leukemia often succumb to infections.

Table E5.1 Cancer risk and detection *(continued).*

5.3 Diagnosis and Treatment

WWW **Media Activity 5.2** Earlier Detection for Ovarian Cancer?

Many cancers can be detected by the excess production of proteins that are normally produced by a particular cell type. For example, prostate cancer can be diagnosed by the presence of high levels of a prostate specific protein in the blood. The high CA125 protein level measured in Nicole's blood led her physician to suspect she had a tumor forming on her remaining ovary. Once a tumor was suspected, Nicole's physician wanted to perform a *biopsy*.

Biopsy and Surgery

A **biopsy** is the surgical removal of some cells, tissue, or fluid used to determine whether they are cancerous or not. When viewed under a microscope, benign tumors consist of orderly growths of cells that resemble the cells of the tissue from which they were taken. Malignant cells do not resemble other cells found in the same tissue—they are dividing so rapidly that they do not have time to produce all the proteins necessary to build normal cells. This leads to the disorderly appearance seen under a microscope (Figure 5.11).

A needle biopsy is usually performed if the cancer is located on or close to the surface of the body. For example, breast lumps are often biopsied with a needle to determine whether the lump contains fluid and is a noncancerous cyst, or whether it contains abnormal cells and is a tumor. When a cancer is diagnosed, surgery is often performed to remove as much as possible without damaging neighboring organs and tissues.

In Nicole's case, getting at the ovary to find tissue for a biopsy required the use of a surgical instrument called a *laparoscope*. For this operation, the surgeon inserted a small light and a scalpel-like instrument through a tiny incision above her navel.

Nicole's surgeon preferred to use the laparoscope since he knew Nicole would find it much easier to recover from laparoscopic surgery than she had from the surgery to remove her other, cystic ovary. Laparoscopy was not possible when Nicole had her other ovary removed—the cystic ovary had grown so large that her surgeon had to make an abdominal incision to remove it.

A laparoscope has a small camera attached that projects images from the ovary onto a monitor the surgeon views during surgery. These images showed that Nicole's tumor was a different shape, color, and texture than the rest of her ovary. It also appeared that the tumor was not confined to the surface of the ovary; in fact, it appeared to have spread deeply into her ovary. Nicole's surgeon decided to shave off only the affected portion of the ovary and leave as much intact as possible, with the hope that the remaining ovarian tissue might

(a) Normal ovarian tissue

(b) Benign ovarian tumor

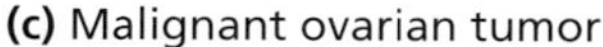

(c) Malignant ovarian tumor

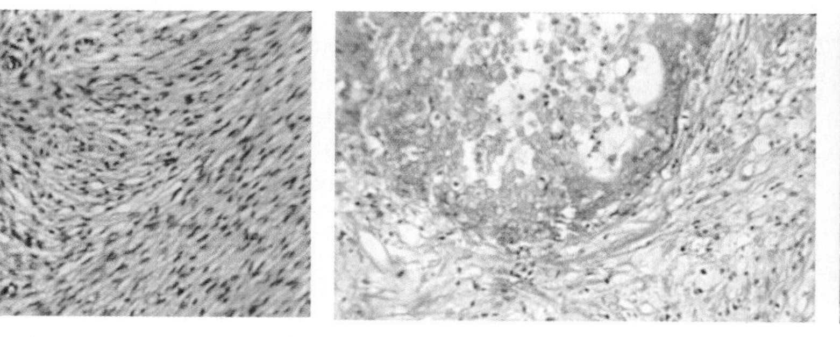

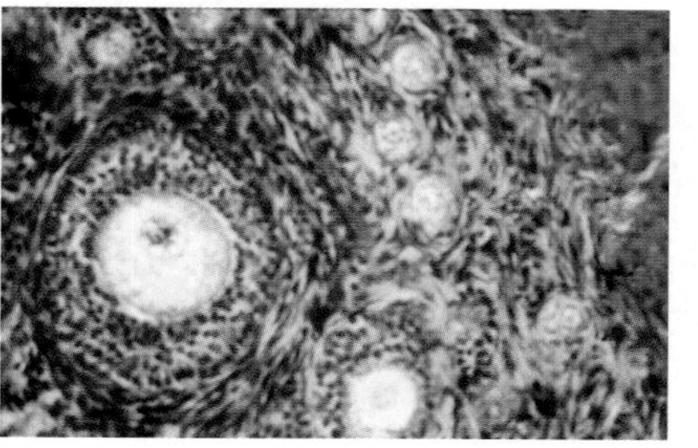

Figure 5.11 Biopsy. When stained and viewed under a microscope, cancer cells have a disorderly appearance. At higher magnifications, it is possible to see tumor cells that were in the process of mitosis when the slide was prepared.

still be able to produce egg cells. He then sent the tissue to a laboratory so that the pathologist could examine it. Unfortunately, when the pathologist looked through the microscope this time, she saw the disordered appearance characteristic of cancer cells. She found cells at every stage of mitosis—Nicole's ovary was cancerous.

Luckily for Nicole, her ovarian cancer was diagnosed very early. Regrettably, this is not the case for most women with ovarian cancer, many of whom are diagnosed when symptoms become severe, leading them to see their physician. The symptoms of ovarian cancer tend to be vague and slow to develop and include abdominal swelling, pain, bloating, gas, constipation, indigestion, menstrual disorders, and fatigue—many women simply overlook this type of discomfort. The difficulty of diagnosis is compounded by the fact that there are no routine screening tests available. For instance, CA125 levels are only checked when ovarian cancer is suspected because these levels vary from individual to individual and are dependent upon the phase of the woman's menstrual cycle. By the time symptoms become severe enough for women to see their doctors, a cancer may have grown quite large and metastasized, making it much more difficult to treat.

Even though her cancer was caught early, Nicole's physician was concerned that some of her cancerous ovarian cells may have spread through blood vessels or lymph ducts on or near the ovaries, or into her abdominal cavity, so he started Nicole on chemotherapy after her surgery.

WWW
Media Activity 5.3 **Alternative Cancer Treatments: What Really Works?**

Chemotherapy and Radiation

Chemotherapy is the injection of chemicals into the bloodstream, selectively killing dividing cells. A variety of chemotherapeutic agents act in different ways to interrupt cell division. For example, some block DNA replication, others prevent mitosis from occurring. One of the drugs Nicole was given was *Cisplatin*. This chemotherapeutic agent disrupts DNA synthesis, and therefore prevents cell division. Another drug given to cancer patients is Taxol. Taxol is a chemical isolated from the bark of Pacific Yew trees. This chemotherapeutic agent prevents microtubules from shortening during cell division, thereby halting mitosis (Figure 5.12). Unfortunately, normal cells that are also dividing are affected by these treatments as well.

Several hours after each chemotherapy treatment Nicole became nauseous; she often had diarrhea and vomited for a day or so after her treatments. This happened because the cells that line the intestines undergo cell division often, and are also affected by chemotherapy.

Midway through her chemotherapy treatments Nicole lost most of her hair because the cells that produce the proteins found in hair divide rapidly, so they were also killed by the chemotherapy chemicals.

Nicole's treatments consisted of many different chemotherapeutic agents, spread over many months because cancer cells can become resistant to the drugs being used against them. Cells that are resistant to one drug proliferate when the chemotherapeutic agent clears away the other cells that compete for space and nutrients. Cells with a preexisting resistance to the drugs are "selected for" and produce more daughter cells with the same resistant characteristics, requiring the use of more than one chemotherapeutic agent. This practice increases the likelihood that cells resistant to one drug will be killed by another drug (Chapter 9).

Cancer patients often undergo radiation treatments as well as chemotherapy. **Radiation** therapy uses high-energy particles to injure or destroys cells by damaging their DNA, making it impossible for these cells to continue to grow and divide. A typical course of radiation involves a series of 10 to 20 treatments performed after surgical removal of the tumor. Radiation therapy is usually only used when cancers are located close to the surface of the body, because it

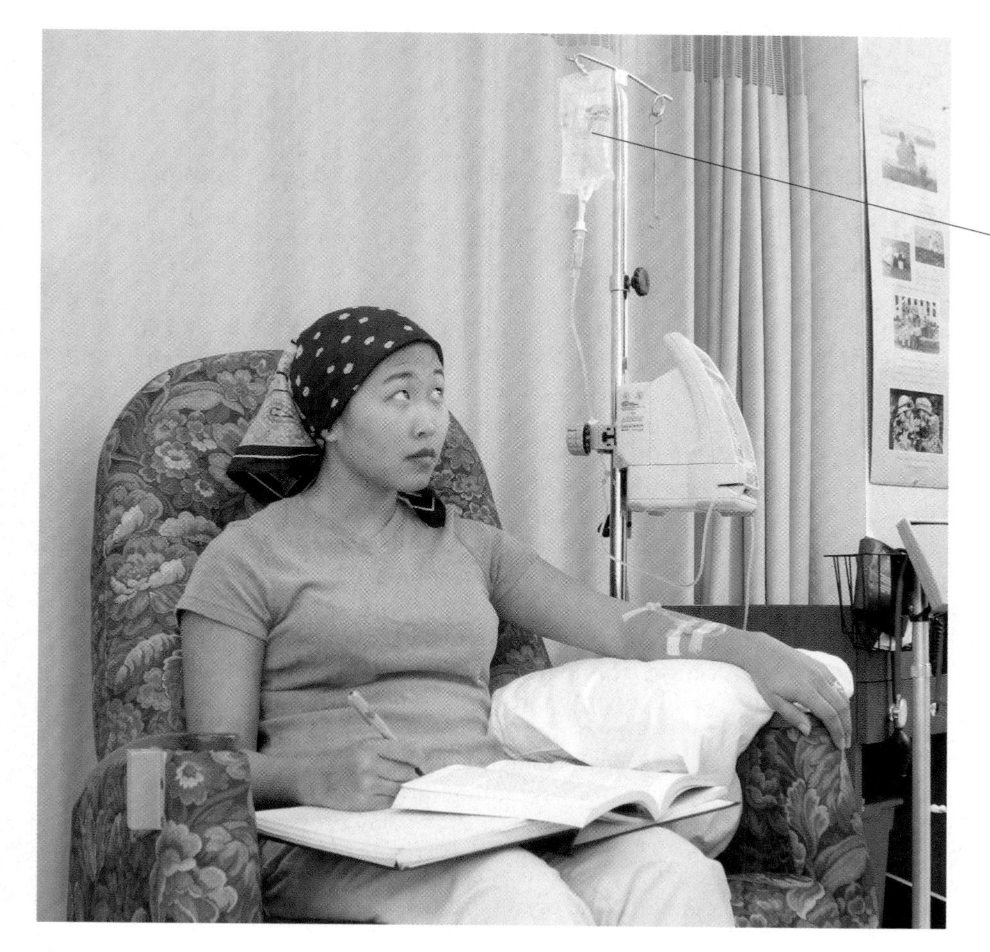

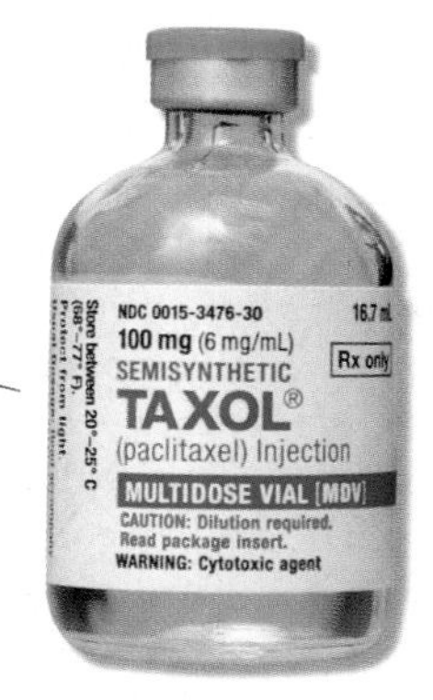

Taxol contains a substance isolated from Yew trees that prevents microtubules from shortening during mitosis, thereby preventing cell division.

Figure 5.12 Taxol. The chemotherapeutic agent Taxol disrupts microtubules, thereby preventing cell division. It is administered through an IV needle.

is difficult to focus a beam of radiation on internal organs such as an ovary. Therefore, Nicole's physician recommended chemotherapy only.

Nicole's chemotherapy treatments took place at the local hospital on Wednesdays and Fridays. She usually had a friend drive her to the hospital very early in the morning, returning later in the day to pick her up. The drugs were administered through an intravenous needle (IV) into a vein in her arm. During the hour or so while she was undergoing chemotherapy, Nicole usually studied for her classes. She remained a full-time student during her treatments and was taking mostly advanced biology and psychology courses. She had considered taking a semester off, but decided she did not want to delay her graduation or allow the cancer to take anything more away from her.

Nicole did not mind the actual chemotherapy treatments that much. The hospital personnel were kind to her, and she got some studying done. It was the aftermath of these treatments that she hated. Most days during her chemotherapy regimen, she was so exhausted that she did not get out of bed until after her morning classes, and the day after her treatments she often slept until late afternoon. Then she would get up and try to get some work done or make some phone calls and go back to bed early in the evening.

Nicole tried to remain upbeat and worked hard to believe that she would get better. It helped her endure the chemotherapy when she could convince herself that it would end soon, and she could have her old life back. However,

Essay 5.2 Experimental Cancer Therapies

Cancers that are not responding to conventional treatments may require experimental therapies. Experimental cancer therapies may prove beneficial, but are considered to be long shots because they are still in the development stages. Typically, people undergoing these therapies participate in controlled experiments that serve to test the effectiveness of various treatments. These controlled experiments are called *clinical trials*.

An experimental therapy called *radio-immunotherapy* is scheduled to undergo clinical trial soon. The goal of this therapy is to deliver radioactive substances directly to tumors without affecting other tissues, and involves attaching radioactive substances to proteins that target cancer cells. These proteins, called *antibodies,* are made by the immune system in response to the presence of foreign substances. Scientists have been able to produce antibodies that bind to substances found on tumor cells. When the radioactive antibodies find the cancer cells, they adhere to them and deliver a lethal dose of radiation.

Other scientists are working on preparing treatments called *cancer vaccines.* While we traditionally think of vaccines as preventing disease, cancer vaccines help stimulate the immune system to more quickly recognize and remove existing cancer cells.

The vaccine is made of substances contained by the tumor. Tumor cells are removed from the patient and grown in the laboratory. They are then treated to make sure that they can no longer divide. The treated cells are injected into the patient along with other substances that are known to stimulate the immune system. The presence of so many "killed" cancer cells in the bloodstream, along with the chemicals that stimulate the immune system, may help the immune system more quickly determine which cells are cancerous and which are not.

Most other cancer-research strategies focus on understanding the genes that control the cell cycle. Many cancer therapies are likely to come from research targeting the specific mutations that cause uncontrollable cell growth. Thus, cancer treatment will change from killing all dividing cells, to killing only those cells with specific characteristics.

Another area for cancer research, one that has shown some promise in mice, is looking at methods of preventing a tumor from obtaining its own blood supply. Chemicals called *angiogenesis inhibitors* may prove useful in depriving human cancers of the nutrients required for growth.

there were days during her chemotherapy that she became very depressed and was upset that all her friends got to be normal college students, but she had to have cancer. She also worried that her ovary would not recover from the surgery and chemotherapy, which meant that she would never be able to have children. Nicole had always assumed that she would have children some day, although she did not currently have a huge desire to experience pregnancy, but she wondered if her feelings would change. Even though she was not planning to marry any time soon, she also wondered how her future husband would feel if she were not able to become pregnant.

After six weeks of chemotherapy, Nicole's CA125 levels started to drop. After another two months of chemotherapy, the levels were down to their normal, precancer level. If Nicole has normal CA125 levels for five years, she will be considered to be in remission. After 10 years of normal CA125 levels, she will be considered cured of her cancer. Because Nicole's cancer responded to chemotherapy, she was spared from having to undergo any other, more experimental treatments (see Essay 5.2).

Nicole recently graduated from college with her degrees in biology and psychology. She plans to attend medical school after taking a year off to travel. Her experience with cancer has convinced her that she wants to be a physician. Nicole hopes that her illness will help her understand her patient's ordeal, something she might not have been capable of had she not had cancer.

CHAPTER REVIEW

Summary

- Unregulated cell division can lead to the formation of a tumor. Benign tumors stay in one place and do not prevent surrounding organs from functioning. Malignant tumors are those that are invasive or those that metastasize to surrounding tissues, starting new cancers.
- The cell cycle includes all the events that occur as one cell gives rise to daughter cells.
- Interphase consists of two gap phases of the cell cycle (G_1 and G_2) during which the cell prepares to enter mitosis, and the S (synthesis) phase. The S phase of interphase occurs between G_1 and G_2. It is during the S phase that the DNA is duplicated. The two duplicated copies, sister chromatids, are attached to each other at a region called the centromere.
- During mitosis the sister chromatids are separated from each other into daughter cells. During prophase, the duplicated DNA condenses into chromosomes. At metaphase these duplicated chromosomes align across the middle of the cell. At anaphase the sister chromatids are separated from each other and placed at opposite poles of the cells. At telophase a nuclear envelope reforms around the chromosomes lying at each pole.
- Cytokinesis is the last phase of the cell cycle. During cytokinesis, the cytoplasm is divided into two portions, one for each daughter cell.
- Unregulated cell division occurs when there are mutations to the genes encoding proteins that regulate the cell cycle. Proto-oncogenes are normal versions of cell-cycle control genes. Mutant cell-cycle control genes are called oncogenes. Oncogenes override cell cycle control checkpoints, designed to regulate cell division. Mutant tumor suppressor genes allow cells with damaged DNA to keep dividing. Changes to genes involved in angiogenesis, contact inhibition, anchorage dependence, and cell mortality are required for a tumor to progress toward malignancy. Mutations can be inherited or induced by exposure to carcinogens.
- Cancer treatment usually begins with a biopsy to determine whether a growth is cancerous, followed by removal of the growth by surgery. Chemotherapy and radiation therapy are attempts to kill any remaining cancer cells. The therapies used depend on the type and extent of the cancer under treatment, and often cause unpleasant side effects.

Key Terms

anaphase p. 115
anchorage dependence p. 119
angiogenesis p. 119
benign p. 111
biopsy p. 125
cancer p. 111
carcinogen p. 121
cell cycle p. 113
cell division p. 111
centrioles p. 115
checkpoints p. 116
chemotherapy p. 126
chromosomes p. 113
contact inhibition p. 119
cytokinesis p. 113
growth factors p. 116
immortal p. 120
interphase p. 113
malignant p. 111
metaphase p. 115
metastasis p. 111
microtubules p. 115
mitosis p. 113
mutations p. 116
oncogenes p. 116
prophase p. 115
proto-oncogenes p. 116
radiation p. 126
risk factors p. 121
sister chromatids p. 113
telophase p. 115
tumor p. 111
tumor suppressors p. 117

Learning the Basics

1. How many chromatids would be present at metaphase in a cell with four chromosomes?
2. List two types of genes that, when mutated, can increase the likelihood of cancer occurring.
3. What properties of cancer cells make them more likely to metastasize?
4. What property of cancer cells do chemotherapeutic agents attempt to exploit?

5. What structure helps move chromosomes during cell division?

6. A cell that begins mitosis with 46 chromosomes produces daughter cells with _____.
   a. 13 chromosomes
   b. 23 chromosomes
   c. 26 chromosomes
   d. 46 chromosomes

7. All of the following are characteristics of telophase during mitosis except _____.
   a. the start of cytokinesis
   b. the nuclear envelope reforming
   c. each chromosome being made of two chromatids
   d. the chromosomes decondensing
   e. microtubules disappearing

8. DNA replication occurs _____.
   a. between G_1 and G_2 of interphase
   b. during G_2
   c. during prophase of mitosis
   d. between metaphase and anaphase

9. The centromere is a region in which _____.
   a. sister chromatids are attached to each other
   b. metaphase chromosomes align
   c. the tips of chromosomes are found
   d. the nucleus is located

10. Proto-oncogenes _____.
    a. are mutant genes some people inherit
    b. are normal genes that encode cell-cycle control proteins
    c. can become oncogenes if a person smokes cigarettes
    d. are proteins that fail to suppress tumor formation
    e. b and c are true

11. At metaphase of mitosis _____.
    a. the chromosomes are condensed and found in the poles
    b. the chromosomes are composed of one sister chromatid
    c. cytokinesis begins
    d. the chromosomes are composed of two sister chromatids, and are lined up along the equator of the cell

12. Sister chromatids _____.
    a. are two different chromosomes attached to each other
    b. are exact copies of one chromosome that are attached to each other
    c. arise from the centrioles
    d. are broken down by Taxol
    e. are chromosomes that carry different genes

Analyzing and Applying the Basics

1. Would a skin-cell mutation your father obtained from using tanning beds make you more likely to get cancer? Why or why not?

2. Assume that you have graduated and become a pharmaceutical researcher attempting to cure cancer. If you could design a drug to disrupt any biological process you chose, what would you try to disrupt and why?

3. Why are some cancers treated with radiation therapy, and others with chemotherapy?

4. In mice, substances that inhibit the formation of new blood vessels, called angiogenesis inhibitors, have been found to decrease mortality from cancer. Why might angiogenesis inhibitors lead to decreased cancer mortality?

5. What steps are required for a benign tumor to become malignant?

Connecting the Science

1. If your friend tells you that he is going to try to prevent cancer in himself by taking nutritional supplements, what advice might you give him?

2. Why might it be good to know if your family has a history of certain cancers?

3. Should society be forced to pay the medical bills of smokers when the cancer risk from smoking is so evident? Explain your reasoning.

4. If you could be tested today to find out whether or not you might develop cancer later in life, would you want to have this information?

5. Are there changes you could make in your own life to decrease your odds of getting cancer?

Media Activities

Media Activity 5.1 Cell Division and Cancer
Estimated Time: 10 minutes
Review the cell cycle and the process of cell division, then review the effects of mutations on control of the cell cycle and the development of cancer.

Media Activity 5.2 Earlier Detection for Ovarian Cancer?
Estimated Time: 15 minutes
Investigate an alternative or experimental cancer treatment. Learn about a particular treatment or prevention method, and then use your knowledge of "good science" to assess the validity of the claims made.

Media Activity 5.3 Alternative Cancer Treatments: What Really Works?
Estimated Time: 15 minutes
Exciting recent research indicates that a potential new test has been developed for the routine and early detection of ovarian cancer. Explore the significance of the test, how it works, the disadvantages, and possible alternatives.

CHAPTER 6

DNA Detective

DNA Structure and Replication, Meiosis

The Romanov family ruled Russia until their overthrow, exile, and 1918 execution.

The fall of communist Soviet Union (shown here by the topple of the statue of Lenin) prompted the desire for a proper burial of the Romanov family.

People believed that bones found in a grave in Ekaterinburg were those of the slain Romanovs.

On the night of July 16, 1918, the Czar of Russia, Nicholas Romanov II, his wife Alexandra, their five children, and four family servants were executed in a small room in the basement of the home in which they were exiled. These murders ended three centuries of Romanov family rule over Imperial Russia.

Prior to his execution, Czar Nicholas II had relinquished his power over the monarchy by abdicating for both himself and on behalf of his only son Alexis, then 13 years old. The Czar had hoped that these abdications would protect his son, the heir to the throne, as well as his whole family from harm.

The political climate in Russia at this time was explosive. During the summer of 1914, Russia and other European countries became involved in World War I. This war proved to be a disaster for the Imperial government. Russia faced severe food shortages, and the poverty of most people in Russia provided a stark contrast to the luxurious lives being lived by their leaders—the Russian people felt a great deal of resentment toward their Czar. This sentiment helped spark the Russian Revolution in 1917. Troops attacked the Romanov family palace and kept the Romanovs under guard until they were exiled to Ekaterinburg, Siberia.

As a result of the Russian Revolution, the Bolshevik army under the leadership of Vladimir Lenin took control of the government. Ridding the country

Scientists made use of DNA evidence to confirm that the bones buried in this shallow grave belonged to the Romanovs.

of any last vestige of Romanov rule became a priority of Lenin and his army. Not only would this make way for a Communist regime, it would serve to garner support among the peasantry, many of whom felt that the exiled Romanovs and their opulent lifestyle had come to represent all that was wrong with Imperialist Russia. Accordingly, Lenin ordered the family's execution.

Shortly after midnight on the night of their execution, the family was awakened and asked to dress for a photograph. The Romanovs were told that there was a rumor that they had escaped to England, and a photograph was required to prove that they were still in Russia. Nicholas, Alexandra, and their children Olga, Tatiana, Maria, Anastasia, and Alexis, along with the family physician, cook, maid, and valet, were escorted to a dark room in the basement. Once the family had arrived and been seated for what they thought would be a photograph, a soldier read a short statement indicating that they were to be killed. Eleven armed men stormed into the basement, and each shot his assigned victim.

After the murder, the bodies of the Romanovs and their servants were loaded into a truck and driven to a remote, wooded area in Ekaterinburg. Historical accounts differ about whether the bodies were dumped down a mine shaft, later to be removed, or were immediately buried. There is also some disagreement regarding the burial of two of the people who were executed. Some reports indicate that all 11 people were buried together, and two of them were badly decomposed by acid placed on the ground of the burial site, or that two of the bodies were burned to ash. Other reports indicate that two members of the family were buried separately. Some people even believe that two victims actually escaped the execution. In any case, the bodies of at least nine people were buried in a shallow grave, where they lay undisturbed until 1991.

Not only the bodies remained buried—for decades, details of the family's murder were hidden in the Communist party archives in Moscow. However, after the dissolution of the Soviet Union, post-communist leaders allowed the bones to be exhumed so that they could be given a proper burial. This exhumation took on intense political meaning because the people of Russia hoped to do more than just give the family a church burial. The event took on the symbolic significance of laying to rest the brutality of the communist regime that took power after the murder of the Romanov family.

Since all that remained of these bodies at the time of their exhumation was a pile of bones, it was difficult to know for certain that these were actually the remains of the royal family. However, there was a great deal of circumstantial evidence pointing toward that conclusion. The size of the bones seemed to indicate that they belonged to six adults and three children. Investigators electronically superimposed the photographs of the skulls on archived photographs of the family. They compared the skeletons' measurements with clothing known to have belonged to the family. Five of the bodies had gold, porcelain, and platinum dental work, which was only available to aristocrats. These and other forensic data were consistent with the hypothesis that these bodies could be those of the Czar, the Czarina, three of their five children, and the four servants.

However, at this point the scientists had only shown that these skeletons *might* be the Romanovs—they had not yet shown with any degree of certainty that these bodies *did* belong to the slain royals. The new Russian leaders did not want to make a mistake when symbolically burying a former regime.

Concrete proof was necessary because there was so much at stake politically. To solve the mystery of who was buried in the Ekaterinburg grave, scientists turned to the chromosomes and DNA isolated from the bones exhumed from the grave.

6.1 Chromosomes and DNA

Chromosomes are condensed forms of DNA found in a cell's nucleus. **DNA** (**deoxyribonucleic acid**) is the molecule of heredity. It is passed from parents to offspring. DNA stores the instructions necessary for making all the proteins required by a cell. There are hundreds, or in some cases thousands, of these protein-building instructions (*genes*) located along the length of any one of the 46 human chromosomes.

Chromosomes

The 46 human chromosomes can be arranged into pairs with 22 pairs of nonsex chromosomes or **autosomes**, and one pair of **sex chromosomes** (the X and Y chromosomes)—the sex chromosomes comprise the 23rd pair. Human males have an X and a Y chromosome, while females have two X chromosomes.

A **karyotype** is a picture of a person's chromosomes. It is usually prepared from the chromosomes that have been removed from white blood cells. (Only white blood cells are used because red blood cells have no DNA.) White blood cells are separated from the rest of the blood and treated with a drug called *colchicine*. Treatment with colchicine prevents the microtubule proteins from functioning properly. Recall from Chapter 5 that microtubules are structures that help move chromosomes around during cell division. When the microtubules are not able to shorten and pull the chromosomes to the poles of the cell, cell division is halted. Colchicine-treated cells therefore contain the duplicated chromosomes stalled at metaphase of mitosis. Thus the chromosomes visible in a karyotype are in the shape of an X, instead of being linear, as is the case with unduplicated chromosomes.

To finish preparing a karyotype, colchicine-treated cells are placed in a solution that causes them to swell. The swollen cells are then dropped onto a stained slide. As the cells hit the slide, they break open and the chromosomes spread out. Using a camera attached to a microscope, scientists take a picture of the stained chromosomes. It is then possible to cut the individual chromosomes out of the photograph and arrange them in pairs (Figure 6.1).

Each chromosome is paired with a mate that is the same size and shape, with its centromere in the same position. These pairs of chromosomes are called **homologous pairs**. Each member of a homologous pair of chromosomes also has the same genes, although not necessarily the same versions of those genes. (Recall from Chapter 4 that different versions of a gene are called *alleles*.) For example, the presence of attached earlobes is genetically determined. For this gene, detached earlobes are dominant, and attached earlobes are recessive. If a person has one copy of the dominant allele, they will have detached earlobes. A person with two recessive alleles of this gene will have attached earlobes. The dominant and recessive versions of the earlobe attachment gene are different alleles of that gene.

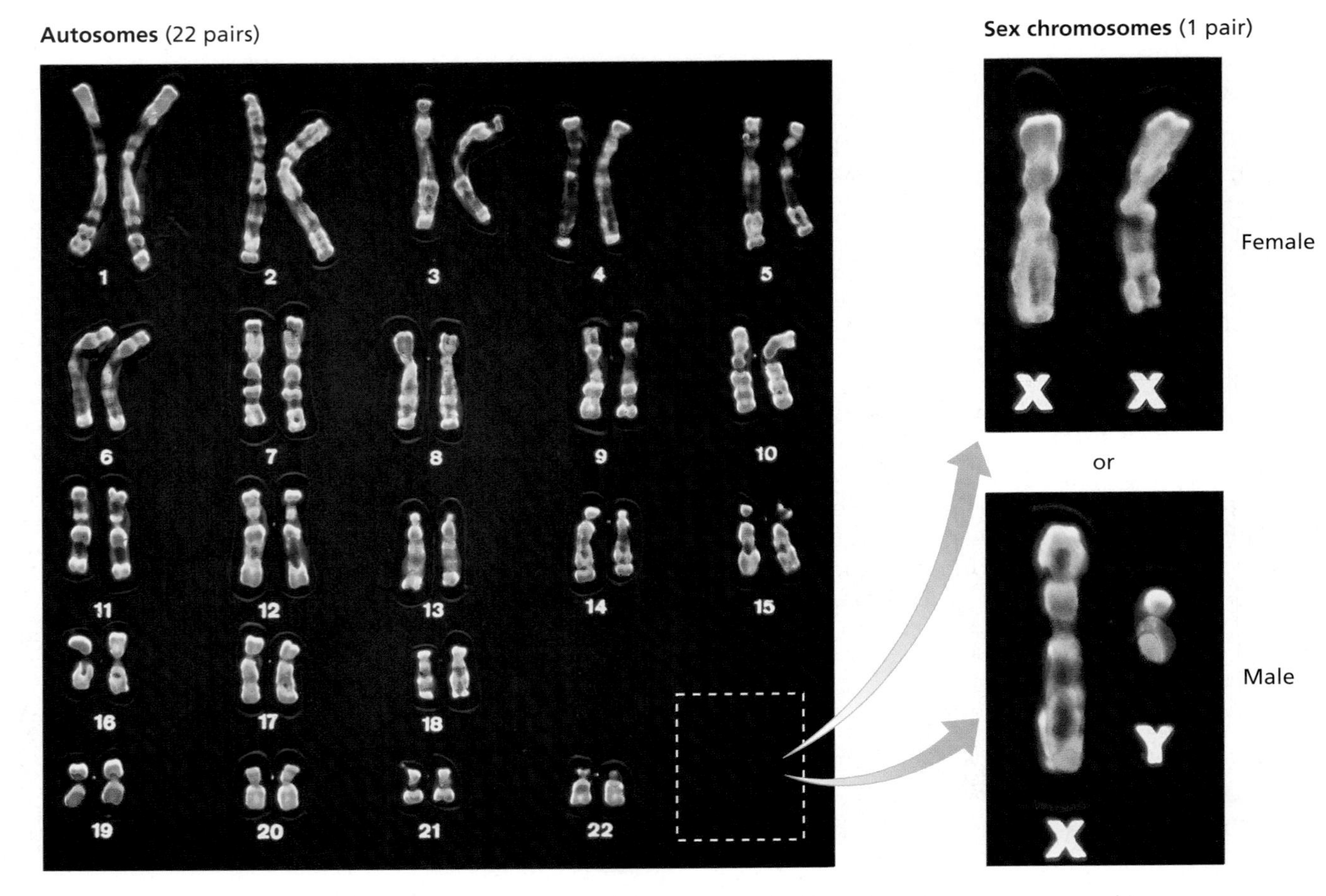

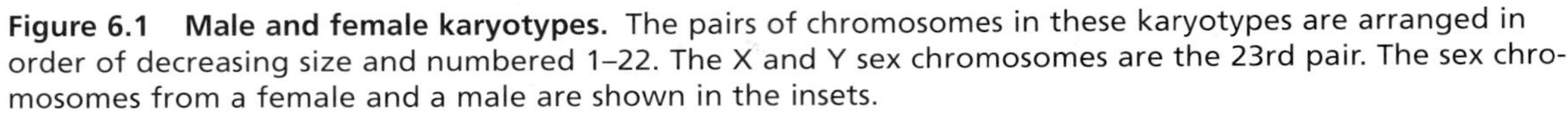

Figure 6.1 Male and female karyotypes. The pairs of chromosomes in these karyotypes are arranged in order of decreasing size and numbered 1–22. The X and Y sex chromosomes are the 23rd pair. The sex chromosomes from a female and a male are shown in the insets.

After pairing the 44 autosomes in the karyotype, the sex chromosomes are paired and the sex of the individual can be determined. If there are two X chromosomes, the individual is female. If there is one X chromosome and one Y chromosome, the individual is male.

Since all that was left of the Romanovs was a pile of scattered bones, it was not possible to perform karyotype analysis to determine the sex of the buried individuals. Even though bone is made of cells, the cells of the buried bones were no longer undergoing cell division, so isolating chromosomes at metaphase was not possible.

Scientists can sometimes determine the sex of an individual on the basis of pelvic structure. Women usually have smaller pelvises with wider pelvic inlets to accommodate the passage of a child through the birth canal. Russian scientists had determined that all three of the smaller skeletons and two of the adult-sized skeletons were probably female (and four of the adult skeletons were male). However, the pelvises had decayed and an unequivocal analysis was impossible.

Scientists had to turn to the DNA itself to perform the more sophisticated analyses required to determine who was buried in the grave.

DNA Structure

Figure 6.2 shows the three-dimensional structure of a DNA molecule and zooms inward to the chemical structure. You can see that DNA is composed of two strands winding around each other to form a double helix. Each strand of the

www
Media Activity 6.2 Examining the 3-D Structure of DNA

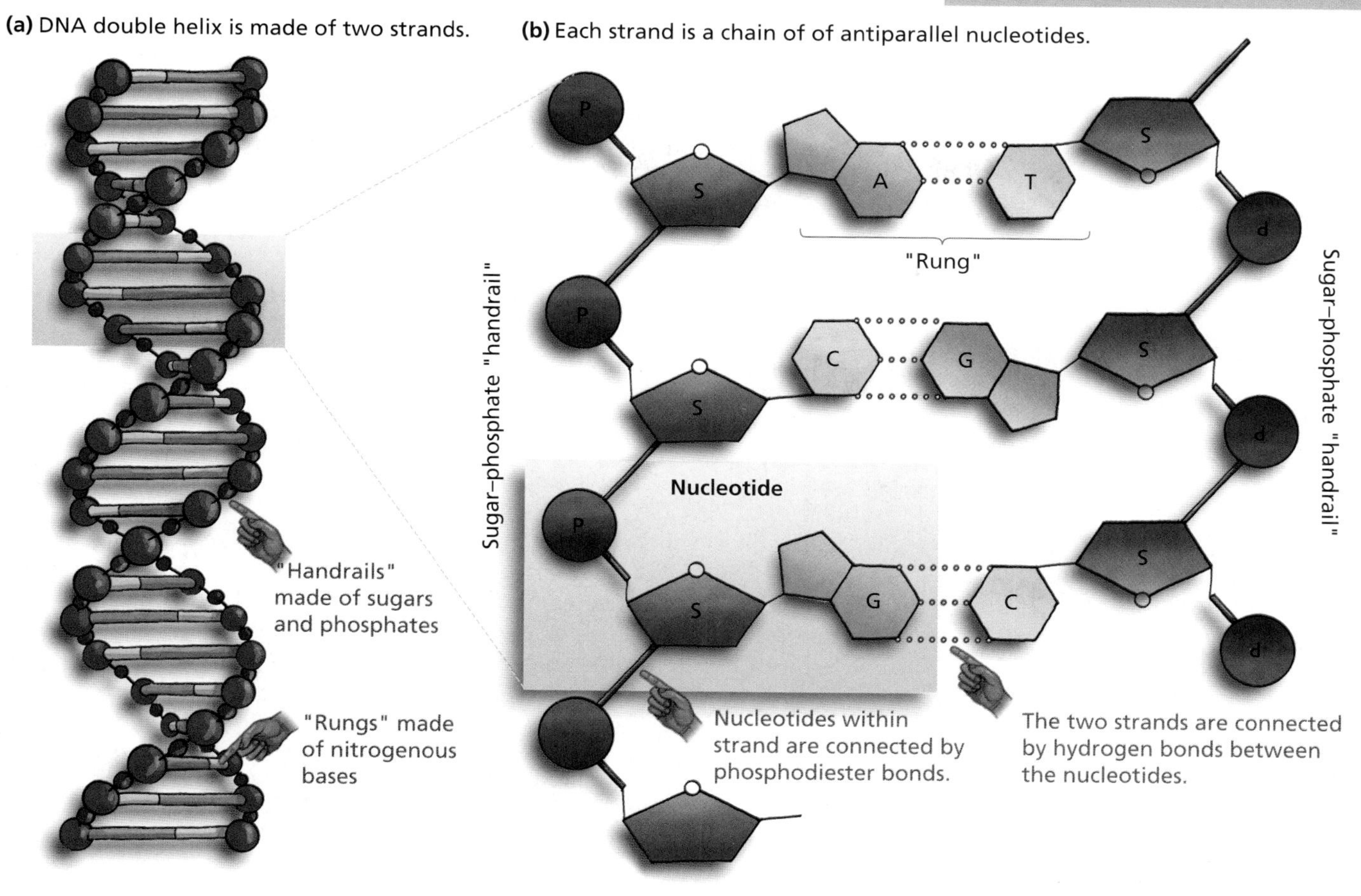

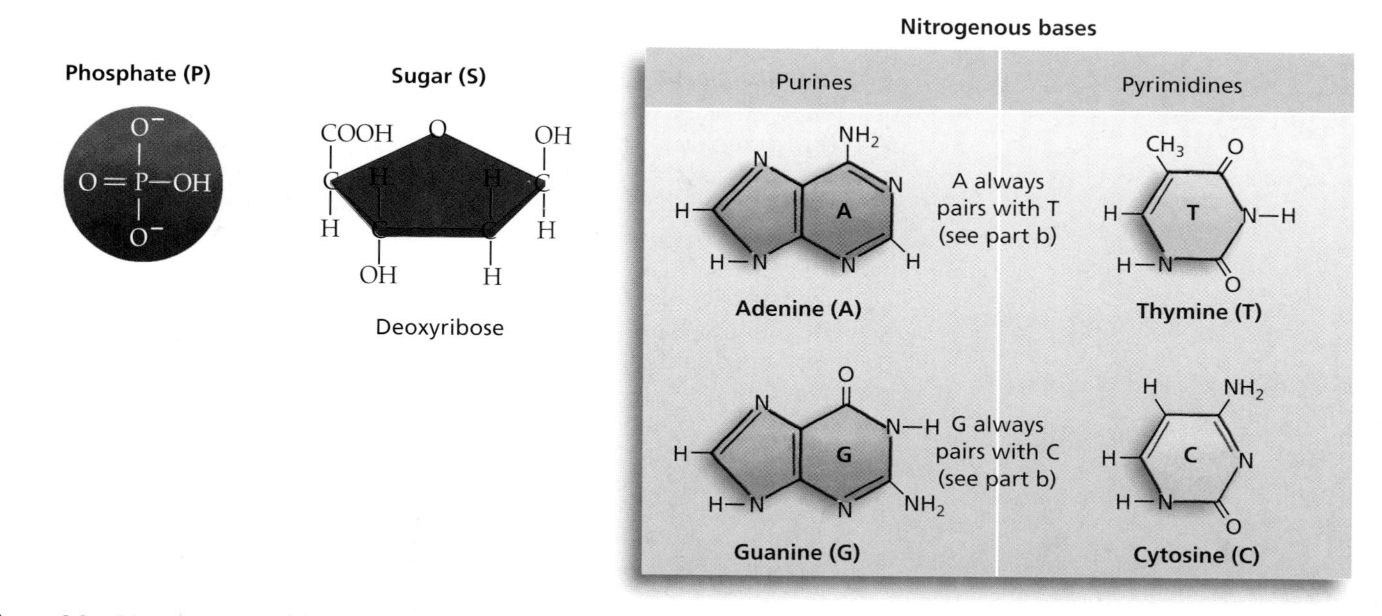

Figure 6.2 DNA structure. (a) DNA is a double-helical structure composed of sugars, phosphates, and nitrogenous bases. (b) Each strand of the helix is composed of repeating units of sugars and phosphates, making the sugar-phosphate backbone, and nitrogenous bases. (c) A phosphate, a sugar, and a nitrogenous base comprise the structure of a nucleotide. Adenine and guanine are purines, which have a two-ring structure, and cytosine and thymine are pyrimidines, with a single-ring structure.

helix consists of a long series of chemical building blocks called **nucleotides**. A nucleotide is made up of a sugar, a phosphate, and a nitrogen-containing base. The sugar in DNA is the 5-carbon sugar **deoxyribose**. The nitrogen-containing bases, or **nitrogenous bases**, of DNA have one of four different chemical structures, each with a different name: **adenine, guanine, thymine,** and **cytosine**. Nucleotides are joined to each other along the length of the helix by a type of chemical bond formed by the hydroxyl (OH) and phosphate (P) groups—called a **phosphodiester bond**.

Nitrogenous bases form hydrogen bonds with each other across the width of the helix. Adenine (A) on one strand always pairs with a thymine (T) on the opposite strand. Likewise, guanine (G) always pairs with cytosine (C). The term *complementary* is sometimes used to describe these pairings. For example, A is complementary to T, and C is complementary to G. Therefore, the order of nucleotides on one strand of the DNA helix predicts the order of nucleotides on the other strand. Thus, if one strand of the DNA molecule is composed of nucleotides AACGATCCG, we know that the order of nucleotides on the other strand is TTGCTAGGC.

As a result of this **base-pairing rule** (A pairs with T; G pairs with C), the width of the DNA helix is uniform. There are no bulges or dimples in the structure of the DNA helix since A and G, called *purines*, are structures composed of two rings while C and T are single-ring structures called *pyrimidine*. A purine always pairs with a pyrimidine and vice versa, so there are always a total of three rings across the width of the helix. Nitrogenous bases are held together across the width of the helix by weak bonds called *hydrogen bonds*. A:T base pairs have two hydrogen bonds holding them together. G:C pairs have three hydrogen bonds holding them together.

Each strand of the helix thus consists of a series of sugars and phosphates alternating along the length of the helix, the **sugar-phosphate backbone**. The strands of the helix align so the nucleotides face "up" on one side of the helix and "down" on the other side of the helix. For this reason the two strands of the helix are said to be *anti-parallel*.

The overall structure of a DNA molecule can be likened to a rope ladder that is twisted, with the sides of the ladder (the hand rails) composed of sugars and phosphates (the sugar-phosphate backbone) and the rungs of the ladder composed of the nitrogenous-base sequences A,C, G, and T. When the DNA is copied, or replicated, the ladder is split up the middle of the rungs, and base pairs are added according to the base-pairing rule.

As we learned in Chapter 4, every individual who is not an identical twin has a unique set of genes, and therefore a unique sequence of nucleotides, which make up genes.

Scientists used the fact that different individuals have distinct nucleotide sequences to test their hypothesis that the bones buried in the Ekaterinburg grave belonged to the Romanov family. The scientists had to answer the following questions:

1. Which of the bones from the pile were actually different bones from the same individuals?
2. Which of the adult bones could have been from the Romanovs, and which bones could have belonged to their servants?
3. Which two Romanov children were missing from the grave?
4. Are these bones actually from the Romanovs, not some other related set of individuals?

All of these questions were answered using a very powerful technique that takes advantage of differences in DNA sequence, called **DNA fingerprinting**.

Essay 6.1 The Many Uses of DNA Fingerprinting

DNA fingerprinting can be used to establish parentage, convict a criminal, exonerate the innocent, and identify the dead. Courts around the world have allowed DNA fingerprinting to be used as evidence in hundreds of murder and rape cases. Many people have been released from prison on the basis of DNA that was present on items that were admitted into evidence and saved in case of an appeal. The United States Army collects DNA samples from all enlisted personnel to facilitate identification of those who are killed during war. Even private citizens can purchase kits that help establish parentage; cells scraped from inside the cheek of purported parents and children can be sent away for accurate and inexpensive DNA analysis by private companies.

The real power of this technique is that it can provide positive identification with great accuracy. More conventional methodologies, such as blood typing, can only exclude people who do not match the blood type of the person in question, and are not as useful to confirm a positive identification.

The odds of any two people sharing the same DNA fingerprint approaches zero. This idea is especially important to understand when DNA fingerprinting is used in court cases. Juries must recognize that the chance of another person coincidentally having the same DNA fingerprint as the accused is very low.

If DNA evidence has been found at the scene of the crime—from semen, hair, tissue, bone marrow, saliva, or urine—the probability of another person having the same pattern can be calculated by multiplying together the frequency with which each visible band occurs in the general population. Scientists use probes (single stranded, radioactively labelled DNA sequences that are complementary to known DNA sequences) for which the frequency in the population is well established. For example, assume that bands produced in a DNA fingerprint occur in the general population with the following frequencies:

band 1 = 2%	band 2 = 1%
band 3 = 5%	band 4 = 1.5%
band 5 = 1.2%	band 6 = 3%

The probability that this combination of bands would be found in two different individuals in the general population is the product of all six frequencies—that is, 5.4×10^{-11}. In other words, the odds of someone other than the accused having the same DNA fingerprint as the accused is 1:5,400,000,000,000 or, one in five trillion-four hundred billion. Since there are only six billion people in the world, it is virtually impossible for two people to have the same DNA fingerprint.

6.2 DNA Fingerprinting

DNA fingerprinting can be used when it is necessary to unambiguously identify people in the same manner that traditional fingerprinting has been used in the past. Essay 6.1 describes some of the many uses of DNA fingerprinting.

To begin this process, the DNA is broken apart with enzymes that cleave, or cut, the DNA at specific nucleotide sequences (Figure 6.3). These enzymes are called **restriction enzymes**, and they act like highly specific molecular scissors. Individual restriction enzymes only cut DNA at specific nucleotide sequences. Because each individual has distinct nucleotide sequences, cutting different people's DNA with the same enzymes releases fragments of different sizes.

The sizes of these fragments are determined by placing them in a gelatin-like substance called an *agarose gel*. When an electric current is applied, the gel impedes the progress of the larger DNA fragments more than the smaller ones. The current also pulls the negatively charged DNA molecules toward the bottom (positively charged) edge of the gel, further facilitating the size-based separation.

The DNA fragments are then lifted out of the gel by placing a special filter paper on top of the gel. The DNA is wicked up through the gel, and attaches to the filter paper. The filter is then treated with chemicals that break the hydrogen bonds between the two strands of the DNA helix. The resulting

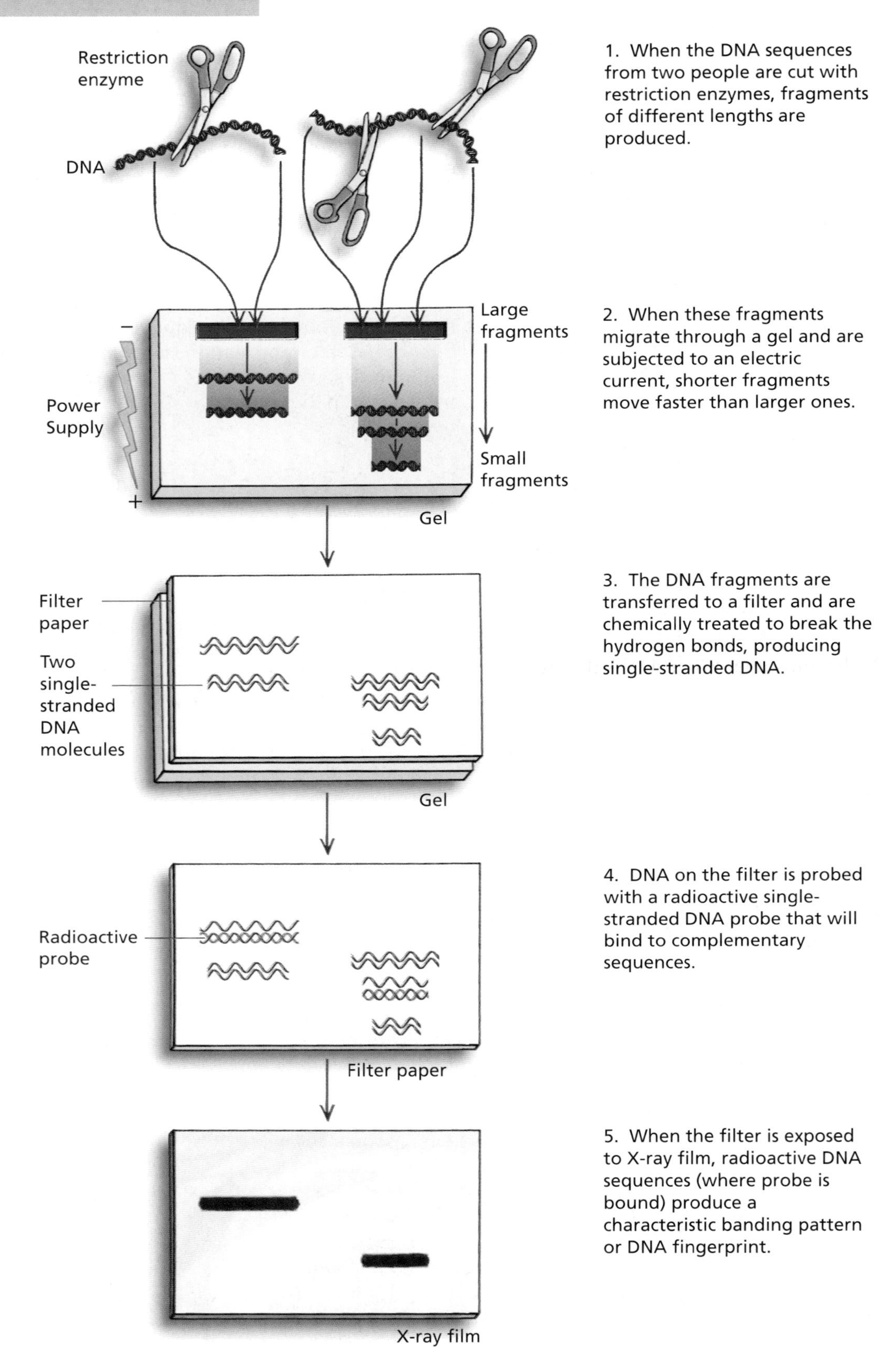

1. When the DNA sequences from two people are cut with restriction enzymes, fragments of different lengths are produced.

2. When these fragments migrate through a gel and are subjected to an electric current, shorter fragments move faster than larger ones.

3. The DNA fragments are transferred to a filter and are chemically treated to break the hydrogen bonds, producing single-stranded DNA.

4. DNA on the filter is probed with a radioactive single-stranded DNA probe that will bind to complementary sequences.

5. When the filter is exposed to X-ray film, radioactive DNA sequences (where probe is bound) produce a characteristic banding pattern or DNA fingerprint.

Figure 6.3 DNA fingerprinting. This illustration outlines the process of preparing DNA fingerprints from two different people.

single-stranded DNA now resembles a ladder that has been sliced up the middle, each rung having been cut in half.

The single-stranded DNA is mixed with specially prepared single-stranded DNA that has been radioactively labeled, called a *probe*. The probe is synthesized so that the phosphate molecules in its sugar-phosphate backbone are the radioactive form of phosphorus, ^{32}P. The probe DNA is designed so that it will base pair with a DNA sequence of the individual being fingerprinted. The probe DNA finds sequences that are complementary and binds to them, forming a double helix.

Once it is bound to the labeled probe, the presence of radioactive DNA can be detected on photographic film. In the same manner that X-ray film records the shape of bones when your body is exposed to X-rays, the film used in DNA fingerprinting records the location of DNA when the radiation emitted from radioactive DNA molecules bombards the film. A piece of X-ray film placed over the filter paper shows the locations of the radioactive DNA as a series of bands. Different individuals have different bands, since the probe binds to differently sized pieces of DNA in each individual. The specific banding pattern that is produced makes up the actual fingerprint. Figure 6.4 shows a hypothetical DNA fingerprint from 20 bone fragments.

In 1992, a team of Russian and English scientists used DNA fingerprinting to determine which of the bones discovered at Ekaterinburg belonged to the same skeleton. Their fingerprinting analysis confirmed that the pile of decomposed bones in the Ekaterinburg grave belonged to nine different individuals.

Since karyotype and pelvic bone analyses were difficult to perform on the decayed bones, scientists also probed the DNA for sequences known to be present only on the Y chromosome. When DNA that was isolated from the children's remains was probed with a Y-specific probe, it became clear that, if these bones did belong to the Romanovs, one of the two missing children was Alexis, the Romanovs' only son.

Once scientists had established that there were bones from nine different people buried in the grave, they tried to determine which bones might belong to the adult Romanovs and which belonged to the servants. For the answer to these questions, scientists took advantage of the fact that Romanov family members would have more DNA sequences in common with each other than they would with the servants. This is because the Czar and Czarina each passed half of their DNA to each of their children. The process by which parents pass half of their DNA to their children is called **meiosis**.

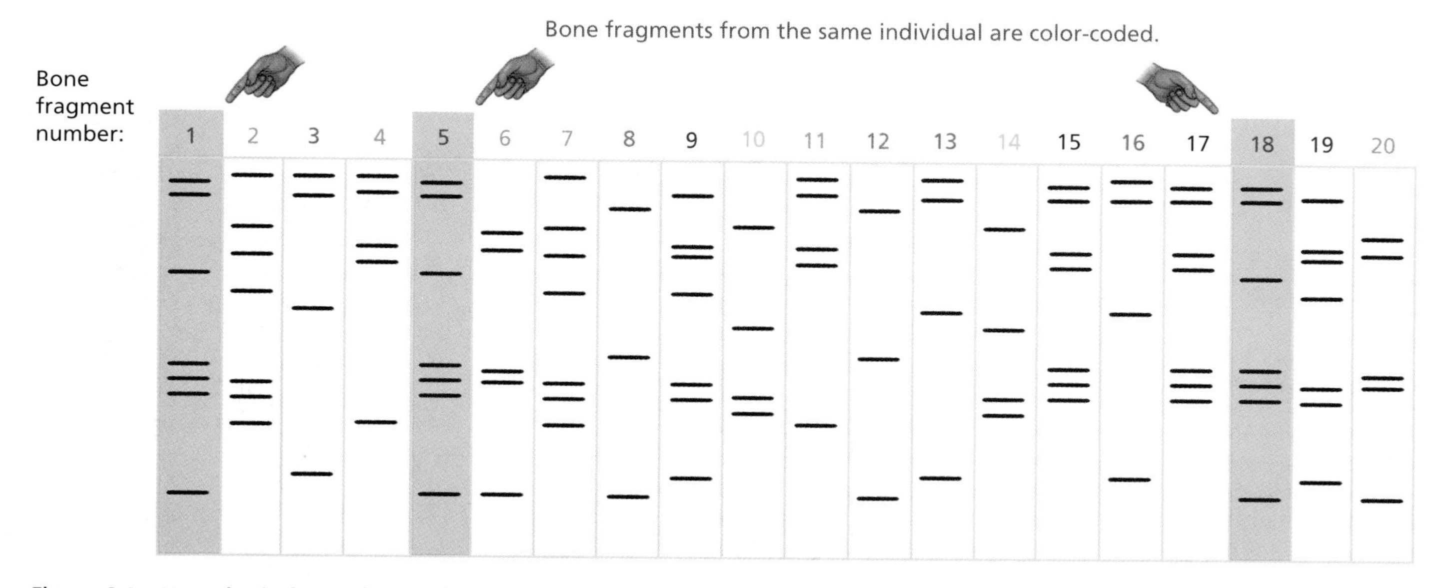

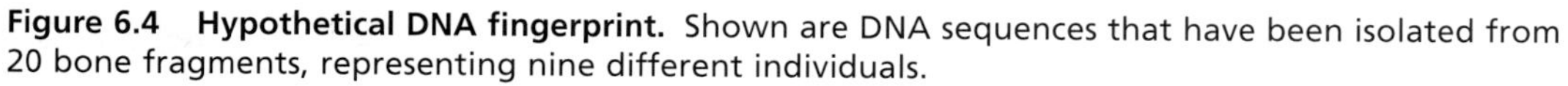
Figure 6.4 Hypothetical DNA fingerprint. Shown are DNA sequences that have been isolated from 20 bone fragments, representing nine different individuals.

6.3 How DNA Passes from Parents to Their Children

Meiosis is a type of nuclear division that occurs only in the cells of the testes and ovaries. During meiosis, chromosomes, including the DNA they carry, are copied and split into specialized sex cells called **gametes**. Human male gametes are called *sperm cells*, and human female gametes are called *egg cells*.

Both sperm and egg cells contain half as many chromosomes as other body cells. Because human body cells have 46 chromosomes, meiosis produces gametes with 23 chromosomes. When an egg cell and a sperm cell combine their chromosomes at fertilization, the developing embryo will then have the required 46 chromosomes.

The placement of chromosomes into gametes is not random—that is, meiosis does not simply place any 23 of the 46 chromosomes into a gamete—meiosis apportions chromosomes in a very specific manner. Recall from Section 6.1 that the 46 chromosomes in human body cells are actually 23 different pairs of chromosomes. These homologous pairs of chromosomes have the same size and shape and the same position of centromere. In addition, homologous pairs of chromosomes carry the same genes along their length, but not necessarily the same alleles of those genes (Figure 6.5).

Meiosis separates the members of a homologous pair from each other. Once meiosis is completed, there is one copy of each chromosome (1–23) in every gamete. When only one member of each homologous pair is present in a cell, we say that that cell is **haploid** (**n**)—both egg cells and sperm cells are haploid. After the sperm and egg fuse, the fertilized egg cell has two sets of chromosomes and is said to be **diploid** (**2n**). All body cells (nongametes) in humans contain homologous pairs of chromosomes and are therefore diploid (Figure 6.6). For a diploid cell in the testes or ovary to become haploid, it must go through the meiotic cell cycle.

The Meiotic Cell Cycle

The meiotic cell cycle includes all of the events that occur as a cell undergoes meiosis to produce daughter cells. This includes interphase, followed by two phases in which divisions of the nucleus take place called *meiosis I* and *meiosis II* (Figure 6.7),

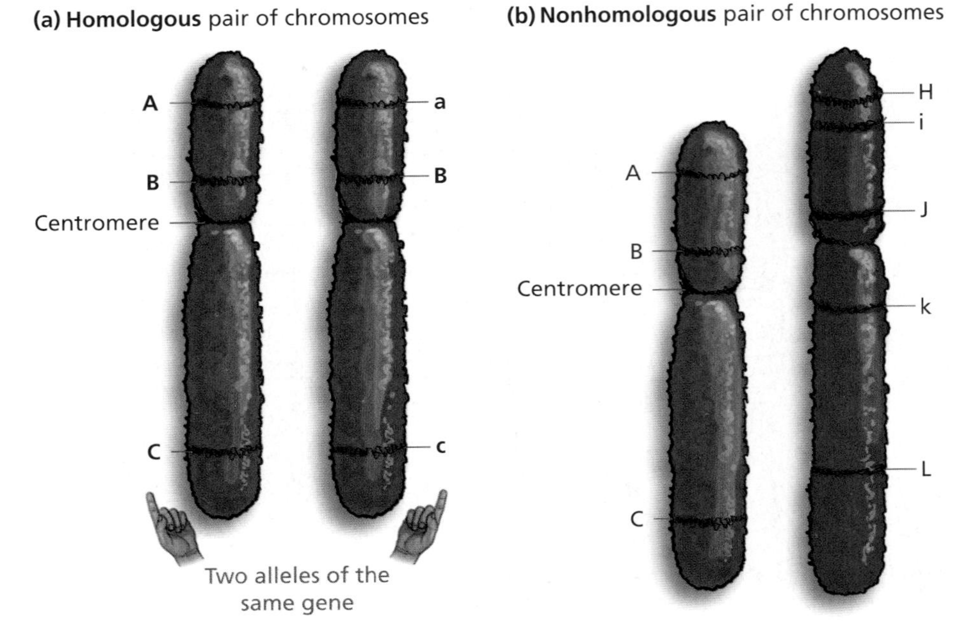

Figure 6.5 Homologous and nonhomologous pairs of chromosomes. (a) Homologous pairs of chromosomes have the same genes (shown here as A, B, and C) but may have different alleles. The dominant allele is represented by an uppercase letter and the recessive allele with the same letter in lowercase. Note that the chromosomes of this pair each have the same size, shape, and position of centromere. (b) The nonhomologous pair of chromosomes at the right are different sizes and shapes and carry different genes.

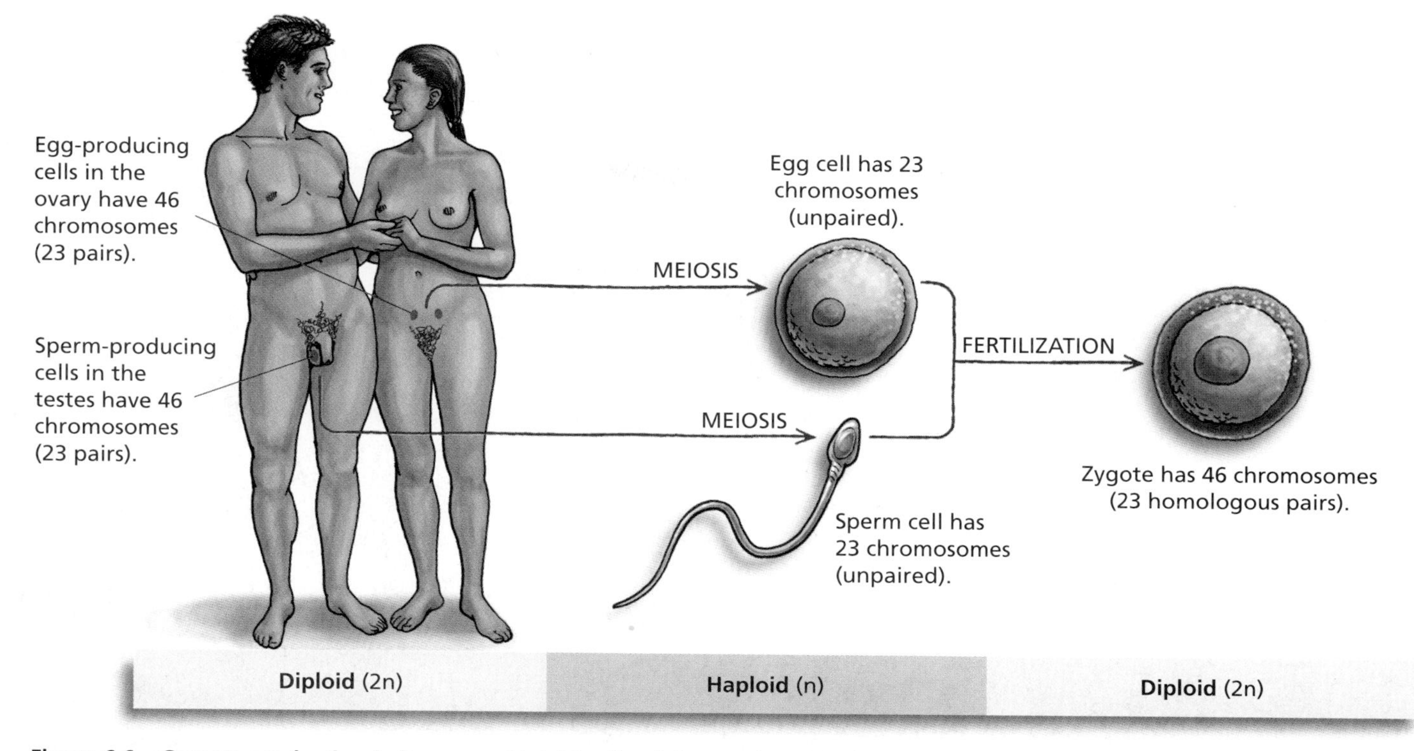

Figure 6.6 Gamete production in humans. Diploid cells of the ovaries and testes undergo meiosis and produce haploid gametes. At fertilization, the diploid condition is restored.

followed by cytokinesis. Meiosis I separates the members of a homologous pair from each other. Meiosis II separates the chromatids from each other. Cytokinesis divides the cytoplasm roughly equally into the resulting daughter cells.

Interphase The interphase that precedes meiosis consists of G_1, S, and G_2. This interphase is similar in most respects to the interphase that precedes mitosis. The centrioles from which the microtubules will originate are present. The G phases are times of cell growth and preparation for division. The S phase is when DNA replication occurs (compare Figure 5.4 with Figure 6.7).

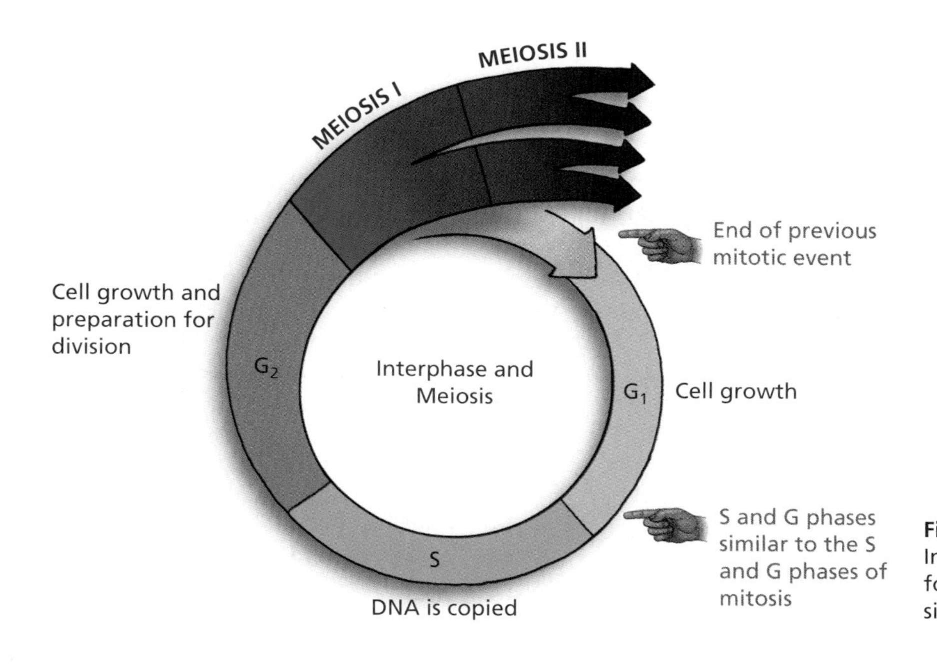

Figure 6.7 Interphase and meiosis. Interphase consists of G_1, S, and G_2 and is followed by two rounds of nuclear division, meiosis I and meiosis II.

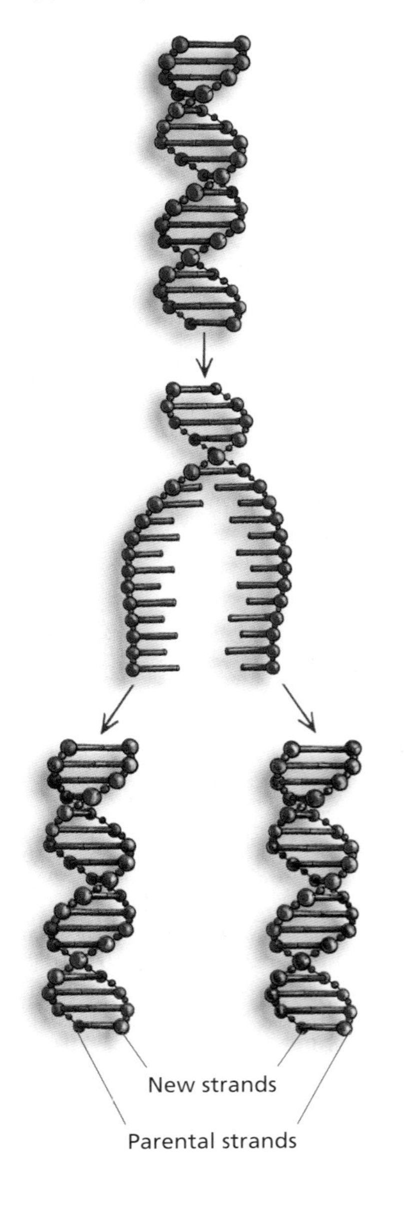

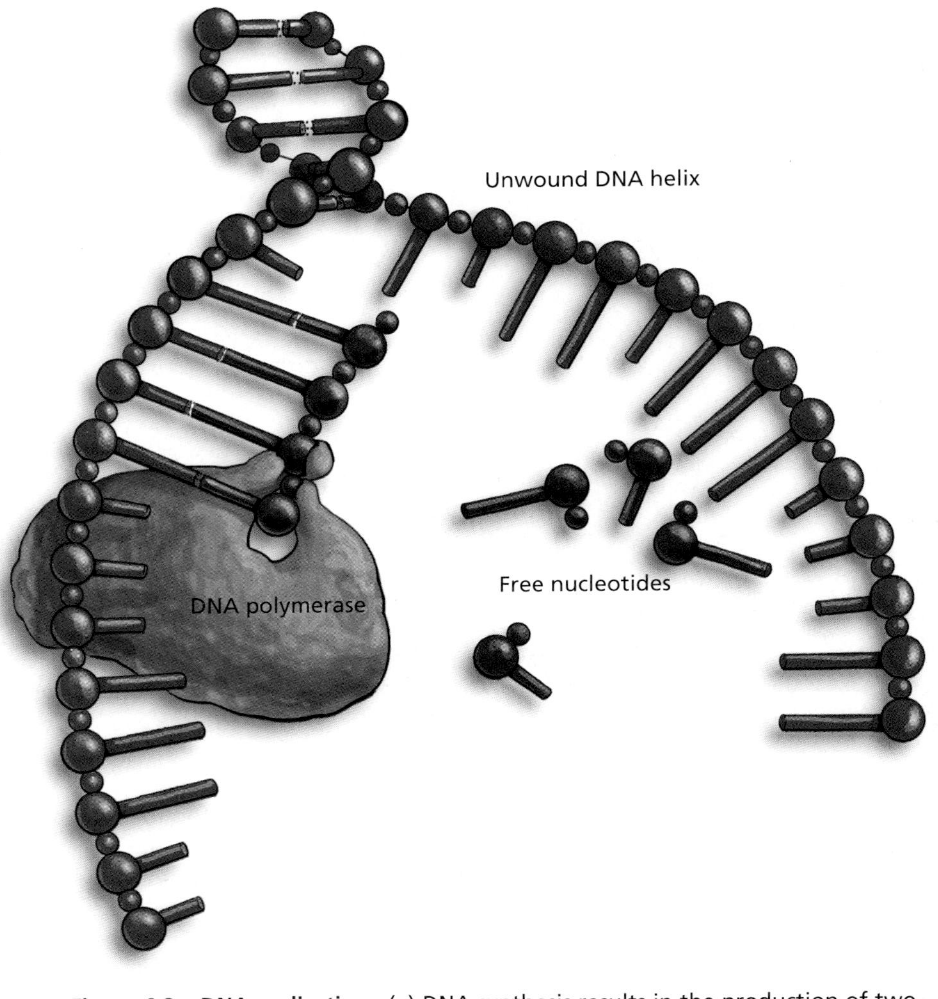

Figure 6.8 DNA replication. (a) DNA synthesis results in the production of two identical daughter DNA molecules from one parent molecule. Each daughter DNA molecule is composed of $\frac{1}{2}$ parental DNA and $\frac{1}{2}$ newly synthesized DNA. (b) The DNA polymerase enzyme moves along the unwound helix, tying together adjacent nucleotides on the newly forming daughter strand.

DNA Replication When a DNA molecule is to be replicated, or copied, it is split up the middle of the helix, and new nucleotides are added to each side of the original parent molecule, maintaining the A:T and G:C pairings. This results in two daughter DNA molecules, each composed of one strand of parental nucleotides and one newly synthesized strand (Figure 6.8a).

To replicate the DNA, an enzyme that facilitates DNA synthesis is required. This enzyme, **DNA polymerase**, moves along the length of the unwound helix and helps bind incoming nucleotides to each other on the newly forming daughter strand (Figure 6.8b). Nucleotides on the daughter strand are complementary to those across from them on on the parent strand. When free nucleotides floating in the nuclear fluid have an affinity for each other (A for T and G for C), they bind to each other across the width (rungs) of the helix. The DNA polymerase catalyzes the formation of the phosphodiester bond along the length (handrails)

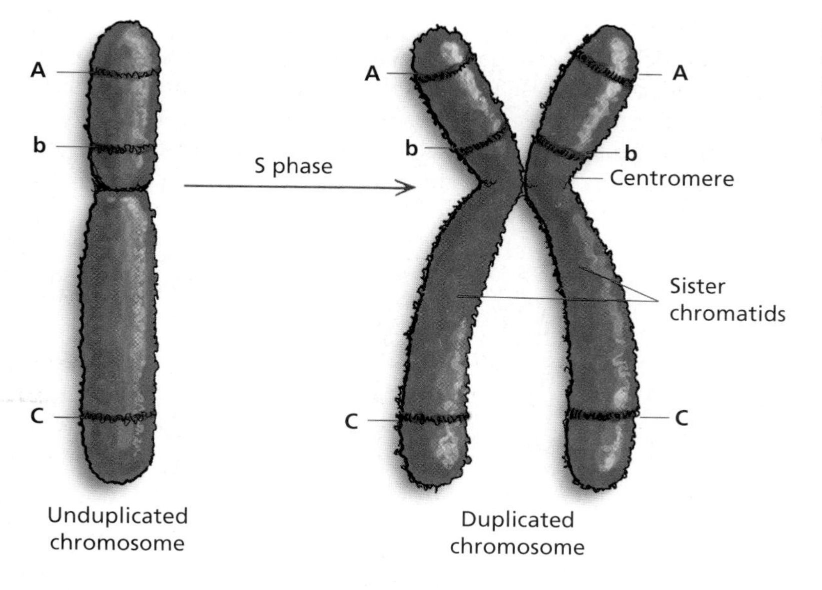

Figure 6.9 Unduplicated and duplicated chromosomes. Unconensed chromosomes are not linear, but are drawn this way here to illustrate that duplicated chromosomes are exact copies of the unduplicated chromosome and are attached at the centromere.

of the helix. The two complementary nitrogenous bases (A,C,G, and T) are joined, and the DNA polymerase advances along the parental DNA strand to the next unpaired nucleotide.

Just as happens in mitosis, when an entire chromosome has been replicated, the duplicated copies are identical to each other, attached at the centromere, and called *sister chromatids* (Figure 6.9).

Once the cell's DNA has been duplicated, it can enter the meiotic phase of the cell cycle, consisting of two cell divisions called meiosis I and meiosis II.

Meiosis I The first meiotic division, meiosis I, consists of prophase I, metaphase I, anaphase I, and telophase I (Figure 6.10).

During prophase I of meiosis, the nuclear envelope starts to break down, and the microtubules begin to assemble. The previously replicated chromosomes condense, so they can be moved around the cell without becoming tangled; now the chromosomes are linear and can be viewed under a microscope. At this time, the homologous pairs of chromosomes can exchange genetic information, a process called **crossing over**.

At metaphase I the homologues line up at the equator, or middle, of the cell. Microtubules bind to the metaphase chromosomes near the centromere. The arrangement of homologous pairs is haphazard with regard to which member of a homologous pair faces which pole. This is called **random alignment**. A detailed description of crossing over and random alignment, along with their impact on genetic diversity, is given below.

At anaphase I the homologues are separated from each other by shortening of the microtubules, and at telophase I the nuclear envelopes reform around the chromosomes. The DNA is then partitioned into each of the two daughter cells by cytokinesis. Because each daughter cell contains only one copy of each type of chromosome, at this point the cells are haploid. Now both of these daughter cells are ready to undergo meiosis II.

Meiosis II Meiosis II consists of prophase II, anaphase II, metaphase II, and telophase II. This second meiotic division is virtually identical to mitosis. At prophase II of meiosis, the cell is readying for another round of division, and the microtubules are lengthening again. At metaphase II the chromosomes align across the equator in much the same manner as they do during mitosis—not as pairs, as was the case with metaphase I. At anaphase II the sister chromatids are separated from each other and move to opposite poles of the cell. At telophase II the separated chromosomes are each enclosed in their own nucleus. Thus, when meiosis works properly, the daughter cells (gametes) will contain one member of each original homologous pair.

www
Media Activity 6.1 Meiosis

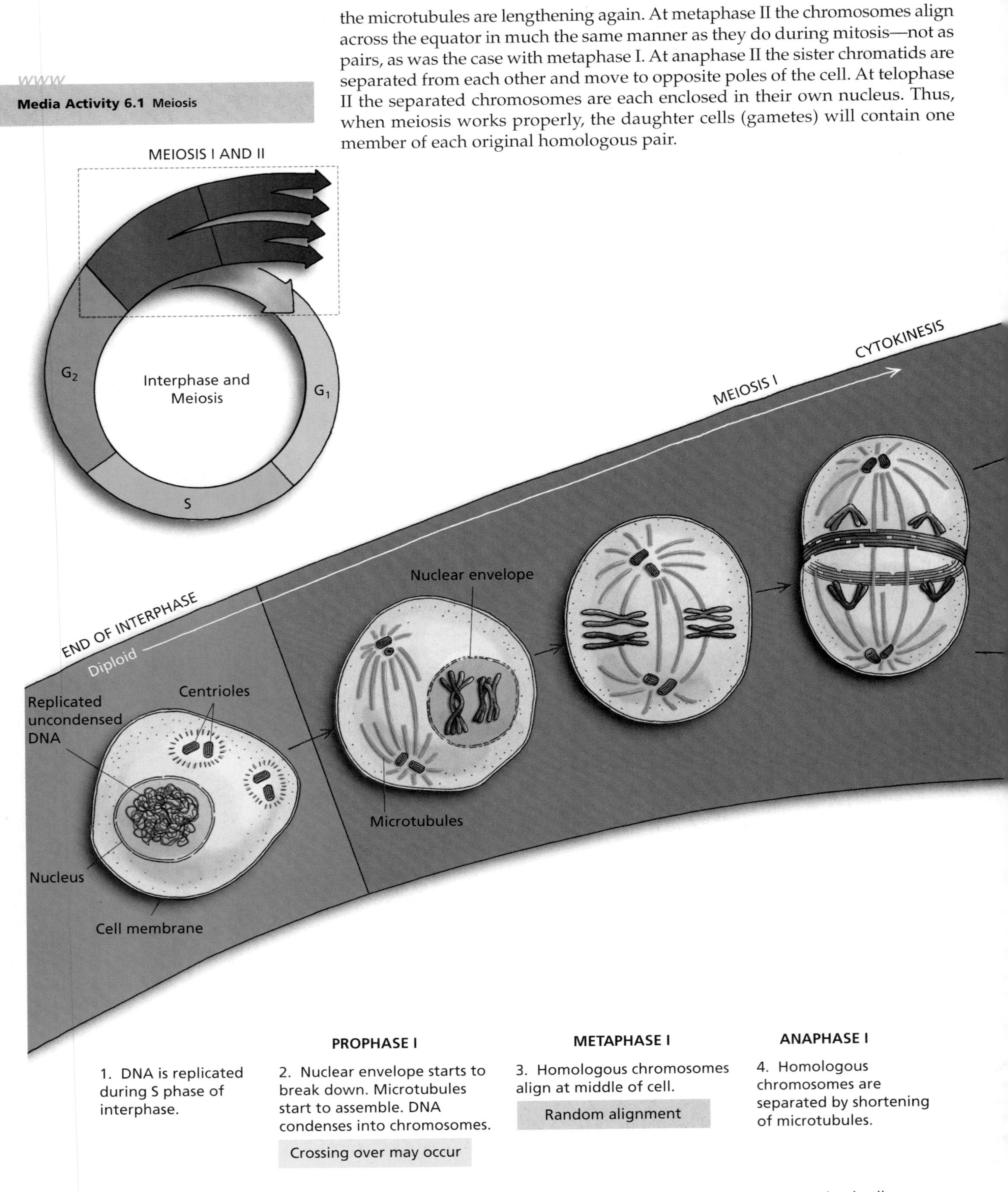
MEIOSIS I AND II
G2
Interphase and
Meiosis
G1
S
END OF INTERPHASE
Diploid
MEIOSIS I
CYTOKINESIS
Replicated
uncondensed
DNA
Centrioles
Nucleus
Cell membrane
Nuclear envelope
Microtubules
PROPHASE I
METAPHASE I
ANAPHASE I
1. DNA is replicated during S phase of interphase.
2. Nuclear envelope starts to break down. Microtubules start to assemble. DNA condenses into chromosomes.
Crossing over may occur
3. Homologous chromosomes align at middle of cell.
Random alignment
4. Homologous chromosomes are separated by shortening of microtubules.

Figure 6.10 The cell cycle. This diagram illustrates interphase, meiosis I and meiosis II, and cytokinesis in an animal cell.

In this fashion, half of a person's genes are physically placed into each gamete; thus, children carry one-half of each parent's genes. This characteristic was exploited in the Romanov case when investigators used DNA fingerprints prepared from the bone fragments of all nine individuals to determine their relationships.

Each band in a DNA fingerprint made from a child must be present in at least one parent. By comparing DNA fingerprints made from the smaller skeletons, scientists were able to determine which of the six adult skeletons

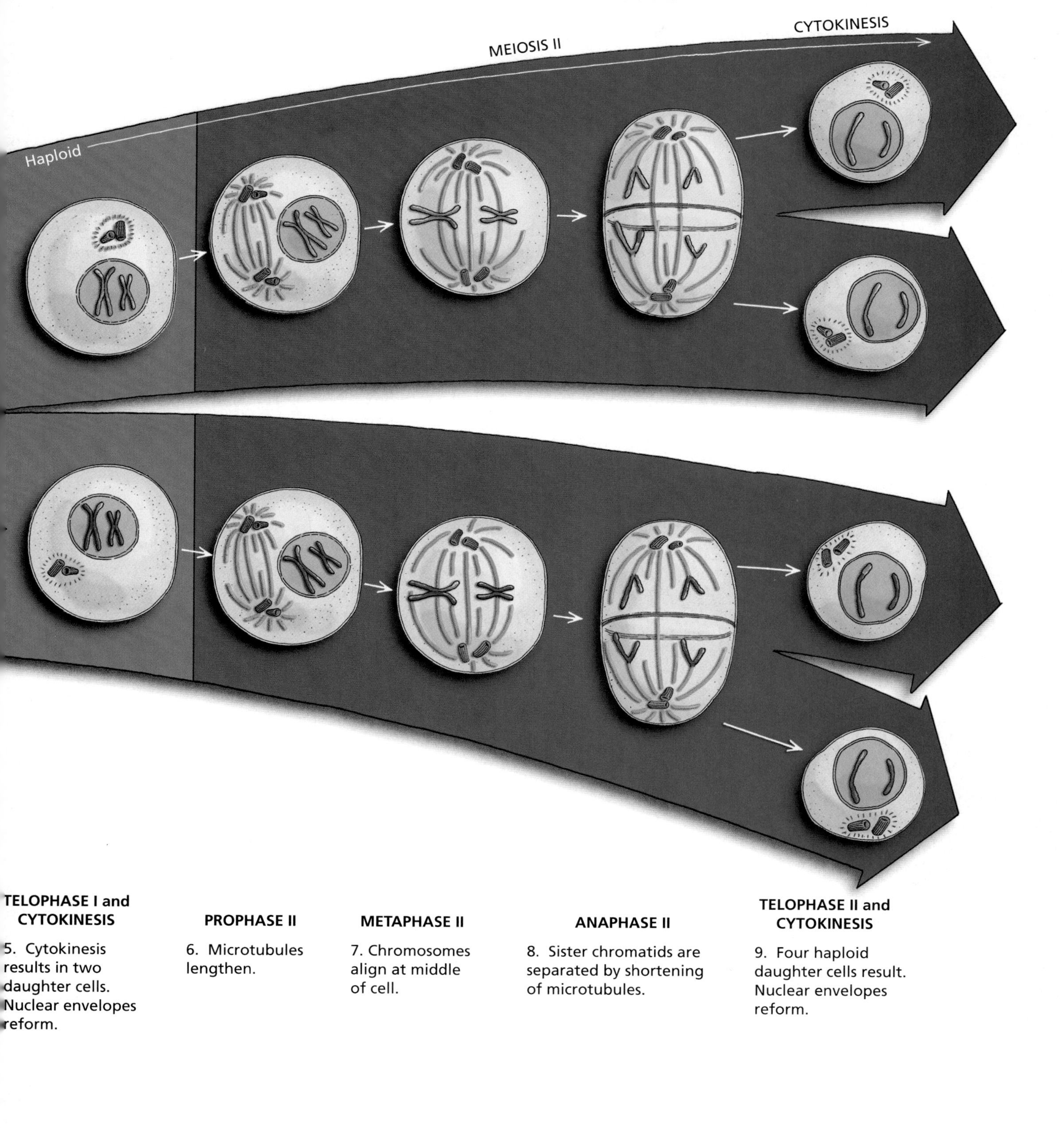

TELOPHASE I and CYTOKINESIS

5. Cytokinesis results in two daughter cells. Nuclear envelopes reform.

PROPHASE II

6. Microtubules lengthen.

METAPHASE II

7. Chromosomes align at middle of cell.

ANAPHASE II

8. Sister chromatids are separated by shortening of microtubules.

TELOPHASE II and CYTOKINESIS

9. Four haploid daughter cells result. Nuclear envelopes reform.

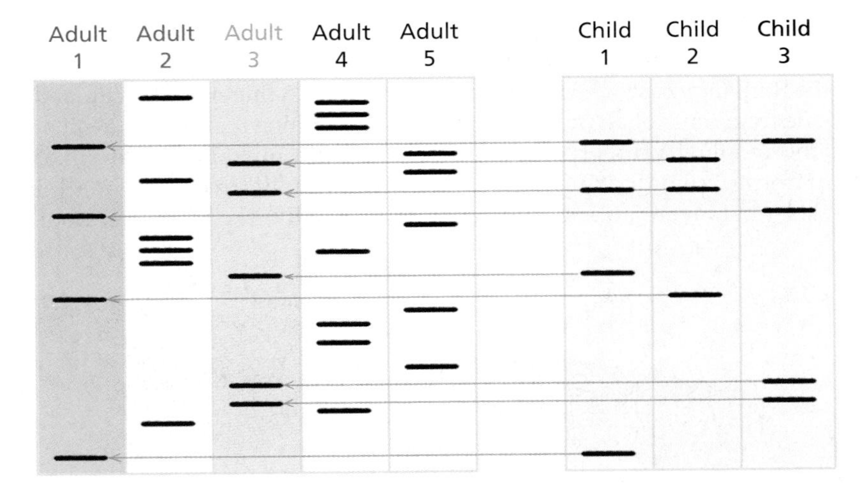

Figure 6.11 Hypothetical fingerprint of adult- and child-sized skeletons. Hypothetical DNA fingerprint made from bone cells found in the Ekaterinburg grave. From the results of this fingerprint, it is evident that children 1, 2, and 3 are the offspring of adults 1 and 3. Note that each band from each child has a corresponding band in either adult 1 or adult 3. The remaining DNA from adults does not match any of the children, so these adults are not the parents of any of these children.

could have been the Czar and Czarina. Figure 6.11 shows a hypothetical DNA fingerprint that illustrates how the banding patterns produced can be used to determine which of the bones belonged to the parents of the smaller skeletons and which bones may have belonged to the servants.

Until the DNA fingerprint evidence became available, the mystery of the two missing skeletons of the Romanov children, Alexis and one daughter, allowed many pretenders to the throne to make claims that could not be disputed. Numerous people from all over the world had alleged themselves to be either a Romanov who had escaped execution or their descendant.

The most compelling of these claims was made by a young woman who was rescued from a canal in Berlin, Germany, two years after the murders. This young woman suffered from amnesia and was cared for in a mental hospital where the staff named her Anna Anderson (Figure 6.12). She later came to believe that she was Anastasia Romanov, a claim she made until her death in 1984. The 1956 Hollywood film *Anastasia*, starring Ingrid Bergman, made Anna Anderson's claim seem plausible, and another, more recent, animated version of the escaped-princess story, also titled *Anastasia*, has many young viewers convinced that Anna Anderson was indeed the Romanov heiress.

Since the sex-typing analysis showed only that one daughter was missing from the grave, but not which daughter, scientists again relied on fingerprinting data. DNA fingerprinting had been done in the early 1990s on intestinal tissue removed during a surgery performed prior to her death; it showed that Anna Anderson was not related to anyone buried in the Ekaterinburg grave—therefore she could not be Anastasia Romanov.

How can scientists know that Anna Anderson's DNA fingerprint could not have been produced from the DNA of a child of the Czar and Czarina? The answer has to do with the way gametes are produced. There is a limited but huge number of possible gametes any person can produce.

The limitation in types of gametes a person can produce arises because all people carry only two alleles of every gene. Therefore, variation for that particular gene is limited to those two alleles. For example, let us assume that both the Czar and Czarina had attached earlobes. This is inherited as a recessive trait so, if true, they would both have carried two copies of the recessive allele and could only pass the recessive allele to their children. Neither parent could have given their children the dominant allele because they did not carry it.

The enormous number of gametes that an individual can produce is due to two events that occur during meiosis I—crossing over and random alignment of the homologues. Both of these processes greatly increase the number of different kinds of gametes an individual can produce, and therefore increase the variation in individuals that can be produced when gametes combine. Thus, both of these processes increase genetic diversity.

Figure 6.12 Anna Anderson. This woman claimed she was Anastasia Romanov.

Crossing Over and Random Alignment

Crossing over occurs during prophase I of meiosis and involves the exchange of portions of chromosomes from one member of a homologous pair with the other member. Crossing over is believed to occur several times on each homologous pair during each occurrence of meiosis.

To illustrate crossing over, let us consider an example using genes involved in the production of red hair and freckles. These two genes are on the same chromosome and are called **linked genes**. Linked genes move together on the same chromosome to a gamete, and they can undergo crossing over.

If a person, say a cousin of the Romanovs, has red hair and freckles, the chromosomes may appear as shown in Figure 6.13. It is possible for this person to produce four different types of gametes with these two genes. Two types of gametes would result if no crossing over occurred between these genes—the gamete containing the red hair and freckle chromosome, and the gamete containing the non-red hair and non-freckle chromosome. Two additional types of gametes could be produced if crossing over did occur—one type containing the

www
Media Activity 6.1 Meiosis

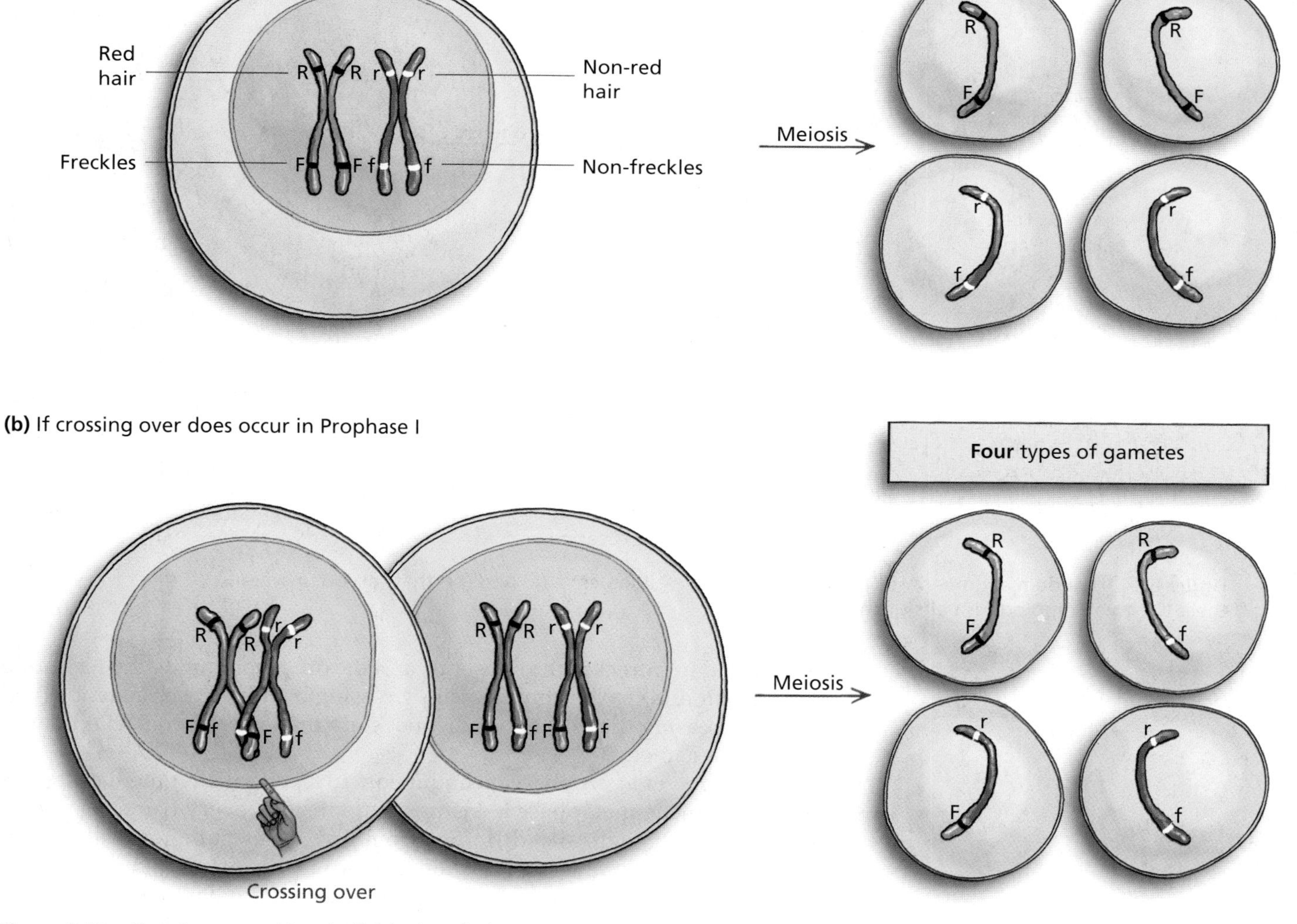

Figure 6.13 Crossing over. If an individual with the above arrangement of alleles undergoes meiosis, he or she can produce (a) two different types of gametes for these two genes if crossing over does not occur; or (b) four different types of gametes for these two genes if crossing over does occur.

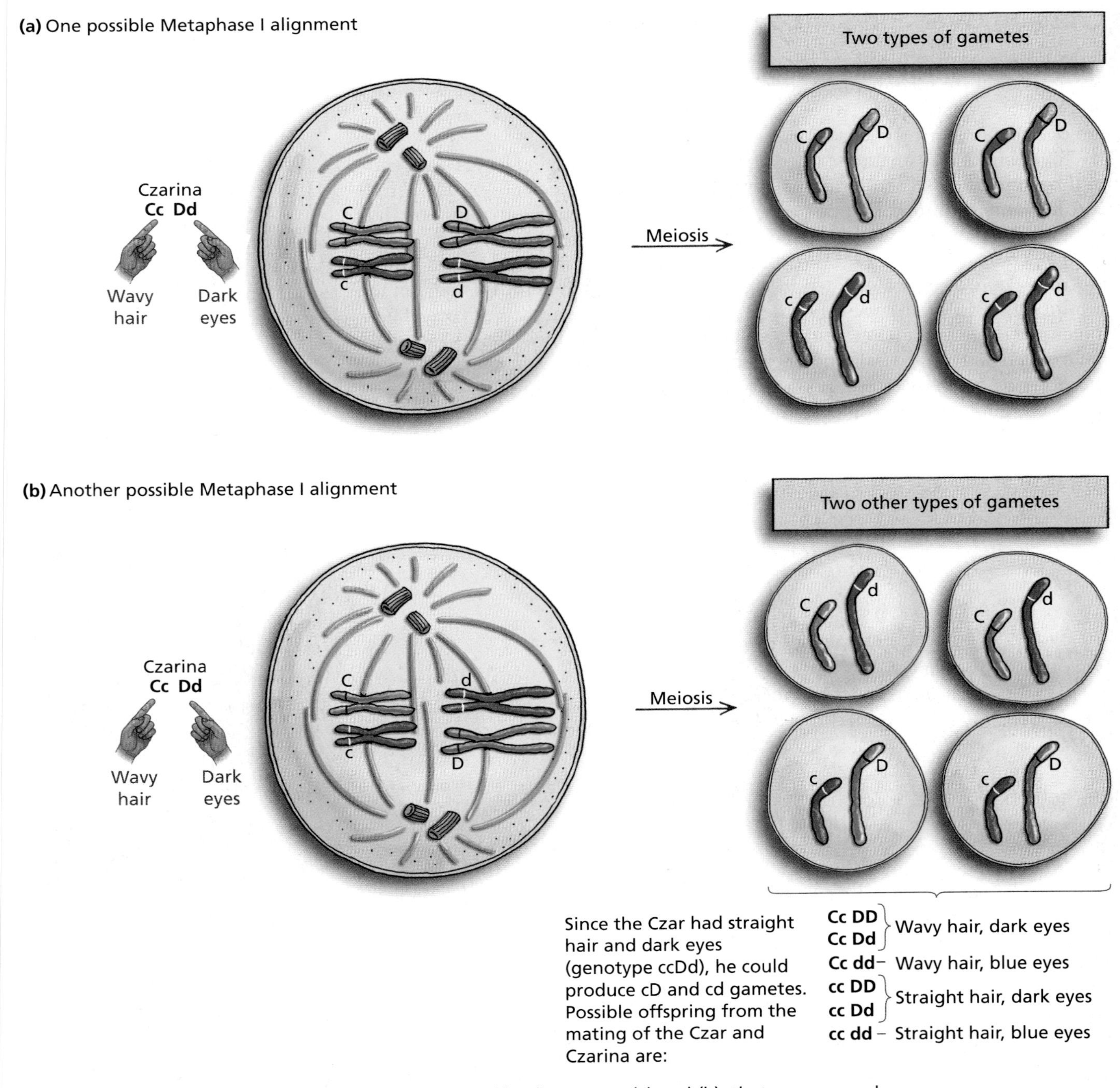

Figure 6.14 Random alignment. There are two possible alignments, (a) and (b), that can occur when there are two homologous pairs of chromosomes.

red hair and non-freckle chromosome, and the other containing the reciprocal non-red hair and freckle chromosome. Crossing over increases genetic diversity by increasing the number of distinct combinations of genes that may be present in a gamete.

Crossing over at prophase I, combined with random alignment of the homologues during metaphase I, ensures that virtually every gamete a person makes will be genetically different from any other gametes that person makes. Random alignment of the homologues is the reason alleles of a gene undergo *independent assortment* (Chapter 4), which we will illustrate in the following example.

The Czar and Czarina were able to produce five genetically distinct children (and could have produced millions more) partly due to events occurring during meiosis. Figure 6.14 diagrams a random alignment with just two chromosomes that carry some of the genes for hair texture and eye color. Curly hair (CC) is dominant over wavy (Cc) or straight hair (cc), and darkly pigmented eyes (DD or Dd) are dominant over blue eyes (dd). As you learned in Chapter 4, eye color is determined by three different genes. Because there are color photos of the Romanovs that allow us to determine their eye color and hair texture, we will use these traits to illustrate independent assortment. However, to simplify things, we will follow the path of only one of the three eye-color genes. The Czar had straight hair and dark eyes (ccDd), while the Czarina had wavy hair and dark eyes (CcDd). Since these genes are located on different chromosomes, together the royal couple could produce children with wavy hair and brown eyes (Tatiana and Maria), wavy hair and blue eyes (Anastasia and Olga), straight hair and blue eyes (Alexis), or straight hair and brown eyes.

The above example shows the genetic diversity than can be generated when only two genes on two different chromosomes are considered. The diversity generated from all 23 pairs of chromosomes is immense. This point can be illustrated using an analogy involving your classmates' shoes.

A pair of shoes is analogous to a homologous pair of chromosomes because the two shoes are similar in size, shape, and style, but are not exactly similar since they fit left and right feet (Figure 6.15). If you ask 23 students to take off their shoes and place them in a row across the front of the classroom, and they arrange their shoes so that the left shoe is on the left, and the right shoe is on the right, the students could then separate all of the left shoes from the right shoes, just as meiosis separates homologous chromosomes. This would produce one pile (gametes) containing all left shoes, and another pile containing all right shoes. Different piles of shoes would result if the very first pair of shoes was reversed so that the left shoe and right shoe exchange places, but the other 22 pairs of shoes stayed as they were. When the shoes are separated this time, one pile would have 22 right shoes and one left shoe, and the other pile would have 22 left shoes and one right shoe. The students could continue making different combinations of left and right shoes for a very long time, because there are 2^{23} (over 8 million) possible ways to line up these pairs of shoes.

The same is true of chromosomes. Consider that each of your parents was able to produce over 8 million genetically different gametes (not even considering crossing over) other than the ones that combined to make you. Hence, the odds of you receiving your particular combination of chromosomes is 1 in 8 million $\times$ 1 in 8 million—or 1 in 64 trillion. Remarkably, together your parents could have made over 64 trillion genetically different children, and you are only one of the possibilities. When you add the diversity that results from crossing over, this number becomes unimaginable. Thus, the Romanov parents could have produced millions upon millions of genetically different children.

In light of this, how was it possible for scientists to rule out Anna Anderson's claim to be a Romanov? Since each child can only receive genes that their parents carry, each band of the DNA fingerprint that is found in a child must also be present in one of the parents. This was not the case when Anna Anderson's DNA fingerprint was compared to those of the Romanov parents—Anna Anderson had DNA sequences in her fingerprint that she could not have received from either the Czar or Czarina.

Thus far scientists had answered three of the questions posed in Section 6.1: They had determined that there were nine different individuals buried in the Ekaterinburg grave; that, if these were indeed the Romanovs, the only son, Alexis, was missing from the grave along with one unspecified daughter; and two of the adult skeletons were the parents of the three children. The last question was still to be answered. How would the scientists show that these were bones from the Romanov family, not just some other set of related individuals? They again used DNA fingerprinting evidence, in conjunction with charts that outline family trees, called *pedigrees*.

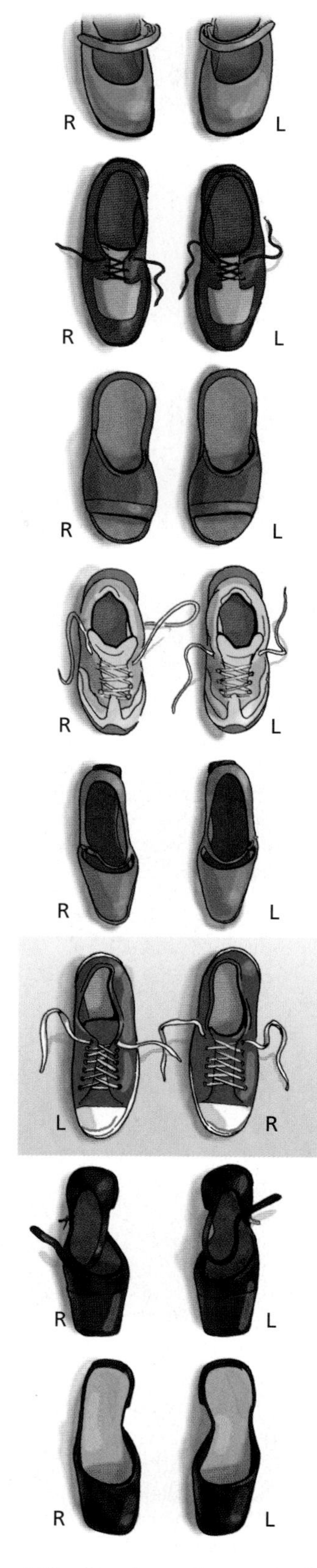

Figure 6.15 Random alignment of shoes. The diversity of left and right shoes is demonstrated if these shoes are separated into piles. Reversing the locations of left and right shoes increases the diversity within each pile of shoes.

6.4 Pedigrees

A **pedigree** is a family tree that follows the inheritance of a genetic trait for many generations. Information was available about the Romanovs' ancestors because they were royalty. In addition, the Romanov family tree had been mapped in detail by scientists who were interested in the inheritance of a rare genetic blood-clotting disorder called *hemophilia*. A person with hemophilia cannot produce the protein called clotting factor VIII. When this protein is absent, blood does not clot well and affected individuals bleed excessively, even from small cuts.

The Czarina's brother had this disease, as did two of her nephews. Historical records indicate that Alexis Romanov, the heir to the throne, had hemophilia and was so ill at the time of his death that his father had to carry him to his execution.

The hemophilia trait is linked to the X chromosome. Genes linked to the X (or Y) chromosome are called **sex-linked traits** because biological sex is inherited along with the X- or Y-linked gene. Most sex-linked traits are located on the X chromosome because it is much larger than the Y chromosome, which carries very little genetic information. Figure 6.16 outlines the inheritance of biological sex and the X-linked hemophilia gene. Since males have only one X chromosome,

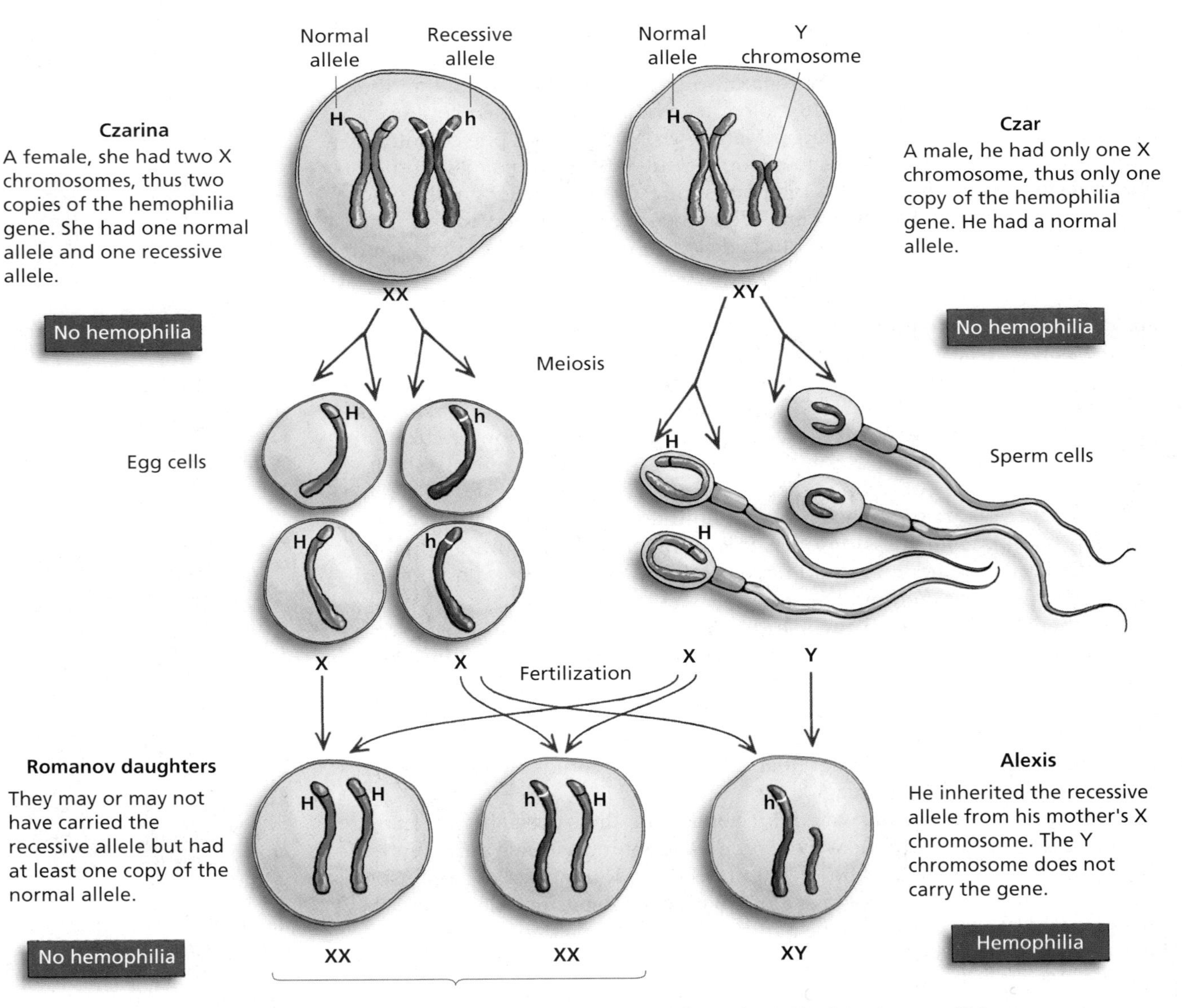

Figure 6.16 Sex determination and sex-linked genes. This illustration shows both the inheritance of biological sex and inheritance of the X-linked hemophilia trait.

they are much more likely to inherit X-linked traits because they have no other X chromosome with a normal allele to provide the missing protein. When recessively inherited traits are located on autosomes (non-sex chromosomes) there are two chances for the normal allele to be present and provide the required protein. For example, human females have two X chromosomes and require two recessive alleles to display the hemophilia phenotype.

Hemophilia was common among European royal families but rare amongst the rest of the population. This was because members of the royal families intermarried so as to preserve the royal bloodlines. Mating between related persons is called *inbreeding*. The offspring of inbred matings tend to have more health problems than offspring produced from unrelated individuals because rare, disease-causing alleles—such as those for hemophilia—are more likely to occur in related individuals. When royal cousins married, as was expected of the Romanovs, each member of the couple may have inherited the same rare recessive allele from the same grandparent. In fact, any time two related individuals mate, the chances that they both carry the same rare recessive allele is much greater than it would be if two unrelated individuals mated. Because people with rare genetic diseases are less likely to survive and pass on their disease, the offspring of inbred matings are selected against. Figure 6.17 is a pedigree that includes the Romanov family

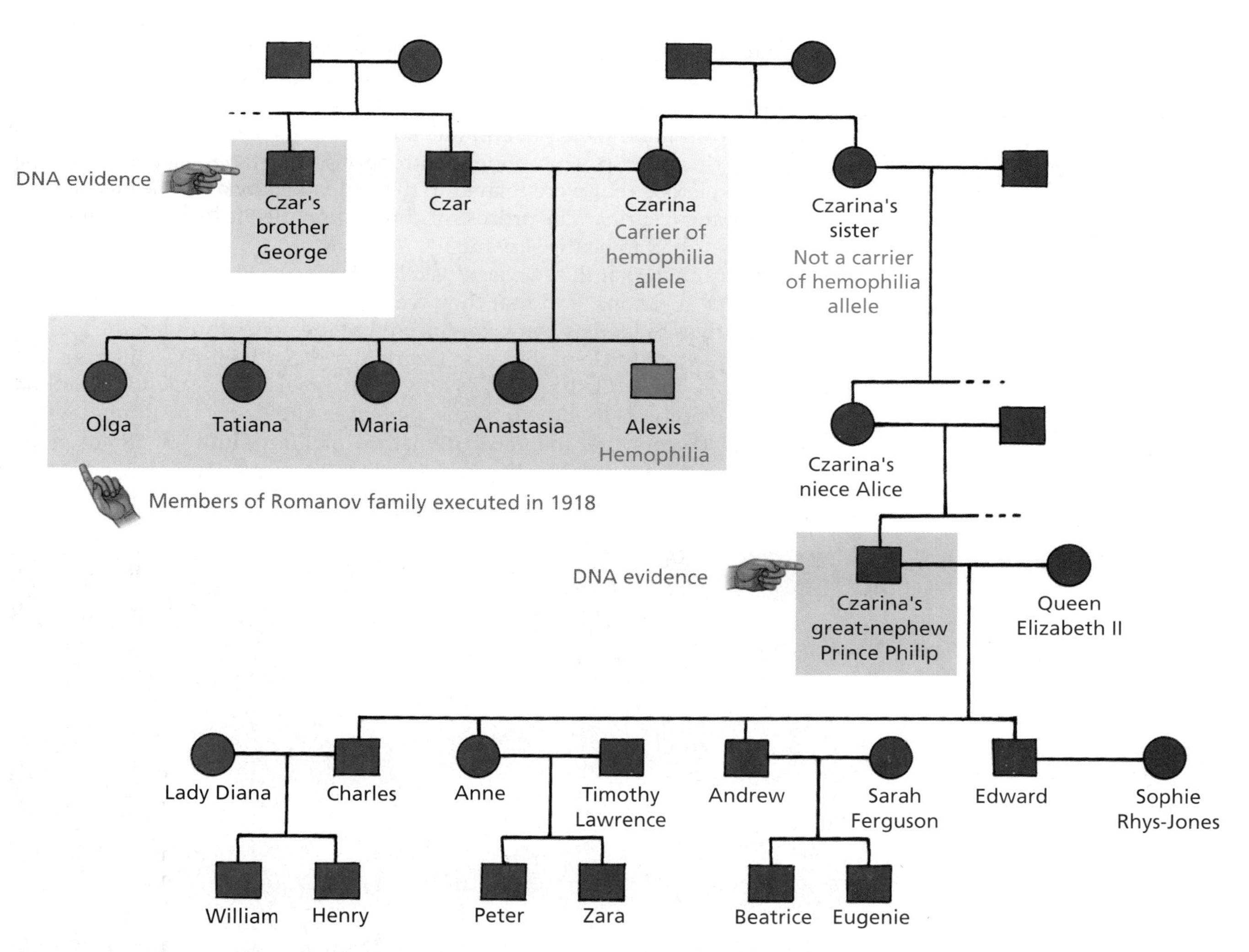

Figure 6.17 Romanov family pedigree. In this pedigree, only the pertinent family members are shown. DNA from the Czar's brother George showed that he was related to the Czar. Note that Prince Philip is the Czarina's great nephew. Prince Philip married Queen Elizabeth II and together they had four children, Charles, Anne, Andrew, and Edward, the current British royal family. Note also that the Czarina's sister does not appear to have been a carrier of the hemophilia allele, since none of her descendants have been affected by the disease.

Figure 6.18 Prince Philip. Husband of Queen Elizabeth II.

showing which individuals were affected with hemophilia. Note their connection to the current British royal family.

The next steps were crucial to conclusively establish the identity of the skeletons. DNA testing was performed on England's Prince Philip (Figure 6.18), who is a great-nephew of Czarina Alexandra. In addition, Nicholas II's deceased brother George was exhumed, and his DNA was tested as well. The DNA testing performed on these individuals showed that George was genetically related to one adult male skeleton, and one adult female skeleton was genetically related to Prince Philip. This evidence strongly supported the hypothesis that the adult skeletons that were related to the child skeletons were indeed those of the Czar and Czarina. The process of elimination dictates that the remaining four skeletons had to be the servants.

Having shown that the two adult skeletons belonged to the parents of the three smaller skeletons, and that they were genetically related to known Romanov relatives, scientists were convinced that the bones found in the Ekaterinburg grave were those of the Czar, Czarina, three of their five children, and their servants. Table 6.1 summarizes how scientists used the scientific method to test the hypothesis that the remains were indeed those of the Romanovs.

In 1998, 80 years after their execution, the Romanov family finally received a church burial (Figure 6.19). The people of post-communist Russia had symbolically laid to rest this part of their country's political history.

Figure 6.19 A church burial for the Romanovs. In 1998, the remains of the Romanov family found in Ekaterinburg were laid to rest.

Hypothesis: The bones found in the Ekaterinburg grave belonged to the Romanov family and their servants.	
Test	**Description of Results**
Analyze teeth	Royalty
Measure skeletons	6 adults and 3 children
Sex typing	One male child missing from grave
DNA fingerprinting	Children in grave are related to two adults in grave
DNA fingerprinting	Claims to be one of the missing Romanov children or their descendants disproved
DNA fingerprinting	The buried Romanovs are related to known Romanovs
Conclusion: When you look at each result individually, the evidence is less compelling than when you look at all the evidence together. As a whole, the evidence strongly supports the hypothesis that it was indeed the Romanovs who were buried in the Ekaterinburg grave.	

Table 6.1 The scientific method. A summary of tests and the conclusions that were drawn from them.

CHAPTER REVIEW

Summary

- Chromosomes and the DNA they carry are found in the nuclei of all cells.
- Karyotypes can be used to study chromosomes. Homologous pairs are lined up and photographed. The complement of sex chromosomes (XX or XY) are used to determine the biological sex of an individual.
- DNA is a double-stranded, helical molecule. Each strand of the helix is composed of nucleotides joined to each other: a sugar, a phosphate group, and a nitrogenous base (A, C, G, or T). Nitrogenous bases pair with each other across the width of the helix: A with T, and C with G.
- Every individual has their own unique DNA sequence.
- DNA fingerprinting is a technique that is used to show the relatedness of individuals based on similarities in their DNA sequences.
- Related individuals share similar DNA sequences because parents pass DNA to children via the process of meiosis.
- Meiosis is the type of cell division that occurs in cells that will produce gametes. Gametes are haploid and contain one member of each homologous pair from the parent cell.
- Prior to meiosis, DNA is replicated. When the two strands of the helix are pulled apart, each strand acts as a template for the synthesis of a new DNA strand. The DNA molecule splits up the middle and complementary base pairs are added to each parental strand of the helix.
- Meiosis I separates the members of a homologous pair from each other. Meiosis II separates the sister chromatids from each other.
- Crossing over and random alignment greatly increase the number of different kinds of gametes any individual can produce.
- Pedigrees are tools that scientists use to study the transmission of genetic traits among related individuals.

Key Terms

adenine p. 138
autosomes p. 135
base-pairing rule p. 138
chromosomes p. 135
crossing over p. 145
cytosine p. 138
deoxyribonucleic acid (DNA) p. 135
deoxyribose p. 138
diploid p. 142
DNA fingerprinting p. 138
DNA polymerase p. 144
gametes p. 142
guanine p. 138
haploid p. 142
homologous pairs p. 135
karyotype p. 135
linked genes p. 149
meiosis p. 141
nitrogenous base p. 138
nucleotides p. 138
pedigree p. 152
phosphodiester bond p. 138
random alignment p. 145
restriction enzymes p. 139
sex chromosomes p. 135
sex-linked trait p. 152
sugar-phosphate backbone p. 138
thymine p. 138

Learning the Basics

1. What enzyme facilitates DNA synthesis?

2. Describe the technique of DNA fingerprinting.

3. Describe the structure and function of the probe used in DNA fingerprinting.

4. State whether the chromosomal conditions in Figure 6.20 are haploid or diploid.

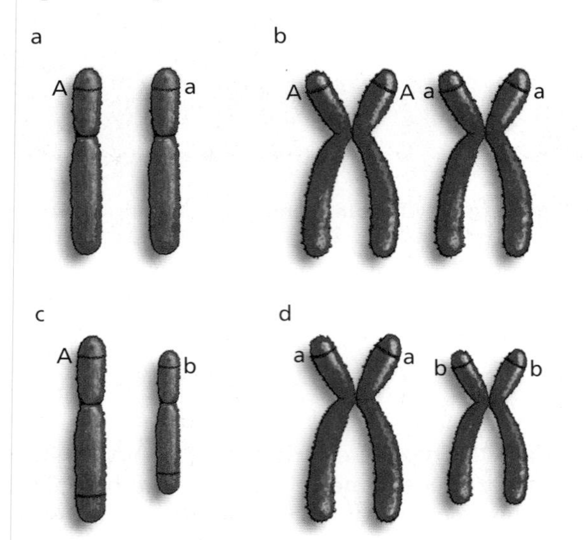

Figure 6.20

5. What three components make up a nucleotide?

6. Which of the following is *not* part of the procedure used to make a DNA fingerprint?
   a. DNA is treated with restriction enzymes.
   b. Cut DNA is placed in a gel to separate the various fragments by size.
   c. The genes that encode fingerprint patterns are cloned into bacteria.
   d. DNA from blood, semen, vaginal fluids, or hair root cells are used for analysis.
   e. An electrical current is used to separate DNA fragments.

7. Which of the following is consistent with this DNA fingerprint?

A	B	C	D
—			
	—	—	
—			—
—		—	—
	—	—	—
	—		—
—			—

   a. B is the child of A and C.
   b. C is the child of A and B.
   c. D is the child of B and C.
   d. A is the child of B and C.
   e. A is the child of C and D.

8. If you cut one linear segment of DNA with one restriction enzyme for which there are two cutting sites, _____ restriction fragments would be generated.
   a. 1
   b. 2
   c. 3
   d. 4
   e. 5

9. To bind to DNA, a probe must be _____.
   a. double stranded
   b. single stranded
   c. radioactive
   d. fluorescent

10. The function of restriction enzymes is to _____.
    a. add new nucleotides to a growing DNA strand
    b. produce double-stranded DNA
    c. join nucleotides together
    d. cleave nucleic acids at specific sites
    e. repair breaks in the sugar-phosphate backbone

11. After telophase I of meiosis, each daughter cell is _____.
    a. diploid, and the chromosomes are composed of one double-stranded DNA molecule
    b. diploid, and the chromosomes are composed of two sister chromatids
    c. haploid, and the chromosomes are composed of one double-stranded DNA molecule
    d. haploid, and the chromosomes are composed of two sister chromatids.

12. The pedigree in Figure 6.21 illustrates the inheritance of hemophilia (sex-linked recessive) in the royal family. What is the genotype of individual II-5 (Alexis)?
    a. X^HX^H
    b. X^HX^h
    c. X^hX^h
    d. X^HY
    e. X^hY

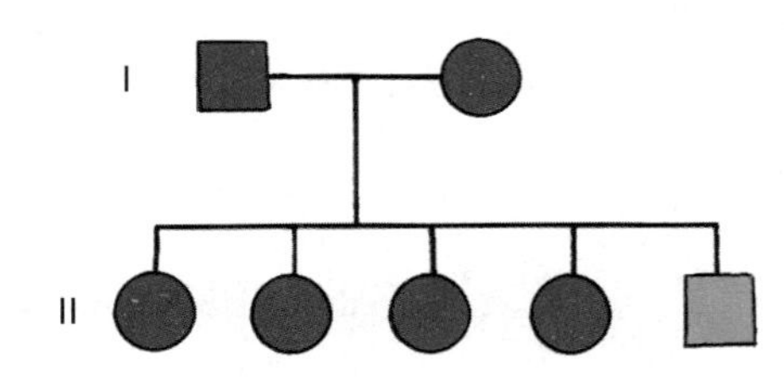

Figure 6.21

Analyzing and Applying the Basics

1. Make a rough sketch of the structure of a DNA molecule, and label each of the following: a nucleotide, a sugar, a nitrogenous base, a purine, a pyrimidine, a hydrogen bond, and a phosphodiester bond.

2. How is it possible that one person can produce millions of different types of gametes?

3. Draw a DNA fingerprint that might be generated by two sisters and their parents.

4. What is the order of nucleotides on the strand opposite this strand of a DNA molecule: ACTTTGACGGTAT?

5. Draw a pedigree of a mating between first cousins.

Connecting the Science

1. Should DNA fingerprinting evidence be required in cases of disputed parentage? Why or why not?

2. Science helped solve the riddle of who was buried in the Ekaterinburg grave, leading to a church burial for some of the Romanov family. In this manner, science played a role in helping the people of Russia come to terms with the brutal communist regime that followed the deaths of the royal family. Can you think of other examples where science has been used to help answer a question with great social implications?

3. Some of the Czar's living relatives have refused to allow their DNA to be tested. Why might they have made that decision?

4. Do you think that the routine use of DNA fingerprinting might be an infringement of privacy?

Media Activities

Media Activity 6.1 Meiosis

Estimated Time: 10 minutes

This visual overview of meiosis as a dynamic process starts with where it occurs in humans and follows the process to the resulting gametes. Includes an explanation of how meiosis produces a wealth of genetic diversity.

Media Activity 6.2 Examining the 3-D Structure of DNA

Estimated Time: 5 minutes

One of the remarkable attributes of DNA is its highly conserved three-dimensional double helix. Web-based modeling programs will allow you to view and study the structure of DNA from your computer.

Media Activity 6.3 Using DNA Fingerprinting

Estimated Time: 10 minutes

In this activity from PBS, students will first read a short introduction on DNA fingerprinting, and then solve the crime of the licked lollipop.

CHAPTER 7

Genetic Engineering

Gene Expression, Genetically Modified Organisms

Scientists who manipulate genes are often depicted as mad scientists.

7.1 Genetic Engineers

7.2 Genetic Engineers Can Use Bacteria to Synthesize Human Proteins

7.3 Genetic Engineers Can Modify Foods

7.4 Genetic Engineers Can Modify Humans

Can scientists bring extinct species back to life as seen in the movie *Jurassic Park*?

Will we some day be able to clone ourselves as in the movie *Multiplicity*?

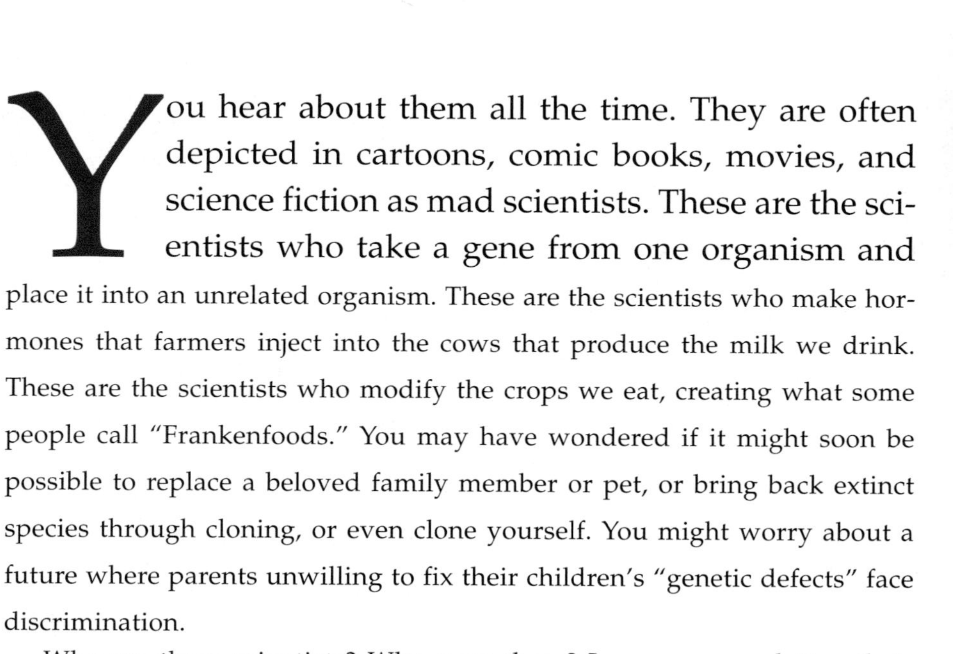

You hear about them all the time. They are often depicted in cartoons, comic books, movies, and science fiction as mad scientists. These are the scientists who take a gene from one organism and place it into an unrelated organism. These are the scientists who make hormones that farmers inject into the cows that produce the milk we drink. These are the scientists who modify the crops we eat, creating what some people call "Frankenfoods." You may have wondered if it might soon be possible to replace a beloved family member or pet, or bring back extinct species through cloning, or even clone yourself. You might worry about a future where parents unwilling to fix their children's "genetic defects" face discrimination.

Who are these scientists? Who pays them? Is anyone regulating their work? Is anyone trying to determine if it is unhealthy to eat these modified foods, whether genetically modified plants will cause environmental problems, or if genetically modified animals are less healthy than their counterparts?

With all kinds of unreliable information coming from diverse sources, it is often hard to separate fact from fiction. To help you sort this out, let us first

Will it be possible to give organisms superhuman powers as in *Spiderman*?

look at the scientists who are involved in manipulating genes, and then learn how they do what they do and about the regulations affecting their work. Finally, we will examine the real prospects and perils of genetic engineering.

7.1 Genetic Engineers

Genetic engineers are scientists who manipulate genes. They make their living working at colleges and universities, for the government, and for private companies. Most of them have had extensive training in genetics. The manipulations genetic engineers perform include changing the gene itself, changing how the gene is regulated (turned on or off), or moving the gene from one organism to another.

The training for the typical genetic engineer involves many years of schooling. After completing an undergraduate degree, some will obtain a master's degree, which takes two to three years and requires course work as well as a thesis research project. If the student does not continue past the master's level, he or she will probably work in a laboratory under the supervision of a more senior scientist.

Students who want to continue their education can apply to graduate schools with Ph.D. (Doctor of Philosophy) programs. Scientists holding a Ph.D. have the title of "Doctor" because they have a doctorate in their chosen field [a medical doctor (M.D.) has a doctorate in medicine]. A Ph.D. program involves more course work and an expanded research component; the results of this research must also be published in a peer-reviewed journal. Most Ph.D. scientists have gone to school five or more years after earning their undergraduate degree.

In scientific fields, graduate students generally get paid a small salary and have their tuition waived by the university. In exchange for tuition and salary, students work as teaching assistants overseeing undergraduate laboratory courses. If your biology course has a laboratory component, you may have had experience with a teaching assistant who is a graduate student in biology.

Most colleges and universities, especially the larger ones, expect faculty members to combine teaching with research. In this way, college professors not only pass information to the next generation; they also add to the knowledge base of their field.

For scientists in academia, obtaining money from granting agencies is imperative for funding their research. Scientists working at public universities have their academic-year salary paid by tax money allocated by the state government in exchange for teaching undergraduates. To support their research programs, scientists apply for grants by justifying the importance of their work and outlining the methods they plan to use to answer the scientific questions they have.

Grant money is provided on a competitive basis by various public and private granting agencies. Public agencies are supported by tax dollars and include the National Institutes of Health (NIH) and the National Science Foundation (NSF). Money from public agencies is subject to governmental control. Private agencies, such as charities and foundations, also award grants.

The federal government employs many biologists—for example, the National Cancer Institute (NCI) employs genetic engineers. Scientists who work in private industry are usually paid by the company they work for, either from profits earned on the products produced, or by private investors in the corporation. Genetic engineers who choose careers in the private sector do not have the opportunity to teach and therefore focus entirely on their research programs. They are likely to be involved in producing genetically modified crops, pharmaceuticals, or laboratory equipment and supplies.

Independent research laboratories are not directly connected with industry or universities. Genetic engineers working at these labs receive money to support their research from endowment funds that have been donated by individuals who wish to see their wealth used for the advancement of science, or from granting agencies. Jackson Laboratory in Bar Harbor, Maine, is famous for genetic studies, particularly on mice. The Mayo Clinic and Foundation in Rochester, Minnesota, is a hospital and world-renowned research center. Salk Institute in San Diego, California, employs scientists working on cancer and immunology research.

Genetic engineers in academia, government, and industry are involved in many different research projects. These projects vary from trying to produce a protein in the laboratory, to changing the genetic characteristics of crop plants, to trying to understand how human genes interact. One of the first genetic engineering projects to seize the attention of the public was the genetic engineering of a protein called *bovine (cow) growth hormone*, or *BGH*.

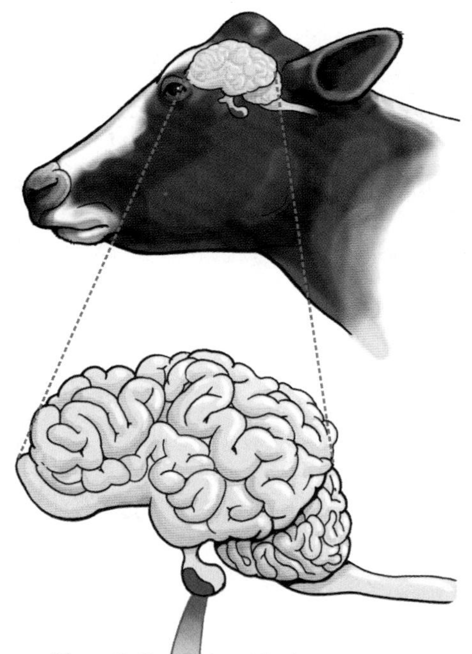

The pituitary gland is the natural source of bovine growth hormone (BGH)

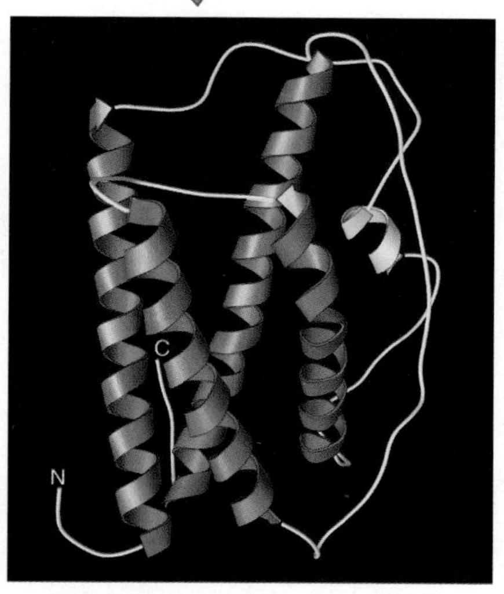

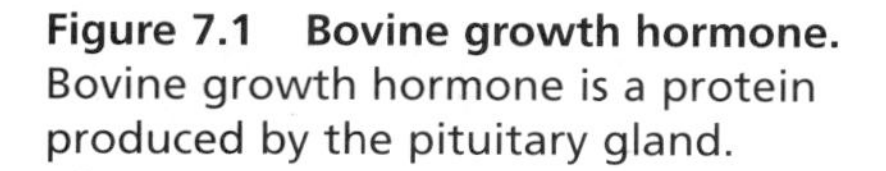

Figure 7.1 Bovine growth hormone. Bovine growth hormone is a protein produced by the pituitary gland.

7.2 Genetic Engineers Can Use Bacteria to Synthesize Human Proteins

During the early 1980s, genetic engineers at the Monsanto Corporation began to produce **recombinant** bovine growth hormone (*r*BGH) in their laboratories. Recombinant (*r*) bovine growth hormone is a protein that has been made by manipulating the **DNA** sequence (gene) that carries the instructions for, or encodes, the growth hormone protein so it can be produced in the laboratory. Hormones are substances secreted from specialized glands. Hormones travel through the bloodstream to affect their target organs. Growth hormone acts on many different organs to increase the overall size of the body.

Before the advent of genetic technologies, growth hormone was procured from the pituitary glands of slaughtered cows and then injected into live cows (Figure 7.1). The same technique has been used to obtain human growth hormone from the pituitary glands of human cadavers. When the human growth hormone is injected into humans who have a condition called *pituitary dwarfism*, their size increases. However, harvesting the growth hormone from the pituitary glands of cows and humans is laborious, and many cadavers are necessary to obtain small amounts of the protein.

Genetic engineers at Monsanto realized that they could produce large quantities of bovine growth hormone in the laboratory, inject dairy cows, and increase milk production, completely bypassing the less efficient surgical isolation of BGH. These scientists understood that if they were successful, Monsanto would stand to make a healthy profit from the dairy farmers who would buy the engineered growth hormone to increase milk yield. Let us examine how this recombinant protein is produced.

Producing *r*BGH

The first step in the production of the *r*BGH protein is to transfer the BGH gene from the nucleus of a cow cell into a bacterial cell. Bacteria with the BGH gene will then serve as factories to produce millions of copies of this gene and its protein product—making many copies of a gene is called **cloning** the gene.

Cloning a Gene Using Bacterial Cells The following steps are involved in moving a BGH gene into a bacterial cell (Figure 7.2):

1. The gene is sliced out of the cow chromosome on which it resides by exposing the cow DNA to enzymes that cut DNA. These enzymes are

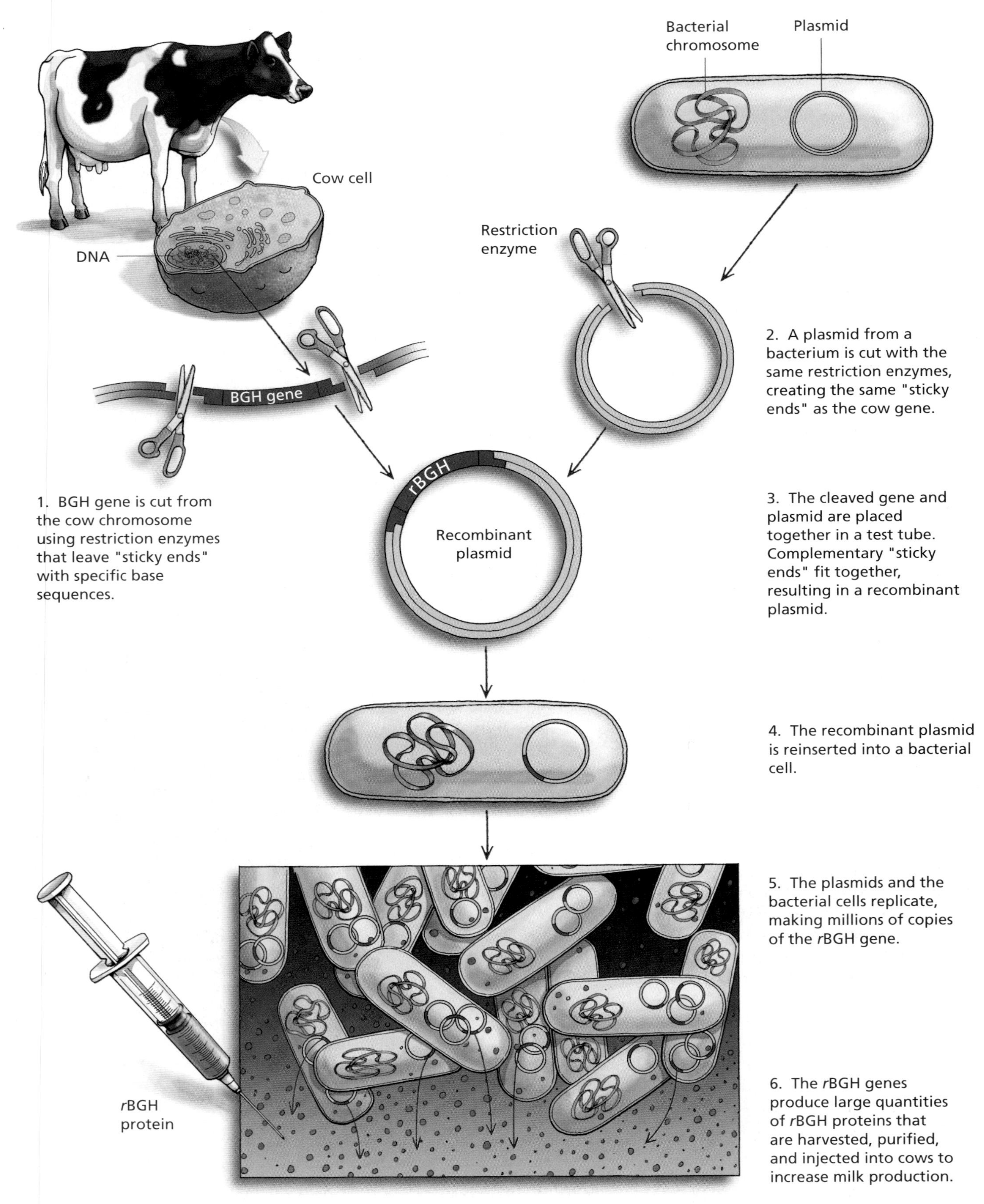

Bacterial chromosome
Plasmid
Cow cell
DNA
Restriction enzyme
BGH gene
rBGH
Recombinant plasmid
rBGH protein
1. BGH gene is cut from the cow chromosome using restriction enzymes that leave "sticky ends" with specific base sequences.
2. A plasmid from a bacterium is cut with the same restriction enzymes, creating the same "sticky ends" as the cow gene.
3. The cleaved gene and plasmid are placed together in a test tube. Complementary "sticky ends" fit together, resulting in a recombinant plasmid.
4. The recombinant plasmid is reinserted into a bacterial cell.
5. The plasmids and the bacterial cells replicate, making millions of copies of the *r*BGH gene.
6. The *r*BGH genes produce large quantities of *r*BGH proteins that are harvested, purified, and injected into cows to increase milk production.

Figure 7.2 Cloning genes using bacteria. Bacteria can be used as factories for the production of human or other animal proteins.

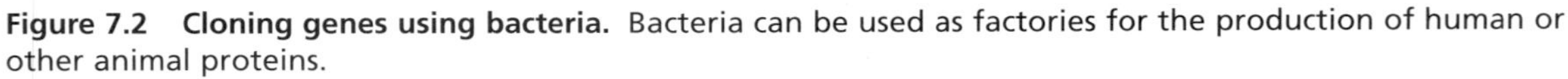

www

Media Activity 7.1A Creation of a Recombinant Plasmid

called **restriction enzymes** and act like highly specific molecular scissors. Individual restriction enzymes only cut DNA at specific sequences, such as:

Note that the bottom middle sequence is the reverse of the top sequence.

Many restriction enzymes cut the DNA in a staggered pattern, leaving "sticky ends," such as:

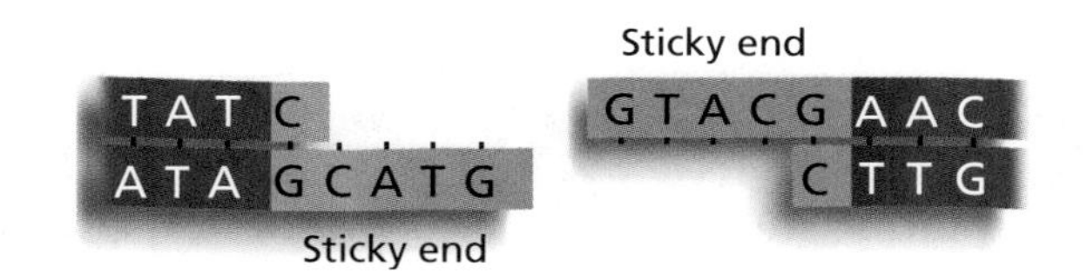

The unpaired bases form bonds with any complementary bases with which they come in contact. The enzyme selected by the scientist cuts on both ends of the BGH gene but not inside the gene.

Since different individual restriction enzymes cut DNA only at specific points, scientists need some information about the entire suite of genes present in a particular organism, the **genome**, to determine which restriction enzyme to use.

2. Once the gene is removed from the cow genome it is inserted into a bacterial structure called a **plasmid**. A plasmid is a circular piece of DNA that normally exists separate from the bacterial chromosome and can replicate itself. Think of the plasmid as a ferry that carries the gene into the bacterial cell where it can be replicated. In order to incorporate the BGH gene into the plasmid, the plasmid is also cut with the same restriction enzyme used to cut the gene. Cutting both the plasmid and gene with the same enzyme allows the "sticky ends" that are generated to base-pair with each other. As we discussed in Chapter 6, a base pair forms when A and T bind to each other and when G and C bind with each other. When the cut plasmid and the cut gene are placed together in a test tube they reform into a circular plasmid with the extra gene incorporated.

   The bacterial plasmid has now been genetically engineered to carry a cow gene. At this juncture, the BGH gene is referred to as the *r*BGH gene with the *r* indicating that this product is genetically engineered, or recombinant, DNA. It is called a *recombinant gene* because it has been removed from its original location in the cow genome and recombined with the plasmid DNA.

3. The recombinant plasmid is now inserted into a bacterial cell. Bacteria can be treated so their cell membranes become porous; when they are placed into a suspension of plasmids, the bacterial cells allow the plasmids back into the cytoplasm of the cell. Once inside the cell, the plasmids replicate, themselves, as does the bacterial cell, making thousands of copies of the *r*BGH gene. Using this procedure, scientists can grow large amounts of bacteria in a liquid that provides them with all the nutrients they need for replication.

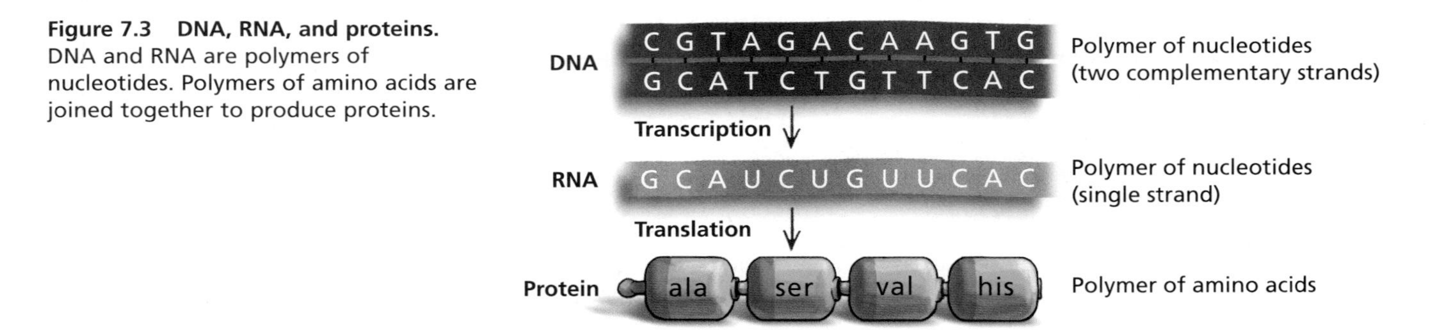

Figure 7.3 DNA, RNA, and proteins. DNA and RNA are polymers of nucleotides. Polymers of amino acids are joined together to produce proteins.

Once scientists successfully clone the BGH gene into bacterial cells, they allow the bacteria to produce the protein encoded by the gene. Producing a protein, or **protein synthesis**, involves using a gene to direct which amino acids will be joined together to produce a specific protein.

Protein Synthesis Understanding protein synthesis requires that you remember a few things from previous chapters. First, a *gene* is a sequence of DNA that encodes a protein. Second, DNA is a polymer of *nucleotides* (A, C, G, and T) that make chemical bonds with each other based on their complementarity (A:T, and C:G). You will see shortly that RNA is also a polymer of nucleotides, but it is usually a single strand. Finally, proteins are polymers of *amino acids* (Figure 7.3).

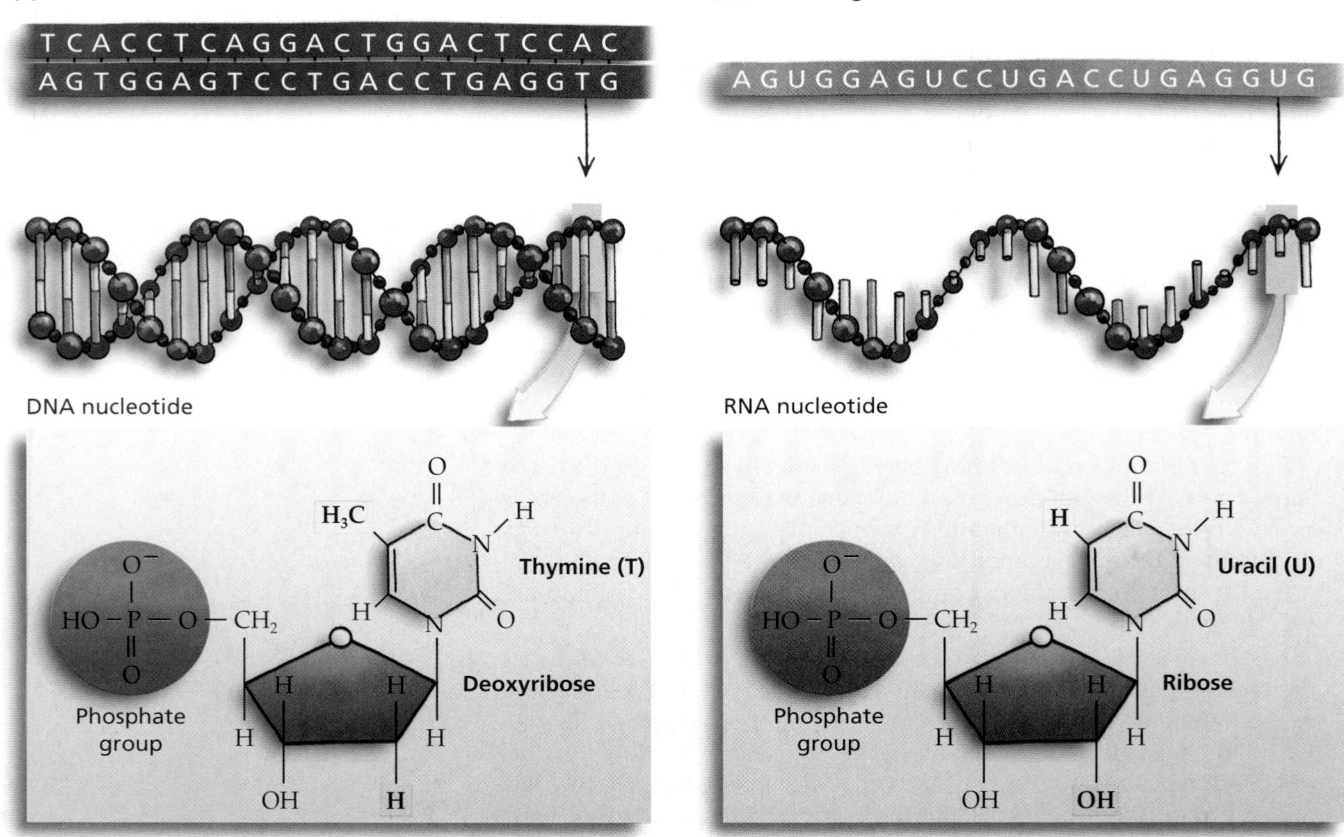

Figure 7.4 RNA and DNA. RNA and DNA are both nucleic acids. The building-block subunits of nucleic acids are the nucleotides. A nucleotide is a sugar, a phosphate, and a nitrogenous base. (a) The sugar in DNA is deoxyribose. (b) The sugar in RNA is ribose. Note that the nitrogenous base T in DNA is replaced with U in RNA.

It is important to understand the difference between *deoxyribonucleic acid* (DNA) and **ribonucleic acid (RNA)**. Recall from Chapter 6 that a nucleotide is composed of a sugar, a phosphate, and a nitrogen-containing base. For DNA nucleotides the sugar is *deoxyribose*, and the nitrogenous bases are A, C, G, and T. The nucleotides that join together to produce RNA are composed of the sugar **ribose**, a phosphate, and the nitrogenous bases A, C, G, and **U (uracil)**—there are no thymines (T) in RNA because U replaces them. In addition, RNA is usually single stranded, not double stranded like DNA (Figure 7.4). RNA is produced by using DNA as a template when a protein needs to be made. RNA nucleotides are able to make base pairs with DNA nucleotides: C and G make a base pair, and U pairs with A.

The process of protein synthesis is also referred to as **gene expression**, since proteins are synthesized when the genes that encode them are turned on. Proteins are only synthesized when a particular cell needs them.

Gene expression involves two main steps. The first step involves producing a copy of the DNA gene sequence, called **transcription**. This copy is synthesized by an enzyme called **RNA polymerase**. During transcription, the RNA polymerase rides along one strand of the DNA helix, along the nucleotides that comprise the gene. The RNA polymerase then ties together complementary RNA nucleotides from the surrounding nuclear fluid to produce a single-stranded RNA molecule that is a complementary copy of the gene. This complementary RNA copy of the DNA gene is called **messenger RNA (mRNA)**, since it carries the message of the gene that is to be expressed (Figure 7.5).

The second step from gene to protein requires that the mRNA be used to produce the actual protein encoded by the gene via a process called **translation**.

Translation occurs on structures called **ribosomes** that we first discussed in Chapter 2 (Figure 7.6a). The mRNA is threaded between the large and small subunits of the ribosome. As it moves through the ribosome, small sequences of mRNA are exposed. These sequences of mRNA are three nucleotides long and encode an amino acid; they are called **codons**.

When an mRNA codon reaches the reading site on the ribosome, structures called **transfer RNA (tRNA)** bind to it (Figure 7.6b). Transfer RNAs bind to codons through interactions between the RNA nucleotides at the base of the tRNA, a region called the **anticodon**, and the mRNA codon. The anticodon on

WWW
Media Activity 7.1C Transcription

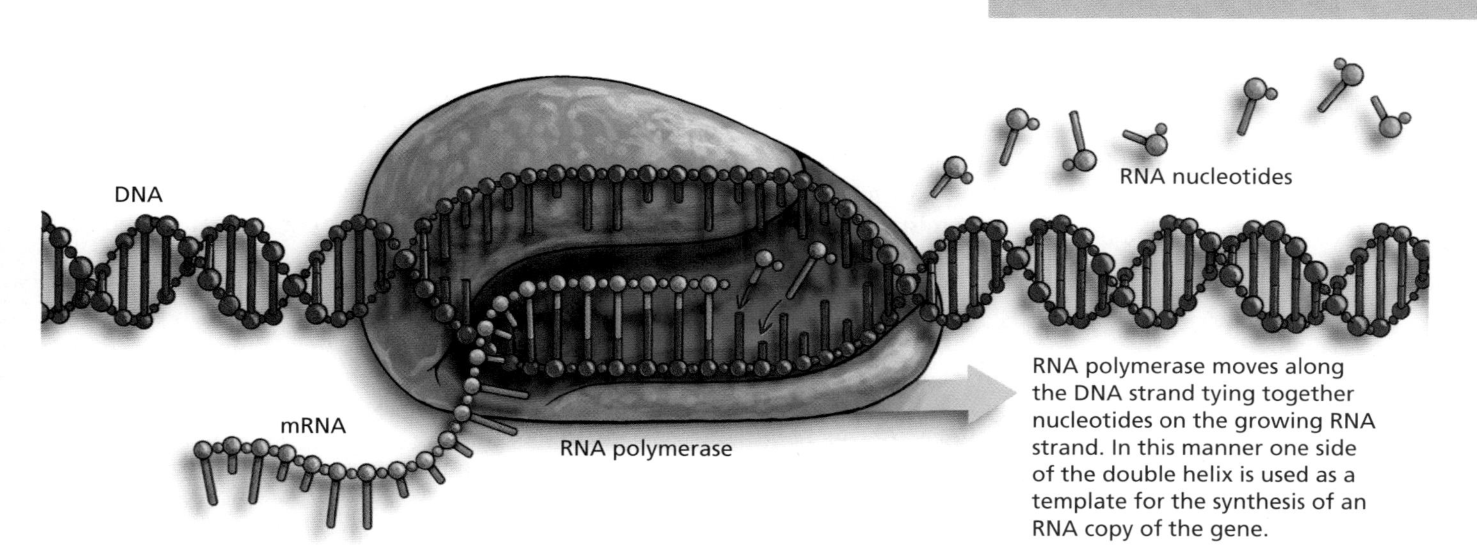

RNA nucleotides
DNA
mRNA
RNA polymerase
RNA polymerase moves along the DNA strand tying together nucleotides on the growing RNA strand. In this manner one side of the double helix is used as a template for the synthesis of an RNA copy of the gene.

Figure 7.5 Transcription. The enzyme RNA polymerase ties together nucleotides within the growing RNA strand as they bind to their complementary base on the DNA. Only when a complementary base pair is made between DNA and RNA does the polymerase add an RNA nucleotide to the growing strand. Complementary bases are formed via hydrogen bonding of A with U and G with C. When the RNA polymerase reaches the end of the gene, the mRNA transcript is released.

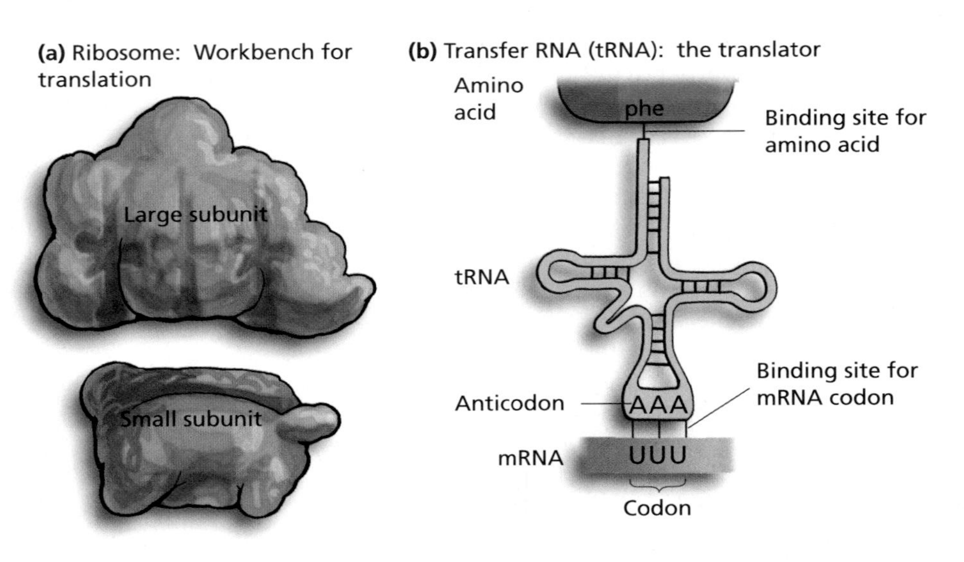

Figure 7.6 Translation. (a) The large and the small subunits of a ribosome work together to provide a "workbench" on which translation can occur. (b) Each tRNA has a specific amino acid bound to it and a specific anticodon that will base pair with the mRNA codon which specifies the required amino acid.

a particular tRNA binds to the complementary mRNA codon. Thus the codon calls for the incorporation of a specific amino acid (Figure 7.6c). When a tRNA anticodon binds to an mRNA codon, the ribosome adds the amino acid the tRNA is carrying to the growing chain of amino acids that will eventually constitute the finished protein. In this manner, the sequence of bases in the DNA dictates which amino acids will be joined together to produce a protein. Protein synthesis ends when a codon that does not call for an amino acid, called a **stop codon**, moves through the ribosome (Figure 7.6d). When a stop codon is present in the ribosome, no new amino acid can be added and the growing protein is released. Once released, the protein folds up on itself and moves to where it is required in the cell.

This is how cells determine which amino acid sequence a gene encodes. Scientists determine the sequence of amino acids a gene calls for by looking at a chart called a **genetic code** (Table 7.1). The genetic code shows which mRNA codons code for which amino acids. If scientists know the DNA sequence, they can use the base-pairing rules to determine the mRNA sequence that would be produced by transcription. A scientist can then use the genetic code to predict the amino acid sequence of the protein that the gene encodes.

In Chapter 4 we discussed how mutations, or changes to the DNA sequence, can produce different alleles. Therefore, these mutations result in the production of a protein different from the one originally called for. If this protein does not have the same amino-acid composition, it may not be able to perform the same job as the original protein (Figure 7.7). There are also cases in which a mutation will not have an effect on a protein. This may occur when amino acids with similar chemical properties are substituted for each other. Different genes result in the production of distinct proteins because different sequences of DNA bases code for the incorporation of different amino acids into each protein.

To help you understand protein synthesis, let us consider its similarity to an everyday process such as baking a cake. To bake a cake, you would consult a recipe book (genome) for the specific recipe (gene) for your cake (protein). You must copy the recipe (mRNA) out of the book so the original recipe (gene) does not become stained or damaged. The original recipe (gene) is left in the book on a shelf (nucleus) so that you can make another copy when you need it. The original recipe (gene) can be used over and over again. The copy of the recipe (mRNA) is placed on the kitchen counter (ribosome) while you assemble the ingredients (amino acids). The ingredients (amino acids) for your cake (protein) include flour, sugar, butter, milk, and eggs. The ingredients are measured in measuring spoons and cups (tRNAs) that are dedicated to one specific ingredient. Like the amino acids that are combined in different orders to produce a

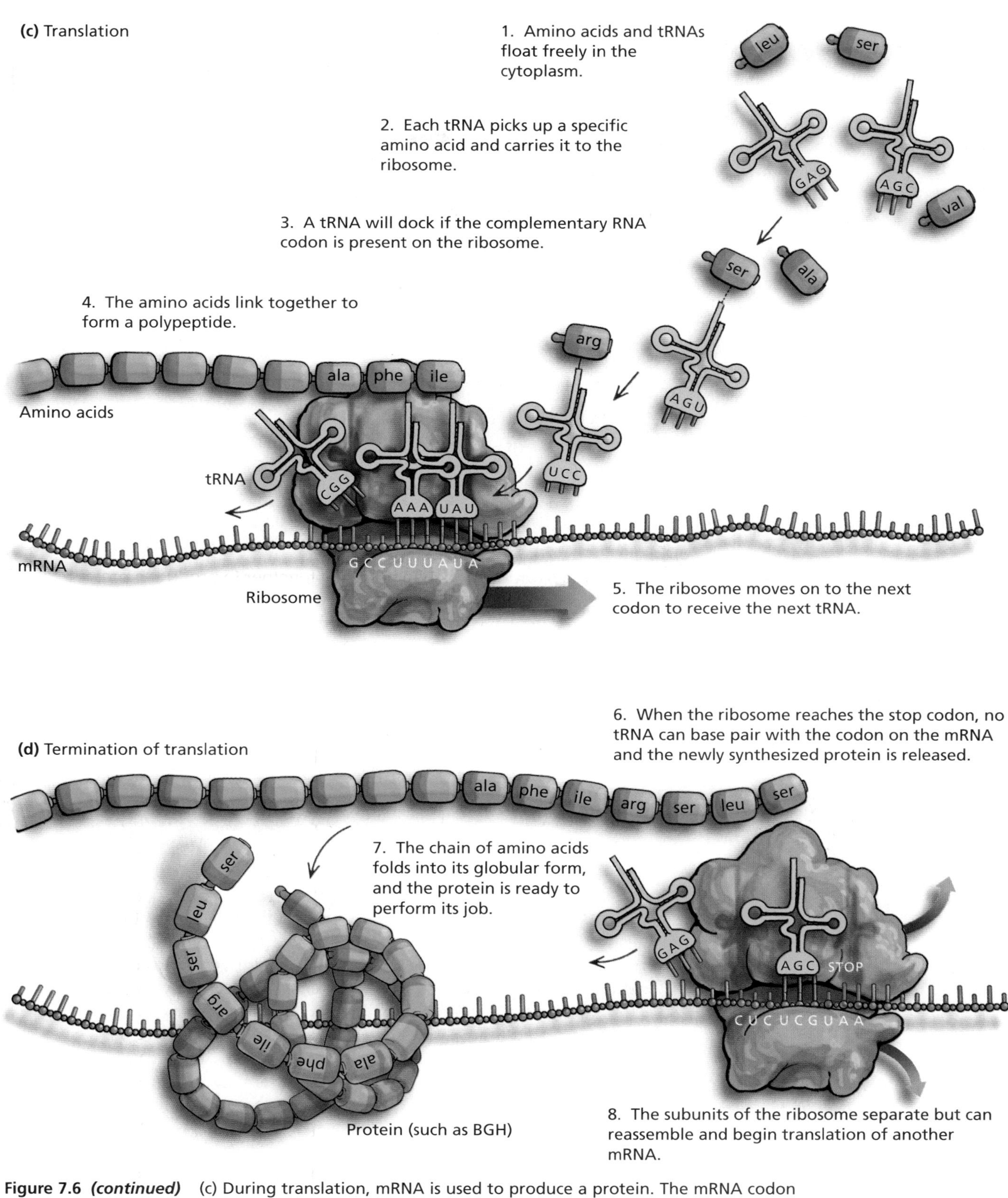

(c) Translation
1. Amino acids and tRNAs float freely in the cytoplasm.
leu
ser
2. Each tRNA picks up a specific amino acid and carries it to the ribosome.
GAG
AGC
val
3. A tRNA will dock if the complementary RNA codon is present on the ribosome.
ser
ala
4. The amino acids link together to form a polypeptide.
arg
AGU
ala
phe
ile
Amino acids
tRNA
CGG
AAA
UAU
UCC
mRNA
GCCUUUAUA
Ribosome
5. The ribosome moves on to the next codon to receive the next tRNA.
(d) Termination of translation
6. When the ribosome reaches the stop codon, no tRNA can base pair with the codon on the mRNA and the newly synthesized protein is released.
ala
phe
ile
arg
ser
leu
ser
7. The chain of amino acids folds into its globular form, and the protein is ready to perform its job.
GAG
AGC
STOP
CUCUCGUAA
Protein (such as BGH)
8. The subunits of the ribosome separate but can reassemble and begin translation of another mRNA.

Figure 7.6 ***(continued)*** (c) During translation, mRNA is used to produce a protein. The mRNA codon that is exposed in the ribosome binds to its complementary tRNA molecule, which carries the amino acid coded for by the DNA gene. When many amino acids are joined together, the required protein is produced. (d) When the translation machinery reaches a stop codon, the newly synthesized protein is released into the cytoplasm.

WWW

Media Activity 7.1D Translation

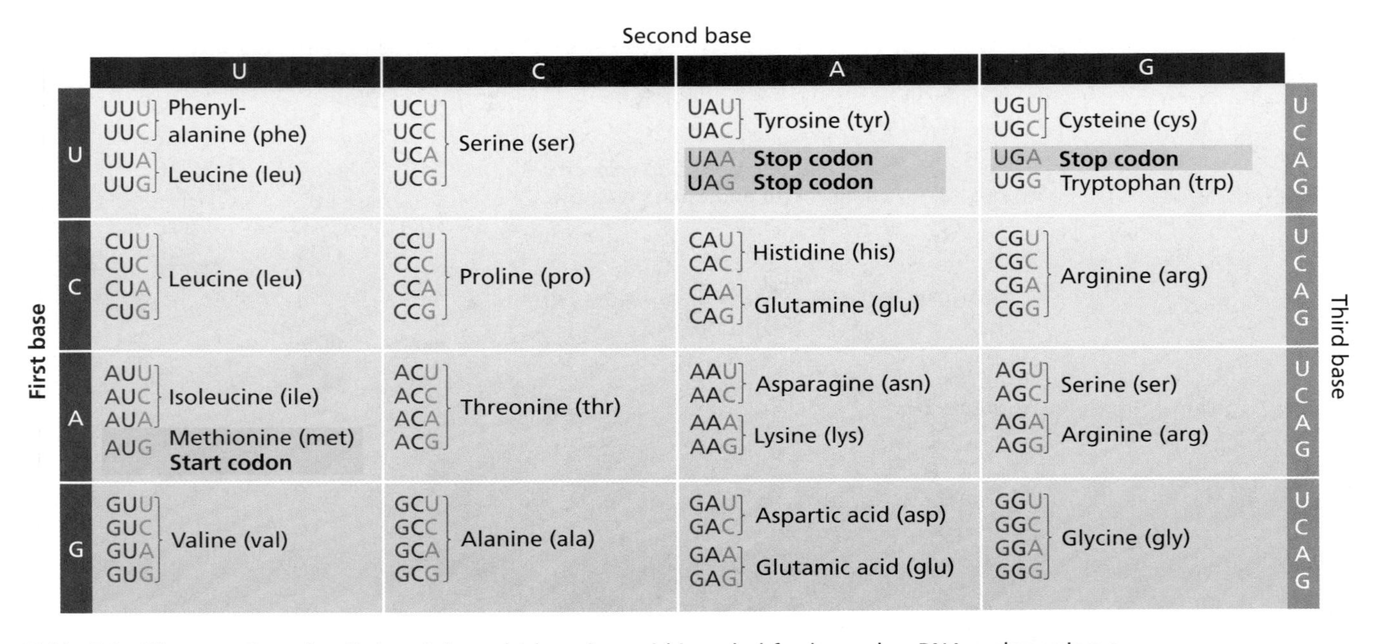

First base	Second base: U	Second base: C	Second base: A	Second base: G	Third base
U	UUU Phenylalanine (phe)	UCU Serine (ser)	UAU Tyrosine (tyr)	UGU Cysteine (cys)	U
U	UUC Phenylalanine (phe)	UCC Serine (ser)	UAC Tyrosine (tyr)	UGC Cysteine (cys)	C
U	UUA Leucine (leu)	UCA Serine (ser)	UAA **Stop codon**	UGA **Stop codon**	A
U	UUG Leucine (leu)	UCG Serine (ser)	UAG **Stop codon**	UGG Tryptophan (trp)	G
C	CUU Leucine (leu)	CCU Proline (pro)	CAU Histidine (his)	CGU Arginine (arg)	U
C	CUC Leucine (leu)	CCC Proline (pro)	CAC Histidine (his)	CGC Arginine (arg)	C
C	CUA Leucine (leu)	CCA Proline (pro)	CAA Glutamine (glu)	CGA Arginine (arg)	A
C	CUG Leucine (leu)	CCG Proline (pro)	CAG Glutamine (glu)	CGG Arginine (arg)	G
A	AUU Isoleucine (ile)	ACU Threonine (thr)	AAU Asparagine (asn)	AGU Serine (ser)	U
A	AUC Isoleucine (ile)	ACC Threonine (thr)	AAC Asparagine (asn)	AGC Serine (ser)	C
A	AUA Isoleucine (ile)	ACA Threonine (thr)	AAA Lysine (lys)	AGA Arginine (arg)	A
A	AUG Methionine (met) **Start codon**	ACG Threonine (thr)	AAG Lysine (lys)	AGG Arginine (arg)	G
G	GUU Valine (val)	GCU Alanine (ala)	GAU Aspartic acid (asp)	GGU Glycine (gly)	U
G	GUC Valine (val)	GCC Alanine (ala)	GAC Aspartic acid (asp)	GGC Glycine (gly)	C
G	GUA Valine (val)	GCA Alanine (ala)	GAA Glutamic acid (glu)	GGA Glycine (gly)	A
G	GUG Valine (val)	GCG Alanine (ala)	GAG Glutamic acid (glu)	GGG Glycine (gly)	G

Table 7.1 The genetic code. Determining which amino acid is coded for by each mRNA codon using a chart called the *genetic code:* Look at the left-hand side of the chart for the first-base nucleotide in the codon; there are four rows, one for each possible RNA nucleotide, A, C, G, or U. By then looking at the intersection of the second-base columns at the top of the chart and the first-base rows, you can narrow your search for the codon to four different codons. Finally, the third-base nucleotide in the codon on the right-hand side of the chart determines the amino acid a given mRNA codon codes for. Note the three codons, UAA, UAG, and UGA, that do not code for an amino acid; these are stop codons. The codon AUG is a start codon, found at the beginning of most protein-coding sequences.

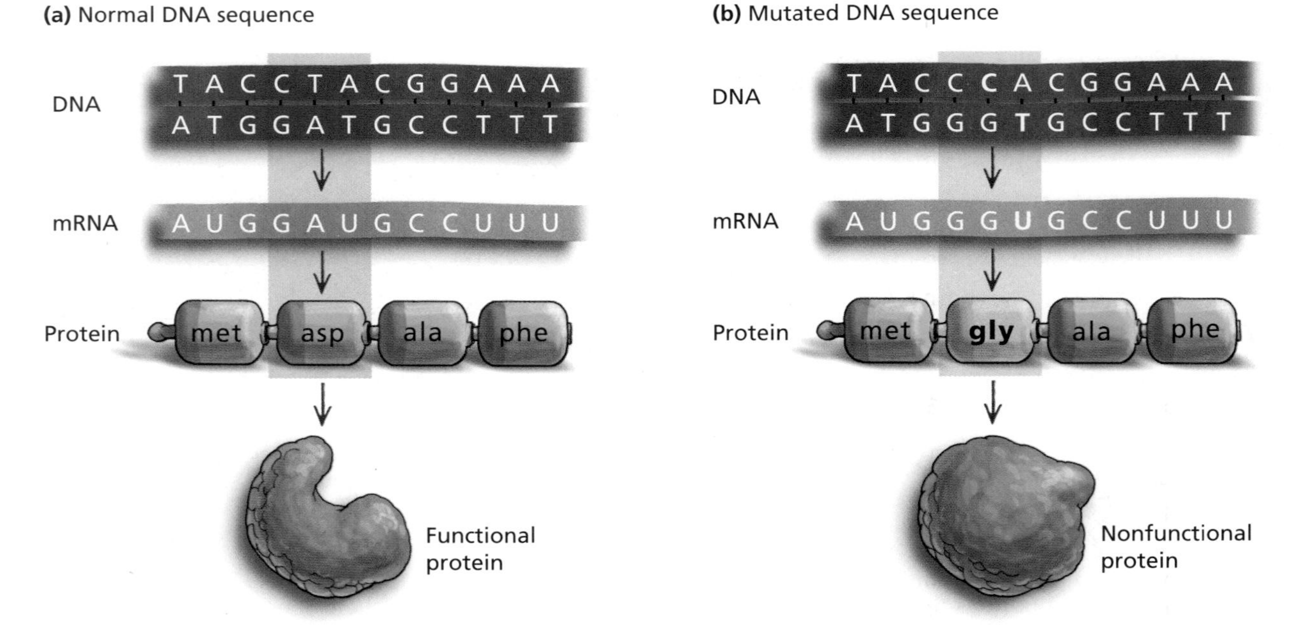

Figure 7.7 Mutation. A single nucleotide change from the normal sequence (a) to the mutated sequence (b) results in the incorporation of a different amino acid. If the substituted amino acid has different chemical properties than the original amino acid, the protein may assume a different shape and thus lose its ability to perform its job.

specific protein, the ingredients in a cake can be used many ways to produce a variety of foods. The ingredients (amino acids) are always added according to the instructions specified by the original recipe (gene).

The *r*BGH protein was synthesized inside bacterial cells. Scientists at Monsanto engineered the bacteria so that they could synthesize the *r*BGH protein by placing the growth hormone gene from cows into a bacterial plasmid. The plasmid was placed back into the bacterial cells, which then transcribed and translated the protein. The scientists were then able to break open the bacterial cells, isolate the BGH protein, and inject it into cows (Figure 7.8).

Figure 7.8 Cow receiving *r*BGH injection. Injections of *r*BGH are used to increase milk production in cows.

Close to one-third of all dairy cows in the United States now undergo daily injections with recombinant bovine growth hormone. These injections increase the volume of milk each cow produces by around 20%.

Prior to marketing the recombinant protein to dairy farmers, the Monsanto Corporation had to demonstrate that its product would not be harmful to cows or the humans that consume cows' milk. This involved obtaining approval from the United States Food and Drug Administration (FDA).

FDA Regulations

The FDA is the governmental organization charged with ensuring the safety of all domestic and imported foods and food ingredients (except for meat and poultry, which are regulated by the United States Department of Agriculture). The manufacturer of any new food that is not **Generally Recognized As Safe (GRAS)** must obtain FDA approval before marketing its product. Adding substances to foods also requires FDA approval, unless the additive is GRAS.

According to both the FDA and Monsanto, there is no detectable difference between milk from treated and untreated cows and no way to distinguish between the two. Even if there were increased levels of *r*BGH in the milk of treated cows, there should be no effect on the humans consuming the milk because we drink the milk, we do not inject it. Drinking the milk ensures that any protein in it will be digested by the body in the same manner as any other protein that is present in food. Therefore, milk from *r*BGH treated cows was deemed safe for human consumption by the FDA in 1993.

In addition, since the milk from treated and untreated cows is indistinguishable, the FDA does not require that milk obtained from *r*BGH-treated cows be labeled in any manner. Vermont is the only state that requires labeling of *r*BGH-treated milk. However, many distributors of milk from untreated cows label their milk as "hormone free," even though there is no evidence of the hormone in milk from treated cows (Figure 7.9).

Figure 7.9 Hormone-free milk. Some dairy companies label their milk as hormone free, although there is no proof that *r*BGH is harmful to humans.

It is not unusual that the majority of this work was performed for a corporation (Monsanto), not a university. There are some fundamental differences between the types of research performed for industry versus that performed at universities and colleges.

Basic versus Applied Research

Scientists in academia often seek answers to questions for which there is no profit motive. Research for which there is not necessarily a commercial application is called **basic research**. This research is largely funded by taxpayers through the NIH or NSF. The premise behind basic research is that we cannot always predict which kinds of scientific understanding will be valuable to society in the future. For instance, basic-research scientists might study transcription or translation simply to better understand the process. Genetic engineers may spend their entire careers trying to understand the conditions under which a particular protein is synthesized.

Funding for basic research is important because we do not know where the next piece of invaluable information will come from. When scientists first began studying the genes in the single-celled eukaryote *Saccharomyces cerevisiae* (bakers' yeast) they probably had no idea that most of the genes present in this yeast were also present in humans. Today, scientists manipulate the environmental conditions of yeast to better understand how genes are regulated in humans. Likewise, scientists interested in studying the diversity of tropical plants (Chapter 3) assisted in the development of many pharmaceutical agents, including some of the anticancer agents described in Chapter 5.

Scientists in industry typically seek to answer questions that will have an immediate and profitable application, like the production of *r*BGH. This more **applied research** is important for scientists in industry because new products and improvements to existing ones increase profitability, which in turn determines the success or failure of the business. One application of applied research that has proven to be very lucrative has been the genetic engineering of crop plants.

www
Media Activity 7.2 **Bioengineering New Products**

7.3 Genetic Engineers Can Modify Foods

Whether you realize it or not, you have been eating genetically modified foods for some time now. This may lead you to wonder why and how plants are genetically modified, whether eating them is bad for your health, or whether growing them is bad for the environment.

Why Are Crop Plants Genetically Modified?

Crop plants are genetically modified to increase their shelf life, yield, and nutritive value. The first genetically engineered fresh produce, tomatoes, were available in American grocery stores in 1994. These tomatoes were engineered to soften and ripen more slowly. The longer ripening time meant that tomatoes would stay on the vine longer, thus making them taste better. The slower ripening also increased the amount of time the tomatoes could be left on grocery store shelves without becoming overripe and mushy. An enzyme called *pectinase* mediates the ripening process in some produce, including tomatoes. This enzyme breaks down pectin, a naturally occurring substance found in plant cells. When the enzyme pectinase is active, it helps break down the pectin and the produce softens.

In tomatoes, genetic engineers inserted a gene that produces an mRNA transcript complementary to the mRNA produced by transcription of a pectinase gene. In double-stranded DNA, the strand that codes for a protein is called the **sense** strand, and its complement is called the **antisense** strand. When the antisense version of the pectinase gene is transcribed, it produces an mRNA that is complementary to the mRNA from the normally transcribed (sense strand) of the pectinase gene. When the mRNA from the genetically engineered antisense gene base pairs with its naturally occurring pectinase complement, ripening is slowed (Figure 7.10). Binding the antisense and sense mRNAs leaves less of the sense pectinase mRNA available for translation. Thus, less of the pectinase enzyme is produced and ripening occurs more slowly.

Improving the yield of crop plants has been the driving force behind the vast majority of genetic engineering. Yield can be increased when plants are engineered to be resistant to pesticides and herbicides, drought, and freezing. For example, a gene from an Arctic fish has been transferred into a strawberry to help prevent frost damage.

Many people believe that improving farmers' yields may help decrease world hunger problems. Others argue that, since there is already enough food

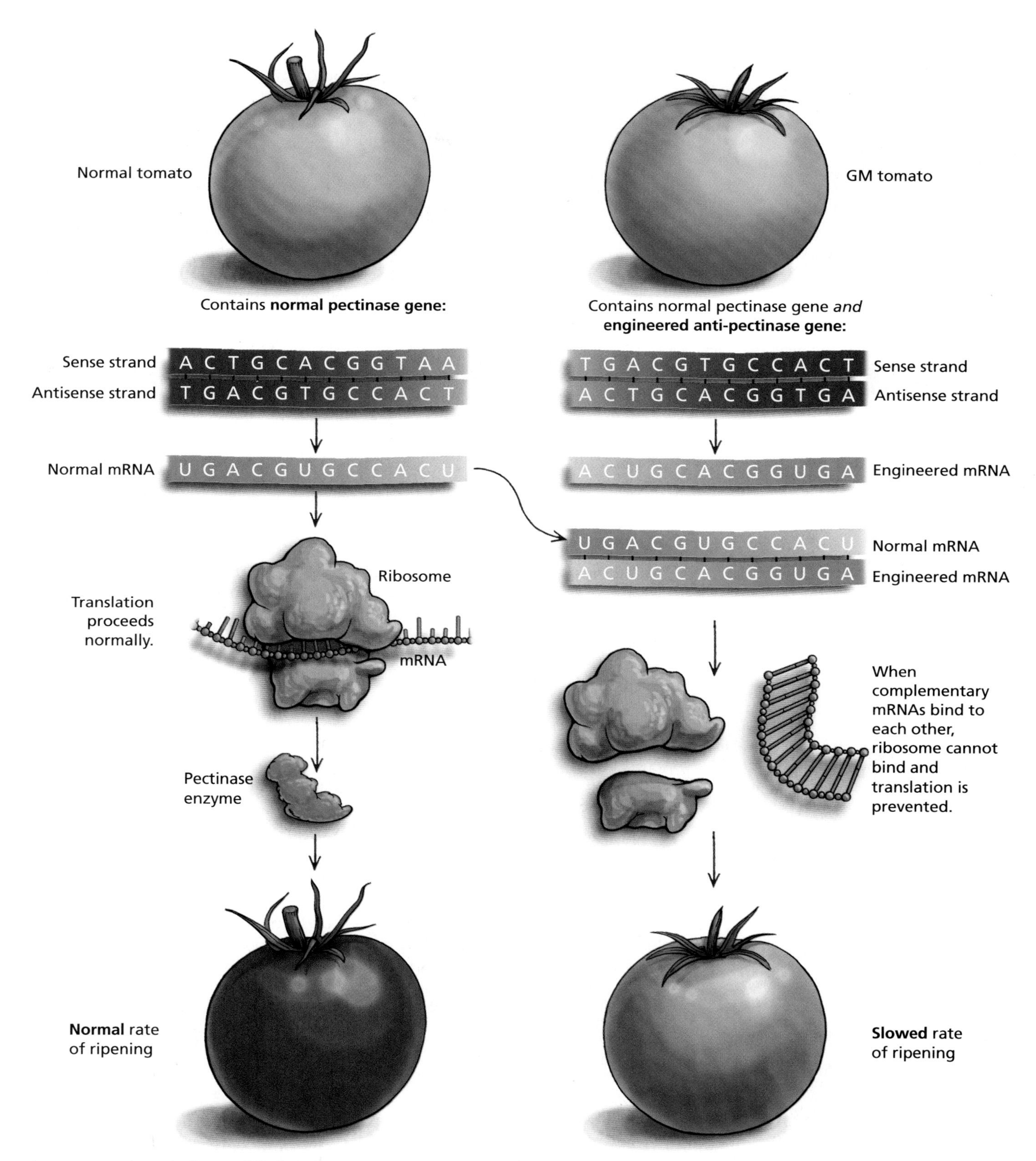

Figure 7.10 Genetically modified tomato. Genetically modified tomatoes produce mRNA that decreases the effects of the pectinase gene. When the normal, sense version of the pectinase gene is transcribed and translated, ripening occurs. When the pectinase mRNA is bound to the engineered antisense version, translation does not occur, and ripening is slowed.

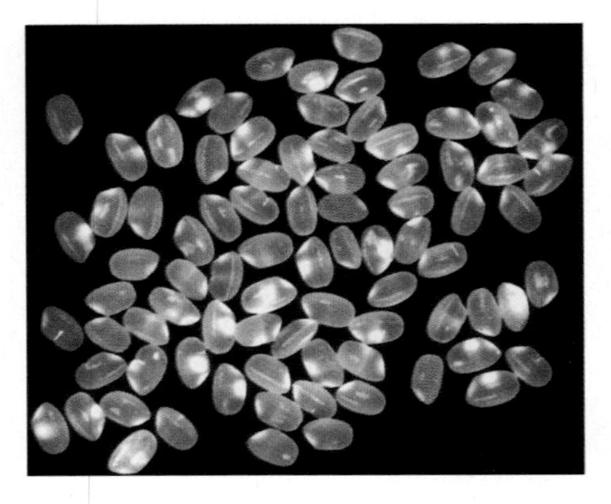

Figure 7.11 Golden Rice. Golden Rice has been genetically engineered to produce more β-carotene. The increased concentration of β-carotene makes the rice look more gold in color than nonmodified rice.

being produced to feed the entire population, it might make more sense to use less technological approaches to feeding the hungry. Significant numbers of people around the world are malnourished, hungry, or starving, not due to a shortage of food but because access to food is tied to access to money or land. However, as the population increases, it may some day be imperative to increase the yield of crop plants in order to feed all the world's people.

Genetic engineers may also be able to increase the nutritive value of crops. Some genetic engineers have increased the amount of β-carotene in rice, a staple food for many of the world's people. Scientists hope the engineered rice will help decrease the number of people who become blind in underdeveloped nations because cells require β-carotene in order to synthesize vitamin A, a vitamin required for vision. Therefore, eating this genetically modified rice, called Golden Rice, increases a person's ability to synthesize vitamin A (Figure 7.11).

How Are Crops Genetically Modified?

To modify crop plants, the gene must be able to gain access to the plant cell, which means it must be able to move through the plant's rigid, outer cell wall. The "ferry" for moving genes into flowering plants is a naturally occurring plasmid of the bacterium *Agrobacterium tumefaciens.* In nature, this bacterium infects plants and causes tumors called *galls* (Figure 7.12a). The tumors are induced by a plasmid, called **Ti plasmid** (for *T*umor *i*nducing).

Genes from different organisms can be inserted into the Ti plasmid by using the same restriction enzyme to cut the Ti plasmid and the gene, and then connecting the plasmid and the gene together and reinserting it into the bacterium. *A. tumefaciens*, with the recombinant Ti plasmid, is then used to infect plant cells. During infection the recombinant plasmid is transferred into the host plant cell (Figure 7.12b). For genetic engineering purposes, scientists use only the portion of a plasmid that does not cause tumor formation.

Moving genes into other agricultural crops such as corn, barley, and rice can also be accomplished by using a device called a **gene gun**. A gene gun shoots tungsten-coated pellets covered with foreign DNA into plant cells (Figure 7.13). A small percentage of these DNA genes may be incorporated into the plant's genome. The gene gun is often used by companies that do not want to pay licensing fees to Monsanto, holder of the *A. tumefaciens* patent.

When a gene from one organism is incorporated into the genome of another organism, a **transgenic organism** is produced. A transgenic organism is commonly referred to as a **genetically modified organism** or **GMO**.

Many people have raised concerns about genetically modified (GM) crop plants. One concern is that large corporations that own many farms, called *agribusiness* corporations, profiting from GM crop production will put owners of family farms out of business. Other concerns focus on the impact of GMOs on human health and the environment (Figure 7.14).

Figure 7.12 Genetically modifying plants. (a) Plants infected by *Agrobacterium tumefaciens* in nature show evidence of the infection by producing tumors called *galls.* (b) The Ti plasmid from *A. tumefaciens* serves as a shuttle for incorporating genes into plant cells. It can be engineered to carry a gene from a different organism and not produce galls. The recombinant plasmid is then used to infect developing plant cells, producing a genetically modified plant. When the plant cell reproduces, it may pass on the engineered gene to its offspring.

(a) Gall caused by *A tumefaciens*

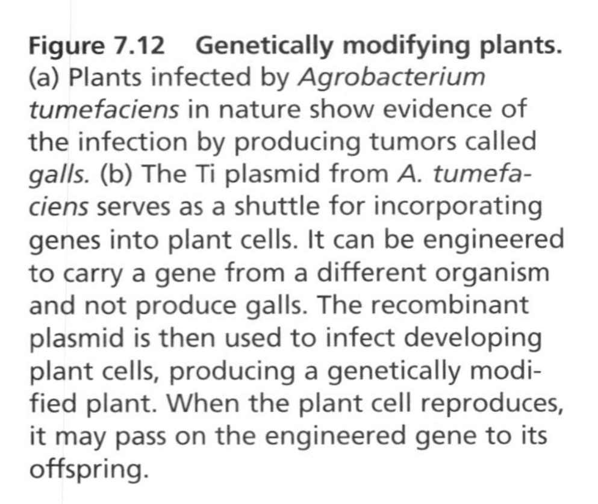

(b) Using the Ti plasmid

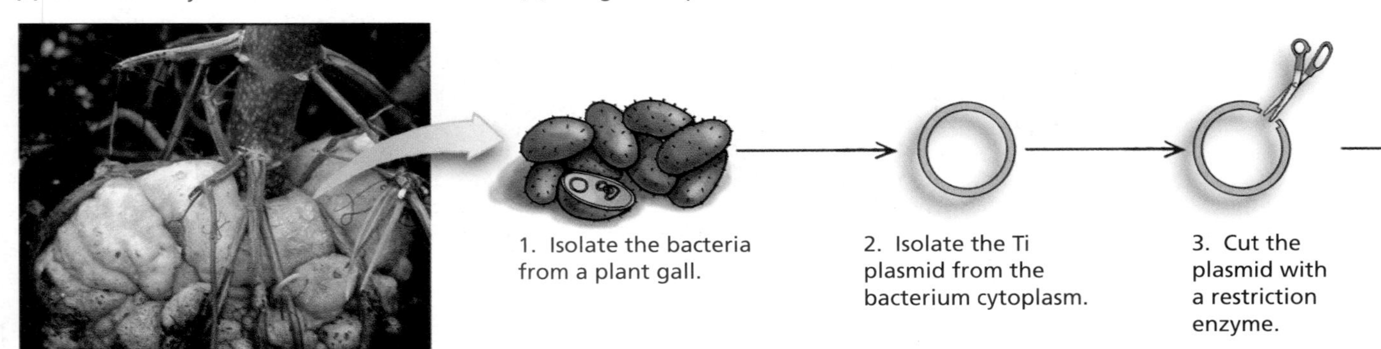

1. Isolate the bacteria from a plant gall.
2. Isolate the Ti plasmid from the bacterium cytoplasm.
3. Cut the plasmid with a restriction enzyme.

(a) Gene gun

(b) How the gene gun works

Gun

Shockwaves

"Bullet"

Microscopic particles coated with gene of interest are "shot" into plant cells.

Plant cells in culture

Figure 7.13 Gene gun. (a) A gene gun shoots a plastic bullet loaded with tiny metal pellets coated with DNA into a plant cell. The bullet shells are prevented from leaving the gun, but the DNA-covered pellets penetrate the cell wall, cell membrane, and nuclear membrane of some cells (b).

Figure 7.14 Protesters at a World Trade Organization meeting in Seattle. These people are concerned about how GMOs may affect humans and the environment.

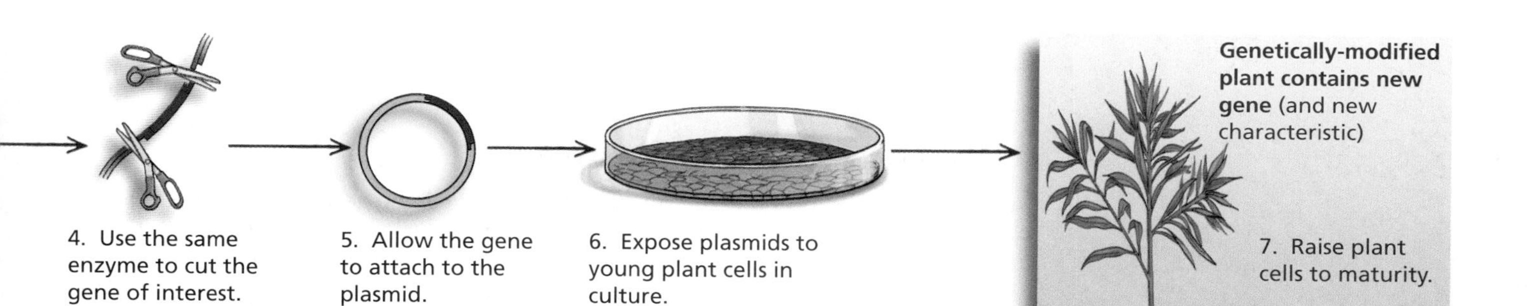

4. Use the same enzyme to cut the gene of interest.
5. Allow the gene to attach to the plasmid.
6. Expose plasmids to young plant cells in culture.
Genetically-modified plant contains new gene (and new characteristic)
7. Raise plant cells to maturity.

Figure 7.15 "No GMO" labelling. Many food manufacturers and consumers consider the use of unmodified foods to be a selling point for their products.

GMOs and Health

Concerns about the potential negative health effects of consuming GM crops have led some citizens to fight for legislation requiring that modified foods be labeled so consumers can make informed decisions about what foods they choose to eat. The manufacturers of GM crops argue that labeling foods is expensive and will be viewed by consumers as a warning, even in the absence of any proven risk. They believe that this will decrease sales and curtail further innovation.

While the labeling controversy rages, the rate at which genetically modified foods floods the market increases. Most of the corn and soy used in cattle feed is genetically engineered, as is much of the canola oil, squash, and potatoes that humans consume. Even processed foods such as bread and pasta contain grains that have been engineered. Products that do not contain GMOs are often labeled to promote that fact (Figure 7.15).

Genetically modified crop plants must be approved by the Environmental Protection Agency (EPA) prior to their release into the environment. The FDA becomes involved in testing the GM crop only when the food the gene comes from has never been tested, or when there is reason to be concerned that the newly inserted gene may encode a protein that will prove to be a toxin or allergen.

If the gene being shuffled from one organism to another is not known to be toxic or cause an allergic reaction, the FDA considers it to be substantially equivalent to the foods from which they were derived, that is, GRAS. If a modified crop contains a gene derived from a food that has been shown to cause a toxic or allergic reaction in humans, it must undergo testing prior to being marketed.

This method of determining potential hazard worked well in the case of a modified soybean that carried a gene from the Brazil nut. This engineering was done in an effort to increase the protein content of soybeans. Since Brazil nuts were known to cause allergic reactions in some people, the modified beans were tested and did indeed cause an allergic reaction in susceptible people. The product was withdrawn and no one was harmed.

Proponents of genetic engineering cite this as an example of the efficacy of the FDA rules. Opponents of genetically modifying foods wonder whether it will always be possible to predict which foods to test. They point out that it is possible for a protein encoded by a gene with no history of toxicity or allergenicity, to interact with substances in its new environment in unpredictable ways.

In terms of toxicity, scientists focus on the protein produced by the modified plant and not the actual gene that is inserted. This is because the gene itself will be digested and broken down into its component nucleotides when it is eaten, and therefore will not be transcribed and translated inside human cells.

Allergy is a serious problem for the close to 8% of Americans that experience allergic reactions to foods. Symptoms of food allergy range from mild upset to sudden death. Genetic engineers must be vigilant about testing foods with known allergens; a person who knows to avoid peanuts may not know to avoid a food that has been genetically modified to contain a peanut gene.

Concern about GM foods is not limited to their consumption. Many people are also concerned about the effects of GM crop plants on the environment.

GM Crops and the Environment

Aside from Golden Rice, most crops that have been genetically modified have been modified to increase their yield. For centuries, farmers have tried to increase yields by killing the pests that damage crops and by controlling the growth of weeds that compete for nutrients, rain, and sunlight. In the United States, farmers typically spray chemical pesticides and herbicides directly on

Figure 7.16 Crop dusting. Pesticides and herbicides are sprayed on crops to improve yields and are found on the surface of many crop plants. The chemicals also leach into the soil, where they can contaminate the groundwater.

to their fields (Figure 7.16). This practice concerns people worried about the health effects of eating foods that have been chemically treated. In addition, both pesticides and herbicides leach through the soil and contaminate the groundwater.

To help decrease farmer's reliance on pesticides, agribusiness companies have engineered plants that are genetically resistant to pests. For example, corn plants have been engineered to kill the European corn borer (Figure 7.17a). To do this, scientists transferred a gene from the soil bacterium *Bacillus thuringiensis* (Bt) into corn. The Bt gene encodes proteins that are lethal to corn borers but not to humans (Figure 7.17b). The idea of using this bacterium for pest control

(a) Corn plants have been engineered to kill the insects that eat them.

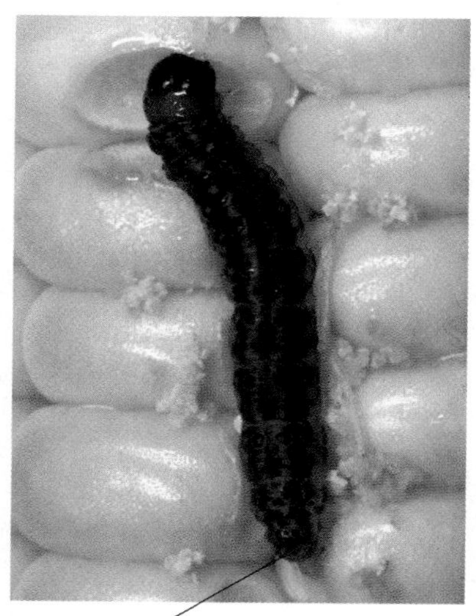
Corn borer

(b) How it works:

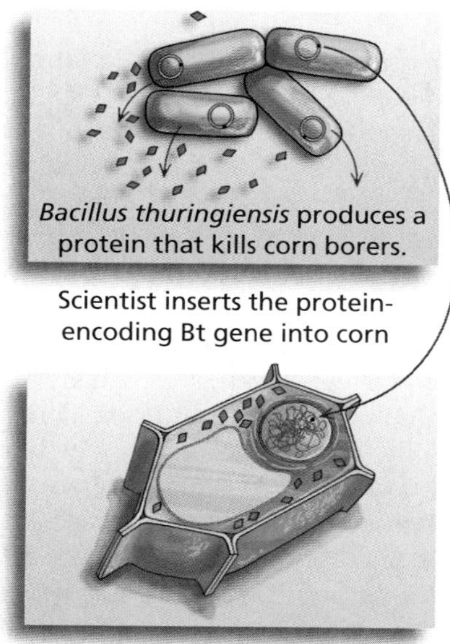
Bacillus thuringiensis produces a protein that kills corn borers.
Scientist inserts the protein-encoding Bt gene into corn
Corn cells produce protein that kills corn borers.

(c) Pollen from Bt corn might unintentionally kill butterfly larvae.

Monarch butterfly catterpillar
Milkweed (common on edges of corn fields)

Figure 7.17 The European corn borer. (a) The European corn borer damages corn and decreases yields. (b) A gene present in the bacterium *Bacillus thuringiensis* produces a protein that is toxic to the corn borer. When this gene is inserted into corn DNA, the plant produces the protein that kills the corn borer, thereby providing resistance to the pest. (c) Some researchers have shown that Bt corn may be harmful to other organisms.

actually came from organic farmers, who have sprayed unengineered *B. thuringiensis* on crop plants for many years. Genetically modified Bt corn has proven to be so successful at resisting the corn borer that close to one-half of all corn currently grown in the United States is engineered with this gene.

Shortly after the arrival of Bt corn, concern arose about its impact on organisms in the surrounding areas. One laboratory study showed that milkweed, a plant commonly found on the edges of cornfields that had been dusted with pollen from Bt corn, was lethal to Monarch butterfly caterpillars, for which milkweed is the only source of food (Figure 7.17c). This research was performed in a laboratory and has been difficult to duplicate on farmers' fields, but there may still be cause for concern about how GM crops will affect other organisms.

Modified corn also caused controversy in 1996 when Bt corn was found in Kraft Taco Bell™ taco shells. Since corn with high levels of Bt had not yet been approved for human consumption, there was a massive recall of the product.

Critics of Bt corn point out that it is only a matter of time before corn borers develop resistance to Bt corn, which will require the development of new varieties of genetically engineered corn. This is true of pesticides applied to crops as well—pests develop resistance because application of a pesticide does not always kill all of the targeted organisms. The few that have preexisting resistance genes and are not susceptible survive and produce resistant offspring. Eventually, widespread resistance develops and a new pesticide must be developed and applied.

This problem is particularly vexing for the organic farmers who were the first to use *B. thuringiensis* for controlling the corn borer, but who did so in a targeted way. When a farmer's chemical overspray drifts to the farms of nearby organic farmers, the organic farmer has lost a powerful tool when the bacterium is killed, and must find another method of controlling this pest.

The continued need for the development of new pesticides in farming is paralleled by farmers' reliance on herbicides. Herbicide-resistant crop plants, such as Round-Up Ready™ soybeans, have been engineered to be resistant to Round-Up™ herbicide, used to control weeds in soybean fields. Farmers can now spray their fields of genetically engineered soybeans with herbicides that will kill everything but the crop plant. Some people worry that this resistance gene will allow farmers to spray more herbicide on their crops, since there is no chance of killing the GM plant, thereby exposing consumers to even more herbicide.

There is also concern that GM crop plants may transfer engineered genes from modified crop plants to their wild or weedy relatives. Wind, rain, birds, and bees carry genetically modified pollen to related plants near fields containing GM crops (or even to farms where no GM crops are being grown). Many cultivated crops have retained the ability to interbreed with their wild relatives; in these cases, genes from farm crops can mix with genes from the wild crops. While this is unlikely to happen with corn or soybeans, which do not have weedy relatives, it has already been seen with canola and is likely to happen with squash and rice. Thus, the herbicide is rendered ineffective since both the crop plant and its weedy relative share the same resistance gene. It may become impossible to determine whether weed plants surrounding fields of engineered crops have been pollinated with pollen containing the modified gene, and there could be unintended consequences for the ecology of the surrounding environment. Also, if pollen from GM crop plants drifts to farms that are not growing modified crops, it becomes impossible to determine whether a crop plant has engineered genes or not. This would be disastrous in the event of a recall.

Genetic manipulation could lead to decreasing variation within a species, and this too can have evolutionary consequences. Most GM corn, in addition to carrying the Bt resistance gene, has also been selectively bred to mature all at once, produce uniform ears, and have a particular nutrient profile. If an unforeseen disease or pest were to sweep through the area containing this corn variety, the disease would probably devastate a large portion of the crop.

Most, but not all, of the genetic engineering that occurs to produce crop plants resistant to pesticides and herbicides is performed by private companies and is designed to maximize profits. For example, Round-Up Ready soybeans are purchased by farmers who then apply Round-Up herbicide; both the GM soybean and herbicide are sold by Monsanto. Some day the techniques pioneered by agribusiness firms may be used to help solve the problem of world hunger, but this has not been the case to date.

7.4 Genetic Engineers Can Modify Humans

WWW **Media Activity 7.3** Gene Therapy Using Stem Cells

Some genetic engineers are attempting to modify humans. These modifications may one day include replacing defective or nonfunctional alleles of a gene with a functional copy of the gene. If this happens, it might be possible for physicians to diagnose genetic defects in early embryos and fix them, allowing the embryo to develop into a disease-free adult. Recent developments that have led to a much better understanding of the human genome may make this scenario more likely.

The Human Genome Project

The **Human Genome Project** involves sequencing, or determining the nucleotide-base sequence (A,C,G, or T), of the entire human genome and the location of each of the 30,000–60,000 human genes. In 1990, the Office of Health and Environmental Research of the United States Department of Energy (DOE), along with the NIH and scientists from around the world, undertook this project. At the time, scientists involved in the project proposed to have a complete accounting of all the genes present in humans by the year 2005. However, the race to complete the sequencing of the human genome sped up drastically due to technological advances and the involvement of a private company named Celera. At stake were the rights to patent the gene sequences (see Essay 7.1). Initially, Celera wanted to retain the rights to the DNA sequences, but government scientists were making sequences available to the public. Eventually, the two groups worked together to publish a working draft of the human genome in 2001.

The scientists involved in this multinational effort also sequenced the genomes of the mouse, the fruit fly, a roundworm, bakers' yeast, and a common intestinal bacterium. Scientists thought it was important to sequence the genomes of organisms other than humans because these **model organisms** are easy to manipulate in genetic studies, and because important genes are often conserved from one organism to another. In fact 90% of human genes are also present in mice, 50% in fruit flies, and 31% in bakers' yeast. Therefore, understanding how a certain gene functions in a model organism helps us understand how the same gene functions in humans.

To sequence the human genome, scientists isolated DNA from white blood cells. They then cleaved the chromosomes into more manageable sizes using restriction enzymes, cloned them into plasmids, and determined the base sequence using automated DNA sequencers (Figure 7.18). These machines distinguish between nucleotides on the basis of structural differences in the nitrogenous bases. Sequence information was then uploaded to the Internet and scientists working on this, or any other project, could search for regions of sequence information that overlapped with known sequences. Using overlapping regions, scientists in laboratories all over the world worked together to patch together DNA-sequence information. In this manner, scientists sequenced

Figure 7.18 DNA sequencing. These machines determine the sequence of nucleotides present in a DNA sample.

Essay 7.1 Patenting

Patents provide protection to people who have invested a lot of time, energy, and money in creating a new product. To patent an object, the inventor has to prove that it has never been made before, involves an inventive step that was not obvious, and that the object serves some useful purpose. Until 1980, living organisms were considered unpatentable because they were considered to be discoveries of nature. This changed when the United States Supreme Court ruled that a genetically engineered bacterium capable of digesting oil could be patented, since it was created by humans. (They were designed to help clean up oil spills.) This ruling paved the way for a company to engineer one gene of a plant and patent it, even though selective breeding over thousands of years actually produced the cultivated plant.

Once a gene is patented, people who want to use it must pay a licensing fee to the person or organization that registered the patent. It is now common practice for scientists to patent gene sequences in the hope that they may one day prove valuable. It is even possible to patent animals—a biotechnology company has patented a genetically engineered strain of mice for use in cancer research. Proponents of patenting live animals argue that those used in research will help researchers find cancer treatments more quickly. Companies that make the large initial investments that are required to produce GMOs argue that the only way to recoup some of these costs is to charge the people who want to use these animals.

Animals are not only being used as research subjects. A pharmaceutical firm recently patented a sheep, named Tracey (Figure E7.1), that can produce the human blood clotting factors absent in people with hemophilia. The proteins are secreted in her milk and can be harvested for human use. Animals like Tracey, called *pharm animals*, are touted as the next great advancement in the production of proteins with pharmacological value to humans.

Figure E7.1 Tracey.

entire chromosomes by "walking" from one end of a chromosome to the other (Figure 7.19). DNA-sequence information obtained by the Human Genome Project may some day enable medical doctors to take a blood sample from a patient and determine which genetic diseases are likely to affect them.

Many people worry about having these types of tests performed because personal information may get back to their insurance companies or employers, but there is a positive side to having all this information available. Once the genetic basis of a disease has been worked out—that is, how the gene of a healthy person differs from the gene of a person with a genetic disease—the information will be used to provide treatments or cures.

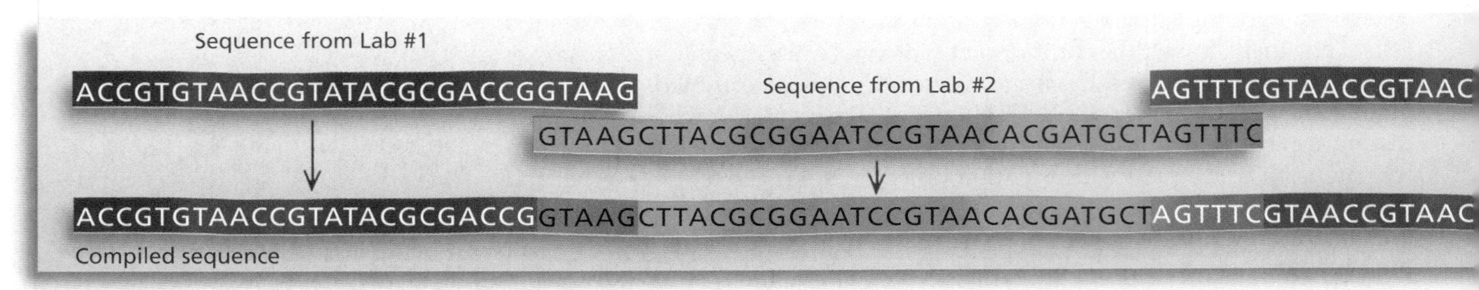

Figure 7.19 Chromosome walking. Scientists can determine the location of genes on chromosomes using DNA sequence information uploaded to the Internet from laboratories around the world. They search for areas that overlap and fill in the gaps, much like assembling a jigsaw puzzle.

Using Genetic Engineering to Cure Human Diseases

Scientists who try to replace defective human genes (or their protein products) with functional genes are performing **gene therapy**. Gene therapy may some day enable scientists to fix genetic diseases in an embryo. To do so, the scientist would supply the embryo with a normal version of a defective gene; this so-called *germ line gene therapy* would ensure that the embryo and any cells produced by cell division would replicate the new, functional version of the gene. Thus, most of the cells would have the corrected version of the gene—when these genetically modified individuals have children, they will pass on the corrected version of the gene. If scientists can fix genetic defects in early embryos, some genetic diseases could be wiped out altogether (see Essay 7.2).

Another type of gene therapy, called *somatic cell gene therapy*, can be performed on body cells to fix or replace the defective protein in only the affected cells. Using this method, scientists introduce a functional version of a defective gene into an affected individual cell in the laboratory, allow the cell to reproduce, and then place the copies of the cell bearing the corrected gene into the diseased person.

This treatment may seem like science fiction, but it is likely that this method of treating genetic diseases will be considered a normal procedure in the not-too-distant future. In fact, genetic engineers already have successfully treated a genetic disorder called **Severe Combined Immunodeficiency Disorder** (**SCID**), a disease caused by a genetic mutation resulting in the absence of an important enzyme, giving the individual a severely weakened immune system. Since their immune systems are compromised, people with SCID are incapable of fighting off any infection, and they often suffer severe brain damage from the high temperatures associated with unabated infection. Any exposure to infection can kill or disable someone with SCID, so most patients are kept inside their homes and often live inside protective bubbles that separate them from everyone, even family members.

To devise a successful treatment for SCID, or any disease treated with gene therapy, scientists had to overcome two major obstacles: getting the therapeutic gene to the right place, and making sure it is expressed in the right manner.

Delivery of the Functional Gene to the Correct Location Proteins break down easily and are difficult to deliver to the proper cells, so it is more effective to replace a defective gene than to continually replace a defective protein. Delivering a normal copy of a defective gene only to the cell type that requires it is a difficult task. SCID, the disorder that has been treated successfully, was chosen in part because defective immune-system cells could be removed from the body, treated, and returned to the body.

Immune-system cells that require the enzyme missing in SCID patients circulate in the bloodstream. Blood removed from a child with SCID is infected with nonpathogenic (non-disease-causing) versions of the virus that causes the common cold. This virus is first engineered to carry a normal copy of the

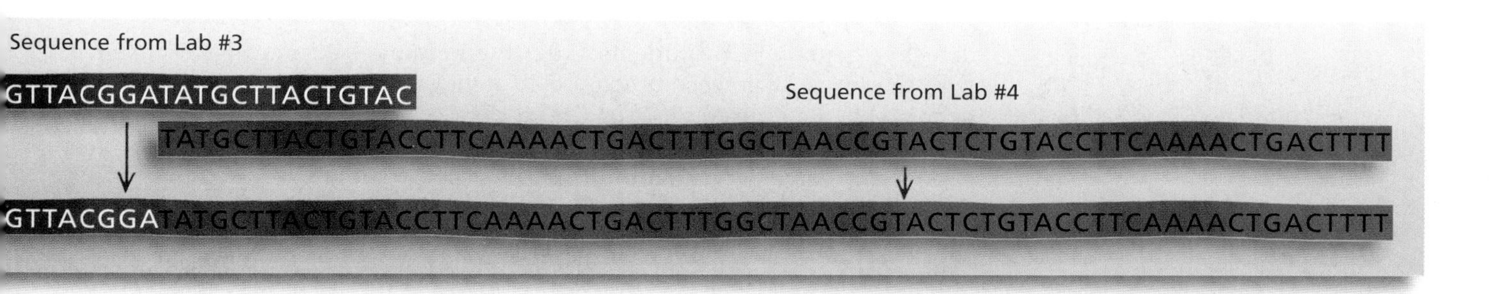

Essay 7.2 Stem Cells

Stem cells are cells that grow and divide without limitation and, given the proper signals, can become any other type of cell.

Some stem cells are embryonic in origin. As a human embryo grows, the early cells start dividing and forming different, specialized cells such as heart cells, bone cells, and muscle cells. Once formed, specialized nonstem cells can only divide to produce replicas of themselves. They cannot backtrack and become a different type of cell.

Embryonic stem cells retain the ability to become virtually any cell type (Figure E7.2). If the cells are harvested from an early embryo (about 5–7 days after conception) and nudged in a particular direction in the laboratory, they can be directed to become a particular tissue or organ.

Tissues and organs grown from stem cells in the laboratory may some day be used to replace organs damaged in accidents or organs that are gradually failing due to *degenerative diseases*. Degenerative diseases start with the slow breakdown of an organ and progress to organ failure. Additionally, when one organ is not working properly, other organs are also affected. Degenerative diseases include stroke, diabetes, liver and lung diseases, heart disease, and Alzheimer's disease.

Stem cells could provide healthy tissue to replace those damaged by spinal cord injury or burns. New heart muscle could be produced to replace that damaged during a heart attack. A diabetic could have a new pancreas, and people suffering from osteoarthritis could have replacement cartilage to cushion their joints. Thousands of people waiting for organ transplants might be saved if new organs were grown in the lab.

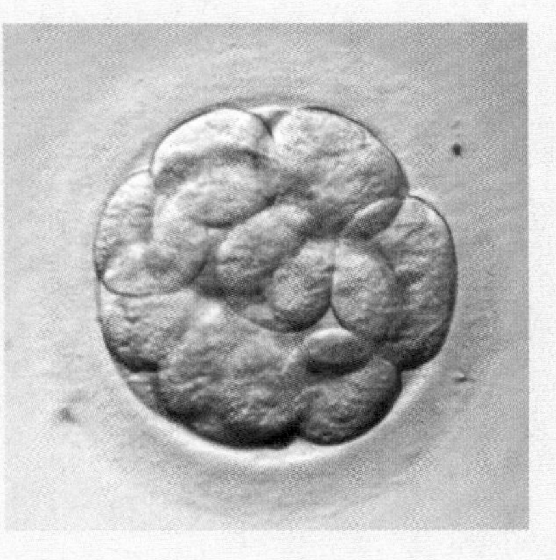

Figure E7.2 Early human embryos in a petri dish.

One problem with stem-cell research is that the embryos are destroyed when the stem cells are removed—and many people object to the destruction of early embryos. Currently, the federal government will fund research using leftover embryos from fertility treatments, but will not support research using embryos created solely for research purposes. This ban only applies to federally funded research projects, which means that in the United States, research on embryos can only be performed by genetic engineers who obtain grants from nongovernmental sources unless they have access to the limited numbers of embryos created during fertility treatments.

In vitro (Latin, meaning "in glass") fertilization procedures often result in the production of excess embryos because a large number of egg cells are harvested from a woman who wishes to become pregnant. These egg cells are then mixed with her partner's sperm in a petri dish, resulting in the production of many fertilized eggs that grow into embryos. A few of the embryos are then implanted into the woman's uterus. The remaining embryos are stored so that more attempts can be made if pregnancy does not result or if the couple desires more children. When the couple achieves the desired number of pregnancies, the remaining embryos can, with the couple's consent, be used for stem-cell research.

A solution to the ethical dilemma presented by the use of embryonic stem cells seems to be on the horizon. Scientists have recently discovered that many adult tissues also contain stem cells. Recent studies published in peer-reviewed literature suggest that most adult tissues have stem cells, that these cells can be driven to become other cell types, and that they can be grown indefinitely in the laboratory. Based on success in animal models, there is even evidence that adult stem cells will help cure diseases. In fact, scientists have used stem cells from adult tissues to repair damage in animals due to heart attack, stroke, diabetes, and spinal cord injury.

defective gene in SCID patients. After the immune-system cells are infected with the virus, these recombinant cells, which now bear copies of the functional gene, are returned to the SCID patient (Figure 7.20a).

In 1990, a four-year-old girl named Ashi DiSilva (Figure 7.20b) was the first patient to receive gene therapy for SCID. Ashi's parents were willing to face the unknown risks to their daughter because they were already far too familiar with the risks of SCID—the couple's two other children also had SCID and were severely disabled. Ashi is now a healthy adult with an immune system that is able to fight off most infections.

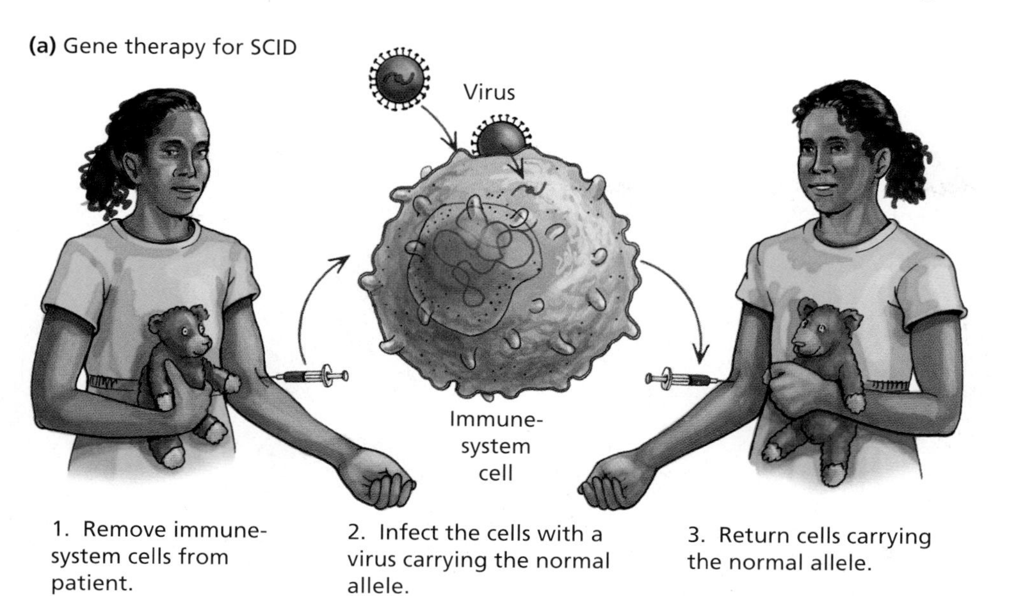
(a) Gene therapy for SCID
Virus
Immune-system cell
1. Remove immune-system cells from patient.
2. Infect the cells with a virus carrying the normal allele.
3. Return cells carrying the normal allele.

(b) SCID survivor

Figure 7.20 Gene therapy in an SCID patient. (a) A virus carrying the normal gene is allowed to infect immune-system cells that have been removed from a person with SCID. The virus inserts the normal copy of the gene into some of the cells, and these cells are then injected into the SCID patient. (b) Ashi DiSilva, the first gene therapy patient. (Photo ©1995 Jessica Boyatt)

However, Ashi must continue to receive treatments, because blood cells, whether genetically engineered or not, have limited life spans. When most of Ashi's engineered blood cells have broken down, she must be treated again. Thus, she undergoes this gene therapy a few times each year. Since Ashi's treatment, many other SCID patients have been successfully treated and live normal lives. Unfortunately, Ashi's gene therapy does not prevent her from passing on the defective allele to her biological children because this therapy is not "fixing" the allele in her ovaries.

Even though Ashi will need lifelong treatment and could pass the defective allele to her children, she is lucky that her genetic disease was amenable to gene therapy. There are not many genetic diseases for which the defective cells can be removed from the body, treated, and reintroduced to the body. Nor are there many genetic diseases for which contributions from many genes and the environment are not a factor.

For gene therapy to be successful in curing more genetic diseases, it is necessary for scientists to not only deliver the gene to the correct location, they must also make sure the gene is turned off and on at the proper times. In other words, the expression of the gene must be regulated.

Regulating Gene Expression Scientists will only succeed with gene therapy if they can learn to regulate the expression of a gene once it is located in the proper place. Recall from Chapter 4 that each cell in your body, except sperm or egg cells, has the same complement of genes you inherited from your parents. However, different genes are expressed in different cells at different times. Heart cells differ from eye cells because each cell type expresses only a small percentage of its genes. For gene therapy to work, scientists must learn how to turn the right genes on in the right cell type at the right time.

With *r*BGH, farmers can regulate how much protein to inject into the bloodstream of a cow, but a gene inserted into the genome must respond to environmental cues telling it when and when not to produce a protein.

Gene expression is most commonly *regulated*—turned off or on modulated more subtly—by controlling the rate of transcription. Adjacent to an actual gene is a sequence of nucleotides called the **promoter**, which functions in helping to

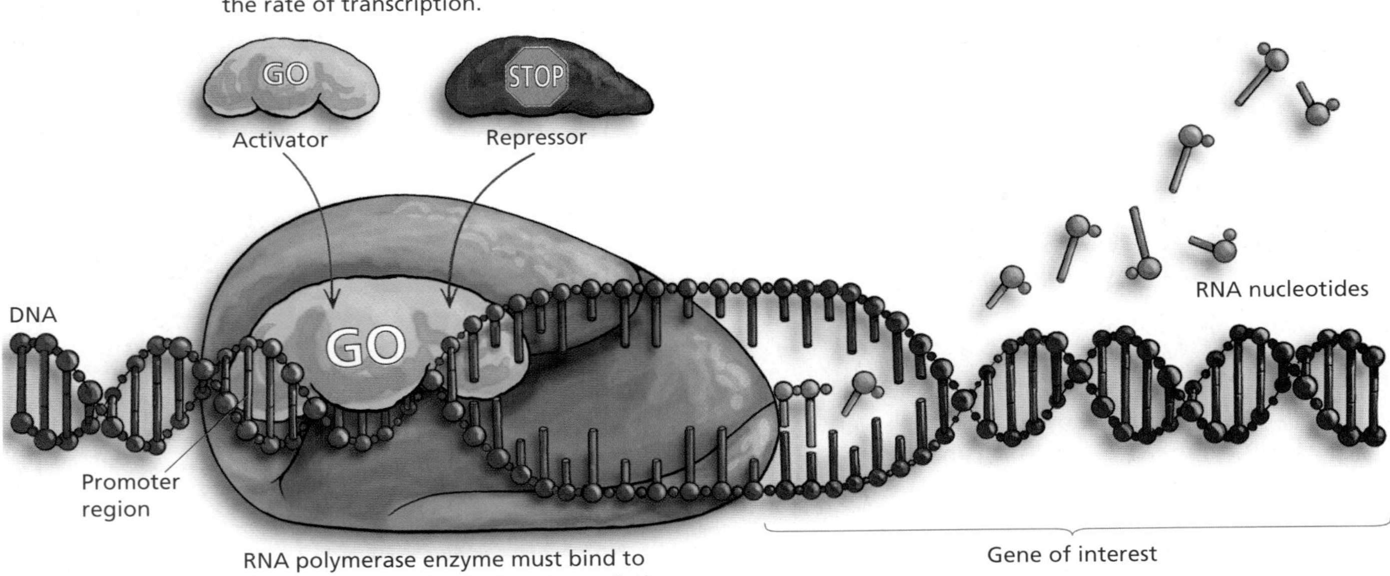
Various proteins activate or repress the rate of polymerase binding to the promoter, affecting the rate of transcription.
GO
STOP
Activator
Repressor
DNA
GO
RNA nucleotides
Promoter region
RNA polymerase enzyme must bind to the promoter region before transcription of the gene can begin.
Gene of interest

Figure 7.21 Regulation of gene expression. When a gene needs to produce the protein it encodes, the RNA polymerase enzyme binds to the promoter adjacent to the gene to begin transcription. Other proteins, called activators or repressors, increase or decrease the rate of polymerase binding, thereby influencing the overall rate of protein synthesis.

WWW
Media Activity 7.1B Regulation of gene expression

regulate gene expression. When a cell requires a particular protein, the RNA polymerase enzyme binds to the promoter for that particular gene and transcribes the gene. Other proteins in the cell can activate or repress the rate at which the polymerase binds to initiate transcription as well (Figure 7.21).

The rate at which the polymerase binds to the promoter is also affected by substances that are present in the cell. For example, the presence of alcohol in a liver cell might result in increased transcription of a gene involved in the breakdown of alcohol.

Once genetic engineers find better methods for delivering gene sequences to the required locations and can regulate their expression, gene therapy will be far more effective. However, gene therapy is an attempt to modify only one or a few genes. A far more controversial type of genetic engineering involves making an exact copy of an entire organism, a process called *cloning*.

It May Soon Be Possible to Clone Humans

Human cloning occurs commonly in nature via the spontaneous production of identical twins. These clones arise when an embryo subdivides itself into two separate embryos early in development—this is not the type of cloning that many people find objectionable. People are more likely to be upset by cloning that involves selecting which traits an individual will possess. Natural cloning of an early embryo to make identical twins does not allow any more selection for specific traits than does fertilization. However, in the future it may be possible to select adult humans who possess desired traits and clone them.

Cloning offspring from adults with desirable traits has been successfully performed on cattle, goats, mice, cats, pigs, rabbits, and sheep. In fact, the animal that brought cloning to the attention of the public was a ewe named Dolly.

Dolly was cloned when Scottish scientists took cells from the mammary gland of an adult female sheep and fused it with an egg cell that had previously

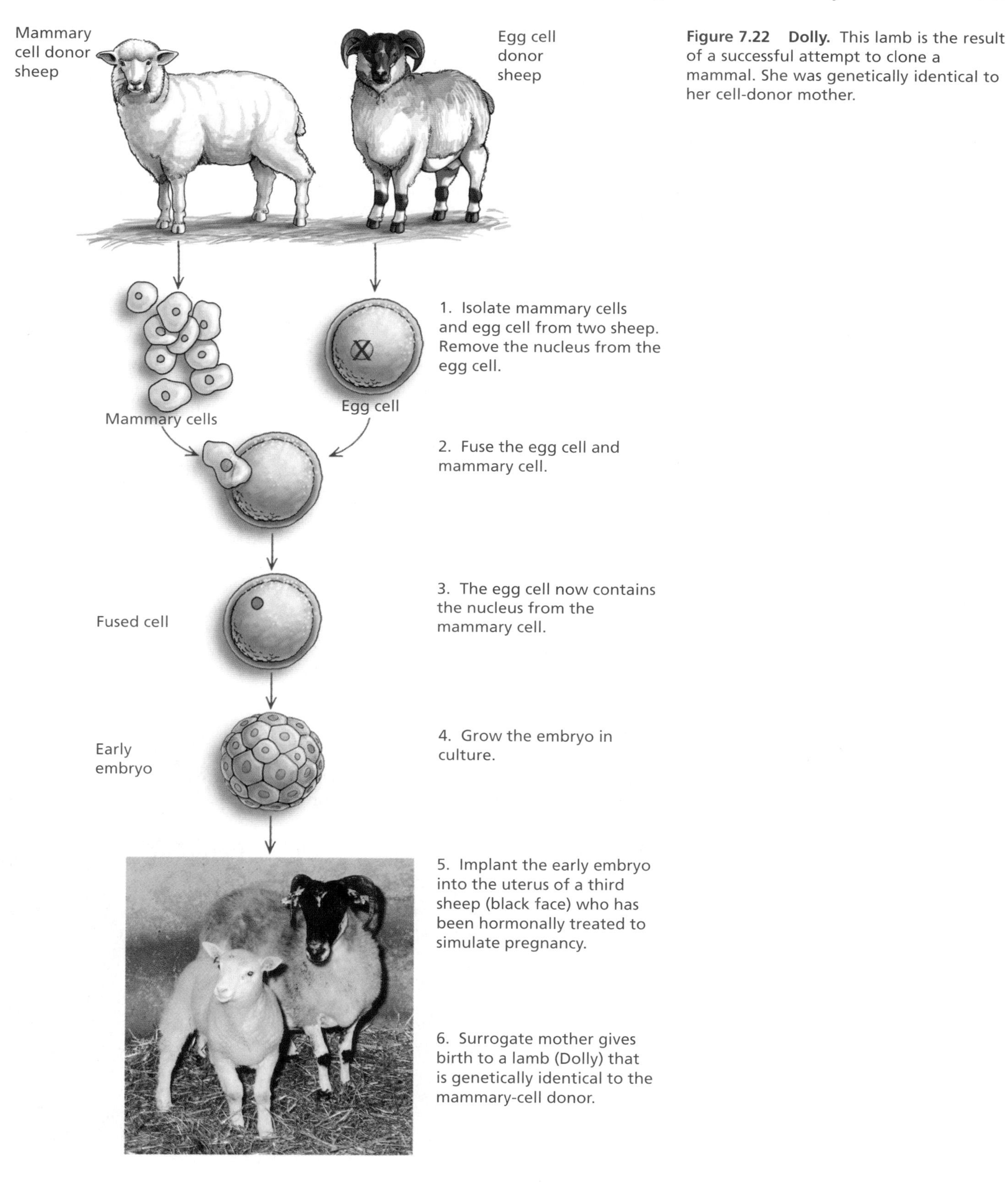

Mammary cell donor sheep
Egg cell donor sheep
Mammary cells
Egg cell
1. Isolate mammary cells and egg cell from two sheep. Remove the nucleus from the egg cell.
2. Fuse the egg cell and mammary cell.
Fused cell
3. The egg cell now contains the nucleus from the mammary cell.
Early embryo
4. Grow the embryo in culture.
5. Implant the early embryo into the uterus of a third sheep (black face) who has been hormonally treated to simulate pregnancy.
6. Surrogate mother gives birth to a lamb (Dolly) that is genetically identical to the mammary-cell donor.

Figure 7.22 Dolly. This lamb is the result of a successful attempt to clone a mammal. She was genetically identical to her cell-donor mother.

had its nucleus removed. Treated egg cells were then placed in the uterus of an adult ewe that had been hormonally treated to support pregnancy. Scientists had to try many, many times before this **nuclear transfer** technique worked. In all, 277 embryos were constructed before one was able to develop into a live lamb (Figure 7.22). Dolly was born in 1997.

The research that led to Dolly's birth was designed to provide a method of ensuring that cloned livestock would have the genetic traits that made them most beneficial to farmers. Sheep that produced the most high-quality wool and cattle that produced the best beef would be cloned. This is more efficient than allowing two prize animals to breed because each animal would only give half its genes to the offspring. Therefore, there is no guarantee that the offspring of two prize animals will have the desired traits.

There is no equivalent reason for cloning humans, and no one knows if nuclear transfer will work in humans—or if cloning is safe. If Dolly is a representative example, cloning animals may not be safe. In 2003, Dolly was put to sleep to relieve her from the discomfort of arthritis and a progressive lung disease, conditions usually found only in older sheep. The fact that she developed these conditions has lead scientists to question whether Dolly's chromosomes were older than Dolly herself. Dolly was cloned from a six-year-old sheep, and she died at age six. This would make her chromosomes closer to twelve years old—approximately the age non-cloned sheep would develop these conditions. Chromosomes may "age" due to the shortening of their tips, or telomeres (discussed in Chapter 5), that occurs with each round of cell division.

The debate about human cloning mimics the larger debate about genetic engineering. As a society, we need to determine whether the potential for good outweighs the potential harm for each application of these technologies (Figure 7.23). When it comes to human cloning, the potential for abuse could be substantial. Will adults clone themselves and raise their clones as children? Will clones of ill people be used as organ donors? Will clones of physically fit individuals be used to fight wars? Will clones be considered people or property? Because no one knows for sure if human cloning is even possible, these questions remain unanswerable.

Some reasons why the work of genetic engineers is important	Some reasons why the work of genetic engineers is controversial
• GM animals and crops may make farms more productive. • GM crops may be made to taste better, last longer, or contain more nutrients. • Genetic engineers hope to cure diseases and save lives.	• GM crops encourage agribusiness, which may close down some small farms. • GM animals and crops may cause health problems in consumers. • GM crops might have unexpected adverse effects on the environment. • Present research might lead to the unethical genetic modification of humans.

Figure 7.23 Are genetic engineers doing more good than harm? This chart lists some of the pros and cons of genetic engineering.

CHAPTER REVIEW

Summary

- Bovine growth hormone is a protein produced by the pituitary glands of cows. To increase the quantity of milk a cow produces, additional growth hormone is injected into the cow.
- Modern genetic engineering techniques enable scientists to produce recombinant BGH in the lab by placing the gene for growth hormone into plasmids, which then clone the gene by making millions of copies of it as they replicate themselves inside their bacterial hosts.
- Bacteria can then express the gene by making an mRNA copy (transcription) and translating the mRNA into a protein. Translation takes place on ribosomes. When an mRNA codon is present in the ribosome binding site, a tRNA carrying the called-for amino acid is able to bind to the mRNA present at the ribosome and its amino acid is added to the growing polypeptide chain. The final protein product (growth hormone) is then isolated and injected into cows.
- Crop foods may be genetically modified to increase their yield, shelf life, and nutritive content.
- A plasmid or gene gun is used to insert a particular gene into plant cells.
- There is concern that GM foods may negatively affect the environment or people who consume them.
- Information about genes obtained from the Human Genome Project can be used to help replace genes that are defective or missing in people with genetic diseases. Gene therapy is still considered experimental, but may hold tremendous promise once scientists figure out how to target genes to the right locations and express them in the proper amounts.
- Gene expression is regulated at sequences called promoters. The RNA polymerase enzyme binds to the promoter more or less efficiently to increase or decrease transcription of a given gene.
- Cloning animals with desirable agricultural traits has occurred. It may someday be possible to clone humans, but it is unclear if these humans would be healthy.

Key Terms

anticodon p. 165
antisense p. 170
applied research p. 170
basic research p. 169
cloning p. 161
codons p. 165
DNA p. 161
gene expression p. 165
gene gun p. 172
generally recognized as safe (GRAS) p. 169
gene therapy p. 179
genetic code p. 166
genetic engineer p. 160
genetically modified organism (GMO) p. 172
genome p. 163
Human Genome Project p. 177
messenger RNA (mRNA) p. 165
model organisms p. 177
nuclear transfer p. 183
plasmid p. 163
promoter p. 181
protein synthesis p. 164
recombinant p. 161
restriction enzymes p. 163
ribonucleic acid (RNA) p. 165
ribose p. 165
ribosome p. 165
RNA polymerase p. 165
sense p. 170
Severe Combined Immunodeficiency Disorder (SCID), p. 179
stop codon p. 166
Ti plasmid p. 172
transcription p. 165
transfer RNA (tRNA) p. 165
transgenic p. 172
translation p. 165
uracil p. 165

Learning the Basics

1. List the order of nucleotides on the mRNA that would be transcribed from the following DNA sequence: CGATTACTTA
2. Using the genetic code (Table 7.1), list the order of amino acids encoded by the following mRNA nucleotides: CAACGCAUUUUG
3. Why are plasmids used when cloning genes using bacteria?
4. Describe all the subcellular structures that participate in translation.
5. What would happen to the expression of a gene if its promoter were blocked?
6. Transcription _____.
   a. synthesizes new daughter DNA molecules from an existing DNA molecule
   b. makes an RNA copy of a gene that is to be translated
   c. pairs thymines (T) with adenines (A)
   d. occurs on ribosomes

7. Transfer RNA (tRNA) _____.
   a. carries monosaccharides
   b. is made of messenger RNA
   c. has an anticodon region, which is complementary to the mRNA codon
   d. is the site of protein synthesis

8. If a cell needs a particular protein _____.
   a. the gene that codes for that protein is transcribed and then translated
   b. a hormonal message is sent to the brain, telling the brain to produce this protein
   c. DNA synthesis results in the production of the protein
   d. mitosis occurs

9. During the process of transcription _____.
   a. DNA is used to make DNA
   b. DNA is used to make RNA
   c. DNA is used to make proteins
   d. RNA is used to make proteins

10. Translation is the conversion of _____.
   a. DNA into RNA
   b. RNA into DNA
   c. DNA into protein
   d. RNA into protein

11. If the sequence of nucleotides in a gene was GGCCTTAA, the mRNA sequence would be _____.
   a. CCGGUUAA
   b. CCGGAATT
   c. CCGGAAUU
   d. AATTCCTT

12. An anticodon is a part of _____.
   a. DNA
   b. tRNA
   c. mRNA
   d. ribosome
   e. amino-acid-activating enzyme

13. The RNA polymerase enzyme binds to _____, initiating transcription.
   a. amino acids
   b. the tRNA
   c. the promoter sequence
   d. the ribosome

14. A particular triplet of bases in the coding sequence of DNA is TGA. The anticodon on the tRNA that binds the mRNA codon is _____.
   a. TGA
   b. UGA
   c. UCU
   d. ACU

15. _____ is not involved in the process of translation.
   a. mRNA
   b. tRNA
   c. ribosomes
   d. DNA

16. RNA and DNA are similar because _____.
   a. they are both double-stranded helices
   b. uracil is found in both of them
   c. both contain the sugar deoxyribose
   d. both consist of a sugar, a phosphate, and a base

Analyzing and Applying the Basics

1. Describe the steps you would take to clone a gene from humans using bacteria.

2. How could a mutation (change in DNA) to a promoter region alter gene expression?

3. Why are the cells that comprise the various tissues of your body different, even though they all contain the same genes?

4. Why are Ti plasmids and gene guns used to insert genes into plant cells?

5. Why are some genetic defects more likely to be cured by gene therapy than others?

Connecting the Science

1. Who should decide whether human cloning is ethical?

2. The first "test-tube baby," Louise Brown, was born over 30 years ago. Sperm from her father was combined with an egg cell from her mother. The fertilized egg cell was then placed into the uterus of her mother for the period of gestation. At the time of her conception, many people were very concerned about the ethics of scientists performing these *in vitro* fertilizations. Do you think human cloning will eventually be as commonplace as *in vitro* fertilizations are now? Why or why not?

3. Taxpayers support basic research with state and federal funding. Do you think it is important to financially support basic as well as applied research?

4. Do you think it is acceptable to grow genetically modified foods if health risks turn out to be low but environmental effects are high?

Media Activities

Media Activity 7.1 Making Recombinant Bovine Growth Hormone
Estimated Time: 5 minutes
Explore the processes of creating a recombinant plasmid, the regulation of gene expression, transcription of the recombinant gene into mRNA, and translation of the mRNA into a protein.

Media Activity 7.2 Bioengineering New Products
Estimated Time: 15 minutes
There are many ways of getting bioengineered products besides using bacteria. Explore the possibilities of using mouse urine, goat's milk, and chicken eggs as possible biological "factories" for the production of recombinant products.

Media Activity 7.3 Gene Therapy Using Stem Cells
Estimated Time: 15 minutes
There are many controversies associated with the use of gene therapy and stem cells, and unexpected problems have arisen. Explore the hope it offers, as well as the frightening possibilities associated with use of genetically altered stem cells.

CHAPTER 8

Where Did We Come From?

The Evidence for Evolution

Should public-school students be taught about alternative hypotheses to evolution?

One idea about the origin of humans: Special creation.

Another idea: Evolution.

In August of 1999, the Kansas State Board of Education adopted new statewide educational standards for primary and secondary students. Conspicuously missing from the list of topics young Kansans were required to know before graduating was a portion of the subject of biological evolution. Specifically, the standards excluded the theory that describes the descent of modern species from extinct ancestors. "We are only being honest with our students," said school board member Steve Abrams. "Evolution is only a theory."

The school board's action in Kansas was not an isolated event. Elected officials in several states have debated about requiring their students to learn the theory of evolution. In early 2002, the Ohio state school board considered requiring the teaching of "alternatives" to evolution, in addition to the theory of evolution, in public school curriculums. In fact, the debate about whether children should learn about evolution is nearly as old as the theory itself.

Probably the subject of evolution would not cause so much controversy, in Kansas and elsewhere, if it did not address fundamental questions of human existence: Who are we and where do we come from? Many religious traditions include as part of their beliefs the understanding that humans were designed by an intelligent supernatural being. In contrast to this belief, called *special creation*, evolutionary theory argues that humans arose through natural processes,

Why do biologists insist that only evolution be taught?

and they are the descendants of ancient apes. On the surface, these two ideas appear to be competing hypotheses about the origins of humanity. However, U.S. federal courts have consistently ruled that religious beliefs about human origins have no place in American public-school science classrooms.

If special creation is not allowed to be presented in public schools because it is religion, the theory of evolution should not be presented either, reasoned a majority of Kansas state school board members in 1999. After all, they argued, it seems that the theory of evolution represents a religious belief as well, just one that *rejects* any action of a higher power.

Did the school board decision in Kansas represent a revolution in science education—the recognition that scientific theories are as much a matter of faith as religious doctrine is? Or, as others have said, does it represent a dangerous trend in U.S. education—one that sets religious belief on equal footing with scientific understanding? This chapter will examine these questions by exploring the theory of evolution and the origin of humans as a matter of science.

WWW
Media Activity 8.1 Microevolution in Bacteria

8.1 What Is Evolution?

"Evolution" really has two different meanings to biologists. The term can refer to either a process or an organizing principle, that is, a theory. Generally, the word *evolution* means change, and the *process* of biological evolution is derived from this definition. That is, **biological evolution** is a change in the features of individuals in a biological population that occurs over the course of generations.

For example, consider the species of organism commonly known as head lice. A **species** (Latin, meaning "kind") consists of a group of individuals that can regularly breed together and is generally distinct from other species in appearance or behavior. Most species are physically subdivided into smaller groups, **populations**, that are somewhat independent of other populations—that is, most individuals in a population breed with other individuals in the same population. As some parents of small children have discovered, some populations of head lice in the United States have become immune to the pesticide permethrin, found in over-the-counter delousing shampoos. Initially, lice infections were readily controlled through treatment with these products; however, over time, populations of lice changed to become less susceptible to the effects of these chemicals. The evolution of resistance can occur rapidly—a study in Israel demonstrated that populations of head lice were four times less susceptible to permethrin only 30 months (40 lice generations) after it came into widespread use in that country.

Note that in this example, *individual* head lice did not "evolve" or change; each individual louse is either susceptible to, or resistant to, permethrin. Instead, the population as a whole changed from one in which most lice were susceptible to one in which most lice were resistant to the pesticide. This change in the characteristics of the population took many generations. According to the definition above, the lice have evolved.

In general, the fact that the process of evolution occurs is not questioned. Evolutionary change in biological populations, such as the development of pesticide resistance in insects and antibiotic resistance in bacteria, has been observed multiple times. Changes that occur in the characteristics of populations are referred to as **microevolution**. Chapter 9 explores the consequences of microevolution resulting from our efforts to combat certain human diseases.

Few members of the Kansas state school board, or their supporters, disagree that populations of organisms can change over time. The *theory* of evolution is much more controversial in society than the process. Some of the controversy is generated by the use of the word *theory*. When people use the word in everyday conversation, they often are referring to a tentative explanation with little systematic support. A sports fan might have a theory about why her team is losing, or a gardener might have a theory about why his roses fail to bloom. Usually, these ideas amount to a "best guess" regarding the cause of some phenomenon. A **scientific theory** is much more substantial—it is a body of scientifically acceptable general principles that help explain how the world works. Scientific theories are supported by numerous lines of evidence and have stood up to repeated experimental tests. For instance, the theory of gravity explains the motion of the planets; atomic theory explains the interactions between chemical elements and molecules; and the theory of relativity explains the relationship between mass and energy, which led directly to the development of the atomic bomb. The theory of evolution is an organizing principle for understanding how species originate and why they have the characteristics they exhibit. The **theory of evolution** can thus be stated:

> All organisms present on Earth today are descendants of a single common ancestor, and all organisms represent the product of millions of years of microevolution.

In other words, modern animals, plants, fungi, bacteria, and other living things are related to each other and have been diverging from their common ancestor by various processes since the origin of life on this planet. The origin of modern species from a common ancestor is what most nonscientists think of when they hear the word *evolution*. This part of the theory of evolution, called the theory of common descent, is illustrated in Figure 8.1. However, most biologists conceive of the theory of evolution as encompassing both the theory of common descent and the processes by which evolutionary change occurs.

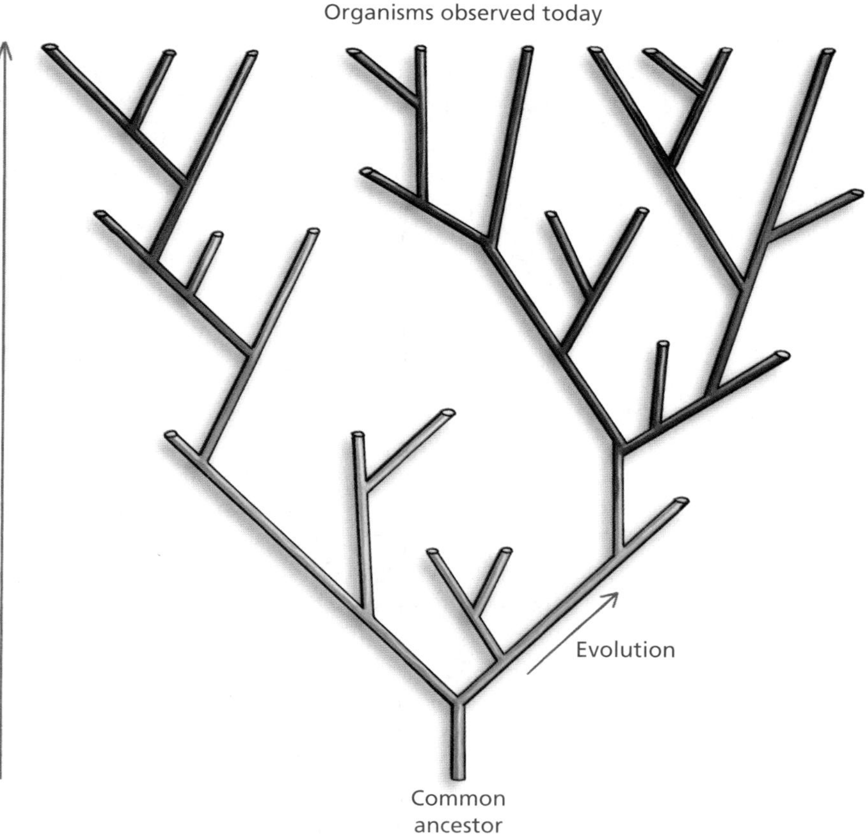

Figure 8.1 The theory of common descent. All modern organisms descended from a single common ancestor.

Figure 8.2 Charles Darwin. As a young naturalist, Darwin conceived of and developed the theory of evolution. He spent nearly 25 years collecting data and building evidence for his hypothesis before publishing it.

8.2 Charles Darwin and the Theory of Evolution

Charles Darwin is credited with bringing the theory of evolution into the mainstream of modern science (Figure 8.2). The youngest son of a wealthy physician, Darwin had spent much of his early life as a lackluster student. After dropping out of medical school at Edinburgh University and at the urging of his father, Darwin entered Cambridge University to study for the ministry. Darwin barely made it through his classes, but did strike up friendships with several members of the scientific community at the college. One of his closest companions at Cambridge was Professor John Henslow, an influential botanist. It was Henslow who secured Darwin his first "job" after graduation. In 1831, at age 22, he set out on what would become his life-defining journey—the voyage of the *HMS Beagle*.

The *Beagle's* mission was to chart the coasts and harbors of South America and included a naturalist for "collecting, observing, and noting anything worthy to be noted in natural history." Henslow recommended Darwin for this unpaid position after two other candidates had turned it down.

The hypothesis that organisms, and even the very rock of Earth, had changed over time was not new when Darwin embarked on his voyage. The first modern evolutionist, Jean Baptiste Lamarck, had published his theory of evolution in 1809. Lamarck's theory of evolution was similar to Darwin's in many respects, but his scientific contemporaries were not willing to consider hypotheses that proposed that species changed through time. Not only were Lamarck's contemporaries unconvinced by his proposed mechanism for species change—that traits developed over the lifetime of an individual were passed on to its offspring—they were also unwilling to question the hypothesis that Earth and its organisms were created by a divine designer.

The most influential stop Darwin made on his journey was at a small archipelago of volcanic islands off the coast of Ecuador. The Galapagos Islands were not very appealing to Darwin at first—they seemed harsh and nearly lifeless. However, during the month the *Beagle* spent sailing the islands, Darwin collected an astonishing variety of organisms. Many of the birds and reptiles he observed appeared to be unique to each island—so that, while all islands had populations of tortoises, the type of tortoise found on one island was different from the types found on other islands (Figure 8.3). Darwin reflected on these and other observations after his return to England and concluded that the different species of tortoises on the different islands arose through evolution from a single ancestral tortoise species.

Darwin outlined the evidence for his conclusion that life had evolved in a book titled, *On the Origin of Species by Means of Natural Selection, or the Preservation of Favoured Races in the Struggle for Life*. The main point of the text was to put forward a hypothesis about how species come about—in other words, the process of evolution. He called this process *natural selection*. Darwin's ideas about the process of evolution are discussed in detail in Chapter 9. However, Darwin devoted the last several chapters of *The Origin of Species* to describing the evidence for what was then the *hypothesis* of **common descent**. A hypothesis is a tentative explanation for an observation. Darwin argued that the best explanation for observations of the relationships among modern organisms, and for the existence of fossils of extinct organisms, was that all organisms had descended from a single common ancestor in the distant past.

Darwin knew that the hypothesis of common descent was radical, and that other scientists, including Lamarck and his grandfather Erasmus Darwin, had been vilified for proposing it. This knowledge prevented him from sharing his ideas with all but a few close friends. Darwin was finally spurred into action in 1858, after receiving a letter from Alfred Russel Wallace. With the letter was a manuscript detailing a mechanism for evolutionary change identical to the one

(a) Tortoise from a wet environment.

(b) Tortoise from a dry environment.

Figure 8.3 Giant tortoises of the Galapagos. The species of giant tortoises on the Galapagos Islands from different environments look very different. Individuals with dome-shaped shells (a) are common in areas with abundant vegetation, while those with flatter shells (b) are found where vegetation is less abundant. Darwin felt that the existence of relatively similar, but obviously different, tortoises on these islands was best explained by descent from a common ancestor.

Darwin had hypothesized. Darwin had excerpts of his and Wallace's work presented in July of 1858 at a scientific meeting in London, and the next year he published *The Origin of Species*.

Darwin's fear of being rejected by his scientific colleagues caused him to carefully document and research support for the hypothesis of common descent. By the time *The Origin of Species* was published in 1859, the evidence he had accumulated was overwhelming. In fact, the evidence put forth in the book was so complete, and from so many different areas of biology, that it no longer appeared to be a tentative explanation. In response, scientists began to refer to this inference as the *theory* of common descent.

8.3 Evaluating the Evidence for Evolution

The theory that modern species evolved from extinct ancestors is no longer in question among scientists. Indeed, most biologists would agree that evolution is a scientific law—really the equivalent of a fact. However, the acceptance of the theory of evolution, chiefly the theory of common descent, in the scientific community is in great contrast to the feelings of the nonscientist public. Opinion polls in the United States conducted over the past several decades have consistently indicated that nearly 50% of Americans do not believe that modern species arose from a common ancestor.

Let us explore the statement "evolution is a fact" more closely. When *The Origin of Species* was published, most Europeans believed that **special creation** explained how organisms came into being. According to this belief, God created organisms during the six days of creation described in the first book of the Bible, Genesis. This belief also states that organisms, including humans, have not changed significantly since this beginning. According to biblical scholars, the Genesis story indicates that creation also occurred fairly recently, within the last 10,000 years. Before Darwin, this story was scientists' best explanation for where living beings came from, and it helped explain the wonderful diversity and complexity of life. The belief in special creation grew

Essay 8.1 Origin Stories

Most of us are familiar with the Judeo-Christian origin story. However, there are hundreds of creation stories found among people all around the world. Here is a small sample:

Creation from a Primordial Being (Norse)

Heat from Muspell, a fiery area to the south, met with the cold from icy Ginnungagap in the north and created the frost giant Ymir. A man and woman were born from his armpits, and one of his legs mated with the other to make a son. This began a race of frost ogres. Some melting ice became the cow Audhumla, whose teats gave rise to rivers of milk. The man Buri appeared from a block of ice that Audhumla licked. His descendents included the gods Odin, Vili, and Ve. They killed Ymir, and blood from his wounds caused a flood that killed all people except the giant Bergelmir and his family. The three gods turned Ymir's body into the earth and his blood into the surrounding seas. His bones and teeth became mountains and rocks, his skull became the sky, and his brains became clouds. The gods made the sun, moon, and stars out of sparks from Muspell. Then they made a man and woman (Ask and Embla) from two fallen trees. Odin gave them life, Vili gave them intelligence, and Ve gave them speech, sight, and hearing.

Earth Diver (Huron)

In the beginning, there was only a wide sea. A divine woman fell from the upper world. Two loons saw her falling and caught her to keep her from drowning. They called for help from other animals. One of the animals to come was tortoise, and the woman climbed onto his back. The animals decided the woman should have earth to live on. The tortoise told them all to dive to the bottom of the sea to bring up some earth. Many tried but failed. Finally toad dived; he came back exhausted and almost dead, but with mud in his mouth. Tortoise gave it to the woman, who spread it around the tortoise's shell. It extended on all sides, forming a country. The woman was pregnant with twins, Tijuskeha and Tawiskarong. Tawiskarong, the evil one, was not born in the usual manner, instead he broke through his mother's side, killing her. Her body was buried, and from it came plants and trees. Tijuskeha created useful animals, and Tawiskarong created fierce and monstrous ones. The two brothers eventually dueled; Tijuskeha prevailed and killed his brother, but Tawiskarong's spirit appeared and said he had gone to the far west and that all men would also go to the west when they died.

A Creation Story (Hindu)

There are many creation stories in the Hindu tradition. Hinduism teaches that there are times that the universe takes form and times that it dissolves back into nothing.

Before time there was no heaven or earth. A vast dark ocean washed upon the shores of nothingness. A giant cobra floated in the waters, with the Lord Vishnu asleep in its coils. From far below, a humming sound, Om, began. This sound soon filled the emptiness with energy. As dawn broke, from Vishnu's navel grew a beautiful lotus flower. In the middle of the flower sat Vishnu's servant, Brahma. Vishnu ordered Brahma to create the world, and then he vanished. Brahma split the lotus flower into three parts. He formed one part into the heavens, another into the earth, and from the third he fashioned the sky. Brahma then created trees and plants of all kinds. Next he created animals, insects, birds, and fish.

from belief in Christianity in Europe. This same creation story is shared by other major religions as well, including Judaism and Islam. Interestingly, if another set of beliefs were common in Europe at the time of the development of science, the hypothesis about the origins of living species might have been quite different. Essay 8.1 summarizes the creation stories of some other religious traditions.

Consider special creation as an alternative to the theory of common descent for explaining how modern organisms came to be. Because the idea of special creation requires the action of a supernatural entity—an all-powerful creator—it is not itself a scientific hypothesis. As discussed in Chapter 1, in order for a hypothesis to be testable by science, it must be able to be evaluated through observations of the material universe. Since a supernatural creator is not observable or measurable, there is no way to determine the existence or predict the actions of such an entity. However, the belief of special creation does provide

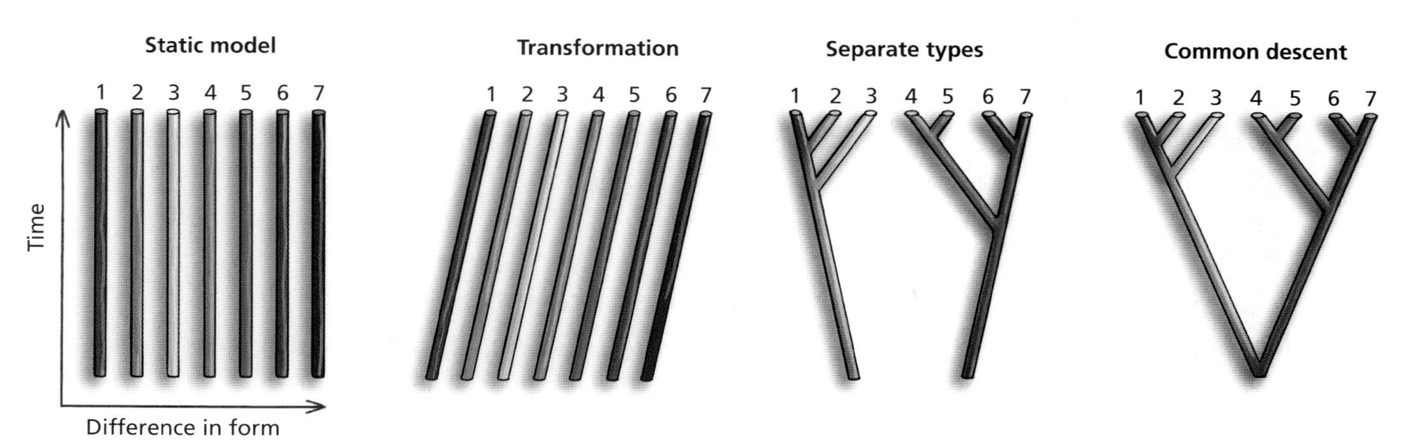

Figure 8.4 Four hypotheses about the origin of modern organisms. A graphical representation of the four hypotheses.

some scientifically testable hypotheses. For instance, that organisms came into being within the last 10,000 years, and they have not changed substantially since their creation. We can call these hypotheses about the origin and relationship among living organisms *the static model*.

There are also several intermediate hypotheses between the static model and evolution. One intermediate hypothesis is that organisms were created, perhaps millions of years ago, and that changes have occurred, but brand-new species have not arisen; we will call this the *transformation* hypothesis. Another intermediate hypothesis is that types of organisms arose separately and since their origin have diversified into numerous species; we will call this hypothesis *separate types*. These hypotheses and the theory of common descent are summarized in Figure 8.4. Polls of the American public indicate that many people feel that these four alternatives—static model, transformation, separate types, and common descent—are all equally reasonable explanations for the origin of biological diversity.

Based on the feelings of many Americans, perhaps the Kansas State Board of Education made a reasonable choice when it decided to remove the theory of evolution (and thus the theory of common descent) from the state's science standards. Rather than bowing to the arguments of scientists who teach only one of many ideas about the origin of modern organisms, the school board elected to allow school districts to suppress discussion of the topic altogether. In Ohio, perhaps the state school board was trying to be fair when they considered requiring the discussion of other hypotheses in addition to the theory of common descent.

But what about all those scientists who insist that the theory of common descent is fact? Why would so many scientists maintain this position? You shall see that the three alternative hypotheses are not equivalent to the theory of common descent. To understand why, we must evaluate the observations that help us test these hypotheses. We will do this as we address one of the most controversial questions underlying this debate: Are humans really related to apes?

Any zookeeper will tell you that the primate house is the most popular exhibit in the park. People love chimps, apes, and monkeys. It is easy to see why—primates are curious, playful, and agile. In short, they are fun to watch. But there is something else that drives our fascination with these wonderful animals: We see ourselves reflected in them. The forward placement of their eyes and their reduced noses appear humanlike; they have hands with fingernails instead of paws with claws; some can stand and walk on two legs for short periods; they

Figure 8.5 Are humans related to apes? Biologists contend that apes and humans are similar in appearance and behavior because we share a common ancestor.

can finely manipulate objects with their fingers and opposable thumbs; they exhibit extensive parental care; and even their social relations are similar to ours—they tickle, caress, kiss, and pout (Figure 8.5).

Why are primates, particularly the great apes (gorillas, chimpanzees, and bonobos) so similar to humans? Scientists contend that it is because humans and apes are descendents of a common biological ancestor.

The Biological Classification of Humans

Humans have long recognized our similarities with the apes. Cultures with close contact with these animals often gave them names that reflect this similarity—such as *orangutan*, a Malay word that is translated as "person of the forest." The Greek naturalist Aristotle, whose 2,000-year-old writings form some of the basis for modern Western science, organized the living world into a linear chain from what he perceived as the lowest form to the highest forms and placed the great apes a step below humans.

As the modern scientific community was developing in the sixteenth and seventeenth centuries, a number of different methods for organizing biological diversity developed. Many of these **classification systems** grouped organisms by similarities in habitat, diet, or behavior; some of these classifications placed humans with the great apes, others did not.

Into the classification debate stepped Carolus Linnaeus, a Swedish physician and botanist. Linnaeus gave all species of organisms a two-part, or *binomial*, name in Latin, which was the common language of science at the time. These Latin names typically contained information about the species' traits—for instance, *Acer saccarhum* is Latin for "maple tree that produces sugar," the tree commonly known as the sugar maple, while *Acer rubrum* is Latin for "red maple."

In addition to the binomial naming system, Linnaeus developed a new way to organize living organisms according to shared physical similarities. His classification system was arranged hierarchically—organisms that shared many traits were placed in the same narrow classification, while those that shared fewer, broader traits, were placed in more comprehensive categories. The hierarchy took the following form, from broadest to narrowest groupings:

Kingdom
Phylum (or Division)
Class
Order
Genus
Species

Thus, for example, all organisms that were able to move under their own power, at least for part of their lives, and relied on other organisms for food were placed in the Kingdom Animalia. Within that kingdom, all organisms with backbones (or another skeletal structure called a *notochord*) were placed in the same phylum, Chordata, and all chordates that possess fur and produce milk for their offspring were placed in the Class Mammalia, the mammals. Humans are mammals, as are dogs, lions, dolphins, and monkeys. The scientific name of a species contains information about its classification as well—for instance, humans, *Homo sapiens*, belong to the genus *Homo*.

Other scientists quickly adopted the logical and orderly Linnaean system of classification, and it became the standard practice for organizing biological diversity. Later scientists added a new level, family, placed between order and genus (Figure 8.6). Even more recently, biologists have added "sub" and "super" levels

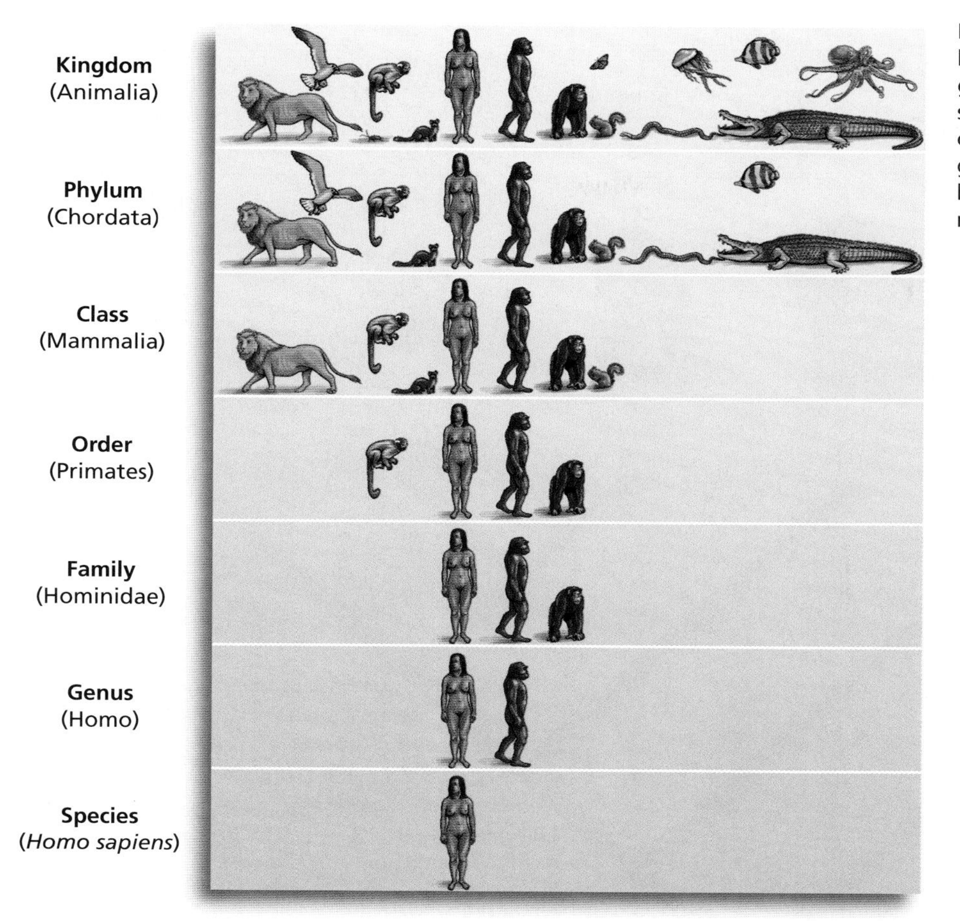

Figure 8.6 The Linnaean classification of humans. Linnaeus's classification system groups organisms into progressively smaller categories. All organisms within a category share basic characteristics—as the groups become narrower toward the bottom of the figure, the organisms look more and more similar.

between these categories as well—such as superfamily between family and order. The recent research discussed in Chapter 3 describes another classification level above kingdom—the domain.

Using this classification system, Linnaeus placed humans, monkeys, and apes in the same order, which he called Primates, because humans have forward-facing eyes and coordinated hands like other primates. The modern classification of humans reflects only refinements of Linnaeus's ideas. Among living primates, humans are most similar to apes. Humans and apes share a number of characteristics, including relatively large brains, more erect posture, lack of a tail, and increased flexibility of the thumb. Scientists place humans and apes in the same family, Hominidae. Humans and the African great apes (gorillas, chimpanzees, and bonobos) share even more characteristics, including elongated skulls, short canine teeth, and reduced hairiness and are placed together in the same *sub*family Homininae.

Linnaeus himself did not believe that evolution occurred—his purpose in developing a biological classification was to determine what he called "God's plan" for the living universe. However, Darwin used Linnaeus's classification as a major facet of his argument for the theory of evolution. Darwin argued that Linneaus's system was an effective way to organize biological diversity *because it reflected the underlying biological relationships among living organisms*. Darwin noted that the levels in Linnaean classification could be interpreted as different degrees of relationship. In other words, all species in the same family share a relatively recent common ancestor, while all families in the same class share a more distant common ancestor. The relationship among species implied by Linnaean

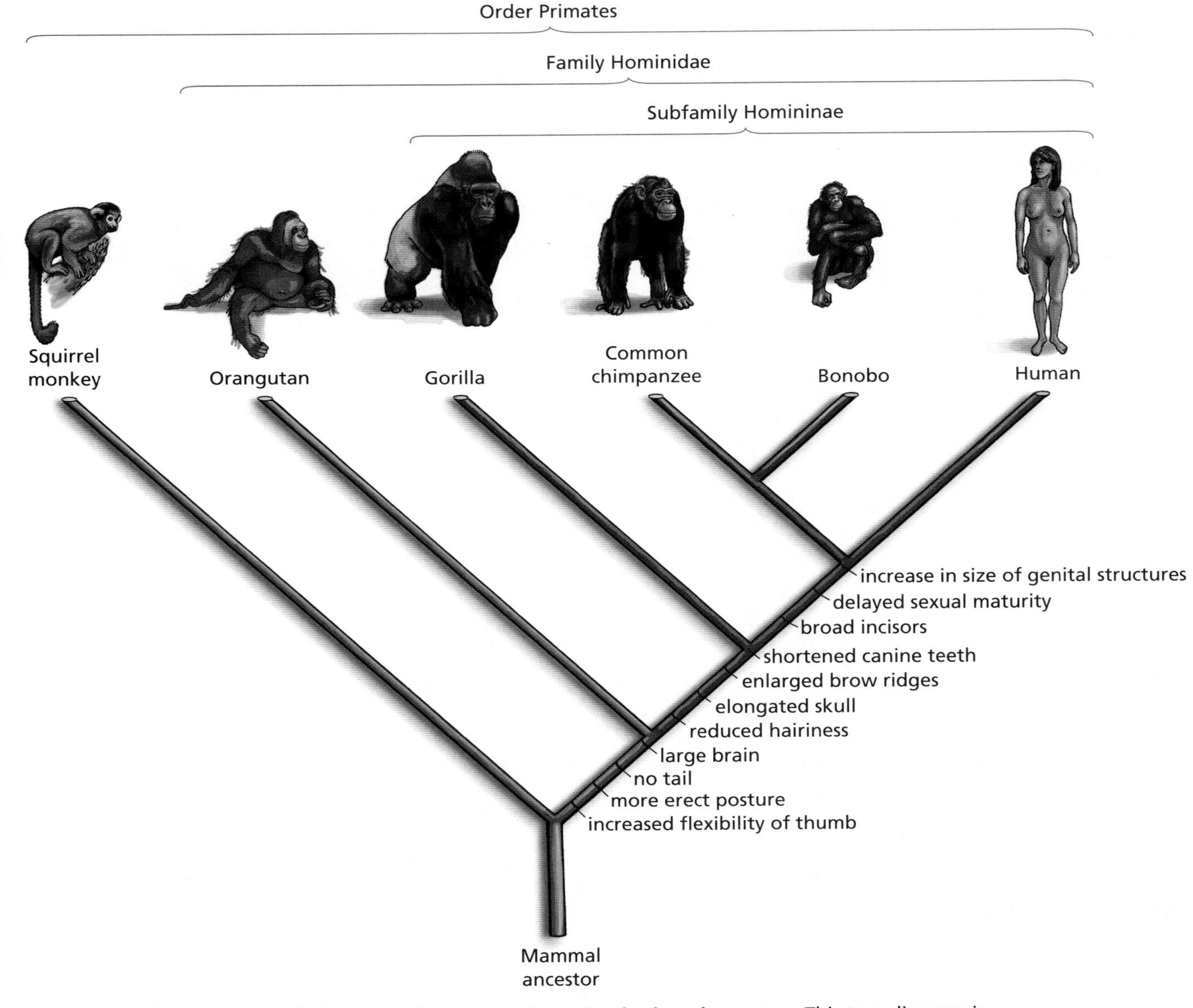

Figure 8.7 Shared characteristics among humans and apes imply shared ancestry. This tree diagram is suggested by the current classification of humans and apes. Characteristics noted on the side of the evolutionary tree are shared by all of the species above that point. The common ancestor of the Hominidae family must have had all of the characteristics shared by the modern members of the family.

classification is typically illustrated with a tree diagram. If we draw the classification of humans, great apes, and other primates as a tree diagram (Figure 8.7), we can see why Darwin concluded that humans and modern apes evolved from the same ancestor.

Does Classification Reflect a Biological Relationship Between Humans and Apes?

The fact that different organisms share many traits does not necessarily indicate that they share biological ancestry. Linnaeus's classification alone does not support any of the four competing hypotheses about the origin of modern organisms. However, the tree of relationships *implied* by Linnaeus's classification forms a

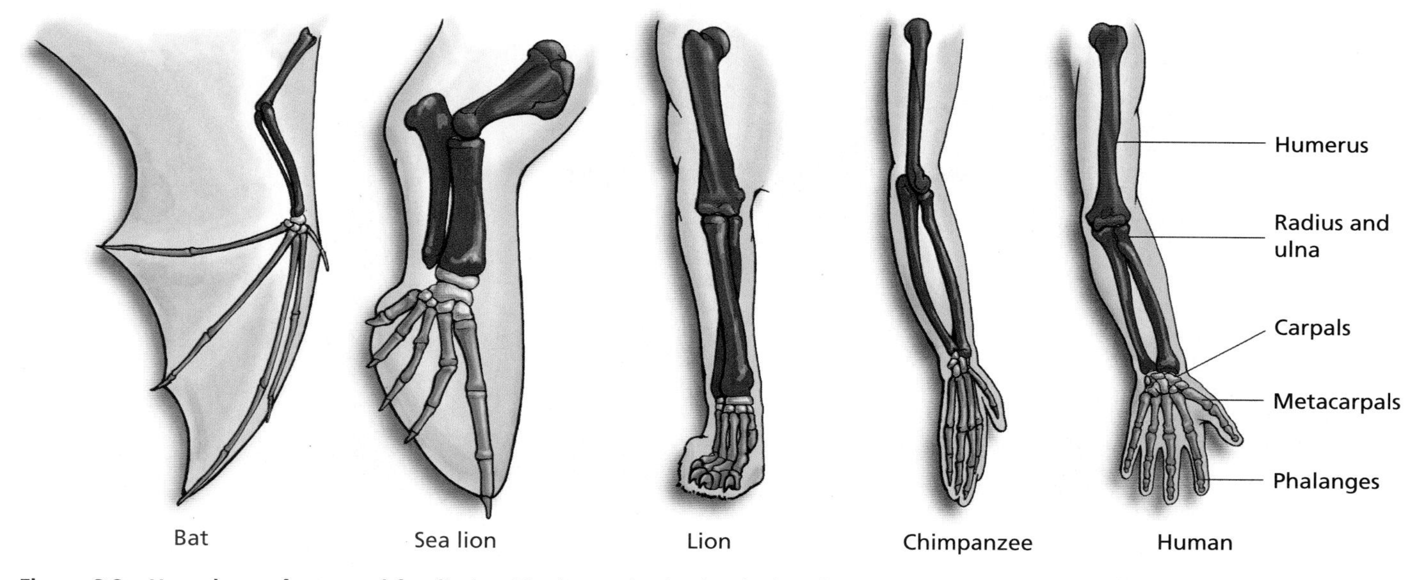

Figure 8.8 Homology of mammal forelimbs. The bones in the forelimbs of these mammals are very similar, despite the fact that they are used for very different functions. Equivalent bones in each organism are shaded the same color. The similarity in basic bone structure may be evidence of shared origin.

hypothesis that can be tested. If modern species represent the descendants of ancestors that also gave rise to other species, we should be able to observe other, less obvious similarities between them in anatomy and genetic material. And if Darwin is right, we should be able to identify extinct organisms that show a transition between a common ancestor and its modern descendants.

WWW
Media Activity 8.2 Evolution of Whales

The Anatomical Evidence Figure 8.8 illustrates the concept of **homology**, similarity in characteristics resulting from common ancestry. Each of the mammal forelimbs pictured has a very different function—bat wings are used for flight, sea lion flippers for swimming, lion legs for running, and human arms for grasping and throwing. However, each of these limbs shares a common set of bones that are in the same relationship to each other, even if they are quite different in size and extent. The most likely explanation for the similarity in the underlying structure of these limbs is that each species inherited the basic structure from the same common ancestor, and the process of evolution led to their unique modification in each group.

Some critics of the theory of evolution counter that a skilled designer—that is, a supernatural creator—could have simply produced these different forelimbs using the same creative base plan. However, some similarities among species appear to be less logical than one would expect from an intelligent designer. For instance, humans contain a number of **vestigial traits**, which appear to have been modified from functional traits found in an ancestor. These traits either do not function in humans or have a function that is highly modified from that of other descendants of the same ancestor. In other words, vestigial traits represent a vestige, or remainder, of our biological heritage. Figure 8.9 provides two examples of vestigial traits in humans. Great apes and humans have a tailbone like other primates, yet neither great apes nor humans have a tail. Additionally, all mammals possess tiny muscles called *arrector pili* at the base of each hair. When the arrector pili contract under conditions of emotional stress or cold temperatures, the hair is elevated. In furry mammals, the arrector pili help to increase the perceived size of the animal, and they increase the insulating value of the hair coat. In humans, the same conditions only produce goosebumps, which provide neither benefit.

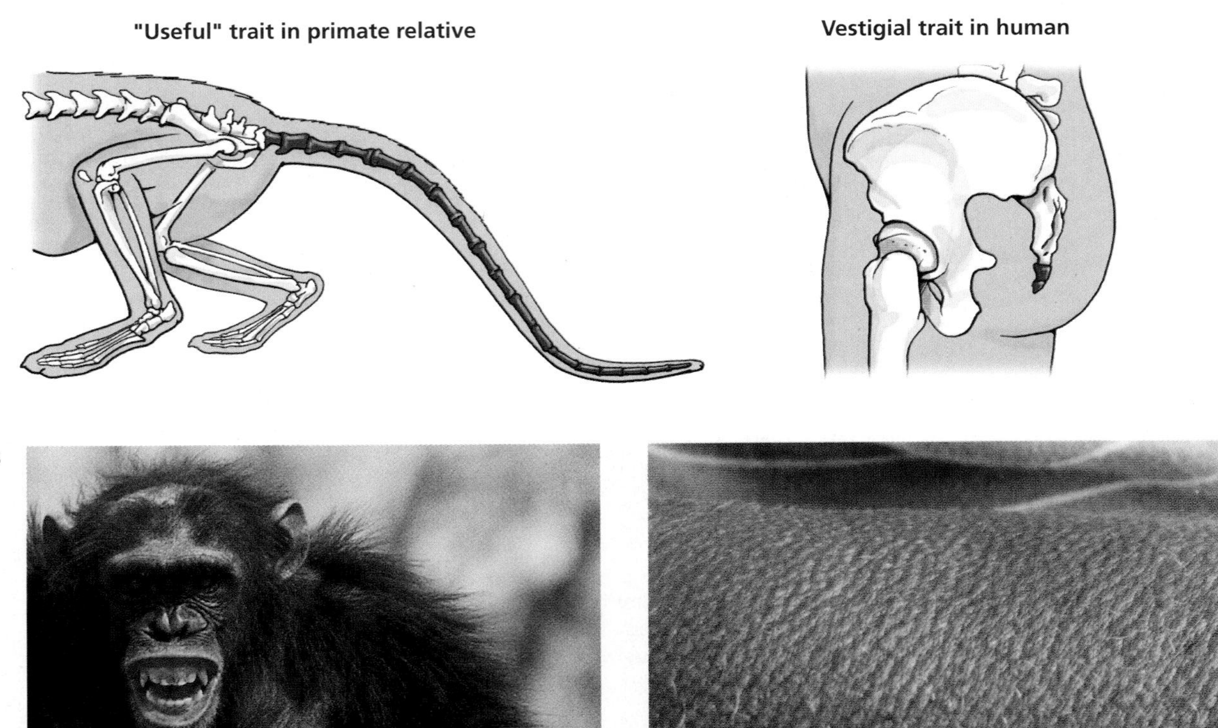

Figure 8.9 Vestigial traits reflect our evolutionary heritage. (a) Humans and other great apes do not have tails, but they do have a vestigial tailbone, which corresponds to the functional tailbone of a monkey. (b) The ability to elevate their fur helps many mammals seem bigger and provides increased insulation in cold conditions. The vestiges of this trait in humans appear as goosebumps, which arise under similar conditions of cold and intense emotion but serve no known function.

Darwin maintained that the hypothesis of evolution provided a better explanation for vestigial structures than the hypothesis of special creation. Because Linnaeus's classification system showed that similarities among species occurred in both useful and vestigial traits, Darwin argued that the Linnaean system must be a reflection of shared ancestry. A useless trait such as goosebumps is better explained as the result of inheritance from our biological ancestors than poor design.

The Biochemical Evidence Since Darwin's time, scientists' understanding of the nature of biological inheritance has expanded immensely. Scientists now understand that differences among individuals arise in large part from differences in their genes. It stands to reason that differences among species must also derive from differences in their genes. If the hypothesis of common descent is correct, species that appear to be closely related must have more similar genes than species that are more distantly related.

The most direct way to measure the overall similarity of two species' genes is to evaluate how similar their DNA is. Recall from Chapter 6 that DNA molecules carry genetic information in the sequence of chemical bases making up linear structures. A single gene on a DNA molecule may be made up of a few hundred

(a) Comparing DNA sequences

(b) Similarity to human DNA sequences

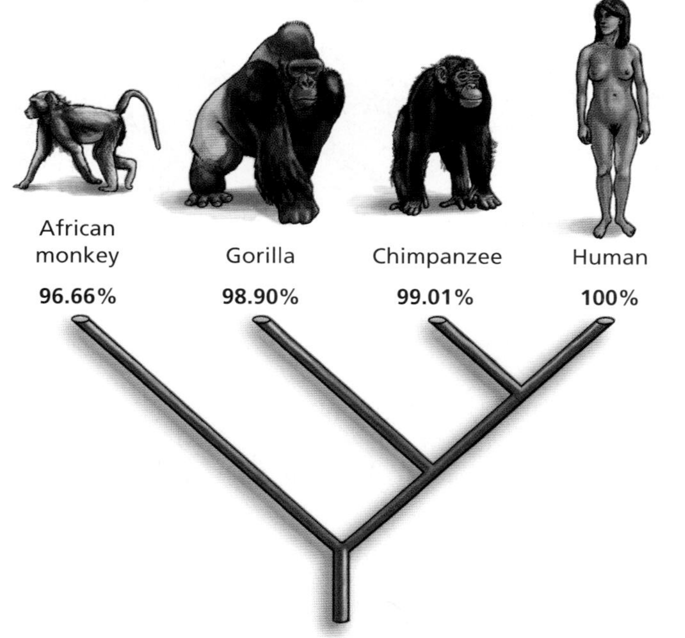

Figure 8.10 Similar organisms have similar DNA sequences. (a) The order of DNA bases contains information about the traits of an organism. We can compare the DNA of different organisms by looking at similarities and differences in the DNA sequences for various genes. (b) Species that appear to be more similar to humans have more similar DNA sequences for the same genes compared to species that are less similar.

to a few thousand bases—this sequence contains information about the structure of a protein, the physical product of this stored information.

Many genes are found in nearly all living organisms. For instance, genes that code for the proteins that help neatly store DNA inside cells are found in algae, fungi, fruit flies, humans, and all other organisms that contain linear chromosomes. Among organisms that share many aspects of structure and function, such as mammals, many genes are shared. However, the *sequences* of these genes are not identical. If we compare the sequence of DNA bases in the same gene found in two different mammals, we find that the more similar their classification, the more similar their genes are (Figure 8.10a). In other words, if classification indicates that two mammals share a recent common ancestor, their DNA sequences are more similar than two mammals that share a more ancient common ancestor.

A comparison of the sequences of dozens of genes that are found in humans and other primates demonstrates this pattern (Figure 8.10b). The DNA sequences of these genes in humans and chimpanzees are 99.01% similar, while humans and gorillas are identical over 98.9% of their length. More distantly related primates are less similar to humans in DNA sequence. This pattern of similarity in DNA sequence exactly matches the biological relationships implied by physical similarity.

At first, this result may not seem especially surprising. If genes are like instructions, you would expect the instructions for building a human and a chimpanzee to be more similar than the instructions to build a human and a monkey. After all, humans and chimpanzees have many more similarities than humans and monkeys. However, remember that the genes being compared perform the same function in all of these species. For example, one of genes in this DNA analysis is *BRCA1*, a gene associated in humans with an increased risk of breast

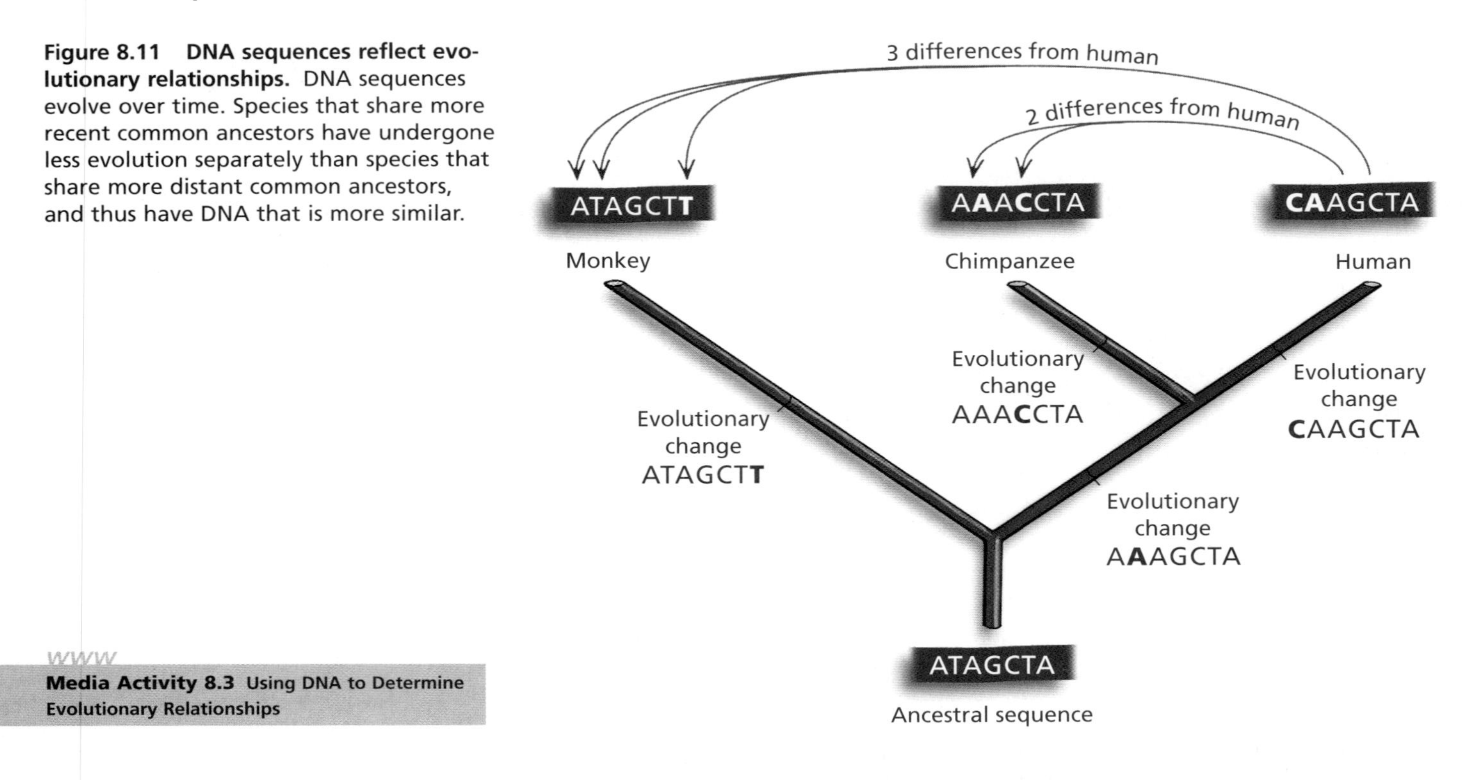

Figure 8.11 DNA sequences reflect evolutionary relationships. DNA sequences evolve over time. Species that share more recent common ancestors have undergone less evolution separately than species that share more distant common ancestors, and thus have DNA that is more similar.

www
Media Activity 8.3 Using DNA to Determine Evolutionary Relationships

and ovarian cancer, but which has the general function of helping repair damage to DNA in all organisms. Given this very basic function, there is no reason to expect that differences in the gene sequences should conform to any particular pattern—if organisms show no biological relationship. But the *BRCA1* gene of humans *is* more like the *BRCA1* gene of chimpanzees than the *BRCA1* gene of monkeys. The best explanation for this observation is that humans and chimpanzees share a more recent common ancestor with each other than either species does with monkeys (Figure 8.11).

The differences in DNA sequence between humans and chimpanzees also allow us to estimate when these two species diverged from their common ancestor. This estimate is based on a *molecular clock* that has been derived from observations of DNA sequence differences in a number of species groups. The principle behind a molecular clock is that the rate of change in DNA sequences, due to the accumulation of mutations within a species, seems to be relatively constant. According to the molecular clock hypothesis, the amount of time required to generate a 0.99% difference in overall DNA sequence (the difference between modern humans and modern chimpanzees) is approximately five to six million years.

Does the Fossil Record Demonstrate a Relationship Between Humans and Apes?

Observations of anatomical and genetic similarities among modern organisms provide good evidence to support the theory of evolution. However, as with nearly all evidence in science, it allows us to infer the accuracy of the hypothesis but does not prove the hypothesis correct. The European scientific community in the nineteenth century was convinced by Darwin's arguments about the origin of similarities and embraced the theory of common descent as the best explanation for the origin of species. Since Darwin's time, scientists have accumulated additional indirect evidence, such as the DNA

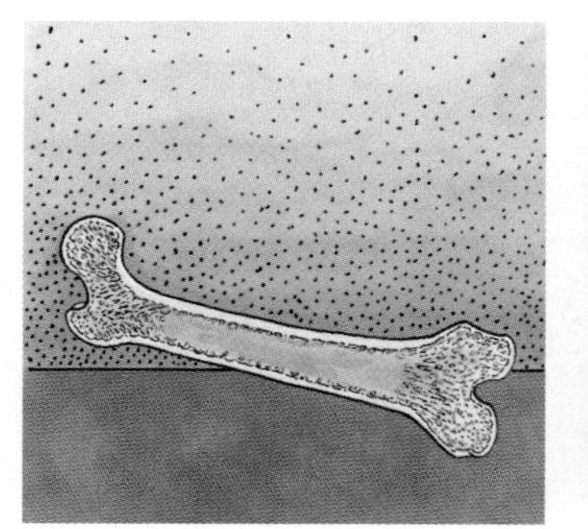

1. An organism is rapidly buried in water, mud, sand, or volcanic ash. The tissues begin to decompose very slowly.

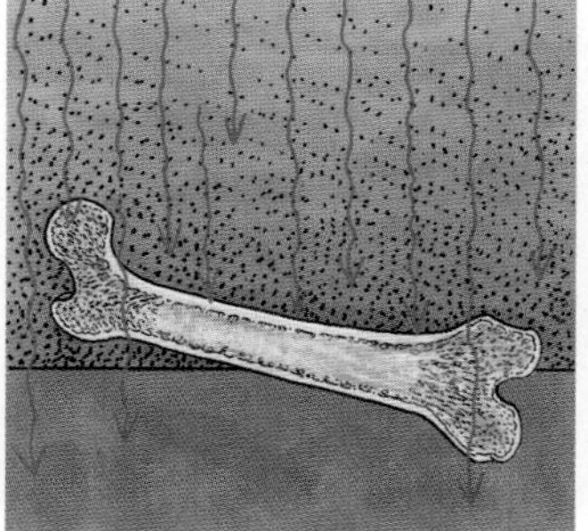

2. Water seeping through the sediment picks up minerals from the soil and deposits them in the spaces left by the decaying tissue.

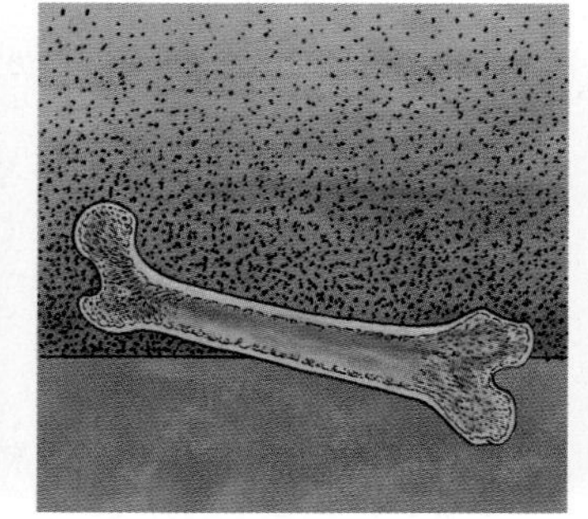

3. After thousands of years, most or all of the original tissue is replaced by very hard minerals, resulting in a rock model of the original bone.

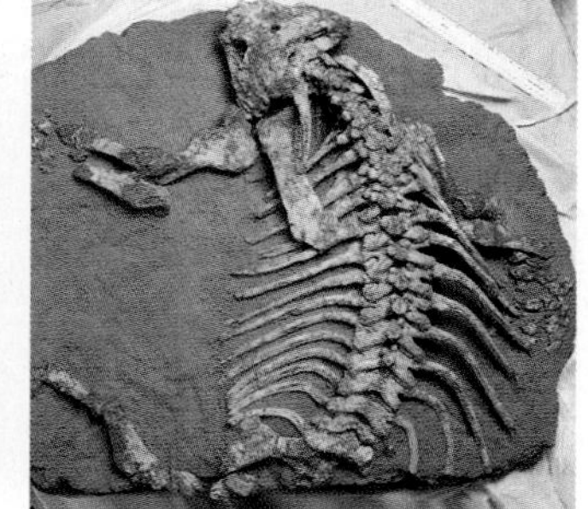

4. When erosion or human disturbance removes the overlying sediment, the fossil is exposed (as shown here looking from above).

Figure 8.12 Fossilization. When bones fossilize, the material that makes up much of their substance slowly decays and is replaced by minerals from water seeping through the sediments they are buried in. Eventually what remains is a rock "model" of the original bone.

sequence similarity discussed above, to support this theory. But the direct evidence of the genealogy of life provided by the fossil record is even more convincing evidence for the theory of common descent.

The Fossil Evidence **Fossils** are the remains left in soil or rock of plants or animals that once existed. Most fossils of large animals are rocks that have formed as the organic material in bone decomposed and the spaces left behind were filled by minerals (Figure 8.12). The process of fossilization requires special conditions because most organisms quickly decompose after death or are scavenged by other organisms; to form fossils, dead organisms have to be protected from these processes. Fossils are thus more likely to form when an organism is rapidly buried by water, mud, or volcanic ash. Fortunately for scientists looking for fossils of *hominins*, humans and human ancestors, these organisms were likely to live near water.

Fossils were well known by Darwin's time. In the early 1800s, *paleontologists*, scientists who search for, describe, and study ancient organisms, were beginning to describe the large fossil remnants of dinosaurs, and the first complete skeleton of a dinosaur was found in southern New Jersey in 1858, one year before *The Origin of Species* was published. In later books, Darwin noted that convincing evidence for human relationships to apes would come from the fossils of human ancestors, which he predicted would be found in Africa, the home of modern chimpanzees.

One key difference between humans and other apes is our mode of locomotion. While chimpanzees and gorillas use all four limbs to move, humans are *bipedal*—that is, they walk upright on only two limbs. The reason bipedalism evolved is unclear. Some scientists, including Darwin, hypothesized that the upright gait was an advantage that freed the hands for tool use. More recently, scientists have hypothesized that an upright posture allowed early hominins to reach fruit in small trees and shrubs. Most likely, bipedalism evolved because it had many benefits to our ancestors.

Whatever its origin, bipedalism evolved through several anatomical changes: The face is now placed on the same plane as the back, instead of at a right angle to it; thus the *foramen magnum*, the hole in the skull that allows the passage of the spinal cord, is found on the back of the skull in other apes, but

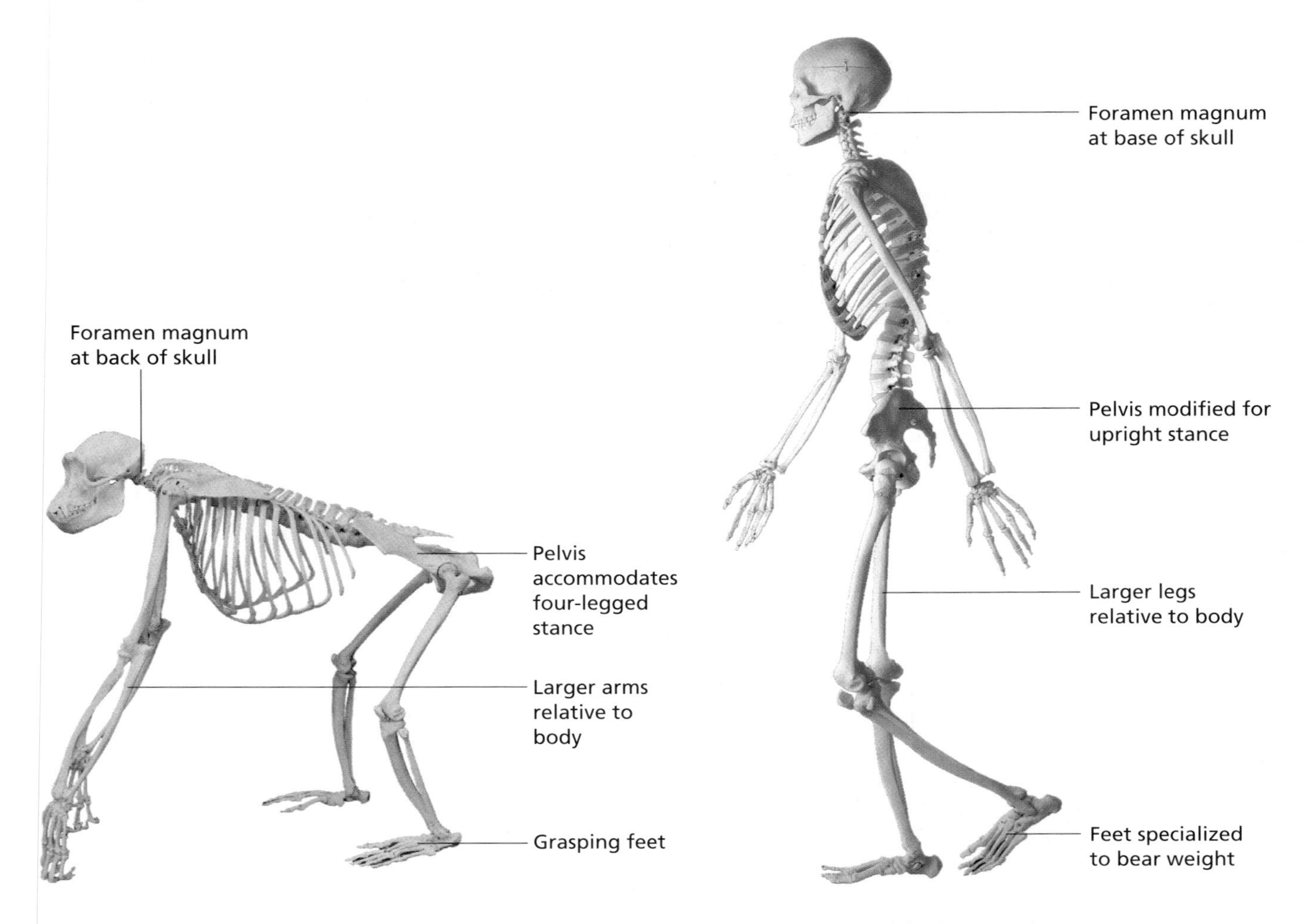

Figure 8.13 Anatomical differences between humans and chimpanzees. Humans are bipedal animals, while chimpanzees typically travel on all fours. The evolution of bipedalism in hominins resulted from several anatomical changes. If any of these features are present in a fossil primate, the fossil is classified as a hominin.

at the base of the skull in humans. In addition, in bipedal apes, the structures of the pelvis and knee are modified for an upright stance, the foot is changed from being grasping to weight bearing, and the lower limbs are elongated relative to the front limbs. Thus, a variety of skeletal features can provide clues about the locomotion of a fossil apelike creature (Figure 8.13).

The first fossil hominins were not found in Africa, but in Europe. A fossil of *Homo neanderthalensis* (Neanderthal man) was discovered by Johann Fuhlrott in 1856 in a small cave in the Neander Valley in Germany. Fuhlrott recognized his find as a primitive human, but the German scientific establishment rejected his interpretation, incorrectly claiming that it was a modern human. Eugene Dubois found fossils of manlike creatures now called *Homo erectus* (standing man) in Java in 1891. It was not until 1924 that the first *African* hominin fossil was found, the Taung child, described by Raymond Dart. This fossil was later placed in the species *Australopithecus africanus*. Paleontologists continue to discover new hominin fossils in southern and eastern Africa, including the famous "Lucy," a remarkably complete skeleton of the species *Australopithecus afarensis*, discovered by paleontologist Donald Johanson in 1974 (Figure 8.14). Lucy's fossil skeleton included a large section of her pelvis, clearly indicating that she walked upright.

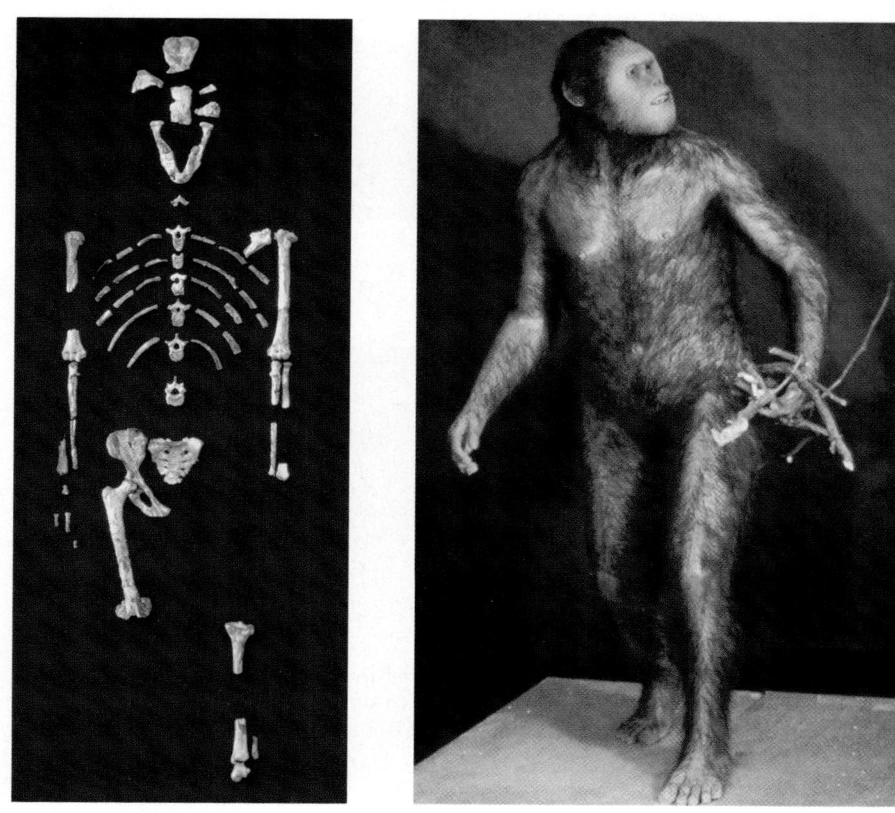

Figure 8.14 A fossil ancestor. (a) Lucy is still the most complete fossil of *Australopithecus afarensis* ever found. Her pelvis and knee joint provide evidence that she walked on two legs. (b) This artist's conception of what Lucy looked like in life is based on her fossil remains as well as other fossils of the same species.

When Did Fossil Hominins Live? The date an ancient fossil organism lived can be determined by estimating the age of the rock that surrounds the fossil. **Radiometric dating** relies on a natural process that results in change in particular *chemical elements*, the basic building blocks of matter. This process is called *radioactive decay* and results as radioactive elements in the rock spontaneously break down into different elements, called "daughter products." Each radioactive element decays at its own unique rate. The rate of decay is measured by the element's *half-life*: the amount of time required for ½ of the amount of the element originally present to decay into the daughter product.

When rock is newly formed from cooled magma, the liquid underlying Earth's crust, it contains a fixed amount of any radioactive element. When the magma hardens, this radioactive element becomes trapped within the resulting rock. As the element decays over time, the amount of the original element in the rock declines and, correspondingly, the amount of daughter product increases. By determining the ratio of radioactive element to daughter product in a rock sample and knowing the half-life of the radioactive element, scientists can estimate the number of years that have passed since the rock formed (Figure 8.15). Some critics of the theory of common descent note that different scientists may calculate different dates for the same fossil and that all of these fossil dates represent estimates with a certain degree of potential error. This uncertainty about the exact age of a fossil occurs because the layers of rock containing fossils did not form from magma, meaning that most fossils cannot be dated directly. Instead, fossils are found in rocks formed from the sediments that initially buried the organism. When fossils are found in sedimentary rock between layers of magma-formed rocks, the fossils are assumed to be intermediate in age to the magma-formed rocks. Even though fossils cannot be

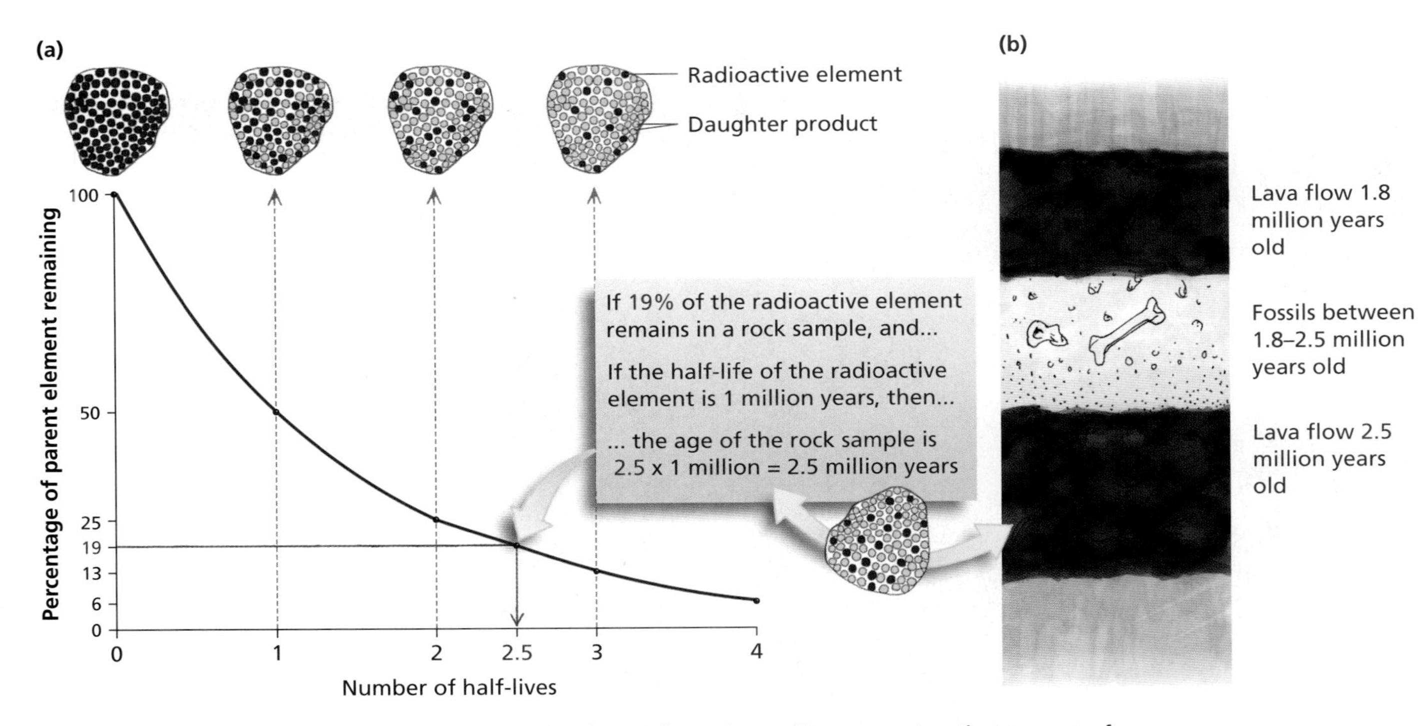

Figure 8.15 Radiometric dating. (a) The age of rocks can be estimated by measuring the amount of radioactive material (designated by dark purple circles) with a known half-life and the amount of daughter material (designated by light blue circles) in a sample of rock. (b) The age of a fossil can be estimated when it is found between two layers of magma-formed rock.

aged with perfect accuracy, the age of particular fossil species inferred from the age of surrounding rocks are always within the same general range. Fossil dating helps place a timeline on the historical record of living organisms.

Scientists have used radiometric dating to estimate the age of the Earth and the time of origin of various groups of organisms. Using this technique, scientists have also determined that the most ancient hominin fossil, the species *Ardepithecus ramidus*, is 5.2 to 5.8 million years old. An even older fossil species, *Orrorin tugenensis*, the famous "millennium man," was recently described as a six-million-year-old human ancestor, but many scientists are reserving judgment about this fossil specimen until more are found. These very early fossils probably represent hominins that are quite similar to the common ancestor of humans and chimpanzees.

Is There a Missing Link? As the number of described hominin fossils has increased, a tentative genealogy of humans has emerged. The fossil species can be arranged in a tentative pedigree from most ancient to most modern species by determining the age of a fossil and the anatomical similarities among organisms (Figure 8.16). What this pedigree indicates is that modern humans are the last remaining branch of a once diverse group of hominins. But does the pedigree provide convincing evidence that modern humans evolved from a common ancestor with other apes?

The common ancestor of humans and chimpanzees is often called the "missing link." Because the common ancestor of humans and chimpanzees has not been identified, some critics of evolution say that the biological relationship between apes and humans remains unproven.

However, finding the fossilized common ancestor between chimpanzees and humans, or between any two species for that matter, is extremely difficult, if not impossible. In order to identify a common ancestor, the evolutionary history of

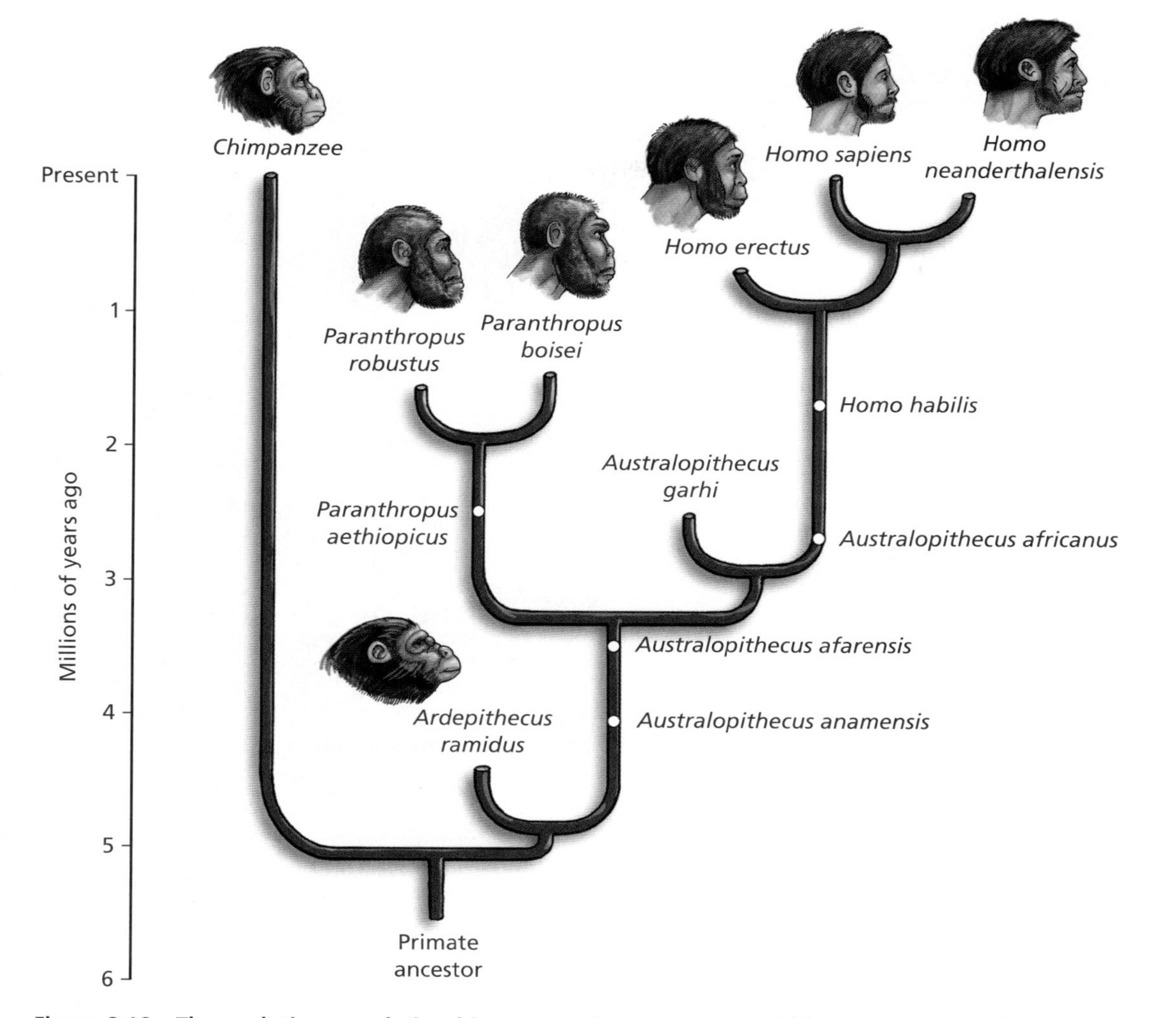

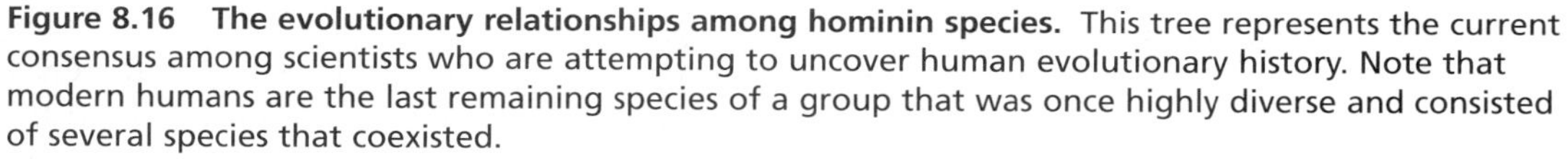
Figure 8.16 The evolutionary relationships among hominin species. This tree represents the current consensus among scientists who are attempting to uncover human evolutionary history. Note that modern humans are the last remaining species of a group that was once highly diverse and consisted of several species that coexisted.

both species since they diverged must be clear. Like humans, modern chimpanzees have been evolving over the five million years since they diverged from humans—in other words, a missing link would not look like a modern chimpanzee with some human features or a cross between the two species (Figure 8.17). While the history of hominins is becoming clearer as new fossils are being identified, much less work has been done on the evolutionary history of chimpanzees.

Furthermore, if we examine the theory of common descent more closely, we can see that accepting this theory does not *require* the identification of a missing link. The theory of evolution is also supported by evidence of intermediate forms between a modern organism and its ancestors. This evidence should be much easier to locate, and it is the type of evidence provided by the hominin fossil record (review Figures 8.14 and 8.16). Besides being bipedal, humans differ from other apes in having a relatively large brain, a flatter face, and a more extensive culture. The oldest hominins are bipedal but are otherwise similar to other apes in skull shape, brain size, and probable lifestyle. More

Figure 8.17 The common ancestor of humans and chimpanzees? Although in popular representations the missing link between humans and chimpanzees is pictured as in this satirical image of Charles Darwin, both humans and chimpanzees have been evolving separately for at least five million years, and their common ancestor is unlikely to resemble either modern species.

modern hominins show greater similarity to modern humans, with flattened faces and increased brain size (Figure 8.18). Even younger fossil finds indicate the existence of symbolic culture and extensive tool use, trademarks of modern humans.

The ancient hominin–modern human transition is not the only fossil record that supports the theory of common descent. Examples of well-described transitions include one between ancient reptiles and mammals, and another between ancient and modern horses (Figure 8.19); many others have been found as well.

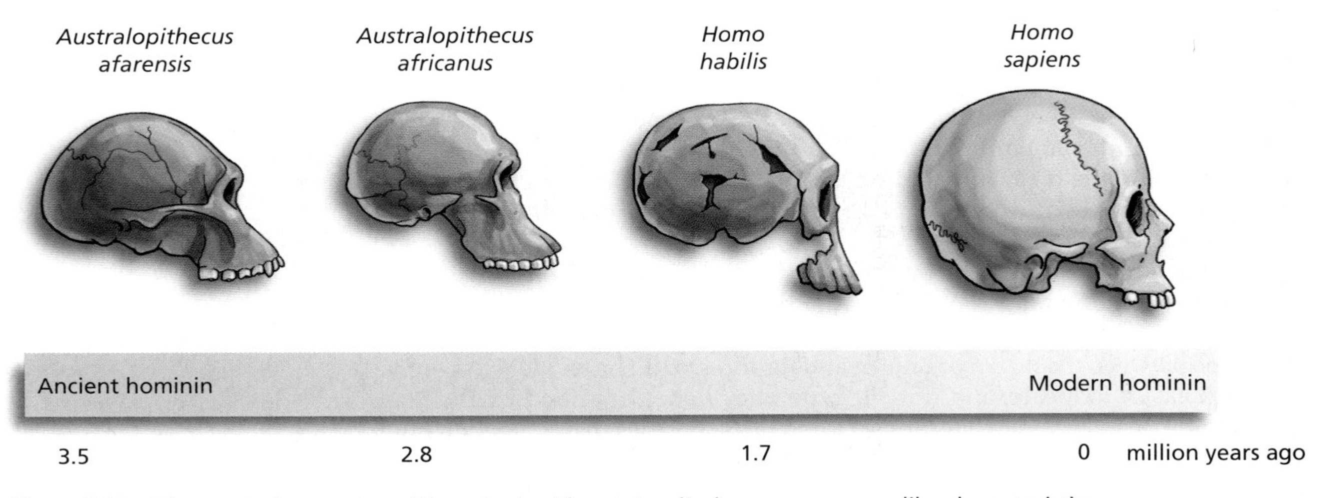

Figure 8.18 The ape-to-human transition. Ancient hominins display numerous apelike characteristics, including a large jaw, small brain case, and receding forehead. More recent hominins have a reduced jaw, larger brain case, and smaller brow ridge, much like modern humans.

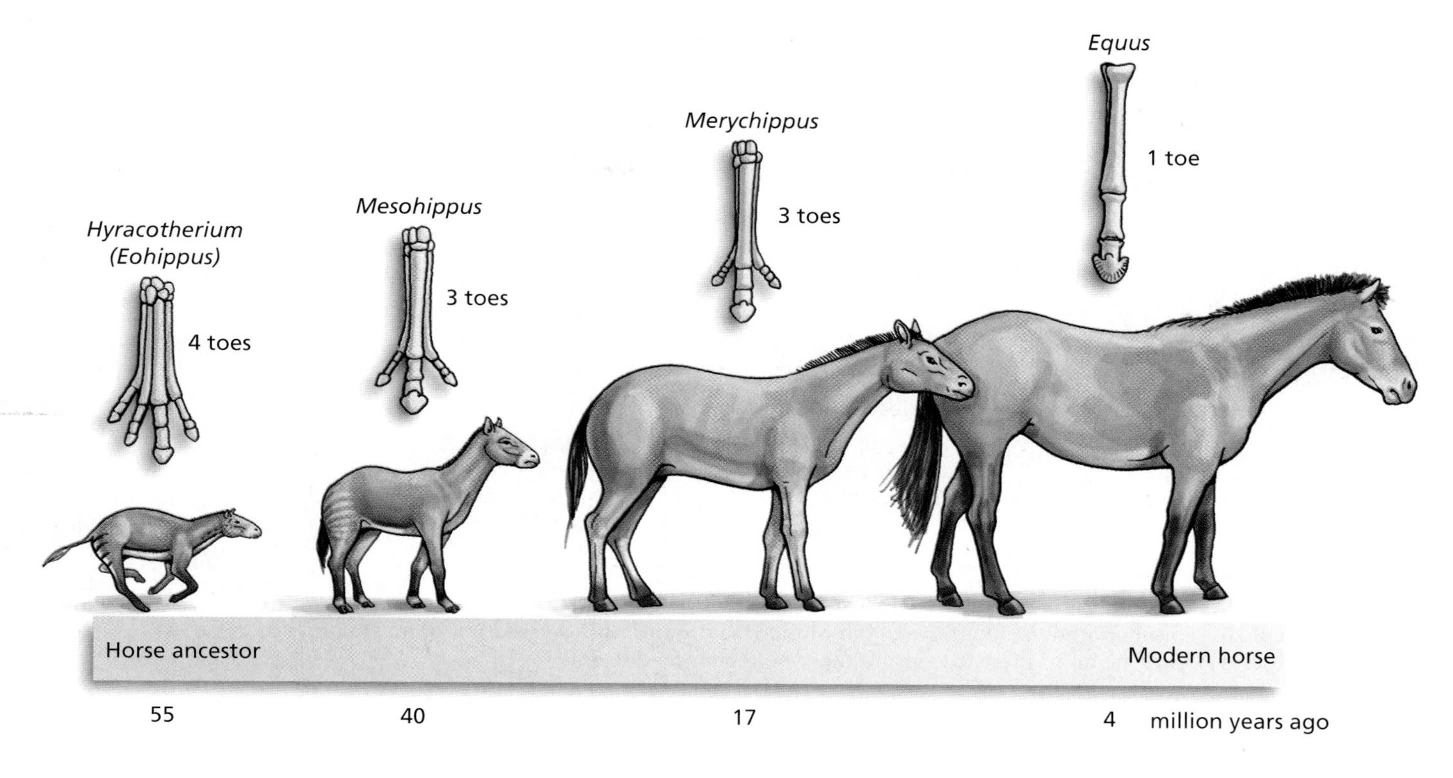

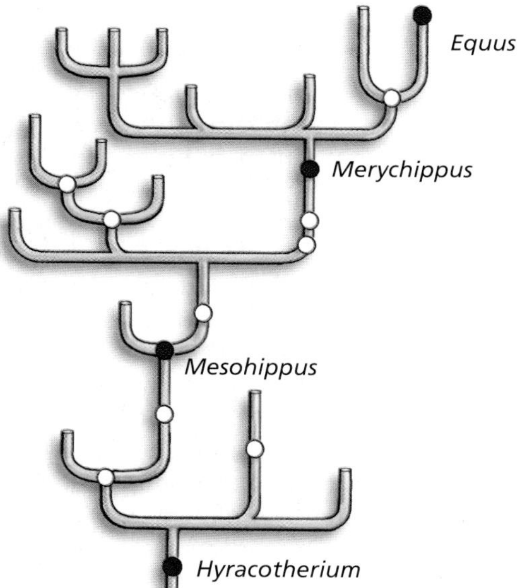

Figure 8.19 The fossil record of horses. Horse fossils provide a fairly complete sequence of evolutionary change from small, catlike animals with five toes to the modern horse, with one massive toe.

8.4 Evaluating the Hypotheses

Now we return to the four competing hypotheses: static model, transformation, separate types, and common descent. Do the observations described in the previous section allow us to reject any of these hypotheses? Figure 8.20 summarizes our findings.

The physical evidence we have discussed thus far allows us to clearly reject only one of the hypotheses—the static model. Radiometric dating indicates that Earth is far older than 10,000 years, and the fossil record provides

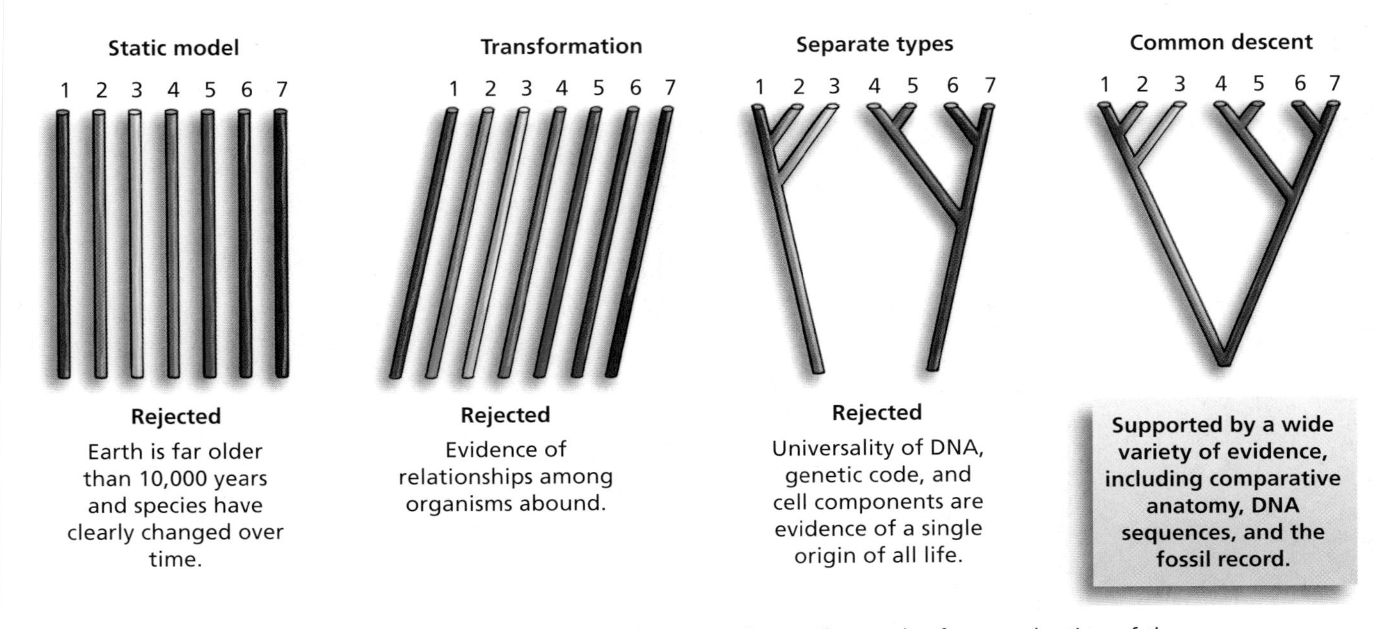
Static model
1 2 3 4 5 6 7
Rejected
Earth is far older than 10,000 years and species have clearly changed over time.
Transformation
1 2 3 4 5 6 7
Rejected
Evidence of relationships among organisms abound.
Separate types
1 2 3 4 5 6 7
Rejected
Universality of DNA, genetic code, and cell components are evidence of a single origin of all life.
Common descent
1 2 3 4 5 6 7
Supported by a wide variety of evidence, including comparative anatomy, DNA sequences, and the fossil record.

Figure 8.20 Four hypotheses about the origin of modern organisms. The result of our evaluation of these hypotheses is to reject all of them except the theory of common descent.

unambiguous evidence that the species that have inhabited this planet have changed over time.

Of the remaining three hypotheses, transformation is the poorest explanation of the observations. If organisms arose separately, and each changed on its own path, there is no reason to expect that different species would share structures—especially if these structures are vestigial in some of the organisms. There is also no reason to expect similarities among species in DNA sequence. The hypothesis of transformation predicts that we will find little evidence of biological relationship among living organisms. As our observations have indicated, evidence of relationship abounds.

Both the hypothesis of common descent and the hypothesis of separate types contain a process by which we can explain observations of relationship. That is, both hypothesize that modern species are descendants of common ancestors. The difference between the two theories is that common descent hypothesizes a single common ancestor for all living things, while separate types hypothesizes that ancestors of different groups arose separately and then gave rise to different types of organisms. Separate types seems more reasonable than common descent to many people. It seems impossible that organisms as different as pine trees, mildew, ladybugs, and humans share a common ancestor. However, several observations indicate that these disparate organisms are all related.

The most compelling evidence for the single origin of all life is the universality of both DNA and of the relationship between DNA and proteins. As you learned in Chapter 7, genes from bacteria can be transferred to plants, and the plants will make a functional bacterial protein. This is only possible because both bacteria and plants translate genetic material into functional proteins in an identical manner. If bacteria and plants arose separately, we could not expect them to translate genetic information identically.

The fact that organisms as different as pine trees, mildew, ladybugs, and humans contain cells with nearly all the same components, processes of cell division, and biochemistry is also evidence of shared ancestry. A mitochondrion could have many different possible structural forms and still perform the same function; the fact that the mitochondria in a plant cell and an animal cell are essentially identical implies that both groups of organisms received these mitochondria from a common ancestor.

Pine trees, mildew, ladybugs, and humans *are* very different. Proponents of the hypothesis of separate types argue that the differences among these organisms could not have evolved in the time since they shared a common ancestor. (See Essay 8.2 to explore hypotheses about the origin of the common ancestor of all life.) But the length of time pine trees, mildew, ladybugs, and humans have been diverging is immense—at least 1,000,000,000 years. The remaining basic similarities among all living organisms serve as evidence of their ancient relationship (Figure 8.21).

Scientists favor the theory of common descent because it is the best explanation for how modern organisms came about. The theory of evolution—including the theory of common descent—is robust, meaning that it is a good explanation for a variety of observations and well supported by a wide variety of evidence from anatomy, geology, molecular biology, and genetics. The theory of common descent is no more tentative than atomic theory; few scientists disagree with the models that describe the basic structures of atoms, and few disagree that the evidence for the theory of common descent is overwhelming. Most scientists would say that both of these theories are so well supported that we can call them fact.

Evolutionary theory helps us understand the functions of human genes, comprehend the interactions among species, and predict the consequences of

Figure 8.21 The unity and diversity of life. The theory of evolution, including the theory of common descent, provides the best explanation for how organisms as distinct as pine trees, humans, mildew, and ladybugs can look very different while sharing a genetic code and many aspects of cell structure and cell division.

Essay 8.2 The Origin of Life

The origin of the first living cell is an active research question in biology. There are two scientific hypotheses regarding the source of this first organism:

Hypothesis 1: The common ancestor arose on another planet and was imported to Earth.

Hypothesis 2: The common ancestor arose on Earth through natural processes from nonliving materials.

To test hypothesis one, scientists look for life on more distant planets. Astronomers can use a tool called spectroscopy to analyze light reflecting from the surface of these planets for the chemical "signature" of life. Spectroscopy allows researchers to determine if a planet has oxygen and methane in its atmosphere, two gases that are only maintained by the action of living organisms. None of the planets nearest Earth display this chemical signature.

At least one meteorite has been described as containing evidence of life—the famous "Martian meteorite," ALH84001,0, found on Antarctic ice in 1984 (Figure E8.2). Some scientists studying this meteorite argued in a 1996 paper that it contained convincing evidence of bacterial life, including fossilized cells, and crystals and other chemicals only known from living organisms. Other scientists argue that these apparent "cells" are really structures created by the impact of the meteorite with boiling hot water—perhaps the event that jettisoned the rock from Mars.

Hypothesis 2, the most well-studied hypothesis about the evolution of life, states that the common ancestor arose on Earth through natural processes from nonliving materials. According to this hypothesis, the process can be broken down into three basic steps.

1. Nonbiological processes assembled the simple molecules that were present early in the history of the solar system into more complex molecules.
2. These molecules then assembled themselves into chains that could store information and/or drive chemical reactions.
3. Collections of these complex molecules were assembled into a self-replicating "cell," with a membrane and energy source. This cell fed on other complex molecules.

There is some experimental evidence to support all three steps of Hypothesis 2. First, Stanley Miller, a young graduate student working in the lab of his mentor Harold Urey in 1953, developed a laboratory apparatus that attempted to recreate conditions on early Earth. After allowing the apparatus, which contained very simple molecules and an energy source, to "run" for one week, Miller found that complex molecules had formed spontaneously. These molecules included the building blocks of proteins and sugars. His results and others support step 1 of this hypothesis. More recent experiments have demonstrated that these building-block chemicals can be induced to form long chains when put in contact with hot sand, clay, or rock. Long chains of DNA and RNA nucleotides, and of amino acids have been created via these methods, providing some experimental support for step 2. Finally, in the early 1980s, two teams of scientists demonstrated that an information-carrying molecule, RNA, could also potentially copy itself, so at least part of step 3 has experimental support.

(a) Martian meteorite

(b) "Cell-like" fossils inside the Martian meteorite

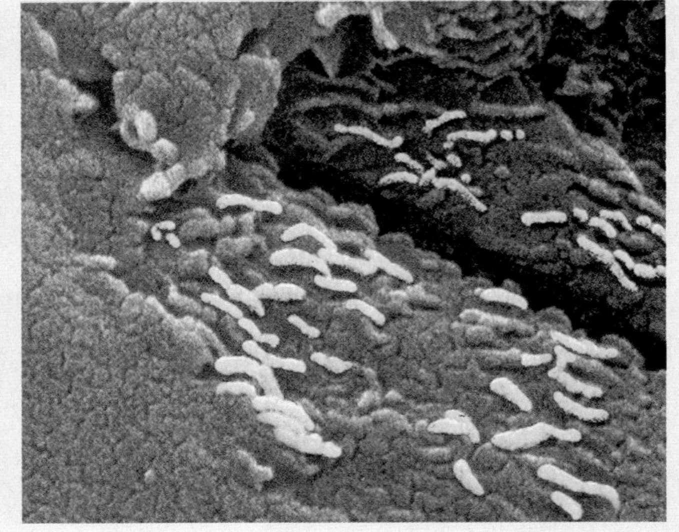

Figure E8.2 The Martian meteorite, ALH840001.

a changing global environment for species. Describing evolution as "*just* a theory" vastly understates the importance of evolutionary theory as a foundation of modern biology. Despite the continued controversies that arise periodically around the country, the citizens of Kansas finally realized this. They decided that students who do not have a grasp of this fundamental biological principle would lack an appreciation of the basic unity and diversity of life, and would fail to understand the implications of evolutionary history and change on the natural world and ourselves. (We will explore why evolution is important for understanding and treating human disease in Chapter 9.) In the elections of 2000, Kansans voted the majority of state school board members who promoted the "evolutionless" science standards out of office. The new school board moved quickly to repeal the old standards and replaced them with standards emphasizing the centrality of evolution to the understanding of biology.

CHAPTER REVIEW

Summary

- The process of evolution is the change that occurs in the characteristics of organisms in a population over time.
- The theory of evolution, as described by Charles Darwin, is that all modern organisms are related to each other and arose from a single common ancestor.
- Linnaeus classified humans in the same order with apes and monkeys based on his observations of physical similarities between these organisms. Darwin argued that Linnaeus's classification was strong support for the theory of common descent.
- The existence of vestigial structures is difficult to explain except via the theory of common descent.
- Modern data on similarities of DNA sequences among organisms also indicate a close biological relationship between humans and apes.
- The fossil record indicates that humanlike animals appeared about five million years ago.
- As predicted by the theory of common descent, ancient hominins have more apelike characteristics than more modern hominins.
- Shared characteristics of all life, especially the universality of DNA and the relationship between DNA and proteins, provide evidence that all organisms on Earth descended from a single common ancestor.

Key Terms

biological evolution p. 190
classification system p. 196
common descent p. 192
fossils p. 203
homology p. 199
microevolution p. 190
populations p. 190
radiometric dating p. 205
scientific theory p. 191
special creation p. 193
species p. 190
theory of evolution p. 191
vestigial traits p. 199

Learning the Basics

1. Describe the theory of common descent.
2. What observations did Charles Darwin make on the Galapagos Islands that helped convince him that evolution occurs?
3. What is a vestigial structure and how does the existence of these structures support the theory of common descent?
4. What information can we gain from fossils, and how does the fossil record provide support for the theory of common descent?

5. What is a "missing link" and why are they difficult to locate and identify?

6. The process of biological evolution _____.
   a. is not supported by scientific evidence
   b. results in a change in the features of individuals in a population
   c. takes place over the course of generations
   d. b and c are correct
   e. a, b, and c are correct

7. In science, a theory is a(n) _____.
   a. educated guess
   b. inference based on a lack of scientific evidence
   c. idea with little experimental support
   d. a body of scientifically acceptable general principles
   e. statement of fact

8. The theory of common descent states that all modern organisms _____.
   a. can change in response to environmental change
   b. descended from a single common ancestor
   c. descended from one of many ancestors that originally arose on Earth
   d. have not evolved
   e. can be arranged in a hierarchy from "least evolved" to "most evolved"

9. Darwin's observations on the voyage of the *HMS Beagle* convinced him that _____.
   a. new species can arise via evolution
   b. modern species are related to each other
   c. similar species share a common ancestor
   d. species can change over time
   e. all of the above

10. Most biologists would agree that the statement, "all modern organisms derive from a single common ancestor," is _____.
    a. a tentative hypothesis
    b. a fact
    c. one of several equally likely hypotheses about the origin of modern organisms
    d. a statement of faith
    e. probably incorrect

11. Which of the following lists places the classification levels in order from broadest grouping to narrowest grouping?
    a. family, phylum, genus, order
    b. phylum, family, genus, class
    c. order, genus, species, phylum
    d. kingdom, order, genus, species
    e. class, phylum, family, order

12. Which of the following is a vestigial trait in humans?
    a. goosebumps
    b. forearms
    c. opposable thumbs
    d. foramen magnum
    e. body hair

13. The DNA sequence for the same gene found in several species of mammal _____.
    a. is identical among all species
    b. is equally different between all pairs of mammal species
    c. is more similar between closely related species than between distantly related species
    d. provides evidence for the hypothesis of common descent
    e. more than one of the above is correct

14. What characteristics of a fossil can paleontologists use to determine whether the fossil is a part of the human evolutionary lineage?
    a. the position of the foramen magnum
    b. the structure of the pelvis
    c. the structure of the foot
    d. a and c are correct
    e. a, b, and c are correct

15. The fossil record of hominins _____.
    a. does not indicate relationship between humans and apes, because a missing link has not been found
    b. dates back at least five million years
    c. indicates that bipedal apes first evolved in Africa
    d. b and c are correct
    e. a, b, and c are correct

Analyzing and Applying the Basics

1. The classification system devised by Linnaeus can be "rewritten" in the form of an evolutionary tree. Draw a tree that illustrates the relationship among these flowering species, given their classification (note that "subclass" is a grouping between class and order):

   Pasture rose (*Rosa carolina*, family Rosaceae, order Rosales, subclass Rosidae)

   Live forever (*Sedum purpureum*, family Crassulaceae, order Rosales, subclass Rosidae)

   Spring avens (*Geum vernum*, family Rosaceae, order Rosales, subclass Rosidae)

   Spring vetch (*Vicia lathyroides*, family Fabaceae, order Fabales, subclass Rosidae)

   Multiflora rose (*Rosa multiflora*, family Rosaceae, order Rosales, subclass Rosidae)

2. DNA is not the only biochemical that is used to test for evolutionary relationships among organisms. Proteins can also be used, and the sequences of their building blocks (called amino acids) can be compared in much the same way DNA sequences are compared. *Cytochrome c* is a protein found in nearly all living organisms—it functions in the transformation of energy within cells. The percent difference in amino acid sequence between humans and other organisms is summarized in the following table:

Organism	Percent Difference from Human *Cytochrome c* Sequence
Chimpanzee	0.0%
Mouse	8.7%
Donkey	10.6%
Carp	21.4%
Yeast	32.7%
Corn	33.3%
Green Algae	43.4%

Draw the evolutionary tree implied by this data that illustrates the relationship between humans and the other organisms listed.

3. The following couplet was written by Ralph Waldo Emerson.

Striving to become a man, the worm
Mounts through all the spires of form.

According to your understanding of the theory of common descent, is this an accurate reflection of how evolution occurs? Explain.

4. Whales and dolphins are sea-dwelling mammals that evolved from land-dwelling ancestors. Describe two pieces of evidence that would help support this hypothesis.

5. Some critics of evolution argue that the theory is not scientific because it is not falsifiable. Are there observations or data that would cause scientists to consider rejecting the theory of common descent? Is the theory of evolution falsifiable?

Connecting the Science

1. Humans and chimpanzees are more similar to each other genetically than many very similar-looking species of fruit fly are to each other. What does this similarity imply regarding the usefulness of chimpanzees as "stand-ins" for humans during scientific research? What do you think it implies in terms of our moral obligations to these animals?

2. Creationists have argued that if students learn that humans descended from animals and are, in fact, a type of animal, these impressionable youngsters will take this fact as permission to act on their "animal instincts." What do you think of this claim?

3. Many high school biology teachers do not teach the theory of evolution, or cover it only briefly, because it is controversial in society. Is the theory of evolution an essential piece of science education? Is there any harm done to students who do not learn the theory of evolution in high school? Is there any harm done to society if a large segment of the public knows little about this theory?

Media Activities

Media Activity 8.1 Microevolution in Bacteria
Estimated Time: 5 minutes
Examine the results of natural selection in a population of *Mycobacterium tuberculosis*, the bacterium that causes the disease tuberculosis.

Media Activity 8.2 Evolution of Whales
Estimated Time: 30 minutes
Scientists have made some exciting new discoveries about the evolution of whales.

Media Activity 8.3 Using DNA to Determine Evolutionary Relationships
Estimated Time: 20 minutes
In addition to anatomical evidence, scientists can now use DNA sequences to trace evolution. This article reports the findings that present-day populations in Africa are descendants of the ancestral human population.

CHAPTER 9

Evolving a Cure for AIDS

Natural Selection

At the height of his pro basketball career, Magic Johnson retired in November 1991 after he learned that he had tested positive for HIV.

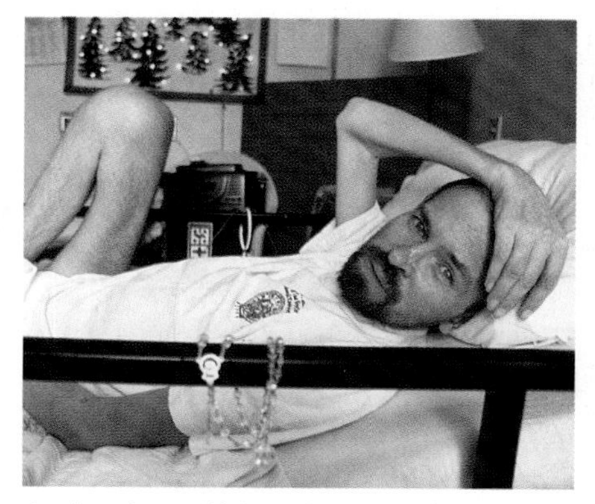

At the time of his retirement, fans expected that Magic would meet the fate of HIV patients such as this man—death from AIDS.

In late 1991, basketball fans around the world were hit with devastating news. Earvin "Magic" Johnson, one of the greatest basketball guards ever to play the game, was retiring at age 32, two years after being named the NBA's most valuable player for the third time in 10 seasons. In his relatively short career, Magic broke the record for the most career assists and led the Los Angeles Lakers to five NBA championships. Why was this talented, popular, and successful athlete with a new wife and a baby on the way leaving the game just as he was reaching his physical prime? He had learned only days before that he was infected with HIV, the virus that causes AIDS.

More than 10 years after his retirement, Magic is healthy and successful. Why?

In 1991, a diagnosis of HIV infection was considered to be a death sentence. At that time, the typical length of time between the diagnosis of HIV and death from AIDS was eight to ten years. Magic Johnson's fans and other NBA players steeled themselves to watch the terrible decline that always occurred with the onset of AIDS.

Fast forward to November 2002, 11 years after Magic's diagnosis. The fit and muscular 43-year-old NBA Hall of Fame member is now part owner of the Los Angeles Lakers and head of a company that owns dozens of movie theaters, coffee shops, and a fast-food franchise. In the time since his diagnosis,

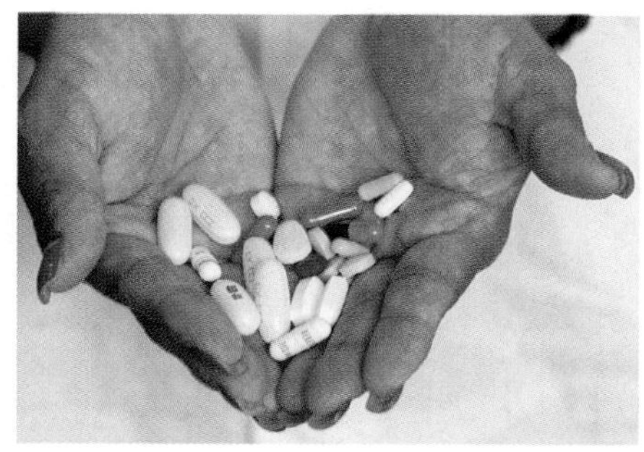

Because modern anti-AIDS therapy has disabled a powerful tool of HIV—evolution.

Magic has won an Olympic gold medal, made two comebacks as an NBA player, coached his former team, celebrated two more Laker NBA championships, and hosted his own late-night television show. He is about as successful as any former sports star, and just about as healthy. Now many of his friends joke that he will be hit by a bus before he dies of AIDS.

Magic Johnson's survival in spite of his HIV infection is partly a testament to the huge effort government scientists have invested to control AIDS. The time since the identification of this new disease and the first drug treatments was less than a decade. Although anti-HIV drugs have been available since 1987, five years before Magic's announcement, most people did not remain healthy for long after they started using these drugs. The failure of these early treatment strategies was the result of a single factor—evolution—and the success of current treatments depends on the understanding and management of this powerful process.

In this chapter, we will explore how the process of evolution has shaped HIV and governed our methods for controlling this killer virus.

9.1 AIDS and HIV

Acquired immune deficiency syndrome, or **AIDS**, was first described in 1981 after dozens of young gay men in New York City and San Francisco were diagnosed with illnesses rarely seen in healthy young people. The susceptibility to these illnesses appeared to be *acquired* (that is, caused by exposure to some factor) because it was seen suddenly in large numbers of individuals.

AIDS Is a Disease of the Immune System

The increased susceptibility to illness in these men resulted from a decline in their immune-system function. As is discussed in detail in Chapter 11, the role of the immune system is to maintain the integrity of the body. The cells of the immune system constantly patrol the tissues and organs of the body for anything that is not clearly produced by the body—that is, anything that is "non-self." Upon encountering non-self entities, the immune system acts to eliminate it—this is known as an *immune response*. A non-self substance, object, or organism typically has a unique chemical signature that causes an immune response. This signature is called an *antigen*. Immune-system cells called *lymphocytes* respond to antigens. Lymphocytes carry proteins on their cell membranes, called *receptors*, which recognize and are attracted to particular antigens (Figure 9.1). The binding of an antigen to an antigen receptor on a lymphocyte starts the immune response. Our bodies can make 100 trillion to 1 quintillion *different types* of antigen receptors, and about 100 million distinct receptors are present on lymphocytes in our bloodstream at any given time. The incredible diversity in antigen receptors is responsible for the immune system's ability to respond to a huge range of non-self objects.

The virus that causes AIDS primarily kills or disables a particular class of lymphocytes called **T4 cells**. T4 cells are also known as *helper T cells* because they serve as the directors of the immune system's response to an antigen. Thus, loss of T4 cells causes *immune deficiency*—that is, affected individuals experience diseases that are normally controlled by healthy immune systems. These include infections by organisms commonly found on our bodies in low levels, such as *Pneumocystis carinii*, a fungus that is found in nearly everyone's

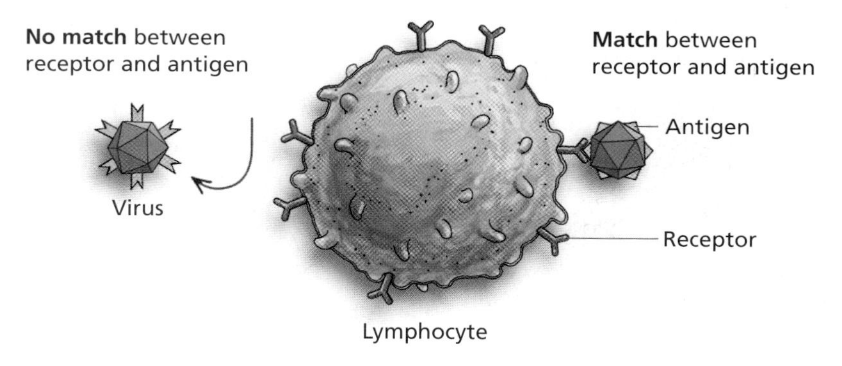

Figure 9.1 Lymphocytes recognize particular antigens. This lymphocyte has receptors for the antigens that are present on one virus but not on another. When an antigen receptor binds to an antigen, a cascade of events causes an immune-system response and often the elimination of the antigen.

lungs by age 30. In healthy people, *P. carinii* is held in check by the immune system, but in AIDS patients, this organism often causes pneumonia and extensive lung damage. Diseases like *P. carinii* pneumonia are called **opportunistic infections** because they only occur when the opportunity arises due to a weakened immune system.

Because individuals with weakened immune systems may have more than one opportunistic infection, each with its own signs and symptoms, there is no single disease that is *always* associated with AIDS. This is why the disease is called a *syndrome*—a group of signs and symptoms indicating that an individual has AIDS. Primary among those signs is the depletion of T4 cells.

HIV Causes AIDS

Within months of the initial reports of this new disease, it became clear that AIDS could be transmitted through both sexual intercourse and contact with the blood of affected individuals. By 1983, scientists in France and the United States had identified the factor causing the transmission and symptoms of AIDS—later named the **Human Immunodeficiency Virus**, or **HIV**. (The evidence linking HIV to AIDS is outlined in Essay 9.1.) Worldwide, the majority of HIV transmission is via sexual intercourse without a condom. In the United States, both unprotected sex and the sharing of needles by injection-drug users are primary modes of HIV transmission.

HIV is a simple structure composed of RNA and essential proteins called *enzymes*, all surrounded by a protein envelope and a membrane coat (Figure 9.2). As with all viruses, HIV can only reproduce by forcing the cells of its **host**, the organism it is infecting, to make copies. HIV does this by first binding to a protein on the surface of a cell (the CD4+ receptor), and then releasing its RNA and enzymes into the cell. The majority of cells infected with HIV are T4 cells, but other cells that carry the CD4+ receptor are susceptible to HIV as well. Once inside the cell, the viral RNA is *reverse transcribed* into viral DNA by the action of one of the viral enzymes. Chapter 7 describes that transcription occurs in cells when the information in DNA is rewritten into the language of RNA—reverse transcription is simply the converse of that process. With the help of another viral enzyme, the viral DNA then inserts itself into the cell's *genome*, where it commandeers the cellular machinery for copying genetic material and producing proteins. In this way, HIV forces the cell to make copies of the virus. The cell now makes new copies of the viral RNA, translates the genes on the viral DNA into the proteins that make up the coat, enzymes, and membrane surface proteins, and assembles new viruses. The newly made copies of the virus then leave the cell by budding off the cell membrane and go on to infect other cells that possess the CD4+ receptor. Infection with HIV usually either disables or kills the host cell. The *life cycle* of HIV is summarized in Figure 9.2, and Chapter 11 provides additional details about the biology of viruses.

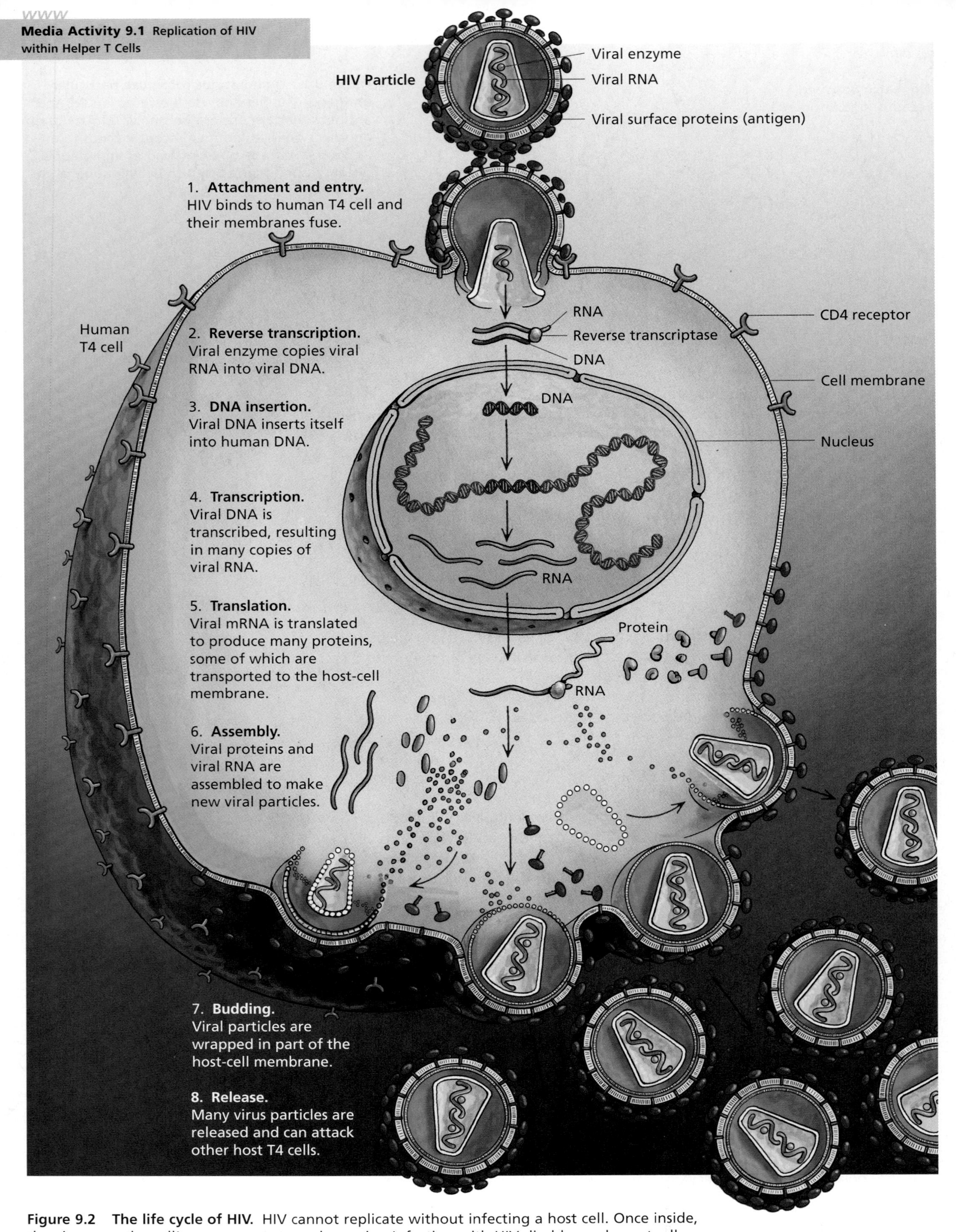
WWW
Media Activity 9.1 Replication of HIV within Helper T Cells
HIV Particle
Viral enzyme
Viral RNA
Viral surface proteins (antigen)
1. Attachment and entry.
HIV binds to human T4 cell and their membranes fuse.
Human T4 cell
2. Reverse transcription.
Viral enzyme copies viral RNA into viral DNA.
RNA
Reverse transcriptase
DNA
CD4 receptor
Cell membrane
3. DNA insertion.
Viral DNA inserts itself into human DNA.
DNA
Nucleus
4. Transcription.
Viral DNA is transcribed, resulting in many copies of viral RNA.
RNA
5. Translation.
Viral mRNA is translated to produce many proteins, some of which are transported to the host-cell membrane.
Protein
RNA
6. Assembly.
Viral proteins and viral RNA are assembled to make new viral particles.
7. Budding.
Viral particles are wrapped in part of the host-cell membrane.
8. Release.
Many virus particles are released and can attack other host T4 cells.

Figure 9.2 The life cycle of HIV. HIV cannot replicate without infecting a host cell. Once inside, the virus uses the cell's components to make copies. Infection with HIV disables and eventually kills the host cell.

Essay 9.1 The Evidence Linking HIV to AIDS

Although a small number of scientists have argued that the link between HIV and AIDS is weak, an enormous number of them agree that the statement "HIV causes AIDS" is a fact. Scientists use Koch's postulates, developed by physician Robert Koch in the nineteenth century, as the litmus test for determining the cause of any epidemic disease. They are summarized as follows:

1. *Association*: The suspected infectious agent is found in all individuals suffering from a particular disease.
2. *Isolation*: The supposed infectious agent can be grown outside the host in a pure culture (without any other microorganisms).
3. *Transmission:* Transfer of the suspected pathogen to an uninfected host produces the disease in the new host.
4. *Isolation from new victim:* The same pathogen must be found in the newly infected host.

Does HIV fulfill Koch's postulates as the cause of AIDS? Let us examine the evidence for each assumption.

Association: Numerous studies from around the world show that virtually all AIDS patients are HIV-seropositive; that is, they carry antibodies that indicate HIV infection.

Isolation: Modern laboratory techniques have allowed the isolation of HIV from virtually all AIDS patients, as well as in almost all HIV-seropositive individuals. In addition, researchers have documented the presence of HIV genes in both of these groups of patients.

Transmission: HIV does not appear to cause AIDS in other animals, so transmission is difficult to demonstrate—we cannot expose an animal to the virus and see if it develops AIDS. However, this postulate has been fulfilled by a series of tragic incidents. In one case, three laboratory workers with no other risk factors developed AIDS after accidental exposure to concentrated HIV at work. In another case, transmission of HIV from a Florida dentist to six patients was documented by genetic analysis of the virus isolated from both the dentist and the patients. The dentist and four of the patients developed AIDS and died. Finally, the Centers for Disease Control has received reports of 57 documented, occupationally acquired HIV infections among health-care workers, of whom 26 have developed AIDS in the absence of other risk factors.

Isolation from new victim: In the case of the three laboratory workers described above, HIV was isolated from each infected individual and its RNA sequence was examined; the HIV proved to be the virus the workers had handled. In the case of the Florida dentist, HIV isolated from his infected patients had very similar RNA sequences, indicating that all of them were infected with the same virus strain.

In short, the link between HIV and AIDS has been firmly established using the standard set by Koch's postulates, and this relationship has been accepted by the vast majority of medical scientists.

The Course of HIV Infection

The early symptoms of HIV infection resemble the flu in about 70% of infected individuals (there are no noticeable symptoms in the remaining 30%). These symptoms occur because the HIV present in the bloodstream is destroying large numbers of T4 cells, thus interfering with normal immune responses. Most people infected with HIV begin to control the virus within six to 12 weeks, and therefore recover from these flu-like symptoms. This seeming recovery from the infection is due to the actions of the immune system.

Among the pool of immune-system cells, particular lymphocytes recognize antigens that are present on HIV particles. *T lymphocytes* that have an HIV-antigen receptor are stimulated to reproduce so that a large number of cells are available to patrol the bloodstream for signs of the virus. (Note that the CD4+ receptor, which HIV binds to in order to enter T4 cells, is not an antigen receptor.) *B lymphocytes* that have an HIV-antigen receptor are also stimulated to divide and will produce anti-HIV **antibodies**, proteins that bind to and help

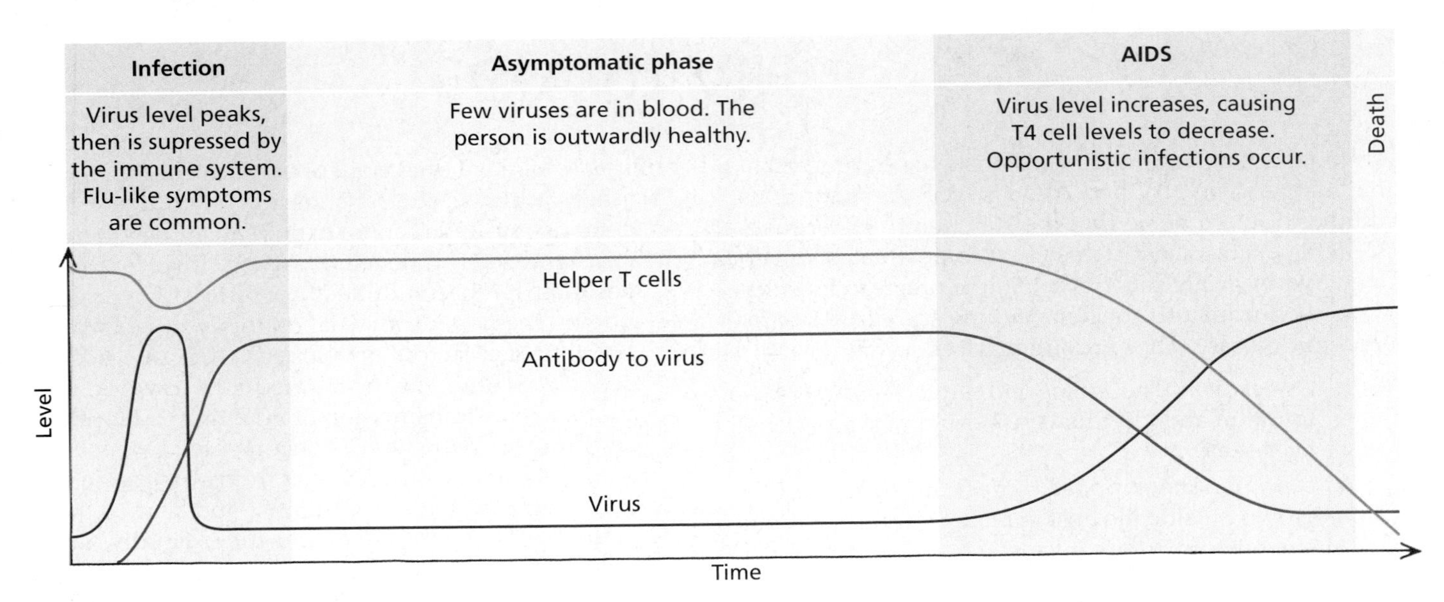

Figure 9.3 The typical course of HIV infection. This graph illustrates the change in HIV levels in the blood, the level of antibodies present, and the level of T4 cells over time. After initial infection, most patients produce enough antibody to control virus levels for months or years. Eventually, however, nearly all HIV-infected people develop AIDS when the body fails to maintain antibodies to the virus.

eliminate particular antigens. Within three months of initial contact with the virus, 95% of infected individuals have high levels of anti-HIV antibodies circulating in their blood which reduce the level of HIV in the bloodstream.

Once the immune response to HIV is fully developed, the levels of T4 cells rebound. At this point, the infected individual is **asymptomatic** and has low levels of detectable virus in the fluid part of the blood and a mostly normal immune response. The asymptomatic phase of HIV infection may last for 10 or more years.

In nearly all HIV-infected individuals who are not receiving drug treatment, the immune system eventually loses control over the virus. At some point, virus levels begin to increase and T4 cell numbers decline, signaling the onset of AIDS (Figure 9.3). Why does HIV eventually win its battle with the immune system? Primarily because of the evolution of HIV within its host.

9.2 The Evolution of HIV

Understanding why most people infected with HIV eventually fail to control this virus requires an understanding of one of the most important ideas in biology—any population has the potential to adapt to its environment if it is given enough time and a mechanism to generate variation.

WWW

Media Activity 9.2 Natural Selection

The Theory of Natural Selection

In *The Origin of Species*, Charles Darwin put forth two major ideas: the theory of common descent and the theory of natural selection. We discussed the theory of common descent in detail in Chapter 8 and learned that all species living today appear to have descended from a single ancestor that arose in the distant past. Darwin's presentation of this theory was thorough and convincing. Within

(a) Variation in coat color

(b) Variation in blooming time

Figure 9.4 Observation 1: Individuals within populations vary. (a) Gray wolves vary in coat color, even within a single pack of animals. (b) Flowers may vary in blooming time, with some individual plants blooming much earlier than others of the same species.

20 years of his book's publication, most scientists accepted Darwin's principle that all living organisms are related to each other through common descent. However, it was another 60 years before the scientific community accepted Darwin's ideas about *how* the great variety of living organisms had come about. Darwin called the process that causes evolution **natural selection**.

The theory of natural selection is elegantly simple. It is an inference based on four general observations:

1. *Individuals within populations vary.* Observations of groups of humans support this statement—people come in an enormous variety of shapes, sizes, colors, and facial features. It may be less obvious that there is variation in nonhuman populations as well—for example, within a single pack of gray wolves in northern North America, there may be individuals who are black, tawny, or reddish in color (Figure 9.4a); or within a single population of flowers, there are some individuals that bloom earlier than the majority, and some that bloom later (Figure 9.4b). We can add all kinds of less obvious differences to this visible variation—for example, differences in blood type among people. Each different type of individual in a population is termed a *variant*.
2. *Some of the variation among individuals can be passed on to their offspring.* Although Darwin did not understand how it occurred, he observed many examples of the general resemblance between parents and offspring. Farmers regularly take advantage of the inheritance of certain variations. For example, some chickens produce more eggs than others, and their offspring often produce more eggs than the offspring of less productive chickens. This enables a farmer to select only the offspring of the best laying hens as the new flock of egg producers. Pigeon breeders took advantage of the inheritance of variation when they produced fancy birds—pigeons with fan-shaped tails were more likely to produce offspring with fan-shaped tails than pigeons with straight

Figure 9.5 Observation 2: Some of the variation among individuals can be passed on to their offspring. Darwin noted that breeders could create flocks of pigeons with fantastic traits by using only those individuals that displayed these traits as parents of the next generation.

tails (Figure 9.5). Darwin hypothesized that offspring tend to have the same characteristics as their parents in natural populations as well.

For several decades after the publication of *The Origin of Species*, the observation that some variations were inherited was the most controversial part of the theory of natural selection. Since scientists could not adequately explain the origin and inheritance of variation, many were unwilling to accept that natural selection could be a mechanism for evolutionary change. When Gregor Mendel's work on inheritance in pea plants (discussed in Essay 4.1) was rediscovered in the 1900s, the mechanism for this observation became clear.

3. *Populations of organisms produce more offspring than will survive*. This observation is clear to most of us—the trees in the local park make literally millions of seeds every summer, but only a few of those that sprout live for more than a few years. In *The Origin of Species*, Darwin gave a graphic example of the difference between offspring production and survival. In his example he used elephants, animals that live long lives and are very slow breeders. A female elephant does not begin breeding until age 30, and she produces about one calf every 10 years until around age 90. Even at this very low rate of reproduction, Darwin calculated that if all of the descendants of a single pair of elephants survived and lived full, fertile lives, after about 500 years their family would have more than 15 million members (Figure 9.6)! Clearly, only a subset of the elephants born in every generation survives long enough to reproduce.
4. *Survival and reproduction are not random*. In other words, the subset of individuals who survive long enough to reproduce is not an arbitrary group. Some variants in a population have a higher likelihood of survival and reproduction than other variants. The relative survival and reproduction of one variant compared to others in the same population is referred to as its **fitness**. Traits that increase an individual's fitness in a particular environment are called **adaptations**. Individuals with

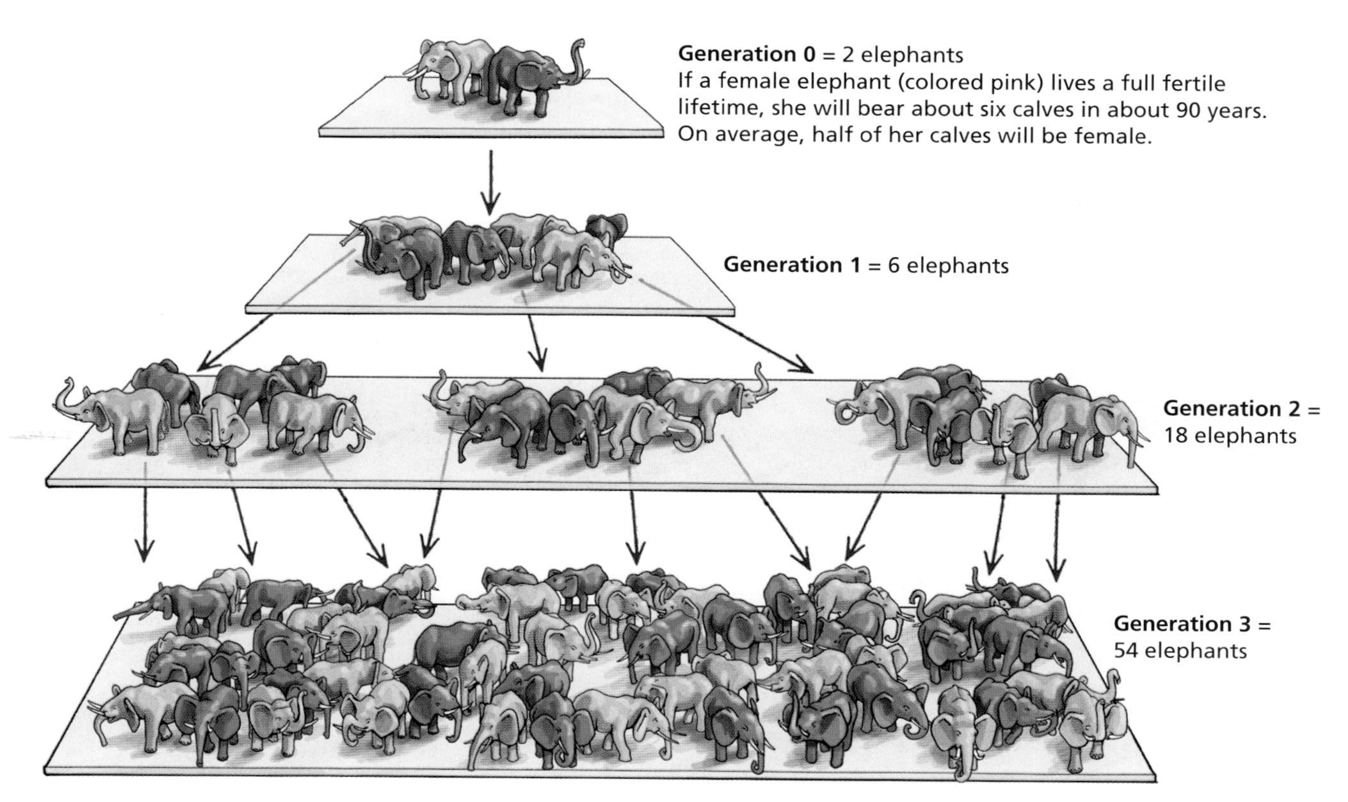
Generation 0 = 2 elephants
If a female elephant (colored pink) lives a full fertile lifetime, she will bear about six calves in about 90 years. On average, half of her calves will be female.
Generation 1 = 6 elephants
Generation 2 = 18 elephants
Generation 3 = 54 elephants

Figure 9.6 Observation 3: Populations of organisms produce more offspring than will survive. Even slow-breeding animals like elephants are capable of producing huge populations relatively quickly.

WWW
Media Activity 9.3 Population Growth and Human Evolution

adaptations to a particular environment are more likely to survive and reproduce than individuals lacking such adaptations.

Darwin referred to the results of differential survival and reproduction as natural selection. Although Darwin used the word *selection*, which implies some active choice, you should note that natural selection is a passive process. Adaptations are "selected for" in the sense that individuals possessing them survive and contribute offspring to the next generation. For example, among the birds called *Darwin's finches* scientists have observed that when rainfall is scarce, a large bill is an adaptation. This is because birds with larger bills are able to crack open large, tough seeds—the only food available during severe droughts. As shown in Figure 9.7, the 300 survivors of a 1977 drought had an average bill depth that was 6% greater than the average bill depth of the original population of 1,300 birds.

Darwin made this inference from these four observations: The result of natural selection is that favorable inherited variations tend to increase in frequency in a population over time, while unfavorable variations tend to be lost. In other words, adaptations become more common in a population as those individuals who possess them contribute larger numbers of their offspring to the succeeding generation. Natural selection results in a change in the traits of individuals in a population over the course of generations—*voilà*, evolution.

Testing Natural Selection Darwin proposed a scientific explanation of how evolution occurs—like all good hypotheses, it needed to be tested. As Darwin noted in *The Origin of Species*, humans have been testing the hypothesis that

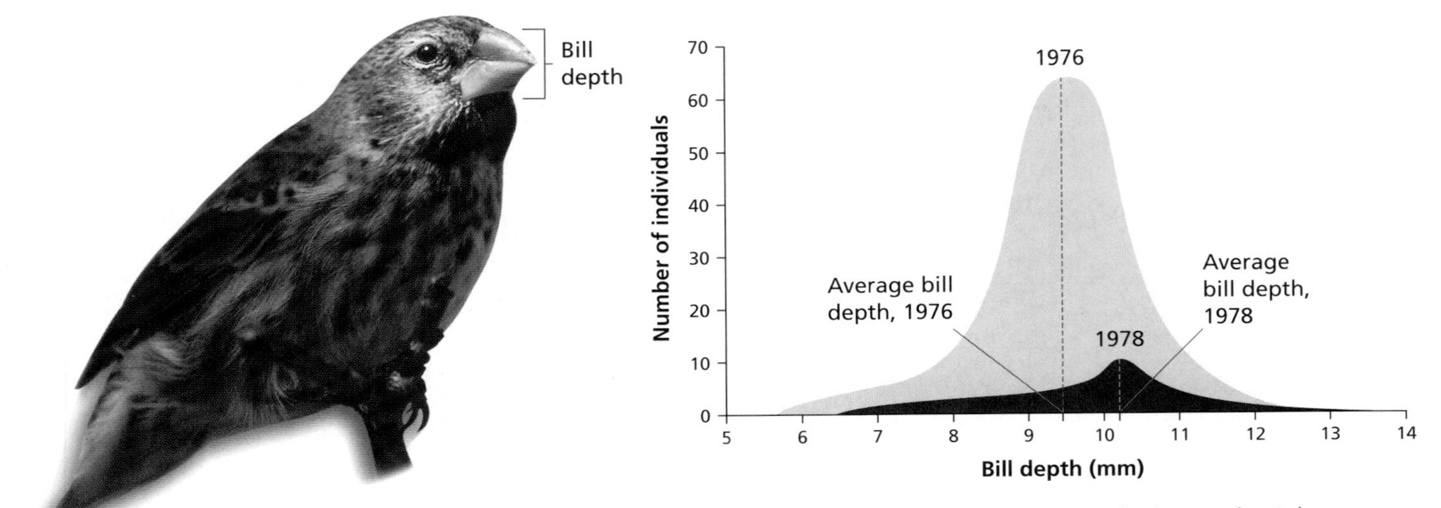

Figure 9.7 Observation 4: Survival and reproduction are not random. Darwin's finches on the Galapagos Islands did not have equal chances of surviving a severe drought. The pale blue curve summarizes bill depth in these birds before the drought. The same population after the drought of 1976 and 1977 (represented by the purple curve) had an average bill depth approximately 0.5 mm greater. This illustrates that during the drought years 1976 and 1977, finches with larger-than-average bills had higher fitness than small-billed birds.

selection causes evolution for thousands of years. By imposing selection on domestic animals and plants, humans have changed the characteristics of populations of these organisms.

Selection imposed by human choice is called **artificial selection**—it is artificial in the sense that humans control the survival and reproduction of individual plants and animals with favorable characteristics to change the characteristics of the population. Pigeon varieties resulted from artificial selection—they evolved through selection by breeders for various traits (Figure 9.8). However, because of the direct intervention of humans on survival and reproduction of these organisms, artificial selection is not exactly equivalent to natural selection.

Scientists have also observed selection occurring among organisms that are exposed to different environments. An example of this kind of experiment is one performed on fruit flies placed in environments containing different concentrations of alcohol. In high concentration, alcohol is toxic to fruit flies—flies exposed to alcohol break it down to make it less toxic. There is variation among fruit flies in the rate at which they detoxify alcohol. In a typical laboratory environment, most flies process alcohol relatively slowly, but about 10% of the population can process alcohol twice as rapidly as the slowest processors.

In their experiment, scientists divided a population of fruit flies into two random groups—initially these two groups had the same percentage of fast and slow alcohol processors. One group of flies was placed in an environment containing typical food sources; the other group was placed in an environment containing the same food spiked with alcohol. After 57 generations, the percentage of fast-processing flies in the normal environment was the same as at the beginning of the experiment—10%. But after the same number of generations, 100% of the flies in the alcohol-spiked environment were the fast-processing variety (Figure 9.9). In other words, the average rate of alcohol processing increased in flies over many generations in the high-alcohol environment; the population had evolved.

The evolution of the fruit flies in this experiment was a result of natural selection. In an environment where alcohol concentrations were high, individuals who were able to process alcohol relatively rapidly had higher fitness. Since they lived longer and were less affected by alcohol, they left more offspring than the slow processors. Thus, in each generation there was a higher frequency of fast-processing individuals than in the previous generation. After many generations, flies that could detoxify alcohol rapidly predominated in the population.

The example of the alcohol-processing flies illustrates two important points. One is that natural selection is situational—only the population of flies in the high alcohol environment evolved a faster rate of alcohol processing. Without a change in the environment, the alcohol-processing rate of the population of flies in the normal environment did not evolve. The second is that natural selection does not cause change in individual flies—flies either can rapidly process alcohol or they cannot. It is the differential survival and reproduction of these types of flies that causes the population to change.

The effects of natural selection have been observed in wild populations as well. A classic example of natural selection in action is the evolution of bill size in Galapagos finches in response to drought. Figure 9.7 illustrated that a nonrandom subset of the finch population survived a 1977 drought; that is, the survivors tended to be those with the largest bills. This resulted in a change in the next generation—the birds hatched in 1978 had an average bill depth 4–5% larger than the pre-drought population. Bill size in this population of birds evolved in response to natural selection.

The Modern Understanding of Natural Selection One barrier to the acceptance of Darwin's theory of natural selection was the lack of understanding of the origins of variation among individuals and the mechanism by which variations were passed to the next generation. Without this understanding, it was difficult to see how natural selection could cause a change in the frequency of particular traits in a population. It was not until scientists began to understand the nature of genes in the early twentieth century that the theory of natural selection was fully accepted by most biologists.

As discussed in Chapter 4, *genes* are segments of genetic material (either *DNA*, or *RNA* in some viruses) that contain information about the structure of molecules called *proteins*. The actions of proteins within an organism help determine its physical traits. Different versions of the same gene are called *alleles*—variation in traits among individuals in a population is often due to variation in the alleles they carry.

We can apply these genetic principles to the fruit flies exposed to a high-alcohol environment. In this population, there are two alleles for the gene that

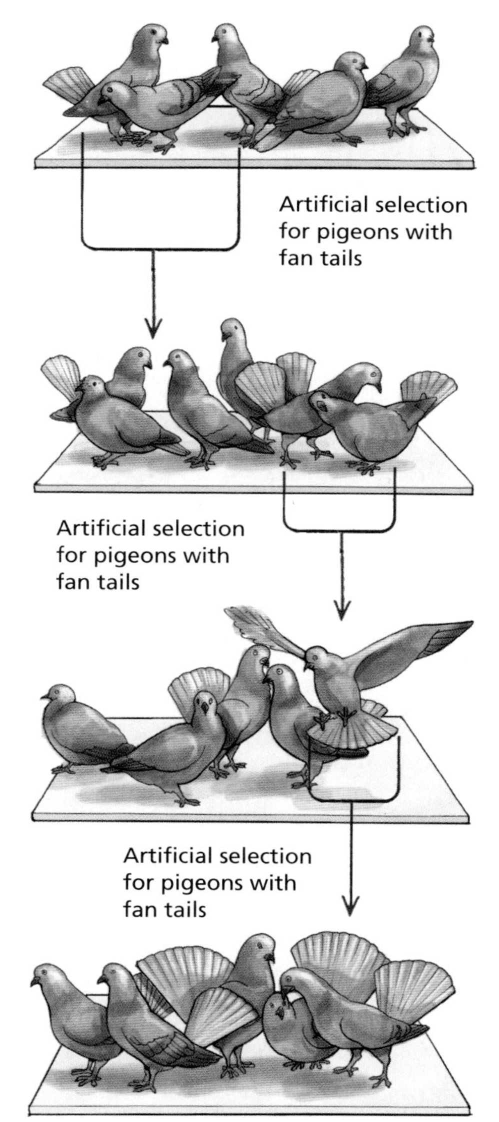

Figure 9.8 Artificial selection can cause evolution. When pigeon breeders selected individuals with certain traits to produce the next generation of birds, they increased the frequency of that trait in the population. Over generations, the trait can become quite exaggerated.

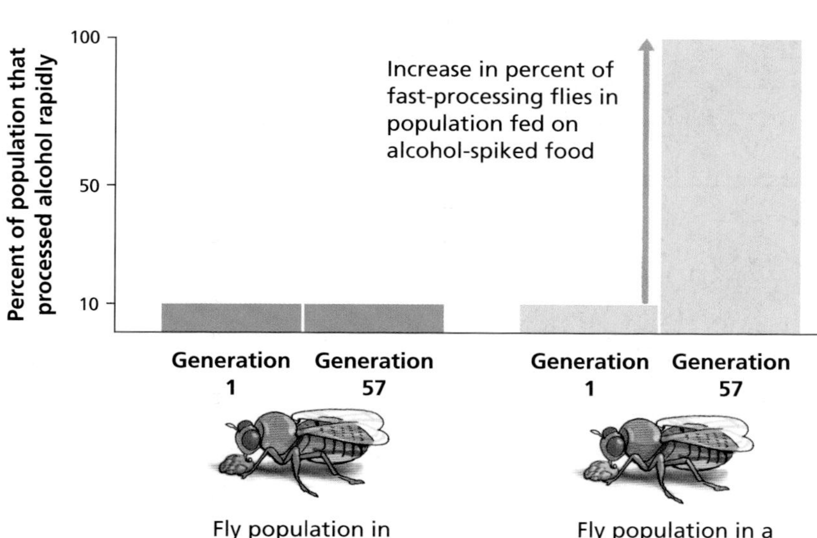

Figure 9.9 An observation of natural selection. When fruit flies are placed in an alcohol-spiked environment, the percentage of flies that can quickly process alcohol increases over many generations as a result of natural selection. In the normal laboratory environment, there is no selection for faster alcohol processing, so the percentage of fast processors does not change.

controls alcohol processing. One allele produces a protein we will term "fast," and the other produces a protein we will call "slow." Flies that can detoxify alcohol rapidly make the fast protein. To make this protein, they must carry two copies of the fast allele. As described in detail in Chapter 4, half of the alleles carried by a parent are passed to their offspring via their eggs or sperm. In the high-alcohol environment, flies with the fast protein had more offspring than flies that produce the slow protein. Since they carry two copies of the fast allele, each of the offspring of a fast processor received a copy of this allele. Therefore, in the next generation a higher percentage of individuals carried the fast allele. We can now describe the evolution of a population as an increase or decrease in the *frequency of an allele* for a particular gene (Figure 9.10).

Understanding the nature of genes also explains the origin of their variations. Different alleles for the same gene arise through *mutation*—changes in the DNA sequence. Mutations occur by chance when DNA is copied before reproduction. If a mutation results in an allele that has a function different from that of the original allele, the resulting variation could become subject to the process of natural selection. The existence of two different alleles for alcohol processing in fruit flies suggests that one of these alleles is a mutated version of the other. In the normal laboratory environment, neither of these alleles appears to have a strong affect on fitness—that the slow processors are more numerous than the

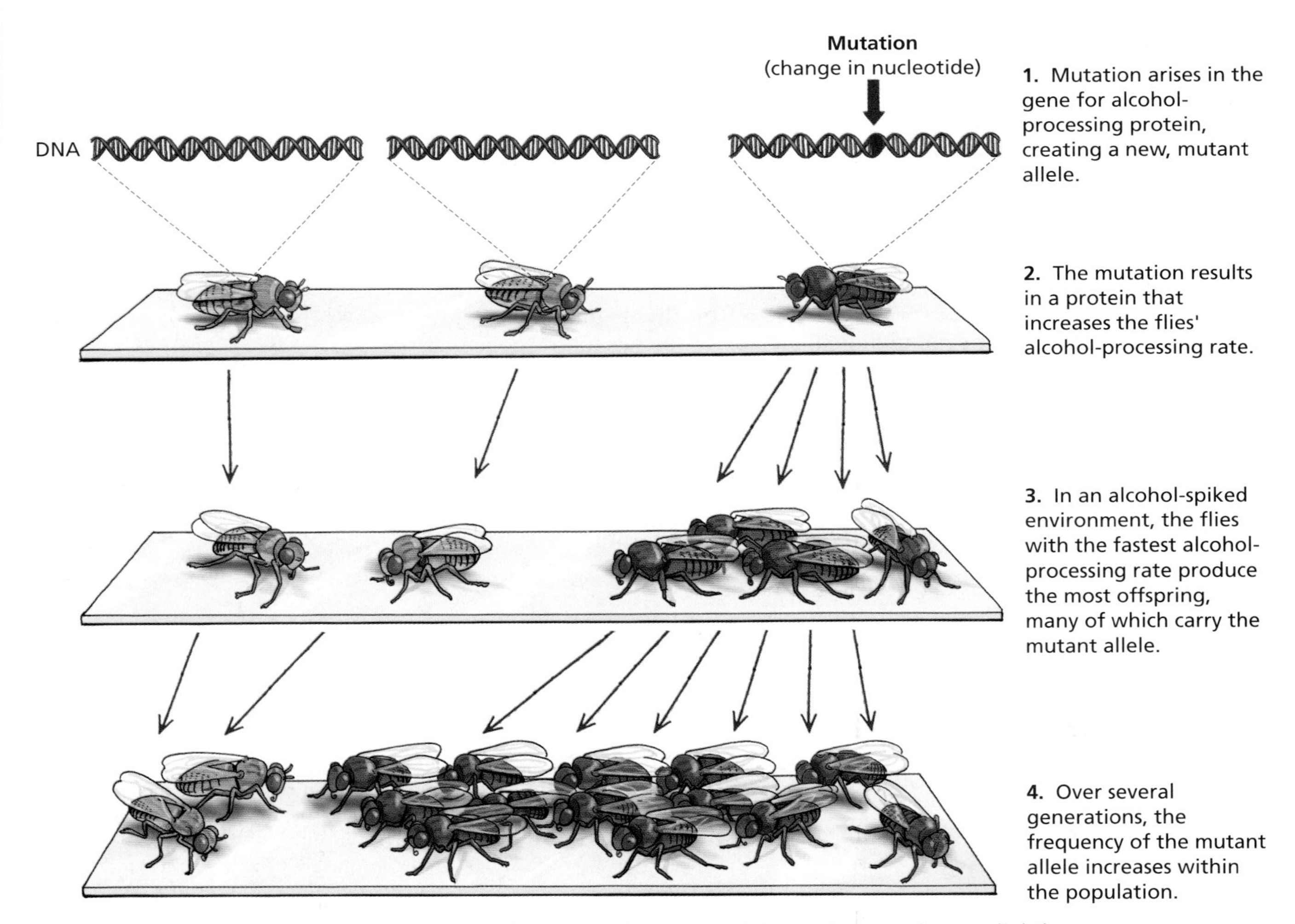

Figure 9.10 Mutation and natural selection. When a gene has mutated, its product may have a slightly different activity. If this new activity leads to increased fitness in individuals carrying the mutated gene, it will become more common in the population through the process of natural selection.

fast processors indicates that there might be a slight disadvantage to carrying the fast protein. However, in a different environment, the mutation resulting in the fast allele is an advantage, and its presence in the population results in the population's evolution. Scientists now understand that the random process of gene mutation generates the raw material—variations—for evolution.

The Natural Selection of HIV

During the asymptomatic period of HIV infection, the numbers of HIV virus particles in an infected person's bloodstream is relatively low. However, the immune response to HIV does not completely eliminate the virus. HIV persists inside immune-system structures called *lymph nodes* where it continues to infect and kill T4 cells. The dying T4 cells release the virus into the bloodstream, where anti-HIV antibodies quickly eliminate them. At the same time, the infected individual is maintaining a high rate of T4 cell production to replace those lost to HIV. In a sense, the virus and the immune system maintain a balance of power during this period.

The population of HIV is not stable during the asymptomatic period, however. HIV particles are constantly being reproduced because they continue to infect cells in the lymph nodes; any time there is reproduction, mutation can occur. As a result, during the asymptomatic period new variants of HIV arise. Some of these HIV variants have mutated antigens. The change in antigens can be great enough that the antibodies that attacked the HIV particles produced early in the infection do not recognize the new variants. Thus, these new variants have higher fitness than the older variants as a result of their longer survival in the bloodstream. As the new HIV antigen variant becomes more common, the host's immune system develops an antibody to it, and HIV again begins to be cleared from the bloodstream—until the next new antigen variant arises through mutation. In other words, the population of HIV inside the host is continually evolving, and the host's immune system is continually trying to "catch up" (Figure 9.11).

Among viruses, HIV has an unusually high rate of mutation. Some scientists estimate that *every single HIV particle produced* has at least one difference from the HIV it arose from. In addition, HIV has an enormously high rate of replication. These two characteristics of HIV result in a population of virus within an asymptomatic host that contains on the order of one billion distinct variants.

HIV's rapid evolution appears to cause the eventual end of the asymptomatic period in an infected person. The immune system is able to produce antibodies to many different HIV antigen variants, but eventually the sheer number of different variants of HIV that the immune system must respond to becomes overwhelming. Finally, one variant arises that escapes immune-system control for a long period, large numbers of T4 cells become infected, and the infected individual becomes increasingly immune deficient. This change initiates the onset of AIDS. The relentless evolution of HIV within an infected person's body eventually exhausts his or her ability to control this deadly virus.

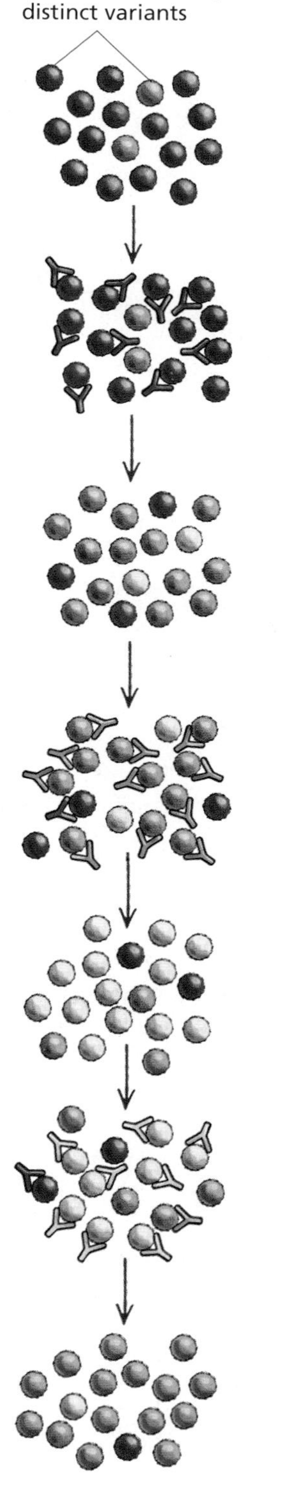
Different colors represent distinct variants
1. Initial infection by HIV particles.
2. Immune response: Most HIV particles are targeted by antibodies and destroyed.
3. Antibody-resistant HIV variants proliferate and new variants arise.
4. Immune response.
5. Antibody-resistant HIV variants proliferate and new variants arise.
6. Immune response.
7. Antibody-resistant HIV variants proliferate.

Figure 9.11 The evolution of HIV. HIV populations evolve in response to changes in the immune system. When the immune system develops a specific response to a strain of HIV, mutants that escape this response proliferate until the immune system develops a response to the mutant strain.

9.3 How Understanding Evolution Can Help Prevent AIDS

Immediately after scientists identified and characterized HIV as the virus that causes AIDS, a search for drugs that would interfere with HIV's ability to replicate began.

One target of anti-HIV drugs is the process of reverse transcription, the rewriting of HIV's genetic information from RNA into DNA. Reverse transcription does not occur in uninfected human cells, and drugs that target this

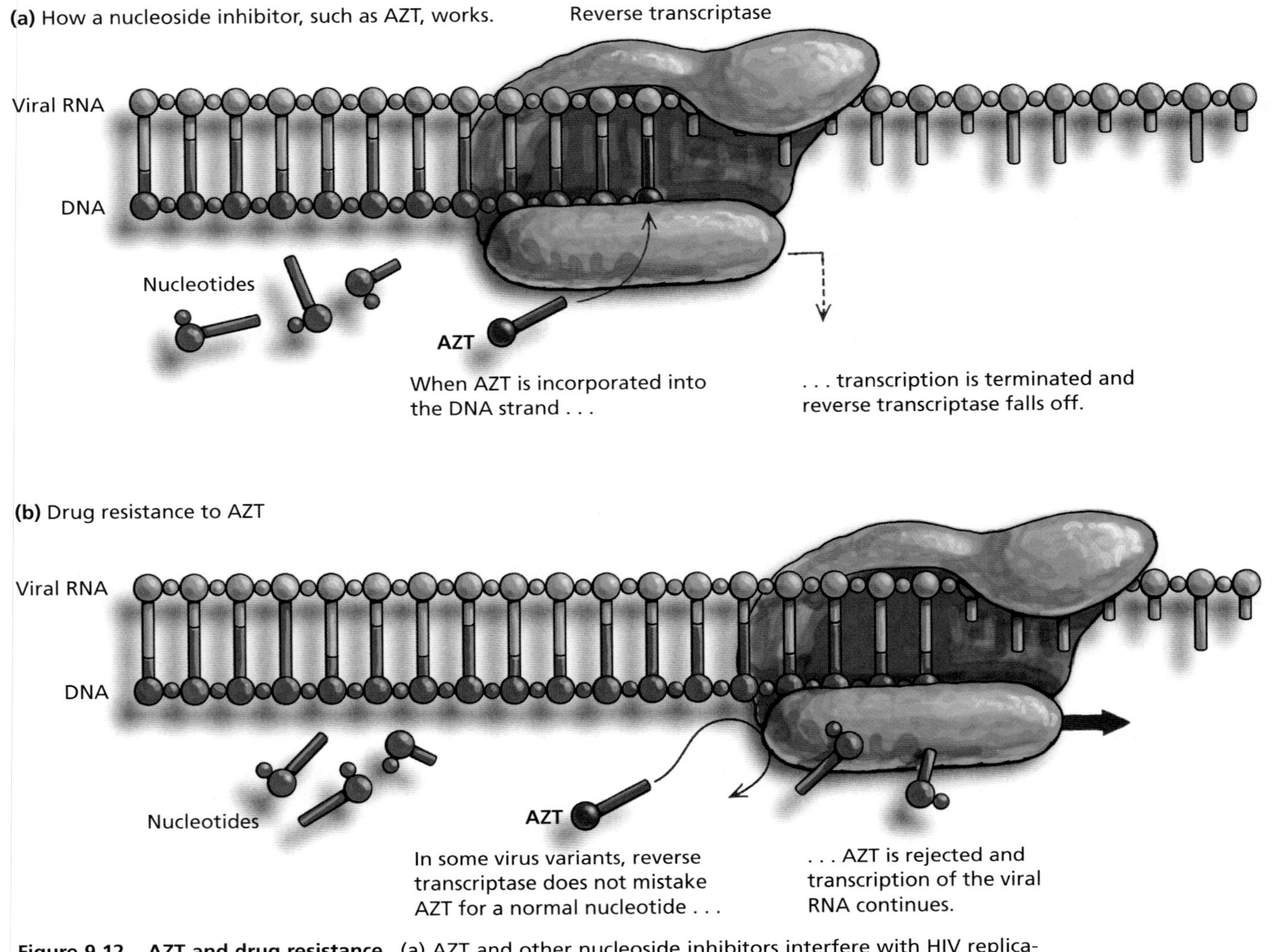
(a) How a nucleoside inhibitor, such as AZT, works.
Reverse transcriptase
Viral RNA
DNA
Nucleotides
AZT
When AZT is incorporated into the DNA strand . . .
. . . transcription is terminated and reverse transcriptase falls off.
(b) Drug resistance to AZT
Viral RNA
DNA
Nucleotides
AZT
In some virus variants, reverse transcriptase does not mistake AZT for a normal nucleotide . . .
. . . AZT is rejected and transcription of the viral RNA continues.

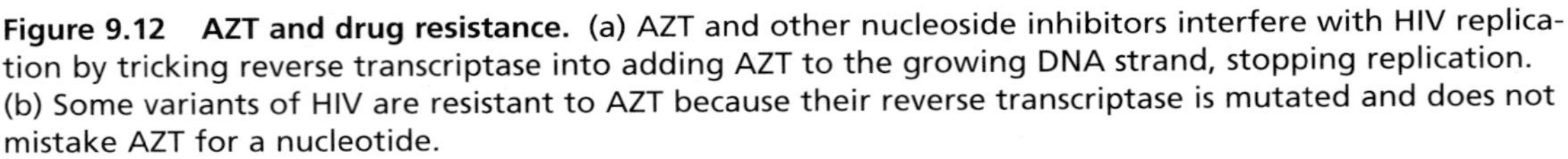
Figure 9.12 AZT and drug resistance. (a) AZT and other nucleoside inhibitors interfere with HIV replication by tricking reverse transcriptase into adding AZT to the growing DNA strand, stopping replication. (b) Some variants of HIV are resistant to AZT because their reverse transcriptase is mutated and does not mistake AZT for a nucleotide.

process have the potential for zeroing in on HIV replication without harming normal functions of the human body. One class of drugs used to inhibit reverse transcription is known as the *nucleoside analogs*. These drugs are similar in structure to one of the four DNA nucleotides described in Chapter 6—A, C, G, and T. Nucleoside analogs inhibit reverse transcription because reverse transcriptase, the enzyme that catalyzes this process, adds one of these analogs to a growing HIV DNA strand in place of the real nucleotide. Once a nucleoside analog is added to a growing DNA strand, replication halts because additional nucleotides cannot be attached to the analog (Figure 9.12a).

One of the first nucleoside analogs approved as treatment for AIDS is known as *AZT*. While it is not free of side effects, some very severe, AZT first appeared to be a wonder drug—nearly eliminating HIV from the blood of patients who had already progressed to AIDS. However, in all cases, AZT failed to keep virus populations low for an extended period. The failure of AZT over time occurred as a result of the evolution of HIV. Among the virus variants present in an infected person, there are some that do not mistake AZT for a normal nucleotide, and never incorporate it into growing HIV DNA

Essay 9.2 Our Evolving Enemies

HIV is not the only *pathogen* that can become resistant to drug treatments. Nearly every pathogen that is controlled by a chemical treatment has the potential to evolve resistance to that control. In the past several decades, doctors have seen a dramatic increase in the number of drug-resistant diseases in a variety of settings.

Gonorrhea

Gonorrhea is a sexually transmitted disease caused by infection with bacteria affecting over 650,000 people in the United States each year—82% of these cases occur in teens and young adults. In the 1980s, gonorrhea became resistant to the antibiotics penicillin and tetracycline. As a result, the Centers for Disease Control recommend second-line antibiotics, such as ciprofloxacin, as the standard treatment for this infection. In the late 1990s, gonorrhea strains that were resistant to these antibiotics began to be detected, notably in Hawaii, where 10% of gonorrhea cases are multiply drug resistant. Unchecked, gonorrhea infections in women can lead to pelvic inflammatory disease, a leading cause of infertility.

MRSA

MRSA is the acronym for methicillin resistant *Staphylococcus aureus. S. aureus* is a common, usually harmless passenger on human skin. Occasionally, it gets inside the body and causes infection. These infections can be as minor as a pimple, or as serious as a lethal blood infection. Elderly people and those with severely weakened immune systems (such as AIDS patients) are most susceptible to so-called staph infections. In many large health care settings, such as hospitals and nursing homes, some strains of *S. aureus* are resistant to methicillin, the first-line antibiotic. Outbreaks of MRSA occur in hospitals and can result in out-of-control infections that are difficult and expensive to treat—one estimate places the cost of MRSA at over $100 million per year. MRSA can be treated with a second-line antibiotic, vancomycin, but strains of MRSA that are less susceptible to vancomycin have recently been identified, causing concern that multiply drug-resistant strains will soon appear.

Tuberculosis

Tuberculosis (TB) is a lung disease caused by bacterial infection. Before antibiotics were available, TB was fatal in about 50% of cases and was a leading cause of death around the world. Once antibiotic therapy became available, the TB cure rate reached 98% and it became extremely uncommon in most developed countries. However, the number of reported cases of TB in the United States and Europe began to increase again during the 1980s and early 1990s, and some strains had become multiply drug-resistant. High levels of drug-resistant TB are now found in Asia and eastern Europe, and surprisingly large numbers of multiply drug-resistant TB cases are being seen in New York City and other large urban centers in the United States.

strands. These variants thus continue to replicate and become the predominant HIV variants in an AZT-treated individual (Figure 9.12b). The HIV population has evolved to become **drug resistant**; that is, not susceptible to the effects of AZT. The evolution of drug resistance in disease-causing organisms is not new—Essay 9.2 describes the disturbing rise of drug-resistant diseases—but the speed at which AZT resistance arose in AIDS patients was an early clue to HIV's amazing capacity to evolve.

Combination Drug Therapy Can Slow HIV Evolution

Since the development of AZT, dozens of new anti-HIV drugs have been developed. In addition to more types of nucleoside analogs, other non-nucleoside analogs that interfere with reverse transcription are now in use, as well as a powerful new class of drugs called *protease inhibitors* that stop HIV replication by interfering with the process that converts inactive viral proteins to active enzymes. However, there are still less than 20 anti-HIV drugs available. When patients take only one of these drugs, HIV develops resistance to it in a short period of time.

Understanding that the rapid rate of HIV evolution decreases the effectiveness of these drugs has led doctors to a new standard of care for the infection.

This standard is the use of **combination drug therapy**, also commonly called *drug cocktail therapy*—a combination of at least two reverse transcription inhibitors and a protease inhibitor. This approach has dramatically decreased the number of AIDS cases and deaths due to AIDS in the United States. The effectiveness of combination therapy is based on the following fact: The greater the number of drugs used, the greater the number of changes required in the virus's genetic material for resistance to develop. The likelihood of a virus arising that is resistant to a single drug is relatively small but still very possible in a patient with one billion different HIV variants. However, the likelihood of a virus arising with resistance to all three drugs in a cocktail is extremely small.

Another key to the effectiveness of combination drug therapy is that the more HIV replication is suppressed, the more slowly new drug-resistant HIV variants can arise. If replication represents the main route by which mutations occur, fewer rounds of replication mean fewer possible mutants. Thus, drug cocktails control HIV populations within people by creating an environment that is difficult to adapt to and by slowing the rate of evolution. Understanding how HIV evolves to defeat the immune system has allowed scientists to devise ways to interfere with this evolutionary process (Figure 9.13).

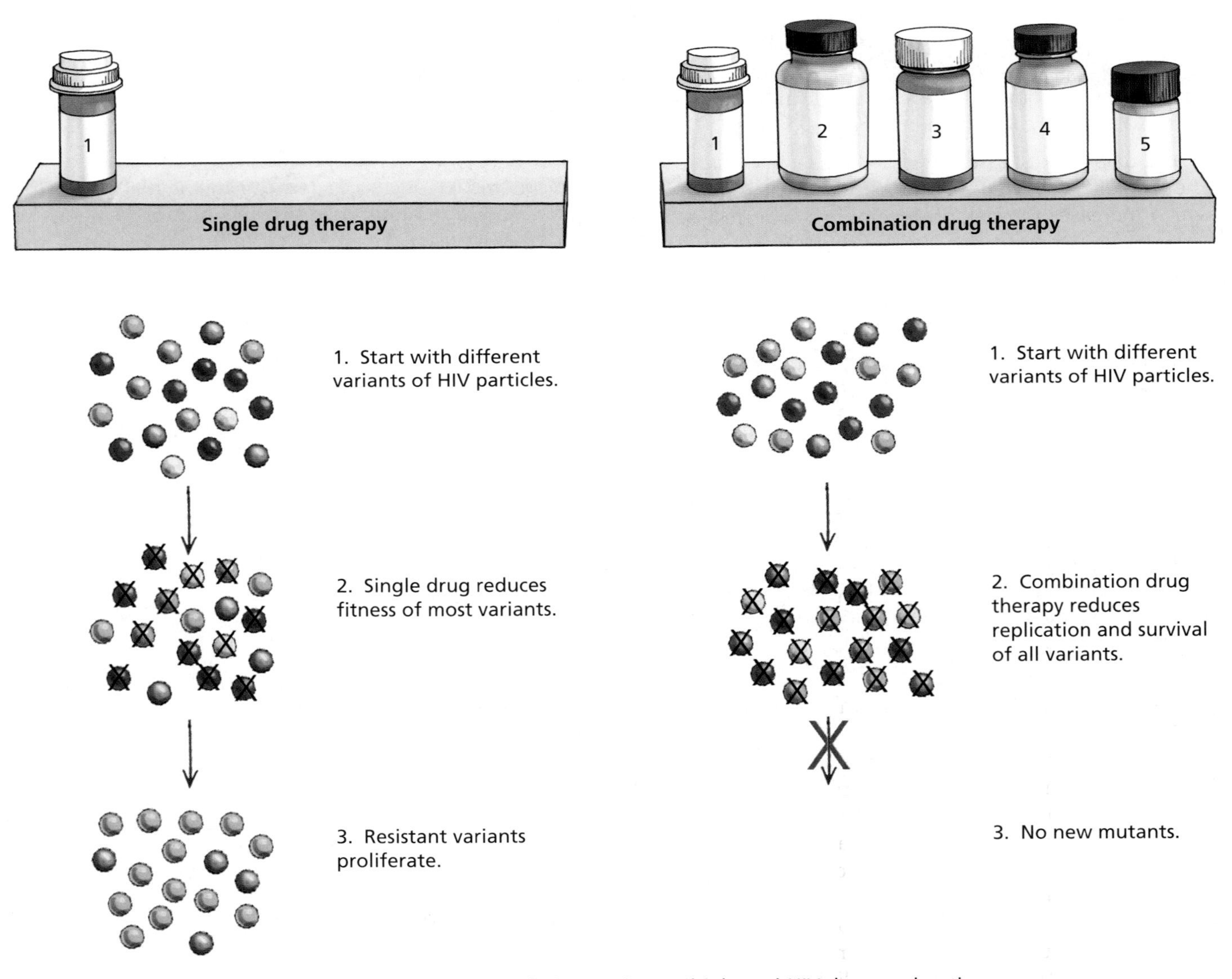
1
Single drug therapy
1
2
3
4
5
Combination drug therapy
1. Start with different variants of HIV particles.
2. Single drug reduces fitness of most variants.
3. Resistant variants proliferate.
1. Start with different variants of HIV particles.
2. Combination drug therapy reduces replication and survival of all variants.
3. No new mutants.

Figure 9.13 Combination drug therapy slows HIV evolution. Using multiple anti-HIV drugs makes the environment much harsher for the virus and decreases the likelihood that a variant with multiple resistances will evolve.

Finally, the process of evolution is only able to work within the limitations of the organism that is experiencing natural selection. Many mutations that confer drug resistance also appear to interfere with HIV's ability to replicate and infect cells. This appears to be the case in 30–40% of HIV infected individuals who host populations resistant to multiple drugs. In these patients, resistance to drug treatment has led to high levels of HIV in the blood. However, these high HIV loads do not cause a decrease in T4 cells, indicating that the drug-resistant variant in their bodies is not a very effective killer of these cells. These patients may live for three or more years with high loads of HIV in their bloodstream before progressing to AIDS—much longer than the majority of people with high loads of non-drug-resistant HIV. In these patients at least, combination drug therapy has led to the selection of less deadly virus variants and has prolonged their lives.

Problems with Combination Drug Therapy

Despite good news of the increased health and prolonged lives of HIV-infected people undergoing combination drug therapy, there are some problems with this approach. Combination drug therapy is expensive, often results in severe and unpleasant side effects, and most importantly, is difficult to follow. Patients may have to take dozens of pills per day, some of which have very different requirements (for instance, some pills must be taken on an empty stomach, while others need to be taken with food or significant amounts of water). All of this effort is to control an infection that initially may not seem to have any symptoms. As a result of the difficulty of combination drug therapy, it is common for individuals to skip doses, or take themselves off the drugs for a period of time. The side effect of these breaks in treatment is an increase in HIV replication, which increases the risk of developing drug-resistant varieties. Patients who do not follow the drug treatment schedule carefully can find themselves with large virus populations that are multiply drug resistant. Even if this resistant virus is less able to infect cells, few people can live with high loads of HIV for long. To control the virus over the long term, individuals with multiply drug-resistant variants must change and perhaps increase the number of drugs in their cocktail—some patients take 10 to 15 different drugs several times per day. Many scientists fear that the rapid rate of HIV evolution will eventually outpace their ability to both develop new drugs and prolong the asymptomatic period of HIV infection.

Perhaps more troublesome than the evolution of HIV within a patient is the potential evolution of the HIV epidemic in response to combination drug therapy. The rate of transmission of HIV has not significantly changed in the United States for a decade—about 40,000 new infections are reported every year. However, many of the HIV strains that are being transmitted today already carry some degree of drug resistance. Currently, between 10–30% of new infections are of drug-resistant HIV. This means that potentially as many as one-third of newly infected people have fewer options for controlling their virus. As the transmission of drug-resistant HIV increases, our ability to control AIDS in the U.S. population will decline. Worse yet, there is some evidence that combination drug therapy has made HIV and AIDS appear to be less of a threat, leading to decreased prevention efforts and an upswing in infection rates. Combination drug therapy does not cure HIV infection—at best it is an expensive and long-term commitment to increase an individual's ability to live with this disease. Increases in transmission and drug resistance will erode the benefits of this powerful therapy.

Combination drug therapy is also not available to all of the 42 million HIV-infected individuals around the world. Currently, combination drug therapy costs $1,000 to $2,000 per month. As discussed in Essay 9.3, HIV continues to disproportionately affect the poorest and most vulnerable members of our society and the global community. The gap between the resources of the most-affected

Essay 9.3 The Global Impact of HIV

AIDS and HIV no longer dominate the headlines in the United States—apparently for a good reason. According to the Centers for Disease Control (CDC), the number of AIDS-related deaths in the United States has plummeted from over 51,000 in 1995 to less than 16,000 in 2001, mostly due to the effectiveness of combination drug therapy. This is a remarkable 70% drop, and as a result AIDS no longer ranks as a leading cause of death in the United States. Currently, nearly twice as many Americans die from gunshot wounds as from the results of HIV infection.

However, the overall death rate due to AIDS obscures some of the important details of this epidemic in the United States. Data collected by the CDC indicate that AIDS is now the fifth leading cause of death among Americans aged 25 to 44, the third leading cause of death among Hispanic males in this age group, and the leading cause of death for black males in this age group. Nearly 14% of new HIV infections are to injection-drug users, and 59% of new HIV infections reported in teenagers are in girls, most of them African-American. In the United States, HIV and AIDS is becoming a disease of the impoverished and marginalized; people with the least access to adequate health care.

The inequity of AIDS is even more profound on a global scale (Figure E9.3). According to the United Nations Joint Program on HIV/AIDS (UNAIDS), 90% of all HIV infections occur in the developing world—an astonishing 70% of all HIV infections occur in Sub-Saharan Africa. In impoverished countries such as Zimbabwe and Botswana, *over one-third* of all adults are infected with HIV. The scale of the epidemic is almost impossible to fathom—nearly 30 million people infected, nearly all of whom will die within the decade. Only a tiny fraction of the infected individuals receive anti-HIV drug therapy, and few of them receive basic medical treatment for opportunistic infections. The epidemic in Africa could soon be matched in other impoverished areas of Asia and eastern Europe—in countries whose economies and health-care systems are sure to be seriously stressed by it.

The devastation caused by AIDS in Africa is exemplified by the country of Malawi, whose people face widespread famine as well as high numbers of AIDS deaths and HIV infections. Adverse weather and changes in government policy contributed to the food crisis, but the impact of AIDS has been substantial. A study performed in central Malawi in 2002 indicated that nearly 70% of households had lost laborers due to illness, and 50% of families delayed planting crops to take care of ill family members or the orphaned children of their relatives. The epidemic is also sapping the ability of the government of Malawi to provide agricultural support for these small farmers because many employees of the government are themselves ill, or must care for ill relatives, and also because government resources are stretched thin by the public-health crisis caused by AIDS.

Rays of hope exist in this bleak landscape, however. According to UNAIDS, prevention programs that emphasize the use of condoms and promote abstinence have helped decrease the number of new HIV infections among young women in Ethiopia, Uganda, and South Africa; and several African governments have begun addressing the AIDS crisis at the highest level. In 2001, the United Nations made a strong commitment to slowing and cushioning the impact of the epidemic around the world, but without resources from wealthy countries, the loss of life and the disruption caused by this disease could be without precedent in human history.

populations and the cost of this promising therapy means that AIDS will continue to kill people by the millions for years to come.

In reality, the best "treatment" for AIDS is to avoid becoming infected with HIV at all (Figure 9.14). HIV is a fragile virus that is only transmitted through direct contact with bodily fluids—primarily blood, semen, vaginal fluid, or occasionally to newborns via breast milk. There is no evidence that the virus is spread by tears, sweat, coughing, or sneezing. It is not spread by contact with an infected person's clothes, phone, or toilet seat. It is not transmitted by insect bite. And it is unlikely to be transmitted by kissing (although any kissing that allows the commingling of blood could lead to HIV transmission). HIV is frequently spread through needle sharing among injection-drug users, but the primary mode of HIV transmission is via unprotected sex, including oral sex, with an infected partner. So, what is the best way to avoid HIV infection? Do not use injection drugs, and if you are sexually active, know your partner's HIV status, drug habits, and sexual activities. According to the Centers for Disease Control,

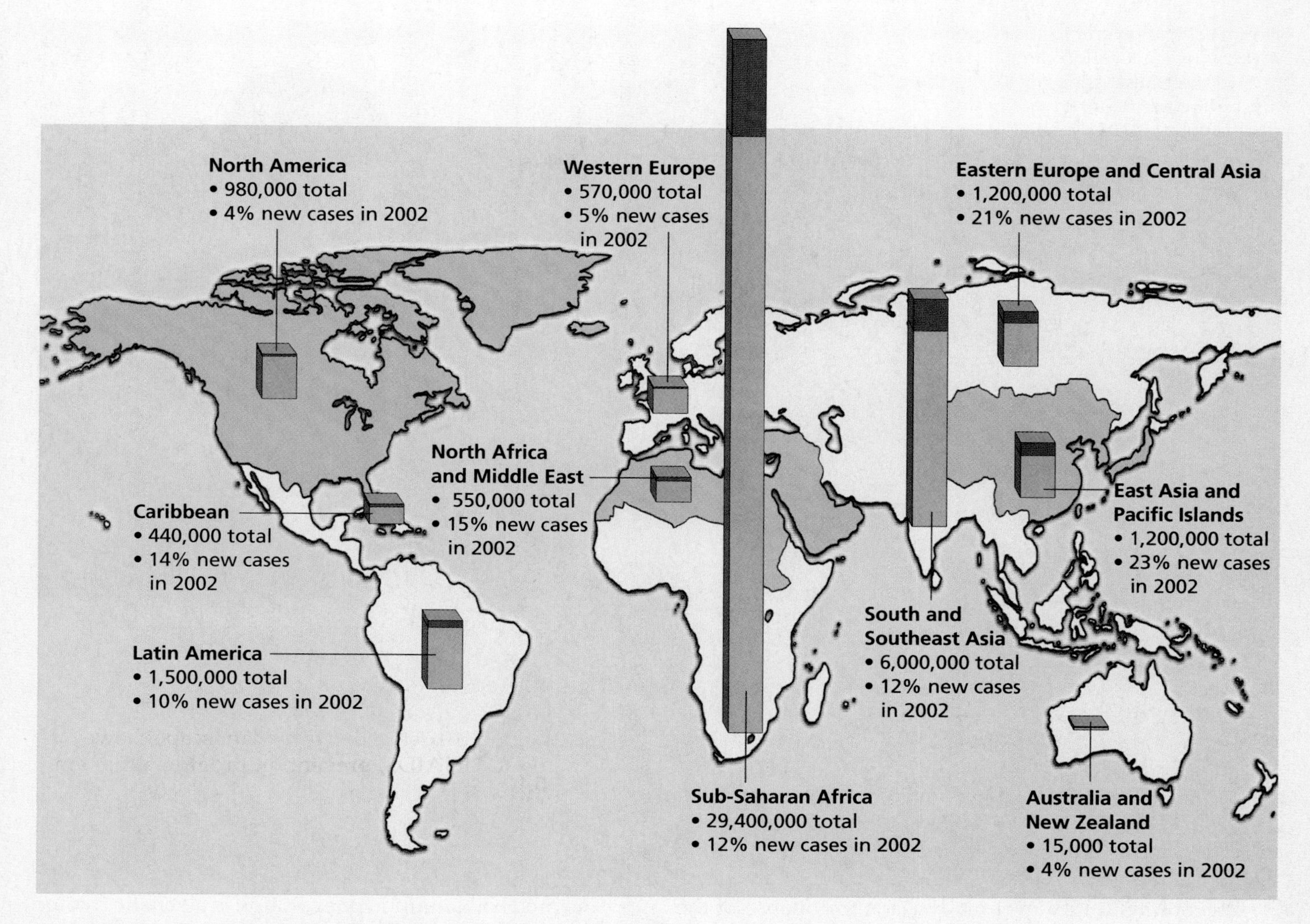

Figure E9.3 The AIDS crisis. HIV infections are not evenly distributed around the world—in general, poorer areas have more cases. In addition, the rate of spread of infections in various areas is not equal. Whereas developed countries have a rate of new infections of 4–5%, areas of Asia have seen a growth in infections of over 20% in the past several years.

about one-quarter of the approximately one million HIV-infected people in the United States do not know that they carry this deadly virus. If your partner might be at risk for HIV infection, practice *safer sex*—that is, use a condom.

Magic's Greatest Trick—Living with HIV

Why has Magic Johnson remained free of AIDS for over 10 years since contracting HIV? Because he has access to the highest-quality medical care, he has the resources and commitment to maintain long-term combination drug therapy, and because these actions have limited the evolution of HIV in his body. Magic is still infected with HIV, and no one knows whether drug therapy will help his body finally eliminate HIV or if the therapy will eventually fail and he will lose the battle with this killer virus. Magic's ability to survive, and even thrive, for more than 10 years since his diagnosis gives us hope that someday HIV will not be a death sentence for anyone.

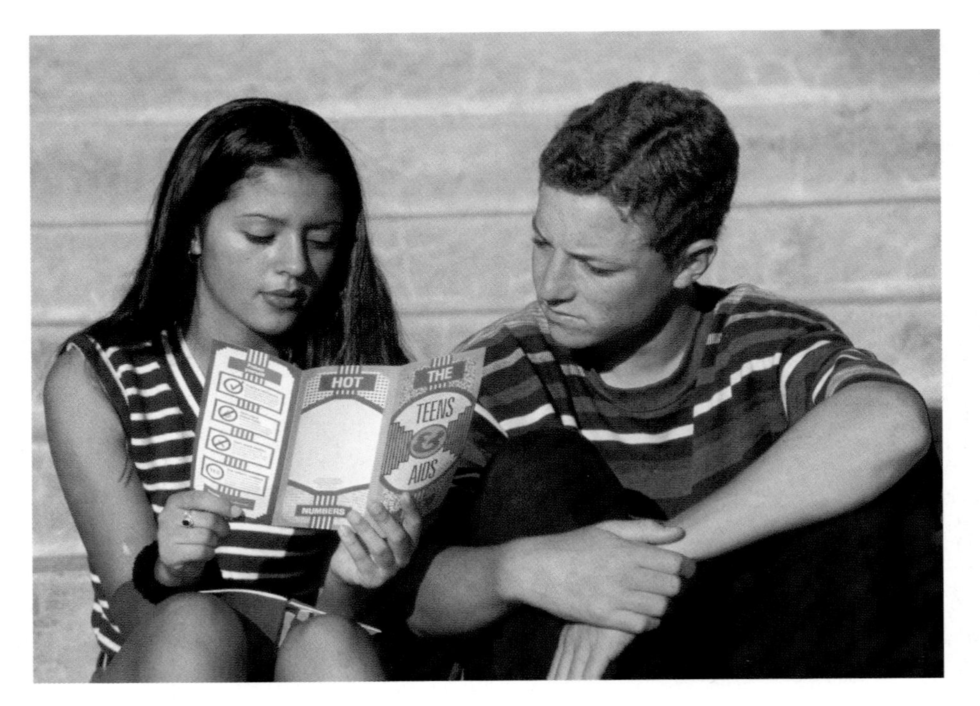

Figure 9.14 Preventing the tragedy of AIDS. Public education campaigns to reduce the transmission of HIV are the most effective means of preventing deaths due to AIDS.

www

Media Activity 9.4 AIDS: Exploring the Myths

CHAPTER REVIEW

Summary

- Infection with HIV eventually leads to collapse of the immune system, resulting in AIDS.
- The immune system of an individual infected with HIV can initially control the virus, but the evolution of HIV leads to the eventual loss of immune-system cells.
- Natural selection is a mechanism for evolutionary change in populations.
- Advantageous traits, called *adaptations,* increase an individual's fitness, his or her chance of survival, and/or reproduction.
- The increased fitness of individuals with particular adaptations causes the adaptation to become more prevalent in a population over generations.
- Selection has been demonstrated to cause evolution.
- Artificial selection, when humans control an organism's fitness, enables the production of different breeds of animals and varieties of plants.
- Populations exposed to environmental changes have been shown to evolve traits that make them better fit to the environment.
- The modern definition of evolution is a genetic change in a population of organisms.
- The traits of an organism are partially determined by alleles.
- Different alleles for a gene arise through the process of mutation.
- Alleles that code for adaptations become more common in a population over generations.
- HIV eventually overwhelms the immune system because it continually adapts to the body's attempts to control it.
- HIV can be controlled longer through the use of combination drug therapy.
- Mutants to multiple anti-HIV drugs are relatively unlikely, and drug therapy suppresses HIV replication, thereby reducing the production of mutant varieties.
- Varieties of HIV that can survive in an environment containing multiple drugs are sometimes less deadly than nonresistant varieties.

- Anti-HIV combination drug therapy is selecting for drug-resistant viruses, both within patients and in the general population.
- Anti-HIV combination drug therapy has major disadvantages, including high cost, difficulty in following the treatment schedule, and severe side effects.
- The best treatment for AIDS is prevention of HIV infection, which is primarily accomplished through safer sex practices.

Key Terms

Acquired Immune Deficiency Syndrome (AIDS) p. 218
adaptation p. 224
antibodies p. 221
artificial selection p. 226
asymptomatic p. 222
combination drug therapy p. 232
drug resistant p. 231
fitness p. 224
host p. 219
Human Immunodeficiency Virus (HIV) p. 219
natural selection p. 223
opportunistic infection p. 219
T4 cell p. 218

Learning the Basics

1. What is the relationship between HIV infection and AIDS?

2. Define fitness as used in the context of evolution and natural selection.

3. What are the four observations that led to Charles Darwin's inference of the theory of natural selection?

4. Define artificial selection, and compare and contrast it with natural selection.

5. Describe how HIV evolves when it is exposed to a drug that interferes with replication.

6. An individual infected with HIV _____.
   a. has AIDS
   b. may eventually develop AIDS
   c. cannot develop an immune response to the virus
   d. can transmit the virus to others by coughing or sneezing
   e. will probably not live very long

7. Which of the following observations is *not* part of the theory of natural selection?
   a. Populations of organisms have more offspring than survive.
   b. There is variation among individuals in a population.
   c. Modern organisms descended from a single common ancestor.
   d. Traits can be passed from parent to offspring.
   e. Some variants in a population have a higher probability of survival and reproduction than other variants.

8. The best definition of evolutionary fitness is _____.
   a. physical health
   b. the ability to attract members of the opposite sex
   c. the ability to adapt to the environment
   d. survival and reproduction relative to other members of the population
   e. overall strength

9. An adaptation is a trait of an organism that increases _____.
   a. its fitness
   b. its ability to survive and replicate
   c. in frequency in a population over many generations
   d. a and b are correct
   e. a, b, and c are correct

10. Natural selection can cause _____.
    a. evolution
    b. survival of the fittest
    c. adaptations to occur via mutation
    d. differences in survival and reproduction
    e. traits to be passed from parents to offspring

11. The heritable differences among organisms are a result of _____.
    a. differences in their DNA
    b. mutation
    c. differences in alleles
    d. a and b are correct
    e. a, b, and c are correct

12. The immune system of an HIV-infected individual ______.
    a. can eliminate HIV entirely
    b. cannot cause HIV populations to decline
    c. causes selection for HIV variants which escape immune-system control
    d. evolves quickly in response to HIV infection
    e. cannot make antibodies to HIV

13. HIV evolves rapidly because it ______.
    a. has a very high rate of reproduction
    b. has a very high mutation rate
    c. can detoxify anti-HIV drugs
    d. a and b are correct
    e. a, b, and c are correct

14. All of the following statements about multiply drug-resistant HIV are true *except* that it ______.
    a. is more common in individuals who are not consistent in taking drugs
    b. is often less able to replicate and infect other cells than nonresistant varieties
    c. can be transmitted to uninfected people
    d. can be cured by taking the patient off anti-HIV drugs
    e. is less likely to arise when virus replication is suppressed

15. HIV is transmitted via ______.
    a. sexual intercourse with an infected person
    b. shaking hands with an infected person
    c. using the same bathroom as an infected person
    d. a bite from an insect that has previously bitten an infected person
    e. all of the above

Analyzing and Applying the Basics

1. The wide variety of dog breeds is a result of artificial selection from wolf ancestors. Use your understanding of artificial selection to describe how a dog breed such as the Chihuahua may have evolved.

2. The striped pattern on zebras' coats is considered to be an adaptation that helps reduce the likelihood of a lion or other predator identifying and preying on an individual animal. The ancestors of zebras were probably unstriped. Using your understanding of the processes of mutation and natural selection, describe how a population of striped zebras might have evolved from a population of unstriped zebras.

3. The evolution of HIV in a patient taking drugs requires an adaptation to very strong natural selection—in other words, viruses without the appropriate adaptation do not survive. However, many adaptations in nature are subtler in their effects, and result in only relatively small increases in reproduction of the variant or its chance of survival, so that individuals without the adaptation do survive and reproduce. Are more subtle adaptations likely to rapidly become prevalent in a population? Explain your answer.

4. Are all features of living organisms adaptations? How could you determine if a trait in an organism is a product of evolution by natural selection?

5. A human generation is approximately 20 years. A typical bacterial generation is about 20 minutes. Which of these two types of organisms is likely to evolve more rapidly? Why?

6. Essay 9.2 discusses several instances of drug resistance having evolved in pathogens that were previously easy to control. Given what you have learned in this chapter about minimizing the development of resistance in HIV, describe policies that might decrease the risk of resistance evolving in other dangerous organisms.

Connecting the Science

1. The theory of natural selection has been applied to human culture in many different realms. For instance, there is a general belief in the United States that "survival of the fittest" determines which businesses are successful and which go bankrupt. How is the selection of "winning" and "losing" companies in our economic system similar to the way natural selection works in biological systems? How is it different?

2. HIV is primarily transmitted via sexual activity. What public policies do you think are most effective at reducing the rate of transmission of HIV/AIDS? Do you think that since HIV infection no longer appears to be a death sentence, it will become transmitted at increasingly higher rates? Why or why not?

3. In developing countries, where most of the population cannot afford combination drug therapy, 95% of worldwide HIV/AIDS cases occur. Does the United States have an obligation to provide people in the developing world with low-cost, effective anti-AIDS therapy? In countries where the needs of daily survival often overshadow the requirement to take the drugs in the proper dosage, drug-resistant strains of HIV may be more likely to develop. What do you think will best help reduce the toll of HIV/AIDS in these regions?

Media Activities

Media Activity 9.1 Replication of HIV within Helper T Cells
Estimated Time: 15 minutes
In this animation you will explore how HIV enters host cells, how it replicates and destroys the host, and why the loss of HIV-infected cells compromises the immune response.

Media Activity 9.2 Natural Selection
Estimated Time: 10 minutes
In studying populations of Alpine Skypilots, scientist Candace Galen proposed that natural selection was responsible for the observed differences. If natural selection is at work, then its four postulates should be true. In this activity, you will explore each of these postulates in the Alpine Skypilot populations.

Media Activity 9.3 Population Growth and Human Evolution
Estimated Time: 15 minutes
In this activity you will use Internet resources to answer a number of questions on human population growth and relate population growth to natural selection.

Media Activity 9.4 AIDS: Exploring the Myths
Estimated Time: 20 minutes
There are many myths and stereotypes surrounding AIDS. You will try to track down the truth by exploring on-line resources and articles.

CHAPTER 10

Who Am I?

Species and Races

How should a woman who has Asian, Black, White, and Native-American grandparents respond to this question?

What race does Indigo belong to?

Do the races on Indigo's census form represent different "basic types" of humans?

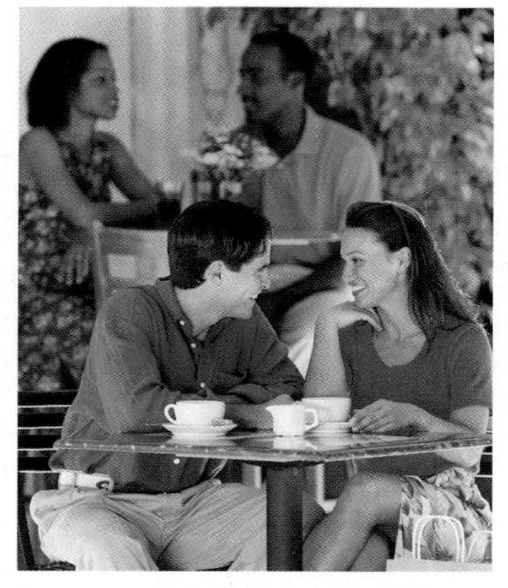

Are we more similar to people of the same race than to people of different races?

Indigo pondered the choices in front of her: White, Black, American Indian or Alaskan Native, Vietnamese, Other Asian. As the daughter of a man with African-American and Choctaw ancestry, and a woman with a White-American father and a Laotian mother, Indigo was not sure what race she should report on the U.S. census form. She called her sister, Star.

"I checked them all," Star explained. "Didn't you read the instructions? It says 'Mark one or more races'."

Indigo was not satisfied. "I did read the instructions!" she exclaimed. "It's just that ... well, it doesn't feel right. I mean, most people assume I'm Black, OK, and we've never even met mom's father's side of the family. I don't really think of myself as White."

"Well, then just check American Indian, Asian, and Black, I guess," offered Star.

Indigo let out an exasperated sigh, "I don't think I belong to any of these races. Maybe they should have a category for 'none of the above'!"

"I'm not sure that would do it either, honey," sympathized Star. "Maybe 'human' would be more appropriate."

Indigo and Star spent a few more minutes on the phone, but after she hung up, Indigo still felt troubled. "Why do I need to specify my racial category?

And what does it mean? If I'm part White and part Black, am I somehow different from each group?"

Indigo's questions reflect those posed by many people over the years. Why do human groups differ from each other in skin color, eye shape, and stature? Do these physical differences reveal underlying basic biological differences among these groups?

10.1 All Humans Belong to the Same Species

In the mid-1700s, the Swedish scientist Carolus Linnaeus began the task of cataloguing all of nature. As described in Chapter 8, Linnaeus developed a classification scheme that grouped organisms according to shared common traits. The primary category in his classification system was **species**, a group whose members have the greatest resemblance. Linnaeus assigned a two-part name to each species—the first part indicates the **genus**, or broader group to which the species belongs, while the second part is specific to the particular species within the genus. Linnaeus coined the binomial name *Homo sapiens* (*Homo* meaning "man," and *sapiens* meaning "knowing or wise") to describe humans. Although Linnaeus recognized the impressive variability among humans, by placing all of us in the same species he acknowledged our basic unity. (Linnaeus did classify humans into different *varieties* within the same species, a point we will return to in Section 10.2.)

Modern biologists have kept Linnaeus's basic classification, although a **subspecies** name, *Homo sapiens sapiens*, has been added to distinguish modern humans from humans that first appeared approximately 250,000 years ago. What does it mean to belong to a species?

The Biological Species Concept

According to the **biological species concept**, a species is defined as a group of individuals that, in nature, can interbreed and produce fertile offspring but do not breed with members of another species. In many ways, species are considered the fundamental units of biology. Like all living units, species grow and reproduce, but unlike cells or individuals, species evolve and can change over the course of many generations. Because traits can only be passed on to individuals of the same species, any trait that occurs in a member of a species can only spread within that species. When the frequency of a trait in a species changes over the course of generations, the species has evolved. The fact that species have the potential to evolve as a unit provides the basis of the biological species concept.

As we discussed in Chapter 9, differences in traits among individuals arise partly from differences in their genes. New *alleles* of a gene occur when the DNA that makes up the gene *mutates*. A particular allele can become more common in a species through the process of evolution. Scientists refer to the sum total of the alleles found in the individuals of a species as its **gene pool**. We can think of a single species as making up an impermeable container for that species' gene pool—a change in the frequency of an allele in a gene pool can only happen within a species.

The spread of an allele throughout a species' gene pool is called **gene flow**. Gene flow cannot occur between different biological species because a pairing between members of these different species fails to produce fertile offspring. This phenomenon, known as **reproductive isolation**, can take two general forms: *prezygotic* (before fertilization) barriers, or *postzygotic* (after fertilization) barriers.

Figure 10.1 A prezygotic barrier to reproduction. (a) Female blue-footed boobies will not mate with males who fail to perform this dance. (b) Male blue-footed boobies will not mate with females who do not engage in the pointing display with them. These behaviors prevent reproduction between unrelated booby species.

Prezygotic barriers to reproduction occur when individuals from different species either do not attempt to mate with each other, or if they do, fail to produce a fertilized egg. For example, many of the songs and displays produced by birds serve as prezygotic barriers. Male blue-footed boobies, sea birds that look almost as goofy as their name implies, perform an elaborate dance for the female before they mate (Figure 10.1a). This dance involves much waggling and displaying of their electric blue feet and differs from the dances performed by males of other, related booby species. A female blue-footed booby will not respond until she has witnessed several rounds of the dance, at which time she will engage the male in a pointing display (Figure 10.1b). In this display, both birds point their bills skyward, drop their wings, and call out their mating song. The male's dance and the pairs' pointing display presumably provide a way for both birds to recognize that they belong to the same booby species. If a female is courted by a male that cannot perform the "Blue-footed Booby Dance," she will not mate with him.

The most common prezygotic barrier between species that *will* mate with each other is an incompatibility between eggs and sperm. For fertilization to occur, a sperm cell must bind to a protein on the surface of an egg cell. If the egg does not recognize the sperm (that is, if the egg does not have a protein that will bind to the sperm), no offspring can be produced. Among animal species that utilize external fertilization and release their sperm and eggs into the environment, such as fish, amphibians, and sponges, this method of reproductive isolation is widespread.

Postzygotic barriers occur when fertilization does occur as a result of mating between two members of different species, but the resulting offspring does not survive or is sterile. Mules are the result of successful mating between a horse and a donkey, two different species. Mules have a well-earned reputation as tough and sturdy farm animals, but they are sterile and cannot produce their own offspring. Most instances of postzygotic barriers are less obvious—the offspring of most *interspecies hybrids*, that is, of parents from two different species, do not survive long after fertilization. Postzygotic barriers are often a result of an incompatibility between the genes of different species. Different species have different genes, placing these genes in combination in an offspring provides incomprehensible information about how to build a body.

Figure 10.2 Reproductive isolation between horses and donkeys. (a) Mules carry one set of horse chromosomes and one set of donkey chromosomes. (b) The sperm or eggs that are produced will not result in living offspring. The mule's sterility is a postzygotic barrier to reproduction between the horse species and the donkey species.

(a) Mule, a cross between a horse and a donkey

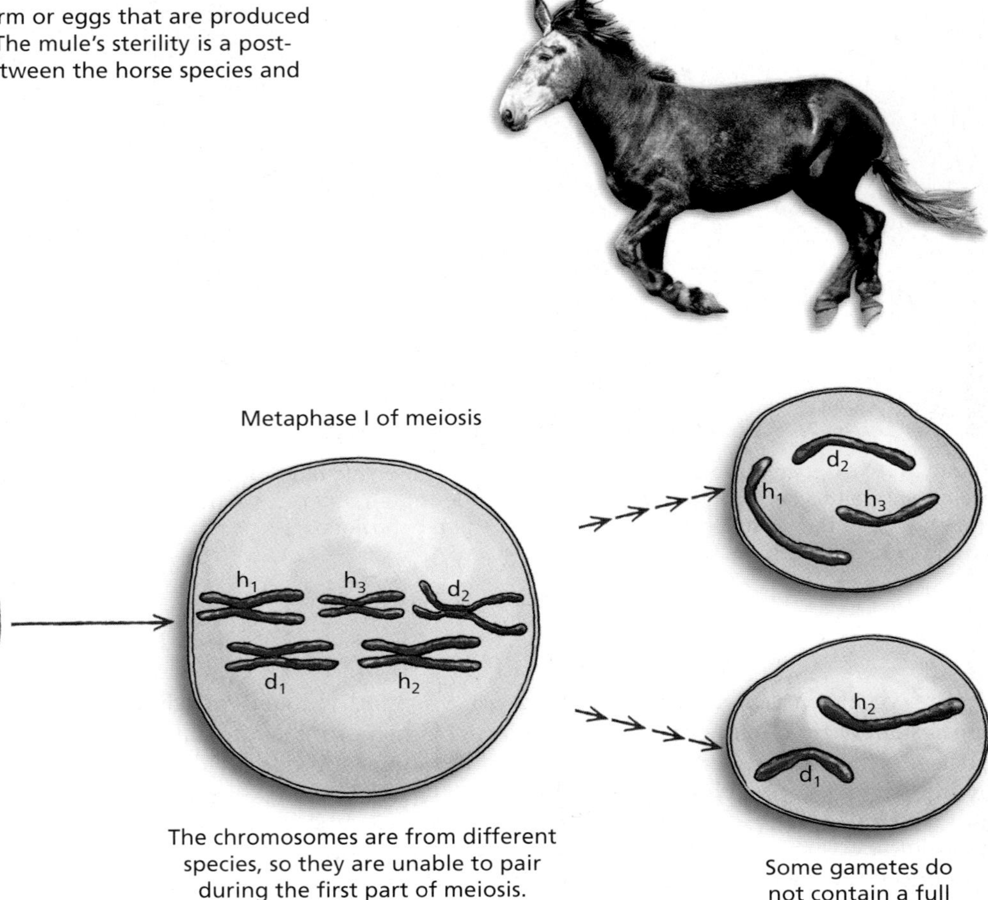

(b) Why mules are sterile

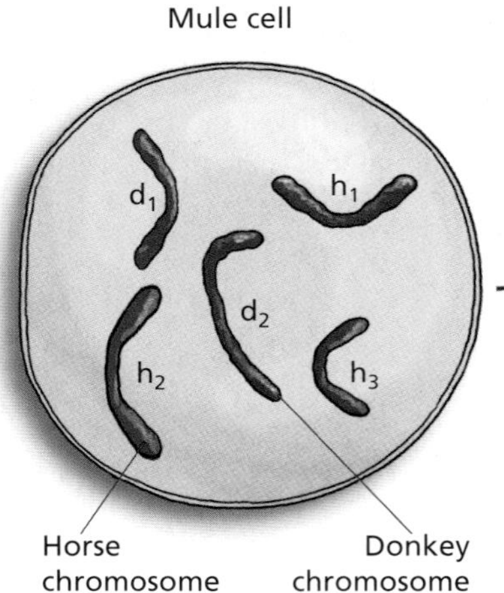

In the case of mules and other sterile hybrids, the genetic incompatibility is not so large that the offspring cannot develop. Instead, the postzygotic barrier of hybrid sterility occurs because these hybrids cannot produce proper sperm or egg cells. Recall from Chapters 4 and 6 that during the production of eggs and sperm compatible genetic sequences called *chromosomes* pair up and are separated during the first cell division of *meiosis*. Because a hybrid forms from the sets of chromosomes from two different species, the chromosomes cannot pair up correctly during this process, and the sperm or eggs that are produced in these animals will have too many or too few chromosomes and cannot lead to normal offspring (Figure 10.2).

Different groups of humans show no evidence of postzygotic barriers to reproduction, and Indigo's diverse ancestry clearly demonstrates that no prezygotic barriers exist that prohibit mating among the races listed on her census form. Thus, all humans belong to the same biological species. To understand the concept of races *within* a species however, we must examine the process of speciation.

The Process of Speciation

According to the theory of common descent discussed in Chapter 8, all modern organisms descended from a common ancestral species. This evolution of one or more species from an ancestral form is called **speciation**, and the process of speciation is often referred to as **macroevolution**.

For one species to give rise to a new species, most biologists agree that three steps are necessary:

1. Isolation of the gene pools of subgroups, or *populations*, of the species;
2. Evolutionary changes in one or both of the isolated populations; and
3. The evolution of reproductive isolation between these populations, preventing any future gene flow.

Recall that gene flow occurs when reproduction is occurring within a species. Now imagine what would happen if two populations of a species became *physically* isolated from each other, so that the movement of individuals between these two populations was impossible. Even without prezygotic or postzygotic barriers to mating between these two populations, gene flow between them would cease.

What is the consequence of eliminating gene flow between two populations? New alleles that arise in one population may not arise in the other, so while a new allele may become common in one population, it may not exist in the other. Even among existing alleles, one may increase in frequency in one population, but not in the other. In this way, each population would be evolving independently. Over time, the traits found in one population begin to differ from the traits found in the other population. In other words, the populations begin to **diverge** (Figure 10.3).

The gene pools of populations may become isolated from each other for several reasons. Often a small population becomes isolated when it migrates to a location far from the main population. This is the case on many oceanic islands, including the Galapagos and Hawaiian Islands. Species on these islands appear to be the descendants of species from the nearest mainland. The original ancestral migrants arrived on the islands by chance. Because it is rare for organisms from the mainland to find their way across hundreds of miles of open ocean to these islands, populations at each site are practically completely isolated from each other (Figure 10.4).

Populations may also be isolated from each other by the intrusion of a geologic barrier. This could be an event as slow as the rise of a mountain range or as rapid as a sudden change in the course of a river. The emergence of the Isthmus of Panama about three million years ago represents one such intrusion event. This land bridge connected the formerly separate continents of South and North America but *divided* the ocean gulf between them. Scientists have described several pairs of species of snapping shrimp on both sides of the isthmus that appear

www
Media Activity 10.1A Populations Diverge if Gene Flow Is Cut Off

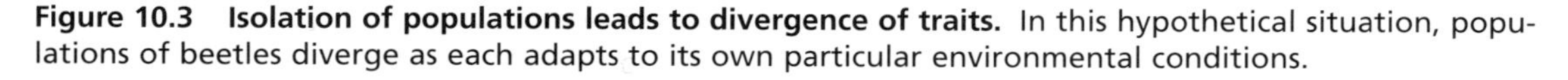
Populations become isolated (no gene flow)
Evolutionary changes accumulate over time, and the populations diverge in their characteristics.

Figure 10.3 Isolation of populations leads to divergence of traits. In this hypothetical situation, populations of beetles diverge as each adapts to its own particular environmental conditions.

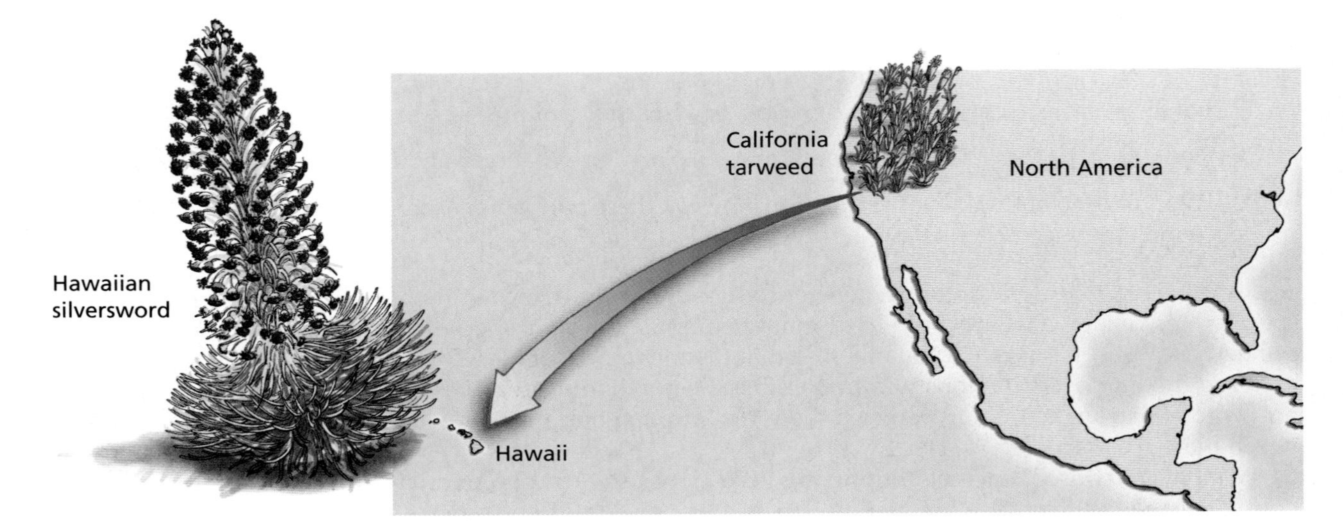

Figure 10.4 Migration leads to speciation. The ancestor of Hawaiian silverswords was the much smaller and less dramatic California tarweed. Tarweed seeds were blown or carried by birds to the Hawaiian Islands, creating an isolated population. With no gene flow between the two populations, Hawaiian silverswords evolved into a very different group of species.

Media Activity 10.1B Migration Leads to Speciation

to have speciated after this event. These shrimp species seem to be related to each other because of similarities in appearance and lifestyle. In each case, one member of each pair is found on the Caribbean side of the land bridge, while the other is found on the Pacific side (Figure 10.5). This geographic pattern indicates that the two species in each pair descended from a single species. These original

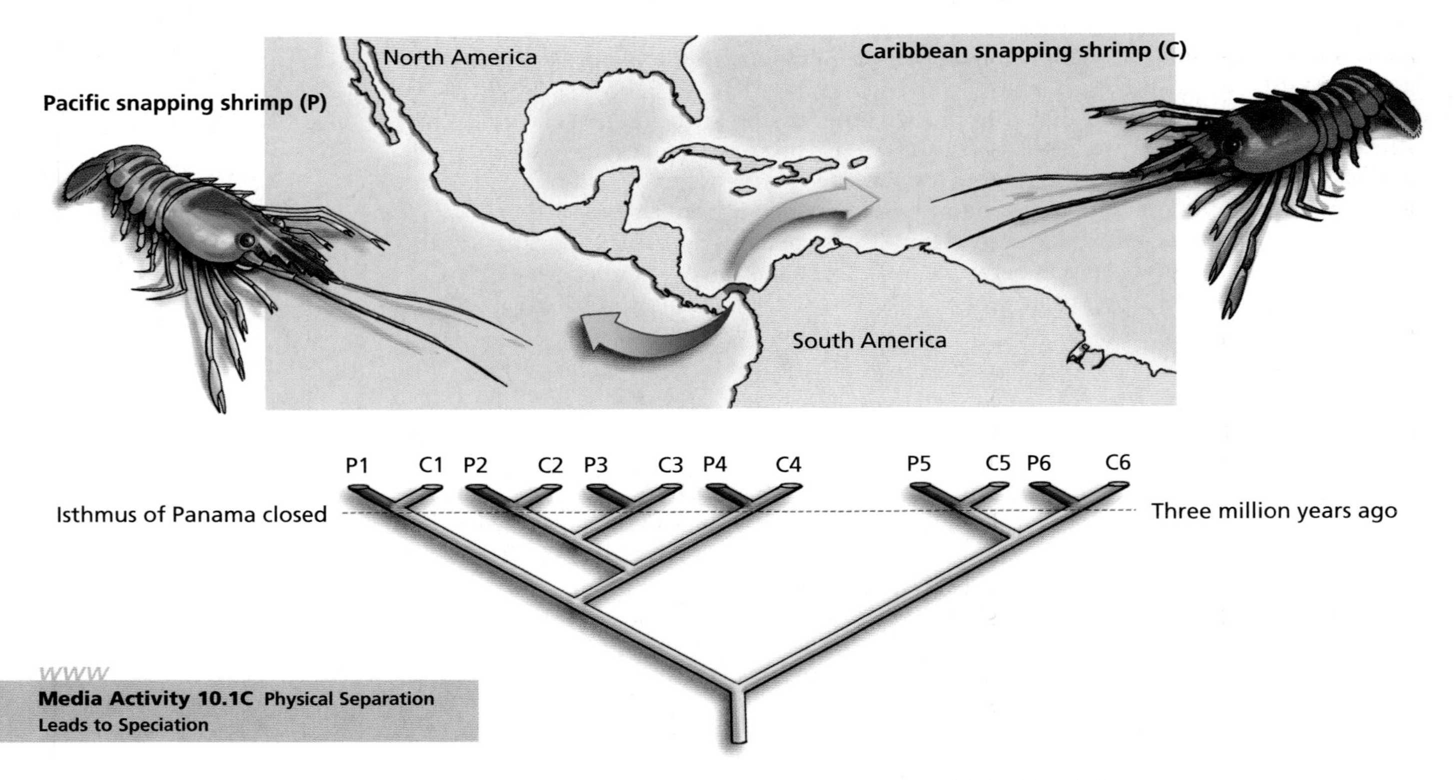

WWW
Media Activity 10.1C Physical Separation Leads to Speciation

Figure 10.5 Physical separation leads to speciation. Species of snapping shrimp on either side of the Isthmus of Panama can be paired according to similarities in appearance and habit. Pairs are numbered for simplicity—the letter "C" before the number indicates the species is found on the Caribbean side of the Isthmus, while a "P" indicates a Pacific species. This pattern indicates that recent speciation in this group occurred after the Isthmus emerged and divided formerly continuous populations of ancestral species.

species were most likely found throughout the gulf before the isthmus arose and were divided into two isolated populations after the land bridge appeared. Once in isolation, the populations diverged into different species.

Isolation of the gene pools of populations may also occur even if the populations are living in physical proximity to each other. This appears to be the case in populations of apple maggot fly, a species that provides one of the clearest examples of macroevolution "in action."

Apple maggot flies are so named because they are notorious pests of apples grown in northeastern North America. However, apple trees are not native to North America—they were first introduced to this continent less than 300 years ago. Apple maggot flies also infest the fruit of hawthorn shrubs, a group of species that *are* native to North America. Apple maggot flies appear to have descended from hawthorn-infesting ancestors that began to use the novel food source of apples after the fruit began to be cultivated in their home range. Apples and hawthorns live in close proximity, and apple maggot flies clearly have the ability to fly between apple orchards and hawthorn shrubs. At first glance, it does not appear that the apple maggot flies that eat apples and those that eat hawthorn fruit are isolated from each other.

However, upon closer inspection, populations of apple maggot flies on apples and those on hawthorns actually have little opportunity for gene flow between them. Flies mate on the fruit where they will lay their eggs, and hawthorns produce fruit approximately one month after apples do. Each population of fly has a strong preference for which fruit it will mate on, and flies that lay eggs on hawthorns develop much more quickly than flies that lay eggs on apples. There appears to be little mixing between the apple-preferring and hawthorn-preferring populations.

Scientists who have examined the gene pools of the two groups of apple maggot flies find that they differ strongly in the frequency of some alleles, so much so that they refer to them as incipient (or newly forming) species. Thus it appears that divergence of two populations can occur even if those populations are in contact with each other, as long as some other factor—in this case the timing of mating and reproduction—is keeping their gene pools isolated (Figure 10.6).

For incipient species that have diverged in isolation to become truly distinct biological species, they must become reproductively isolated. This may happen when enough divergence between the populations has occurred so that individuals of different populations are no longer genetically compatible. There is no hard-and-fast rule about how much divergence is required—sometimes a difference in a single gene can lead to incompatibility, while other times populations demonstrating great physical differences can produce healthy and fertile hybrids (Figure 10.7).

Once reproductive isolation occurs, each species may take radically different evolutionary paths because gene flow between the two species is impossible. Over thousands of generations, species that derived from a common ancestor can accumulate many differences, even completely new genes.

The period between separation of populations and the evolution of reproductive isolation—that is, the development of two new species from a common ancestral species—could be thought of as a period during which biological races of a species may form. Determining if the racial groupings on Indigo's census form came about via this process is our focus in the next section.

www

Media Activity 10.1D Subdivision of the Environment Leads to Speciation

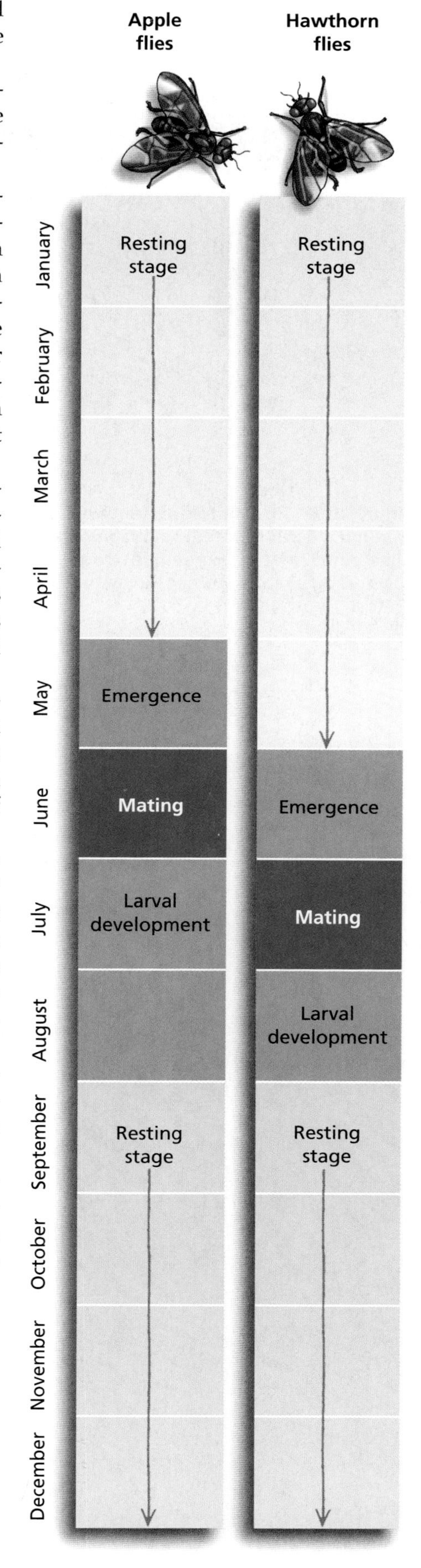

Figure 10.6 Incipient speciation in apple maggot flies. This graph illustrates the life cycle of two populations of the apple maggot fly: one that lives on apple trees, and another that lives on hawthorn shrubs. Note that the mating period for these two populations differs by a month. This results in little gene flow between these two populations.

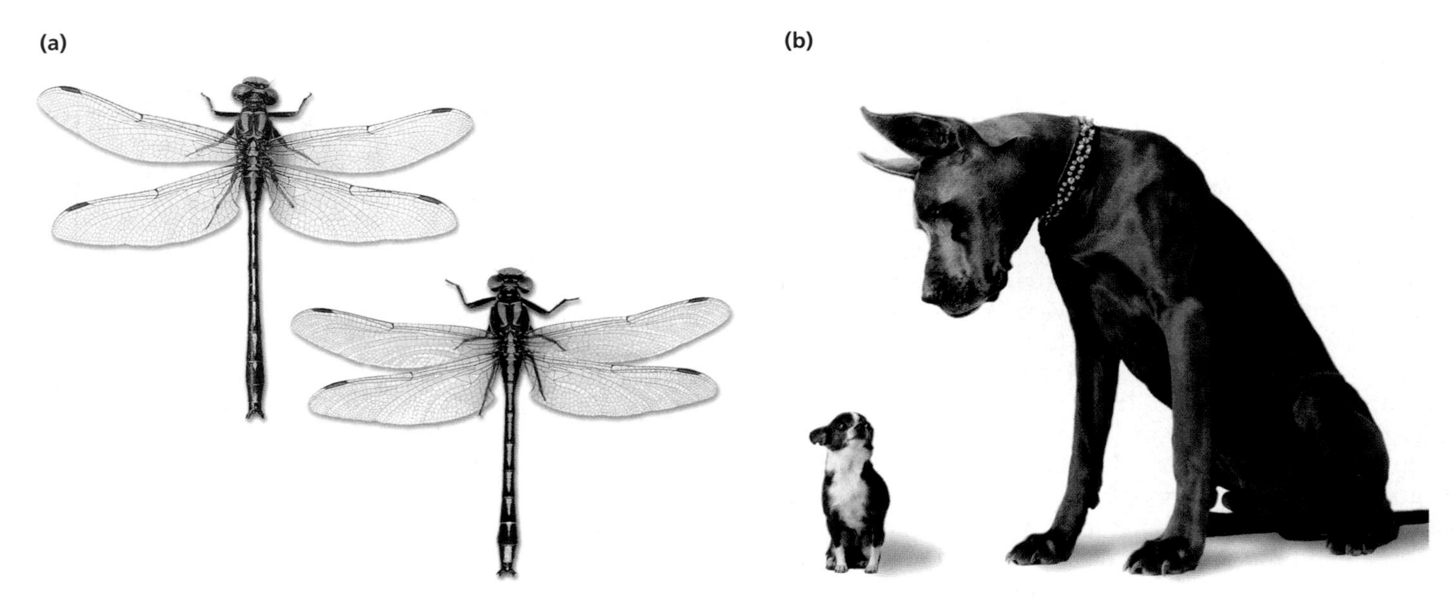

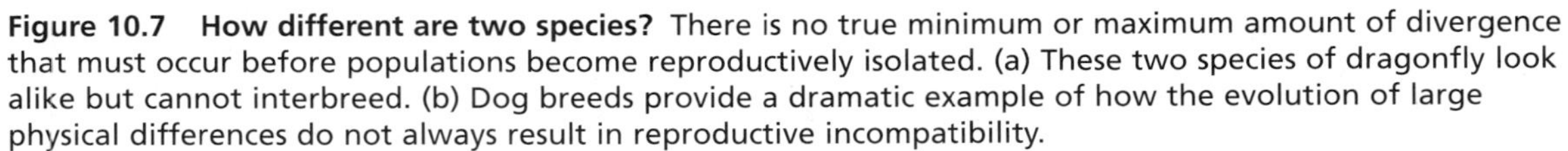

Figure 10.7 How different are two species? There is no true minimum or maximum amount of divergence that must occur before populations become reproductively isolated. (a) These two species of dragonfly look alike but cannot interbreed. (b) Dog breeds provide a dramatic example of how the evolution of large physical differences do not always result in reproductive incompatibility.

10.2 The Race Concept in Biology

Biologists do not agree on a standard definition of "biological race." Populations of birds with slightly different colorations might be called different races by some bird biologists, while other biologists would argue that the contrasts in color are unimportant. However, Indigo's question about *how* the racial group she identifies with matters leads us to a definition of race that does have a specific meaning. What she wants to know is: If she identifies herself as a member of a particular race, does that mean that she is more closely related, and thus biologically more similar, to other members of the same race than she is to members of other races? The definition of **biological race** that addresses this question is the following: Races are populations of a single species that have diverged from each other. These could be populations that are currently in the process of speciating, or they could be populations that will never be reproductively isolated but have little gene flow between them. With little gene flow, evolutionary changes that occur in one race may not occur in a different race. (Note that many scientists prefer the term *subspecies* over *race* for these subgroups within a species.) Indigo's question about whether race matters, at least biologically, can now be restated as: Do the racial groups on the census form represent populations of the human species whose gene pools were isolated until relatively recently?

Humans and the Race Concept

Until the height of the European colonial period in the seventeenth and eighteenth centuries, few cultures distinguished broad groups of humans based on shared physical characteristics. People primarily identified themselves and others as belonging to particular cultural groups, with different customs, diets, and languages. As northern Europeans began to contact people from other parts of the world, being able to set these people "apart" made the process of colonization and subjugation less morally questionable. Thus, when

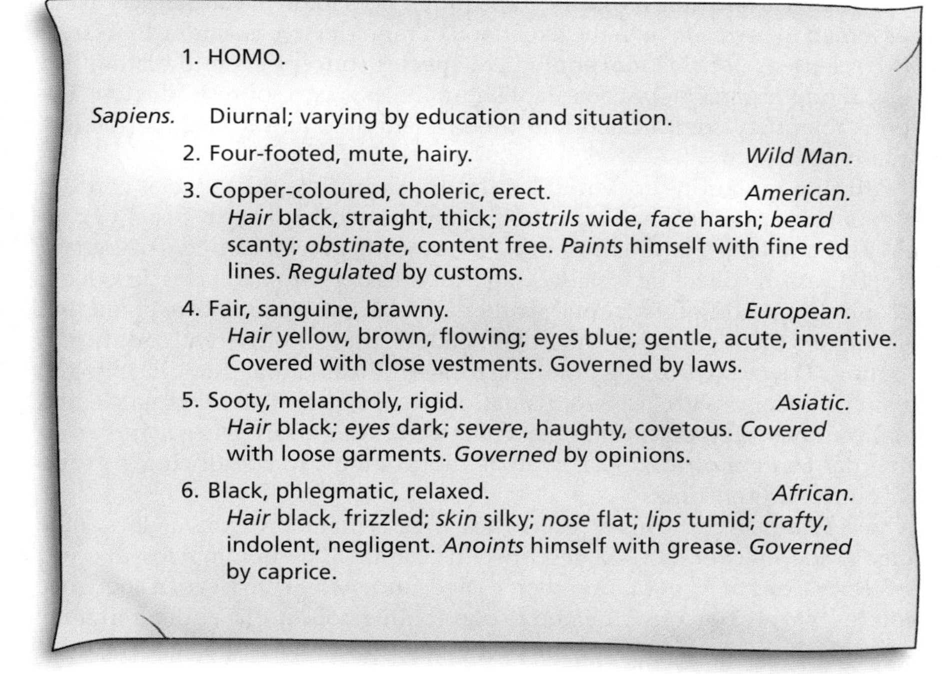
1. HOMO.

Sapiens. Diurnal; varying by education and situation.

2. Four-footed, mute, hairy. *Wild Man.*
3. Copper-coloured, choleric, erect. *American.*
Hair black, straight, thick; *nostrils* wide, *face* harsh; *beard* scanty; *obstinate*, content free. *Paints* himself with fine red lines. *Regulated* by customs.
4. Fair, sanguine, brawny. *European.*
Hair yellow, brown, flowing; eyes blue; gentle, acute, inventive. Covered with close vestments. Governed by laws.
5. Sooty, melancholy, rigid. *Asiatic.*
Hair black; *eyes* dark; *severe*, haughty, covetous. *Covered* with loose garments. *Governed* by opinions.
6. Black, phlegmatic, relaxed. *African.*
Hair black, frizzled; *skin* silky; *nose* flat; *lips* tumid; *crafty*, indolent, negligent. *Anoints* himself with grease. *Governed* by caprice.

Figure 10.8 Linneaus's classification of human variety. Linneaus published this classification of the varieties of humans in the 10th edition of the *Systema Naturae* in 1758. The behavioral characteristics he attributed to each variety reflect a widespread bias among his contemporaries that the European race was superior to other races. Interestingly, one variety of humans recognized by Linneaus—the "Wild Man" in his classification—is the chimpanzee!

Linnaeus classified all humans as a species, he was careful to distinguish definitive *varieties* (what we would now call races) of humans. Linnaeus recognized five races of *Homo sapiens* and set the European race as the superior form (Figure 10.8).

Numerous scientists since Linneaus have also attempted to establish the number of human races. The most common number is six: White, Black, Pacific Islander, Asian, Australian Aborigine, and Native-American; although some scientists have described as many as 26 different races of the human species. The physical characteristics that are used to identify these races are typically skin color, hair texture, and eye, skull, and nose shape.

To answer Indigo's question about the biological meaning of race, we must determine if the physical characteristics Linnaeus and other scientists used to delineate their proposed human races developed when these groups were evolving independently (or mostly independently) of each other. We can test this hypothesis by looking at the fossil record for evidence of isolation during human evolution and by looking at the gene pools of these proposed races for the vestiges of that isolation.

Modern Humans: A History

WWW
Media Activity 10.2 **How Did Human Groups Evolve?**

The ancestors of humans are known only through the fossil record. We cannot delineate these fossil species using the biological species concept. Instead, paleontologists use a more practical definition: A species is defined as a group of individuals that have some reliable physical characteristics distinguishing them from all other species. In other words, individuals in the same species have similar *morphology*—they look alike. The differences in physical characteristics are assumed to correlate with the likelihood of reproductive isolation. In the real world, scientists use this **morphological species concept** even to distinguish among living organisms, since establishing the existence of reproductive isolation is relatively complicated and time consuming, and in fact, impossible for many species.

Natural populations are variable, thus the morphological species concept presents some special challenges for scientists working with fossil organisms. The challenge is illustrated by a dinosaur genus named *Triceratops*. After 80 years of collecting fossils, 30 species of *Triceratops* had been described, each one slightly different from the others. Eventually scientists realized that every one of these "species" had been collected from only two counties in Wyoming. The close proximity of these fossils indicated that they all belonged to a single species with large and small males, large and small females, and juveniles. Paleontologists must use clues about the location, age, and environmental context of fossils, as well as morphology, to convincingly group fossils into different species.

One advantage of the fossil record, however, is that it provides a view of the change in species over time. As described in Chapter 8, the hominin fossil record (the fossil record of humans and their extinct ancestors) consists of a sequence of species that are obviously related to one another, follow one another in time, and are generally interpreted as making up an evolutionary lineage. The morphological differences between hominin species are clear, meaning that reconstructing the movement of human ancestors out of Africa is relatively straightforward.

The immediate predecessor of *Homo sapiens* was *Homo erectus*, a species that first appeared in east Africa about 1.8 million years ago and spread to Asia and Europe over the next 1.65 million years. Fossils identified as early *H. sapiens* appear in Africa approximately 250,000 years ago. The fossil record shows that early humans rapidly replaced *H. erectus* populations in the Eastern Hemisphere.

There is considerable debate among paleontologists about whether *H. sapiens* evolved just once, in Africa (this is called the *out-of-Africa* hypothesis), or throughout the range of *H. erectus* (known as the *multiregional* hypothesis). Even if *H. sapiens* evolved in Africa and then migrated to Europe and Asia, it is unclear whether populations of early humans hybridized with *H. erectus* in different areas of the globe (the *hybridization and assimilation* hypothesis). Because this scientific question is still unresolved, it is difficult to know when the ancestral population of modern humans split into regional populations—it could be anytime from 1.8 million to 150,000 years ago (Figure 10.9).

Most paleontologists favor the out-of-Africa hypothesis—that all modern human populations descended from African ancestors within the last few hundred thousand years. This hypothesis is supported by the close genetic similarity among people from very different geographic regions. Humans have much less genetic diversity (measured by the number of different alleles that have been identified for any gene) than any other great ape, which is an indication that they are a young species that has had little time to evolve many different gene variants. The out-of-Africa hypothesis is also supported by evidence that human populations in Africa are more genetically diverse than other human populations around the world. This observation indicates that African populations are the oldest human populations. If the out-of-Africa hypothesis is correct, the physical differences we see among human populations must have arisen in the last 150,000 to 200,000 years, or about 10,000 human generations. In evolutionary terms, this is not much time. The recent shared ancestry of human groups does not support the hypothesis that the commonly defined human races are very different from each other.

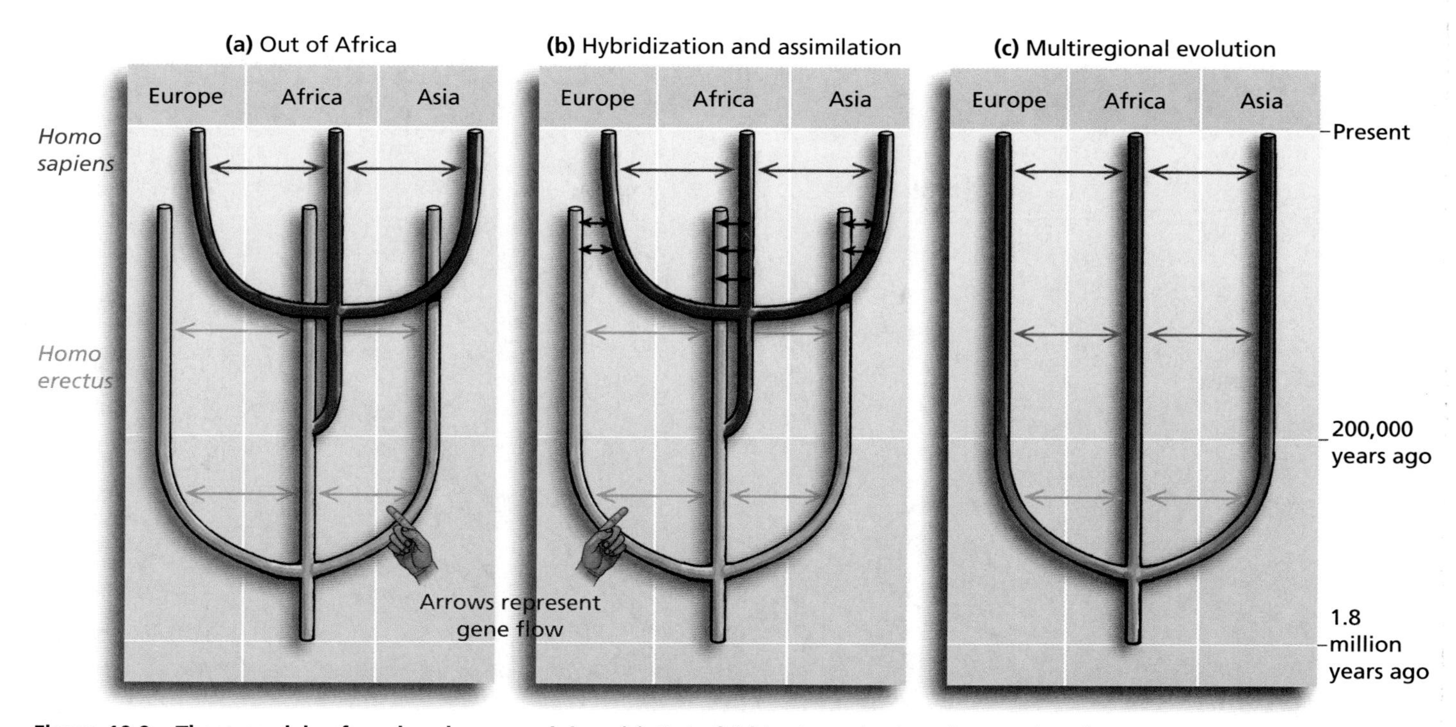

Figure 10.9 Three models of modern human origins. (a) Out of Africa hypothesizes that modern humans arose in Africa and replaced populations of *Homo erectus* that had moved around the globe. (b) Hybridization and assimilation is an intermediate hypothesis that states that *H. sapiens* evolved in Africa and replaced *H. erectus* populations, but that there was some hybridization among local forms of *H. erectus* and incoming *H. sapiens*. (c) Multiregional evolution hypothesizes that *H. sapiens* evolved from *H. erectus* multiple times throughout its range.

Testing the Hypothesis of Human Races

While the evidence discussed thus far indicates that members of the human species are not very different from each other, Indigo's question about the meaning of race is still relevant. After all, even if two races only differ from one another slightly, if the difference is consistent, perhaps it is fair to say that people are biologically more similar to members of their own race than to people of a different race.

If a population represents a biological race, there will be a record in its gene pool of its isolation from other groups. Recall that when populations are isolated from each other, little gene flow occurs between them. If an allele appears in one population, it cannot spread to another, and changes in allele frequency occurring in one population do not necessarily occur in others. Therefore, we can make two predictions to test a hypothesis of biological races. If a race has been isolated from other populations of the species for many generations, it should have (1) some unique alleles, and (2) allele frequencies for some genes that differ from the allele frequencies (for those genes) of other races. The tree diagram in Figure 10.10 illustrates this by comparing two hypothetical races. In the figure, populations A, B, and C are all part of the same race (Race 1), and populations X, Y, and Z are part of a separate race (Race 2). The grid at the bottom of the tree illustrates the allele frequency of four genes in the ancestral population. For instance, there are two alleles for gene 2—one that is very common (allele a), and one that is rare (allele b). The two races described at the top of the tree originated when the ancestral population split, and the two resulting populations became isolated from each other. Not long after this divergence, mutation causes a new allele for gene 2 (allele c) to arise, but only in Race 2. Because the two races

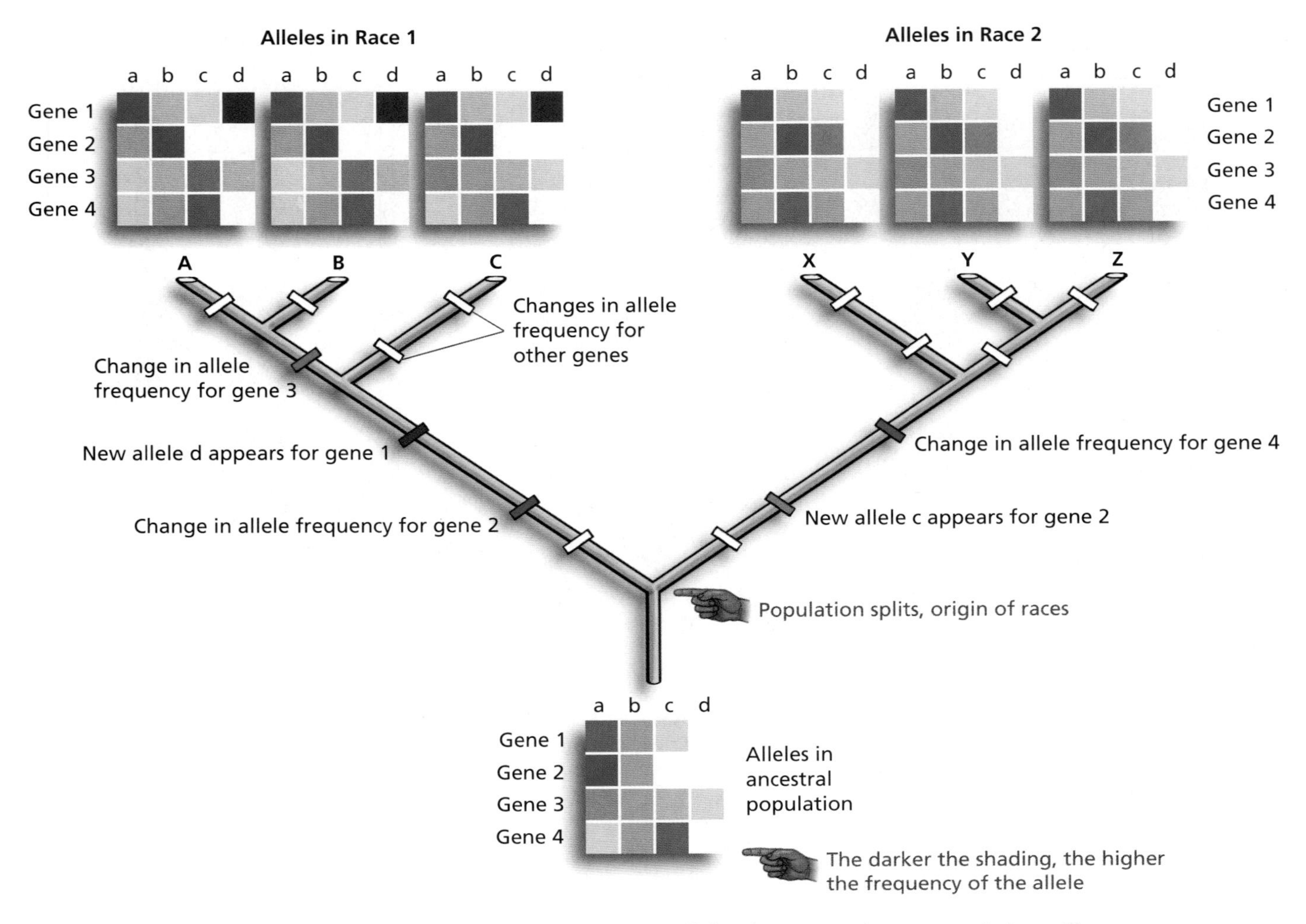

Figure 10.10 Genetic evidence of isolation between races. New alleles that appear in one population will not spread to the other because gene flow is restricted, so each race should have unique alleles. Allele frequencies should also be more similar among populations within a race than between races.

are isolated, this allele does not spread to Race 1. Additionally, because populations X, Y, and Z diverge *after* this allele appears in the race, all of these populations contain individuals who carry this allele. At the same time in Race 1, natural selection is resulting in a change in the allele frequency of gene 2. Perhaps the environment inhabited by Race 1 favors individuals who carry allele b—this results in these individuals having more offspring than individuals who carry only allele a, and thus allele b becomes more common in this race. This evolutionary change occurred before the divergence of populations A, B, and C, so all of these populations have a similar pattern of allele frequency for this gene, but the pattern differs from that in Race 2. In fact, if you compare the allele frequency grids of all the populations, you will notice that populations A, B, and C are similar (although not identical), and populations X, Y, and Z are similar, but population A is very different from populations X, Y, and Z.

Observing a pattern of unique allele frequencies in different populations of the same species is one piece of supporting evidence showing that the populations have been isolated from each other. For example, this type of observation helped scientists establish that the apple-eating and hawthorn-eating populations of apple maggot flies were isolated from each other. In other words, scientists are convinced that these populations of flies are different races.

Recall the six major human races described by many authors: White, Black, Pacific Islander, Asian, Australian Aborigine, and Native American. Do these groups show the predicted pattern of race-specific alleles and unique patterns of allele frequency? In a word, no.

Let us first examine alleles that were once thought to be unique to a race. Sickle-cell anemia is a condition we discussed in Chapter 4. This disease is found in individuals who carry two copies of the sickle-cell allele, an allele that results in red blood cells that deform into a sickle shape under certain conditions. The consequences of these sickling attacks include heart, kidney, lung, and brain damage. Many individuals with sickle-cell anemia do not live past childhood.

Sickle-cell anemia was once thought of as unique to the "Black race." Nearly 10% of African-Americans and 20% of Africans carry one copy of the sickle-cell allele. However, if we examine the distribution of the sickle-cell allele more closely, we see that the pattern is not quite so simple. Just as we can divide the human species into populations that share similarity in skin color and eye shape (the typical races), we can divide these races into smaller populations that share other morphological features—for example, average height or hair color—and cultural and language similarities. When we do this, we find that not all populations classified as Black have a high frequency of the sickle-cell allele. In fact, in populations from southern and north-central Africa, which are traditionally classified as Black, this allele is very rare or absent. Among populations that are classified in the White and Asian races, there are some in which the sickle-cell allele is relatively common, such as among White populations in the Middle East and Asian populations in northeast India. (Figure 10.11). Thus, the sickle-cell allele is not a characteristic of all Black populations, nor is it unique to a supposed "Black race."

Similarly, cystic fibrosis, a disease caused by a recessive allele that results in respiratory and digestive problems and early death, was often thought of as a

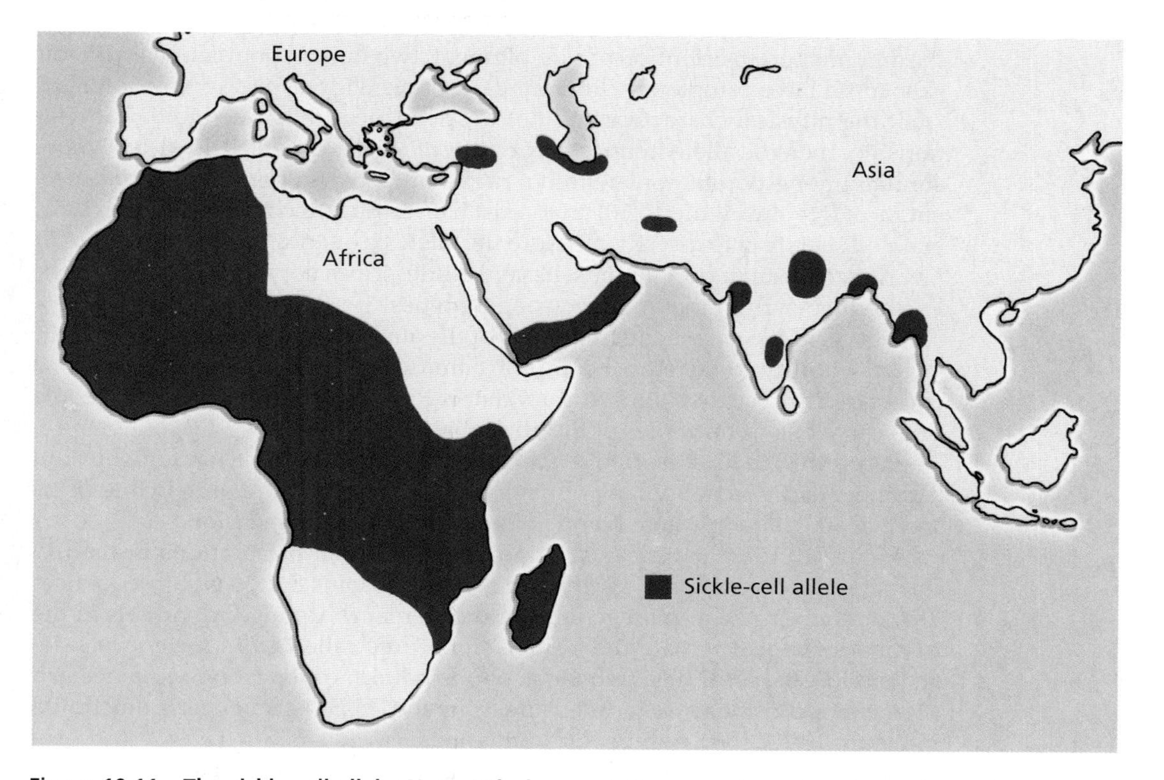

Figure 10.11 The sickle-cell allele: Not a "Black gene." The map illustrates where the sickle-cell allele is found in human populations. Note that it is not found in all African populations and is found in some European and Asian populations.

disease of the "White race." As with sickle-cell anemia, it has become clear that the allele that causes cystic fibrosis is not found in all White populations and is found, in low frequency, in some Black and Asian populations. Thus, the cystic fibrosis allele is not a characteristic of all White populations, nor is it unique to a supposed "White race."

These examples of the sickle-cell allele and the cystic fibrosis allele demonstrate the typical pattern of gene distribution. Scientists have not identified a single allele that *is* found in all (or even most) populations of a commonly described race and *is not* found in other races. The hypothesis that human races represent mostly independent evolutionary groups is not supported by this observation.

What about the second prediction of the hypothesis that human racial groups are biologically independent—that we should observe unique patterns of allele frequency within these different races? If evolution is occurring mostly within racial groups, allele frequencies for various genes should be more similar among populations within a racial group than among populations in different racial groups.

Until the advent of modern techniques allowing scientists to isolate genes and the proteins they produce, there was no way of directly measuring the frequency of alleles for particular genes in a population. However, scientists could evaluate the racial categories already in place and *assume* that genetic differences among these groups reflected their average physical differences. Thus, populations with dark skin were assumed to have a high frequency of "dark skin" alleles, while populations with light skin were assumed to have a low frequency of these alleles. Similar assumptions were made about a range of physical differences—eye shape, skull shape, and hair type all clearly have a genetic basis, and all clearly differ among racial categories. These observations appear to support the hypothesis that different races have unique allele frequencies. However, physical characteristics such as skin color, eye shape, and hair type are each influenced by several different genes, each with a number of different alleles. Because skin color is affected by numerous genes, each of which has an effect on the amount and distribution of skin pigment, two human populations with fair skin could have completely different gene pools with respect to skin color.

If the physical characteristics that describe races illustrate biological relationship, then the allele frequency for *many different* genes should also be more similar among populations within a race than between populations of different races (review Figure 10.10). In the last half-century, modern techniques have enabled scientists to directly measure the allele frequency of different genes in a variety of human populations. Essay 10.1 describes how we can calculate allele frequency from the frequency of genotypes.

Figure 10.12a shows the frequency of the allele that interferes with an individual's ability to taste phenylthiocarbamide (PTC). People who carry two copies of this recessive allele cannot detect PTC, which tastes bitter to people who carry one or no copies of the allele.

Figure 10.12b lists the frequency of one allele for the gene haptoglobin 1 in a number of different human populations. Haptoglobin 1 is a protein that helps "rescue" the blood protein hemoglobin from old, dying red blood cells.

Figure 10.12c illustrates variation among human populations in the frequency of a repeating DNA sequence on chromosome 8. Repeating sequences are common in the human genome and differences among individuals in the number of repeats create the unique signatures called *DNA fingerprints* described in Chapter 6. The frequency of one pattern of repeating sequence in a segment of chromosome 8, called allele 4 of the D8s384 sequence, is illustrated in Figure 10.12c for a number of populations.

You should note that the human populations in Figure 10.12 are listed by increasing frequency of the allele in the population. If the hypothesis that human racial groups have a biological basis is correct, populations from the same racial

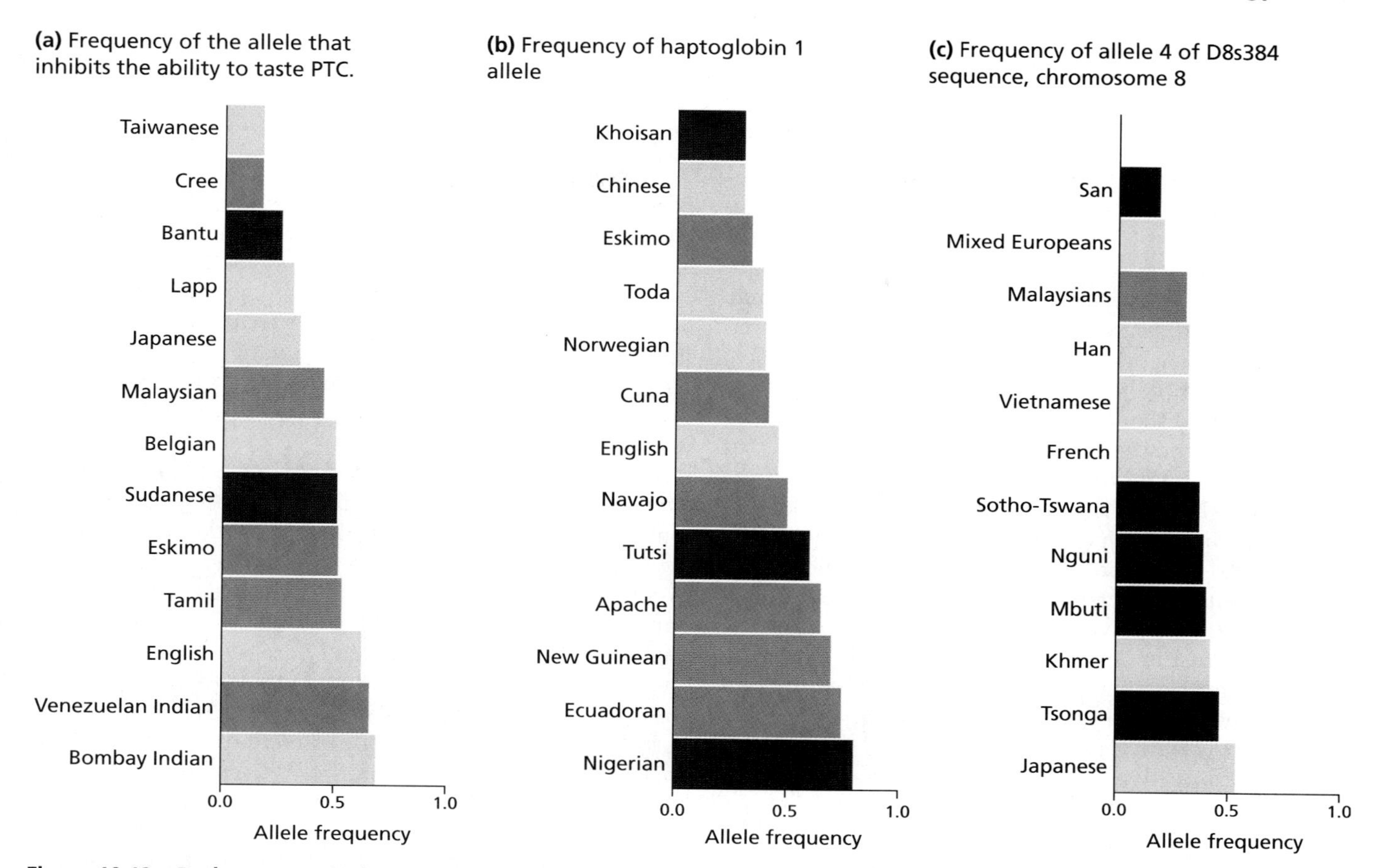

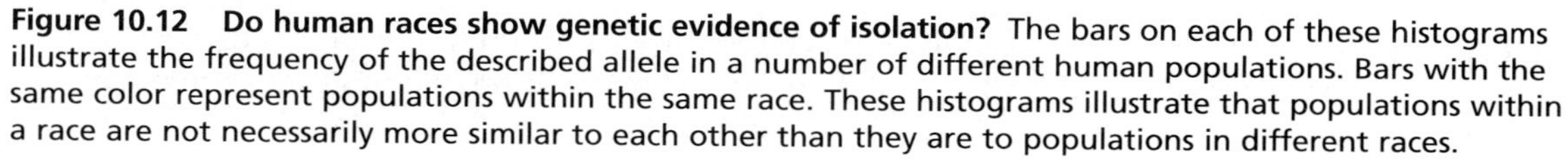

Figure 10.12 Do human races show genetic evidence of isolation? The bars on each of these histograms illustrate the frequency of the described allele in a number of different human populations. Bars with the same color represent populations within the same race. These histograms illustrate that populations within a race are not necessarily more similar to each other than they are to populations in different races.

group should be clustered together on the graphs. The color-coding of each population group in a graph corresponds to the racial category in which they are typically placed. To evaluate this hypothesis, we are using stereotypical colors for the races—pale brown for White, dark brown for Black, yellow for Asian, red for Native American, and medium brown for Pacific Islanders.

What we see in the three graphs is that allele frequencies for these genes are *not* more similar within racial groups than between racial groups. In fact, in two of the three graphs, the populations with the highest and lowest allele frequency belong to the same race—for these genes, there is more variability within a race than average differences among races. These observations do not support the hypothesis that morphological similarity among populations reflects some underlying close biological relationship. The fossil evidence and genetic evidence indicate that the six commonly listed racial groups do *not* represent biological races.

Human Races Have Never Been Truly Isolated

The genetic analysis that caused scientists to reject the hypothesis that human populations within the same races are very similar to each other and consistently different from other races has actually shown that exactly the opposite is true. The evidence that human populations have been "mixing" since modern humans first evolved is contained within the gene pool of human populations.

Essay 10.1 The Hardy-Weinberg Theorem

Reginald Punnett, of Punnett-square fame, was a scientist at Cambridge University during the early 1900s. Punnett is considered to be one of the fathers of modern genetics. His verification of Mendel's work helped establish Cambridge University as a center of genetic research. Among Punnett's accomplishments was his dissemination of Mendel's work to a wide and somewhat skeptical scientific audience.

One of these skeptics was George Udny Yule, a statistician at University College London. Yule argued that inheritance could not work by Mendelian principles—as an example, he used the dominant trait brachydactyly (having six fingers). Since this allele is dominant over the five-fingered condition, Yule asserted, we would expect it to eventually become more common. He based this assertion on the observation that a cross between heterozygotes results in three-fourths of the offspring expressing the dominant trait and one-fourth expressing the recessive trait—thus you should get three brachydactyls for every five-fingered person.

Punnett intuitively knew that Yule's assertion was false, but he did not have the mathematical background to prove his intuition. For help, he turned to another Cambridge scientist named Godfrey Hardy. Hardy was a renowned "pure mathematician" whose teaching revolutionized mathematics education. Legend has it that Hardy wrote the mathematical proof on his shirt cuff during a dinner party, and he felt that it was so simple it fell below his standards of publication. He did eventually publish it as a letter to the editor in the journal *Science* at nearly the same time as Wilhelm Weinberg's identical proof was published in a German journal. Hardy's letter was his only contribution to the field of biology—in his famous autobiography, *A Mathematician's Apology*, it receives no mention.

The Hardy-Weinberg theorem holds that allele frequencies will remain stable in populations that are large in size, randomly mating, and experiencing no migration or natural selection. Subsequent geneticists used this theorem as a baseline for predicting how allele frequencies would change if any of the assumptions of the theorem were violated. In other words, the Hardy-Weinberg theorem enables scientists to quantify the effect of evolutionary change on allele frequencies. Today, the Hardy-Weinberg theorem forms the basis of the modern science of population genetics.

In the simplest case, the Hardy-Weinberg theorem (which we abbreviate to Hardy-Weinberg) describes the relationship between allele frequency and genotype frequency for a gene with two alleles in a stable population. Hardy-Weinberg labels the frequency of these two alleles p and q.

Imagine that we know the frequency of alleles for a particular gene in a population; let us say 70% of the alleles in the population are dominant (A), and 30% are recessive (a). Thus, $p = 0.7$ and $q = 0.3$. Each gamete produced by members of the population carries one copy of the gene, so 70% of the gametes produced by this entire population will carry the dominant allele and 30% will carry the recessive allele. The frequency of gametes produced of each type is equal to the frequency of alleles of each type (Figure E10.1a).

For the purposes of Hardy-Weinberg, we assume that every member of the population has an equal chance of mating with any member of the opposite sex. In other words, there is no relationship between the alleles an individual carries for that gene and the alleles of her or his partner. The fertilizations that occur in this situation are analogous to the result of a lottery drawing. In this analogy, we can imagine individuals in a population each contributing an equal number of gametes to a "bucket." Fertilizations result when one gamete drawn from the sperm bucket fuses with another drawn from the egg bucket. Since the frequency of gametes carrying the dominant allele in the bucket is equal to the frequency of the dominant allele in the population, the chance of drawing an egg carrying the dominant allele is 70%.

In Figure E10.1b, a modified Punnett square illustrates the relationship between allele frequency in a population and genotype frequency in a stable population. On the horizontal axis of the square, we place the two types of gametes that can be produced by females in the population (*A* and *a*), while on the vertical axis we place the two types of gametes that can be produced by males. In addition, on each axis is an indication of the frequency of these types of egg and sperm in the population: 0.7 for *A* eggs and *A* sperm, 0.3 for *a* eggs and *a* sperm. Used like the typical Punnett square in Chapter 4, the grid of the square also shows the frequency of each genotype in this population. The frequency of the *AA* genotype in the next generation will be equal to the frequency of *A* sperm being drawn (0.7) times the frequency that *A* eggs will be drawn (0.7), or 0.49. This calculation can be repeated for each genotype. The frequency of the *AA* genotype is $p \times p(= p^2)$, the *aa* genotype $q \times q(= q^2)$, and the *Aa* genotype $p \times q \times 2(= 2pq)$, because an *Aa* offspring can be produced by an *A* sperm and an *a* egg, or an *a* sperm and an *A* egg (E10.1b). Yule was proven wrong: The dominant/recessive relationship among alleles does not determine the frequency of genotypes in a population. Hardy and Weinberg mathematically proved that the frequency of genotypes in one generation of a population depends upon the frequency of genotypes in the previous generation of the same population. The dominant trait brachydactyly is rare in human populations because the allele for the trait is in very low frequency—in the absence of natural selection, it should remain rare.

Scientists rarely have information about allele frequency—however, they often have information about genotype frequency. When scientists know the frequency of a phenotype produced by a recessive allele, they know the frequency of that genotype. They can then use Hardy-Weinberg to calculate the allele frequency in a population. For instance, if the frequency of individuals with sickle-cell anemia is one in 100 births (0.01), we know that q^2—the frequency of homozygous recessive individuals in the population—is equal to 0.01. Therefore q is simply the square root of this number, or 0.1.

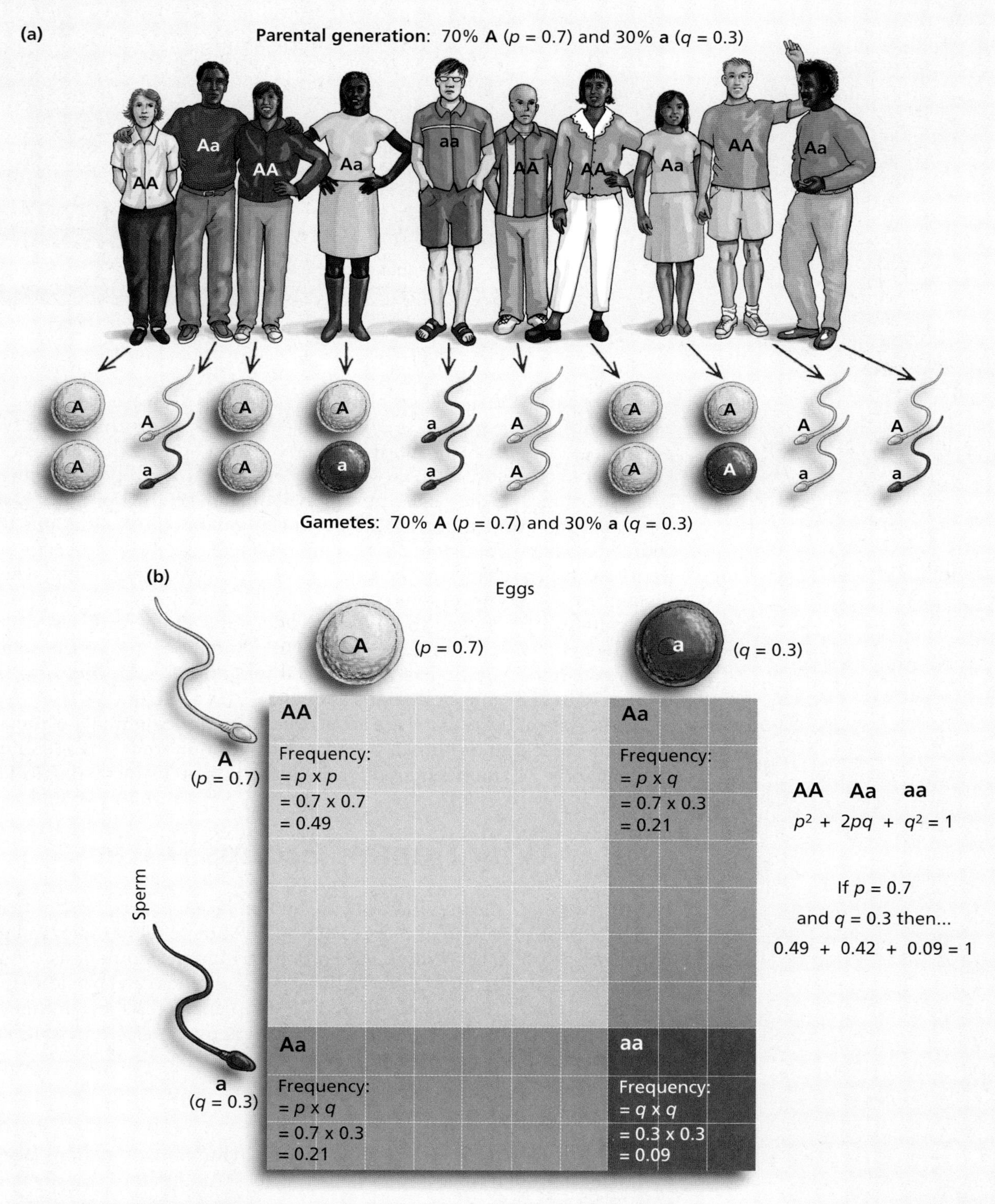

Figure E10.1 The relationship between allele frequency and gamete frequency.

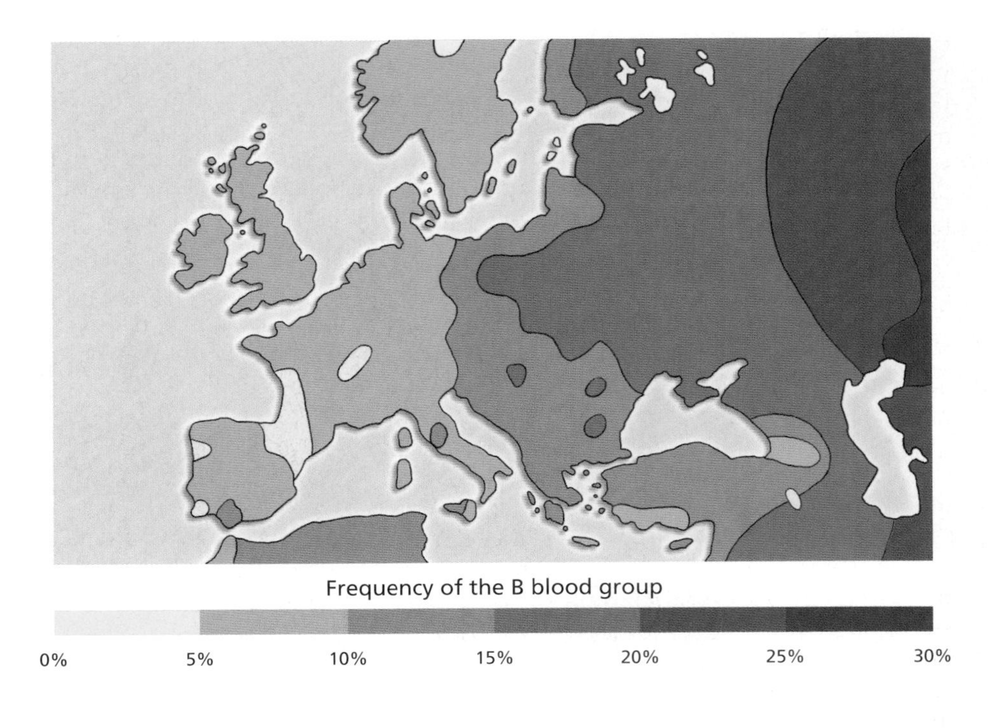

Figure 10.13 The map of blood types indicates mixing between populations. The frequency of the type B blood group in Europe declines from east to west across the continent. This pattern reflects the movement of alleles from Asian populations into European populations over the past 2,000 years.

For instance, the frequency of the B blood group decreases from east to west across Europe (Figure 10.13). The I^B allele that codes for this blood type apparently evolved in Asia, and the pattern of blood group distribution seen in Figure 10.13 corresponds to movement of Asians into Europe beginning about 2,000 years ago. As the Asian immigrants mixed with the European residents, their alleles became a part of the European gene pool. Populations that encountered a large number of Asian immigrants (that is, those closest to Asia) experienced a large change in their gene pools, while populations that were more distant from Asia encountered a more "dilute" immigrant gene pool made up of the offspring between the Asian invaders and their European neighbors. Other genetic analyses have indicated that populations that practiced agriculture arose in the Middle East, migrated throughout Europe and Asia, and interbred with resident populations about 10,000 years ago

These data indicate that there are no clear boundaries within the human gene pool. Interbreeding of human populations over hundreds of generations has prevented the isolation required for the formation of distinct biological races.

10.3 Why Human Groups Differ

As you learned in the previous section, human races such as those indicated on Indigo's census form do not represent mostly independent biological groups. However, as is clear to Indigo, and all of us, human populations differ from each other in many traits. Different human populations can be readily grouped together on the basis of some shared physical characteristics. In this section, we explore what is known about why populations share certain traits and differ in others.

Natural Selection

Recall the distribution of the sickle-cell allele in human populations (see Figure 10.11). It is found in some populations of at least three of the typically described races. The frequency of the sickle-cell allele in these populations is much higher than scientists would predict if its only effect was to cause a life-threatening disease when it is *homozygous*. If causing disease was its only effect, individuals who carry the allele would have lower *fitness* (that is, fewer surviving offspring) than individuals who do not carry the allele, because at least some of the carriers' offspring would have sickle-cell disease. *Natural selection*, the process described in Chapter 9 that results in an increase in the frequency of successful alleles and a decrease in unsuccessful ones, should cause the sickle-cell allele to become rare in a population. The reason that the sickle-cell allele is common in certain populations has to do with the advantage it provides to *heterozygotes*, individuals who carry one copy of the sickle-cell allele, in particular environments.

The sickle-cell allele has the highest frequencies in populations at high risk for malaria. Malaria is caused by a parasitic organism that spends part of its life cycle feeding on red blood cells, eventually killing the cells. Because their red blood cells are depleted, people with severe malaria suffer from *anemia*, which often results in death. When individuals carry a single copy of the sickle-cell allele, their blood cells deform when infected by a malaria parasite; these deformed cells quickly die, reducing the parasite's ability to reproduce and infect more red blood cells.

The sickle-cell allele reduces the likelihood of severe malaria; so natural selection has caused it to increase in frequency in susceptible populations. The protection the sickle-cell allele provides to heterozygote carriers is demonstrated by the overlap between the distribution of malaria and the distribution of sickle-cell anemia (Figure 10.14). The allele for sickle-cell disease is not associated with a particular racial category; it is associated with populations that live in particular environments.

Another physical trait that has been affected by natural selection is nose form. In some populations, most individuals have broad, flattened noses; in others, most people have long, narrow noses. The pattern of nose shape in populations generally correlates to climate factors—populations in dry climates

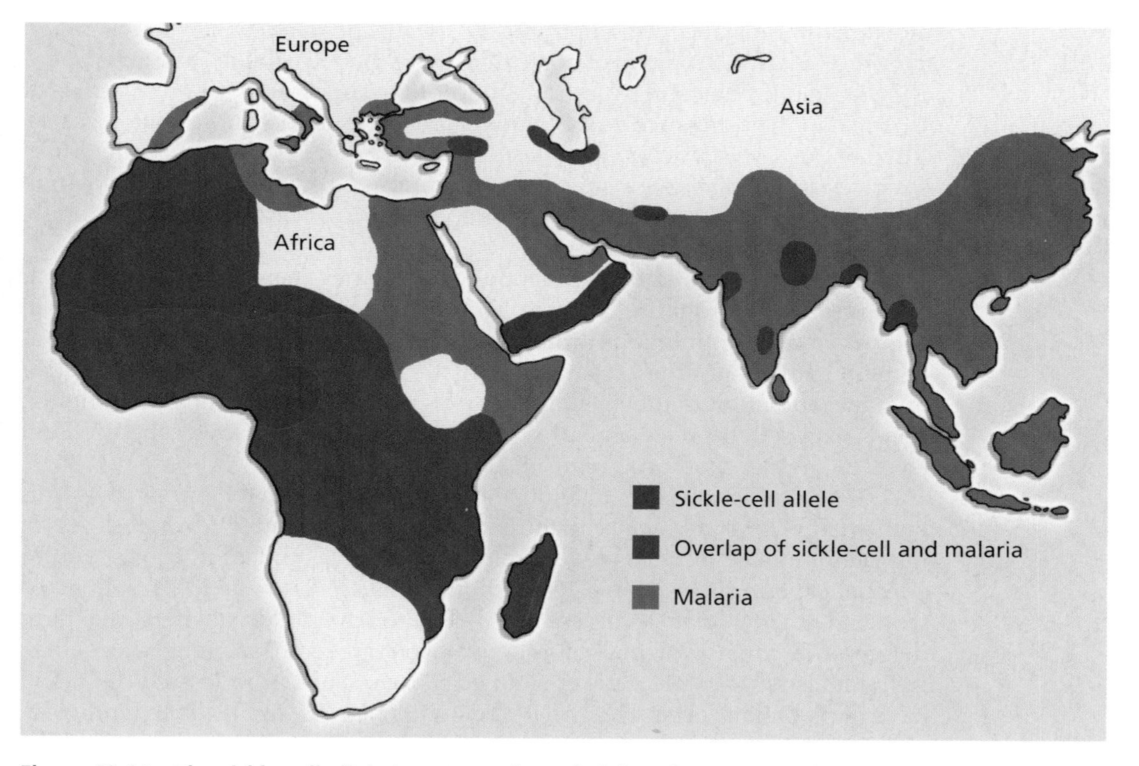

Figure 10.14 The sickle-cell allele is common in malarial environments. This map shows the distributions of the sickle-cell allele and malaria in human populations. The overlap is one piece of evidence that the sickle-cell allele is in high frequency in certain populations because it provides protection from malaria.

Figure 10.15 Nose shape is affected by natural selection. (a) Long, narrow noses are more common among populations in cold, dry environments; (b) broad, flattened noses are more common in warm, wet environments.

(a) Ethiopian with a narrow nose

(b) Bantu with a broad nose

tend to have narrower noses than populations in moist climates. Long, narrow noses appear to increase the fitness of individuals in dry environments, serving to increase the water content of inhaled air before it reaches the lungs. A narrower nose has a greater internal surface area—exposing inhaled air to more moisture. For instance, among tropical Africans, people who live at drier high altitudes have much narrower noses than those who live in lush rainforest areas (Figure 10.15). Interestingly, our preconception puts these two populations of Africans in the same race and explains differences in their nose shape as a result of natural selection, but we place White and Black populations into different races and explain their skin color differences as evidence of long isolation from each other. However, like nose shape, skin color is a trait that is strongly influenced by natural selection.

Traits that are shared by populations because they share environmental conditions rather than sharing ancestry are termed *convergent*. The pattern of skin color in human populations around the globe appears to be the result of **convergent evolution**. When scientists compare the average skin color in a native human population to the level of ultraviolet (UV) light to which that population is exposed, they see a nearly perfect correlation—the lower the UV-light level, the lighter the skin (Figure 10.16).

UV light is light energy in a range that is not visible to the human eye. This high-energy light interferes with the body's ability to maintain adequate levels of the vitamin folate. Folate is required for proper development in babies and for adequate sperm production. Men with low folate levels have low fertility, and women with low folate levels are more likely to have children with severe birth defects; individuals who maintain adequate folate levels have higher fitness than individuals who do not. Darker-skinned individuals absorb less UV light and have higher folate levels in high-UV environments than light-skinned individuals. Therefore, in environments where UV-light levels are high, dark skin is favored by natural selection—that is, dark skin is an *adaptation* in these environments.

Human populations in low-UV environments face a different challenge. The absorption of UV light is essential for the synthesis of vitamin D. Among its many functions, vitamin D is crucial for the proper development of bones.

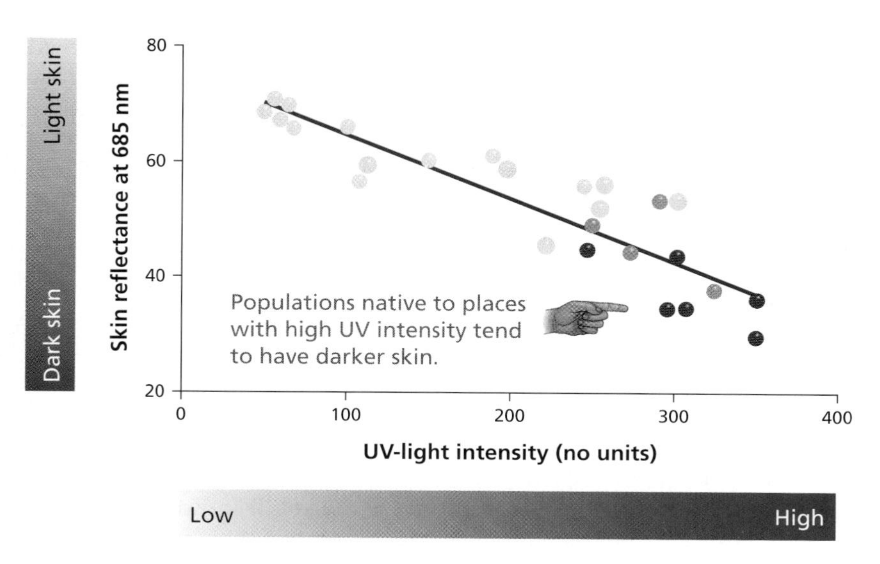

Figure 10.16 There is a strong correlation between skin color and UV-light exposure. The average reflectance of skin in various native human populations is correlated to the average UV radiation these populations experience. Reflectance is an indication of color: Higher reflectance indicates lighter skin. The color of the dots on the graph specifies the racial category of each population.

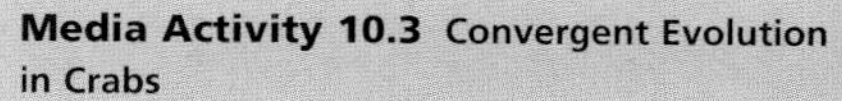

Women are especially harmed by low vitamin D levels—inadequate development of the pelvic bones can make safely giving birth impossible. There is no risk of not making enough vitamin D when UV-light levels are high, regardless of skin color. However, in areas where levels of UV light are low, individuals with lighter skin are able to maximize their absorption of what is available, and thus have higher levels of vitamin D. In these environments, light skin has been favored by natural selection (Figure 10.17). An exception to this pattern is for populations living in low-UV environments but which have high levels of vitamin D in their diet, so being able to make vitamin D is less important.

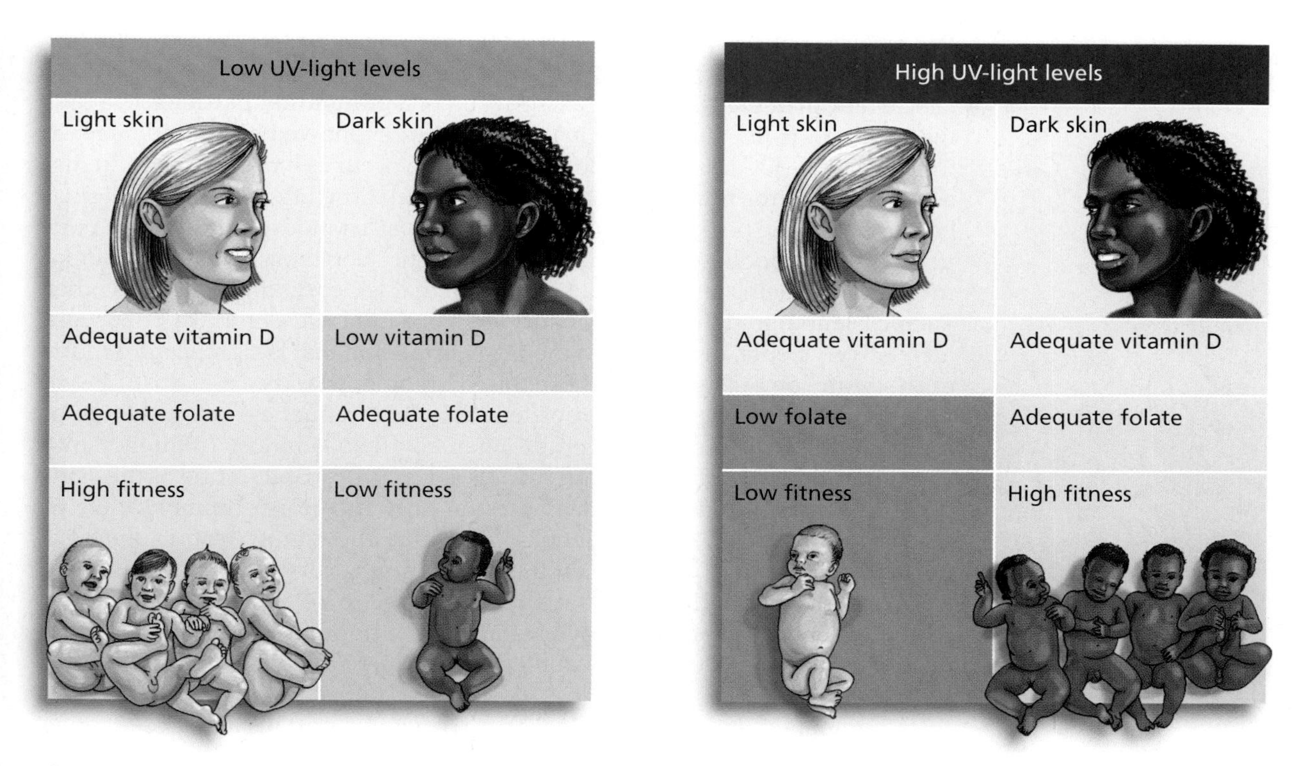

Figure 10.17 The relationship between UV-light levels, folate, vitamin D, and skin color. Populations in regions where UV-light levels are high experience selection for darker, UV resistant skin. Populations in regions where UV-light levels are low experience selection for lighter, UV "transparent" skin.

Because UV light has important effects on human physiology, it has served as a mechanism for natural selection on skin color in human populations. Where UV-light levels are high, dark skin is an adaptation and populations become dark skinned. Where UV-light levels are low, light skin is usually an adaptation and populations evolve to become light skinned. The pattern of skin color in human populations is a result of the convergence of different populations in similar environments, not evidence of separate races of humans.

Genetic Drift

As we have seen, differences among populations may arise through the effect of natural selection in various environments. However, differences may also arise through chance processes. A change in allele frequency that occurs as a result of chance is called **genetic drift**. Human populations tend to travel and colonize new areas, so we seem to be especially prone to evolution via genetic drift.

A common cause of genetic drift occurs when a small sample of a larger population establishes a new population. The gene pool of this sample is rarely an exact model of the source population's gene pool. The difference between a subset of a population and the population as a whole is called *sampling error*. As we discussed in Chapter 1, sampling error is more severe for smaller subsets of a population—such as those that typically found new settlements. This type of sampling error is often referred to as the **founder effect** (Figure 10.18a).

Genetic diseases that are at unusually high levels in certain populations appear to result from the founder effect. For example, the Amish of Pennsylvania are descended from a population of 200 German founders established approximately 200 years ago. Ellis-van Creveld syndrome, a recessive disease that causes dwarfism (among other effects), is 5,000 times more common in the Pennsylvania Amish population than in other German-American populations. This difference is a result of a single founder in that original population who carried this very rare recessive allele. Since the Pennsylvania Amish usually marry others within their small religious community, this allele has stayed at a high level—one in eight Pennsylvania Amish are carriers of the Ellis-van Creveld allele, compared to less than one in 100 non-Amish German-Americans.

Genetic drift may also occur as the result of a **population bottleneck**, a dramatic but short-lived reduction in population size followed by a rapid increase in population (Figure 10.18b). Bottlenecks often occur as a result of natural disasters. As with the founder effect, the new population differs from the original because the gene pool of the survivors is not an exact model of the source population's gene pool. A sixteenth-century bottleneck on the island of Puka Puka in the South Pacific resulted in a population that is clearly different from other Pacific Island populations: The 17 survivors of a tidal wave on Puka Puka were all relatively petite, resulting in a modern population that is significantly shorter in stature, on average, than those found on other islands.

Even without a population bottleneck, allele frequencies may change in a population due to chance events. When an allele is in low frequency in a small population, only a few individuals carry a copy of it. If one of these individuals fails to reproduce, or passes on only the more common allele to surviving offspring, the frequency of the rare allele may drop in the next generation (Figure 10.18c). When the population is very small, there is a relatively high probability that a rare allele will fail to be passed on to the next generation because of chance events; and if the population is small enough, even relatively high-frequency alleles may be lost after a few generations by this process.

A population that illustrates the effects of genetic drift in small populations is the Hutterites, a religious sect with communities in South Dakota and Canada. Modern Hutterite populations trace their ancestry to 442 people who migrated from Russia to North America between 1874 and 1877. Hutterites tend

(a) Founder effect: A small sample of a large population establishes a new population.

Frequency of red allele is low in original population.

Several of the travelers happen to carry the red allele.

Frequency of red allele much higher in new population.

(b) Population bottleneck: A dramatic but short-lived reduction in population size.

Frequency of red allele is low in original population.

Many survivors of tidal wave happen to carry red allele.

Frequency of red allele much higher in new population.

(c) Chance events in small populations: The carrier of a rare allele does not reproduce.

Frequency of red allele is low in original population.

The only person with red allele happens to fall out of tree and die.

Red allele is lost.

Figure 10.18 The effects of genetic drift. A population may contain a different set of alleles because (a) its founders were not representative of the original population; (b) a short-lived drop in population size caused a change in allele frequency in one of the populations; (c) one population is so small that low-frequency alleles are lost by chance.

to marry other members of their sect, so the gene pool of this population is small and isolated from other populations. Genetic drift in this population over the last century has resulted in a near absence of type B blood among the Hutterites compared to a frequency of 15–30% in Europeans and other European migrants in North America.

Humans are a highly mobile species and have been founding new populations for millennia. Most early human populations were also probably quite small. These factors make human populations especially susceptible to the effects of genetic drift, and have contributed to differences among modern human groups.

Assortative Mating and Sexual Selection

Some differences between human populations may be attributed to the ways in which people choose their mates. Individuals usually prefer to marry someone who is like themselves, a process called **assortative mating**—in other words, tall men marry tall women, and light-skinned women marry light-skinned men. When two populations differ in obvious physical characteristics, the number of matings between them may be small if the traits of one population are considered unattractive to members of the other population. Assortative mating tends to maintain and even emphasize physical differences between populations.

In addition to the tendency to mate assortatively, men and women within a population may have preferences for particular physical features in their mates. When a trait influences the likelihood of mating, that trait is under the influence of a form of natural selection called **sexual selection**. Sexual selection is responsible for many of the differences between males and females within a species. For instance, the enormous tail on a male peacock resulted from female peahens choosing mates with showier tails. In humans, there is some evidence that the difference in overall body size between men and women is a result of sexual selection—namely, a widespread female preference for larger males. In both cases, female choice for particular mates has led to the evolution of the population.

Some scientists have hypothesized that a trait common in the Khoikhoi people of South Africa evolved as a result of sexual selection. Women in this population store large amounts of fatty tissue in their buttocks and upper thighs, giving them a body shape that is considerably different from other African populations. Men in these populations prefer women with this body shape and appear to have caused selection for this trait in the population. The morbid fascination that many Europeans had for this body type and one consequence of that fascination are the subject of Essay 10.2. It is possible that many physical features that are unique to particular human populations evolved as a result of sexual selection.

10.4 The Meaning of Differences Among Human Populations

The discussion in this chapter may still leave Indigo unsatisfied. Scientific data indicate that the racial categories on her census form are biologically meaningless. Races that were once thought of as unitary groups have been revealed to be hugely diverse collections of populations. Two unrelated individuals of the "Black race" are no more biologically similar than a Black person and a White person. Yet everywhere she looks, Indigo sees evidence that the racial categories on her census form matter to people—from the existence of her college's Black Student Association to the heated discussions in her American Experience class about immigration policies in the United States.

Part of the disparity between what science has revealed and what our common experience tells us about the reality of race comes from the fact that racial categories are *socially* meaningful. In the United States, we all learn that skin color, eye shape, and hair type are the primary physical characteristics that denote meaningful differences among groups. These physical characteristics have this significance because of the history of European colonization, slavery, immigration, and Native-American oppression. In other words, race is a *social construct*—a product of history and learned attitudes. The construction of racial groups allowed some "races" to justify unethical and inhumane treatment of other "races." Thus, human races were described

Essay 10.2 The Hottentot Venus

The Khoikhoi were a cattle-herding tribe in what is now South Africa before the era of European colonialism. Their Europeanized name was apparently applied by early Dutch settlers—the Dutch word *hotteren-totteren* means to stammer or stutter, and the "Hottentots" had a distinctive language containing a series of consonants that sound like clicks to the untrained ear.

Some of the physical features of the Khoikhoi people are very different from European features, and they fascinated Victorian society in Europe. The low point in this fascination is exemplified by the life of Sara Baartman. Sara was a slave in Cape Town when she was introduced to British Marine Sergeant William Dunlop in 1810. Dunlop purchased the 20-year-old Sara and brought her to England—he hoped that Europeans would pay for a chance to glimpse her unique genitals and large buttocks.

Baartman was advertised as "The Hottentot Venus" and paraded naked before huge audiences in London. She was an instant sensation, and her appearance and exhibition also helped cement ideas about the "animal nature" of Black Africans in the European consciousness. After four years in England, Baartman was moved to Paris, where she was exhibited as part of a traveling circus. In 1815, abandoned by the circus owner, Baartman was forced into prostitution to survive. She died at age 25, an alcoholic and possibly suffering from syphilis and tuberculosis. The French anatomist Georges Cuvier made a plaster cast of Baartman's body, preserved her genitals in formaldehyde, and handed her remains over to the Musée de l'Homme (Museum of Mankind), where they were displayed publicly until surprisingly recently—1974.

South Africa first officially requested the return of Baartman's remains in 1994, when president Nelson Mandela brought the issue to the attention of French president François Mitterand during a state visit to South Africa. The French government did not respond and various Khoikhoi groups began campaigning for Baartman's return. The Musée de l'Homme asserted its ownership of the remains and cited the interests of unspecified "scientific research" in response to requests from the South Africans.

Continued pressure and public attention focused on Sara Baartman's case has led to resolution of this historical wrong. Diana Ferrus, a Khoikhoi writer, created a poetic tribute to Baartman while studying in Europe in 1998. Her haunting poem came to the attention of Nicolas About, a French senator. He wrote to Ferrus assuring her that he would press for Baartman's case. "They wanted to pass her off as something monstrous. But where in this affair is the true monstrosity?" About asked during the senate hearing on the bill he sponsored to return Baartman's remains to South Africa. Finally, in January of 2002, the French legislature voted overwhelmingly to repatriate Sara Baartman's remains to South Africa. Sara finally returned home in April 2002 (Figure E10.2).

Figure E10.2 Sara Baartman's homecoming. Khoikhoi leaders accompany the remains of Sara Baartman to a burial ceremony in South Africa, August 2002.

in the seventeenth century primarily to support **racism**, the idea that some groups of people are naturally superior to others. The United States government collects information about race on the census form as part of its effort to measure and ameliorate the lingering effects of historical state-supported racism, but the Census Bureau acknowledges that the races with which people identify "should not be interpreted as being primarily biological or genetic. . . ."

Figure 10.19 Race is a social construct. Recent Western history has resulted in a society that emphasizes similarity of skin color as evidence of relationship while minimizing the importance of variability in a host of other traits, such as height and weight, within a race.

It may be easier to see that racial categories are socially constructed if you imagine what might have happened if Western history had followed a different path. If the origin of American slaves had been from around the Mediterranean Sea, we might now identify racial groups on the basis of some other physical difference besides skin color—perhaps height, weight, or the presence of thick facial hair. Alternatively, compare the racial groupings in modern North America to those in modern Rwanda, where individuals are identified with different racial groups (Hutu and Tutsi) based on physical stature only. This classification reflects the differential social status attained by the typically taller Tutsi tribe and the typically shorter Hutu tribe under European colonization in the nineteenth century. In the United States, we would classify Hutu and Tutsi in the same "Black race"—an assignment that many members of these two groups would vigorously reject. In every society, children learn from birth which physical differences among people are significant in distinguishing "us" from "them." Even if a child is never explicitly taught racial categories, the fact that many communities are highly segregated into racial enclaves provides a lesson about which physical characteristics mark someone as "different from me."

When socially constructed racial categories are considered biologically meaningful, they become traps that are extremely difficult for individuals to escape. The most important insight that has come from studies of human diversity is that grouping human populations on the basis of skin color and eye shape is as arbitrary as grouping them on the basis of height and weight (Figure 10.19). However, arbitrary groupings are not necessarily bad—we all group *ourselves* by arbitrary categories: Christian or Muslim, blue-collar or white-collar worker, cat person or dog person. Even if the racial categories on the census form were part of a racist system, when people identify themselves as members of a particular race, they are acknowledging a shared history with others who also identify themselves as members of that race. This self-identification can be important for realizing individual and group goals of equality and self-determination, as well as continuing the fight against the real and serious vestiges of state-supported racism. The biological evidence tells Indigo that she is able to choose her racial category based on her own history and relationships—and she should feel free to choose "none of the above" if she desires.

CHAPTER REVIEW

Summary

- All humans belong to the same biological species, *Homo sapiens sapiens.*
- Species are groups of individuals that can interbreed and produce fertile offspring.
- Species are reproductively isolated from each other, thus separating the gene pools of species so that changes in the allele frequency of a gene occurs within a species.
- Reproductive isolation is maintained by prezygotic or postzygotic factors.
- Speciation occurs when populations of a species become isolated from each other. These populations diverge from each other, and reproductive isolation between the populations evolves.
- Biological races are populations of a single species that have diverged from each other.
- The fossil record provides evidence that the modern human species is approximately 200,000 years old, which is not much time for human races to have evolved.
- Modern human groups do not show evidence that they have been isolated from each other: There are no alleles that are unique to a particular "race," and populations that are similar in skin color do not demonstrate other genetic evidence of relationship, including similar allele frequencies for a number of genes.
- Genetic evidence indicates that human groups have been mixing for thousands of years.
- Similarities among human populations may evolve as a result of natural selection. The sickle-cell allele is more common in populations where malaria incidence is high, and light skin is more common in areas where the UV-light level is low; both adaptations are a result of natural selection in these environments.
- Human populations may show differences due to genetic drift.
- Assortative mating or sexual selection may create differences among human populations.
- Race in the human species is a social construct that is based on shared history and self-identity.

Key Terms

assortative mating p. 264
biological race p. 248
biological species concept p. 242
convergent evolution p. 260
diverge p. 245
founder effect p. 262
gene flow p. 242
gene pool p. 242
genetic drift p. 262
genus p. 242
macroevolution p. 244
morphological species concept p. 250
population bottleneck p. 262
racism p. 265
reproductive isolation p. 242
sexual selection p. 264
speciation p. 244
species p. 242
subspecies p. 242

Learning the Basics

1. Describe the three steps of speciation.
2. Can speciation occur when populations are not physically isolated from each other? How?
3. Describe how biological races form, and how they can be identified using genetic clues.
4. How is allele frequency calculated?
5. Describe three ways that evolution can occur via genetic drift.
6. Which of the following is an example of a prezygotic barrier to reproduction?
   a. A female mammal is unable to carry a hybrid offspring to term.
   b. Hybrid plants produce only sterile pollen.
   c. A hybrid between two bird species sings a song that is not recognized by either species.
   d. A male fly of one species performs a "wing-waving" display that does not convince a female of another species to mate with him.
   e. A hybrid embryo is not able to complete development.

7. According to the most accepted scientific hypothesis about the origin of two new species from a single common ancestor, most new species arise when _____.
   a. many mutations occur
   b. populations of the ancestral species are isolated from each other
   c. there is no natural selection
   d. the Creator decides that two new species would be preferable to the old one
   e. the ancestral species decides to evolve

8. For two populations of organisms to be considered separate biological species, they must be _____.
   a. reproductively isolated from each other
   b. unable to produce living offspring
   c. physically very different from each other
   d. a and c are correct
   e. a, b, and c are correct

9. Which of the following alleles is unique to a particular human race?
   a. The sickle-cell allele to the Black race.
   b. The cystic fibrosis allele to the White race.
   c. The haptoglobin 1 allele to the Native-American race.
   d. The type B blood group allele to the Asian race.
   e. None of the above.

10. The statement that "human populations classified in the same race appear to be more genetically similar than human populations placed in different races" is _____.
    a. true
    b. false

11. Differences among human populations can arise through _____.
    a. natural selection
    b. genetic drift
    c. the founder effect
    d. a and c are correct
    e. a, b, and c are correct

12. Similarity in skin color among different human populations appears to be primarily the result of _____.
    a. natural selection
    b. convergence
    c. shared ancestry
    d. a and b are correct
    e. a, b, and c are correct

13. The tendency for individuals to choose mates who are like themselves is called _____.
    a. natural selection
    b. sexual selection
    c. assortative mating
    d. the founder effect
    e. random mating

14. Imagine that the shipwreck survivors on *Gilligan's Island* never made it back to civilization. After many generations, explorers find the descendants of the original seven survivors. (Recall that the seven castaways included four men and three women, and that one of the women—Lovey Howell—was past childbearing age.) Interestingly, 60% of the individuals in the descendent population have spectacular red hair, like Ginger's. The percentage of redheads on the island greatly exceeds the percentage in the mainland population (the source) from which the original seven castaways came. The difference between the island population and the source population is probably due to _____.
    a. shared common ancestors
    b. the founder effect
    c. a higher percentage of redheads in the castaway population than in the source population
    d. b and c are correct
    e. a, b, and c are correct

15. When you identify yourself as a member of a particular human group based on shared customs, language, and recent history, you are using the _____.
    a. biological definition of race
    b. social construct definition of race
    c. psychological definition of race
    d. incorrect definition of race
    e. morphological definition of race

Analyzing and Applying the Basics

1. Spotted owls in the northwestern United States are listed as threatened by the federal government, while those found in the Southwest are not considered at risk of extinction. The spotted owls in these two different areas are very similar in appearance and behavior, including mating and territorial calls. Great distances separate them, and individuals from the two populations rarely come in contact with each other. Should these two populations be considered different races or species? What information would you need to test your answer?

2. The Hawaiian Islands are an archipelago formed by a series of volcanoes. These islands formed in sequence—the island chain runs roughly east to west, and the age of the islands increases in this direction. There are hundreds of species of fruit fly found on the Hawaiian Islands. A particular species usually lives on only one island, and each island has a unique set of species. All of these species probably evolved from a small number of immigrants that arrived on the islands over one million years ago. Scientists have debated whether the

unique set of fruit fly species on each island consists of descendants of a single immigrant species to that island, or whether flies have "island hopped"—that is, immigrants from a species that evolved on one island moved to another island where they formed a new species. What predictions would you make about patterns of allele frequency in different species if all the species on an island are most closely related to each other? What prediction would you make if species were "island hoppers"?

3. The frequency of phenylketoneuria (PKU) in Irish populations is one in every 7,000 births, while the frequency in urban British populations is 1 in 18,000, and only 1 in 36,000 in Scandinavian populations. Give two reasons why this allele, which results in severe mental retardation in *homozygous* individuals, may be found in different frequencies in these populations. (If you have read Essay 10.1, you should also be able to use Hardy-Weinberg to calculate the allele frequency of this allele in each population.)

4. Medical researchers have often excluded particular racial groups from their studies to minimize variability among the subjects of their studies. Given the biological understanding of race, does this policy make sense?

5. Populations experiencing genetic drift gradually lose genetic diversity over time. Genetic drift is a serious problem in captive populations of endangered animals, such as those kept in zoos. To combat the effects of genetic drift, zoos "swap" animals for breeding purposes. How does this practice reduce the speed of genetic drift?

Connecting the Science

1. In the last paragraph of this chapter, we state that, "When socially constructed racial categories are considered biologically meaningful, they become traps that are extremely difficult for individuals to escape." What does this mean? How are current racial categories different from other arbitrary groupings of humans? Is there a way out of this trap?

2. Black people in the United States have higher rates of hypertension (high blood pressure), heart disease, and stroke than White people. Is this difference likely to be biological? How could you test your hypothesis?

3. The only information that was collected from *every resident* of the United States in the 2000 census was name, place of residence, sex, age, and race. (Note that some residents received a "long form" with many additional questions, but everyone had to answer these five basic questions.) Do you think it is important to collect race data from all citizens? Why or why not? Is there some other piece of information that you think is more useful to the government?

Media Activities

Media Activity 10.1 How Can New Species Be Formed?
Estimated Time: 10 minutes
Explore the various ways in which the gene pool of an ancestral population can be separated into several isolated gene pools, resulting in speciation—the evolution of the ancestral population into several different species.

Media Activity 10.2 How Did Human Groups Evolve?
Estimated Time: 15–20 minutes
This activity allows you to examine the data used to support the out-of-Africa hypothesis of modern human origins and compares that data to what you might expect if the competing multiregional evolution hypothesis were correct.

Media Activity 10.3 Convergent Evolution in Crabs
Estimated Time: 5–10 minutes
Usually, animals that look similar are thought to be related. Similar appearance, however, can also be due to similar selection pressures operating on an animal. This article explores the crab body shape, demonstrating that although different kinds of crabs do often look quite similar, they do not seem to be closely related to each other.

CHAPTER 11

Will Mad Cow Disease Become an Epidemic?

Immune System, Bacteria, and Viruses

Scientists are trying to prevent the spread of mad cow disease.

11.1 Infectious Agents

11.2 Epidemics

11.3 The Body's Response to Infection: The Immune System

11.4 Preventing the Spread of Prion Diseases

The brains of affected individuals look spongy.

In deer, this disease is called *chronic wasting disease*.

This disease can also be spread to humans.

In the late 1980s, many dairy cows in Great Britain started behaving in an uncharacteristic manner. Dairy farmers noticed that some of their cows were uncoordinated and seemed apprehensive. These cows would shake and tremble and rub parts of their bodies against walls or fences. Farmers watched with alarm as their cows staggered around, giving the impression that they were mad or irritated, leading to the name "mad cow disease." Several months after the onset of symptoms, the cows inevitably died. Autopsies revealed that the brains of the "mad" cows had holes in them characteristic of a class of diseases known as *spongiform encephalopathies*.

An encephalopathy is a *pathology*, or disease, of the brain. The diseased cows' brains resembled porous, natural sponges filled with holes. Mad cow disease is a type of spongiform encephalopathy that only affects cows, so it is also called *bovine spongiform encephalopathy*, or *BSE*.

Spongiform encephalopathies have been around for many years. They have been diagnosed in sheep, whose skin becomes so itchy that they are compelled to scrape off their wool (hair) on fences, leading the disease to be called *scrapie*. When the disease is present in elk and deer, it is called *chronic wasting disease*, due to the emaciated appearance of affected animals. In some regions of the United States, deer hunting and the consumption of deer meat has decreased dramatically due to concern about eating meat from animals with this disease.

Humans have also been affected by spongiform encephalopathies. An obscure disease called *Kuru* has long been known to affect natives of the eastern highlands of New Guinea. The disease occurs mostly in women and young children as a result of the tribal custom of honoring the dead by eating their brains. Affected individuals lose coordination and often become demented. They, too, inevitably die.

A long-recognized but previously very rare condition called Creutzfeldt-Jakob Disease (CJD) is another spongiform encephalopathy that affects humans. Like the diseased cows, affected humans became very agitated, dizzy, and short-tempered. They experienced short-term memory loss, lack of coordination, and slurred speech. As was the case with cows, this disease is lethal in all the affected humans. Typically, the disease affects only the elderly. However, a recent and alarming trend has been an increase in the rate of diagnosis of this disease in young British patients.

On average, these young patients lived just over a year after they were diagnosed, and when examined at autopsy, their brains looked more similar to those of BSE-infected cattle than CJD-infected humans. The increased number of people diagnosed with the disease, the structure of their brains after death, and the lower age of those infected led scientists to believe that BSE was, in some way, being transmitted from infected cows to these individuals. This new, transmissible form of the disease was named *new-variant CJD (nvCJD)*.

Scientists and doctors, veterinarians, hunters, and patients and their families all over the world are concerned about the preventing the spread of these diseases. The first step toward understanding this disease and preventing its spread in cows and humans required that scientists determine what was causing it.

11.1 Infectious Agents

Infectious diseases in humans typically result when an *infectious agent* gains access to the body and uses the body's resources for its own purposes. Infectious diseases differ from genetic diseases in that they are usually caused by organisms such as bacteria and viruses, rather than being caused by malfunctioning genes—although malfunctioning genes can make an organism more susceptible to infection.

Disease-causing organisms are called **pathogens**. When a pathogen can be spread from one organism to another, it is said to be **contagious**. When it finds a tissue inside the body that will support its growth, it becomes **infectious**. Some organisms can cause an infection in an individual but not be contagious if the infected individual cannot pass on the infection. Organisms that can only be seen when viewed under a microscope are called *microscopic organisms* or **microbes**. Bacteria and viruses are the most common infectious microbes. Table 11.1 lists the various pathogens and their effects.

Bacteria

Bacteria (Table 11.1) are a diverse group of single-celled organisms. They are tiny and numerous. In fact, there are more bacteria in your mouth than there are humans on Earth. These rod-shaped (bacilli), spherical (cocci), or spiral-shaped (spirochetes) cells are prokaryotic, and therefore do not contain organelles that are surrounded by membranes, such as the endoplasmic reticulum or the Golgi complex (see Chapter 2). Prokaryotes also do not have a nucleus. Unlike their eukaryotic counterparts, which contain their DNA in a nucleus, prokaryotes have DNA that is coiled up inside the *nucleoid region*. It is typically a double-stranded circular molecule. In addition to the large DNA chromosome, bacteria may also contain small, circular extra-chromosomal DNA (DNA that is separate from the chromosome) called *plasmids*. In Chapter 7 you learned that plasmids are used by scientists during genetic engineering. Some of these plasmids also carry genes that are resistant to antibiotics, which allow bacteria to resist the drugs that have been designed to kill them. Essay 11.1 addresses the growing problem of antibiotic resistance in pathogenic bacteria.

Most bacterial cells are surrounded by a **cell wall** that provides rigidity and protection; it is composed of carbohydrate and protein molecules. The cell walls of many bacteria are surrounded by a gelatinous **capsule**, which helps the bacteria attach to cells within tissues they will infect. Bacteria also may have external **flagella** to aid in motility, and *pili*, which help some bacterial cells pass genes to each other.

Bacteria reproduce by a process called **binary fission**. When a bacterial cell divides by binary fission, the single, circular chromosome attached to the plasma membrane located inside the cell wall is copied. The copy is attached to another site on the plasma membrane, and the membrane between the attachment sites grows and separates the two copies of the original chromosome until it eventually produces two separate daughter cells.

Bacteria can reproduce rapidly under favorable conditions, doubling their population every 20 minutes or so. For example, a single *Salmonella* bacterium in a chicken salad sandwich will give rise to over 33 million bacterial cells after eight hours when kept at room temperature. This exponential growth occurs because one cell gives rise to two, and those two yield four, the four divide to become eight, then 16, and so on (Figure 11.1).

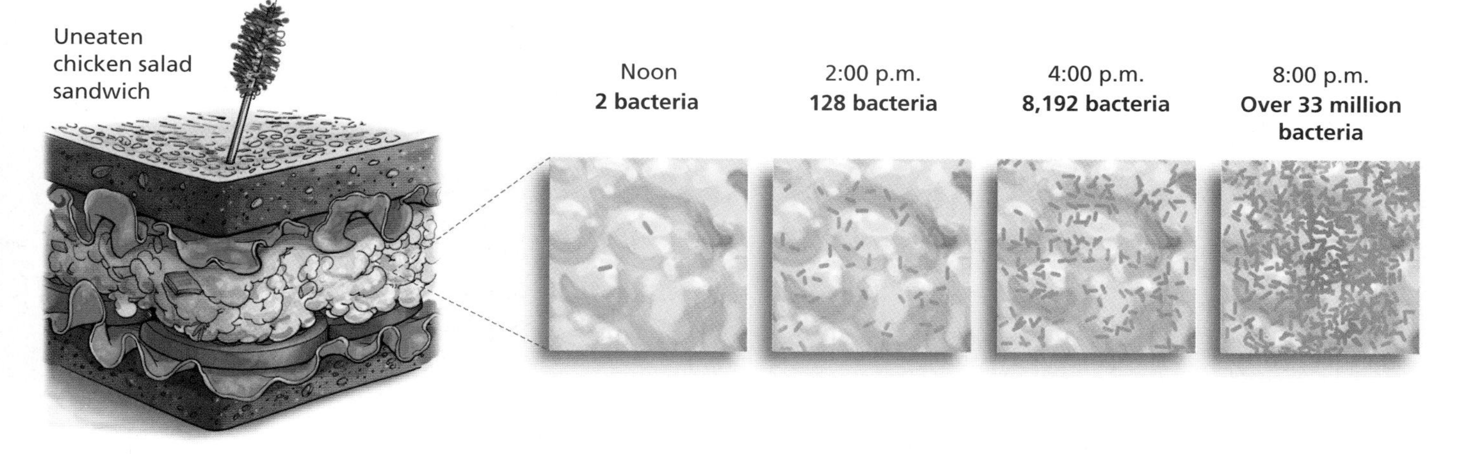

Figure 11.1 Exponential growth of bacteria. *Salmonella*, the bacterium that causes food poisoning, reproduces every 20 minutes. If a sandwich with two bacteria is left out for 20 minutes, each bacterial cell will make a copy of itself, yielding four bacteria. After 40 minutes there will be eight bacteria; within eight hours there will be 33 million bacteria.

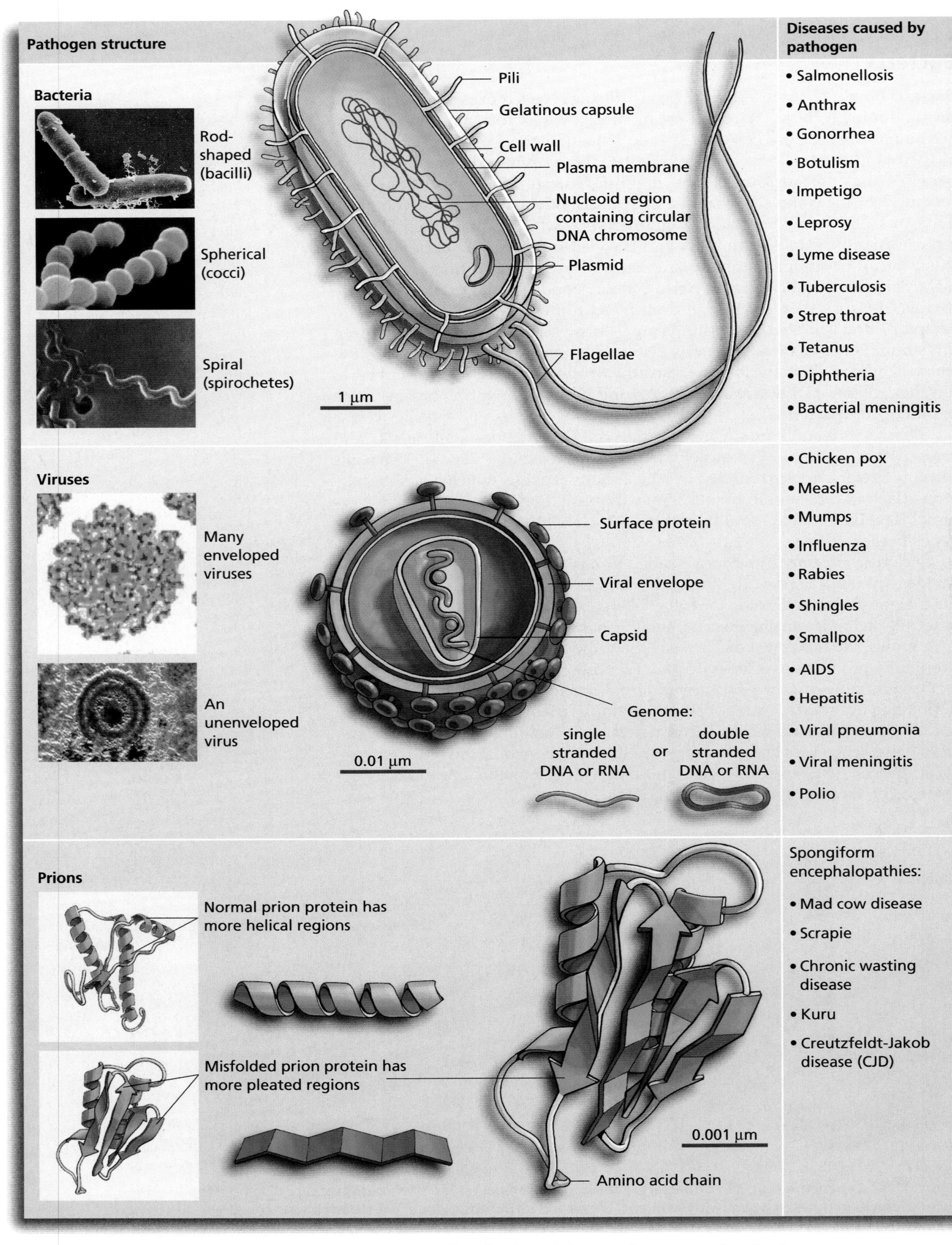

Table 11.1 Bacteria, viruses, and prions. The structure and replication of these pathogens are described.

Mode of reproduction or duplication

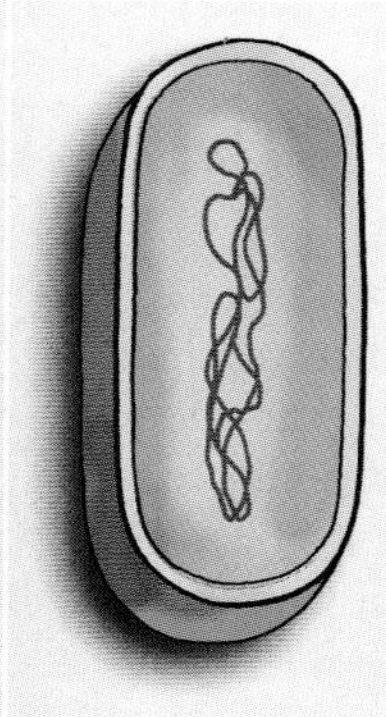

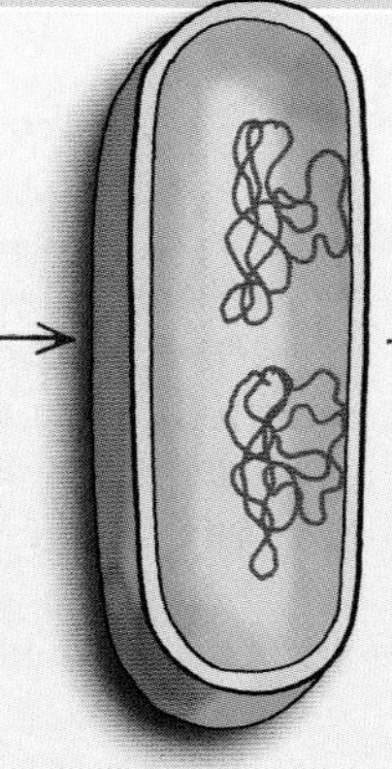

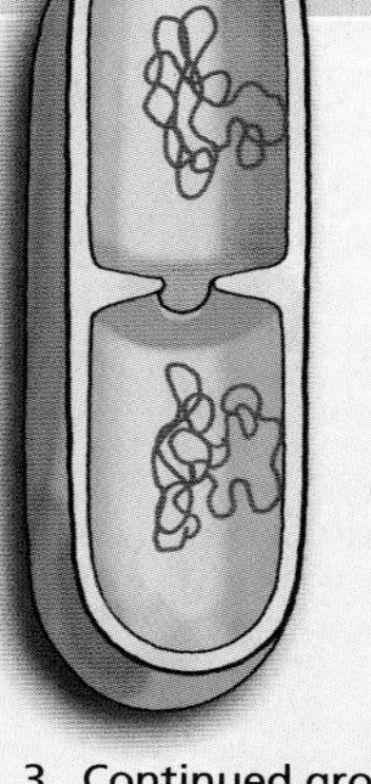

1. Bacterium starts with one copy of circular DNA chromosome in nuclear region.

2. The circular chromosome is copied and each copy is attached to the plasma membrane.

3. Continued growth separates the two chromosomes. The plasma membrane grows inward in the middle and a new cell wall is constructed.

4. Daughter cells separate. Chromosomes return to nucleoid region.

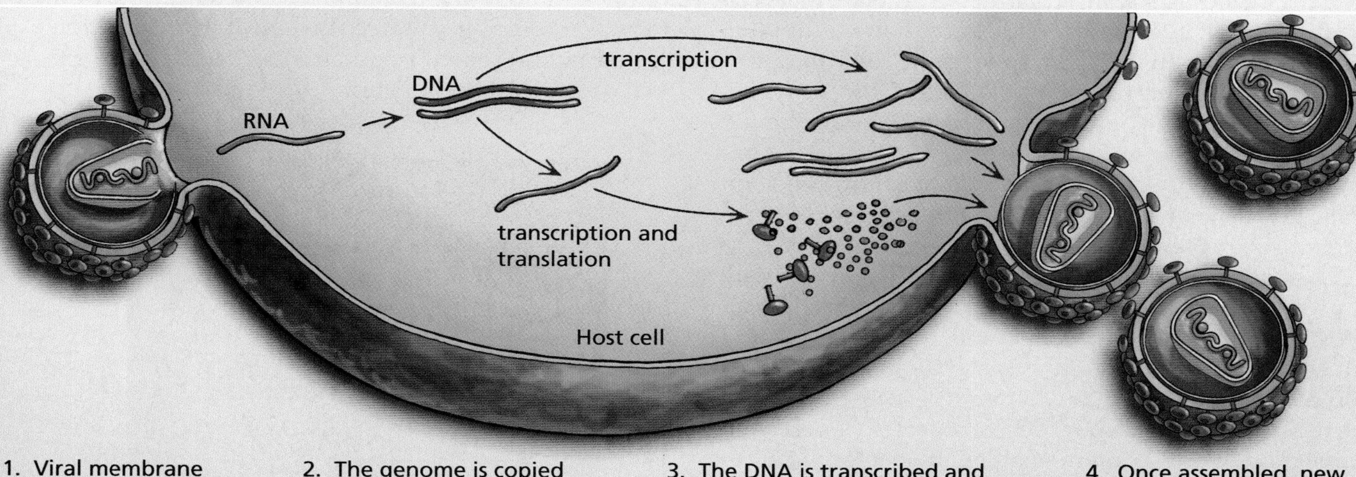

1. Viral membrane fuses with host cell, the capsid is removed, and the genome enters.

2. The genome is copied and used to synthesize many copies of double stranded DNA.

3. The DNA is transcribed and translated by the host cell. More viral proteins and copies of the viral genome are produced.

4. Once assembled, new viruses leave the host cell and infect other cells.

Misfolded prion

Normal prion

1. A misfolded prion protein searches for a normal one and refolds it into the mutant form.

2. The number of misfolded prions is now doubled. Each refolds a normal prion protein into the mutant form.

3. The number of misfolded prions continues to double, eventually causing host cell to burst and spread infection to other cells.

WWW
Media Activity 11.1A Structure and Reproduction of Infectious Agents

Essay 11.1 Antibiotic-resistant Bacteria

Antibiotics are medications that are prescribed for people with bacterial infections. When you were a child and had an ear infection, your parents probably gave you spoonfuls of the pink medicine (Figure E11.1), and now when you have a sinus infection your doctor prescribes other antibiotics. These infections are easily treated and require fairly little inconvenience besides a trip to the doctor and pharmacy. Unfortunately, the days of easily treatable bacterial infections may soon be over because the bacteria that cause many diseases—including tuberculosis, malaria, ear infections, and gonorrhea—have become resistant to the antibiotics developed to cure them.

Antibiotic resistance arises as a result of natural selection. There is variation within any population of individuals, including within a population of bacterial cells. Some bacteria, even before exposure to the antibiotic, carry genes that enable them to resist it. You can develop a drug-resistant infection by selecting for the resistant bacteria within your own body or by contracting the resistant bacteria from someone else.

Tetracycline is an antibiotic used to kill the bacteria that cause acne. It works to prevent translation, the production of proteins, in prokaryotes. Since the workbenches of translation, the ribosomes, differ structurally in prokaryotes and eukaryotes, this antibiotic—like all effective antibiotics—selectively kills bacterial cells and not eukaryotic host cells. One bacterial cell in a population of millions may have a different DNA sequence in the region of DNA that encodes for a protein involved in translation. If this protein enables the bacterial cell to withstand tetracycline treatment, it will survive and pass on the resistant sequence to its offspring. In this manner, those bacteria that are not killed are selected for, and resistant infections develop.

The problem of resistance in bacteria seems to have originated from medical and agricultural overuse of antibiotics. Some patients ask their doctors to give them antibiotics for a cold, cough, or the flu—all of which are caused by viruses. For example, tetracycline would not be able to kill a viral infection because the virus is using your eukaryotic ribosomes for translation and is thus unaffected by the treatment. Adding to the problem of medical overuse is that people often discontinue their prescribed antibiotics when they feel better. Not finishing all of the prescribed medication allows resistant bacteria to proliferate and the infection will come back, resistant to that antibiotic, in a few weeks, and another course of medication will be required.

The use of antibiotics in agriculture is also on the rise. Animals such as cows and chickens are given antibiotic drugs in their feed to prevent them from becoming ill. Animals who are fighting infections do not gain as much weight and must then be sold for less money. Antibiotic-treated animals may harbor resistant bacteria, and eating undercooked meat from an animal infected by a resistant population of bacteria can give you a drug-resistant infection.

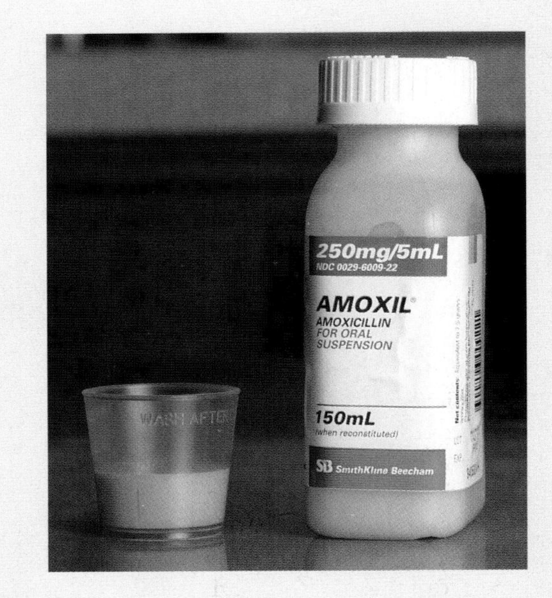

Figure E11.1 Amoxycillin is a familiar antibiotic.

Salmonella reproduces quite quickly in food because the bacteria have access to all the nutrients they need, and they reproduce even more quickly at room temperature than when kept in a refrigerator.

When an infection occurs in your body, the rapidly growing bacteria use your cells' nutrients to support their growth, effectively preventing your cells from functioning properly. However, most cases of disease-causing bacterial infection result from more than just the large numbers of bacteria in your system. The symptoms of a disease arise as the result of the effects of biological molecules secreted by the bacterial cells, called **toxins**, that block cellular processes.

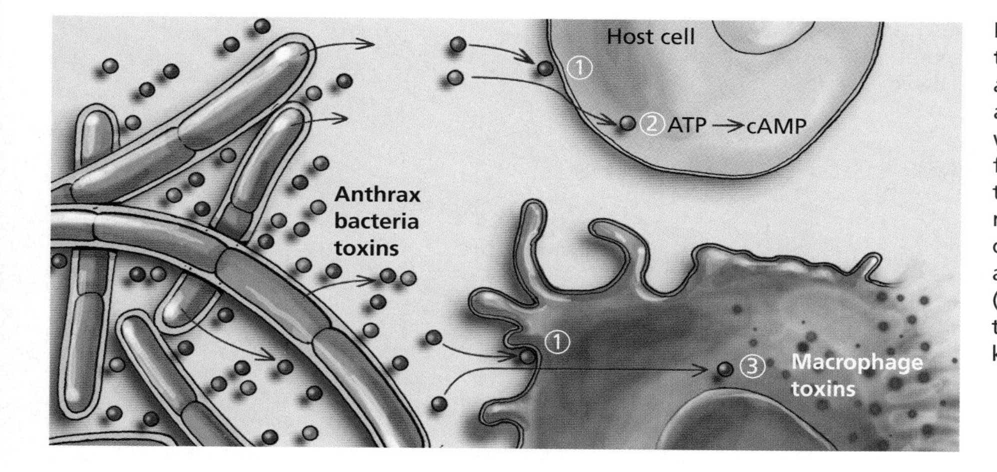

Figure 11.2 Bacteria may secrete toxins that damage host cells. *Bacillus anthracis*, the bacterium that causes anthrax, secretes a trio of toxins that work together to destroy host cells. The first toxin (labelled 1) helps the other two toxins pass through the host cell's plasma membrane; the second toxin (labelled 2) converts ATP into cAMP, resulting in the accumulation of fluid; and the third toxin (labelled 3) causes the host's immune cells to swell and release their own toxins, killing the host cell.

Anthrax is an infectious disease caused by the bacterium *Bacillus anthracis*. This microorganism lives mainly in soil and produces structures called **spores** that can be lethal when inhaled. This disease is rarely contracted by humans—weapons-grade anthrax requires very difficult and very expensive processing to make the spores airborne. Anthrax is an example of a disease that is infectious without being contagious because infected individuals do not pass the disease to other people. When anthrax is inhaled or lodged in the skin, the disease results from the secretion of a trio of toxins (Figure 11.2).

The first toxin, called the *protective factor*, inserts itself into the human host's cell membrane and allows the other two toxins to gain access to the cell's interior. The second toxin, called the *edema factor*, removes the two terminal phosphates from an adenosine triphosphate molecule (ATP), producing cyclic AMP (cAMP). Cyclic AMP relays the messages sent to cells by hormones—for instance, a rise in cAMP in response to the hormone adrenaline causes an increased heart rate. When cAMP levels are not properly regulated, cells are unable to appropriately respond to hormonal signals. This can influence water balance and leads to the accumulation of fluids inside cells, a condition called *edema*.

The third toxin released during an anthrax infection is called the *lethal factor*. When this toxin gains access to white blood cells called *macrophages*, it causes them to swell and release the toxic chemicals they normally use to kill bacteria. The toxic chemicals released by macrophages kill the host organism unless antibiotics have been administered soon after exposure—this is why anyone suspected of having come in contact with anthrax spores is given antibiotics even before infection is confirmed (Figure 11.3).

Viruses

www
Media Activity 11.2 The Revenge of the Chickenpox: The Virus Strikes Back

Viruses (Table 11.1) are not considered to be living organisms because they cannot replicate (copy) themselves without the aid of a host cell. Hosts for viral infections include many organisms other than humans, such as bacteria and plants. Viruses lack the enzymes for metabolism and contain no ribosomes, and therefore cannot make their own proteins. They also lack cytoplasm and membrane-surrounded organelles. Viruses are really nothing more than packets of nucleic acid (DNA or RNA) surrounded by a protein coat.

The genetic material, or *genome*, of a virus can be DNA or RNA; it can be double stranded or single stranded; it can be linear or circular. For example, the herpes virus has a double-stranded DNA genome, and the polio virus has a single-stranded RNA genome. The genes of a virus can code for the production of all the proteins required to produce more viruses.

Figure 11.3 The anthrax threat. In 2002, investigators wore protective clothing to prevent anthrax spores from landing on their skin and a respirator to avoid inhaling spores.

WWW
Media Activity 11.3 New Antidote for the Anthrax Bacterium

The protein coat surrounding a virus is called its **capsid**. Many of the viruses that infect animals develop an additional layer called the **viral envelope**. The envelope is derived from the cell membrane of the host cell and may contain some additional proteins encoded by the viral genome.

Infection by an enveloped virus occurs when the virus gains access to the cell by fusing its envelope with the host's cell membrane. An unenveloped virus uses its capsid proteins to bind to receptor proteins in the plasma membrane of a host cell. Some capsid proteins function as enzymes that digest holes in the plasma membrane, thereby allowing the viral genome to enter into cells. Once inside the host cell, the capsid is removed.

Whether the virus is enveloped or not, after the genome enters the host cell, the infection continues when the virus makes copies of itself. First the genome is copied, then the virus uses the host-cell ribosomes and amino acids to make viral proteins for building new capsids and synthesizing some of the envelope proteins. Once assembled, the new virus exits the cell, leaving behind some viral proteins in the host's cell membrane, and moves to another cell to spread the infection.

The genome, composed of DNA or RNA, is replicated when the virus uses the host cell's enzymes and nucleotides to produce new nucleic acids. Viruses with DNA genomes use the host cell's DNA polymerase, the DNA-copying enzyme, to replicate their own DNA. As explained in Chapter 9, RNA viruses, such as HIV, copy their genomes with the help of a virally encoded enzyme called **reverse transcriptase** (Figure 11.4). Recall that transcription is the general process of copying DNA to make RNA—reverse transcriptase transcribes RNA into DNA. When a virus has an RNA molecule as its genome, the RNA must be converted to DNA, because DNA—whether produced by copying a

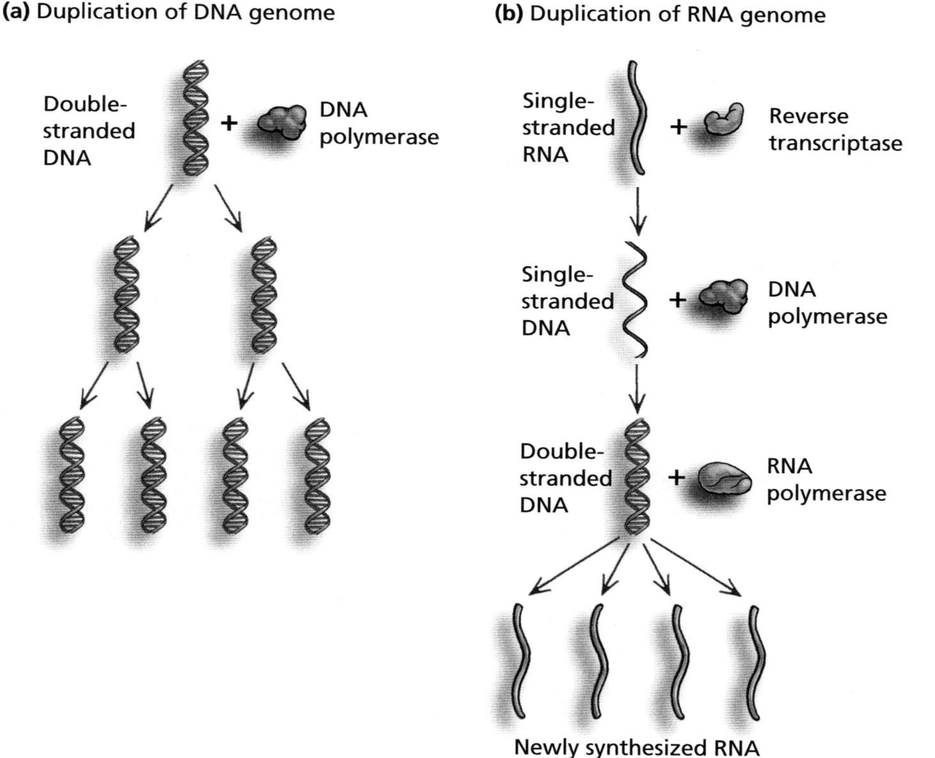

Figure 11.4 Duplicating viral genomes. (a) DNA is duplicated when the DNA polymerase enzyme uses each strand of the double helix as a template for producing a complementary daughter strand. (b) RNA is duplicated when the enzyme reverse transcriptase uses DNA nucleotides to produce single-stranded DNA from RNA. The DNA polymerase then uses the single DNA strand as a template for the production of the other side of the DNA helix. Copies of the original RNA molecule are then produced when RNA polymerase uses the double-stranded DNA as a template.

DNA genome or by reverse transcription of an RNA genome—is transcribed and translated to produce viral capsid and envelope proteins.

Some viruses can insert a DNA copy of their genome into the host cell's genome. These viruses (such as the virus that causes herpes) can become **latent**, or dormant, for long periods of time. Herpes blisters do not return when the virus is latent but do return when the virus becomes active again. When the viral genome is activated, the virus uses the inserted copy of its genome to direct the synthesis of new viruses for infection.

Bacteria and viruses are the most common infectious agents. However, the spongiform encephalopathies are not caused by *either* of these two organisms. Scientists were able to determine this because the infectious agent causing these diseases does not respond to the treatments that usually kill bacteria and viruses—such as heat, chemicals, and radiation. Instead, the spongiform encephalopathies seem to be caused by a novel class of infectious agents called *prions*.

Prions

After a cell synthesizes a protein, the protein folds into its characteristic shape. If a protein is folded wrong, it can no longer perform its job properly. A **prion** (Table 11.1) is a normally occurring protein produced by brain cells that, when misfolded, causes spongiform encephalopathy.

The term *prion* is a shortened form of the term *proteinaceous infectious particle*. Normal prions are present in the brains of all mammals that have been studied. When highly magnified, the normal shape of a prion resembles a coil. The misfolded version of this protein resembles a sheet of paper that has been accordion-folded lengthwise several times when viewed from one end.

The normal role of the prion in the brain is not yet clear, but experiments in mice lacking the prion gene, and therefore unable to make the normal version of the protein, indicate that it may protect mammals against dementia and the other degenerative disorders associated with aging. The very rare CJD seen in elderly people is believed to either arise spontaneously when a prion is mistakenly misfolded or caused by a mutation to the gene that encodes the prion protein in humans, leading to the production of a misfolded prion protein.

In contrast, the newly emerging form of the disease that affects younger people, nvCJD, results when an individual is *infected* by misfolded prion proteins. Remarkably, the misfolded protein searches out properly folded proteins and refolds them into the mutant, disease-causing version. This is very unusual behavior for a protein—no other known protein has this capability.

Over time, the nerve cells in the brain become clogged with the misfolded prions, causing them to improperly transmit normal impulses and eventually cease to function altogether. Ultimately the cells burst, freeing their misfolded prions to find and refold normal prions in other nerve cells. Finally, the brain becomes riddled with empty spaces formerly occupied by normal cells, producing the spongelike character of the diseased brain.

Unlike viruses and bacteria, prions have no RNA or DNA, which were once thought to be necessary for any infectious agent to multiply. Furthermore, although most proteins within cells are easily broken down, prions resist degradation, allowing these rogue disease agents to propagate relentlessly.

All of the infectious agents that had been studied so far had their own genomes, and many researchers did not believe that proteins themselves could be infectious. Dr. Stanley Prusiner, a neurologist at the University of California, San Francisco, was the first to systematically study prions in the early 1980s and the first to link them with disease (Figure 11.5). Although many scientists remained skeptical about his findings, Prusiner was awarded the 1997 Nobel Prize in Medicine or Physiology for his research. However, understanding the nature of prions as infectious agents is only the first step toward preventing spongiform encephalopathies from becoming widespread in humans.

Figure 11.5 Dr. Stanley Prusiner. Dr. Prusiner, left, won the 1997 Nobel prize for his work with prions.

11.2 Epidemics

WWW **Media Activity 11.4** To Vaccinate or Not? What Would Happen if We Stopped?

An **epidemic** is a contagious disease that spreads rapidly and extensively among a population. An **epidemiologist** is a scientist who attempts to determine who is prone to a particular disease, where risk of the disease is highest, and when the disease is most likely to occur. Epidemiologists try to answer these questions by determining what the victims of the disease have in common. By identifying what factors increase the risk of a disease, epidemiologists help formulate public health policy. When it is difficult to pinpoint what is causing a disease or how it is spread, deadly epidemics can result (see Essay 11.2). For an epidemic to occur, the infectious agent must cause disease and must be transmissible from one organism to another.

Transmission of Infectious Agents

Infectious diseases are transmitted via contact with the disease-causing organism. This can include contact with infected fluids such as blood, saliva, and semen, as well as exposures facilitated by an intermediate host. Transmission also occurs via inhalation or ingestion of the disease organism or its toxins (Table 11.2).

Exposure to Infected Body Fluids As you learned in Chapter 9, AIDS is caused by exposure to infected blood. Hepatitis is another disease caused by contact with infected blood. There are at least six different hepatitis viruses, but the most dangerous is the virus known as hepatitis C. Exposure to hepatitis C leads to chronic liver diseases such as cirrhosis (irreversible, potentially fatal scarring of the liver), liver cancer, and liver failure. Hepatitis C ranks second to alcoholism as a major cause of liver disease and is the leading reason for liver transplants in the United States.

It is estimated that about 3% of the world's population carries this virus, but they are unaware of it because there are often no symptoms until the liver damage is severe. However, people who donate blood may find out that they have the disease before it causes serious damage, since blood banks

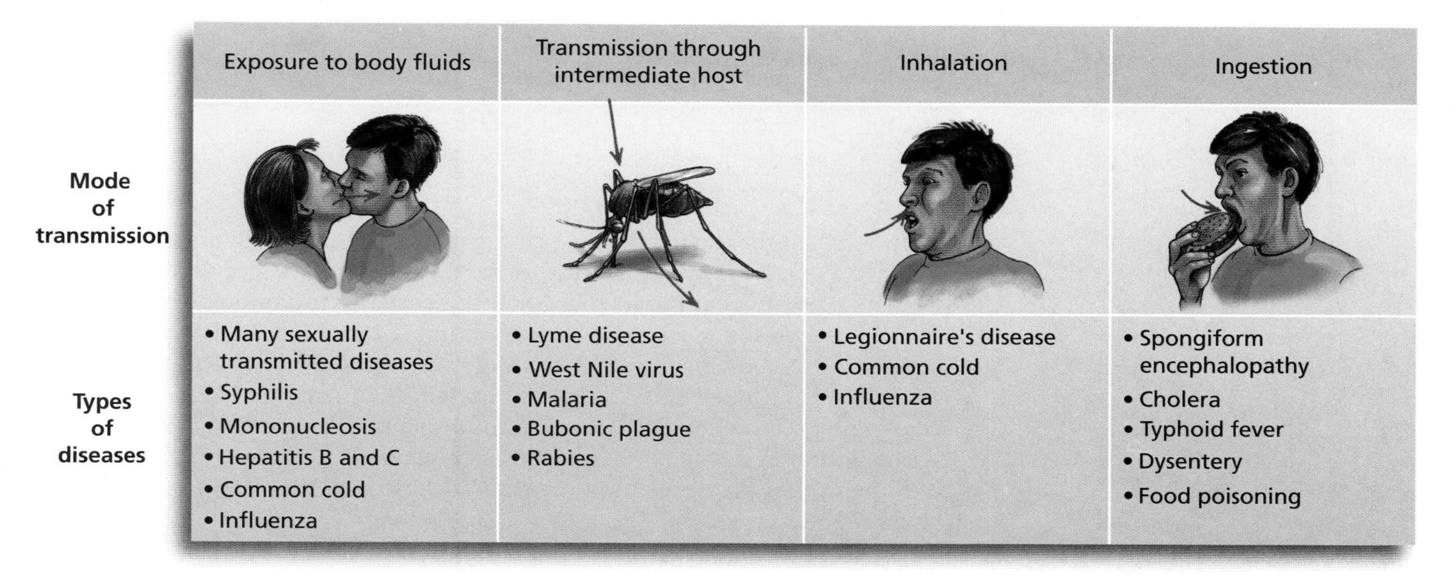

	Exposure to body fluids	Transmission through intermediate host	Inhalation	Ingestion
Mode of transmission				
Types of diseases	• Many sexually transmitted diseases • Syphilis • Mononucleosis • Hepatitis B and C • Common cold • Influenza	• Lyme disease • West Nile virus • Malaria • Bubonic plague • Rabies	• Legionnaire's disease • Common cold • Influenza	• Spongiform encephalopathy • Cholera • Typhoid fever • Dysentery • Food poisoning

Table 11.2 Transmission of infectious agents. Infectious agents are transmitted four ways, allowing pathogenic organisms to gain access to a host.

Essay 11.2 Epidemics: The Plague and Polio

The Plague

The plague that devastated Europe during the Middle Ages became an epidemic because people did not know what was causing it or how it was transmitted. We now know that the plague—or *black death* as it came to be known—was caused by the bacterium *Yersinia pestis*, a parasite of rats transmitted by fleas that pick it up from the rat's blood. *Y. pestis* multiplies in the flea's gut and is spread to humans when the fleas bite them. The risk of epidemic arose when people on farms and in cities al-

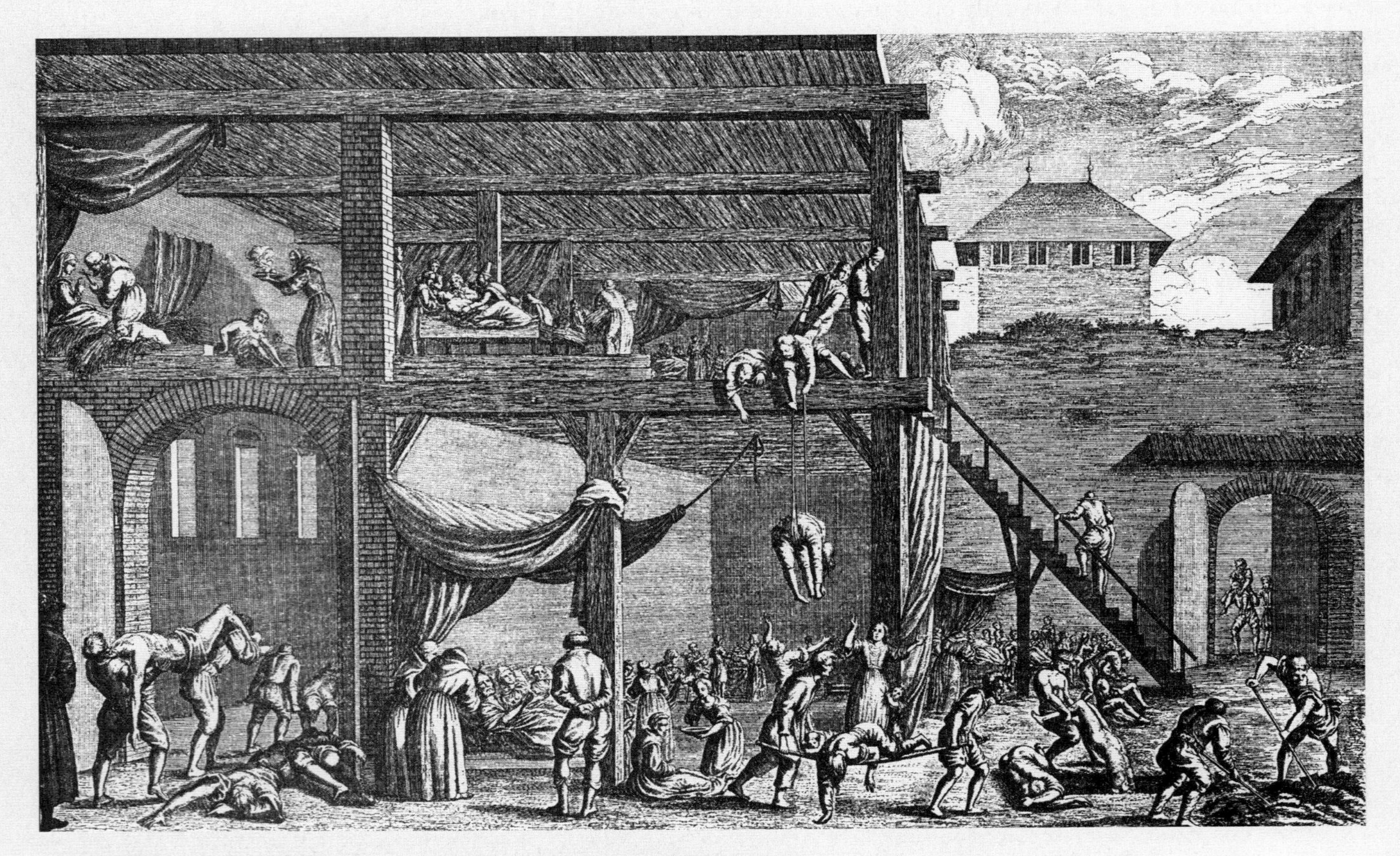

Figure E11.2A The black death killed many people in the Middle Ages.

routinely test for the hepatitis C virus and inform donors who test positive. Prior to 1990, there were no tests for hepatitis C and people who had transfusions prior to that are at risk for the disease. Intravenous-drug use and possibly some skin-piercing practices, notably tattooing, body piercing, and acupuncture, may also be contributing to the spread of hepatitis.

Saliva is another body fluid that transmits infectious disease. Mononucleosis is spread when an infected individual kisses or shares a drinking glass with an uninfected person. This disease is caused by a virus that infects cells of the immune system—the symptoms include sore throat, fever, and excessive fatigue.

The exchange of semen and vaginal fluids facilitates the spread of sexually transmitted diseases (STDs). Table 11.3 describes the origin, symptoms, treatment, and incidence of most STDs.

lowed rats to flourish in barns, in alleys full of discarded food, and on ships. The plague is still with us today, just not in epidemic proportions.

The sickness began with a fever, followed a few days later by swollen lymph glands in the armpits, neck, and groin. The bacteria also infected the nervous system, causing lethargy and hallucinations. Ulcers and open sores covered the bodies of the infected. Sleepless and unable to bear the touch of clothing, some victims walked the streets naked and died. At least half of the people that were infected died within five days. Ultimately, close to 95% of the victims died (Figure E11.2A).

While it was unclear what was causing the disease, it was recognized that plague often followed the docking of a new ship. Consequently, the policy of *quarantining*, or isolating, sailors and passengers arriving in port cities for 40 days was instituted. However, this did not prevent the spread of the plague since rats could scurry up and down mooring ropes and were carried around the world.

Panic caused by a lack of understanding about the cause and transmission of infectious disease is not limited to the distant past. Fear and panic reigned in the 1950s in the United States when an epidemic of polio struck.

Polio

Polio infected people of all ages, but it crippled and killed mostly children. Any time a child came down with a fever and sore throat, parents braced themselves for the telltale signs of polio—a stiff neck and aching muscles—which often led to paralysis. The paralysis was fatal if it spread to the muscles that control breathing and swallowing (Figure E11.2B).

Scientists and citizens alike knew about infectious diseases. There was some understanding that this disease was probably caused by a microbe—but how it was transmitted, and what the microbe was, were unknown. In an effort to prevent the virus from spreading, children were not allowed to go to playgrounds, movie theaters, or libraries. Many families sent their children to the presumed safety of the country, away from crowded urban areas.

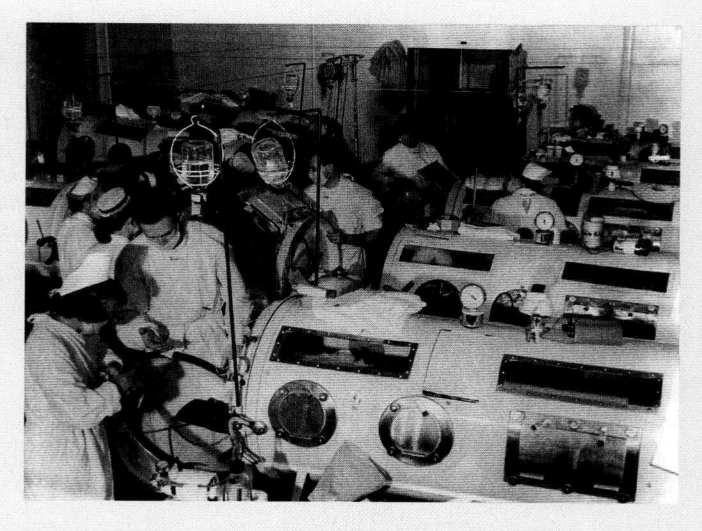

Figure E11.2B Polio victims with extensive paralysis were placed in "iron lungs" to help them breathe.

Eventually, scientists discovered that polio was caused by a virus that was transmitted via oral–fecal contact. People pass it on if they fail to wash their hands after a bowel movement and then touch food or other people's mouths.

When children have some exposure to the virus in infancy, slight or no illness develops and the immune system is stimulated to prevent illness from later exposure. When sanitation is limited, infection and immunity are acquired early in life, and epidemics are less likely. As sanitation and hygiene improves in economically developed countries, infection is delayed and older children and young adults are susceptible to the virus. Purified water and pasteurized milk further decreased the odds of babies being infected with a small dose that would stimulate an immune response and protect them against later infections with a larger dose of virus.

Once the polio virus was isolated in the laboratory, vaccines could be prepared. Today, babies receive the vaccine, which generates the same immunity as getting the childhood disease itself, and the disease has nearly disappeared.

Transmission Through an Intermediate Host Transmission of microorganisms can occur through an intermediate organism, such as an insect. Lyme disease, caused by the bacterium *Borrelia burgdorferi*, is transmitted via the bite of an infected tick. Not all types of ticks carry the bacteria. In the eastern United States, the deer tick is usually responsible for transmitting the disease. In the western states the western black-legged tick is the culprit. A rash with a clear center that resembles a bull's eye may, but does not always, appear any time from one day to one month after the infected tick's bite. Symptoms of the disease include numbness and pain in the arms or legs, paralysis of the facial muscles, fever, fatigue, stiff neck, and headaches. There are now antibiotics available that can cure this disease when it is diagnosed early.

The West Nile virus uses the mosquito as its intermediate host. This virus causes encephalitis (swelling of the brain) in humans and can kill certain species

Bacterial pathogens	Mode of infection	Symptoms	Treatment and prevention	Incidence
Chlamydia Caused by the bacterium *Chlamydia trachomatis*	Chlamydia spreads easily during sexual contact. This bacterium infects the urethra, cervix, uterus, the oviducts of women, and the urethra of men.	Symptoms include pelvic pain and fluid discharge. Untreated chlamydia can lead to pelvic inflammatory disease (see below) and infertility.	Antibiotics are effective. Condoms prevent transmission.	Chlamydia is the most common bacterial STD. Many of the approximately 3 million cases that occur each year in the U.S. are undiagnosed, because there may be no symptoms until many years after infection.
Gonorrhea Caused by the bacterium *Neisseria gonorrhoeae*	Gonorrhea, sometimes referred to as "the clap," is most often transmitted by sexual contact with an infected partner but may also be spread by infected bodily fluids to newborns during birth or to infants by contact.	Symptoms may include a thick discharge from the penis or vagina. However, many people experience no symptoms. In fact, 80% of gonorrhea cases are asymptomatic. Untreated gonorrhea can cause infertility in women if bacteria spread to the oviducts and cause pelvic inflammatory disease (see below).	Because this STD is bacterial, it can be treated with antibiotics. Condoms prevent transmission.	Over half a million cases of gonorrhea are diagnosed in the U.S. each year.
Pelvic Inflammatory Disease (PID) Caused by gonorrhea or chlamydia	Pelvic inflammatory disease (PID) is an infection of the female reproductive tract. The disease usually occurs when sexually transmitted bacteria ascend from the vagina into the uterus and oviducts.	Symptoms, when present, include pelvic pain and difficulty becoming pregnant due to scarring and blockage of reproductive organs caused by the infection.	Antibiotics can kill PID–causing organisms but will not heal damaged reproductive organs. Infertility treatments vary. Condoms prevent transmission.	About 1 million U.S. women are diagnosed with PID each year. About 100,000 of them become infertile as a result of PID, and thousands suffer complications of pregnancy.

Table 11.3 Sexually transmitted diseases (STDs). The origins, symptoms, and prevalence of many STDs are discussed.

WWW
Media Activity 11.5 A New Epidemic of Syphilis?

Viral pathogens	Mode of infection	Symptoms	Treatment and prevention	Incidence
AIDS Caused by the human immunodeficiency virus (HIV)	HIV spreads through oral, anal, and vaginal sex, as well as through blood transfusions and shared intravenous needles.	Over time, HIV infection weakens the immune system so much that infections that are normally easily controlled cause severe damage.	AIDS can be controlled with powerful combination-drug therapies, but cannot be cured. Condoms prevent transmission.	Worldwide, about 42 million people are living with AIDS/HIV. In the United States close to 1 million people are diagnosed with AIDS.
Genital Warts Caused by the human papilloma virus (HPV)	Genital warts usually result from sexual contact (intercourse or oral-genital contact) with an infected partner.	Genital warts appear as growths or bumps on the pubic area, penis, vulva, or vagina. Only some types of HPV cause genital warts. Other types of HPV can cause abnormal cell changes on a woman's cervix, which can result in cervical cancer.	Currently, there is no treatment to cure HPV, although warts can be removed surgically, burned off with lasers, or frozen off with liquid nitrogen. Condoms do not prevent transmission because warts can be present on areas other than those contacted by the condom.	Some studies estimate that the majority of the sexually active population has been exposed to at least one or more of the over 70 different types of HPV.
Hepatitis B Caused by the hepatitis B virus (HBV) or by noninfectious agents such as chemical poisons, drugs, and alcohol	Hepatitis B is transmitted through blood and other bodily fluids. Newborns can get Hepatitis B from their mothers during delivery.	HBV infections cause inflammation and scarring (cirrhosis) of the liver, which may be fatal.	Hepatitis B is preventable through vaccination, and in many states, children are routinely immunized against this disease. A few antiviral drugs are effective for treating chronic HBV infection. Condoms prevent transmission.	Of the six described hepatitis viruses, Hepatitis B is most commonly transmitted sexually. Approximately 200,000 cases are diagnosed annually in the United States and 4,000–5,000 people die from HBV infection.
Herpes Simplex Caused by one of two viruses: herpes simplex type 1 (HSV-1) or herpes simplex type 2 (HSV-2)	Herpes is spread by direct skin-to-skin contact, usually by kissing, or oral, vaginal, or anal intercourse.	Herpes is a common and usually mild infection. It can cause cold sores or fever blisters on the mouth or face. It can also cause similar symptoms in the genital area, known as genital herpes. After initial infection, HSV remains latent for a while, then becomes active again. This results in the occasional and seemingly random appearance of blisters throughout an infected individual's lifetime.	Antiviral medications can lessen the duration and discomfort of herpes outbreaks but do not cure individuals of the virus. Condoms reduce the likelihood of transmission.	Nearly 20% of the population is infected with genital herpes.

Table 11.3 ***(continued).***

Insect, protozoan, and fungal pathogens	Mode of infection	Symptoms	Treatment and prevention	Incidence
Pubic Lice Also known as *crabs*. Caused by the insect *Pediculus pubis*	Pubic lice are transmitted through skin-to-skin contact or contact with an infected bed, towel, or clothing.	The most common symptom of crabs is itching of the pubic area. The itching is caused by an allergic reaction to the bites, and usually starts about five days after the initial infection.	Pubic lice are cured by washing the affected area with a delousing agent.	In the United States, there are an estimated 3 million cases of crabs every year.
Trichomoniasis Caused by the parasitic protozoan (single-celled eukaryote) *Trichomoniasis vaginalis*	Trichomoniasis, sometimes called "trich," is transmitted by vaginal intercourse.	The major symptom of trichomoniasis infection in women is vaginal itching with a frothy yellow-green vaginal discharge. Most men do not have symptoms, but some may experience irritation in their urethra after urination or ejaculation.	Antibiotics are effective. Condoms prevent transmission.	This is an extremely common STD, infecting up to 15% of sexually active women in the U.S. (over 2 million per year).
Yeast Caused by fungi of the genus *Candida*	Yeast are normal inhabitants of the female reproductive tract. They increase in number when a woman is weakened by illness or upset by stress. Antibiotics, taken to treat bacterial infections, can kill vaginal bacteria and allow the yeast to grow, leading to yeast overgrowth. Yeast can also be passed from person to person, such as through sexual intercourse.	Yeast infections are characterized by a thick whitish discharge from the vagina and vaginal itching. The discharge often smells sweet, like baking bread.	Antifungal medicines can cure yeast infections.	Nearly 75% of all adult U.S. women have had at least one genital yeast infection. On rare occasions, men may also experience it.

Table 11.3 ***(continued).***

of birds. Mosquitoes become infected when they feed on infected birds, the virus may then circulate in their blood for a few days. Infected mosquitoes can then transmit the West Nile virus to humans and animals while feeding on their blood. Following transmission by an infected mosquito, West Nile virus multiplies in the person's blood system and eventually reaches the brain.

Inhalation Other pathogens cause disease when they enter the body along with the air we breathe. The influenza virus is the pathogen that causes the flu. The flu is transmitted when a person who has it coughs, sneezes, or speaks and sends the virus into the air and other people then inhale the virus. The flu may also be spread when people touch a surface that has flu viruses on it and then touch their nose or mouth. A person can spread the flu even before he or she feels sick, starting from one day prior to the onset of symptoms until about seven days after symptoms start. These symptoms usually begin suddenly and may include fever, headache, tiredness, a dry cough, sore throat, and body aches.

Ingestion The types of spongiform encephalopathies that are transmissible are spread by ingesting food containing misfolded prions. Scientists believe that the cows with mad cow disease ingested misfolded proteins when they ate the remnants of diseased animals that were present in their feed.

After a cow has been slaughtered, what is left of the carcass is cooked and then ground up, producing a type of cattle feed called meat-and-bone meal. Feed mills buy the meal and make it into a toasted, crunchy breakfast-cereal-like substance that is fed to cows (Figure 11.6). When cows with mad cow disease are unwittingly turned into meat-and-bone meal, their infectious prions are fed to other cows which become infected, thus spreading the BSE to more cows.

Humans become infected with misfolded prions when they eat meat from diseased cows (or from other humans, as is the case with Kuru). Although the pathogen that causes spongiform encephalopathies in cows and humans infects the brain and spinal cord—parts of the cow that are not normally consumed—meat can be contaminated during the rendering process. For example, hamburger containing parts of the spinal cord that were stripped off the carcass may be infectious. If people eat contaminated meat, their immune systems have a very difficult time helping them fight the infection.

Figure 11.6 Meat-and-bone meal cow feed. Waste products from slaughterhouses are used to make a feed additive for cattle.

www
Media Activity 11.1B Defense Against Infectious Agents

11.3 The Body's Response to Infection: The Immune System

Prions enter the body via contaminated food. Therefore, prions have circumvented the first of the body's three lines of defense (Figure 11.7)—the skin and mucous membranes. These external barriers are *nonspecific;* that is, they do not distinguish one pathogen from another.

The second line of defense is also nonspecific. This internal defense system partly consists of white blood cells that indiscriminately attack and ingest invaders. The ingestion of pathogens by cells is called **phagocytosis**. (Phagocytosis is not limited to ingesting invading organisms—it is also used by cells that must use food or materials that are too large to enter through the cell membrane). Phagocytic white blood cells called **macrophages** move throughout the lymphatic fluid, cleaning up dead and damaged cells. To destroy the cells, macrophages extend their long *pseudopodia* (cellular extensions used for eating and moving), grab the invading organism, and engulf it; enzymes inside the macrophage help break the invader apart. Much of the destruction of the offending cells occurs in the lymph nodes, which is why they swell when we are ill with an infection. Swelling and inflammation are signs that this second line of defense has been employed.

The third line of defense is the immune system. The immune system is a *specific* defense system consisting of many millions of white blood cells called **lymphocytes**. Lymphocytes travel throughout the body by moving through spaces between cells and tissues or by transport via the blood and the lymphatic system (Figure 11.8).

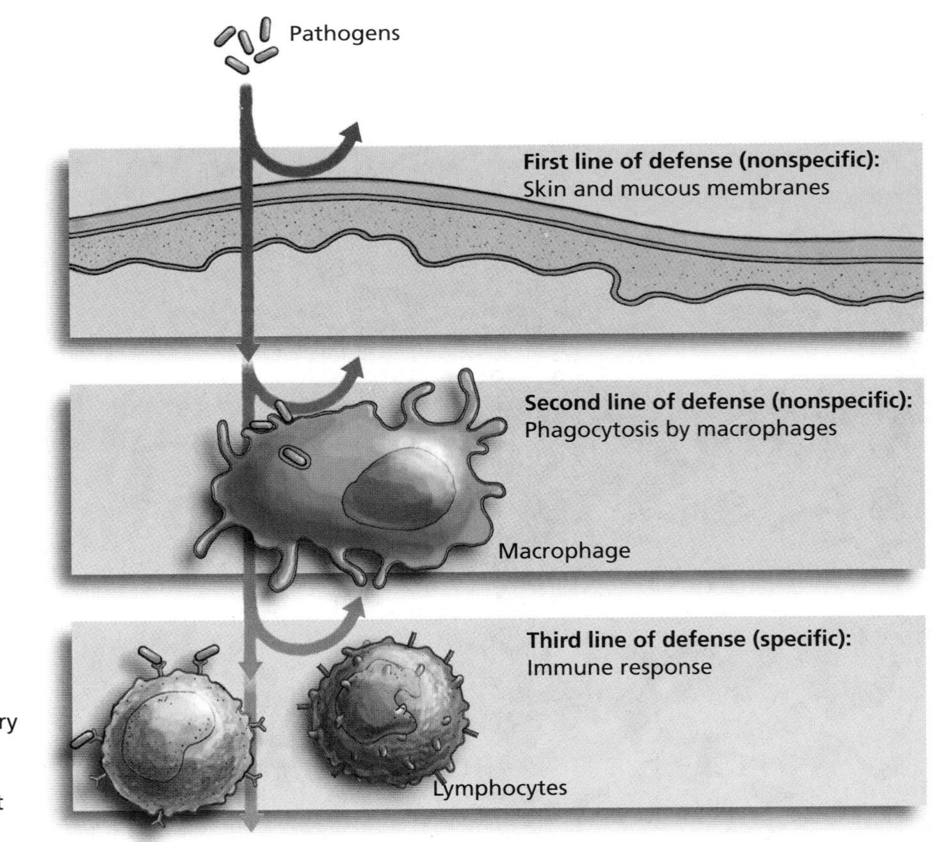

Figure 11.7 The body's defense against infection. The body has three lines of defense, the skin and mucous membranes try to keep infection out, phagocytes attack pathogens that slip through, and lymphocytes target specific pathogens and attempt to contain them.

The specific response generated by the immune system is triggered by proteins and carbohydrates on the outer membranes of bacterial cells or cells that have been infected by a virus. Molecules that are foreign to the host and stimulate the immune system to react are called **antigens**.

In addition to being present on bacteria and cells that have been infected by a virus, antigens are also present on tissues transplanted from one person to another, because every individual's cells have a characteristic set of proteins and sugars on their surfaces. This is the reason people receiving organ transplants are given drugs to suppress their immune systems.

Regardless of the source, when an antigen is present in the body, the production of two types of lymphocytes is enhanced: the **B lymphocytes** (**B cells**) and **T lymphocytes** (**T cells**). Like macrophages, these lymphocytes circulate throughout the blood and lymph system and are concentrated in the spleen and lymph nodes. Because lymphocytes recognize specific antigens, they are said to display *specificity*. B and T cells are able to recognize specific antigens when they have the correct **antigen receptor** on their cell membrane. A single B or T cell bears over 100,000 identical copies of the same receptor for one specific antigen. The antigen receptors that are present on B cells, called **antibodies**, are proteins produced in response to the presence of antigens. The antigen receptors on T cells are called **T-cell receptors**.

The ability of the B and T cells to respond to specific antigens begins before birth. Thus, we are able to respond to infectious agents the first time we are exposed. This ability continues into adulthood because these cells are manufactured, at the rate of about 100 million per day, throughout our lives.

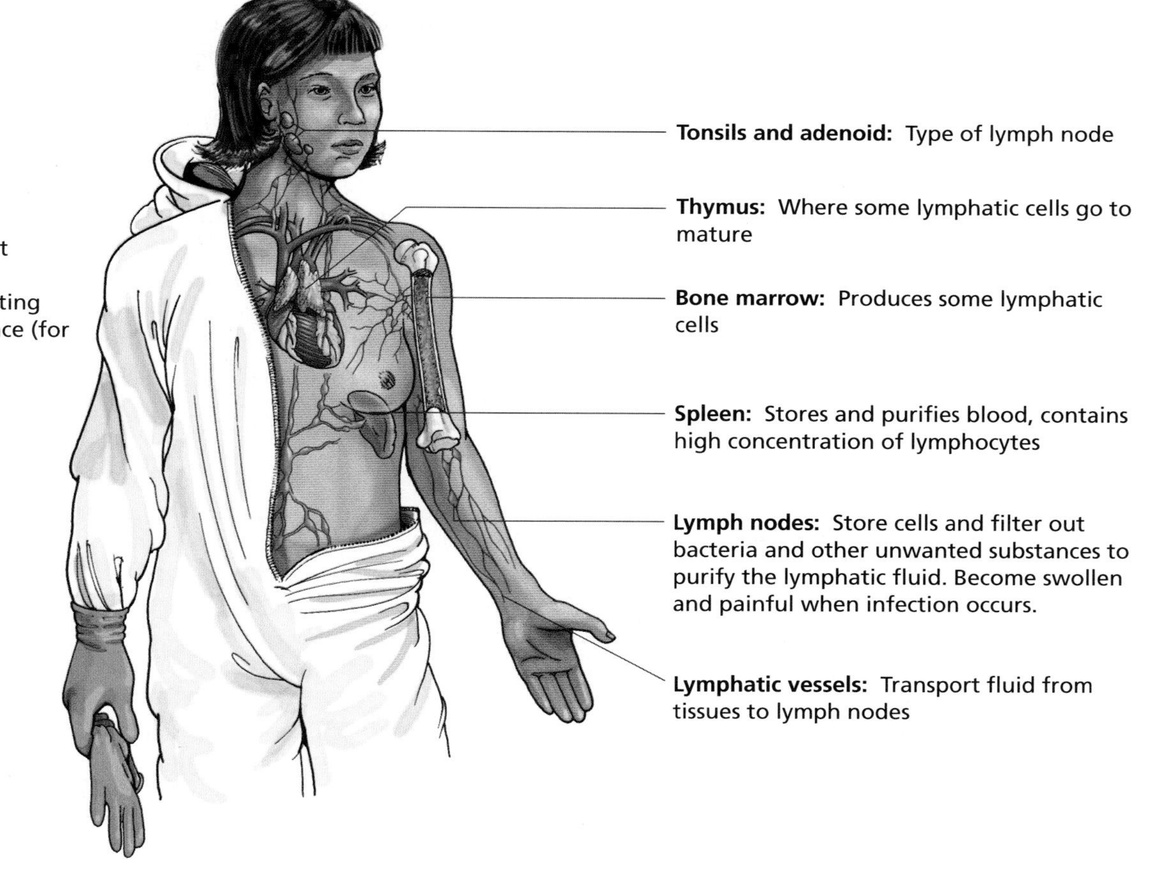
The best way to prevent illness is to keep the pathogens from contacting the body in the first place (for example, in the case of anthrax, by wearing a protective suit).
Tonsils and adenoid: Type of lymph node
Thymus: Where some lymphatic cells go to mature
Bone marrow: Produces some lymphatic cells
Spleen: Stores and purifies blood, contains high concentration of lymphocytes
Lymph nodes: Store cells and filter out bacteria and other unwanted substances to purify the lymphatic fluid. Become swollen and painful when infection occurs.
Lymphatic vessels: Transport fluid from tissues to lymph nodes

Figure 11.8 The lymphatic system. Pathogens trigger a response by the lymphatic system. The various organs of this system work to eliminate infections.

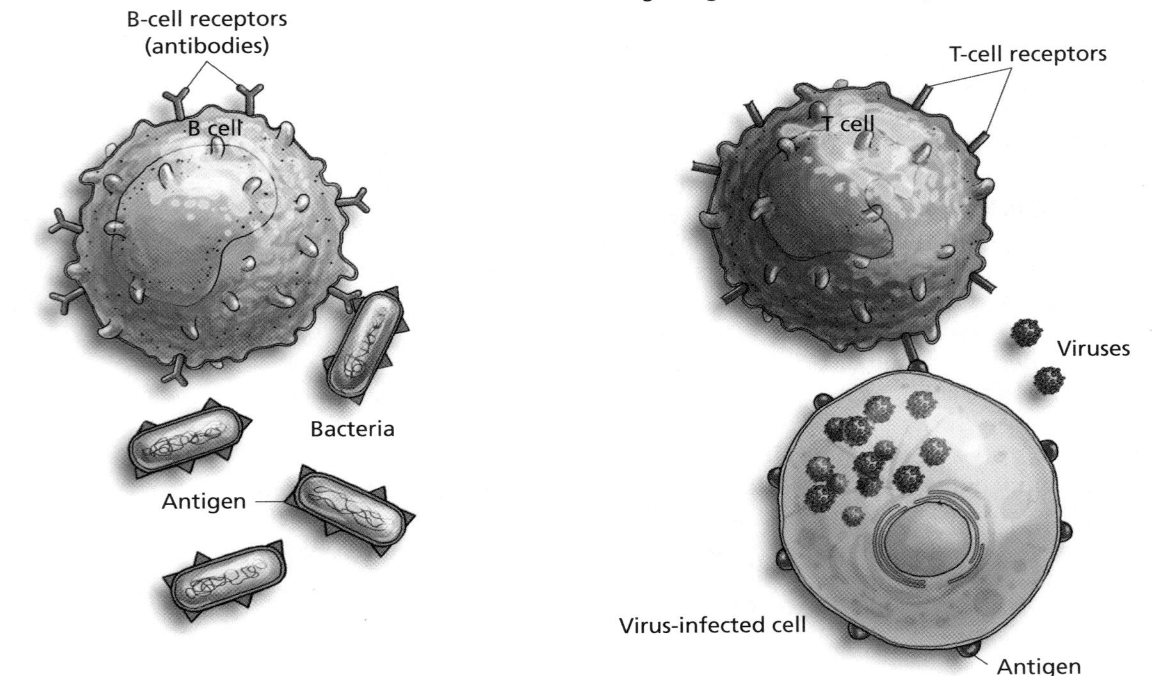
(a) B lymphocytes
B cells recognize and react to small free-living microorganisms such as bacteria and the toxins they produce.
B-cell receptors (antibodies)
B cell
Bacteria
Antigen
(b) T lymphocytes
T cells recognize and react to body cells that have gone awry, such as cancer cells or cells that have been invaded by viruses. T cells also respond to larger organisms, such as fungi and parasitic worms.
T-cell receptors
T cell
Viruses
Virus-infected cell
Antigen

Figure 11.9 B cells and T cells. The antibodies produced by B cells function as antigen receptors. T cells also have antigen receptors. The receptors on the surface of B and T cells help cells recognize invaders.

The ability to respond to an infection, the **immune response**, actually results from the increased production of B and T cells. B and T cells recognize different types of antigens: B cells recognize and react to small, free-living microorganisms such as bacteria and the toxins they produce; T cells recognize and respond to body cells that have gone awry, such as cancer cells or cells that have been invaded by viruses. T cells also respond to transplanted tissues and larger organisms such as fungi and parasitic worms (Figure 11.9).

Making B and T Cells

Lymphocytes are produced from special cells that have the ability to become any other cell type, called *stem cells* (see Essay 7.2). Many parts of the body, including the bone marrow, retain a supply of stem cells that can develop into more specialized cells. Bone-marrow stem cells enable the bone marrow to produce blood cells throughout the lifetime of an individual. Lymphocytes are produced from the stem cells of bone marrow and released into the bloodstream.

Some lymphocytes continue their development in the bone marrow and become B cells. Others take up residence in the thymus gland (see Figure 11.8). The thymus gland, located behind the top of the sternum, stimulates T cells to develop. When immune cells are produced in the **b**one marrow, they are called B cells; when they are produced in the **t**hymus, they are called T cells (Figure 11.10).

During the maturation process, each B and T cell must produce its specific receptor. As B and T cells develop, the DNA in each cell does something very

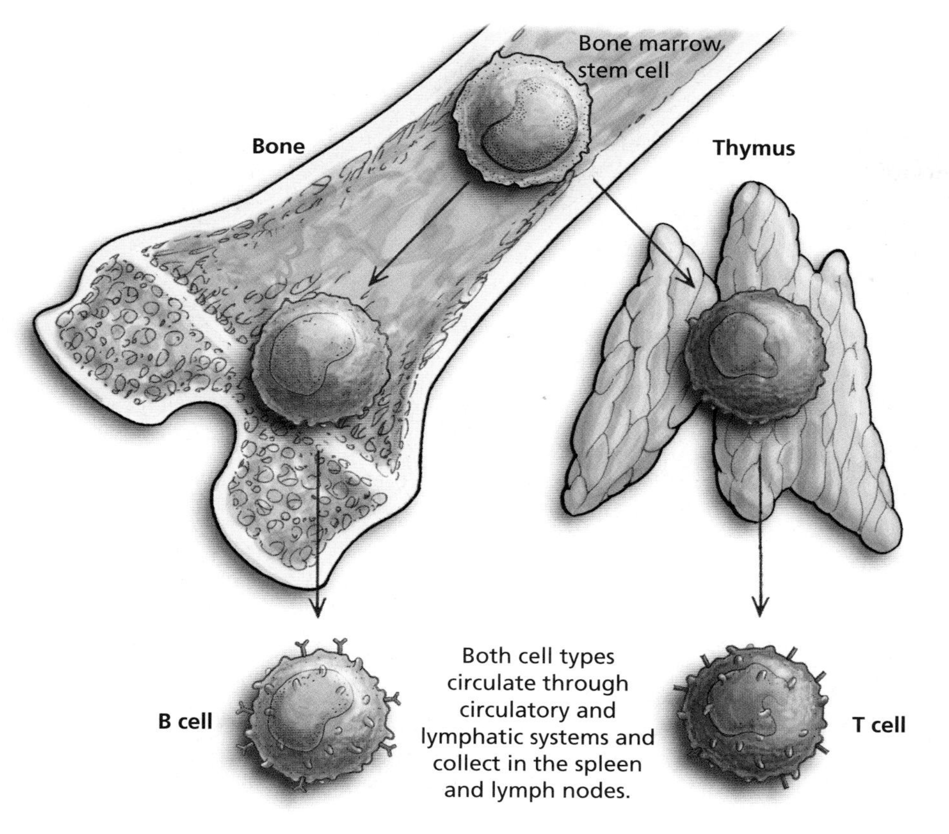

Figure 11.10 Lymphocyte development. Lymphocytes develop from cells in the bone marrow. Lymphocytes that continue their development in the bone become B cells. Those that move to the thymus to continue their development become T cells. Both B and T cells are found in the lymphatic organs.

unusual: The DNA rearranges itself; some portions are cut out, others are shuffled around (Figure 11.11). Each unique arrangement of DNA encodes a different receptor protein. Once synthesized, the proteins move to the surface of the B or T cell and act as antigen receptors.

While B and T cells are maturing, their antigen receptors are tested for potential self-reactivity. The cells of a given individual have characteristic proteins on their surfaces, and developing lymphocytes are tested in the thymus

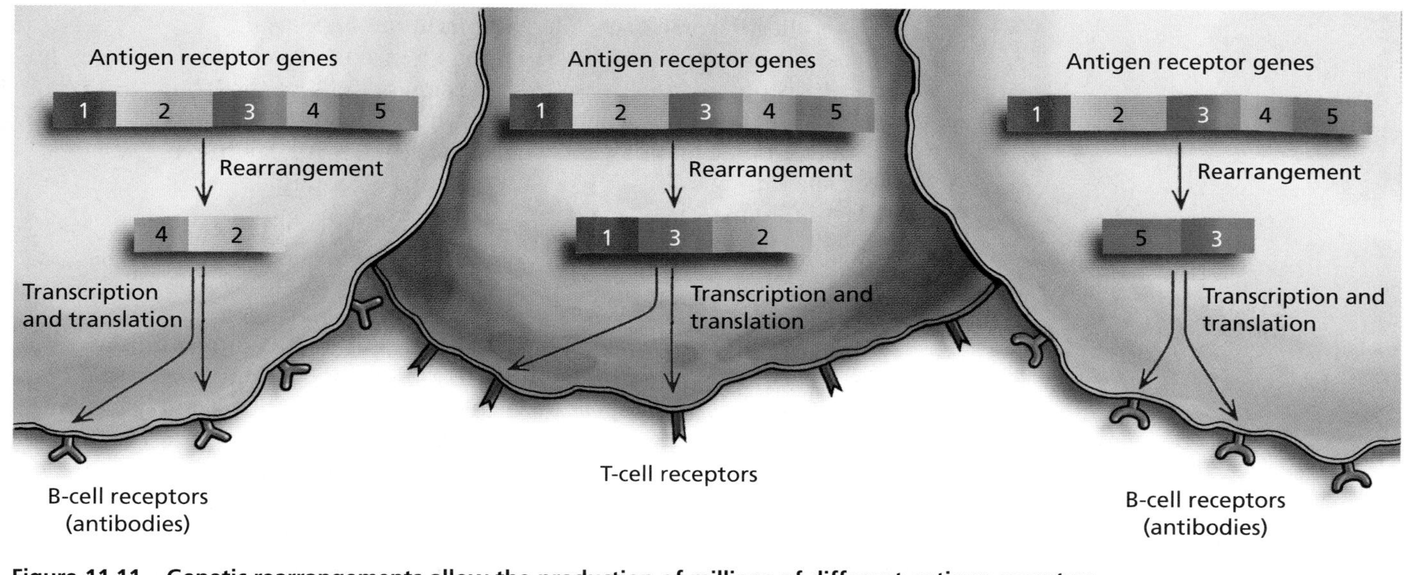

Figure 11.11 Genetic rearrangements allow the production of millions of different antigen receptors. Rearrangements of the genes involved in producing antigen receptors increase the number of antigens to which each of us can respond.

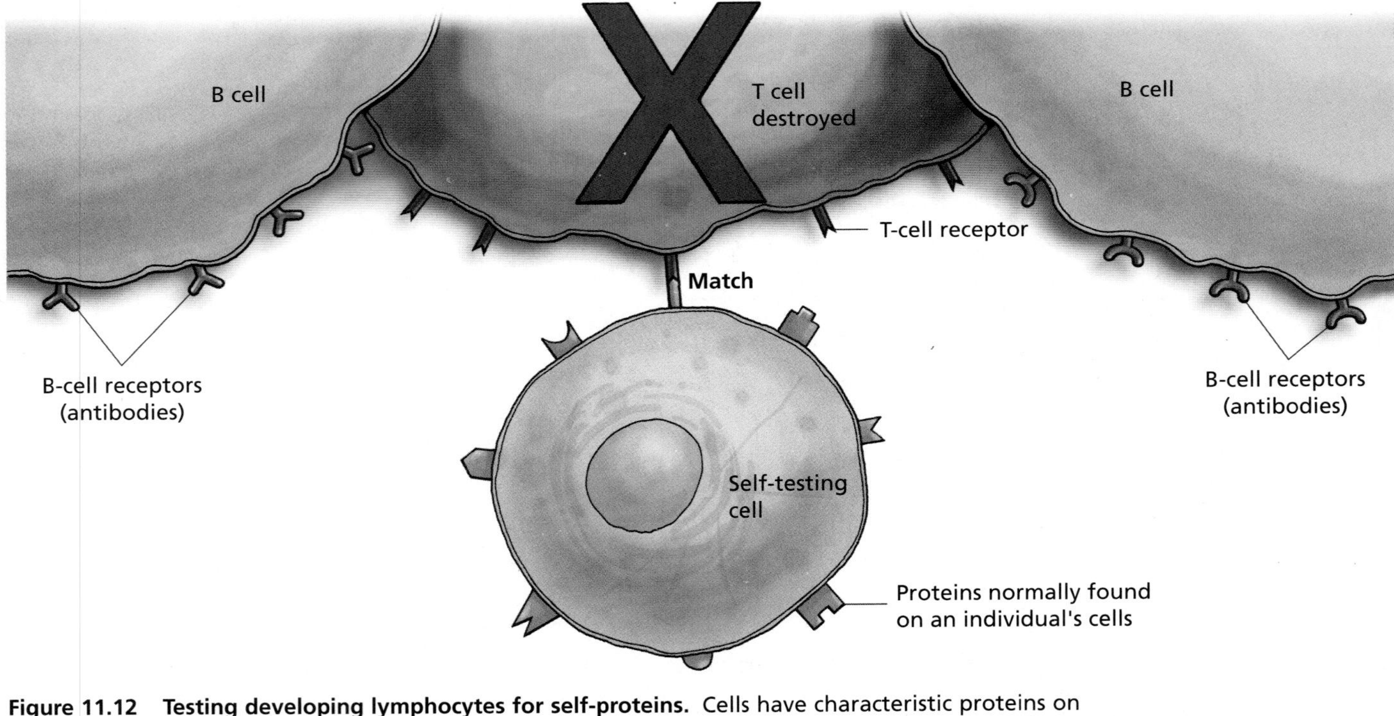

Figure 11.12 Testing developing lymphocytes for self-proteins. Cells have characteristic proteins on their surfaces. Developing lymphocytes are tested to determine whether they are self or non-self. Lymphocytes that bind to antigen receptors on testing cells are destroyed; those that do not bind are allowed to develop.

to determine whether they will bind to self-proteins. Any developing lymphocyte whose antigen receptors bind to self-proteins is eliminated, making an immune response against one's self less likely. Lymphocytes whose receptors do not bind are then allowed to develop to maturity (Figure 11.12). Thus, the body normally has no mature lymphocytes that react against self-proteins, and the immune system exhibits self-tolerance. When this testing fails, the cells of the immune system attack normal body cells. Diseases that result when a person's immune system is attacking itself are called **autoimmune diseases**.

Multiple sclerosis is an autoimmune disease that occurs when T cells specific for a protein on nerve cells attack these cells in the brain. In insulin-dependent diabetes, T cells and B cells attack cells that produce the hormone insulin in the pancreas. Lupus is a disease that occurs when self-antibodies to the nuclei of *all* cells are formed. These anti-nuclear antibodies build up in cells, causing inflammation of many tissues in the body.

WWW
Media Activity 11.6 Gene Therapy Trials for Immune Deficiency

Immune Response

Even though we have a single immune system, it is diversified into two subsystems so we can combat the multitude of infectious agents we encounter in our lifetimes. This diversification is a result of the differing approaches that B and T cells have to ridding the body of infectious agents once they are found. B cells provide a response called **humoral immunity**, while T cells provide a **cell-mediated immunity**.

Humoral Immunity Blood and lymphatic fluid were referred to as "body humours" in medieval times; B cells fostered immunity and were given the name *humoral immunity*. When a B cell encounters an antigen, it immediately makes

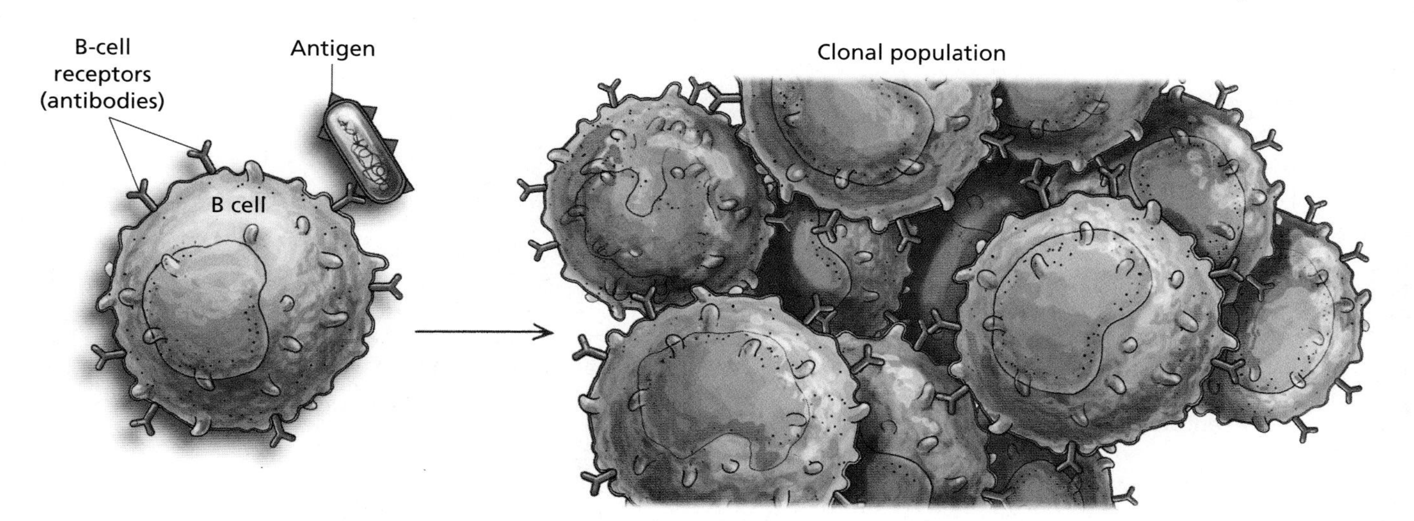

Figure 11.13 Clonal populations. When a lymphocyte binds an antigen, it proliferates to produce many copies of the lymphocyte and its antigen receptor. This strengthens the immune system's ability to rid the body of that infectious agent.

copies of itself, resulting in a population of identical cells able to help fight the infection. This population of cells is called a **clonal population**.

The entire clonal population has the same DNA arrangement, and all the cells in a clonal population carry the same antigen receptor on their membrane. The cells of the clonal population, called **memory cells**, will help the body respond more quickly if the infectious agent is encountered again. Should subsequent infection occur, the large number of memory cells facilitates a quicker immune response (Figure 11.13).

In addition to the memory cells produced when an antibody binds to an antigen, B cells also produce **plasma cells**. Plasma cells secrete antibodies specific to an antigen. The antibodies secreted by plasma cells circulate within body fluids, including tears and saliva; when they encounter an antigen, they bind to it. This antigen–antibody complex then combines with proteins in the blood called **complement proteins**. When complement proteins attach to antibody–antigen complexes on the pathogen's plasma membrane, they cause the cell to break open (Figure 11.14). Antigen–antibody complex binding also increases phagocytosis and the overall ability of the immune system to destroy invaders.

Cell-mediated Immunity T cells also respond to infection by undergoing rapid cell division to produce memory cells and by becoming specialized cells. However, unlike B cells, T cells do not secrete antibodies; instead, they directly attack other cells. Two of these attacking cell types are the *cytotoxic T cells* and *helper T cells*. **Cytotoxic T cells** attack and kill body cells that have become infected with a virus. When a virus infects a body cell, viral proteins are placed on the surface of the host cell. Cytotoxic T cells recognize these proteins as foreign, bind to them, and destroy the entire cell. They do this before the virus has had time to replicate by releasing a chemical that causes the plasma membranes of the target cell to leak.

Helper T cells, also called *T4 cells*, can be thought of as boosters of the immune response. These cells detect invaders and alert both the B and T cells that infection is occurring. Without helper T cells, there can be almost no immune response. Helper T cells also secrete a substance that greatly increases the level of cytotoxic T cell response. The AIDS virus, HIV, infects helper T cells, thus crippling the body's ability to respond to any infection (see Chapter 9).

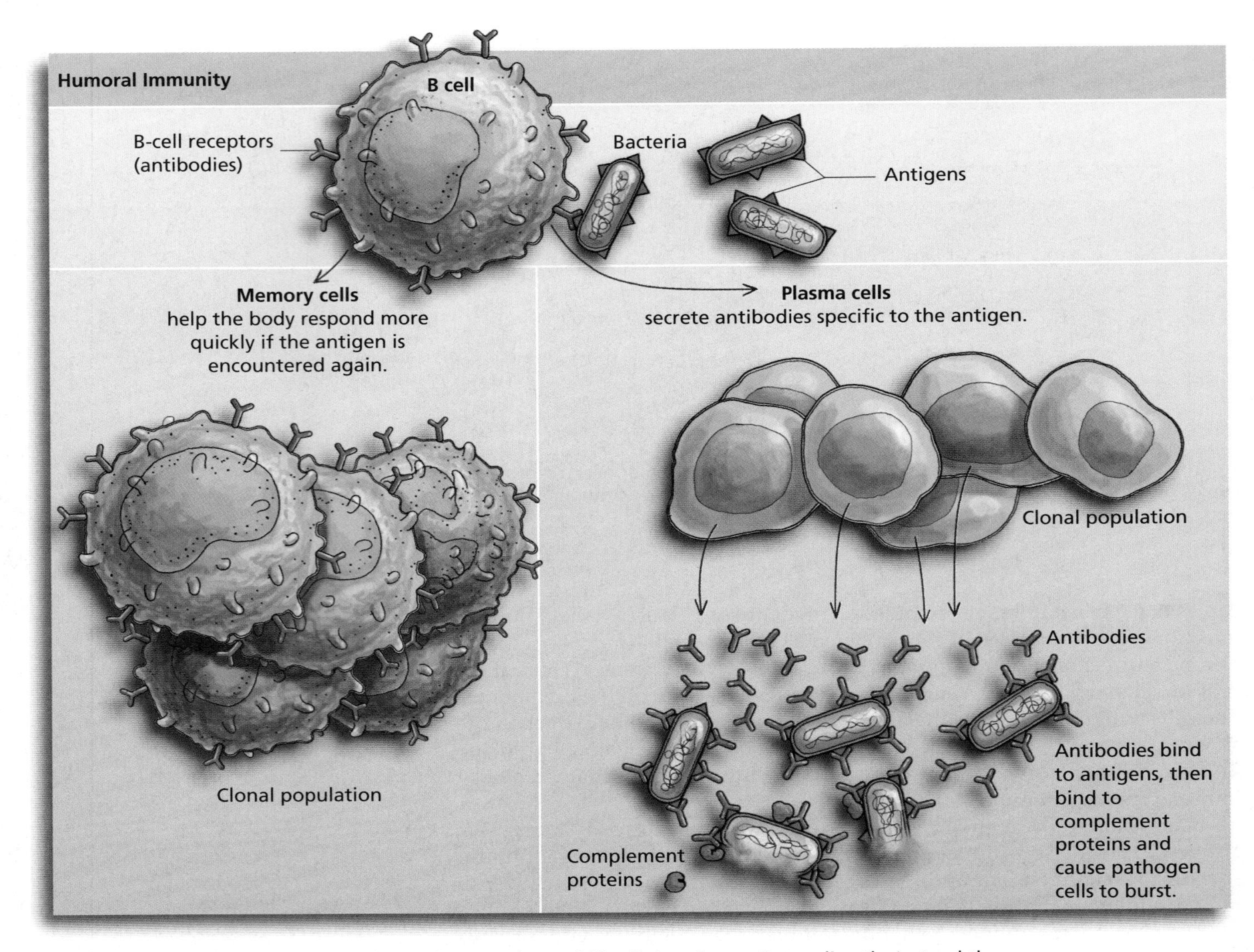

Figure 11.14 Humoral immunity. B lymphocytes do not kill cells bearing antigens directly. Instead they make and secrete antibodies specific to antigens. The antibodies circulate in the blood and lymph. The antibody–antigen complex binds with complement proteins in the blood, and together the antibody and complement can burst bacterial plasma membranes.

Macrophages also help fight infection by transferring an antigen to their own plasma membranes, thus alerting the immune system, via the T cell, to a foreign antigen present in the body. When the macrophage presents an antigen to a T cell that has the correct receptor, the T cell replicates itself to produce more memory cells, more cytotoxic T cells, and more helper T cells. Figure 11.15 summarizes the cell-mediated immune response.

There Is No Immune Response to Prions

People die from infectious diseases when their immune systems have not made enough B or T cells to fight off the infection. In some cases, infectious agents carry antigens for which no amount of DNA rearrangement can yield the proper receptor. In other cases, one group of people may be able to rearrange their DNA to produce the right receptors while others do not. This partly explains why some people do not become ill when exposed to an infectious disease, while others die from the same exposure.

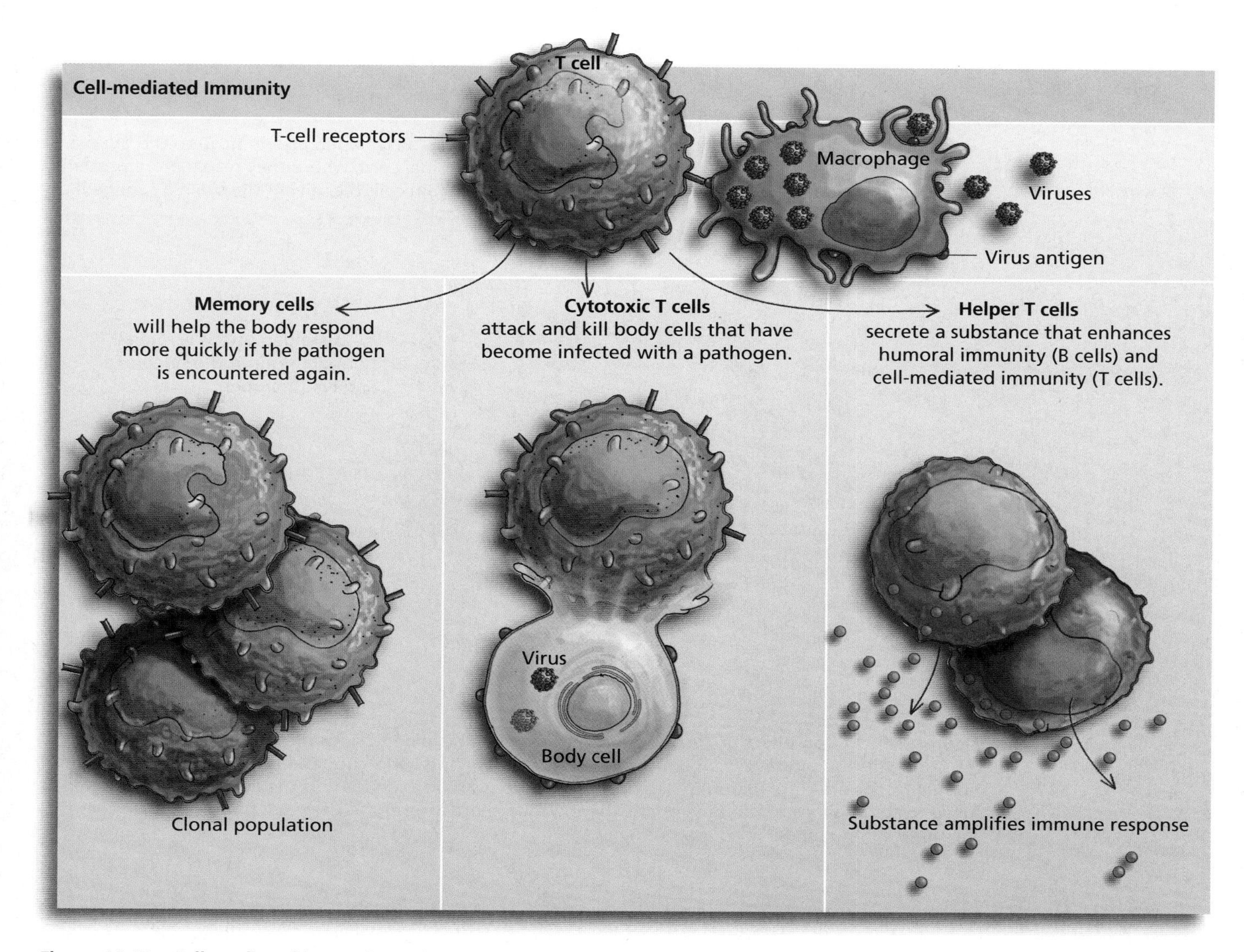

Figure 11.15 Cell-mediated immunity. T lymphocytes divide to produce different populations of cells: (1) Memory cells carry the specific antigen receptor; (2) cytotoxic T cells attack and kill cells; and (3) helper T cells boost the immune response.

Sometimes an infectious agent is effectively combated by the immune system, only to reemerge in a form that is newly pathogenic. Pathogens are under evolutionary pressure to change in order to evade the immune system (see Essay 9.2). Flu vaccines must be given every year because the flu virus rearranges its DNA sequences rapidly. This results in different proteins being encoded and placed on the surface of the virus, thereby preventing existing memory cells from recognizing a virus they have already met. Because the proteins on the surface of a flu virus change so quickly, a vaccine that was prepared to protect you from last year's flu virus will not necessarily work against this year's flu virus.

In the case of the spongiform encephalopathies, the immune system does not mount a response to misfolded prions at all, because they are just refolded versions of the normal prion protein. It seems that the refolded version of the protein is similar enough to the normal version that the immune system recognizes them both as self.

After misfolded bovine prions are ingested by humans, they move to the spleen and lymph nodes and refold all the prions there. From there, the refolded

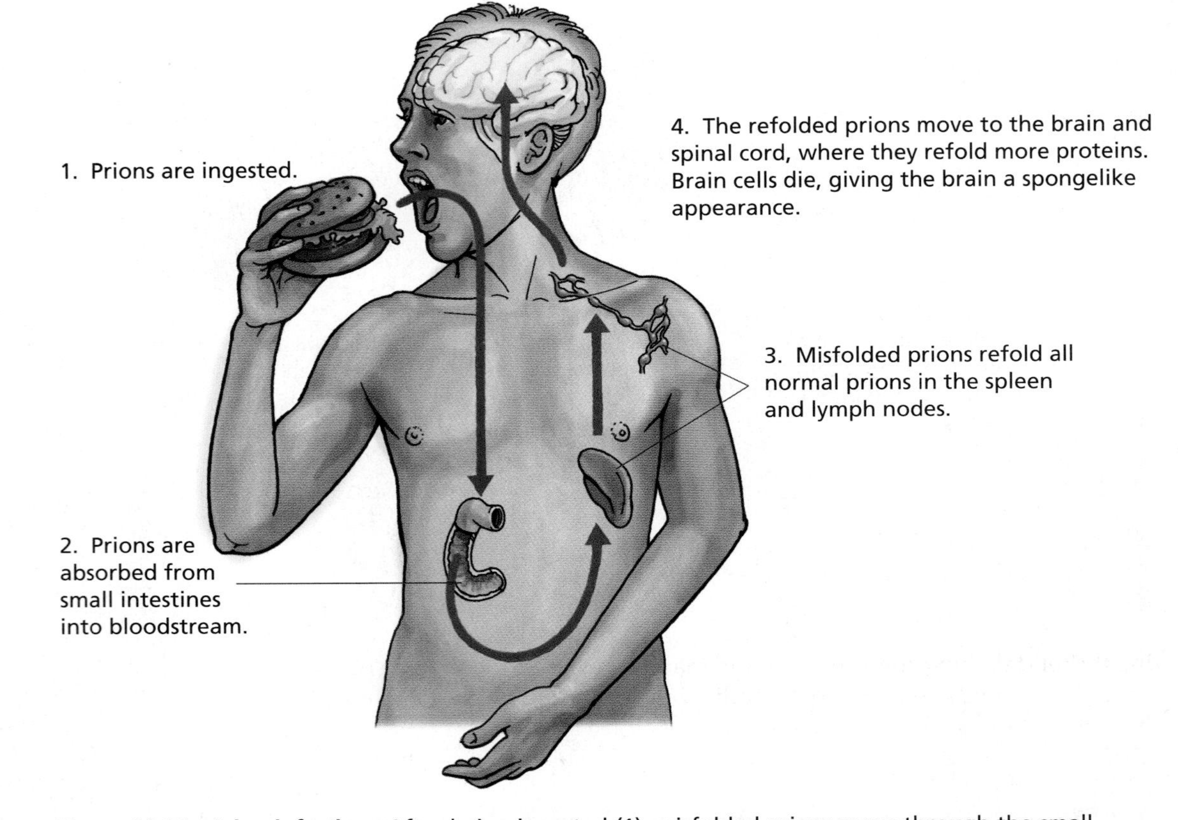

1. Prions are ingested.
4. The refolded prions move to the brain and spinal cord, where they refold more proteins. Brain cells die, giving the brain a spongelike appearance.
3. Misfolded prions refold all normal prions in the spleen and lymph nodes.
2. Prions are absorbed from small intestines into bloodstream.

Figure 11.16 Prion infection. After being ingested (1), misfolded prions move through the small intestine into the blood (2) and gain access to the spleen and lymph nodes. Inside these lymph organs, normal prions are refolded until all of the host's prions are refolded. All of the misfolded prions then move to the brain to refold host prions there. When there are very few normal prions left, the cells of the brain can no longer function normally.

proteins move to the brain and spinal cord, where they refold even more normal proteins (Figure 11.16). The lack of normal proteins damages the brain and eventually causes the patient to die.

Prions are proteins, not bacteria or viruses, and conventional treatments for infectious diseases will not prevent or cure the spongiform encephalopathies. Bacterial infections are usually treated with antibiotics, which have no effect on prions. Many viral and bacterial infections can be prevented through the use of vaccines given in an attempt to bolster the immune system prior to actual exposure to a pathogen. Vaccines are made of components of the disease-causing organisms—such as proteins from the plasma membrane of a bacterial cell, parts of a virus, or a whole virus that has been inactivated. The immune system responds to the introduced vaccine's challenge by producing the clonal population of memory cells that will be prepared for a real infection by the virus, should it occur. Some vaccines require multiple doses before a sufficient response is generated; others require booster shots to maintain protection.

An anti-prion agent would have to be able to determine which prions were properly folded and which were not. This makes treating spongiform encephalopathies very difficult. Because heat and radiation—generally effective in destroying bacteria and viruses—do not work against prions either, scientists hope to discover how to stabilize the normal prion protein so it will maintain its normal shape in brain cells. Until this happens, epidemiologists and other scientists are focusing on preventing the spread of the disease.

11.4 Preventing the Spread of Prion Diseases

WWW
Media Activity 11.7 Prions in Other Folding Protein Diseases?

Epidemiologists estimate that the number of people who had eaten products made from infected cows before they were removed from the food supply could be as high as 100,000. Sadly for these people, mostly citizens of the United Kingdom, there will probably be no cure. Fortunately, they will probably not infect those who take care of them, since prions are infectious but not contagious—there has never been a reported case of transmission between humans.

The fact that new variant CJD is only spread via ingestion of prion-contaminated foods makes containing the spread of the disease less difficult, but preventing an epidemic in the United States still presents a formidable challenge.

Efforts aimed at controlling the spread of the disease have focused on preventing contaminated animals and foods from entering other countries. In the United Kingdom (abbreviated UK, includes England, Scotland, Wales, and Northern Ireland), officials have slaughtered large numbers of cattle (Figure 11.17), and all of the cattle from farms where an infection has been diagnosed are destroyed. Veterinarians have received intensive training to increase their ability to recognize early signs of BSE. It is hoped that this will help prevent the slaughter (for meat) of diseased cows that are in the early stages of the disease but not yet showing the tell-tale symptoms. Finally, and maybe most importantly, the British government banned the practice of feeding cattle the meat-and-bone meal made from other ruminants (cloven-hoofed and horned animals such as cows, sheep, goats, and deer). These animals have stomachs that are divided into four compartments and chew a cud consisting of regurgitated, partially digested food. Except for mink, cats, and humans, prion diseases have chiefly infected ruminants. The British government has also implemented measures designed to prevent prion-harboring tissues, such as brains and spinal cords, from contaminating cattle meat during slaughter. Following the institution of these measures, Great Britain has experienced a decrease in the number of cattle with BSE from a peak incidence of 36,680 confirmed cases in 1992 to 2,254 confirmed cases in 1999.

In 1997, the U.S. Department of Agriculture (USDA) banned the import of live ruminants and most ruminant products—including meat, meat-and-bone

Figure 11.17 Elimination of infected cows. Cows diagnosed with mad cow disease are slaughtered and burned, along with all of the rest of the cows on the farm.

meal, offals (offals are waste parts from butchered animals, sometimes called *variety meats*), and glands—from all of Europe to prevent the spread of the disease to the United States. In addition, the Food and Drug Administration (FDA) prohibited the use of virtually all mammalian protein in the manufacture of animal feeds given to ruminants in the United States.

The incubation period, the time period from initial infection to first signs of a disease, for nvCJD can be as long as eight years. This means that people who ate prion-infected meat while in the UK or infected meat imported to the United States before the 1997 ban will probably become ill by 2005.

Almost all of the nvCJD patients worldwide had spent many years in the UK. To date, there has been only one case of human infection in the United States—that person was a UK citizen during the height of the BSE epidemic. The U.S. government has examined brain specimens from close to 15,000 indigenous cows and has found no evidence of prion disease, as diagnosed by spongelike brain tissue.

In light of all these preventative measures, the answer to whether or not mad cow disease—and its human counterpart nvCJD—will become an epidemic in the United States will be decided, in large part, by the level of stringency practiced by the makers of animal feed, the farmers responsible for feeding cows and observing their behavior, and the governmental regulatory agencies charged with protecting human health.

CHAPTER REVIEW

Summary

- Infectious diseases are usually caused by pathogens, such as bacteria and viruses.
- Bacteria are single-celled organisms with DNA genomes that cause disease by using host resources to rapidly reproduce, and by releasing toxins into the host.
- Viruses can only reproduce inside host cells. They are composed of nucleic acid and protein. They cause disease by using host-cell resources, and by destroying host cells as a part of their infectious cycle.
- Prions are novel proteinaceous infectious agents that lack a genome. They cause disease by refolding the host cell's normally occurring prions. Prions are found in high concentrations in the brain and spinal cord. Once misfolded, the host cell's prions are unable to perform their normal functions.
- Misfolded prions can arise spontaneously when a normally occurring prion mistakenly assumes the misfolded shape, or as the result of a mutation to the gene that encodes the prion protein.
- Misfolded prions can also arise as a result of ingesting misfolded proteins from a diseased organism. Once inside the body, the misfolded prions refold normal proteins so that they assume the disease-causing shape.
- An epidemic is a contagious disease that spreads rapidly and extensively among a population. The spread of infectious disease results from physical contact with the disease-causing organism.
- The immune system fights most disease-causing infections.
- B and T cells carry receptors for millions of different antigens. These receptors are produced by rearranging the coding segments of DNA.
- The humoral response of the immune system involves B cells, which divide to produce cells that carry and secrete antibodies when exposed to an antigen.
- The cell-mediated response of the immune system involves T cells, which become specialized into different cell types upon exposure to an antigen. These cells speed up the immune response, destroy virus-infected cells, and boost the response of the B cells.
- Prions do not evoke an immune response because the immune system recognizes them as self.
- Prion diseases cannot be treated by antibiotics or prevented by vaccines and new approaches for dealing with this unique type of infectious disease must be found.

Key Terms

antibody p. 289
antigen p. 289
antigen receptor p. 289
autoimmune disease p. 292
B lymphocyte (B cell) p. 289
bacteria p. 273
binary fission p. 273
capsid p. 278
capsule p. 273
cell-mediated immunity p. 292
cell wall p. 273
clonal population p. 293
complement protein p. 293
contagious p. 272
cytotoxic T cell p. 293
epidemic p. 281
epidemiologist p. 281
flagella p. 273
helper T cell p. 293
humoral immunity p. 292
immune response p. 290
infectious p. 272
latent p. 279
lymphocyte p. 288
macrophage p. 288
memory cell p. 293
microbe p. 272
pathogen p. 272
phagocytosis p. 288
plasma cell p. 293
prion p. 279
reverse transcriptase p. 278
spores p. 277
T-cell receptor p. 289
T lymphocyte (T cell) p. 289
toxin p. 276
viral envelope p. 278
virus p. 277

Learning the Basics

1. Describe the structure of a typical bacterium and a typical virus.

2. Describe how bacteria and viruses reproduce.

3. What roles do B cells and T cells have in the immune response?

4. How do B cells and T cells produce their receptors?

5. How do prions cause spongiform encephalopathies?

6. The immune system _____.
   a. has a gene for each antigen
   b. undergoes genetic rearrangement in response to different antigens
   c. is able to make many antigen receptors by rearranging DNA of immune cells
   d. can always devise an antibody that will bind to an antigen

7. Autoimmune diseases result when _____.
   a. a person's endocrine system malfunctions
   b. liver enzymes malfunction
   c. B cells attack T cells
   d. the immune system fails to differentiate between self and non-self cells

8. Which of the following cell types divides to produce cells that make antibodies?
   a. helper T cells
   b. B cells
   c. cytotoxic T cells
   d. all of the above

9. Helper T cells secrete substances that _____.
   a. help prevent leukemia
   b. prevent bacteria from entering cells
   c. boost B cell and cytotoxic T cell response
   d. inhibit reverse transcriptase
   e. stimulate the thymus to make more B cells

10. Blood cells are produced in the _____.
    a. heart
    b. bone marrow
    c. bloodstream
    d. capillaries

11. The immune system can recognize a virus you have been exposed to once because _____.
    a. you harbor the virus for many years
    b. we have genes to combat every type of virus
    c. a cell that makes receptors to a virus multiplies upon exposure to it and produces memory cells
    d. a copy of the viral genome is inserted into a memory cell

12. Which of the following is a false statement about sexually transmitted diseases?
    a. Chlamydia and gonorrhea, if left untreated, can cause infertility in females.
    b. Genital warts are prevented by the use of a condom.
    c. The virus that causes genital warts can cause cervical cancer.
    d. Yeast infections are caused by a fungus.

Analyzing and Applying the Basics

1. How do genetic diseases and infectious diseases differ?

2. Why do you need a flu shot every year, but only one inoculation against some diseases?

3. Why is your immune system usually more effective at fighting off infection the second time you are exposed to a pathogen?

4. Why might one person die of an infectious disease while another recovers?

5. How do prions differ from other infectious agents?

Connecting the Science

1. Some parents refuse to have their children vaccinated. Should these parents be forced to allow their children to be vaccinated to protect society against the reemergence of once-deadly killers such as polio? Defend your answer.

2. Should people with infectious diseases be quarantined? Why or why not?

3. How might changes in the environment increase the likelihood of the emergence of novel infectious diseases?

4. Edward Jenner intentionally exposed a child to smallpox in order to test his hypothesis that the child had already acquired immunity via exposure to cowpox virus. Why might Jenner have felt this was a necessary risk? Would such treatment be ethical? Why or why not?

Media Activities

Media Activity 11.1 The Immune System and Infectious Agents
Estimated Time: 15 minutes
Explore the diversity of infectious agents, their mode of reproduction or duplication, and how the human body defends against these agents.

Media Activity 11.2 The Revenge of the Chickenpox: The Virus Strikes Back
Estimated Time: 15 minutes
The infectious agent responsible for chickenpox, the *varicella-zoster* virus, sometimes causes shingles in older people, because the virus initially evades the immune system and then hides in the nerves in a latent form. You will explore information on what shingles is, how it occurs, and what can be done to treat the affected patient.

Media Activity 11.3 New Antidote for the Anthrax Bacterium
Estimated Time: 10 minutes
The disease anthrax gained renewed attention when it was utilized as a bioterrorism tool in the United States. New studies on the disease have identified a possible antidote as well as a detection method for the agent.

Media Activity 11.4 To Vaccinate or Not? What Would Happen If We Stopped?
Estimated Time: 10 minutes
What would happen if we stopped a vaccinating program against a disease like measles? What would be the effect on the rates of disease occurrence, the impact on families, and the financial costs of health care needed for infected individuals? This activity explores the potential effects of halting a current vaccination program.

Media Activity 11.5 A New Epidemic of Syphilis?
Estimated Time: 10 minutes
The number of cases of syphilis in the United States had been steadily declining since the 1940s, but there has been a recent increase in numbers of affected individuals. In this activity, you will explore why epidemiologists are concerned about the higher rates of incidence.

Media Activity 11.6 Gene Therapy Trials for Immune Deficiency
Estimated Time: 15 minutes
One strategy to help patients with immune system deficiencies is to use gene therapy. Retroviruses are used to deliver "good" genes into the host cells of a patient to correct their "bad" or missing genes. There are risks associated with these types of therapy, as we will explore in this activity.

Media Activity 11.7 Prions in Other Folding Protein Diseases?
Estimated Time: 15 minutes
Prions have been identified in a variety of spongiform encephalopathies, but now researchers are interested in how the concept of the misfolded protein may apply to other diseases. This activity explores some of these folding protein diseases.

CHAPTER 12

Gender and Athleticism

Developmental Biology, Reproductive Anatomy, and Endocrinology

Many young athletes dream of playing professional sports.

12.1 The Origin of Biological Sex Differences

12.2 Sex Differences that Do Not Affect Athleticism

12.3 Sex Differences that Can Affect Athleticism

12.4 Culture Affects Athleticism

Most professionals get their start playing on paid minor league teams.

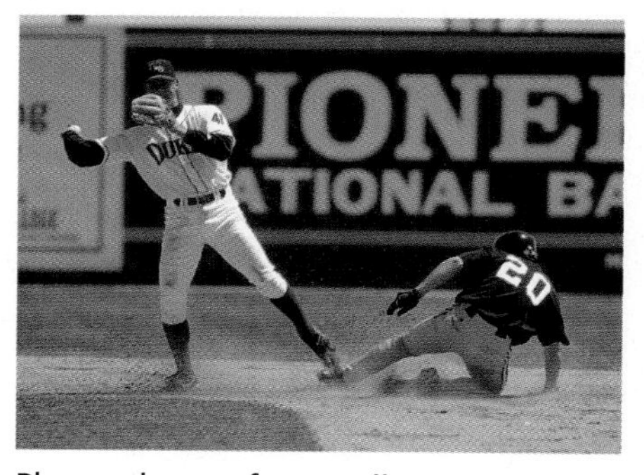

Players that perform well may get a shot at the big leagues.

The scene is one that young athletes dream about on baseball fields all over the world. For one young baseball pitcher, the dream has come true. On a warm Friday evening late in July, the sun is beginning to set over left field in a beautiful, open-air, brick baseball stadium. On the mound, a 23-year-old lefthander stands motionless, squinting to better see the catcher's signals.

The scoreboard above the centerfield wall shows that it is the top of the sixth inning, and the Duluth-Superior Dukes are leading the Sioux Falls Canaries by the score of two to zero. On the field and in each dugout are young, intense, and hopeful baseball players, all of whom have signed professional contracts, each of them dreaming of one day playing major-league baseball.

The Dukes pitcher has only given up three hits this game, but the Canaries are threatening, with one runner on base and the number three hitting center fielder at bat. Two weeks before this game, this pitcher and batter had faced each other during the pitcher's first professional start and the batter had hit a solo home run that put the Canaries ahead for good.

The pitcher glances at the runner on second base and turns to face the batter. In one strong, fluid motion, the pitcher rears back and hurls the ball

Ila Borders is the first woman to win a professional baseball game.

across the inside corner of home plate. "Strike three. The batter is out," yells the home-plate umpire.

The stadium erupts as the fans jump to their feet cheering wildly, and the 5′-10″ pitcher walks off the mound toward the dugout to a standing ovation. For the next half inning, the hometown fans chant the pitcher's name in unison, until the southpaw finally comes out of the dugout to salute the cheering fans. When the pitcher raises the black and purple Dukes baseball cap toward the fans behind home plate, her long hair spills across the back of her uniform, obscuring her name and number.

The pitcher is Ila Borders, the first woman in the starting lineup of a men's professional baseball game and, on that warm July evening, the first woman to win a men's professional baseball game.

This accomplishment was so impressive that *Sports Illustrated* honored her as one of the top 100 female athletes in history, the uniform and glove that served her in this historic victory adorn the prestigious Baseball Hall of Fame in Cooperstown, New York, and national television stations ESPN and CNN broadcast footage of her pitching that night.

Will Ila Borders be the first of many women to play major league baseball? Will women infiltrate professional football and hockey leagues next? Will men and women someday race against each other in track and field or swimming events? Is the current disparity in athletic performance between men and women more about differing opportunities for males and females than about physical differences? Or, are the bodies of women and men simply constructed so differently that they will only rarely be able to compete against each other?

To answer these questions, we need to delve into the study of sex differences—average differences between males and females. They can be biological, such as differences in reproductive anatomy, or cultural, such as differences in dress or hair style.

In this chapter, we will first examine sex differences that are biological in origin, then look at how some of these differences affect sports performance, and finally we will turn our attention to the ways in which gender socialization and biological differences interact to determine athleticism.

WWW
Media Activity 12.1A Hormones and Gamete Production

12.1 The Origin of Biological Sex Differences

Many biological sex differences are caused by the two sexes having different levels of hormones. **Hormones** are substances that travel through the circulatory system and act as signals to turn on genes in their target organs. Hormones that circulate throughout the body, along with the organs and glands that produce them, make up the **endocrine system.**

The Endocrine System

The endocrine system is an internal system of communication involving hormones, the glands that secrete them, and the target cells that respond to them.

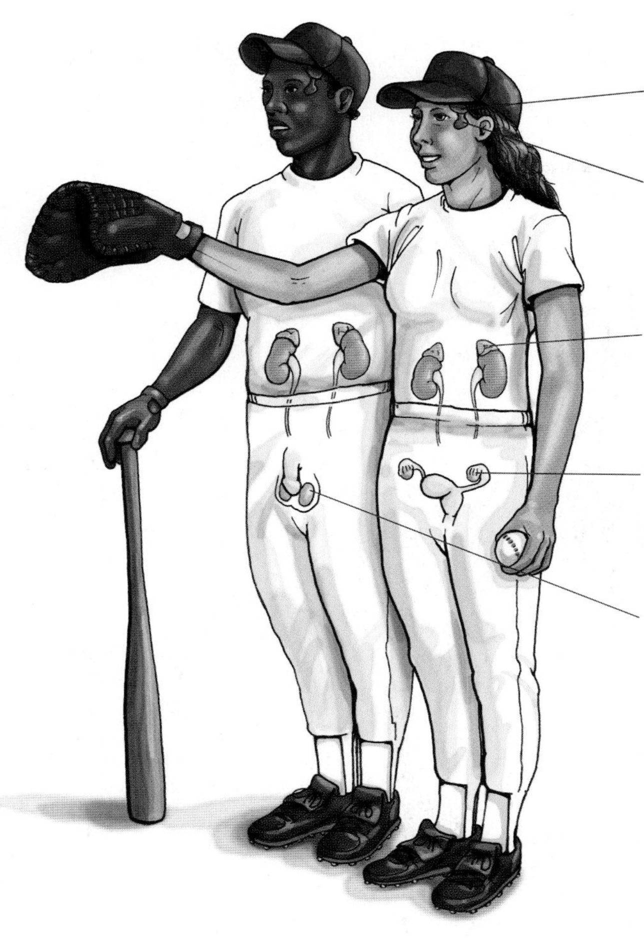
Hypothalamus
Secretes gonadotropin-releasing hormone (GnRH).
Pituitary gland
Responds to GnRH by secreting the pituitary gonodatropins follicle-stimulating hormone (FSH) and luteinizing hormone (LH).
Adrenal glands
Secrete adrenaline, corticosteroids, testosterone (masculinizing hormone), and estrogen (feminizing hormone).
Ovaries
Respond to FSH and LH by secreting **estrogen,** which regulates menstruation, maturation of egg cells, breast development, pregnancy, and menopause.
Testes
Respond to FSH and LH by secreting **testosterone,** which aids in sperm production, increased muscle mass, and voice deepening.

Figure 12.1 Endocrine organs involved in the production of sex differences. There are many organs in the endocrine system. Those involved in the production of sex differences include the hypothalamus and pituitary gland of the brain, the adrenal glands that sit on the kidneys, the testes in males, and the ovaries in females.

Many organs are involved in the endocrine system. Five of these organs are involved in the production of biological sex differences: the *hypothalamus*, the *pituitary gland*, the *adrenal glands*, the *ovaries*, and the *testes* (Figure 12.1).

Hypothalamus and Pituitary Gland The **hypothalamus**, located deep inside the brain, regulates body temperature and affects behaviors such as hunger, thirst, and reproduction. In the reproductive system, the hypothalamus secretes a hormone that stimulates the activities of the **gonads** (testes or ovaries) called **gonadotropin-releasing hormone (GnRH)**. GnRH moves through a complex of veins routed directly to the **pituitary gland** located at the base of the skull. Once it reaches the pituitary gland, GnRH stimulates the synthesis and release of **pituitary gonadotropins**, **follicle-stimulating hormone (FSH)**, and **luteinizing hormone (LH)**. In males, these two hormones are involved in the production of sperm; in females, they help regulate ovulation and menstruation.

Adrenal Glands An adrenal gland sits atop each kidney. These glands secrete the hormone *adrenaline* in response to stress or excitement, but they also secrete **androgens**, masculinizing hormones such as **testosterone**, as well as **estrogens**, which are feminizing hormones. Because estrogen and testosterone are responsible for many of the anatomical sex differences in males and females, these hormones are called **sex hormones**.

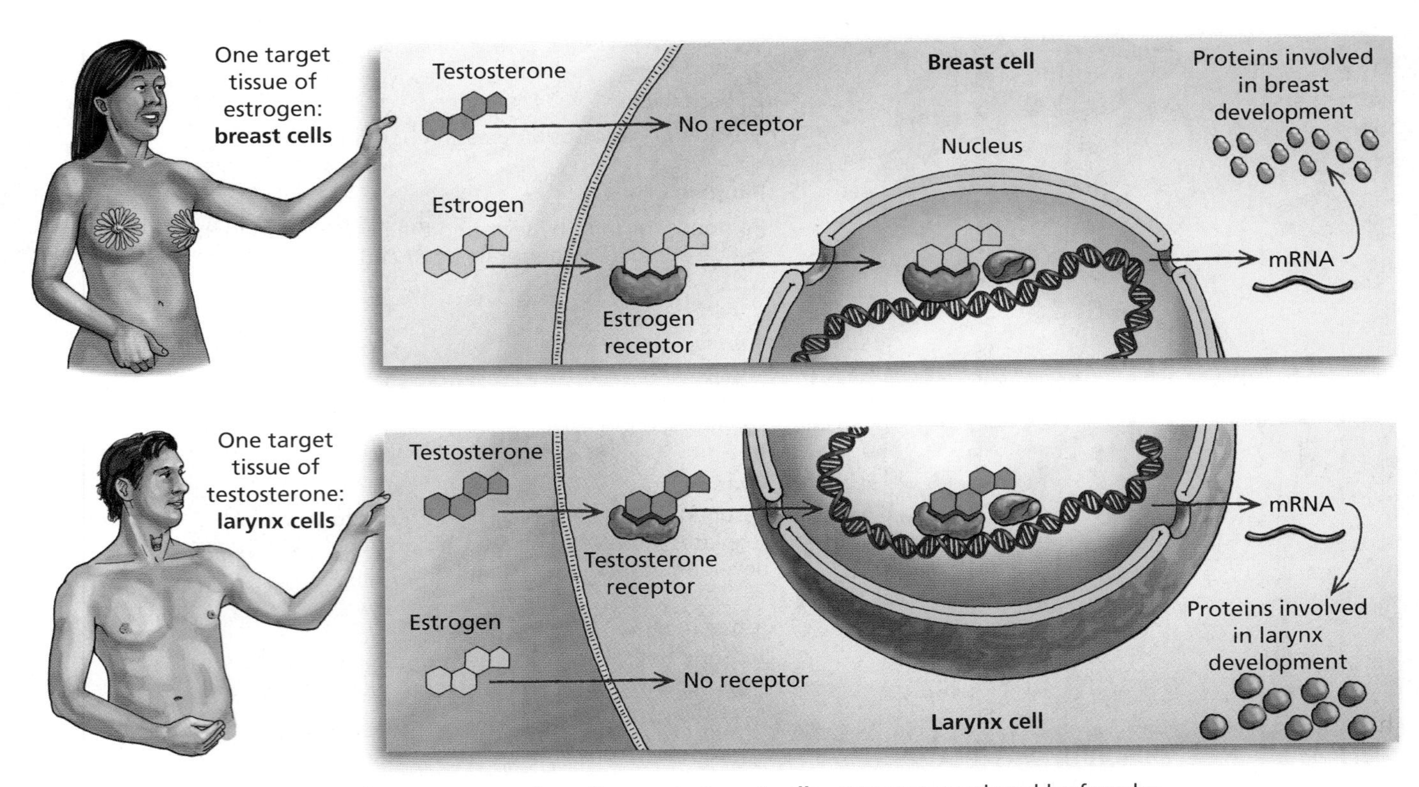

Figure 12.2 Hormones act to increase expression of genes in target cells. Estrogen produced by females enters the cells that compose the breast tissue, where it binds to its receptor. The estrogen and its receptor then bind to DNA to increase the expression (transcription) of genes required for the growth and development of the breast at puberty. Males have lower levels of estrogen and fewer estrogen receptors in their breast tissue so the breasts do not develop to the extent that they do in females. In males, testosterone enters the cells of the larynx and finds receptors in the cells of this tissue that increase expression (transcription) of proteins that change the structure of the larynx, changing the sounds it produces, resulting in voice deepening.

WWW
Media Activity 12.2 Anabolic Steroids: Can They Bridge the Gender Gap?

Sex hormones are *steroid hormones,* because their synthesis requires the steroid cholesterol as a precursor or starting material. Steroid hormones are fat soluble and move across cell membranes readily. However, they act only on *target cells* that contain receptors for them. Once a sex hormone passes into the cell and binds to a receptor, the hormone and receptor move together to the cell nucleus and either enhance or inhibit the expression of different genes (Figure 12.2).

The cells of the outer covering of each adrenal gland synthesize sex hormones from cholesterol. While cholesterol can be synthesized in many body tissues, it is only converted into sex hormones in the adrenal glands, testes, and ovaries. Thus, the adrenal glands of both males and females secrete small amounts of both testosterone and estrogen.

Adrenal glands also secrete *corticosteroids*. Corticosteroids help the body deal with stress and promote the synthesis of glucose from noncarbohydrate sources, such as proteins. Due to corticosteroids, muscle proteins are broken down to provide fuel to the body during intense workouts. Some athletes take steroid hormones to counteract the corticosteroids, which allows them to train harder for longer periods of time. *Anabolic steroids* are hormones that build up muscle. They can be naturally occuring, like testosterone, or synthetic versions of testosterone or its precursors, like androstenedione. The synthetic versions enable the body to bounce back faster after intense training when taken orally or injected, although at huge physical costs—leading to, or causing, sterility, increased aggressiveness, paranoia, heart and liver disease, and kidney damage in both males and females. Female users often have excessive facial and body hair, deepened voices, irregular menstrual cycles, infertility, enlarged genitalia, and diminished breast size. Male users can become bald, have enlarged breasts, and decreased sperm production, leading to infertility. Their use is illegal because of the physical damage they cause.

www
Media Activity 12.3 Infertility Research

Testes The paired testes are oval organs suspended in the scrotum. Testes secrete *testosterone*, the hormone that aids in sperm production, hair thickness and distribution, increased muscle mass, and voice deepening. Beginning at puberty, sperm are produced by the cells of the testes. Sperm production is most efficient at temperatures that are lower than body temperature; therefore the testes are kept outside the body cavity in the scrotum.

Ovaries The paired ovaries are about the size and shape of almonds in the shell. They produce and secrete estrogen. Estrogen regulates many functions in the female body, including menstruation, the maturation of egg cells, breast development, pregnancy, and the cessation of menstruation after reproductive age called *menopause*. Inside the ovaries are all of the cells that can mature into the egg cells that will be ovulated. The production of egg cells begins while a female is *in utero*, pauses at birth, resumes at puberty, and continues until menopause.

By the time the endocrine system begins to produce hormones, a person has already been programmed to secrete either mostly androgens or mostly estrogens. This programming occurs during development inside the mother's uterus.

Sex Differences That Arise During Development

The hypothalamus, pituitary gland, and adrenal glands of both males and females produce the same hormones in similar amounts—it is the hormones secreted by the ovaries and testes that determine biological sex differences. Once a fetus develops either testes or ovaries they produce hormones that influence the development of other structures involved in reproduction.

Sex differences in reproductive organs are the result of the developmental pathways followed by males and females. These pathways lead to different structures for making sperm and egg cells, called gonads; different structures for carrying the sperm and egg cells, called **ducts**; and different external reproductive structures, called **genitalia**.

Gonad Differentiation During the first seven weeks of development, a male and female fetus are indistinguishable unless one looks at their chromosomes. The embryonic gonads are composed of two masses that can become either the paired ovaries of a female or the paired testes of a male. It is only when the cells of the embryonic gonads become specialized during development, or **differentiate**, that development in females and males begins to diverge.

For undifferentiated gonads to develop into testes, a gene on the Y chromosome, the ***SRY* gene**, must be present. The *SRY* (Sex-determining Region of the Y chromosome) gene encodes a protein that causes expression of the genes required for testicular development.

While less is known about ovarian development, it is becoming clear that many genes must be expressed for ovarian development to occur. These genes, some of which are called the ***Gpbox* genes**, seem to act in a similar manner to the *SRY* gene by turning on other genes involved in the production of ovaries—these genes would be expressed only in the absence of the *SRY* gene.

Males, with their one X and one Y chromosome, express the *SRY* gene and not the *Gpbox* genes. Since females have two X chromosomes and no Y chromosome,

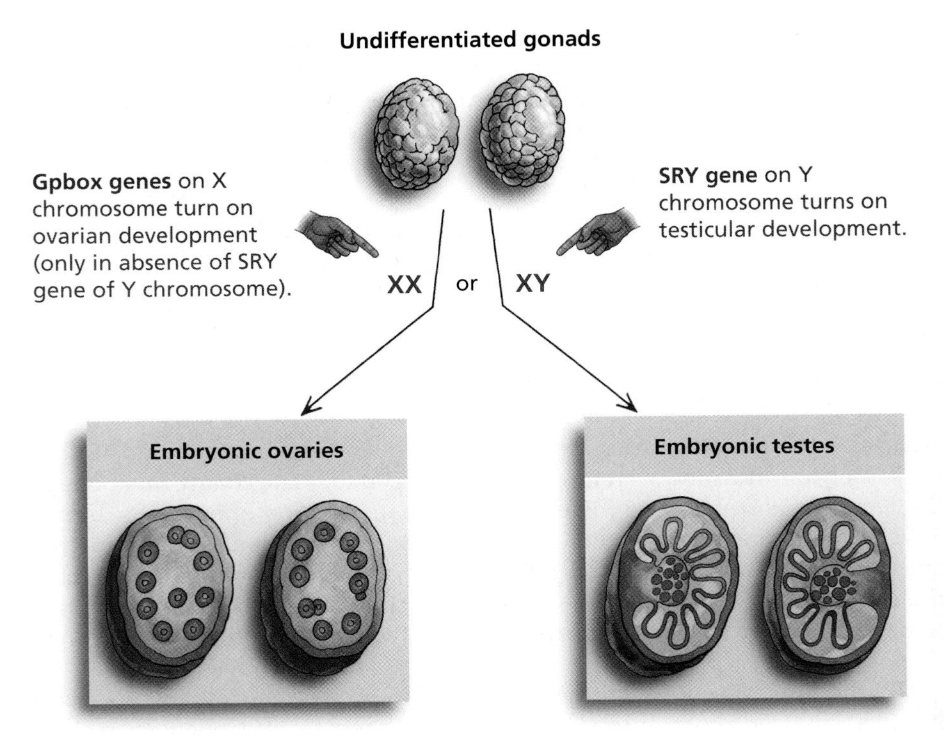
Undifferentiated gonads
Gpbox genes on X chromosome turn on ovarian development (only in absence of SRY gene of Y chromosome).
SRY gene on Y chromosome turns on testicular development.
XX
or
XY
Embryonic ovaries
Embryonic testes

Figure 12.3 Genes involved gonadal differentiation. When the Y-linked *SRY* gene is expressed, testes develop; when the *SRY* gene is absent, *Gpbox* genes are expressed and ovaries develop.

they express the *Gpbox* genes and not the *SRY* gene (Figure 12.3). The expression of these genes leads to the differentiation of the male and female gonads.

Once differentiated, the gonads produce their sex-specific hormones. The cells of the embryonic testes make testosterone, which directs the development of the internal duct systems and formation of external genitalia. Once the embryonic external genitalia have been masculinized, the testes will not secrete testosterone until puberty. The cells of the ovaries do not produce estrogen during development. Estrogen production occurs only after puberty.

Differentiation of the Duct System The ducts of the reproductive system are the structures that carry the sperm and egg cells, or gametes. In males these sperm-carrying ducts include the **epididymis**, **vas deferens**, and **urethra**, in females, egg cells travel through the **oviducts**, **uterus** and **cervix**, and **vagina** (Figure 12.4).

Prior to differentiation, the embryonic duct system consists of two separate sets of tubes that lie side-by-side. For each sex, one duct system stays in place, or persists, and the other degenerates or regresses. These embryonic duct systems are called the **Müllerian duct system** and **Wolffian duct system**.

For male duct development, the Müllerian duct system regresses and the Wolffian duct system persists. Wolffian duct development occurs when testosterone is present. The testes also produce a hormone called **anti-Müllerian hormone** that causes regression of the Müllerian ducts.

Female duct development is stimulated by the presence of gene products from many different genes. When these genes are expressed in the developing ovary, the female pathway is followed, thus preventing testosterone synthesis and secretion. Since the Wolffian duct system requires testosterone to stimulate its development, this structure regresses in females.

External Genitalia Differentiation In male and female embryos, one embryonic structure can be molded into either the penis and scrotum of the male external genitalia or the clitoris and vulva of the female external genitalia.

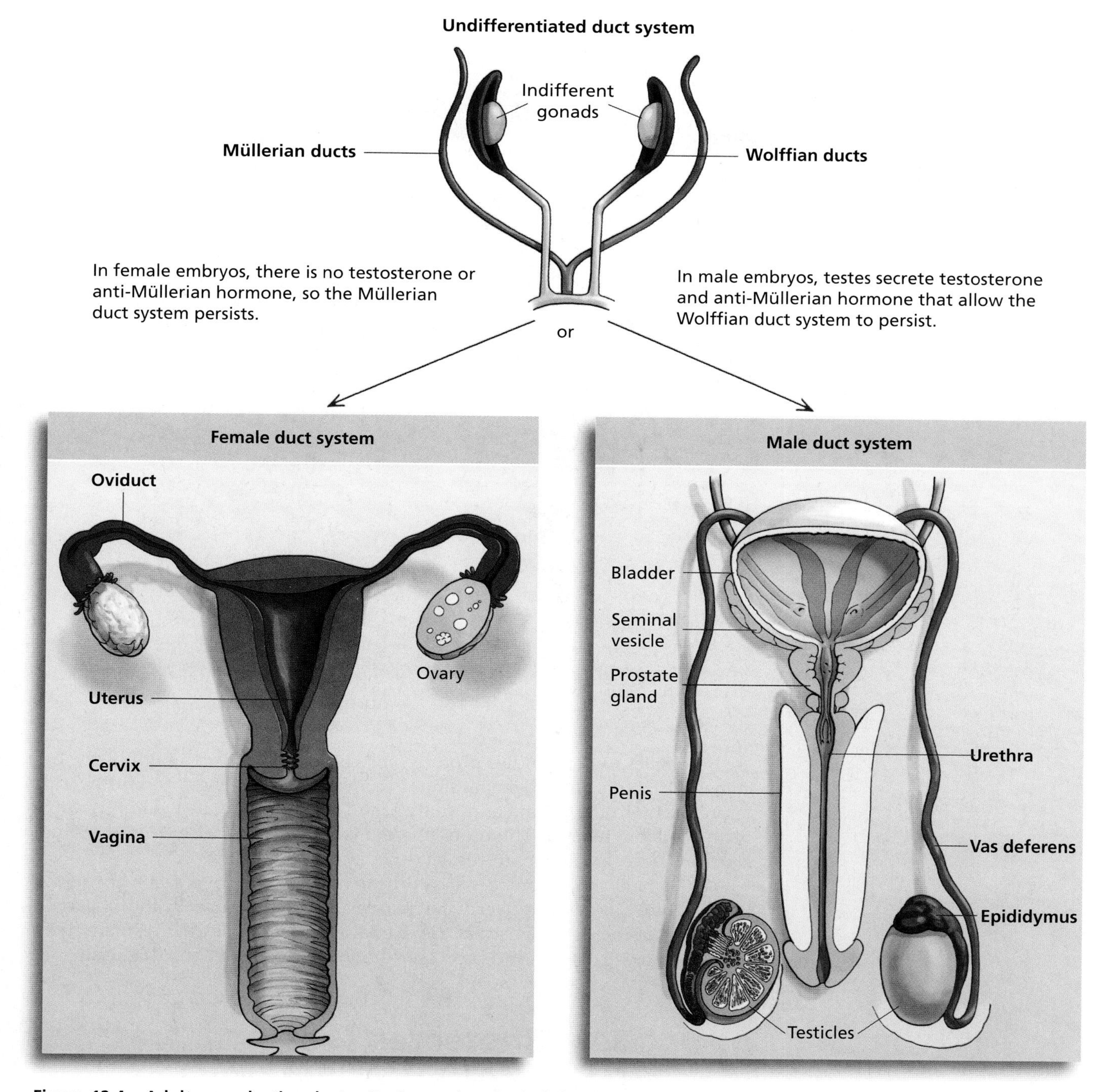

Figure 12.4 Adult reproductive ducts. During embryological development, the two separate duct structures exist side-by-side in the abdomen of the embryo. In an XY male, development of testes and secretion of testosterone and anti-Müllerian hormone allows the Wolffian duct system to persist. In female embryos, there is no testosterone or anti-Müllerian hormone and the Müllerian duct system persists.

(Figure 12.5). Thus, it is said to be a *bipotential* structure; this structure consists of a genital tubercle, a urogenital sinus, and two labioscrotal swellings.

In the presence of dihydroxytestosterone (DHT), an androgen formed from testosterone, the genital tubercle forms the penis, the urogenital sinus fuses to form the urethra, and the labioscrotal swellings fuse to form the scrotum.

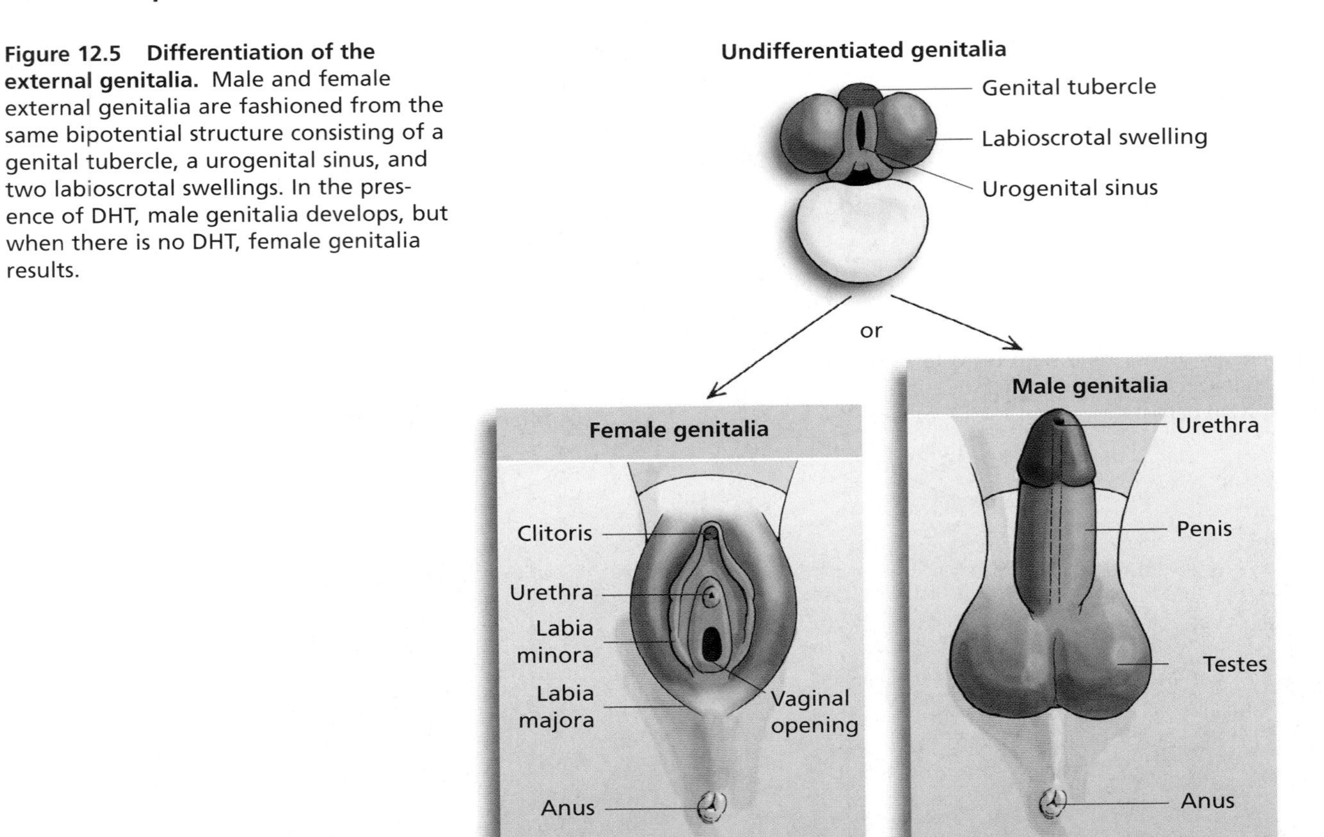

Figure 12.5 Differentiation of the external genitalia. Male and female external genitalia are fashioned from the same bipotential structure consisting of a genital tubercle, a urogenital sinus, and two labioscrotal swellings. In the presence of DHT, male genitalia develops, but when there is no DHT, female genitalia results.

In females, whose bodies produce very little DHT, the genital tubercle becomes the clitoris, the urogenital sinus becomes the labia minora and further differentiates into the vagina and urethra, and the labioscrotal swellings fold to form the labia majora. Therefore, the penis and clitoris arise from the same starting tissue, as do the scrotum and vulva.

While many sex differences arise during development, it is when hormones begin to be expressed at puberty that real sex differences arise. In addition, only a fraction of the hormones that are expressed affect athleticism—many hormones act on structures and processes that have no impact on athleticism.

12.2 Sex Differences That Do Not Affect Athleticism

Puberty marks the beginning of sperm production in males and the beginning of egg-cell maturation and menstruation in females. Boys typically begin puberty around age 13. Puberty in males includes the enlargement of the penis and testes, an overall growth spurt, the growth of muscles and skeleton resulting in wide shoulders and narrow hips, and changes in hair growth including pubic, underarm, chest, and facial hair. In additional, the larynx and vocal cords lengthen and enlarge to produce a deeper voice.

For girls, the first signs of puberty occur around age 11. These signs include breast development, increased fat deposition, a growth spurt, pubic and underarm hair growth, and the commencement of menstruation.

Whether male or female, as the brain matures the hypothalamus begins to secrete GnRH. GnRH secretion causes the pituitary gland to secrete FSH and

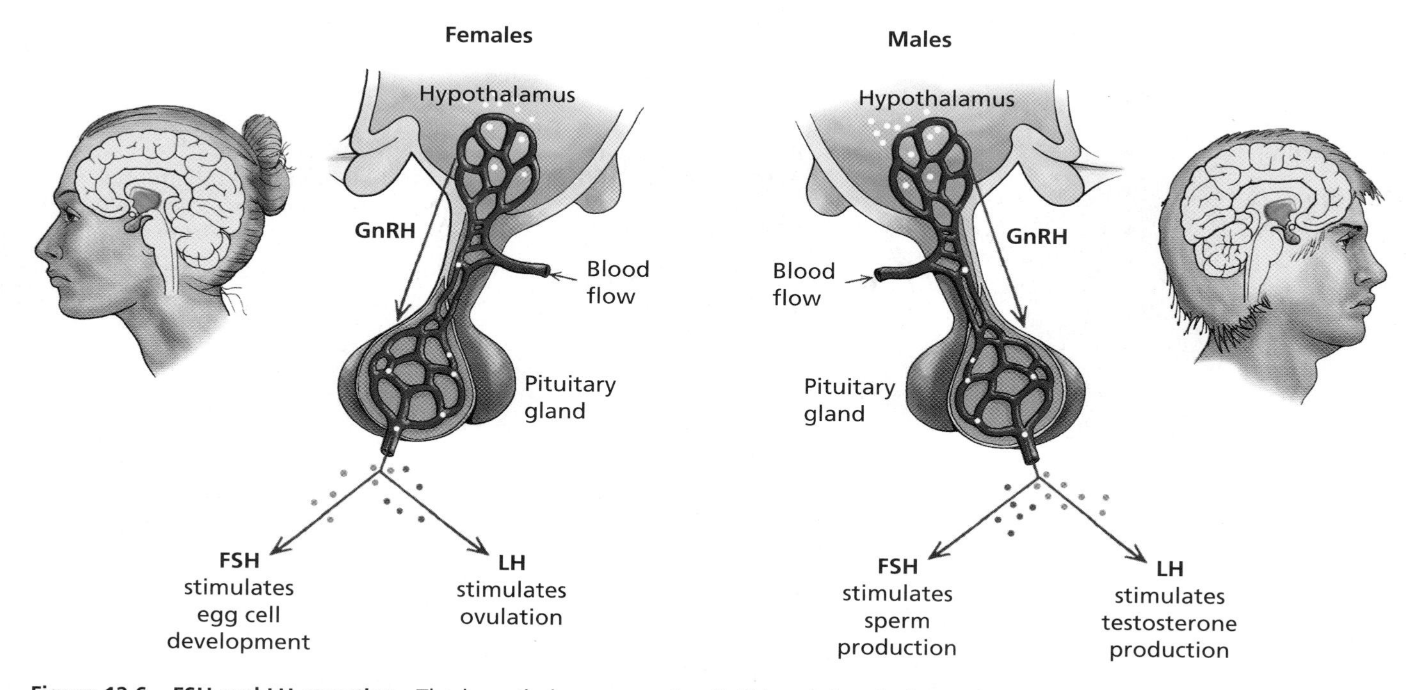

Figure 12.6 FSH and LH secretion. The hypothalamus secretes GnRH and the pituitary gland secretes FSH and LH. FSH and LH act on the testes of a male, resulting in sperm and testosterone production, but in females they stimulate the ovaries to produce egg cells.

LH. In males, FSH stimulates sperm production and LH stimulates testosterone production. In females, FSH stimulates egg-cell development and LH stimulates ovulation (Figure 12.6).

Producing Sperm Cells

Spermatogenesis, the production of sperm, begins to occur in the testes at puberty. The testes consist of many highly coiled tubes surrounded by connective tissues. These tubes are the **seminiferous tubules**, where sperm are formed. The **Leydig cells** that are interspersed between the seminiferous tubules produce testosterone and other androgens (Figure 12.7a).

Cells that will undergo meiosis to produce sperm line the walls of the seminiferous tubules. Recall from Chapter 6 that meiosis is cell division in which one parent cell divides its chromosomes into four daughter cells. The parent cell first duplicates each chromosome and then separates the pairs of chromosomes, called *homologous pairs of chromosomes*, from each other.

During spermatogenesis, each parent cell first duplicates by mitosis (Chapter 5), and then one of the two daughter cells undergoes meiosis. Since each round of mitosis produces two cells, only one of which goes on to perform meiosis, a man will never exhaust the supply of cells that can be used to make sperm cells.

The diploid cell (*diploid* means two sets of chromosomes, or 46 chromosomes in humans) that begins meiosis is called a **primary spermatocyte**. After the first meiotic division, which separates the 23 homologous pairs of chromosomes, the cell is called a **secondary spermatocyte**. Secondary spermatocytes, with their 23 duplicated but unpaired chromosomes, are haploid (*haploid* means one set of chromosomes). These cells then undergo meiosis II to produce haploid cells that no longer have duplicated chromosomes, called **spermatids**.

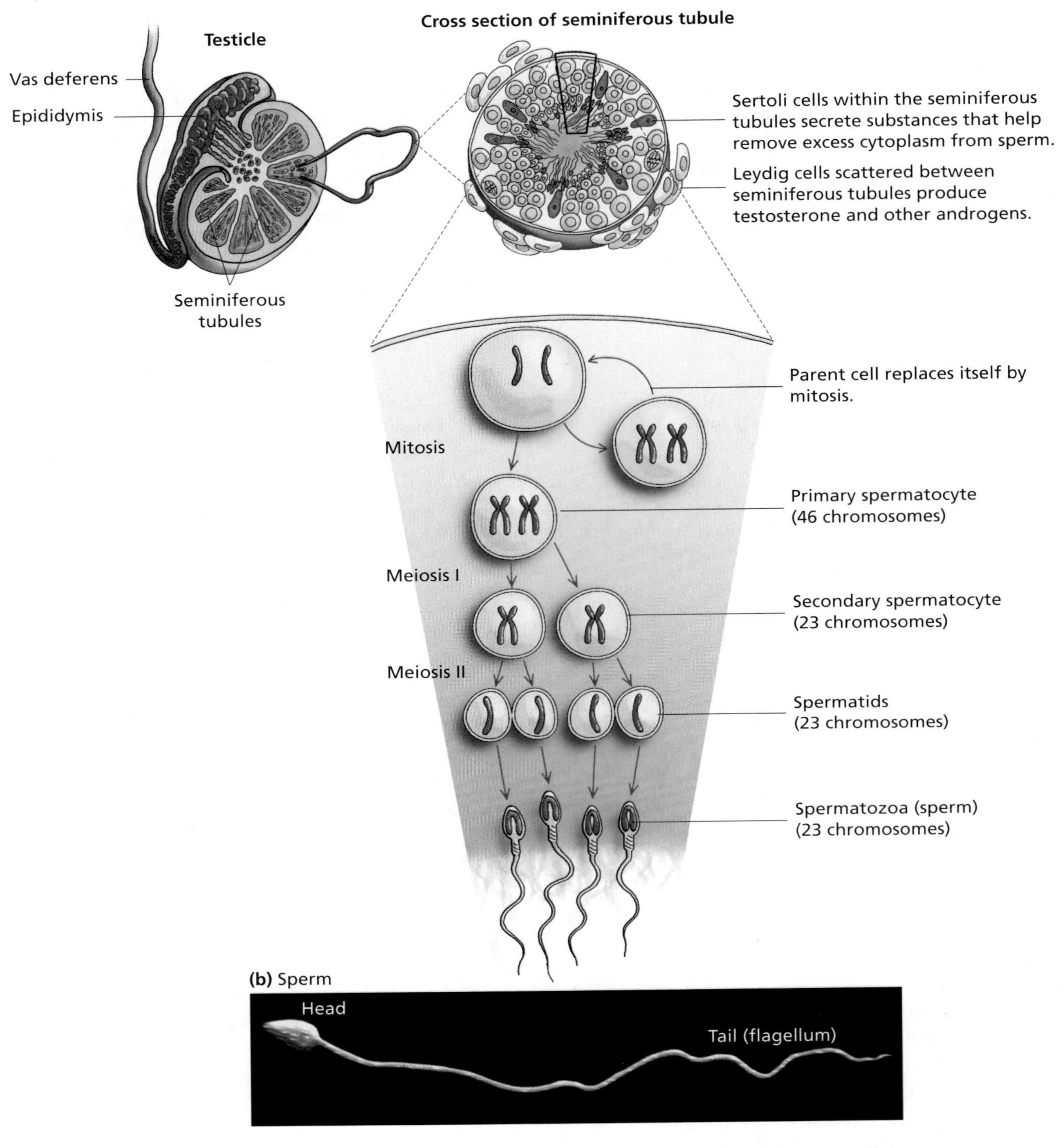

Figure 12.7 Spermatogenesis. (a) Sperm are produced in the seminiferous tubules of each testicle. Cells in the testes first undergo a round of mitosis producing two identical daughter cells with 46 chromosomes. The primary spermatocyte begins meiosis with 46 chromosomes; after meiosis I the two resultant daughter cells are called *secondary spermatocytes.* Meiosis II results in the production of four haploid daughter cells containing 23 chromosomes, called *spermatids*. Spermatids then modify to become mature spermatozoa. (b) At the tip of a spermatozoa is the acrosomal vesicle; the head of the sperm contains the DNA. The mid-piece contains mitochondria to provide energy for the flagellum.

Cells that aid the developing sperm, called **Sertoli cells**, are also located in the seminiferous tubules. These cells secrete the substances required for further sperm development and convert the spermatids into **spermatozoa** (sperm) by removing some of their excess cytoplasm.

The mature sperm is composed of a small head containing the DNA, a midpiece that has mitochondria inside to provide energy for the journey to the oviduct, and a tail (flagellum) to propel the sperm (Figure 12.7b). At the tip of the sperm's head lies the *acrosomal vesicle* derived from the cell's Golgi bodies (recall from Chapter 2 that the Golgi is the cell's internal protein-sorting center). The acrosomal vesicle is full of enzymes that help a sperm cell gain access to the egg cell's nucleus.

After sperm are produced in the seminiferous tubules, they move into a long, coiled tube called the *epididymis*, where the sperm gain the ability to move themselves by using their flagellum. During ejaculation, the sperm are propelled from the epididymis to the vas deferens, a long tube that joins with the ejaculatory duct at a gland called the **seminal vesicle**. The seminal vesicle produces a fructose-rich fluid that helps supply the sperm with energy they will need for propelling themselves. Sperm and these associated fluids are called **semen**. The sperm then move along the ejaculatory duct and pass through the prostate gland located at the base of the bladder. Within the prostate gland, the ejaculatory duct merges with the urethra coming from the urinary bladder. At ejaculation, the urethra carries the semen out of the body through the penis.

It takes about 70 days for a mature sperm to be produced. Males start making sperm at puberty and continue to do so, albeit at a slowing rate, until very old age.

Producing Egg Cells

Oogenesis, the formation and development of female gametes, occurs in the ovaries and results in the production of egg cells (Figure 12.8a). A small percentage of these egg cells will be ovulated, and an even smaller percentage may be fertilized. While spermatogenesis begins at puberty, oogenesis actually begins while the female is still *in utero*, then pauses until puberty when it continues each month until menopause. Amazingly, the egg cell that helped produce you started developing inside your mother while she was developing inside your grandmother's uterus!

Developing egg cells, called **oogonia**, begin meiosis but pause at prophase I, at which stage they are called **primary oocytes**. Each primary oocyte is surrounded by a single layer of flattened cells called **follicle cells**. The primary oocyte surrounded by the follicle cells is called the **primary follicle**. A female produces around 2 million of these potential egg cells prior to her birth, but some of these cells degenerate until she has about 700,000 at birth and around 350,000 by puberty.

When maturation of egg cells commences at puberty, the follicle cells begin dividing, forming layers and secreting estrogen. During the *menstrual cycle*, one primary follicle continues its development from prophase I to metaphase II of meiosis in the ovary. About 12 hours after this **secondary oocyte** forms, it is expelled from the ovary—the release of an egg cell from the ovary is called **ovulation**. The secondary oocyte ovulates around the fourteenth day of the menstrual cycle. It takes about three days for the secondary oocyte, an object about the size of a pin head, to move from the ovary through the oviduct, uterus, and cervix, exiting the body through the vagina. If there are sperm present in the oviduct and the secondary oocyte is fertilized, meiosis resumes from metaphase II. Women are fertile, or able to produce healthy gametes, only a few days per month. A woman can diagnose the fertile time period if she learns to recognize when she is ovulating (Essay 12.1).

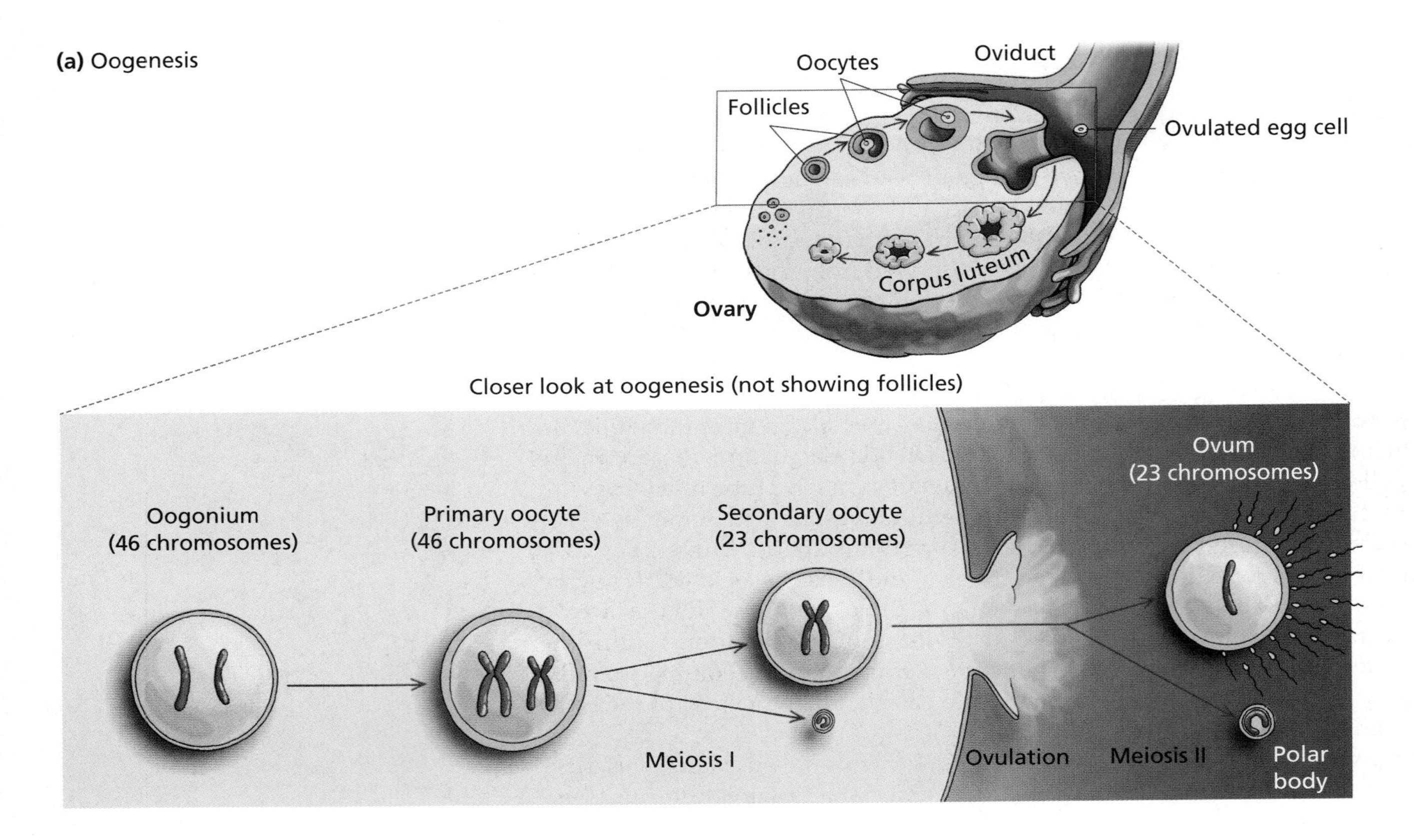

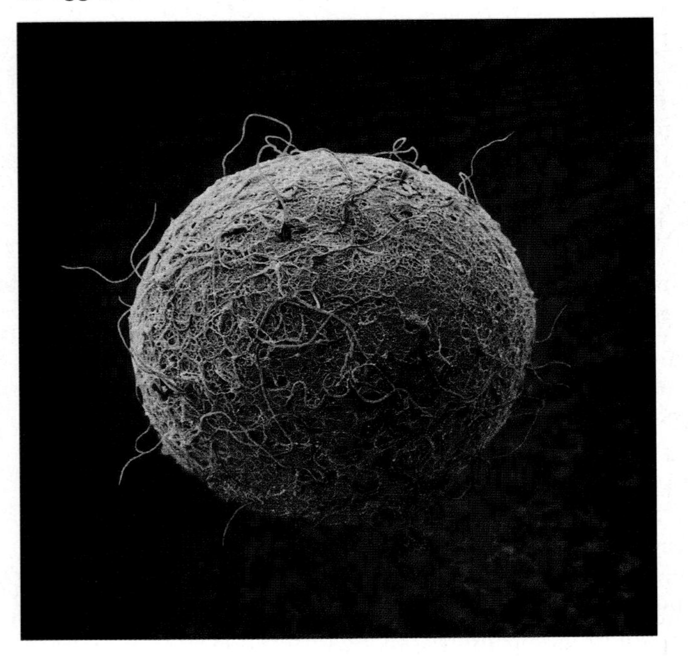

Figure 12.8 Oogenesis. (a) Through meiosis, human females produce egg cells that will be ovulated. From birth to puberty, the process is paused at prophase I. At puberty, one primary oocyte per month moves from prophase I to metaphase II and the secondary oocyte will move through telophase II only if fertilized. Cell division results in the production of polar bodies and one ovum, thus producing one functional gamete, not four as in males. (b) Once fertilized, the resulting cell contains all the materials required to produce a human.

Essay 12.1 Predicting the Fertile Period by Diagnosing Ovulation

Fertility is the ability to produce healthy sperm and egg cells. Men are fertile during their entire reproductive lives—spermatogenesis begins at puberty and continues until death. A woman is only fertile for a few days a month, around the time of ovulation. During this time, a woman can determine if she is fertile—that is, whether or not she has ovulated—in a variety of ways. Some women can feel when they have ovulated: Prostaglandin, the hormone produced by the endometrium in response to the LH surge, stimulates muscle contraction and aids in propelling the egg down the oviducts. The muscular contractions cause lower abdominal pain, known as *mittelschmerz*, in some women.

It is also possible to diagnose ovulation by measuring a woman's body temperature every day. Body temperature increases by a half degree after ovulation, and an egg cell is fertilizable for about 24 hours. By regularly taking her temperature, a woman can determine when she *was* fertile. If a woman tracks her cycle and has regular cycles, she can predict when ovulation is likely to occur in the future.

For many women, the most reliable indicator of fertility is the presence of thin, stringy cervical mucus that has both the consistency and appearance of raw egg white. Secretions from the cervix change in response to the actions of estrogen and progesterone. During the first half of the cycle, prior to ovulation when estrogen levels are high, the cervical mucous increases in abundance, thins, and becomes slick in texture. When the level of progesterone is high after ovulation, during the second half of the cycle, cervical mucous is whitish and gummy. The mucous is produced in the "crypts" or folds that line the cervix, and the parallel strands of mucous function as channels for the sperm to move through, preventing them from becoming stuck in the folds of the cervix (Figure E12.1). Sperm die within a few hours if this type of mucus is not present, but when the mucus is present they live from three to five days. Therefore, a woman is fertile from the beginning of the stringy discharge when the mucus enables sperm to reach the oviducts where they wait for the egg cell. Fertility continues for several days after the stringy discharge because sperm have already gained access to the oviducts and can survive for three to five days. Since the egg breaks down quickly, a woman is fertile five days prior to ovulation and the day of ovulation itself, which corresponds to the time that stringy cervical mucus is present, plus a few more days.

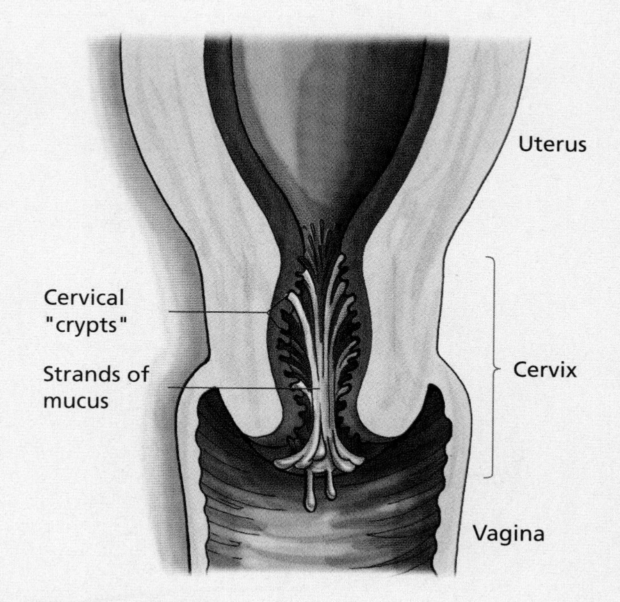

Figure E12.1 Cervix anatomy.

During the meiosis of oogenesis, an interesting type of cellular division occurs. Instead of the equal division of the parent cell into two same-sized daughter cells, the meioses of oogenesis produces one small cell and one large cell. The smaller cell is called a **polar body** and does not contain enough nutrients to undergo any further development. The large cell, called the **ovum**, receives the majority of the cytoplasmic nutrients and organelles and is thus better prepared to proceed with meiosis and, should fertilization occur, early embryonic development. Chromosomes are inherited in equal proportions from both parents, and most of the cytoplasm and its organelles are inherited along with the egg cell (Figure 12.8b).

Menstruation

The term **menstrual cycle** refers specifically to changes that occur in the uterus. The menstrual cycle depends on intricate interrelationships among the brain, ovaries, and lining of the uterus or **endometrium**. During the course of a single menstrual cycle, a woman's body prepares an egg for potential fertilization and her uterus for a potential pregnancy; then her body evaluates whether pregnancy has occurred; if it has not, the uterine lining is excreted and a new cycle begins.

The changes in hormone levels, endometrium condition, and ovaries occurring throughout the 28-day cycle are illustrated in Figure 12.9. Most women's bodies do not adhere precisely to a 28-day cycle—some women have longer cycles, others have shorter cycles, and some women do not have cycles that are regular at all—they can range from 20 to 40 days. The first day of a menstrual cycle is considered to be the first day of actual bleeding.

Like many other biological processes, the menstrual cycle is self-regulating, operating by a mechanism called *feedback*, in which a product of the process

WWW
Media Activity 12.1B Menstral Cycle and Birth Control

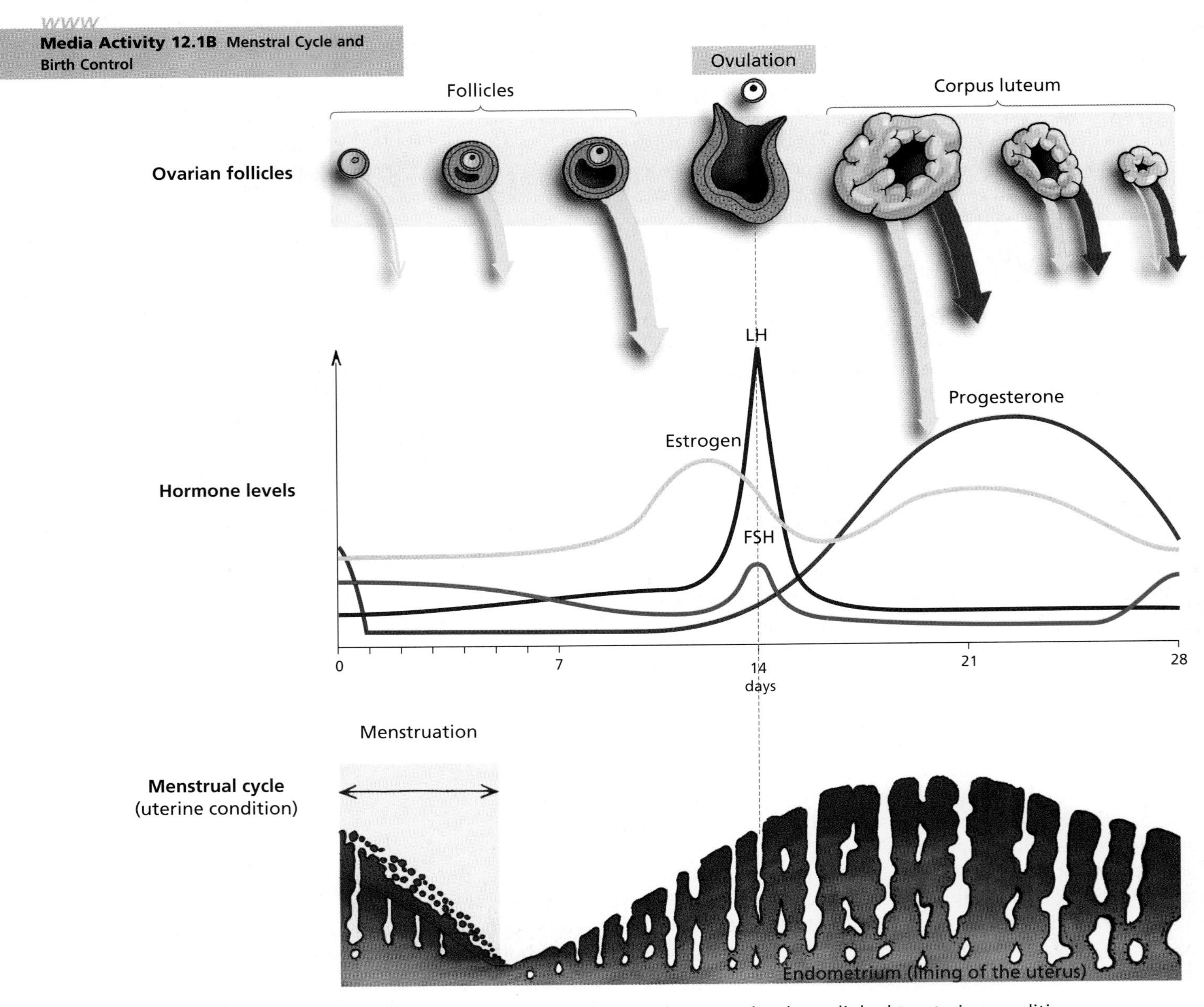

Figure 12.9 The events of the menstrual cycle. Changes in hormone levels are linked to uterine condition and the state of ovarian follicles over the course of the menstrual cycle.

regulates that process. Negative feedback slows or stops a process; positive feedback speeds up the process.

High levels of estrogen provide positive feedback to the hypothalamus, which acts on the pituitary gland to increase the secretion of FSH and LH. Conversely, high levels of the hormone **progesterone** have a negative feedback effect on the hypothalamus, and GnRH secretion is decreased (Figure 12.10).

When a woman is actually menstruating, the uterus is shedding its lining. In the ovary, the primary follicle is beginning to grow under the actions of FSH. As the follicle grows, it produces estrogen. After menstruation has ended, the growing follicle and its estrogen stimulates the cells of the endometrium to grow and divide. Estrogen also acts on breast cells, causing them to swell and leading to breast tenderness.

Once the primary follicle is large enough, it produces enough estrogen to stimulate GnRH release. This leads to a spike in both FSH and LH levels, which lasts about 24 hours. Ovulation occurs 10–12 hours after the LH peak, around 14 days before menstruation in a typical 28-day cycle. (A woman with a 22-day cycle would ovulate around the eighth day.) The primary follicle has receptors for LH, and in response to the LH surge the egg is stimulated to develop. The follicle then literally bursts open, releasing the egg, which is drawn into the oviducts.

After ovulation, a follicle is called the **corpus luteum** (Latin meaning "yellow body," indicating the color of this structure on the ovary). The corpus luteum is a hormone-producing tissue that makes progesterone and estrogen.

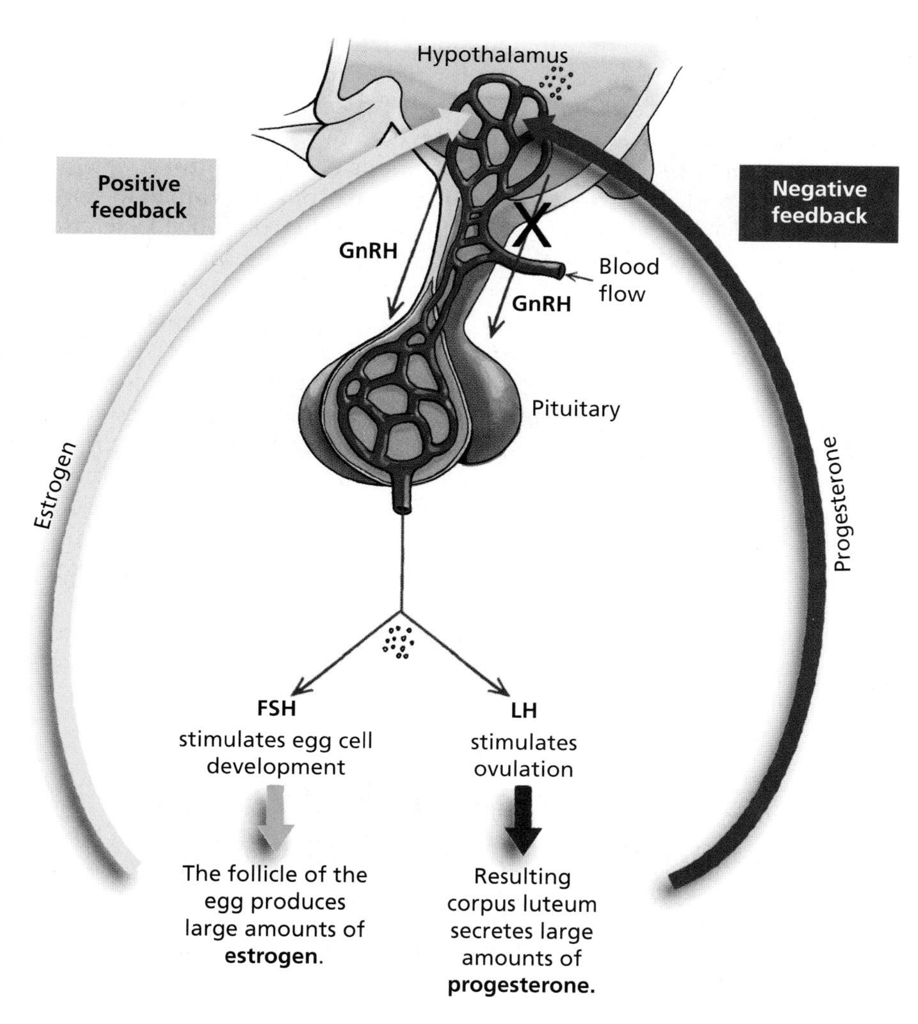

Figure 12.10 Feedback operating during menstruation. High levels of estrogen provide positive feedback to the hypothalamus; conversely, high levels of the hormone progesterone have a negative feedback effect on the hypothalamus. These two feedback loops do not occur simultaneously because follicle development takes place before ovulation and production of the corpus luteum.

Progesterone keeps the secretions from the cervix thick, making it more difficult for sperm to gain access to the egg cell. Progesterone also helps maintain the blood flow to the uterine lining to support early fetal development. In addition, progesterone inhibits continued LH production, so the LH surge is inhibited by the ovarian hormones it stimulates—a good example of negative feedback.

If fertilization does not occur, the corpus luteum degenerates about 12 to 14 days after ovulation. Progesterone and estrogen levels fall because the corpus luteum is no longer secreting them, and the lining of the uterus is shed. The loss of endometrial tissue is an active process and is caused by an increase in the enzyme that breaks down collagen, a fibrous protein that serves as the base for the endometrium. The production and activity of this enzyme is tightly controlled by different levels of estrogen and progesterone.

The loss of estrogen and progesterone also causes the arteries supplying the uterus to spasm, causing menstrual cramps and menstruation begins. The decreasing levels of progesterone also serve to release the hypothalamus from inhibitory control so the LH and FSH levels rise, and the cycle starts over.

If fertilization has occurred, the process of endometrial breakdown is halted. The endometrium and a membrane produced by the developing embryo, the **placenta**, release a hormone that extends the life of the corpus luteum. With the corpus luteum intact, progesterone and estrogen levels remain high, and the endometrium is maintained. In a pregnant woman, the corpus luteum finally disintegrates after about 6 or 7 weeks of pregnancy, when the placenta begins to produce progesterone.

The birth control pill supplies the body with low continuous doses of synthetic versions of estrogen and progesterone. The estrogen present in the pill is at a level low enough to prevent ovulation. The progesterone contained in the pill serves as a backup in case ovulation does occur by making the mucus secreted by the cervix resistant to the ascent of sperm. Low hormone levels also prevent the lining of the uterus from developing enough to support a pregnancy. See Table 12.1 for a list of other methods of birth control.

Pregnancy and menstruation are not good reasons for women to stop exercising. Several studies have shown that women who continue to exercise during pregnancy have shorter labors than those in a sedentary control group. Female athletes also have fewer complications in pregnancy and delivery.

You can see from the description above that a lot of hormonal and physical changes occur during the menstrual cycle, and many people have wondered whether a woman's performance—athletic or mental—is altered during any portion of the cycle. The answer is an unequivocal *no*. Researchers who measured speed and strength among sprint swimmers and weight lifters found the women's performances remained consistent throughout the menstrual cycle. Other research has shown no differences in maximal bicycling and jumping power at different stages of the menstrual cycle. In fact, study after study has shown no loss of physical or mental function occurring during menstruation or prior to menstruation.

While some sex differences, such as differences in reproductive organs, gametogenesis, and the presence of menstruation in females do not affect athleticism, other sex differences can have an impact.

www
Media Activity 12.4 Men's and Women's Health Issues

12.3 Sex Differences That Can Affect Athleticism

Sex differences that influence athleticism include skeletal differences, differences in muscle mass, differences in body-fat storage and utilization, and cardiovascular differences (Figure 12.11).

	Method	Mode of action	Risks	Percent failure*
Abstinence	**Abstinence**	Sperm and egg never have contact.	No associated risks.	0.0
Hormonal methods	**Combination birth control pill**	Synthetic estrogen and progesterone given at continuous doses (versus the cyclic fluctuations of a menstrual cycle) prevent ovulation.	Increased risk of heart disease and fatal blood clots for women over age 35 who smoke. May have slightly increased risk of breast cancer. Decreased risk of uterine and ovarian cancers.	3.0
Hormonal methods	**Minipill**	The minipill contains progesterone only. Since there is no estrogen to combat the effects of progesterone, cervical mucous is thickened, sperm ascent is impeded, and the uterine lining is not prepared to support a pregnancy.	Increased risk of ovarian cysts.	13.2
Hormonal methods	**Depo-Provera**	This method requires the user to have progesterone injections every three months. Same mode of action as the minipill.	Under investigation.	0.3
Hormonal methods	**Norplant**	These progesterone-containing implants have the same mode of action as the minipill.	Abnormal vaginal bleeding.	0.09
Sterilization	**Female sterilization**	Cutting or blocking oviducts prevents sperm and egg contact.	Surgical infection.	<1.0
Sterilization	**Male sterilization**	Cutting each vas deferens prevents sperm from being ejaculated in the semen.	Link to prostate cancer under investigation.	<1.0

*Percent of users per year who have an unintended pregnancy

Table 12.1 Birth-control methods. Many birth-control devices and methods exist. The merits and efficacy of some of them are outlined here.

	Method	Mode of action	Risks	Percent failure*
Barrier methods	**Cervical cap** 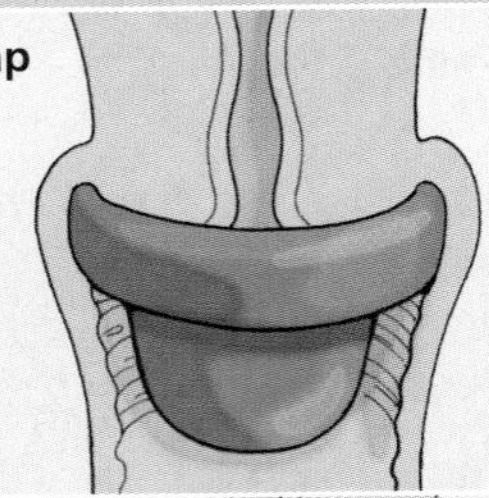	When inserted against cervix before intercourse, prevents sperm and egg contact.	No known risks.	18
	Diaphragm 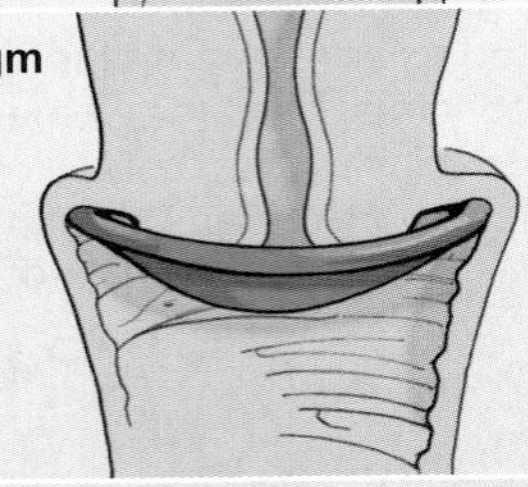	When inserted into vagina before intercourse, prevents sperm and egg contact.	No known risks.	18
	Female condom 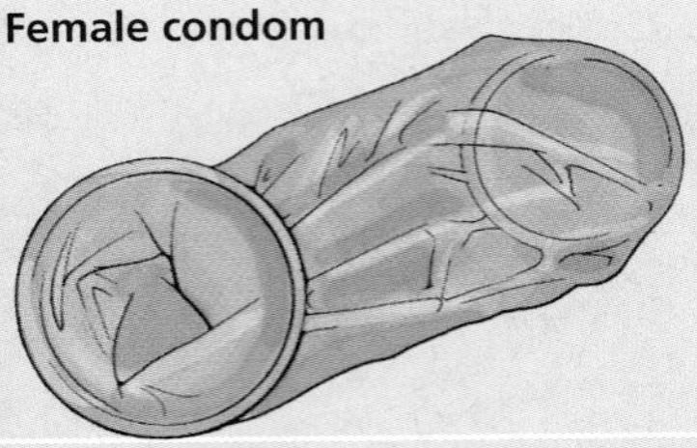	Held against cervix by a flexible ring, the female condom prevents sperm and egg contact.	No known risks.	21
	Male condom 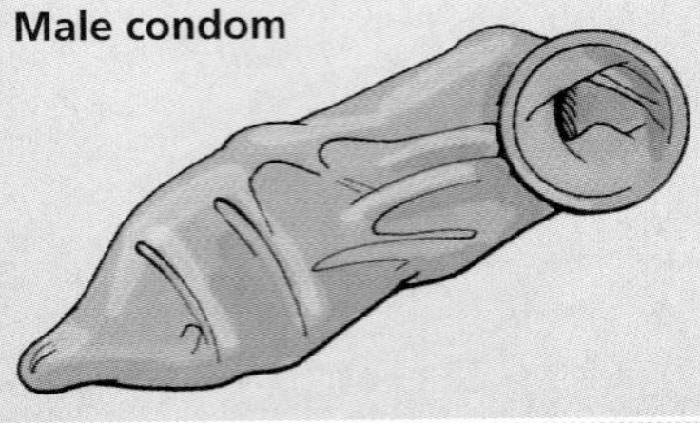	Prevents sperm and egg contact.	No known risks.	12
Other methods	**Spermicides** 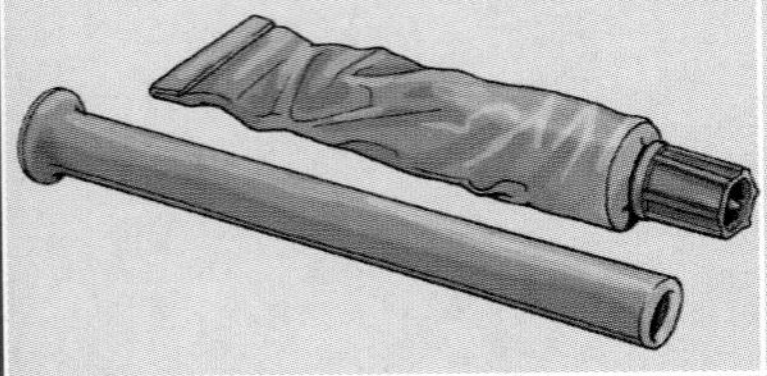	When inserted into vagina 1 hour before intercourse, kills sperm.	No known risks.	21
	Fertility awareness 	Abstinence for the 4 days before and 4 days after predicted time of ovulation.	None.	20
	Intrauterine device 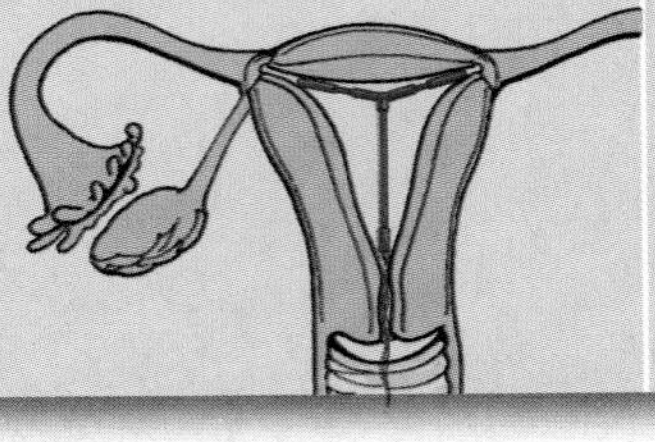	When this small plastic device is inserted into the uterus by a physician, it prevents fertilization and prevents uterus from supporting a pregnancy.	May increase risk of pelvic inflammatory disease (see Chapter 11).	<2

*Percent of users per year who have an unintended pregnancy

Table 12.1 ***(continued)***

www
Media Activity 12.1C Hormones and Sexual Differences

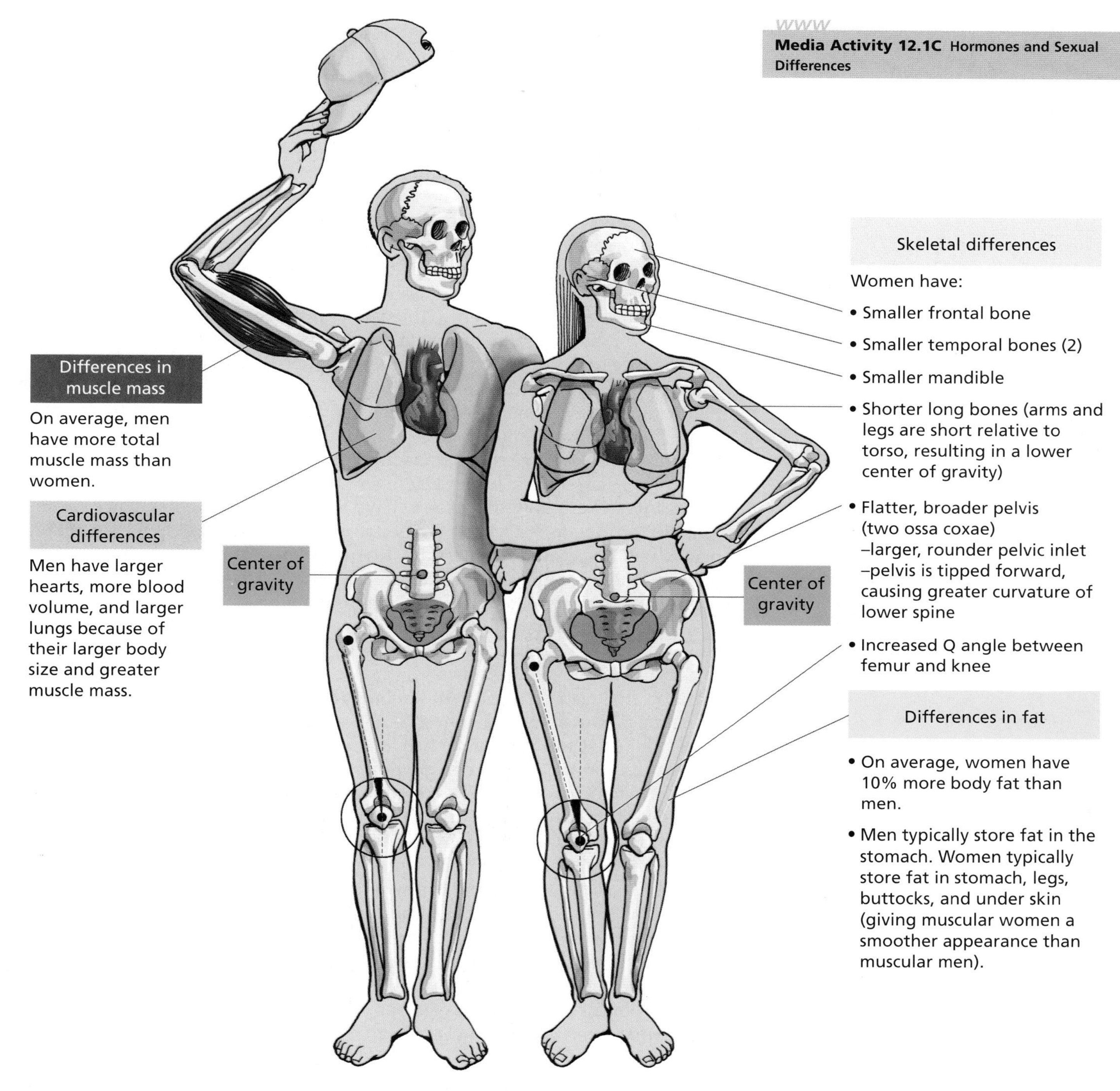

Differences in muscle mass
On average, men have more total muscle mass than women.
Cardiovascular differences
Men have larger hearts, more blood volume, and larger lungs because of their larger body size and greater muscle mass.
Center of gravity
Center of gravity
Skeletal differences
Women have:
• Smaller frontal bone
• Smaller temporal bones (2)
• Smaller mandible
• Shorter long bones (arms and legs are short relative to torso, resulting in a lower center of gravity)
• Flatter, broader pelvis (two ossa coxae)
–larger, rounder pelvic inlet
–pelvis is tipped forward, causing greater curvature of lower spine
• Increased Q angle between femur and knee
Differences in fat
• On average, women have 10% more body fat than men.
• Men typically store fat in the stomach. Women typically store fat in stomach, legs, buttocks, and under skin (giving muscular women a smoother appearance than muscular men).

Figure 12.11 Sex differences that influence athleticism. Human male and female skeletons are very hard to distinguish. Consistent differences are found in the mandible, temporal bone, frontal bone, and the two ossa coxae. The female pelvic inlet is larger and rounder than the male pelvic inlet. Q angles in males are typically smaller than Q angles in females. Men have less body fat and different storage patterns. Men also have larger hearts and more muscle mass.

Skeletal Differences

From infancy to adulthood, humans grow in overall size, but a larger proportion of growth occurs in the extremities (legs and arms) than in the torso. To convince yourself of this, consider how short toddlers' legs look in relationship to their bodies. A final growth spurt occurs during adolescence and puberty, again with a larger proportion of growth occurring in the arms and legs than in the torso. In part because puberty both occurs later and lasts longer in men than in women, the average adult man has longer arms and legs than the average adult woman. Men and women sitting together tend to be closer in height than men and women standing together because of this difference in relative length. Overall, the average man is 15 cm taller than the average woman. Keep in mind however, that shorter bones do not mean weaker bones. Women's bones are neither softer nor more fragile than men's.

The differences between women's and men's skeletons contribute to average differences in success in certain physical activities. Because women generally have long torsos relative to their leg length, their center of gravity (the point on the body where the weight above equals the weight below) is lower than on men. Individuals with a lower center of gravity are better able to maintain balance.

Competitive gymnastics recognizes this difference in center of gravity—women, but not men, compete on the balance beam apparatus. Ice skating, ballet, and other forms of dance also require a lot of balance, and the lower center of gravity may enhance a woman's ability to stay on her feet.

Men, with their generally longer legs and arms, have more difficulty balancing, but they bring more power to activities that rely on the lever action of their extremities. A longer lever transmits more force; therefore, men have faster and stronger slap shots in hockey, stronger kicks in soccer, and faster swings and throws in baseball. Longer legs also have longer stride lengths, which means men will usually be faster than women unless women make up for the difference by having greater muscle mass. Remember that these are average differences and there is a great deal of variability within each sex—a tall woman will naturally have a longer stride than a short man.

Aside from differences in leg and arm length, there are few differences in the human skeleton that affect athleticism. In fact, there are only six bones that differ between same-sized males and females. Four of the bones that differ are found on the head. These are the **mandible**, or jaw bone, which is larger in males; the paired **temporal bones** found near the temple, which have a larger opening in males to allow for connection of thicker muscles to support the large jaw; and the **frontal bone** of the cranium, or forehead, which is generally more rounded in females and has a less pronounced ridge above the eyes. The other two bones that show general differences between men and women are the two **ossa coxae**, which form the bony pelvis.

The differences between the pelvises of men and women evolved in response to different selective pressures. The evolution of upright-walking early humans required that these changes take place. However, the changes to the ossa coxae, combined with our large head size, results in the most difficult birth of any animal. Early human females with rounder pelvic inlets may have been selected for because they were less likely to die during childbirth, since rounder inlets generally permit the easier passage of a child's head through the birth canal.

The round pelvic inlet that is typical of the majority of women is produced by a bony pelvis that is flatter and broader than a man's pelvis, which has an inlet that resembles an elongated oval. A flatter pelvis also requires that the bony pelvis be tipped forward to bring the hipbones to the front of the body, which is maintained by the curvature of the lower spine. The spinal curvature and greater pelvic tilt in individuals with broad pelvises elevates the buttocks and gives a curvy appearance to the profile of women. Conversely, the male pelvis lowers the buttocks and gives a flat appearance to a man's profile.

When you bring your feet together, your *femurs*, or thighbones, extend diagonally to your knees from where they attach at the hips. A broader pelvis means that the femurs are farther away from each other at the point of attachment than femurs attached to a narrower pelvis. Therefore, the **Q angle** formed between the kneecap and femur increases as the broadness of the bony pelvis increases.

An increased Q angle has been thought to be a sex-influenced risk factor for knee injury, since the muscles in the thigh tend to pull the kneecap up and toward the outside of the knee. For instance, knee injuries among basketball players are much more common in women, who generally have greater Q angles, than men, who generally have lesser Q angles. However, many scientists who study sports injuries are quick to point out that social differences in early physical activity and conditioning may have prevented development of the leg-muscle strength that support the knee-stressing movements typical of basketball. To truly discern the effects of greater Q angles, a study must compare men and women who have followed similar conditioning and training regimens.

Most analyses indicate that injuries are sport-specific rather than gender-specific. When men and women at the same level of fitness are compared, women were no more likely to suffer injury.

Differences in Muscle Mass

www

Media Activity 12.5 Muscle Mass Data

While the role of the Q angle may or may not influence the rate of knee injury, one clear advantage the male body has in sports requiring greater strength is that a taller, wider, heavier skeletal frame supports more muscle.

Testosterone released by the testes at puberty results in the increased size of muscle fibers, long cells that contract when stimulated by nerve impulses. Testosterone also increases the total amount of muscle on the body, or muscle mass, because it promotes retention of the dietary nitrogen used to synthesize muscle proteins. Males have more testosterone than females, although the ovaries and adrenal glands do release small quantities of this hormone, and the difference in muscle mass accounts for an overall difference in strength between males and females. In sports that require brute strength, such as football, men have an advantage, but in sports like baseball, where success depends more on agility and hand-eye coordination, sex differences are reduced.

Muscle build-up is a function of testosterone concentration, thus women cannot have bulging muscles, no matter how hard they work out. The muscles of women who exercise become smoother, firmer, and stronger but do not balloon out. Fat deposition also changes the appearance of muscles. Even if a man and a woman had similar amounts of muscle mass, the man would look more muscular because women have a different pattern of body fat deposition. Women have more body fat under their skin, which tends to smooth out the appearance of muscles beneath the skin.

Differences in Body Fat

Differences in body fat arise at puberty when young women begin to store fat in their stomach, buttocks, and legs. This, in addition to skeletal differences, gives women's bodies their curvy shape. Men usually carry their fat in the abdomen. Overall, women have about 10% more body fat then men, which is necessary to maintain fertility. As we discussed in Chapter 2, body fat is required for female fertility because a hormone called *leptin*, secreted by fat cells, tells the brain if there are enough fat stores to support a pregnancy. When a female does not have enough body fat, the hormones that regulate menstruation are blocked and menstruation ceases. Lack of menstruation can be permanent and results in sterility and bone damage. Excessive exercise or starvation that leads to the cessation of menstruation, called *amenorrhea*, causes permanent damage when the estrogen that normally increases prior to ovulation is not produced.

Women may metabolize fat a little differently than men. Females seem to utilize more fat to produce energy than males, who tend to use the body's stored sugars more readily. In women, this has the effect of slowing down glucose metabolism, meaning that more sugars are available for prolonged exercise. This difference may result in a greater tolerance for endurance events in women.

Figure 12.12 Swimmer in streamline position. Fat makes a woman's body smoother and more buoyant.

The increased body fat that provides increased endurance may also be a physical advantage for women in long-distance swimming events—fat increases buoyancy and enables women to maintain the most energy-conserving streamlined position with less effort (Figure 12.12). In addition, fat provides increased insulation, thus slowing the rate of body-heat loss, and stores energy that can be converted to ATP during endurance events.

Women have improved their performance in endurance events more than men in recent years. For example, in 1972—the first year women were officially allowed to compete in marathons—the first-place woman won the Boston Marathon in a time of 3:10:26. That same year, the first man to cross the finish line did so with a time of 2:15:39. In 2002 the first-place woman finished with a time of 2:20:43, a difference of almost 50 minutes; the first-place man finished with a time of 2:09:02, only a six-and-a-half minute difference. In 30 years, women have made substantial athletic gains, bringing the finishing times of men and women closer together. This could simply be due to training; men have been training longer than women have and they once had access to better coaching and practice facilities. Whether or not this trend continues, ultimately making women faster than men at endurance events, remains to be seen—but men still have the advantage when it comes to events requiring a high level of intensity over a shorter period of time due to sex differences in the cardiovascular system.

Cardiovascular Differences

Women have smaller hearts and less blood volume relative to body size than similarly sized men. Men have more blood volume to provide blood to their greater muscle mass. Women also have smaller lungs that work harder to transport oxygen to the cells and tissues that need it. If a man and woman of similar size are running side-by-side at the same speed, the woman's body must work harder to bring oxygen to her tissues.

What Do These Sex Differences Mean? While we can determine average differences between males and females in terms of skeletal size, muscle mass, body fat, and cardiovascular activity, these differences are often of little value in the real world. Average differences hide the fact that there are a wide *range* of values for each of these characteristics influencing athleticism, and the ranges for males and females typically show a great deal of overlap. Consider the example of body fat: The average healthy woman has 22% body fat, and the average healthy man has 14%. This seems like a large difference, but most males and females fall in the same range (Figure 12.13). Healthy women have from 12–32% body fat, and the range is between 3% and 29% for men. The majority of both males and females fall within the 12–29% range.

In fact, it is impossible to determine a person's gender on the basis of their body-fat percentage. If all you knew about a person was that their body-fat percentage was 18%, you would have no way to predict whether they were male or female. In the same way, it would be very difficult to predict a person's body-fat percentage solely on the basis of his or her gender.

The range of body-fat percentages for normal women shows a 20% difference from lowest to highest; for men the range varies by 26%. However, the average difference between men and women is only 7%. As is often the case when we try to study differences between various groups, the differences *within* a group are greater than the differences *between* two groups.

The overlap of ranges also limits assumptions we can make about athletic performance. If we randomly picked 10 men and 10 women from a college course and timed them running 100 meters, we might see a slight difference in average scores between the males and the females. However, the large overlap of ranges would make it impossible to predict a person's gender on the basis of his or her score or predict a person's score on the basis of his or her gender. In addition, if we also timed 10 women from the track team, they would surely be at the fastest end of the distribution—probably much faster than most or all of the randomly picked men. So, average differences can tell us something about athleticism, but they are certainly not the whole story.

To understand the whole story of athleticism and gender, we must evaluate the impact our culture has on exacerbating average differences in athletic performance.

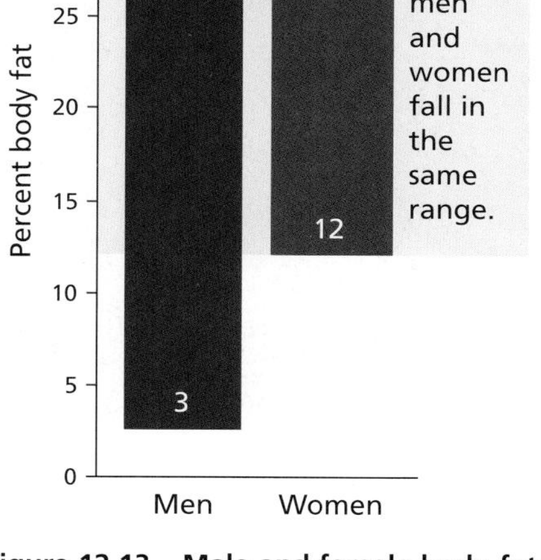

Figure 12.13 Male and female body-fat percentages. This graph shows that there is a lot of overlap in body-fat percentages between the sexes.

12.4 Culture Affects Athleticism

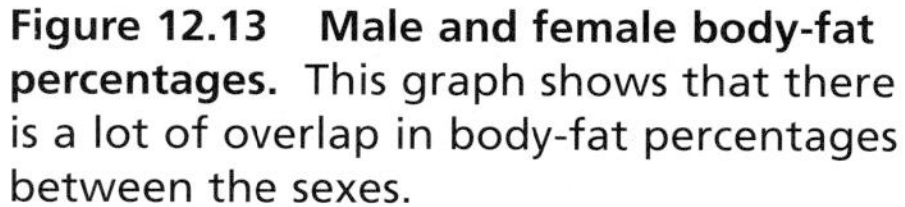

Media Activity 12.6 Current State of Coeducational Teaming

Biological and social differences interact to produce differences in athleticism between the genders. Strong cultural forces often determine the amount of time athletes spend practicing, as well as the amount of support they receive for doing so.

If a boy is struggling with hitting a baseball or throwing a football with a tight spiral, he knows there is the expectation that he is able to do this. His status in the neighborhood and on the school playground may even be determined by his athletic abilities. Conversely, girls who are struggling with developing their athletic skills are aware that there is no expectation that they will become athletes. Unless they are highly motivated, it is often easier for them to quit than to develop a skill.

A girl who wants to participate in sports may end up playing with boys, either in her neighborhood or on mixed-sex teams when she is young (Figure 12.14). While she may not face overt sexism, she is likely to overhear her male teammates chiding each other with taunts such as "You throw like a girl," or "Rally, rally, the pitcher's name is Sally." With these cultural messages, some girls may feel that sports are a male domain and that it is in some way unfeminine to be athletic.

As a young girl, Ila Borders faced many of these hurdles. She was prevented from registering for Little League baseball when she first tried to play at age 10. In college, where she was the first woman to be given a baseball scholarship, she was taunted by her own teammates as well as by members of the opposing team. In a television interview with Mike Wallace of CBS's *60 Minutes*, Borders spoke of how her own teammates would throw baseballs at her when she had her back turned, and how players from opposing teams yelled obscenities at her.

Figure 12.14 Young athletes. Prior to puberty, boys and girls often participate on teams together.

College scholarships and fan support are on the rise for women who have the support and desire to overcome cultural obstacles, and laws have been enacted to support equal participation of males and females in federally funded sports programs.

Title IX is a clause of the Federal Education Amendments that was signed into law in June of 1972. It states that, "No person in the United States shall, on the basis of sex, be excluded from participation in, be denied benefits of, or be subjected to discrimination under any education program or activity receiving federal assistance." For college and university athletic programs, this meant that the proportion of male and female athletes must equal the proportion of males and females comprising the general student population. Schools are considered to be in compliance with Title IX if they can demonstrate that they are making a serious effort to increase women's participation.

Today, more than 100,000 women participate in intercollegiate athletics, a fourfold increase since 1971 (Figure 12.15). Financial support for female athletes, and their social acceptance, has grown tremendously.

Most of us agree that it is important to support all athletes, regardless of their gender. Studies have shown that kids of both sexes who participate in sports are less likely to drop out of high school, smoke, or drink. The celebration of all athletes encourages the critical values learned from sports: teamwork, discipline, pride in accomplishment, drive, and dedication. These values should be instilled in *all* children, regardless of their gender.

Children who exercise are also more likely to continue exercising in adulthood than children who do not exercise. In a culture where one in four adults does not get the recommended 30 minutes of exercise most days of the week, it makes sense to encourage everyone to become more athletic. When good exercise habits are carried into adulthood, there is a decreased risk of heart disease, obesity, diabetes, and many cancers. Additional benefits include lowered

Figure 12.15 College-aged athletes. Participation in college athletics programs has increased for women athletes since the passage of the Title IX amendment.

cholesterol, and preliminary studies suggest that exercise may also decrease anxiety and depression. Sex differences that lead to differences in strength, speed, balance, and endurance are meaningless for recreational athletes. What really matters to most athletes is finding a sport they love so they can keep healthy and active for as long as possible.

CHAPTER REVIEW

Summary

- Many of the biological sex differences between males and females arise as a result of different levels of hormones that circulate in the bloodstream.
- Hormones involved in the production of sex differences are produced by the endocrine glands, including the hypothalamus and pituitary gland of the brain. The hypothalamus of a male or a female produces gonadotropin-releasing hormone, which stimulates the pituitary gland to produce follicle-stimulating hormone and luteinizing hormone. These hormones act on the testes of a male and the ovaries of a female. The adrenal glands also produce hormones, notably the hormone adrenaline along with small amounts of testosterone and estrogen in both males and females.
- Embryonic gonads become either testes or ovaries. Ductal structures exist side-by-side in male and female embryos. In each sex, one structure regresses. Male and female external genitalia are fashioned from the same embryonic material. The expression of sex-specific genes and hormones determines whether a particular embryo becomes male or female.
- At puberty the male gonads secrete testosterone and the female gonads secrete estrogen. Testosterone leads to the production of sperm. Estrogen production leads to ovulation and, along with progesterone, helps regulate the menstrual cycle.
- Sex differences that affect athleticism include skeletal differences—males are taller and women have a lower center of gravity; differences in muscle mass—males have more muscle mass; differences in body fat—females have more body fat; and cardiovascular differences—males have larger hearts that pump more blood and carry more oxygen to their tissues.
- There is a great deal of statistical overlap in athletic abilities between males and females.
- Sex differences in athleticism are influenced by cultural forces that shape our views of male and female athletes.

Key Terms

androgen p. 305
anti-Mullerian hormone p. 308
cervix p. 308
corpus luteum p. 317
differentiate p. 307
ducts p. 307
endocrine system p. 304
endometrium p. 316
epididymis p. 308
estrogen p. 305
follicle cell p. 313
follicle-stimulating hormone (FSH) p. 305
frontal bone p. 322
genitalia p. 307
gonadotropin-releasing hormone (GnRH) p. 305
gonads p. 305
***Gpbox* genes** p. 307
hormones p. 304
hypothalamus p. 305
Leydig cells p. 311
luteinizing hormone (LH) p. 305
mandible p. 322
menstrual cycle p. 316
Müllerian duct system p. 308
oogenesis p. 313
oogonia p. 313
ossa coxae p. 322
oviducts p. 308
ovulation p. 313
ovum p. 315
pituitary gland p. 305
pituitary gonadotropins p. 305
placenta p. 318
polar body p. 315
primary follicle p. 313
primary oocyte p. 313
primary spermatocyte p. 311
progesterone p. 317
Q angle p. 322
secondary oocyte p. 313
secondary spermatocyte p. 311
semen p. 313
seminal vesicle p. 313
seminiferous tubule p. 311
Sertoli cells p. 313
sex hormone p. 305
spermatid p. 311
spermatogenesis p. 311
spermatozoa p. 313
***SRY* gene** p. 307
temporal bone p. 322
testosterone p. 305
urethra p. 308
uterus p. 308
vagina p. 308
vas deferens p. 308
Wolffian duct system p. 308

Learning the Basics

1. List the functions of the endocrine organs involved in producing sex differences.

2. How do the undifferentiated gonads become either the testes or the ovaries?

3. How do the ducts of an embryo become able to carry sperm cells? Egg cells?

4. Why do biological sex differences in athleticism arise at puberty?

5. How do the hormones of the menstrual cycle interact to regulate menstruation and ovulation?

6. The endocrine organ that sits atop a kidney is _____.
   a. the pituitary gland
   b. the hypothalamus
   c. the ovary
   d. the adrenal gland
   e. the testicle

7. The embryonic gonads _____.
   a. differentiate into the penis if androgens are present, or into the clitoris if estrogens are present
   b. become testes when the *SRY* gene is expressed
   c. differentiate at puberty in response to differential hormone production
   d. become testes when LH is present and ovaries when FSH is present

8. The embryonic ducts _____.
   a. arise from a bipotential gonad
   b. arise from paired structures, one of which regresses and the other persists in each embryo
   c. differentiate based on the amount of FSH and LH in the bloodstream
   d. arise when testosterone stimulates the penis to develop in males and the vulva in females
   e. carry sperm and eggs produced *in utero*

9. Gametogenesis _____.
   a. begins at puberty in males and females
   b. requires that the Leydig cells of males produce semen
   c. results in the production of diploid cells from haploid cells
   d. begins at puberty in females
   e. requires that meiosis halve the chromosome number so that sperm and eggs carry half the number of chromosomes

10. Menstruation is regulated such that _____.
    a. increasing estrogen levels have a positive feedback effect on FSH and LH
    b. increasing FSH levels lead to ovulation
    c. as progesterone levels increase, so do FSH and LH levels
    d. ovulation occurs on the fifth day of the cycle
    e. the placenta produces FSH, which stimulates ovulation

Analyzing and Applying the Basics

1. How might a mutation to the *SRY* gene alter development in an XY male?
2. Why does the continuous dose of estrogen provided by the birth-control pill prevent ovulation?
3. Why are most men taller than most women?
4. In what ways do spermatogenesis and oogenesis differ? In what ways are they similar?
5. In what ways do cardiovascular differences between males and females benefit each sex?

Connecting the Science

1. Can you think of other examples of cultural influences that influence our views of biological differences between males and females?
2. Can you think of other situations in which males and females behave differently, but that difference does not necessarily make one sex better than the other?
3. Do you suppose the baseball players Ila Borders pitched against were more motivated to get a hit off of her versus other pitchers?

Media Activities

Media Activity 12.1 Hormones and Sexual Development
Estimated Time: 10 minutes
See how hormones governing the production of eggs and sperm determine sex differences in humans, and the effects those hormones have on physical differences between sexes.

Media Activity 12.2 Anabolic Steroids: Can they Bridge the Gender Gap?
Estimated Time: 35 minutes
What does the data really say about steroid use? Are men and women athletes similarly affected by steroid use? Can the use of steroids make the playing field completely level for coeducational athletics? This activity should answer these, and perhaps many more questions you may have about steroids.

Media Activity 12.3 Infertility Research
Estimated Time: 20 minutes
Many people are concerned that human activity is affecting the rate of fertility in the United States. The human reproductive system involves extensive hormone signaling; it is important for students to be aware that chemicals in the environment may interfere with this signaling, which sometimes causes problems such as infertility.

Media Activity 12.4 Men's and Women's Health Issues
Estimated Time: 25 minutes
This exercise is designed to make students aware of health issues that are unique to men and women, and to investigate whether this uniqueness has an effect on athletic differences.

Media Activity 12.5 Muscle-Mass Data
Estimated Time: 20 minutes
In this activity, you will objectively analyze data about muscle mass and strength to see both the differences and similarities (or overlap) in muscle mass and distribution, and how it may affect athleticism in men and women.

Media Activity 12.6 Current State of Coeducational Teaming
Estimated Time: 35 minutes
This activity is designed to investigate the current state of coeducational sports, and examine what keeps the gender debate alive in our culture.

CHAPTER 13

Attention Deficit Disorder:

Brain Structure and Function

Many schoolchildren are taking the prescription drug . . .

13.1 The Nervous System

13.2 The Brain

13.3 Neurons

13.4 The Environment and ADD

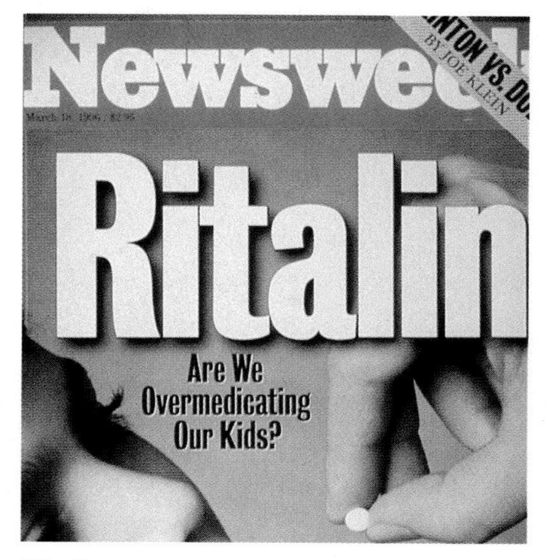

Ritalin . . .

to control Attention Deficit Disorder.

Lunchtime at a typical United States elementary school includes more than crustless peanut butter sandwiches, juice boxes, potato chips, and harried lunchroom aides. For a growing number of children, the lunch period also includes a trip to the nurse's office for a dose of methylphenidate, more commonly known by its trade name, *Ritalin*™. At least 6% of all school-aged boys and 2% of school-aged girls line up to take the drug, aimed at treating a syndrome called *Attention Deficit Disorder*, or *ADD*.

In 1980, ADD first appeared in the third edition of psychiatry's main reference text, the *Diagnostic and Statistical Manual of Mental Disorders* or *DSM-III* published by the American Psychiatric Association. This book guides thousands of mental health professionals and millions of their patients. The editors of the fourth edition, released in 1987, attempted to increase awareness of the hyperactivity component of ADD by renaming the syndrome Attention Deficit Hyperactivity Disorder, or ADHD. However, most people still call the syndrome ADD.

Children are diagnosed as having ADD if they persistently show a combination of traits, including forgetfulness; distractibility; fidgeting; restlessness; impatience; difficulty sustaining attention in work, play, or conversation; or difficulty following instructions and completing tasks. According to

Some college students who have *not* been prescribed the drug use it illegally to pull all-night study sessions.

criteria issued by the American Psychiatric Association, at least six of these traits must be present to a degree that is maladaptive, and these behaviors must cause impairment in two or more settings—for example, at school and at home.

Ritalin is a class of drug known as a *stimulant*. Stimulants temporarily increase activity in an organism. Giving a stimulant to someone who is having trouble concentrating seems illogical. However, Ritalin has a paradoxical effect in that it actually causes people to slow down and pay closer attention. The fact that the drug helps people concentrate has led to its illegal use, sometimes by college students who need to cram for an exam.

Some people argue that ADD is caused by brain malfunction, so it is best to treat ADD with medicines. Others argue that ADD is, if not caused by, at least exacerbated by our fast-paced society and is therefore better treated by changing the environment of the affected child. To comprehend the debate, understanding the biology of the human nervous system is required.

13.1 The Nervous System

Every second, millions of signals make their way to your brain and inform it about what your body is doing and feeling. Your **nervous system** interprets these messages and decides how to respond. Responding to stimuli requires the actions of specialized cells called **neurons**. Neurons are capable of carrying electrical and chemical messages back and forth between your brain and other parts of the body. These electrochemical message-carrying cells are often bundled together, producing structures called **nerves**.

Neurons and nerves are extensively networked throughout your body. They are found in your brain, spinal cord, and sense organs such as your eyes and ears. Nerves connect various organs to each other and link the nervous system with other organ systems (Figure 13.1).

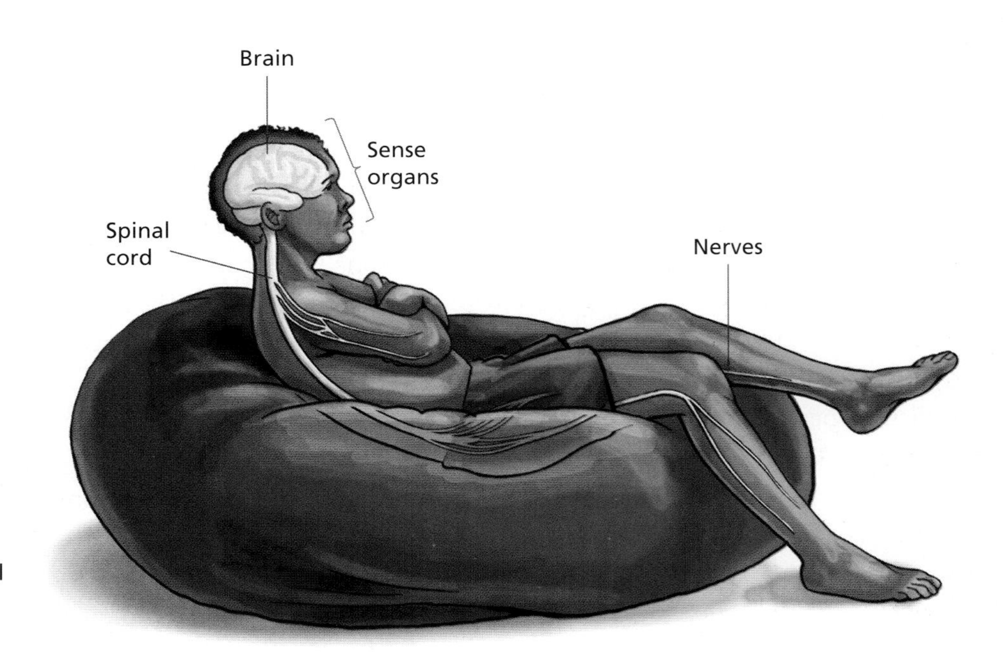

Figure 13.1 The nervous system. The nervous system includes the brain, spinal cord, sense organs, and the nerves that interconnect those organs.

To carry information between parts of the body, the cells of the nervous system pass electrical and chemical signals to each other. Signals are transmitted from one end of a nerve cell to the other end, between nerve cells, and from nerve cells to the cells of tissues, organs, or glands that respond to nerve signals called **effectors**. Information is carried along nerves by electrical changes called **nerve impulses** and transmitted between nerve cells and from nerve cells to the cells they act on by chemical stimuli, called **neurotransmitters**, released from nerve cells.

Neurons can be grouped into three general categories: (1) **sensory neurons** carry information toward the brain; (2) **motor neurons** carry information away from the brain toward effector organs and glands; and (3) **interneurons**, located between sensory and motor neurons within the brain or spinal cord. Regardless of category, their job is to integrate information from various sensory inputs. Most actions involve input from all three sources, with sensory input followed by integration and motor output (Figure 13.2).

Sensory input is detected by *sensory receptors*. Receptors are usually neurons or other cells that communicate with sensory neurons. They detect changes in conditions inside or outside the body. When receptors are stimulated, signals are generated and carried to the brain.

The *general senses* are temperature, pain, touch, pressure, and body position, or proprioception. The sensory receptors for the general senses are scattered throughout the body. The *special senses* are smell, taste, equilibrium, hearing, and vision. The sensory receptors for these five special senses are found in complex sense organs (Table 13.1).

As sensory information is passed to and from your brain, it travels through the main nerve pathway: the **spinal cord**. Your spine, which protects your spinal cord from injury, is made up of 33 separate bones called **vertebrae**.

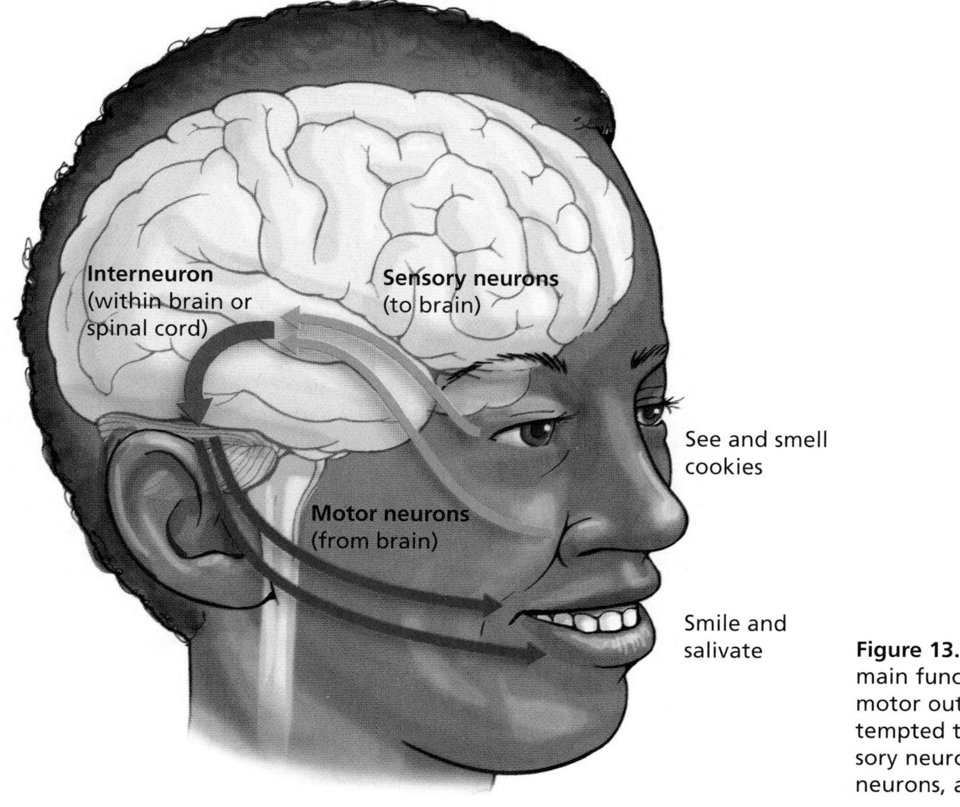

Figure 13.2 Neurons. The nervous system's three main functions are sensory input, integration, and motor output. Most actions, such as being tempted to eat cookies, involve input from sensory neurons, followed by integration, via interneurons, and motor output.

	Sense and its location	Sensory receptors	Description
General senses—throughout body	**Temperature, pain, pressure, touch** 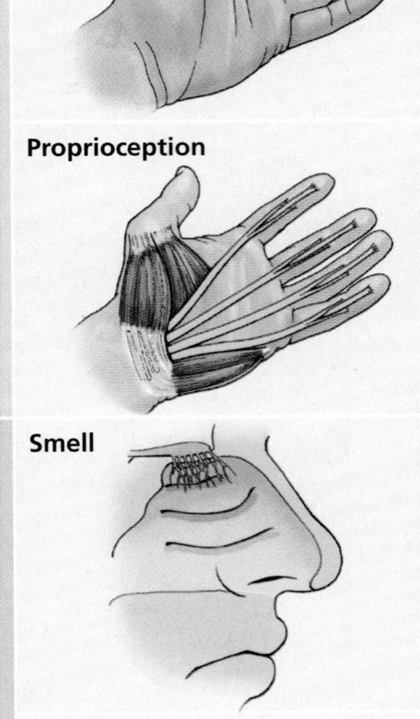	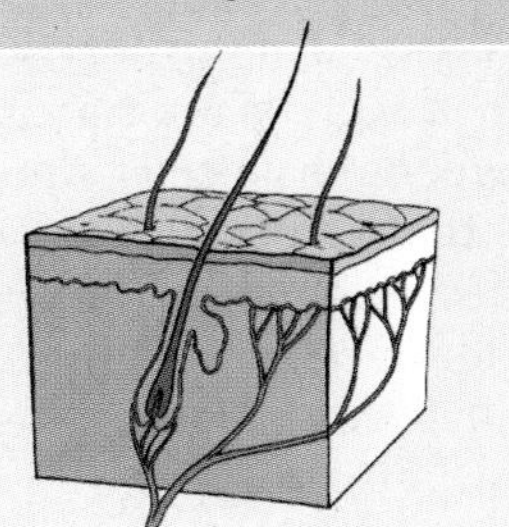	Sensory nerve endings in the skin are sensitive to temperature, pain, pressure, and touch. These receptors are specialized to respond to different levels of sensation (for example, slight cold versus freezing cold).
	Proprioception		Body position, or proprioception, is monitored by neurons that sense the position of joints, the level of tension in joints and ligaments, and the state of muscular contraction. Information gathered by these neurons is relayed to the brain, where it is integrated with balance information from the inner ear.
Special senses—found in complex sense organs	**Smell** 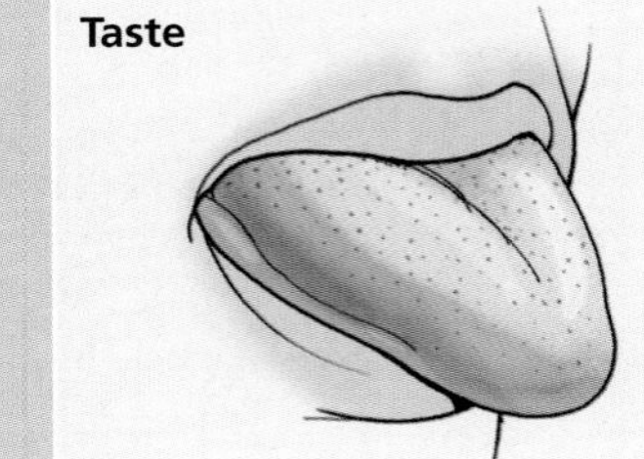		The sense of smell, or olfaction, is provided by the paired olfactory organs located in the nasal cavity. Receptors within the olfactory organs are specialized neurons. Chemicals bind to, and stimulate, receptor molecules on hairlike projections, called *cilia*, that line the nasal cavity. From the olfactory organs, nerve impulses are relayed to the brain.
	Taste 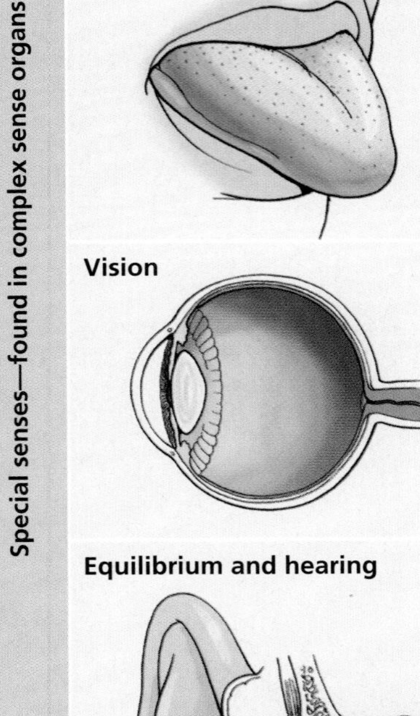		Taste receptors are found along the surface of the tongue, clustered together into taste buds. When food or drink contacts the taste buds, a nerve impulse is sent to the brain. There are four primary taste sensations: sweet, salty, sour, and bitter. Each taste bud has a sensitivity to one of these tastes.
	Vision	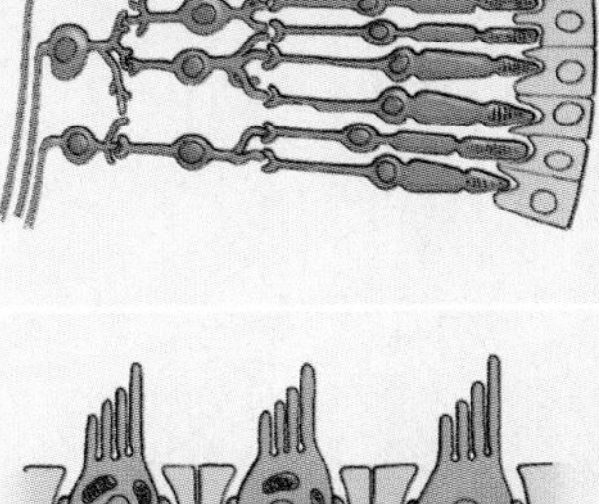	The eyes contain sensory receptors that permit sight. Neuronal receptors in the retina relay sensory information to the brain.
	Equilibrium and hearing	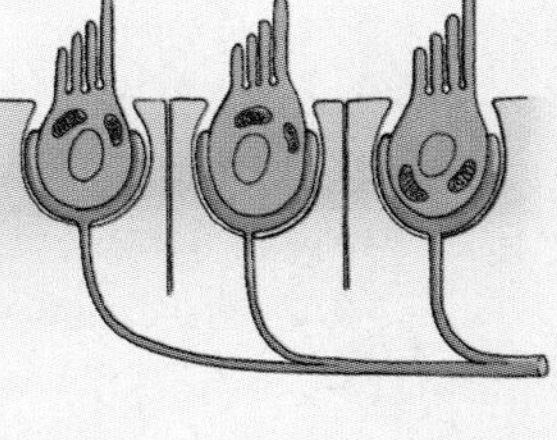	The senses of equilibrium and hearing are provided by the inner ear. Equilibrium determines the position of the body in space by monitoring gravity, acceleration, and rotation. Hearing enables us to detect and interpret sound waves. The receptors for both of these senses are ciliated cells. Movement of the cilia causes the generation of nerve impulses.

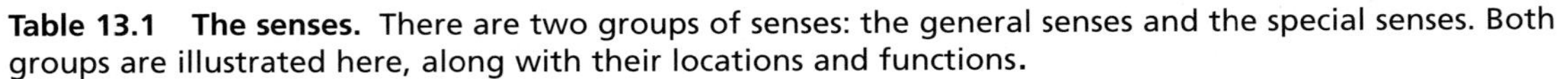

Table 13.1 The senses. There are two groups of senses: the general senses and the special senses. Both groups are illustrated here, along with their locations and functions.

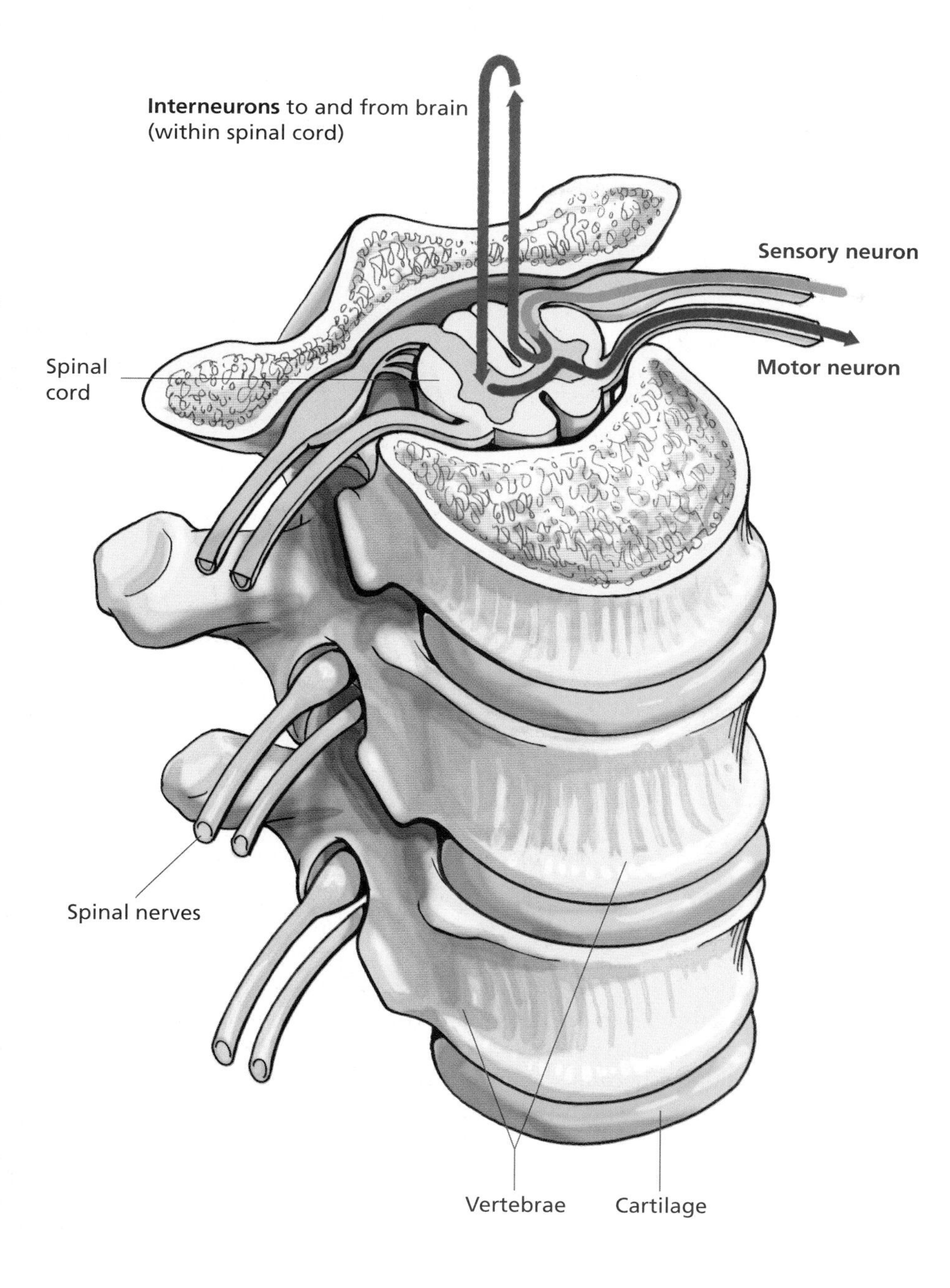

Figure 13.3 Spinal nerves and vertebrae. Spinal nerves branch out between the vertebrae and go to all parts of the body.

Spinal nerves branch out between the vertebrae and go to every part of the body (Figure 13.3).

In addition to transmitting messages to and from the brain, the spinal cord also serves as a reflex center. **Reflexes** are automatic responses to a stimulus. They are prewired in a circuit of neurons, called a **reflex arc**, which often consists of a sensory neuron that receives information from a sensory receptor, an interneuron that passes the information along, and a motor neuron that sends a message to the muscle that needs to respond.

Reflexes make a person react quickly to dangerous stimuli; for instance, the withdrawal reflex occurs when you encounter a dangerous stimulus such as touching something hot. When you touch something hot, sensory neurons from touch receptors send the message to your spinal cord. Within the spinal cord, interneurons send the message to motor neurons to withdraw your hand from the hot surface (Figure 13.4).

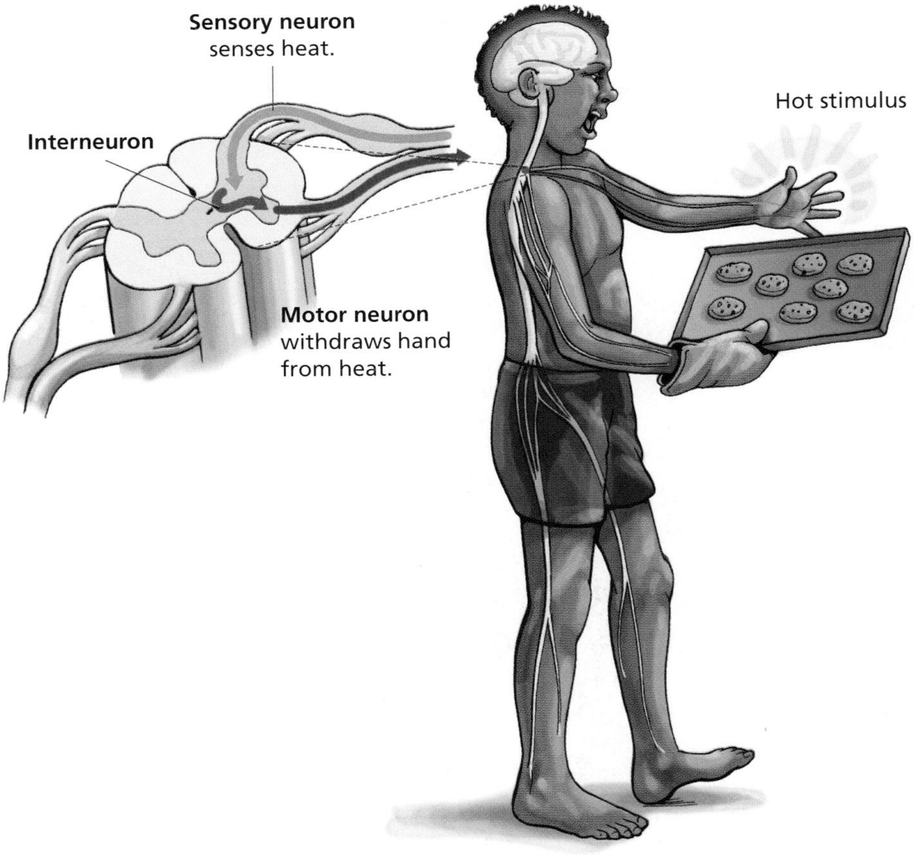

Figure 13.4 A reflex arc. A reflex arc can consist of a sensory receptor, a sensory neuron, an interneuron, a motor neuron, and an effector. Touching a hot baking sheet evokes the withdrawal reflex.

Sensory neuron senses heat.
Hot stimulus
Interneuron
Motor neuron withdraws hand from heat.

While the spinal reflexes are removing your hand from the source of the heat, pain messages are also being sent through your spinal cord to your brain. This takes a little longer because the distance to the brain is longer than the distance to the spinal cord. Therefore, by the time the pain message reaches your brain, your hand has already been removed from the hot surface.

The reflex arc illustrates how the brain, spinal cord, and nerves work together to evoke a response. Even though these parts of the nervous system work together, they are separated into two different anatomical subdivisions. The **central nervous system (CNS)**, consists of your brain and spinal cord and is responsible for interpreting and acting upon information received by the senses. The CNS is the seat of functions such as intelligence, learning, memory, and emotion. The second subdivision of the nervous system, the **peripheral nervous system (PNS)**, includes the network of nerves that radiates out from your brain and spinal cord (Figure 13.5).

WWW
Media Activity 13.1A **Brain and Neuron Structure**

13.2 The Brain

The brain is the region of the body where decisions are reached and where bodily activities are directed and coordinated. About the size of a small cantaloupe, the brain is housed inside the skull, where it sits in a liquid bath, called *cerebrospinal fluid*, that protects and cushions it.

In addition to housing neurons, the brain is composed of other cells called **glial cells**. There are 10 times as many glial cells in the brain as there are neurons.

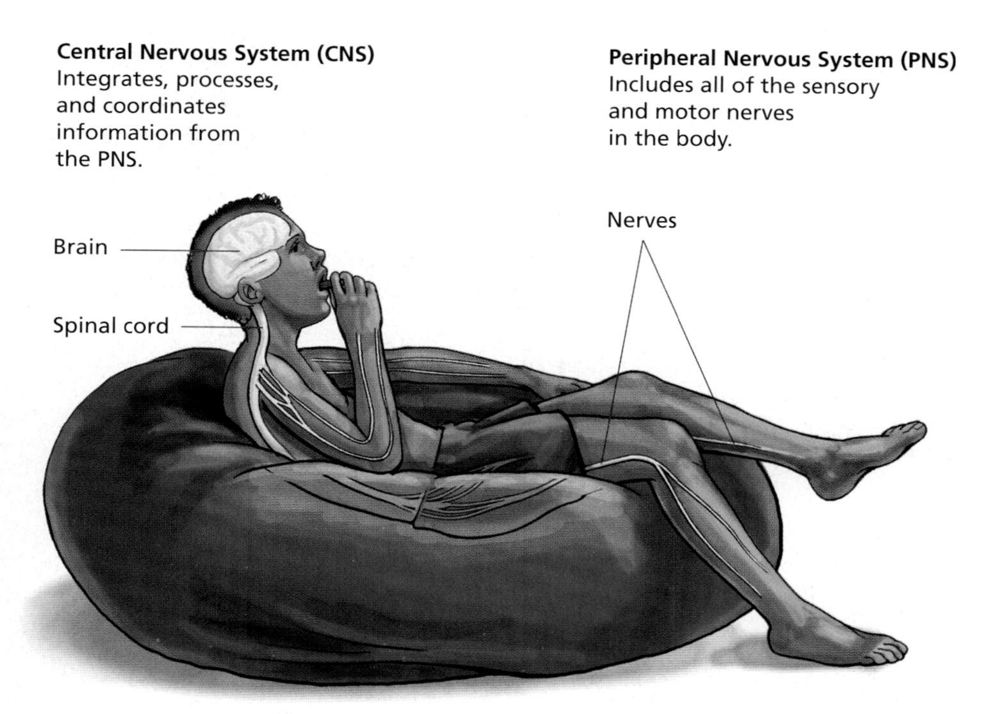
Central Nervous System (CNS)
Integrates, processes, and coordinates information from the PNS.
Peripheral Nervous System (PNS)
Includes all of the sensory and motor nerves in the body.
Brain
Spinal cord
Nerves

Figure 13.5 The central and peripheral nervous systems. The CNS (brain and spinal cord) integrates, processes, and coordinates information from the PNS. The nerves of the peripheral nervous system communicate sensory and motor information to the rest of the body.

In contrast to neurons, glial cells do not carry messages. Instead, they provide support to the neurons by supplying nutrients, helping repair the brain after injury, and attacking invading bacteria.

Structurally, the brain is subdivided into three anatomical regions: the cerebrum, the cerebellum, and the brain stem (Figure 13.6).

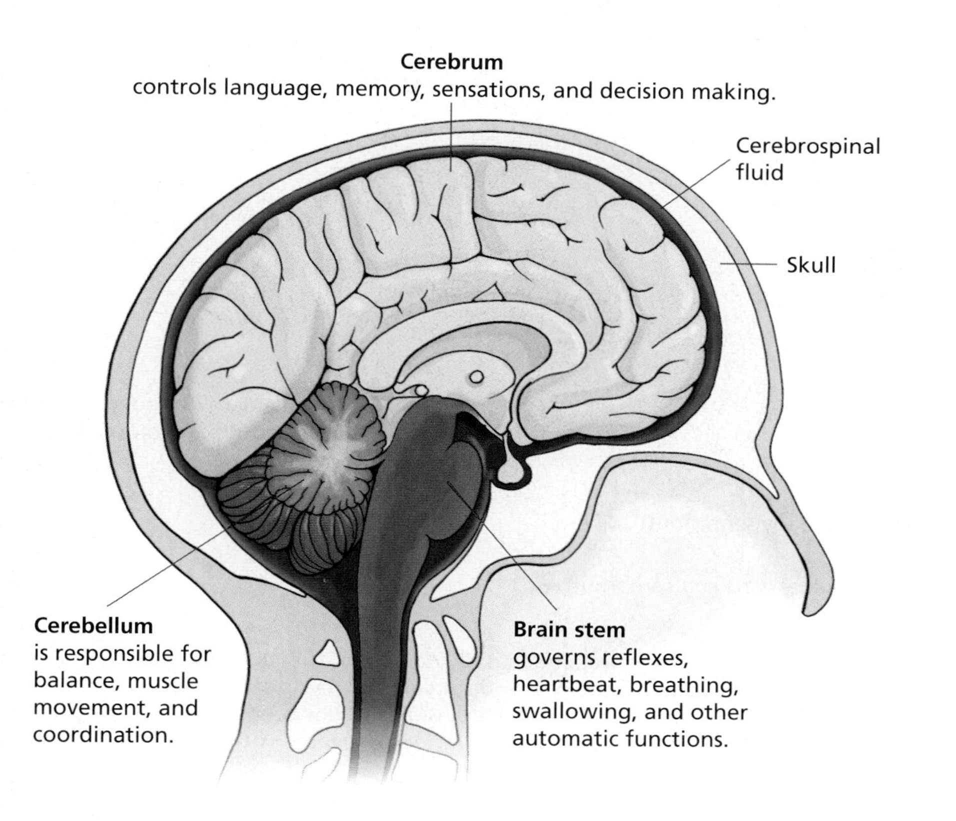
Cerebrum
controls language, memory, sensations, and decision making.
Cerebrospinal fluid
Skull
Cerebellum
is responsible for balance, muscle movement, and coordination.
Brain stem
governs reflexes, heartbeat, breathing, swallowing, and other automatic functions.

Figure 13.6 Anatomy of the brain. The brain has three main parts: the cerebrum, the cerebellum, and the brain stem.

Cerebrum

The **cerebrum** fills the whole upper part of the skull. This is the part of the brain in which language, memory, sensations, and decision making are controlled. The cerebrum has two hemispheres, each of which has four lobes (Figure 13.7). The four lobes are:

1. The *temporal lobe,* which is involved in processing information from the ears and some information from the eyes, as well as memory and emotion.
2. The *occipital lobe,* which processes information from the eyes.
3. The *parietal lobe,* which processes information about touch and is involved in self-awareness.
4. The *frontal lobe,* which processes voluntary muscle movements and is involved in planning and organizing future expressive behavior.

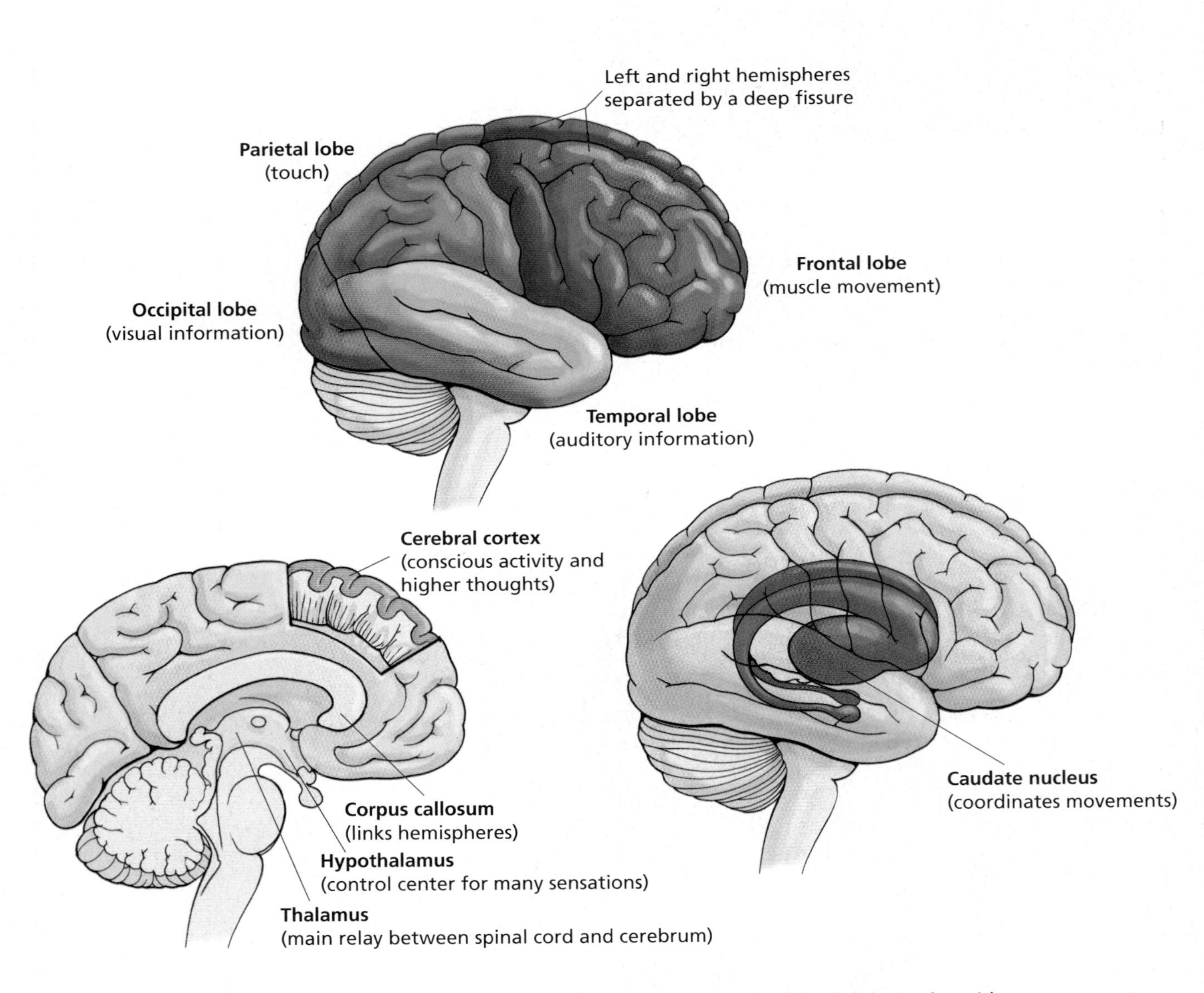

Figure 13.7 Structure of the cerebrum. The highly convoluted cerebral cortex is divided down the midline into left and right hemispheres, and each hemisphere is divided into four lobes: temporal, occipital, parietal, and frontal. The hemispheres are connected by the corpus callosum. Deep inside and between the two cerebral hemispheres are the thalamus and the hypothalamus. A caudate nucleus is located within each cerebral hemisphere.

The deeply wrinkled outer surface of the cerebrum is called the **cerebral cortex**. Wrinkling increases surface area. The cortex contains areas for understanding and generating speech, areas that receive input from the eyes, and areas that receive other sensory information from the body. It also contains areas that allow planning.

The cerebrum and its cortex are divided from front to back into two halves—the right and left cerebral hemispheres—by a deep groove called a *fissure*. At the base of this fissure lies a thick bundle of nerve fibers, the **corpus callosum**, that provides a communication link between the hemispheres. The **caudate nuclei** are paired structures found deep within each cerebral hemisphere. These structures function as part of the pathway that coordinates movement.

Also deep inside the brain, lying between the two cerebral hemispheres, are the **thalamus** and the **hypothalamus** (see Figure 13.7). The thalamus relays information between the spinal cord and the cerebrum. The thalamus is the first region of the brain to receive messages signaling such sensations as pain, pressure, and temperature, which are then relayed to the cerebrum. The cerebrum processes these messages and sends signals to the spinal cord and to neurons in muscles when action is necessary. The hypothalamus is located just under the thalamus and is about the size of a kidney bean. The hypothalamus is the control center for sex drive, pleasure, pain, hunger, thirst, blood pressure, and body temperature. The hypothalamus also releases hormones that regulate the production of sperm and egg cells as well as the menstrual cycle (Chapter 12).

Cerebellum

The **cerebellum** controls balance, muscle movement, and coordination (*cerebellum* means "little brain" in Latin). Since this brain region ensures that muscles contract and relax smoothly, damage to the cerebellum can result in rigidity and in severe cases, jerky motions. The cerebellum looks like a smaller version of the cerebrum (see Figure 13.6). It is tucked beneath the cerebral hemispheres and also has two hemispheres connected to each other by a thick band of nerves. Additional nerves connect the cerebellum to the rest of your brain (Figure 13.8).

Brain Stem

The **brain stem** lies below the thalamus and hypothalamus (see Figure 13.8). It governs reflexes and some spontaneous functions, such as heartbeat, respiration, swallowing, and coughing.

The brain stem is composed of the **midbrain**, **pons**, and **medulla oblongata**. Highest on the brain stem is the midbrain, which adjusts the sensitivity of your eyes to light and of your ears to sound. Below the midbrain is the pons (*pons* means "bridge"). The pons functions as a bridge, allowing messages to travel between the brain and spinal cord. The medulla oblongata is a continuation of the spinal cord. It conveys information between the spinal cord and other parts of the brain.

The functions of the brain are divided between the left and right hemispheres because the nerve fibers cross over each other at the brain stem; thus, the brain's left hemisphere controls the right half of the body, and vice versa. The areas that control speech, reading, and the ability to solve mathematical problems are located in the left hemisphere, while areas that govern spatial perceptions (the ability to understand shape and form) and the centers of musical and artistic creation reside in the right hemisphere.

The **reticular formation** (see Figure 13.8), found in the medulla oblongata, is an intricate network of neurons that radiates toward the cerebral cortex. The reticular formation functions as a filter for sensory input; it analyzes the constant onslaught of sensory information and filters out stimuli that require no response. This prevents the brain from having to react to repetitive, familiar

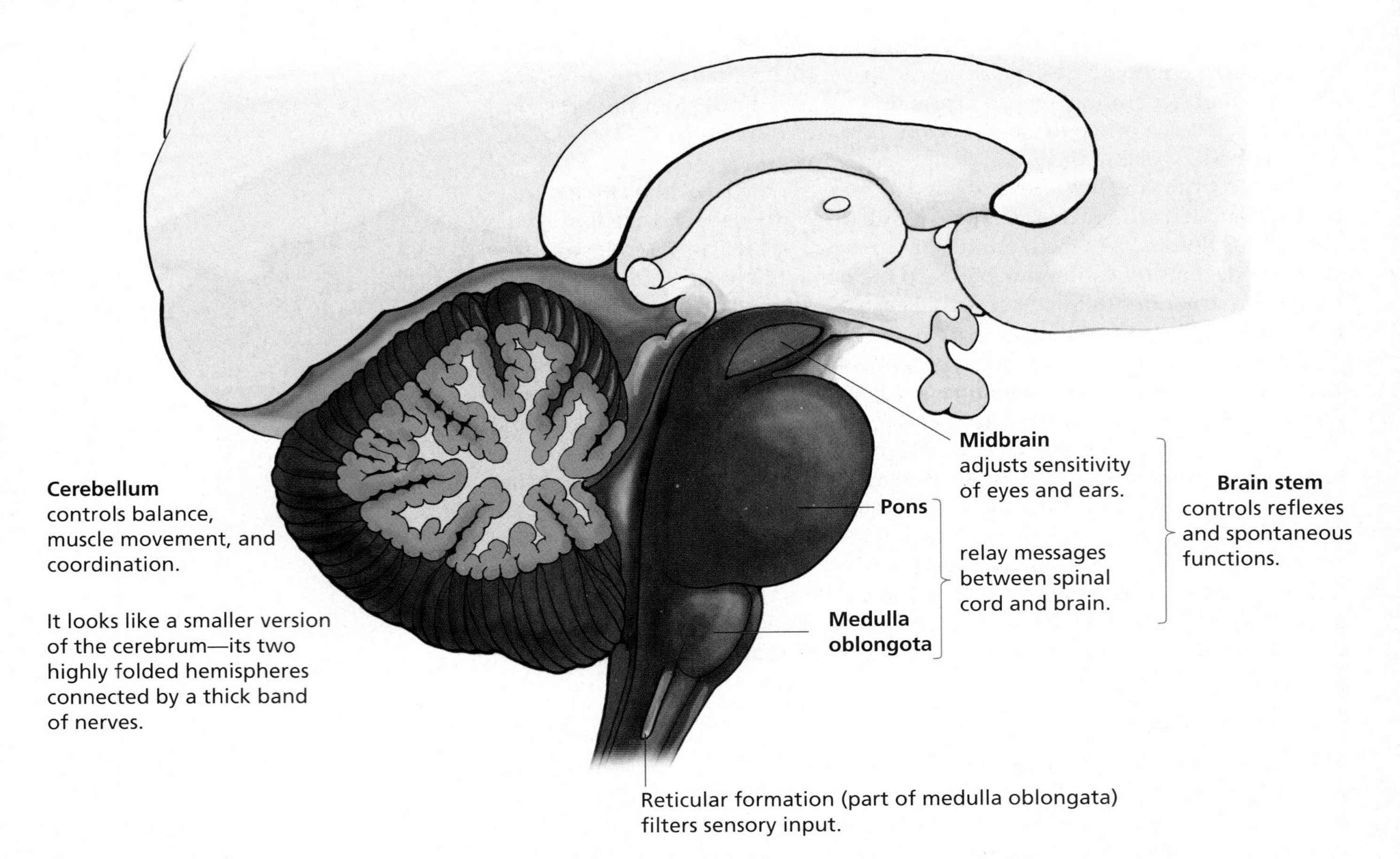

Midbrain
adjusts sensitivity
of eyes and ears.
Brain stem
controls reflexes
and spontaneous
functions.
Cerebellum
controls balance,
muscle movement, and
coordination.
Pons
relay messages
between spinal
cord and brain.
It looks like a smaller version
of the cerebrum—its two
highly folded hemispheres
connected by a thick band
of nerves.
Medulla
oblongota
Reticular formation (part of medulla oblongata)
filters sensory input.

Figure 13.8 The cerebellum and brain stem. The brain stem consists of the midbrain, pons, and medulla oblongata. The reticular formation lies within the medulla oblongata.

stimuli such as the sound of automobile traffic outside your dorm room or the sound of your roommate's breathing while you are asleep.

The reticular formation also functions as an activating center by keeping the cerebral cortex alert. While conscious activity originates in the cerebral cortex, it can only do so if the reticular formation is keeping the cortex alert.

ADD and Brain Structure and Function

Scientists have access to technologies that enable them to view the brain in the same manner that X-rays help physicians check for broken bones. Some **neurobiologists**, biologists who study the nervous system, use these technologies to find physical evidence in the brain structures of people diagnosed with **Attention Deficit Disorder**, or **ADD**. Some neurobiologists believe they have found subtle differences in the structure and function of the brains of people with ADD.

One study, which compared the brains of people with and without ADD, found that the corpus callosum (the band of nerve fibers that connects the two hemispheres of the brain) was slightly smaller in people with ADD. A similar study discovered size differences in the caudate nuclei (involved in coordinating movement)—the caudate nucleus in the right hemisphere of people with ADD was slightly smaller than in people who did not have the disorder.

Other studies have detected functional differences between the brains of people with and without ADD. Some researchers have gathered evidence indicating that the reticular formation is not as effective at filtering out signals and allows too much information to be sent on to the cerebral cortex in people diagnosed as having ADD. One research team observed decreased blood flow through the right caudate nucleus in people with ADD. Another study showed that an area of the cortex metabolizes less glucose in adults with ADD. Glucose metabolism is an indicator of activity—the more glucose metabolized, the more active an area is. This finding could indicate that this area, involved in predicting the consequences of various actions, might be less active than in people who do not have ADD.

It is important to recognize that, if there are brain differences between people with ADD and people who do not have the diagnosis, the differences may have arisen as a result of unique experiences the person has had. In other words, they may reflect inborn biological differences or distinct use patterns in much the same manner that people develop different muscles in response to exercise.

Differences in brain structure and function, whether present at birth or developed later in life, are closely related to how neurons work. By investigating the ways in which messages are transmitted by neurons, researchers have been able to propose additional hypotheses about the causes and effects of ADD.

13.3 Neurons

WWW
Media Activity 13.1A Brain and Neuron Structure

Neurons are highly specialized cells that usually do not divide. Therefore, damage to a neuron cannot be repaired by cell division and often results in permanent impairment. For example, damage to spinal nerves results in lifelong paralysis, because messages can no longer be transmitted from spinal nerves to muscles. Likewise, injury to the brain can result in permanent brain damage if neurons in the brain are harmed.

Neuron Structure

Neurons are composed of branching **dendrites** that radiate from a bulging **cell body** housing the nucleus and organelles and a long, wirelike **axon**, terminating in knobby structures called the **terminal boutons** (Figure 13.9). A nerve impulse usually travels down the axon of one neuron and is transmitted to one of the dendrites of another neuron.

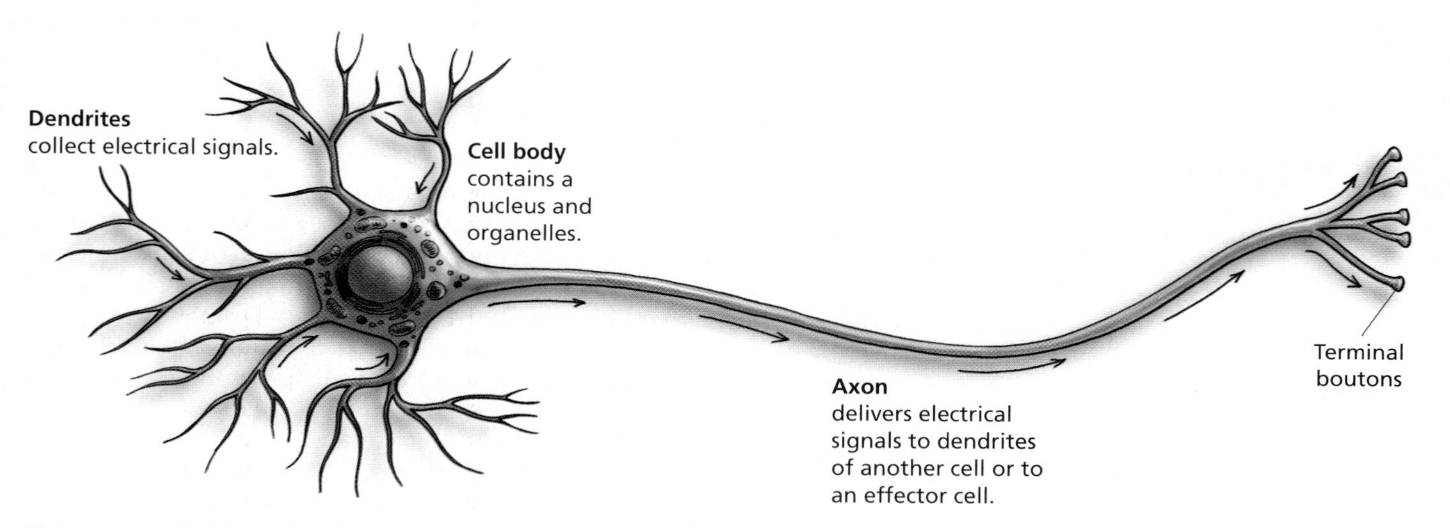
Dendrites
collect electrical signals.
Cell body
contains a nucleus and organelles.
Axon
delivers electrical signals to dendrites of another cell or to an effector cell.
Terminal boutons

Figure 13.9 The structure of a generalized neuron. A neuron consists of the branching dendrites, the cell body, the axon, and terminal boutons. Nerve impulses are propagated in the direction of the arrows.

(a) Axons covered in myelin sheath. **(b)** Nerve impulses travel more quickly on myelinated nerves.

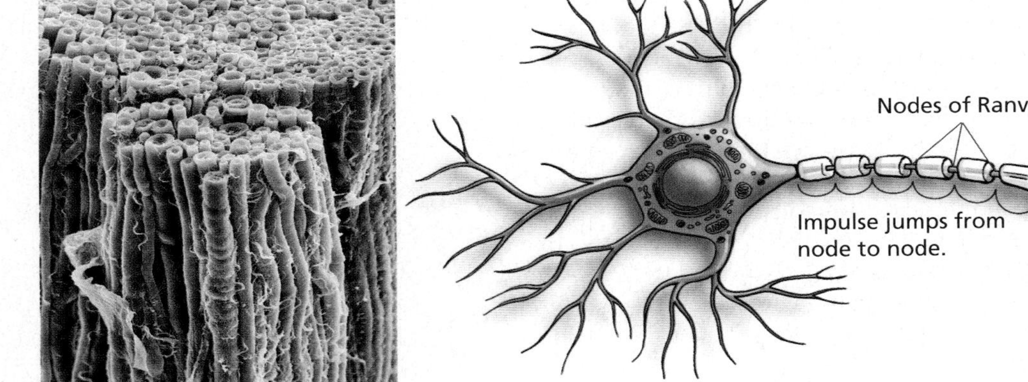
Nodes of Ranvier
Axon
Schwann cell
Impulse jumps from node to node.
Schwann cells collectively form the myelin sheath.

Figure 13.10 Myelination. (a) This photo shows a cut bundle of nerve fibers. Each individual nerve consists of a nerve cell axon (green) covered in an insulating layer of myelin (whitish). (b) Nerve impulses travel more quickly on myelinated nerves by jumping from one node of Ranvier to the next.

The axons of many neurons are coated with a protective layer called the **myelin sheath** (Figure 13.10a). The myelin sheath is formed by supporting cells such as **Schwann cells** (named for the German cell-biologist, Theodore Schwann). The myelin sheath is composed mostly of lipid and has the same appearance as animal fat. Due to its white color, nervous tissue composed of myelinated cells is called **white matter**. The myelin sheath functions like insulation on a wire by preventing sideways message transmission, which increases the speed at which the electrochemical impulse travels down the axon.

Long stretches of the myelin sheath are separated by small interruptions in the Schwann cells, leaving tiny, unmyelinated patches called the **nodes of Ranvier**. Nerve impulses "jump" successively from one node of Ranvier to the next (Figure 13.10b). This jumping transmission of nerve impulses is up to 100 times faster than signal conduction would be on a completely unmyelinated axon—nerve impulses can travel along myelinated axons at a rate of 100 meters per second (greater than 200 miles per hour).

Some neurons are not myelinated and thus transmit nerve impulses more slowly. Unmyelinated axons, combined with dendrites and the cell bodies of other neurons, look gray in cross section and so are called **gray matter** (Figure 13.11).

WWW
Media Activity 13.1B Neuron Function

Neuron Function

Neurons transmit impulses from one part of the body to another. Many kinds of stimuli, including touch, sound, light, taste, temperature, and smell, cause neurons to fire in response. When you touch something, signals from touch sensors travel along sensory nerves from your skin, through your spinal cord, and into your brain. Your brain then sends out messages through your spinal cord to the motor nerves, telling your muscles how to respond. To evoke this response, nerve cells must transmit signals along their length and from one cell to the next.

Generating and Propagating Nerve Impulses Nerve impulses are small *ionic* (involving charged particles) changes that are conducted along the length of the neuron. The inside of a neuron is negatively charged relative to the

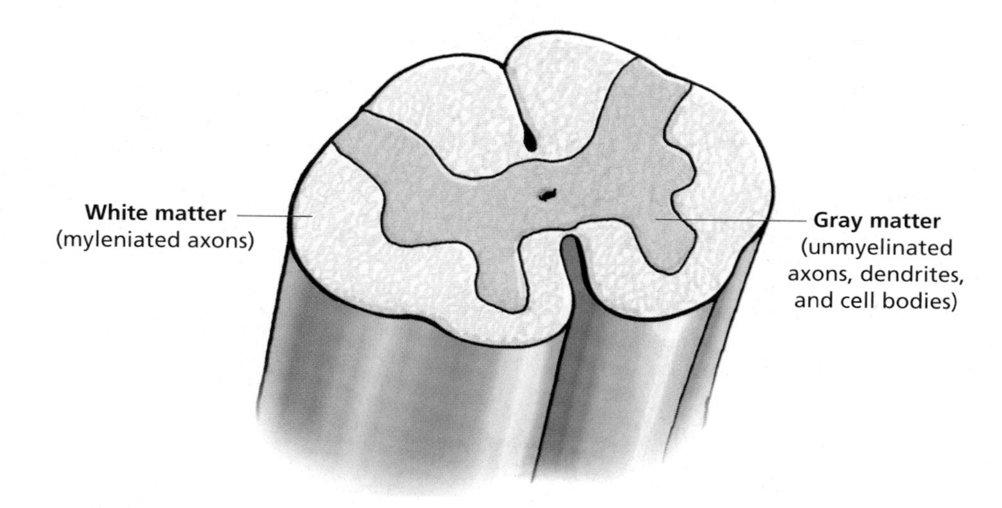

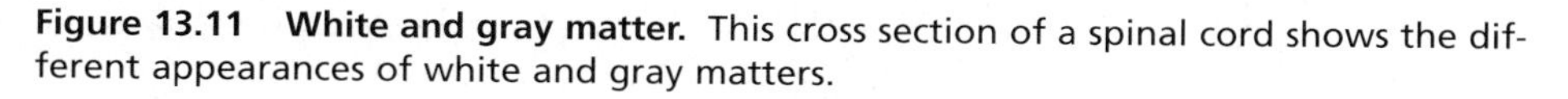

Figure 13.11 White and gray matter. This cross section of a spinal cord shows the different appearances of white and gray matters.

outside, because there are more sodium ions (Na^+) on the outside of the neuron than the inside. The cell membrane of a neuron maintains the charge differential. In Chapter 2 we learned that cell membranes are structures made of lipids and proteins, which surround the cell and regulate which substances are allowed access to a given cell type. A resting neuron maintains the difference in charge by using protein pumps in the cell membrane that physically pump sodium out of the cell. The sodium then leaches into the cell through the pores located in the cell membrane called *sodium channels* (Figure 13.12a).

When a neuron is stimulated, its cell membrane undergoes a rapid change whereby sodium channels open, allowing sodium ions into the cell. A change in charge occurs as the positively charged sodium ions enter (Figure 13.12b). The reduction in the charge difference across the cell membrane is called **depolarization**. Ultimately, a wave of depolarization, called an **action potential** or *nerve impulse*, is propagated to the end of the axon. A domino effect occurs as the positively charged sodium ions spread their charges toward each successive sodium channel, causing each to open and let in more sodium (Figure 13.12c). After the signal passes, Na^+ is again pumped out to restore the charge difference. Once the signal has travelled along the length of the axon, it must be passed to the next neuron or to the effector organ, muscle, or gland.

Synaptic Transmission Since most neurons are not directly connected to each other, the signal must be transmitted to the next neuron across a gap between the two neurons, called the **synapse**. The synapse consists of the terminal boutons of the *presynaptic* neuron, the space between the two adjacent neurons, and the cell membrane of the *postsynaptic* neuron. Saclike structures called **vesicles** are found in the terminal boutons of the presynaptic neuron. Each vesicle in a particular neuron is filled with a specific chemical *neurotransmitter*. When an electrical impulse arrives at the terminal bouton of a nerve cell, neurotransmitters are released. Once released, they diffuse across the synapse and bind to specific receptors on the cell membrane of the postsynaptic neuron. The binding of a neurotransmitter to the postsynaptic cell membrane once again stimulates a rapid change in the uptake of sodium ions by the postsynaptic neuron. The sodium channels open and sodium ions flow inward, causing another depolarization and generating another action potential. In this manner, the nerve

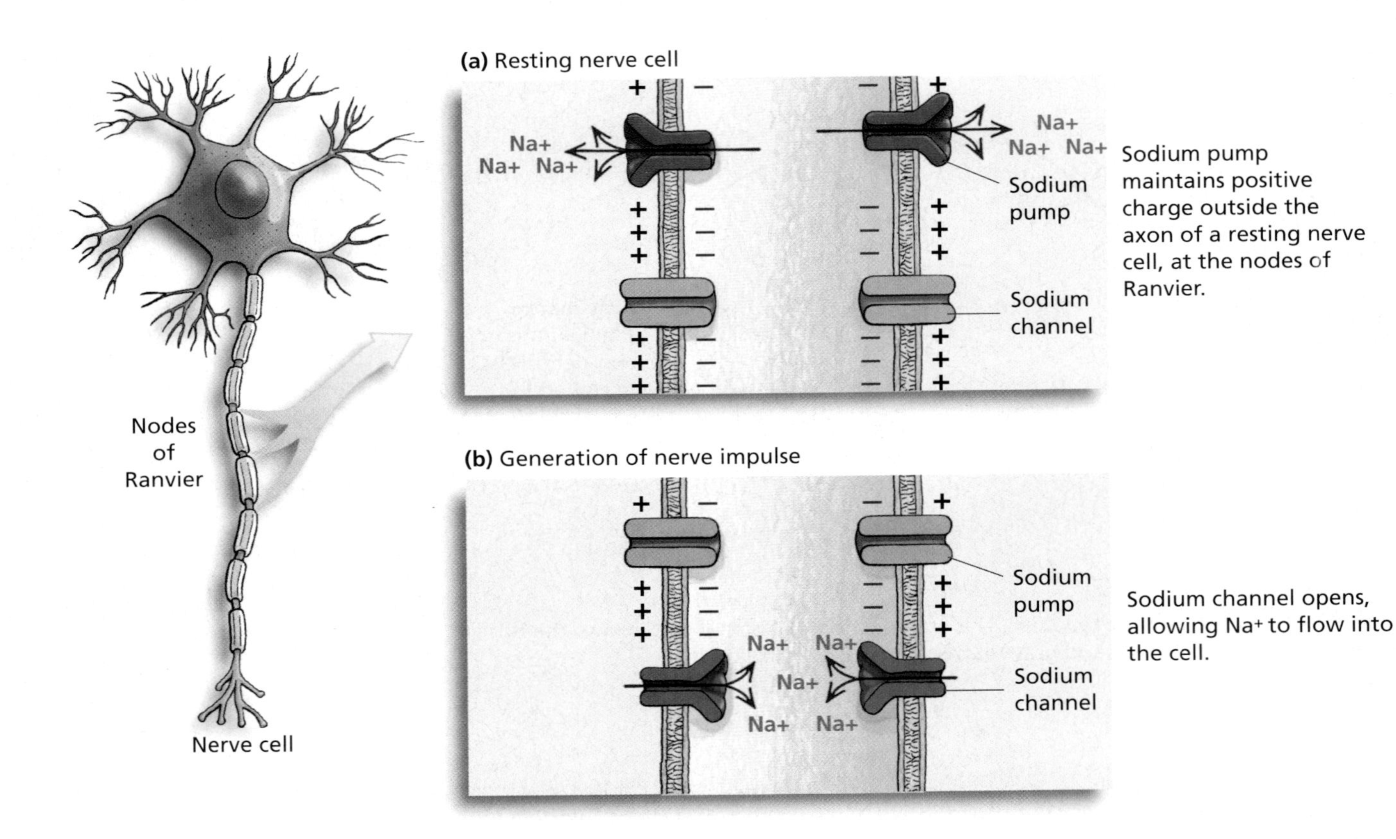

(c) Propogation of nerve impulse

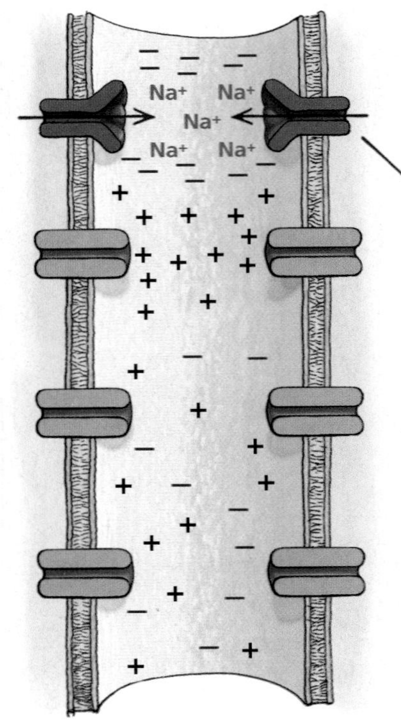

1. Nerve impulse starts with inflow of Na+ ions, which attract negatively charged ions and repulse positively charged ions.

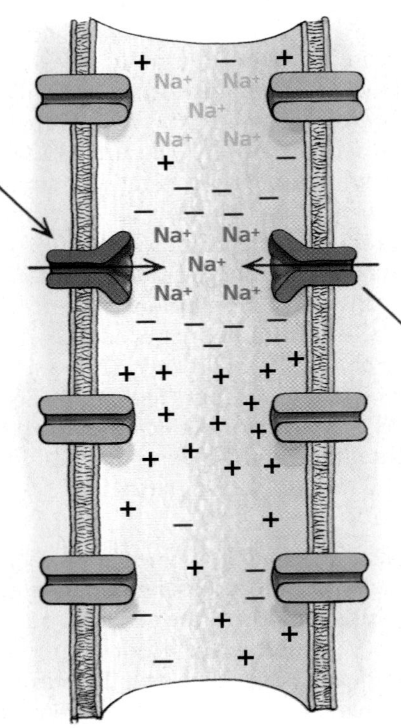

2. The spread of positively charged ions toward the next sodium channel causes it to open, allowing the Na+ to rush in.

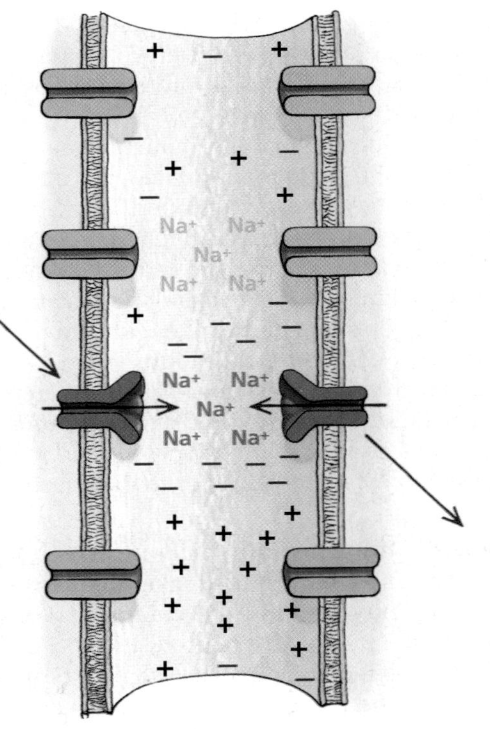

3. The depolarization passes down the axon, propagating the nerve impulse.

Figure 13.12 Generating and propagating a nerve impulse. (a) The outside of a neuron is more positively charged than the inside. This difference in charge is maintained by protein pumps that pump sodium ions (Na^+) out of cells. (b) A nerve impulse starts with an influx of Na^+ ions. (c) The influx of positively charged ions attracts negative charges and repulses other positive charges, which spread away from the channel that allowed the sodium to enter, propagating the depolarization.

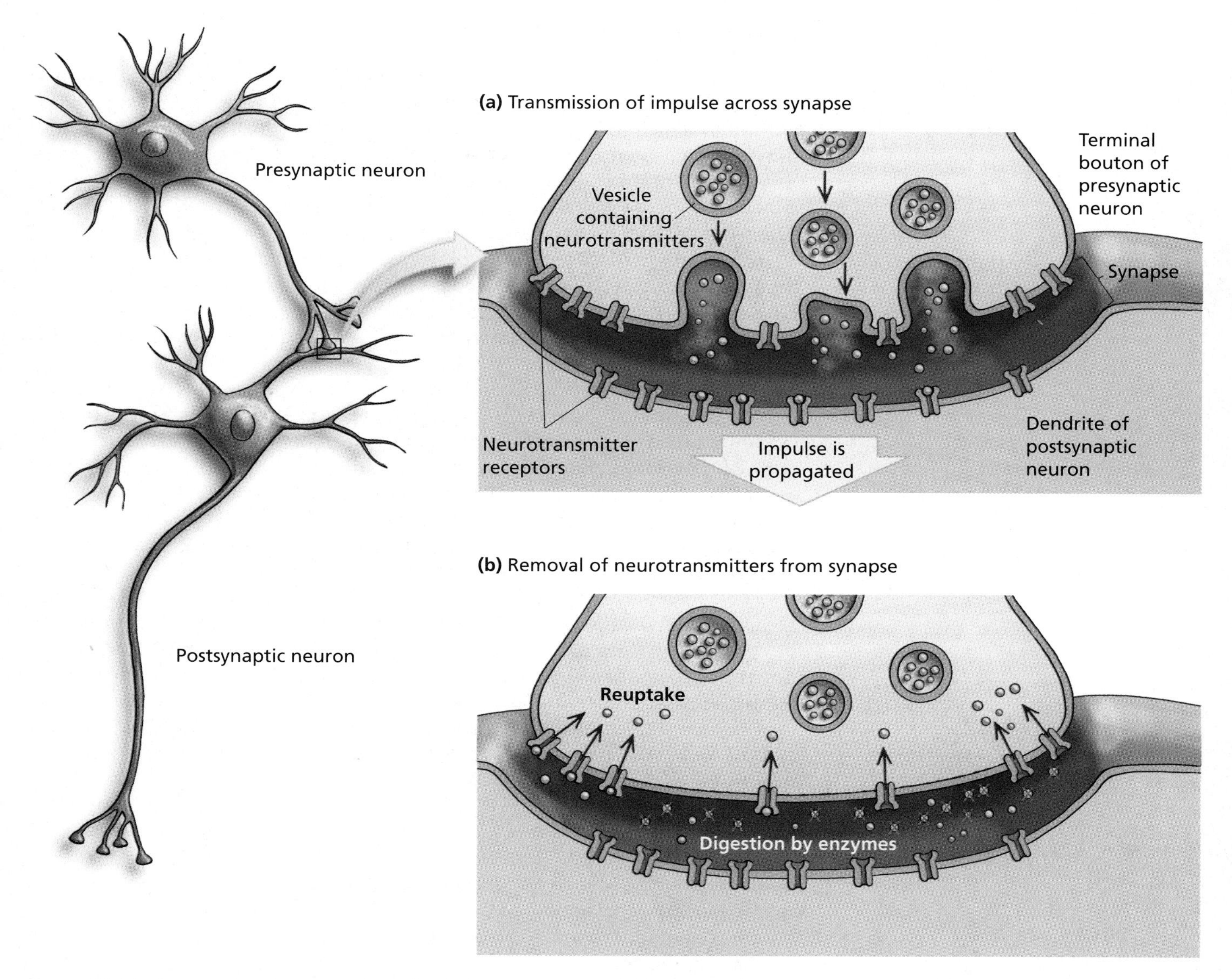

Figure 13.13 Propagating the nerve impulse between neurons. (a) The nerve impulse is transmitted from the terminal bouton of one neuron to the dendrite of the next. Neurotransmitters are released from the presynaptic neuron, travel across the synapse to the postsynaptic neuron, and bind to receptors on dendrites of the postsynaptic neuron. (b) After the neurotransmitter evokes a response, it is removed from the synapse by enzymes present in the synapse that break the neurotransmitter apart or by reuptake by the presynaptic cell.

impulse is propagated from one neuron to the next until the signal reaches the tissue it will affect (Figure 13.13a).

Over a dozen known chemicals function as neurotransmitters. Each type of neuron secretes one type of neurotransmitter, and each cell responding to it has receptors specific to that neurotransmitter.

After the neurotransmitter evokes a response, it is removed from the synapse—some are broken down by enzymes in the synapse; others are reabsorbed by the neuron that secreted them, a process called **reuptake** (Figure 13.13b). Reuptake occurs when neurotransmitter receptors on the presynaptic cell permit the neurotransmitter to reenter the cell. This allows the neuron that released the neurotransmitter to use it again. Both breakdown by enzymes and reuptake by receptors prevent continued stimulation of the postsynaptic cell.

Neurotransmission and ADD

There is a great deal of evidence linking disturbances in neurotransmission to various diseases. For example, *Alzheimer's disease*, a progressive mental deterioration in which there is memory loss along with the loss of control of bodily functions that eventually results in death, is thought to involve impaired function of the neurotransmitter *acetylcholine* in some neurons. Drugs that inhibit the enzyme *cholinesterase*, which breaks down acetylcholine, can temporarily improve mental function but cannot stop this progressive illness.

Depression, a disease that involves feelings of helplessness, despair, and thoughts of suicide, may be caused by or result in decreased levels of the neurotransmitter *serotonin*. Antidepressants blocking the actions of enzymes that degrade serotonin or inhibit its reuptake help alleviate many symptoms of this disease in some people.

Abnormal levels of the neurotransmitter **dopamine** are believed to be involved in producing the symptoms of ADD. This neurotransmitter controls emotions and complex movements. Some researchers hypothesize that people with ADD symptoms may have less dopamine than other people. Decreased dopamine levels may cause ADD symptoms because dopamine suppresses the responsivity of neurons to new inputs or stimuli. Therefore, someone with a low concentration of dopamine may respond impulsively in situations in which pausing to process the input would be more effective.

The cause of decreased dopamine levels in some people with ADD may actually involve an overabundance of dopamine receptors on the presynaptic cell. During reuptake, dopamine receptors on the presynaptic cell remove dopamine from the synapse to prevent continued stimulation of the postsynaptic neuron. Some studies have shown elevated numbers of these receptors in people with ADD. **Ritalin**, the drug usually prescribed to treat symptoms of ADD, may work to decrease the impact of these extra reuptake receptors.

WWW
Media Activity 13.2 Is Ritalin the Only Option for the Treatment of ADD?

Ritalin

Ritalin is thought to increase dopamine's ability to stimulate the postsynaptic cell by blocking the actions of the dopamine reuptake receptor (Figure 13.14). This leaves more dopamine around for a longer period of time, resulting in a decrease in the symptoms of ADD.

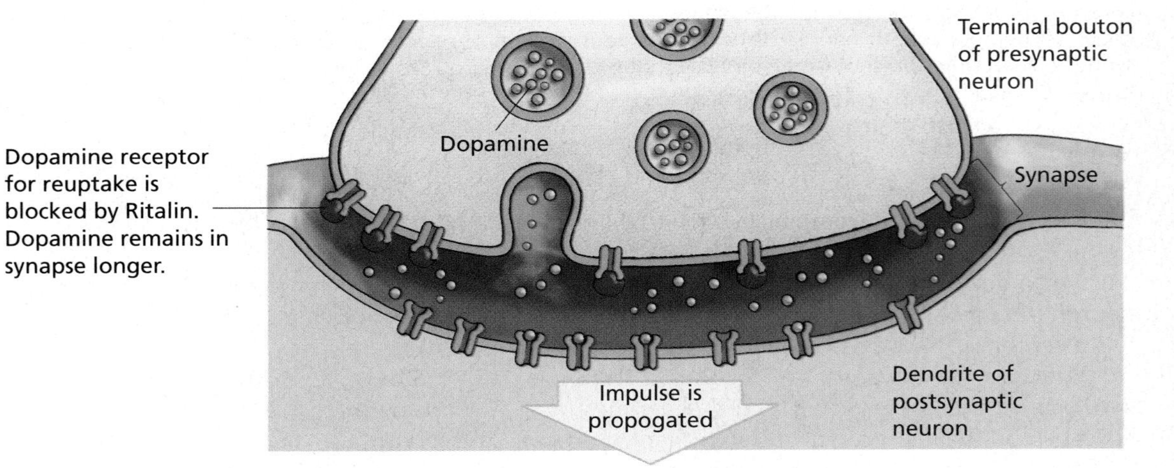

Figure 13.14 The mechanism of Ritalin action. Ritalin blocks dopamine receptors on the presynaptic neuron. This blocks reuptake and thereby allows dopamine access to the postsynaptic neuron for a longer period of time.

Other **stimulants** work through similar mechanisms. In addition to their direct effects on dopamine levels, all stimulants—whether legal like Ritalin obtained by prescription or illegal like cocaine, speed, and other amphetamines—have similar effects on the body. These drugs affect the heart, blood vessels, and lungs on their way to the brain by increasing heart rate, elevating blood pressure, and expanding airways in the lungs.

When stimulants are taken in high doses, the user feels euphoric and has more energy and endurance, a sense of power, and a feeling of mental sharpness. As the drug wears off, heightened fatigue, insomnia, poor concentration, irritability, tearfulness, and depression are common; other effects include personality changes, skin rashes, fever, nausea, and headaches. Abusing stimulants can lead to psychotic episodes, delusions, seizures, hallucinations, and sudden death.

Many of the elementary school students who grow up taking Ritalin to help them with schoolwork are still having their prescriptions filled when they head off to college. Once on campus, these students no longer stand in line for the school nurse to provide the prescribed dose. Instead, they self-administer the stimulant. Some take the prescribed dose, others use the drug only to pull an all-nighter, and still others give pills to their friends.

Due to its availability and status as a prescription drug—not a street drug such as speed or other amphetamines—students may see Ritalin as a less serious drug than other stimulants, and they are often unaware of the dangers of abusing this drug. Recent studies indicate that about 15% of college students report that they have used the drug recreationally.

If you have not been prescribed Ritalin but use it to help focus for an exam, there is no evidence that it actually helps you learn. This is because it is probable that you do not have low dopamine levels to begin with. For Ritalin abusers, the euphoria that is experienced, coupled with the fact that they are more likely to sit in a chair and focus on their schoolwork, may make abusers *believe* that Ritalin is helping them learn. As with any drug, the potential for dangerous side effects should be given serious consideration prior to use. Essay 13.1 describes many recreational drugs and their effects on the body.

The dopamine reuptake theory is not the only hypothesis about the biological cause of ADD symptoms, nor does it explain all cases of ADD. Not all drugs that alleviate ADD symptoms have any impact on dopamine levels, and many times, drugs must be combined with changes in environmental condition for the symptoms of ADD to improve.

13.4 The Environment and ADD

WWW
Media Activity 13.3 Is Our Society Over-medicated?

There is some evidence of biological differences in the brains of people diagnosed with ADD, but many people wonder if environmental factors are exacerbating the ADD situation. In the last decade, the number of people diagnosed with ADD has risen from about 1 million to between 6 million and 11 million. Such a dramatic increase in a very short period of time is not likely due to biology alone. It would take thousands of years for evolution to affect this magnitude of biological change within a population (Chapter 8).

While it may simply be that doctors are getting better at diagnosing ADD cases, it may also be that societal changes are having a negative impact on children. This is an area of active research and debate. A wide variety of people are involved in this debate, including scientists, medical doctors, psychologists, teachers, and parents.

Some people believe that the increased number of people diagnosed with ADD reflects the increased pressures children and adults face in our fast-paced,

Essay 13.1 Recreational Drugs and the Nervous System

Drugs affect each of us differently because of differences in overall metabolism, weight, and gender. The gender difference is clearly illustrated by the way alcohol consumption affects males and females.

It takes the liver one hour to metabolize one drink (a 12-ounce beer, a 6-ounce glass of wine, or a mixed drink with one ounce of alcohol) in a person who weighs approximately 150 pounds. Prior to its metabolism, the alcohol moves from the stomach throughout the body via the circulatory system. Therefore, if a person consumes more than one drink per hour, alcohol accumulates in the body and causes negative effects.

Drug	Mechanism of action	Desired mental effect	Side effect
Alcohol Alcohol is a byproduct of fermentation, the process some organisms use to produce energy when oxygen is not available. When yeast cells break down the sugars present in grapes, wine is produced. When yeast cells ferment barley, beer is produced. The fermentation of juniper berries produces gin.	Alcohol is a depressant. It gains access to cells in the brain by simply diffusing across cell membranes. It does not require the presence of specific receptors. Alcohol is a central nervous system (CNS) depressant because it inhibits neurotransmission in the reticular formation, thus inhibiting the activity of a large variety of neurons in the brain.	Reduced anxiety and a sense of well being, loss of concern for social constraints.	Impaired judgment, slurred speech, unsteady gait, slower reaction times, uncontrollable emotions. Chronic alcohol abuse leads to loss of intellectual ability and liver damage. Alcohol kills nerve cells that cannot be regenerated. As nerve cells die, the brain actually gets smaller. The frontal lobes, where judgment, thought, and reason are centered, are the first to die.
Amphetamines Used legally to treat obesity, asthma, and narcolepsy. Herbal products containing ephedra, a drug originally procured from the Mah huang plant and also synthesized in laboratories, have been marketed as treatments for weight loss and for performance enhancement. Methamphetamine is an illegal amphetamine. A crystalline form of methamphetamine called *ice* is smoked to produce effects similar to crack cocaine.	CNS stimulant that increases bodily activity by increasing the activity of the reticular formation. These drugs mimic the actions of the neurotransmitter norepinephrine, a hormone produced in response to stress. Amphetamines can also block reuptake from the synapse and inhibit the enzyme that normally breaks down norepinephrine resulting in prolonged stimulation of the postsynaptic cell. Can also increase dopamine release.	Small doses make a person feel more energetic, alert, and confident.	Effects wear off quickly, causing the person to suddenly "crash" when neurotransmitter stores are depleted, leading to depression and fatigue. Over time, the brain responds to amphetamines by decreasing the amount of its own neurotransmitters, leaving the user reliant on increasing doses to achieve the desired effect. Prolonged use of amphetamines also results in aggressiveness, delusions, hallucinations, and violent behaviors. Amphetamine use also causes blood vessels to spasm, clots to form, insufficient blood to flow to the heart, accumulation of fluid in the lungs, and death.

Table E13.1 Recreational drugs.

As a group, women metabolize alcohol more slowly than men, in part because their generally smaller bodies have proportionally smaller livers. Also, we learned in Chapter 12 that men have a greater blood volume than women, so the same amount of alcohol in a similarly sized male and female would be more dilute in the male. Finally, although the liver performs most of the alcohol breakdown, the stomach lining contains enzymes that start the process prior to its delivery to the liver through the bloodstream. The average woman has less of this enzyme than the average man. Therefore, women feel the effects of alcohol more quickly than men because more of the alcohol in a drink reaches their bloodstream. Table E13.1 outlines many drugs—how they work and what they do.

Drug	Mechanism of action	Desired mental effect	Side effect
Caffeine A naturally occurring chemical found in plants such as coffee, tea, and cocoa.	Caffeine is a general stimulant, one that affects all cells, not just those of the CNS. Caffeine does not bind receptors; it gains access to cells and, once inside, acts to increase metabolism. It does this by increasing the production of glucose. Increased glucose levels mean that the cell can support increased activity.	Mental alertness, increased energy.	Side effects of caffeine use include insomnia, anxiety, irritability, and increased heart rate.
Cocaine Cocaine, extracted from the leaves of the coca plant, which is grown in the mountains of South America, can be inhaled or injected. A more potent form of cocaine, crack, is smoked.	Cocaine is a stimulant that increases the levels of dopamine and norepinephrine by decreasing reuptake.	A rush of intense pleasure, increased self-confidence, and increased physical vigor.	Increased heart rate and blood pressure, narrowing of blood vessels, dilation of pupils, a rise in body temperature, and reduction of appetite. When cocaine wears off, its effects are followed by a period of deep depression, anxiety, and fatigue. Abuse of this drug can leave a person unable to feel positive feelings without the drug.
Ecstasy or MDMA A white crystalline powder that is primarily ingested in pill or capsule form.	Stimulant and hallucinogenic that acts to prevent serotonin reuptake. It also floods neurons with several other neurotransmitters.	Euphoria, enhanced emotional and mental clarity, increased energy, heightened sensitivity to touch, and enhanced sexual response.	Studies in rats have shown that ecstasy use permanently damages neurons involved in utilizing serotonin, and may also result in permanent memory damage. Short-term problems include confusion, anxiety, paranoia, depression, and sleeplessness that may last for several weeks. When combined with physical exertion, such as dancing, Ecstasy use may lead to severe dehydration and hypothermia, a serious condition characterized by a life-threatening decrease in body temperature.

Table E13.1 ***(continued)***

Drug	Mechanism of action	Desired mental effect	Side effect
Lysergic Acid Diethyl Amide (LSD) and Mushrooms LSD is a derivative of the fungus *Claviceps purpurea*, which grows on rye. LSD is related to the compound psilocybin, found in certain mushrooms.	Hallucinogenic thought to bind to serotonin receptors in the brain, increasing the normal response to serotonin.	Heightened sensory perception and bizarre changes in thought and emotion, hallucinations.	Hallucinations can lead users to dangerous actions. Heavy use leads to permanent brain damage, including impairments of memory and attention span, and can lead to psychosis.
Marijuana Leaves, flowers, and stems of the Indian hemp plant *Cannabis sativa*. Marijuana contains a drug called delta-9-tetrahydrocannabinol (THC).	There are receptors for THC in the portions of the brain that influence mood and pleasure, memory, pain sensations, and appetite. THC is thought to work by increasing dopamine release.	Altered sense of time, an enhanced feeling of closeness to others, and an increased sensitivity to stimuli. Large doses can cause hallucinations.	Marijuana use impairs driving ability by slowing reaction time, reducing coordination, and impairing one's ability to judge time, speed, and distance. Marijuana is also thought to impair short-term memory and to slow learning because it interferes with one's ability to pay attention and to store and acquire information. Since there are receptors for THC on the hypothalamus, the part of the brain that regulates sex-hormone secretion, long-term marijuana use can decrease testosterone production and disrupt menstruation. Long-term use is also associated with loss of motivation.
Nicotine A chemical found in tobacco plants used to make cigarettes. It is just one of the over one thousand chemicals found in cigarette smoke, but one of the most likely to affect the brain.	Nicotine is a stimulant that affects the brain by stimulating neurons in the cerebral cortex that have nicotine receptors to produce acetylcholine, epinephrine, and norepinephrine.	Increased alertness and awareness, appetite suppression, relaxation.	Smoking cigarettes increases the odds of obtaining virtually every type of cancer. Nicotine also causes increased heart rate and blood pressure.
Opiates Derived from the opium poppy. Heroin, morphine, and codeine are opiates. Morphine and codeine can be legally used to control pain.	Hallucinogenic that causes drowsiness. Opiates act by binding to opiate receptors in neurons that control feelings of pleasure. Opiate receptors are thought to have evolved to bind to opiates that are produced by the brain in response to exercise. These opiates are called endorphins.	A quick intense feeling of pleasure, followed by a sense of well-being and drowsiness.	Addiction, poor motor coordination, depression. High doses can cause coma and death. Opiates are thought to change the brain's ability to respond to normal pleasures.

Table E13.1 ***(continued)***

success-driven society. In the United States, the majority of elementary-school children's caretakers work full time—one parent in single-parent families, both parents in two-parent families—according to the 2000 census data. Today's children are involved in more and more time-consuming after-school activities such as sports, dance, music, theater, and art lessons. Some opponents of Ritalin point out that, with so much to pay attention to, it is not surprising that some kids have a hard time staying focused.

Another societal change that has paralleled increases in Ritalin use is the increase in the number of high-school graduates attending college. Between 1973 and 1994 the percentage of high-school graduates that went directly to college rose from 47% to 62%. The increased pressure to attend college means that most kids, regardless of academic ability, must be prepared for the rigors of college study. This places more pressure on teachers to put more kids on a college track, more pressure on parents who know their children will have a hard time making a good living without a college degree, and more pressure on students who have to measure up to these expectations.

Some people involved in the Ritalin debate point out that with more kids facing higher expectations, there will be more kids who need Ritalin because they simply have a hard time coping. Others argue that there have always been people who were more active, distractible, and impulsive than the norm. What *is* new is that these traits are now routinely considered to be problems that require medication.

Deciding whether ADD is caused by biology or environment influences treatment options. If this disease is solely one of neurobiology, then using medication alone to treat it makes sense. If there is an environmental component, then making environmental changes should help many people with ADD. Behavior modification therapies have been shown to improve ADD symptoms in kids who are and are not taking Ritalin. For instance, parents of ADD children are counseled about effective parenting strategies for ADD kids—such as not overscheduling them and having clearly defined and enforced rules. Teachers of ADD students often try to minimize distractions by having them sit close to the front of the room and not near hallways or windows and by providing ADD students with detailed schedules and timelines for assignments.

The success of approaches that combine medical and environmental interventions show that it may be more useful to think of ADD as an imbalance between the brain's inborn tendencies and the demands of a person's environment. Many people worry that if our society accepts the problem as only brain-based, we are much less likely to attempt to change the environment that could actually be exacerbating the problem.

It is easier to accept biological explanations for many problems and treat them with medicines instead of working to change the environment that is contributing to the problem. People take antacids to treat stomach disorders that, in some cases, would be more effectively treated through dietary changes. Likewise, heart disease is medicated when, for many people, dietary changes coupled with exercise would work as well or better. A substantial increase in use of the antidepressant Prozac has also occurred in the last decade. While there are undoubtedly people whose severe depression (like severe ADD) must be medicated, the rapid rise in Prozac use may indicate that some people are in need of dramatic changes in their environment—a person in a horrible situation might well be depressed. Taking Prozac may make the person feel better, but modifying the bad environment would probably be a better way of dealing with the problem.

As you can see, the debate about the relative roles of biology and society in causing ADD is difficult to sort out (Figure 13.15). Current hypotheses suggest that when a child with a biological predisposition toward ADD is living in a stressful environment, there is an increased likelihood of the disease developing.

(a) Biological factors that may influence ADD

	Person without ADD	Person with ADD
Caudate nuclei	Right caudate nucleus larger than left caudate nucleus; normal blood flow.	Two caudate nuclei same size? Decreased blood flow?
Corpus callosum	Normal size	Slightly smaller?
Cerebral cortex	Normal activity	Parts of cortex less active?
Reticular formation	Normal traffic flow	Not as effective at limiting the number of traffic signals to cortex?

(b) Societal factors that may influence ADD

Increase in number of families with two working parents

Increased participation in sports

Increased participation in music, art, and theater lessons

Increased number of children going to college

(c) Biological and societal factors may combine

Child with ADD brain + Stress ⟶ ADD behavior

Figure 13.15 Biological and societal factors that may influence ADD. Many factors influence ADD; these include differences in brain structure and function as well as stress produced by modern society.

CHAPTER REVIEW

Summary

- Many scientists have attempted to determine whether the brains of those diagnosed with ADD differ from the brains of the rest of the population.
- The nervous system consists of the brain and spinal cord (CNS) and the nerves that carry information to and from the brain (PNS).
- The brain is composed of the cerebrum, where most thinking occurs; the cerebellum, in charge of balance and coordination; and the brain stem, which controls many unconscious functions.
- Neurons are specialized cells whose structure consists of the branching dendrites, the cell body, the axon, and terminal boutons.
- Nerves are bundles of neurons that carry impulses to the brain and back to the muscles and organs on which they act.
- Nerve impulses are generated when depolarization of the cell membrane occurs. The electrical impulse is propagated to the ends of the axon, which house chemical neurotransmitters.
- Nerve impulses cause neurotransmitters to be released into the synapse, from where they make their way to receptors on the next neuron. Binding of the neurotransmitter to the next neuron causes depolarization of that neuron's cell membrane, and the nerve impulse is thus propagated until it reaches the muscle or organ it will affect.
- Neurological disorders can be caused by structural abnormalities of the brain and/or by defects in neurotransmission. Some scientists believe ADD is caused by differences in brain structures in ADD individuals or by deficiencies in the activities of the dopamine neurotransmitter.
- Ritalin helps people with ADD by blocking dopamine reuptake receptors on the presynaptic nerve cell so that dopamine has longer access to the receptors on the postsynaptic nerve cell.
- Some people involved in the Ritalin debate wonder whether our fast-paced society is increasing the incidence of people with ADD.

Key Terms

action potential p. 343
Attention Deficit Disorder (ADD) p. 340
axon p. 341
brain stem p. 339
caudate nuclei p. 339
cell body p. 341
central nervous system (CNS) p. 336
cerebellum p. 339
cerebral cortex p. 337
cerebrum p. 338
corpus callosum p. 339
dendrite p. 341
depolarization p. 343
dopamine p. 346
effector p. 333
glial cell p. 336
gray matter p. 342
hypothalamus p. 338
interneuron p. 333
medulla oblongata p. 339
midbrain p. 339
motor neuron p. 333
myelin sheath p. 342
nerve p. 332
nerve impulse p. 333
nervous system p. 332
neurobiologist p. 340
neuron p. 332
neurotransmitter p. 333
nodes of Ranvier p. 342
peripheral nervous system (PNS) p. 336
pons p. 339
reflex p. 335
reflex arc p. 335
reticular formation p. 339
reuptake p. 345
Ritalin p. 346
Schwann cell p. 342
sensory neuron p. 333
spinal cord p. 333
stimulant p. 347
synapse p. 343
terminal bouton p. 341
thalamus p. 338
vertebrae p. 333
vesicle p. 343
white matter p. 342

Learning the Basics

1. How does the CNS differ from the PNS?
2. Describe the three major brain structures and their functions.
3. What is a neurotransmitter?
4. Where are the receptors for neurotransmitters located?

5. How is a nerve impulse transmitted from one neuron to the next?

6. Neurons maintain a negative charge in their cytoplasm relative to outside the cell via the _____.
   a. Krebs' cycle
   b. electron transport chain
   c. dopamine receptor
   d. sodium pump

7. What do Alzheimer's disease and some forms of depression have in common?
   a. Only old people get these diseases.
   b. They are infectious.
   c. They are caused by changes in neurotransmitter levels.
   d. They are caused by environmental conditions.
   e. All of the above.

8. The effects of a neurotransmitter could be increased by _____.
   a. increasing the number of receptors on the postsynaptic cell
   b. preventing reuptake
   c. providing more enzymes involved in the synthesis of the neurotransmitter
   d. inhibiting enzymes involved in breakdown of the neurotransmitter from the synapse
   e. all of the above

9. The reticular formation _____.
   a. is inhibited by Ritalin
   b. is inhibited by alcohol when its neurons are released from regulation
   c. may be less active in kids with ADD
   d. causes cells in the liver and pancreas to secrete more neurotransmitters

10. Which of the following is a false statement about a recreational drug?
   a. LSD is structurally related to serotonin, grows on rye, and can cause permanent impairments of memory, attention span, and abstract thinking.
   b. Opiates are plant-derived narcotics that are structurally similar to some neurotransmitters.
   c. THC and some opiates mimic naturally occurring compounds in the body and bind to their receptors.
   d. Caffeine stimulates most body cells, not just those of the CNS.
   e. The hypothalamus has receptors for alcohol, which is why sex drive and appetite are altered by alcohol use.

Analyzing and Applying the Basics

1. Why might decreased glucose consumption in a brain structure indicate an underlying problem?

2. How does depolarization of a neural membrane occur?

3. How might a genetic defect in sodium pumping ability influence a neuron?

4. How might blocking neurotransmitter breakdown in the synapse affect neural function?

5. Propose a mechanism for decreasing the amount of dopamine present in the synapse, aside from increased dopamine transport back into the presynaptic neuron.

Connecting the Science

1. Some forms of depression are thought to be caused by decreased amounts of the neurotransmitter serotonin. Does this mean that all depression has a biological cause? Why or why not?

2. Suppose that your friend is an engineering major. This friend finds it very difficult to study for the courses in his major because he finds them uninteresting. He talks to you about his idea that taking Ritalin might help him study. What advice would you have for your friend?

3. Why do you think that so many American kids are taking Ritalin?

Media Activities

Media Activity 13.1 Nervous System Communication
Estimated Time: 5 minutes
You will explore the structure and function of nerves and the nervous system, and the factors that influence nervous-system communication in this activity.

Media Activity 13.2 Is Ritalin the Only Option for the Treatment of ADD?
Estimated Time: 15 minutes
Investigate the prescription drug, Ritalin, as well as other options for treating ADD.

Media Activity 13.3 Is Our Society Over-medicated?
Estimated Time: 15 minutes
You will investigate several prescription drugs used to treat depression, and then collect information on how a drug works and how often it is prescribed. You will also explore how simple it is to acquire these medications and nonprescription herbal supplements without a prescription.

CHAPTER 14

Is Earth Experiencing a Biodiversity Crisis?

Ecology and Conservation Biology

The Lost River sucker is endangered . . .

. . . and these farmers are angry.

Who has the right to use this lake for their survival—the farmers or the fish?

In June and July of 2001, angry farmers in the Klamath Basin of Oregon and California repeatedly occupied and disrupted a Bureau of Reclamation facility. The farmers used diamond-bladed chainsaws, pry bars, and blowtorches to open the head gate of an irrigation canal designed to bring water from Upper Klamath Lake to the agricultural fields of the basin. The water the irrigation canal delivers is crucial to the success of crops grown in the high desert of southern Oregon and northern California, and farmers in the Klamath Basin had been able to count on receiving this life-giving water since the early 1900s. The water source was suddenly and completely cut off in April 2001, when the U.S. Fish and Wildlife Service determined that a severe drought and the resulting loss of water from Upper Klamath Lake was decreasing the survival of several species of threatened and endangered fish—including the shortnose and Lost River suckers.

Why should we care about the fate of such a controversial endangered species—or any endangered species?

Many residents of the Klamath Basin are outraged that their homes and livelihoods are threatened by an attempt to preserve some dull-colored and economically unimportant fish. Ty Kliewer, a senior at Oregon State University whose family farms in the Basin, told his senator that he had learned the importance of balancing mathematical and chemical equations in school. "It appears to me that the people who run the Bureau of Reclamation and the U.S. Fish and Wildlife Service slept through those classes," Kliewer said. "The solution lacks balance and we've been left out of the equation." His comment is echoed by thousands of people all over the United States whose jobs have been threatened or eliminated by the government's attempts to protect endangered species.

Why is the federal government so concerned with threatened and endangered species that they are willing to take actions that cause the disruption and displacement of ordinary Americans, such as the Klamath Basin farmers? Why are such drastic measures necessary to save imperiled organisms?

14.1 The Sixth Extinction

The government agencies that stopped water delivery to the Klamath Basin farmers were acting under the authority of the **Endangered Species Act (ESA)**, a law passed in 1973 with the express purpose of protecting and encouraging the population growth of threatened and endangered species. The Lost River and shortnose suckers were once among the most abundant fish in Upper Klamath Lake—at one time they were harvested and canned for human consumption. Now, with populations of fewer than 500 and little reproduction, these fish are in danger of **extinction**—the complete loss of the species. Critically imperiled species such as the Lost River and shortnose suckers are exactly the type of organisms legislators had in mind when they enacted the ESA.

The ESA was passed as a result of the public's concern about the continuing erosion of **biological diversity**, the entire variety of living organisms. Passage of the ESA was prompted in part by the fate of the whooping crane, one of only two cranes native to North America and among the most majestic of all birds. Biologists estimate that there were more than 1,000 whooping cranes alive in the mid-1860s. By 1938—due to the loss of nesting areas and because of human hunting—the bird had disappeared from much of the continent and only two small flocks existed. By 1942, only 16 birds remained in the wild.

Increasing the number of whooping cranes has proven a challenge, because these birds reproduce slowly and require "teachers" to help them learn migratory routes. To bolster the recovery of this species, the Fish and Wildlife Service and other agencies began the laborious process of hand-rearing cranes, teaching them to behave like cranes, training them to follow traditional migration paths (Figure 14.1), and reintroducing the population to its native range. Thanks to these intensive and costly efforts, the population of whooping cranes in the wild now numbers over 300. Congress passed the ESA to stave off the potential for other species to reach the same precarious position once held by whooping cranes.

Unfortunately, the near-extinction of the whooping crane was not a unique event. Bald eagles, peregrine falcons, gray wolves, and woodland caribou—once abundant species—are or recently were critically endangered. One of the most numerous bird species on the planet, the passenger pigeon (Figure 14.2), was driven to extinction in North America a little over 100 years ago.

The extinction of the passenger pigeon was primarily the result of human activity. The conversion of North American forests to agricultural land reduced the nesting and feeding areas of these birds, and unregulated commercial hunting dealt the final blow. The dramatic population declines of eagles, falcons, wolves, caribou, and the Lost River and shortnose suckers have also been caused by humans. The ESA was drafted because humans appeared to be initiating an

Figure 14.1 Saving endangered species. These whooping cranes are following an ultralight airplane, which is leading them from their summer range in Wisconsin to their wintering grounds in Florida. Since there are no longer any wild whoopers that follow this traditional migration route, humans must teach it to these young cranes.

unprecedented rapid rate of species loss. (See Essay 14.1 for a discussion of possibly human-caused extinctions beginning about 15,000 years ago.)

Critics of the ESA argue that the goal of saving all species from extinction is unrealistic. After all, extinction is a natural process—the species living today constitute less than 1% of the species that ever existed—and trying to save rare species, as we have seen in the Klamath Basin, can be detrimental to humans. The next section explores the scientific questions ESA critics seem to pose: How does the rate of extinction today compare to the rate in the past? Is the ESA just postponing the inevitable, natural process of extinction?

Measuring Extinction Rates

If ESA critics are correct that the current rate of species extinctions is "natural," then extinction rates today should be roughly equal to rates in previous eras. The rate of extinction in the past can be estimated by examining the *fossil record*, the physical evidence left by organisms that existed in the past.

Figure 14.2 An extinct species. Passenger pigeons were once the most common bird in North America, but they were driven to extinction by human activity.

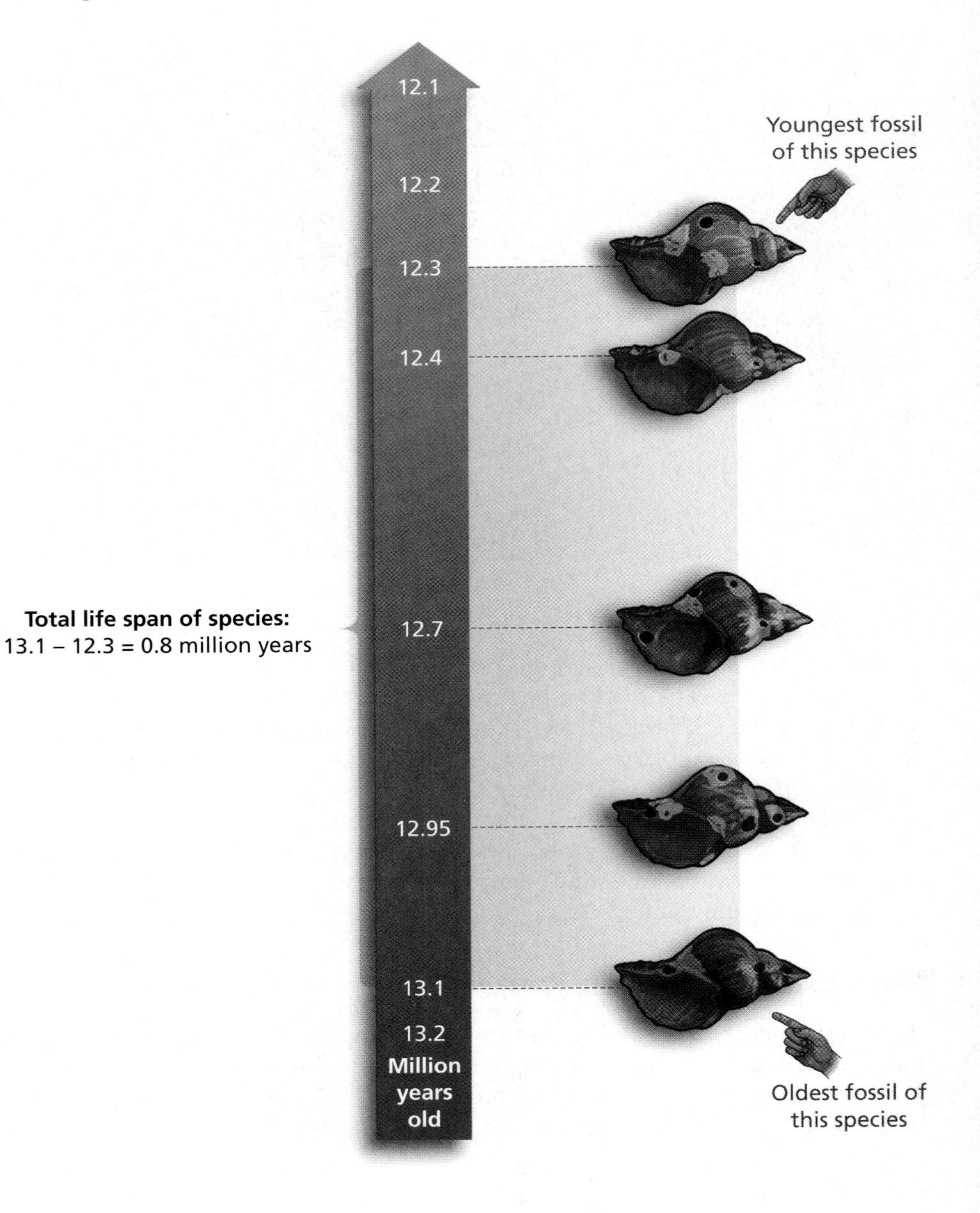

Figure 14.3 Estimating the life span of a species. The ages of these fossil shells are estimated from the age of the rocks they are embedded in via radiometric dating (Chapter 8). Fossils of the same species are arranged on a timeline from oldest to youngest. The difference in age between the oldest and youngest fossil of a species is an estimate of life span.

The span of geologic time in which fossils of an individual species are found represents the life span of that species (Figure 14.3). Biologists have thus estimated that the "average" life span of a species is around 1 million years (although there is tremendous variation), and that the overall rate of extinction is about one species per million per year, or about 0.0001% per year. Some scientists have argued that these estimates are too low, since they are based on observations of the fossil record, a record that may be biased toward long-lived species. However, most biologists agree that the estimates are currently the best approximation of background extinction rates—that is, the constant rate of species loss.

Current rates of extinction are calculated from actual recorded extinctions. This is harder to do than it seems, because extinctions are surprisingly difficult to document. The only way to conclude that a species no longer exists is to exhaustively search all areas where it is likely to have survived. In the absence of a complete search, most conservation organizations have adopted the standard that, to be considered extinct, no individuals of a species must have been seen in the wild for 50 years.

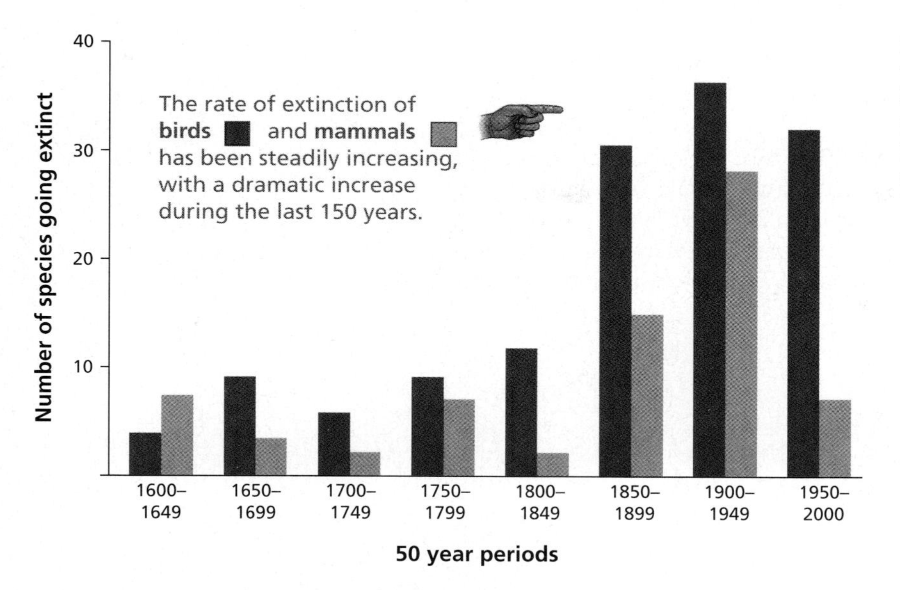

Figure 14.4 Rate of extinction. This graph illustrates the number of species of mammals and birds known to have become extinct since 1600. Note that the rate of extinction in both groups has been generally increasing, with the most dramatic increase occurring within the last 150 years.

A few searches for species of concern give a hint of the recent extinction rate. In peninsular Malaysia, a 4-year search for 266 known species of freshwater fish turned up only 122. In Africa's Lake Victoria, 200 of 300 native fish species have not been seen for years. On the Hawaiian island of Oahu, half of 41 native tree snail species were not found; and in the Tennessee River, 44 of the 68 shallow-water mussel species are missing. However, few of the missing species in any of these searches is officially considered extinct.

The most complete records of documented extinction occur in groups of highly visible organisms, primarily birds and mammals. Since A.D. 1600, 83 mammal species out of an approximate 4,500 identified have become extinct, while 113 of approximately 9,000 known bird species have disappeared. The known extinctions of mammals and birds correspond to a rate of 0.005% per year spread out over the 400 years of these records. Compared to the background rate of extinctions calculated from the fossil record, the current rate of extinction is 50 times higher. If we examine the past 400 years more closely, we see that the extinction rate has actually risen since the start of this historical record (Figure 14.4)—to about 0.01% per year, making the current rate 100 times higher than the calculated background rate.

In addition, there are reasons to expect that the current elevated rate of extinction will continue into the future. The World Conservation Union (known by its French acronym, IUCN), a highly respected global organization made up of and funded by states, government agencies, and nongovernmental organizations from over 140 countries, collects and coordinates data on threats to biodiversity. According to the IUCN's most recent assessment, 11% of all plants, 12% of all bird species, and 24% of all mammal species (the three best-studied groups of organisms) are in danger of extinction, and human activities on the planet pose the greatest threat to most of these species.

Nowhere to Live: Human Causes of Extinction

The dramatic reduction in the numbers of shortnose and Lost River suckers in Upper Klamath Lake is almost entirely the result of human modification of their **habitat**, the place where they live. At one time, 350,000 acres of wetlands protected the quality and regulated the amount of water entering the lake. Most of these wetlands have been drained and converted to irrigated agricultural

Essay 14.1 The Pleistocene Extinctions

During the last great Ice Age, approximately 100,000 to 10,000 years ago, a widespread extinction of large mammals occurred on Earth. Dozens of mammal species in 27 genera were lost, and eight of these genera were in North America—including the wooly mammoth, mastodon, sabertooth cat, and giant ground sloth (Figure E14.1). The majority of extinctions in North America took place at the end of the Pleistocene period, 13,000–10,000 years ago. Curiously, no other group of animals suffered this level of extinction; plants and small animals were largely unaffected.

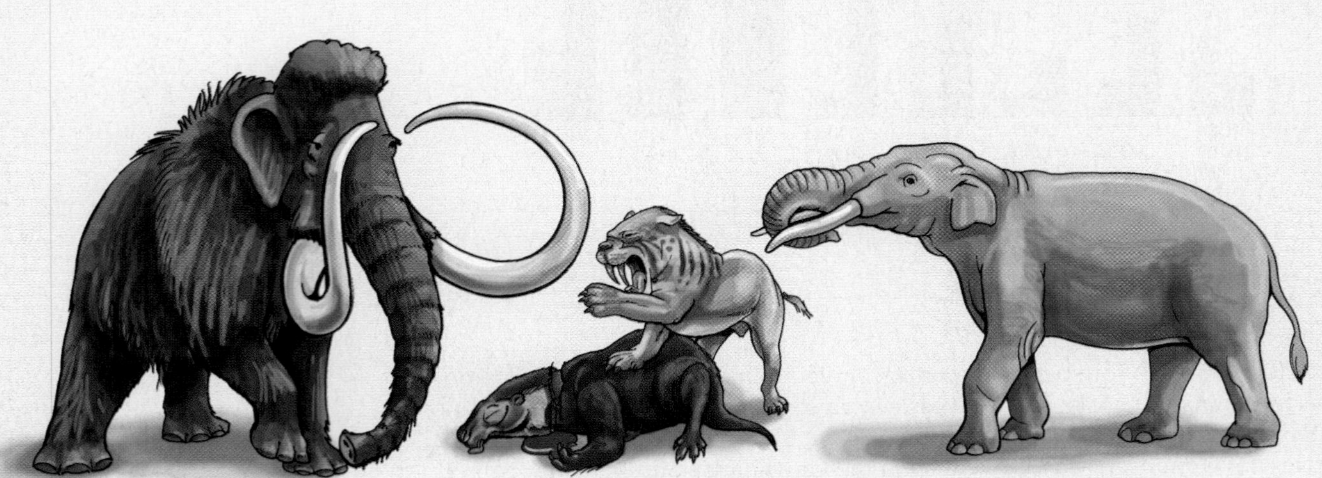

Figure E14.1 Large animals of the Pleistocene.

fields. Alterations of the natural water flow in and out of the lake have reduced the number of offspring produced by suckers by as much as 95%. In addition to the loss of breeding habitat, accumulations of fertilizer from water draining off nearby farms have resulted in massive fish kills in Upper Klamath Lake.

Most other endangered species are in this vulnerable state for similar reasons. The primary cause of species loss is **habitat destruction**—principally, the modification and degradation of natural forests, grasslands, wetlands, and waterways by people during agricultural, industrial, and residential development. An IUCN review of the risks faced by endangered species indicates that 83% of endangered mammals, 89% of endangered birds, and 91% of endangered plants are directly threatened by habitat destruction. As the amount of a *natural landscape* (one that is not strongly modified by humans) declines, the number of species supported by the habitats in these landscapes also falls.

The relationship between the size of a natural area and the number of species it can support follows a general pattern called a **species–area curve**. A species–area curve for reptiles and amphibians on a West Indian archipelago is illustrated in Figure 14.5a. Similar graphs have been generated by studies of different groups of organisms in a variety of habitats, although the actual relationship between habitat area and number of species depends on the type of species that is of interest. The general pattern is that the number of species in an area increases rapidly as the size of the area increases, but the rate of increase slows as the area becomes very large. This "rule of thumb," an approximation derived from the studies, is shown in Figure 14.5b. From the graph, we can estimate that a 10-fold decrease in landscape area will cut the number of species living in the remaining area in half.

The cause of Pleistocene extinctions is the focus of intense research and is a subject of great debate among scientists. The leading hypotheses include overhunting by humans (also called "overkill") and rapid climate change.

The timing of large mammal extinctions in different geographic regions supports the overkill hypothesis. Human populations that were presumably experienced big-game hunters are believed to have crossed the Bering Strait on the land bridge joining Asia to North America approximately 15,000 to 12,000 years ago. This migration shortly precedes the majority of large-mammal extinctions that occurred in North America. Similar patterns are apparent in South America, Australia, and northern Eurasia where mass extinctions occurred shortly after humans moved into each area.

Additional support for the overkill hypothesis is found in the archeological record of these early hunters—spear points have been discovered in association with large-mammal remains at many sites. Large kill sites in ravines and at the bottom of cliffs of wooly mammoths, mastodons, and bison have been identified in several North American locales, indicating that the hunting techniques employed by early humans resulted in the slaughter of thousands of individual animals. The loss of these large grazing animals probably caused the subsequent extinction of other species, including large predators such as the sabertooth cat.

Critics of the overkill hypothesis argue that humans were not numerous enough to drive these abundant animals to extinction in such a short period of time, nor could they have achieved such an effective hunting style. These critics point out that bison remained a mainstay of the Native American diet until European settlement and were incredibly numerous until both the widespread conversion of prairie into agricultural land and indiscriminate sport hunting. These scientists argue that a much more powerful force than humans led to these extinctions—the change in climate that occurred after the retreat of the continental glaciers.

The real challenge of climatic change for large grazers was the subsequent change in vegetation. Records of pollen grains in lake sediments indicate that warmer temperatures at the end of the Ice Age resulted in an increase in forest cover and a decrease in grassland. Supporters of the climate-change hypothesis also argue that the survivors should possess traits that increase their survival in warmer, wetter environments. It is true that the modern descendants of Ice Age survivors (such as bison and beaver) are smaller, as is expected of animals in warmer environments. Perhaps extinct wooly mammoths and mastodons simply did not have genetic traits that would enable their populations to adapt to the new climate conditions. Finally, the fossil record shows that larger animals are always more significantly affected by climate change than smaller animals, which may explain the pattern of loss affecting only the largest mammal species.

It is likely that both climate change and overkill played roles in the extinction of large mammals at the end of the Pleistocene. It also appears that many large animals that survived, but that were negatively affected by climate change, could have been pushed over the brink by human activity.

WWW **Media Activity 14.1** Species–Area Curve

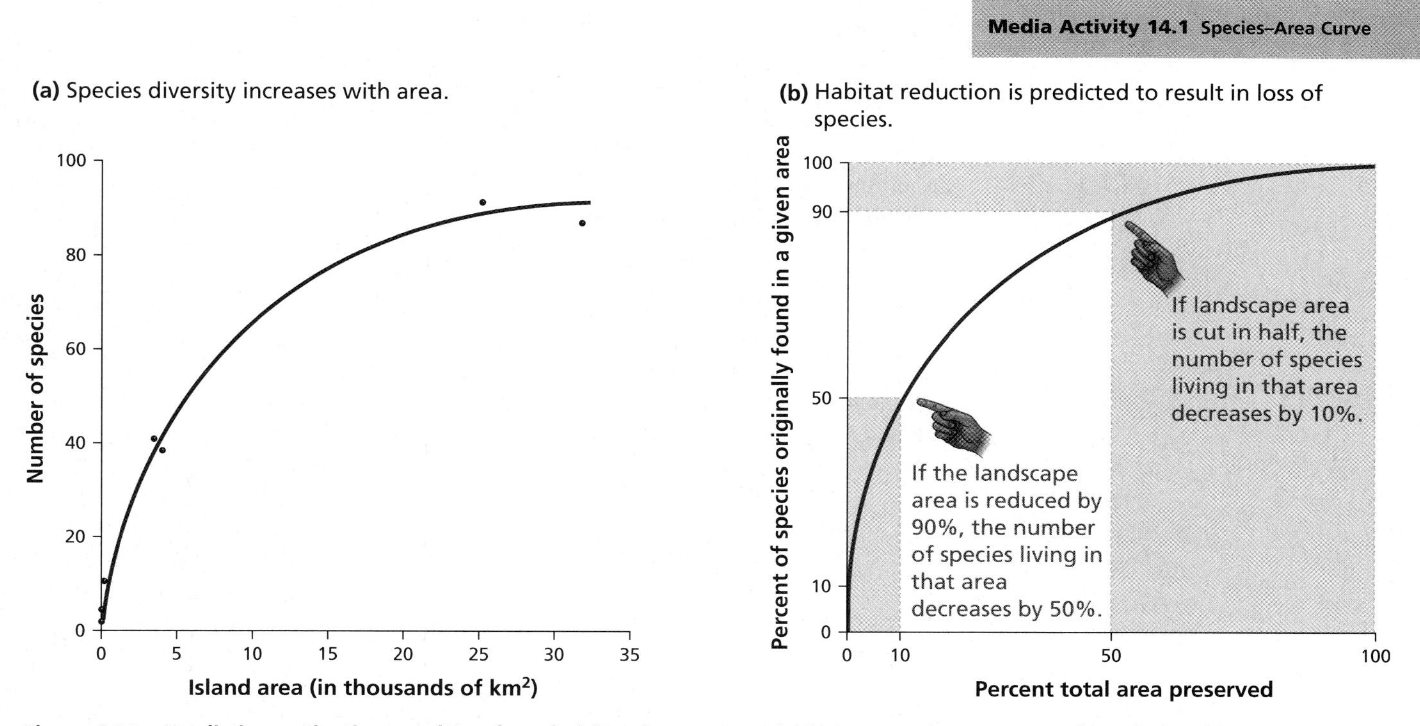

Figure 14.5 Predicting extinction resulting from habitat destruction. (a) This curve demonstrates the relationship between the size of an island in the West Indies and the number of reptiles and amphibians that live there. (b) We use a generalized species–area curve to roughly predict the number of extinctions in an area experiencing habitat loss.

Applying species–area curves to estimate extinction rates requires that we calculate the amount of natural landscape that has been lost in recent decades—a difficult task. Most studies have focused on tropical rainforests, which cover a broad swath of land roughly 20 degrees north and south of the equator. Tropical rainforests contain, by far, the greatest number of species of any *biome*, or major habitat type, on Earth. An early estimate of the rate of habitat destruction in the rainforest was made by biologist Edward O. Wilson, who calculated that about 1% of the tropical rainforest is converted to agricultural use every year. Conservatively estimating the number of species in the rain forest at 5 million, Wilson applied the generalized species–area curve and projected that nearly 20,000 to 30,000 species are lost each year due to rainforest destruction.

More modern studies using images from satellites (Figure 14.6a) indicate that approximately 20,000 square kilometers (about 7,722 square miles, an area the size of the state of Massachusetts) of rainforest in South America's Amazon River basin is cut each year. This is a rate of 2% per year, or twice the rate Wilson estimated. At this rate of habitat destruction, tropical rainforests will be reduced to 10% of their original size within about 35 years. If we apply the species–area curve, the habitat loss translates into the extinction of about 50% of the species that call Amazonian rainforests home. The extinct species in the rainforest would include about 50,000 of all the known 250,000 species of plants, about 1,800 of the known 9,000 species of birds, and about 900 of the 4,500 kinds of mammals in the world.

Of course, habitat destruction is not limited to tropical rainforests. When all of Earth's biomes are evaluated, freshwater lakes and streams, grasslands, and temperate forests are also experiencing high levels of modification. According to the IUCN, if habitat destruction around the world continues at present rates, nearly $\frac{1}{4}$ of *all* living species will be lost within the next 50 years.

Some critics have argued that these estimates of future extinction are too high, because not all groups of species are as sensitive to habitat area as the curve in Figure 14.5b suggests, and many species may still survive and even thrive in human-modified landscapes. Other biologists counter that there are also other threats to species, therefore the rate of species loss should be even higher than these estimates.

Other threats to biodiversity by humans include habitat fragmentation, the introduction of exotic species, overexploitation, and pollution (Figure 14.6b–e). **Habitat fragmentation** occurs when large areas of intact natural habitat are subdivided by human activities. The species remaining in the habitat fragments are more susceptible to extinction because it is more difficult for individuals to move across modified landscapes than across natural landscapes. Habitat fragmentation thus makes it impossible for species to move from an area that is becoming unsuitable because of natural environmental changes to areas that are suitable. Individual organisms within habitat fragments are also more susceptible to being killed or disrupted by humans and human-adapted species (such as domestic cats), because they are in closer proximity to these threats than individuals in unfragmented habitats. For example, grizzly bears need from 200–2,000 square kilometers of habitat to survive a Canadian winter, but the Canadian wilderness is increasingly bisected by roads built for tree harvesting. Every interaction between grizzly bears and humans represents a greater danger to the bears than to humans—for example, of the 136 grizzlies that died in Canada's national parks between 1970 and 1995, 119 were killed by humans. **Exotic species** include domestic cats and organisms introduced by human activity to a region where they had never been found. Exotic species are often dangerous to native species because they have not evolved together—for instance, many birds on oceanic islands such as Hawaii and New Zealand are unable to defend themselves from introduced ground hunters, such as rats. **Overexploitation** encompasses overhunting and overharvesting and occurs when the rate of human destruction or use of a species outpaces its

(a) Habitat destruction

Humans are rapidly destroying tropical rainforests. This 1999 satellite photo illustrates the extent of destruction in an area of Brazilian rainforest, that, until 30 years ago, contained no agricultural lands. The lighter parts of the photo are agricultural fields; the darker regions are intact forest.

(b) Habitat fragmentation

This "island" of tropical forest was created when the surrounding forest was logged. Scientists have documented hundreds of localized extinctions within fragments such as this.

(c) Introduction of exotic species

The introduced brown tree snake is responsible for the extinction of dozens of native bird species on the Pacific island of Guam.

(d) Overexploitation of species

These tiger skins represent a small fraction of the illegal harvest of tigers in Asia, primarily for the Chinese market.

(e) Pollution

Pollution from herbicides appears to be responsible for the increase of deformities in frogs in the midwestern United States and may partially explain the worldwide decline in frog species.

Figure 14.6 The causes of extinction.

ability to reproduce. Humans overexploited passenger pigeons to the point of extinction and decimated populations of gray wolves, sea turtles, and many whale species as well. **Pollution**, the release of poisons, excess nutrients, and other wastes into the environment, poses an additional threat to biodiversity. The massive fish kills in Upper Klamath Lake were caused by fertilizer pollution, and as we discuss in Essay 14.2, global climate change caused by pollution may be the most serious threat to biodiversity yet.

All of these threats indicate that ESA critics who describe modern species extinction rates as "natural" are most likely incorrect. Over the past 400 years, humans have caused the extinction of species at a rate that appears to far exceed past rates, and it is clear that human activities continue to threaten thousands of additional species around the world. In fact, many scientists argue that the Earth is on the brink of a **mass extinction**—a loss of species that is global in scale, affects large numbers of species, and is dramatic in its impact. Earth has experienced five episodes of mass extinction, in which 50–90% of all living species were lost over the course of a few thousand to a few hundred-thousand years (Figure 14.7). Past mass extinctions were probably caused by massive global changes—for instance, changes in sea level brought about by climate fluctuations, or changes in ocean and landform caused by movements of Earth's *tectonic plates*. Many scientists argue that we are now seeing biodiversity's sixth great mass extinction—and the pervasive global change causing this extinction is human activity.

After previous mass extinctions, biological diversity did not reach pre-extinction levels for 5–10 million years. The species that replaced those lost in the mass extinction were also very different. We cannot predict what biodiversity will look like after another mass extinction. Many people who feel a moral responsibility to minimize the human impact on other species and preserve the majority of biological diversity for future generations continue to support actions that preserve species, despite the cost. However, in addition to supporting the ideals of rights for nonhuman species and preservation for the sake of our children and grandchildren, there is a practical reason to prevent the sixth extinction from occurring—the loss of nonhuman species can cause human suffering as well.

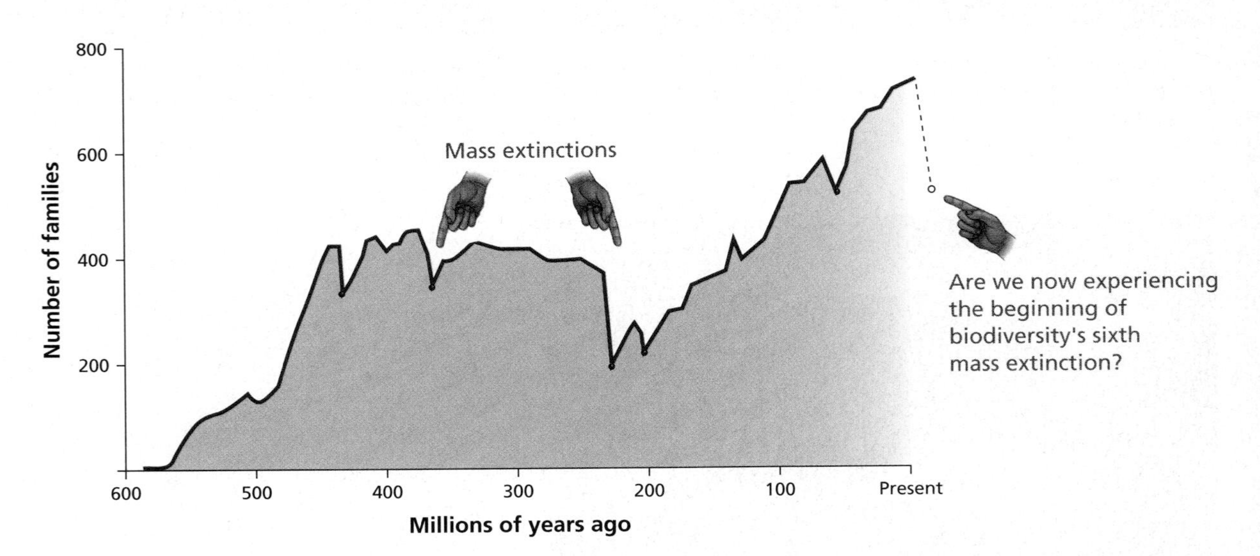

Figure 14.7 Mass extinction. This graph illustrates the general rise in biodiversity over the past 600 million years, as indicated by an increase in the number of marine families present in the fossil record. However, this rise has been punctuated by five mass extinctions, marked with black dots, which resulted in a global decline in biodiversity. The number of species lost during these mass extinction events appears to be even greater than the number of families lost, because families containing many species died out.

14.2 The Consequences of Extinction

Concern over the loss of biological diversity is not simply a matter of ethics or a theoretical issue. The human species and our societies and cultures have evolved with and among the variety of species that exist on our planet, and the loss of these species often results in negative consequences for us.

Loss of Resources

The Lost River and shortnose suckers were once numerous enough to support a fishing and canning industry on the shores of Upper Klamath Lake. Even before the arrival of European settlers, the native people of the area relied heavily upon these fish, which are referred to as "c'wam" and "qapdo" in the Klamath language, as a mainstay of their diet. The loss of these species represents a tremendous impoverishment of wild food sources. The list of species that are harvested directly from natural areas is large and diverse—for example, wood for fuel, shellfish for protein, plants for local medicines, and flowers for perfume. The loss of any of these species affects human populations economically—one estimate places the economic value of wild species in the United States at $87 billion a year, or about 4% of the gross domestic product. Some of the thousands of valuable species known to humans are described in Chapter 3.

Wild relatives of plants and animals that have been domesticated by humans (such as agricultural crops and cattle) are also valuable resources for humans (Figure 14.8). Genes and gene variants that have been "bred out" of domesticated species are often still found in their wild relatives. These genetic resources represent a reservoir of traits that could be reintroduced into agricultural species by breeding or genetic engineering (see Chapter 7). Often, traits that seem unimportant in agricultural crops later prove to be useful, or in fact critical, to the species when it is exposed to new diseases or pests. Agricultural scientists who are attempting to produce better strains of wheat, rice,

(a) Teosinte, ancestor of modern corn (b) Aurochs, ancestor of modern cattle

Figure 14.8 Wild relatives of domesticated crops and animals. (a) Teosinte is the ancestor of modern corn, which was first cultivated in Central America. This species, *Zea diploperennis*, was discovered in a remote Mexican site in 1978. (b) An aurochs, the wild ancestor of modern cattle species, disappeared from its original range in Northern Europe in the 1500s. This modern aurochs was produced after many generations of cross-breeding domesticated cattle and is probably very similar to the extinct wild aurochs.

Essay 14.2 Global Climate Change

Carbon dioxide, a gas produced by burning fossil fuels such as oil, coal, and natural gas, has been steadily accumulating in the atmosphere over the past 150 years (Figure E14.2a). At current rates of production, levels of carbon dioxide in the atmosphere are expected to double by the year 2075.

Scientists who study climate agree that such a large and rapid change in carbon dioxide levels influences weather patterns on Earth. Atmospheric carbon dioxide contributes to the *greenhouse effect*, which keeps Earth relatively warm by preventing the heat generated by the sun shining on its surface from escaping into space (Figure E14.2b). Given this effect, an increase in carbon dioxide in the atmosphere should result in an increase in global temperature. In fact, most computer models of Earth's climate predict that average global temperatures will increase between 3° and 8° F (1.5–4.5° C) by 2075 as a result of increasing carbon dioxide emissions. These models also predict that the warming will not be uniform. Certain areas of the globe, such as at the North and South poles, will warm faster and to a greater degree than other areas, and some regions may even cool slightly. Accompanying this temperature shift will be a change in rain and snowfall—again, a change that varies from region to region. Models predict the future long-term effect of warming on Earth, but there is plenty of evidence that our planet is already warming—from the retreat of alpine glaciers to the gradual upward creep in average yearly temperatures over the past several decades.

Long-term temperature records, such as the data shown in Figure E14.2c that were inferred from ice cores taken from Antarctica, indicate that global temperatures have always fluctuated. However, as you can see on the graph, the concentration of carbon dioxide in the atmosphere is much higher now than at any time in the past 400,000 years. If the change in the level of greenhouse gases is followed by the rapid development of higher global temperatures, species in many habitats will certainly suffer. Melting of glaciers and ice caps will cause sea levels to rise, flooding low-lying coastal areas and some of the most unique and diverse habitats on the planet—tropical oceanic islands. Warmer winters and longer summers will allow some species of plant-eating insects to thrive, potentially causing severe damage to many different environments. Insects that carry disease may also thrive in a warmer climate, threatening the health of human and nonhuman populations alike. If the ice-core record of Figure E14.2c is any guide, it is possible that the current spike in temperature will be followed by rapid global cooling, another blow to the remaining species.

Global warming is already posing a threat to biodiversity. A review published in the journal *Nature* in March 2002 described various species and ecosystems that have been affected by climate change. Many of these species are temperature-sensitive and must move closer to the poles or higher in elevation to find regions with the proper climate. Responses to climate change have been documented—for instance, arctic fox are retreating northward and being replaced by the less cold-hardy red fox; Edith's checkerspot butterfly is now found 124 meters higher in elevation and 92 kilometers north of its range in 1900; and a wide variety of corals have experienced a dramatic increase in the frequency and extent of damage resulting from increased ocean temperatures.

Habitat destruction greatly interferes with a species' ability to respond to global warming. If climate change renders the current habitat of an organism unsuitable for its survival, its only hope of persisting is dispersal to a more appropriate habitat. If the species is slow to disperse, such as a long-lived tree species, or if it cannot cross human-modified landscapes, it may not be able to establish a new population in an appropriate habitat before it becomes extinct in its rapidly changing home. Global climate change is another uncertainty created by humans that threatens the survival of thousands of species on Earth.

and corn look to the wild relatives of these crops as sources of genes for pest resistance and for traits that improve yields in specific environmental conditions. The survival of these wild relatives is crucial for maintaining and expanding crop production. There is value in preserving wild relatives in their natural habitats as well—often the organisms in these communities provide the key to reducing pest damage and disease on the domestic crop. For example, the original habitat of the crop may be home to an insect that is an effective predator of the crop's insect pests.

Other species currently provide, or may provide, future direct benefits to humans for less obvious reasons. Nearly 25% of the drugs prescribed in the United States come from wild sources, and there are potentially thousands of other medically useful organisms that have not been investigated. Modern

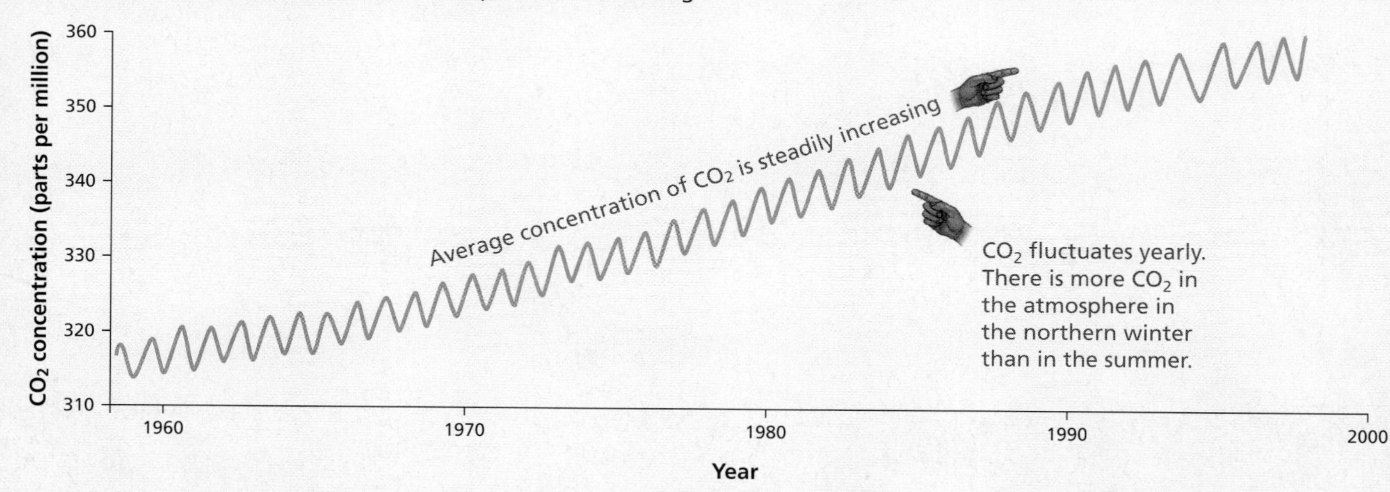

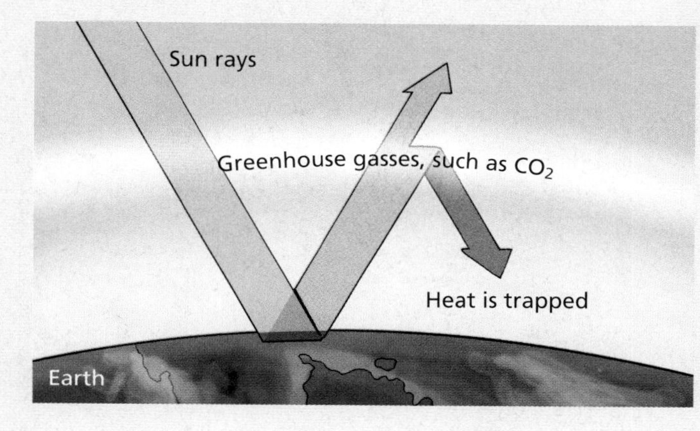

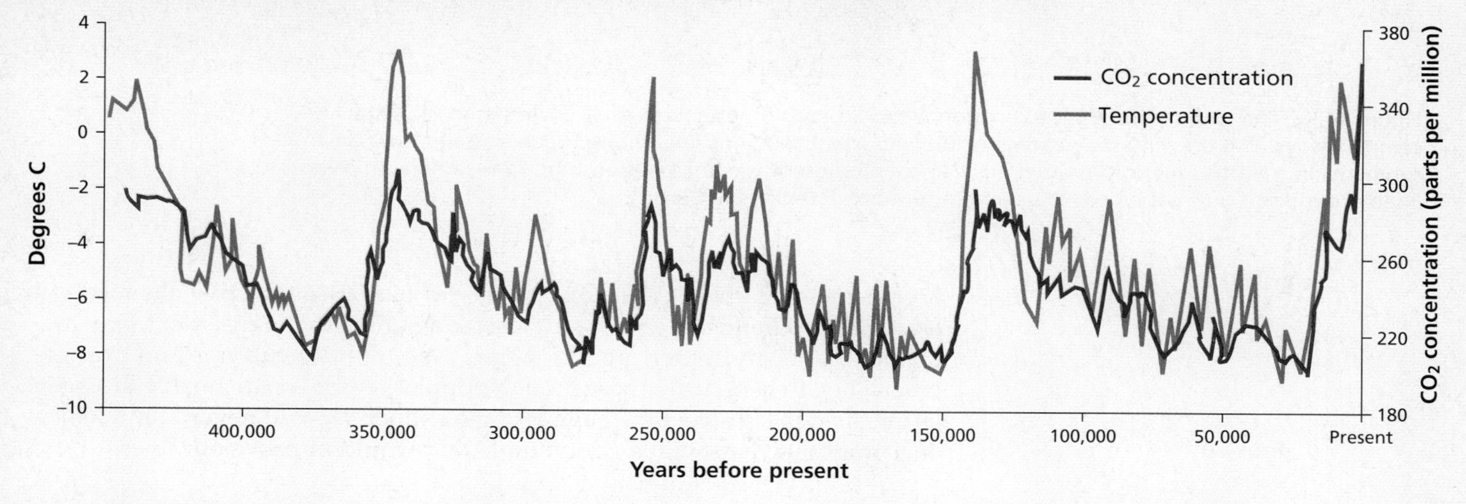

Figure E14.2 (a) The increase of carbon dioxide in the atmosphere near Mauna Loa, Hawaii. (b) The greenhouse effect. (c) Long-term temperature records.

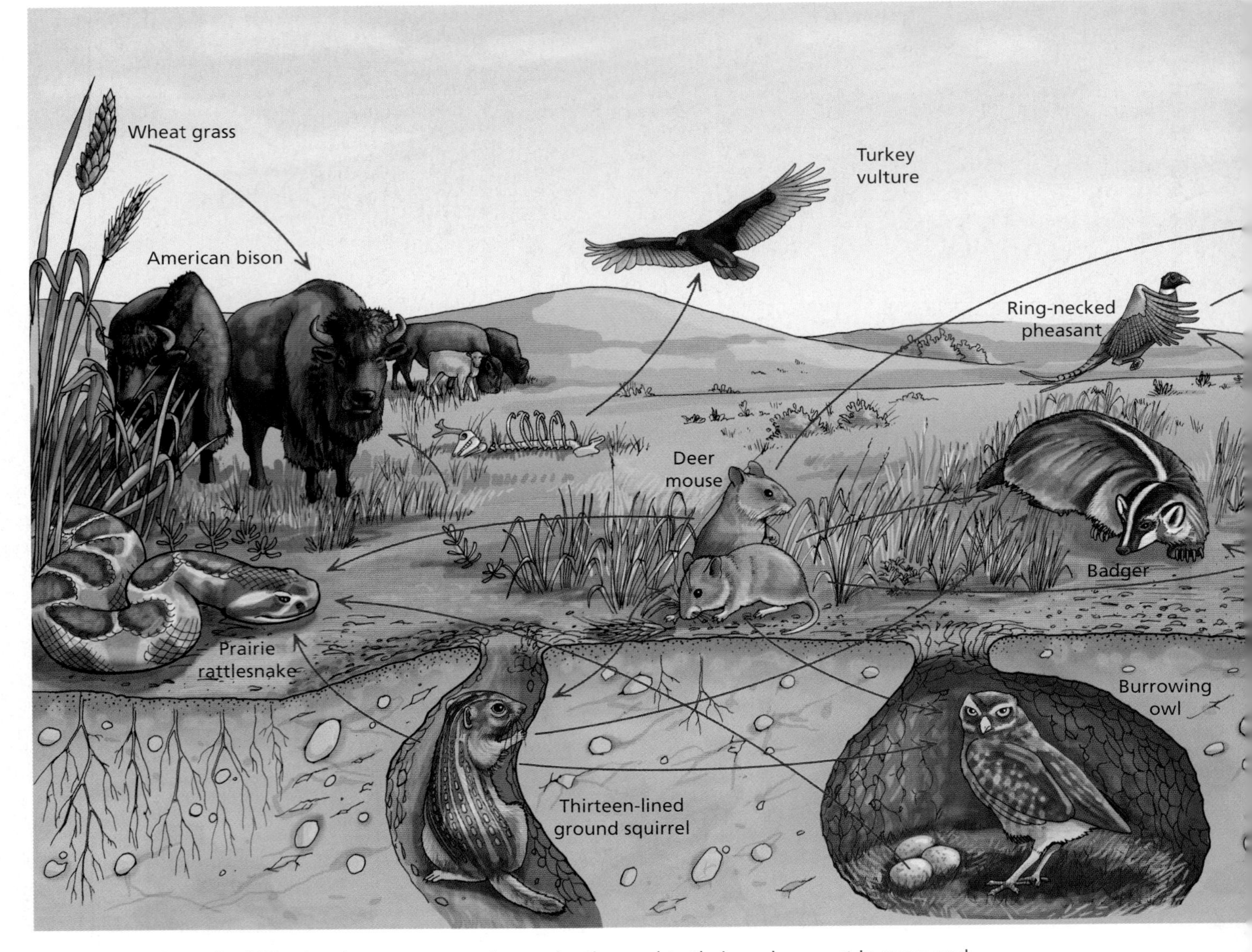

Figure 14.9 The web of life. Species are connected to each other and to their environment in many and complex ways. This drawing shows some of the important relationships among organisms and their environments in a North American prairie. Black arrows represent feeding relationships; for example, thirteen-lined ground squirrels eat wheat grass and in turn are eaten by badgers.

organisms contain an enormous variety of biologically active chemicals; a few of these chemicals are directly useful against human disease. Many medicinal plants and other organisms are used by local cultures and provide crucial health benefits to these human populations. If we are unable to screen living species because they have become extinct, we will never know which ones might have provided compounds that would improve our lives.

Disrupting the Web of Life

Although humans receive a direct benefit from thousands of species, most threatened and endangered species are probably of little or no use to people. Even the Lost River and shortnose suckers, as valuable as they once were to the native people of the Klamath Basin, are not especially missed as a food source—no one has starved simply because these fish are less common.

In reality, most species are beneficial to humans because they are connected to other species and natural processes in an **ecosystem**, which is the sum total of organisms and natural features in a given area. The proper functioning of ecosystems, the natural world's **ecosystem services**, supports human life and our economy. The connection among organisms and natural processes in an ecosystem is often referred to as the "web of life" (Figure 14.9). As with a spider's web, any disruption in one strand of the web of life is felt by other portions of the web. Disruptions caused by the loss of seemingly insignificant species have the potential to cause great damage to human economies.

How Bees Feed the World An interaction between two species that provides benefits to both is called a **mutualism**. Cleaner-fish that remove and consume parasites from the bodies of larger fish and ants that live in the thorns of acacia trees and defend the trees from other insects are examples of mutualists.

Figure 14.10 A mutualism. Honeybees transfer pollen, allowing one plant to "mate" with another plant some distance away.

Benefit to flower:
Its sperm (within the pollen) is carried to the female reproductive structures of another flower, enabling cross-pollination.

Benefit to bee:
Plenty of food in the form of nectar and excess pollen.

However, perhaps the most often-described mutualistic interaction exists between plants and bees (Figure 14.10).

Bees are the primary *pollinators* of many species of flowering plants; that is, they transfer sperm, in the form of pollen grains, from one flower to the female reproductive structures of another flower. The flowering plant benefits from this relationship because insect pollination increases the number and vigor of seeds the plant produces. The bee benefits by collecting excess pollen and the nectar produced by the flower to feed itself and its relatives in the hive.

Wild bees pollinate at least 80% of the $10 billion of agricultural crops in the United States. Thus, populations of wild honeybees have a major and direct impact on approximately $8 billion of our agricultural production and many more billions of dollars of impact around the globe.

Unfortunately, bees have suffered dramatic declines in recent years. According to the U.S. Department of Agriculture, we are facing an "impending pollination crisis," because both wild and domesticated bees are disappearing at alarming rates. These dramatic declines are believed to be the result of *pesticides*, such as chemicals that kill insects, an increased level of bee **parasites** (infectious organisms that cause disease or drain energy from their hosts) caused by poor management of domestic bees, and habitat destruction. The endangerment and extinction of these inconspicuous mutualists of crop plants would be extremely costly to humans.

How Songbirds May Save Forests Wood warblers are a family of North American bird species that are characterized by their small size, colorful summer plumage, and habit of catching and eating insects (Figure 14.11a). Their consumption of another species makes warblers **predators**, and it is in their role as insect predators that they potentially provide benefits to humans.

There are hundreds of millions of individual warblers in the forests of North America during the summer. They spend most of their waking hours catching insects for themselves and their quickly growing offspring; thus, warblers collectively remove literally tons of insects from forest trees and shrubs every year. Most of these insects are predators as well—on plants. By reducing the number of insects in forests, warblers reduce the damage the insects inflict on forest plants. The results of a study that excluded birds from white oak seedlings showed that the trees were about 15% smaller as a result of insect damage over two years when compared to trees where birds were not excluded. Other studies have shown less dramatic benefits.

(a) Black-throated blue warbler, predator of insects.

(b) Forests suffer when insects are unchecked by predators.

Figure 14.11 Birds and forests. (a) The black-throated blue warbler is one of many warbler species native to North American forests. These colorful birds are among the most active predators of plant-eating insects in these forests. (b) Insects can kill many trees, as seen in this photo of a spruce budworm infestation. Warblers and other insect-eating birds probably reduce the number and severity of such insect outbreaks.

Although scientists still disagree about exactly how important warblers and other insect-eating birds are to the survival of trees, most agree that reducing the number of forest pests increases the growth rate of the trees. Harvesting forest trees for paper and lumber production fuels an industry worth over $200 billion dollars in the United States alone. At least some of the wood harvested by the timber industry was produced only because warblers were controlling insects in forests (Figure 14.11b).

Many species of forest-inhabiting warblers appear to be experiencing declines in abundance. The loss of warbler species has several causes, including habitat destruction in their summer and winter habitats and increased predation by human-associated animals such as raccoons and house cats. Although other, less vulnerable birds may increase in number when warblers decline, it is unlikely that warblers' effects on insect pest populations can be completely replaced by less insect-dependent birds. If smaller warbler populations correspond to lower forest growth rates and higher levels of forest disease, these tiny, beautiful birds definitely have an important effect on the human economy.

How an Infected Chicken Could Save a Life When two species of organisms both require the same resources for life, they will be in **competition** for the resources within a habitat. We may imagine lions and hyenas fighting over a freshly killed antelope as a typical example of competition, but most competitive interactions are invisible.

In general, competition limits the size of competing populations. To determine whether two species are in competition, we remove one from an environment; if the population of the other species increases, the two species are competitors. One of the least visible forms of competition occurs among microorganisms. Competitive interactions among microbes may be the most essential factors for maintaining the health of people and ecosystems.

Salmonella enteritidis is a leading (and growing) cause of food-borne illness in the United States—between 2 million and 4 million people in this country will be infected by this bacterium in the coming year, experiencing fever, intestinal cramps, and diarrhea as a result. In about 10% of cases, the infection spreads to the bloodstream—if it is not treated promptly, individuals in whom this happens may die. Nearly 2,000 Americans die as a result of *S. enteritidis* infections every year.

Most *S. enteritidis* infections result from the consumption of raw or improperly cooked poultry products, especially eggs. The U.S. Centers for Disease Control estimate that as many as one in 50 consumers are exposed to eggs contaminated with *S. enteritidis* every year. Most of these eggs have had their shells disinfected and do not look damaged in any way—the bacteria were deposited in the egg by the hen when the egg was forming inside her. Thus, the only way to prevent *S. enteritidis* from contaminating eggs is to keep it out of hens in the first place.

A common way to control *S. enteritidis* is to feed hens *antibiotics*, chemicals that kill bacteria. However, antibiotic use is costly and, like most microbes, *S. enteritidis* strains evolve to become resistant to the effects of the antibiotic. (The evolution of drug resistance is discussed in detail in Chapter 9.) Another way to reduce infections in poultry is to reduce the space available for the bacteria's growth. Most *S. enteritidis* infections originate in an animal's digestive system. If another bacterial species is well-established in a hen's digestive system, *S. enteritidis* has trouble colonizing it. Some poultry producers now establish harmless bacteria in hens' digestive systems, a practice called *competitive exclusion*, to reduce *S. enteritidis* levels in their flocks. This technique involves feeding cultures of bacteria from the intestinal tract of healthy adult chickens to 1-day-old birds to prevent subsequent *S. enteritidis* colonization (Figure 14.12). There is evidence that this practice is working; infections in chickens have dropped nearly 50% in the United Kingdom, where competitive exclusion in poultry is common practice.

The competitive exclusion of *S. enteritidis* from hens' guts by less harmful bacteria mirrors the role of some human-associated bacteria, such as those that normally live in our intestines and genital tracts. For instance, many women who take antibiotics for a bacterial infection develop vaginal yeast infections because the antibiotic kills noninfectious bacteria as well, including species in the genital tract that normally compete with yeast. Competitive exclusion in hen's intestines also mirrors the role of native soil bacteria in maintaining the health and function of soil by balancing all the different species of bacteria that are required for recycling the nutrients within ecosystems. However, the widespread use (and misuse) of antibiotics, the use of pesticides in agricultural settings, and ironically, a "too-clean" environment may reduce the populations of these beneficial bacteria on a local or even global scale. Biologists are a long way from identifying all of the species of bacteria, let alone knowing which are vulnerable to human-caused extinction or identifying their part in maintaining ecosystem health. However, the role of bacteria in preventing *S. enteritidis* poisoning indicates that preserving these competitive interactions is worthwhile.

Biophilia

Species provide benefits to humans directly and as members of functioning ecosystems on which humans rely. In addition, some scientists argue, the diversity of living organisms sustains humans by satisfying a deep psychological need. One of the most prominent scientists to promote this idea is Edward O. Wilson, who calls this instinctive desire to commune with nature *biophilia*.

Wilson contends that people seek natural landscapes because our distant ancestors evolved within such landscapes (Figure 14.13). According to this hypothesis, ancient humans who had a genetic predisposition that drove them to find diverse natural landscapes were more successful than those without this disposition, since more diverse areas provide a wider variety of food, shelter, and tool sources. Wilson claims that we, the modern descendants of successful early humans, have inherited the genetic imprint of our pre-agricultural past.

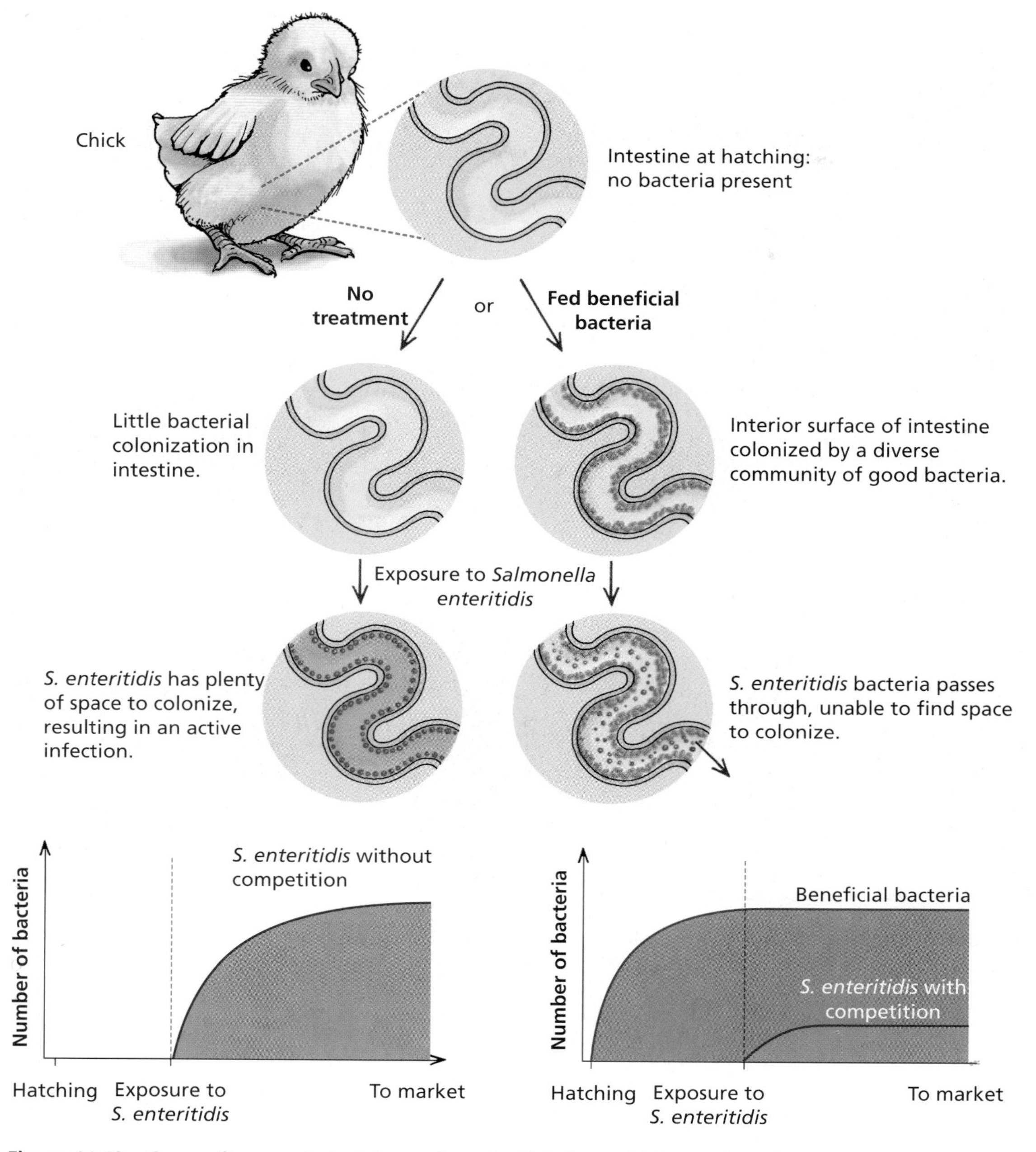

Figure 14.12 Competitors exclude *Salmonella enteritidis* from chickens. If poultry producers feed very young chicks non-disease-causing bacteria, the bacteria take up the space and nutrients in the intestine that would be used by *S. enteritidis*; thus, they will have no site to colonize and increase their population.

While there is as yet no evidence of a "gene" for *biophilia* (in fact, there is considerable debate over whether *any* complex human behavior has a strong genetic component), there is evidence that our experience with nature does have powerful psychological effects. For instance, studies in hospitals indicate that blood pressure drops 10 to 15 points when patients are exposed to serene landscape paintings; and when patients can see trees and other natural scenes from their windows, they need fewer painkillers and recover from illnesses more quickly than patients who are confined to rooms that overlook buildings. Other studies indicate that office workers who have desks facing windows with a view of a natural setting are less stressed and take fewer sick days. While these

Figure 14.13 Is our appreciation of nature innate? Humans evolved in a landscape much like this one in East Africa. Some scientists argue that we have an instinctive need to immerse ourselves in the natural world.

studies are certainly not conclusive, they are intriguing, for they suggest that the continued loss of biological diversity could make human society a less pleasant place to live.

14.3 Saving Species

In the previous sections of this chapter, we have established the potential for a modern mass extinction and described the possibly serious costs of this loss of biodiversity to human populations. Since current elevated extinction rates are largely a result of human activity, reversing the trend of species loss requires political and economic, rather than scientific, decisions. But what can science tell us about how to stop the rapid erosion of biological diversity?

How to Stop a Mass Extinction

In the absence of knowing exactly which species are closest to extinction and where they are located, the most effective way to prevent loss of species is to preserve many habitats. The same species–area curve that Wilson used to estimate

the future rate of extinction also gives us hope for reducing this number. Recall that according to the curve in Figure 14.5b, species diversity declines rather slowly as habitat area declines. Thus, in theory, we can lose 50% of a habitat but still retain 90% of its species. However, this estimate is somewhat optimistic because habitat destruction is not the only threat to biodiversity. In any case, the species–area curve does tell us that if the rate of habitat destruction is slowed or stopped, extinction rates will slow as well.

Given the growing human population (discussed in detail in Chapter 15), it is difficult to imagine completely halting habitat destruction. However, biologist Norman Myers and his collaborators have concluded that 30–50% of all plant, amphibian, reptile, bird, and mammal species are found in just 25 biodiversity "hotspots" that make up less than 2% of Earth's ice-free land area (Figure 14.14). Thus, stopping habitat destruction in the hotspots could greatly reduce the global extinction rate. By focusing immediate conservation efforts on hotspot areas at the greatest risk of losing much of their natural landscape, humans can very quickly prevent the loss of a large number of species. Of course, even preserving these biodiversity hotspots is no easy task—it requires the concerted actions of a diverse community of nations and people, some of whom must also address the immediate concerns of poverty, hunger,

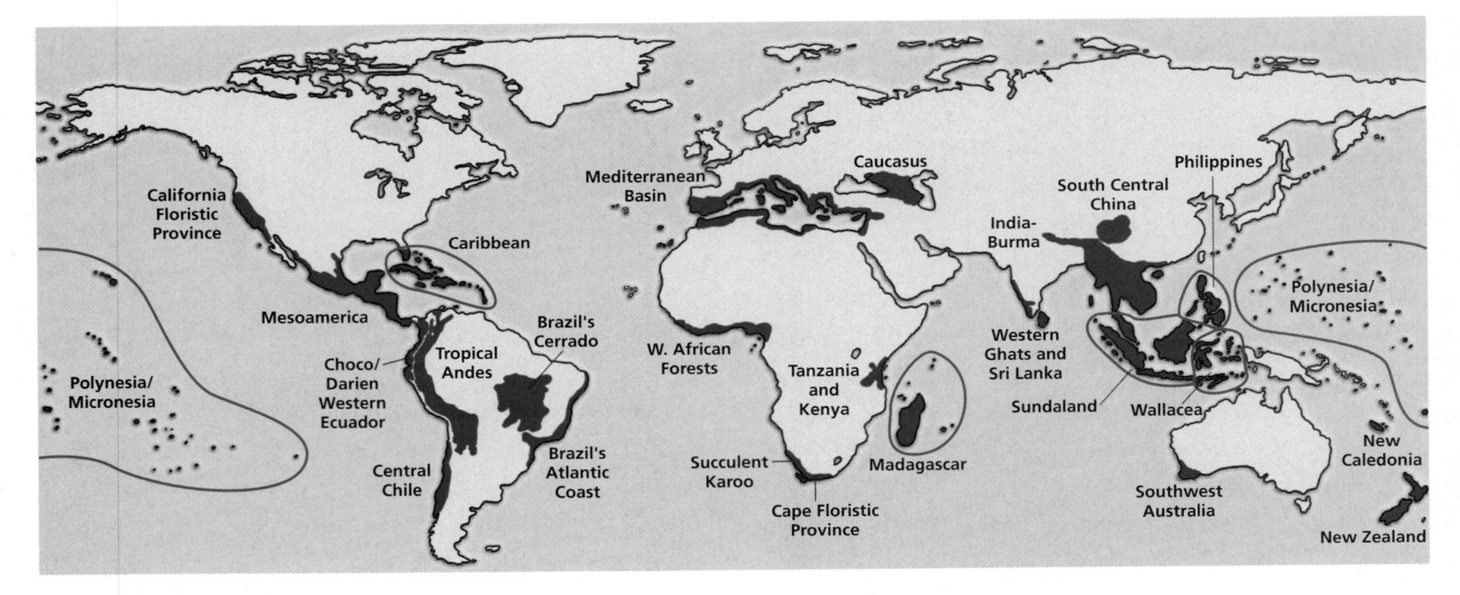

Figure 14.14 Diversity hotspots. This map shows the locations of 25 hotspots that have been identified. Note their uneven distribution around Earth.

WWW
Media Activity 14.2 Conserving Global Biodiversity Hotspots

and disease. It is likely that even with habitat protection, many of the species in these hotspots will become extinct anyway because so little habitat is left for them. In the long term, we must devise ways to preserve biological diversity that include human activity in the landscape. Strategically preserving relatively small amounts of land today may prevent our descendants from identifying this period as the sixth mass extinction.

There are other things everyone can do to reduce the rate of habitat destruction. Reducing the consumption of goods that cause habitat destruction is primary among these. Conversion of land to agricultural production is a major cause of habitat destruction, so reducing our consumption of meat and dairy products (domestic animals require large amounts of crops for feed) is one of the most effective actions we can take. We will explore the effect of agriculture on biodiversity and other aspects of our environment in Chapter 15. Increasing financial support to developing countries, enabling them to take advantage of advances in technology that reduce their use of wood for heating and cooking, may also help slow the rate of habitat destruction. Strategies that can slow the rate of human population growth are another way to avoid a mass extinction.

Although protecting habitat from destruction can reduce extinction rates, for species on the brink of extinction like the shortnose and Lost River suckers, preserving habitat is not enough. These species may only have tiny fragments of suitable habitat remaining, and many have very small populations as well. To save these species, humans must concentrate on their individual needs and causes of endangerment.

WWW
Media Activity 14.3 Biodiversity Inventories for Conservation and Economic Development

One Species at a Time

Habitat destruction is a leading cause of endangerment and extinction, and it follows that the primary requirement for species recovery is restoration of habitat. The ESA requires the Department of Interior to designate *critical habitats* for endangered species, the areas in need of protection for the survival of the species. The amount of critical habitat that is designated depends upon political as well as biological factors. Federal designation of a critical habitat results in restriction of the human activities that can take place there—thus, landowners are usually interested in keeping their lands out of this designation. If landowners are politically powerful, they can exert their influence on elected officials and have profound effects on the *recovery plan* for a particular endangered species.

The biological part of a critical habitat designation includes a study of the habitat requirements of the endangered species and setting a *population goal* for it. The Department of Interior's critical habitat designation has to include enough area to support the recovery population. The designation of critical habitat has an extra benefit as well—protection of this habitat can protect dozens of other less well-known species that may be approaching endangerment.

The recovery plans for both the Lost River and the shortnose suckers sets a short-term goal of one stable population made up of at least 500 individuals for each unique stock of suckers. To understand why at least this many individuals of a species are required to protect the species from extinction, we need to review some of the special problems of small populations.

Growth and Catastrophe in Small Populations A species' *growth rate* is influenced by how long the species takes to reproduce, how often it reproduces, the number of offspring produced each time, and the death rate of individuals under ideal conditions. (Calculation of growth rate is discussed in Chapter 15.) For instance, species that reproduce slowly take longer to grow in number than species that reproduce quickly. Thus the growth rate of an endangered species influences how rapidly it can attain a target population size. Shortnose and Lost River suckers have relatively high growth rates and will meet their population goals quickly if the environment is ideal (Figure 14.15a). For more slow-growing species, such as the California condor (Figure 14.15b), populations may take decades to

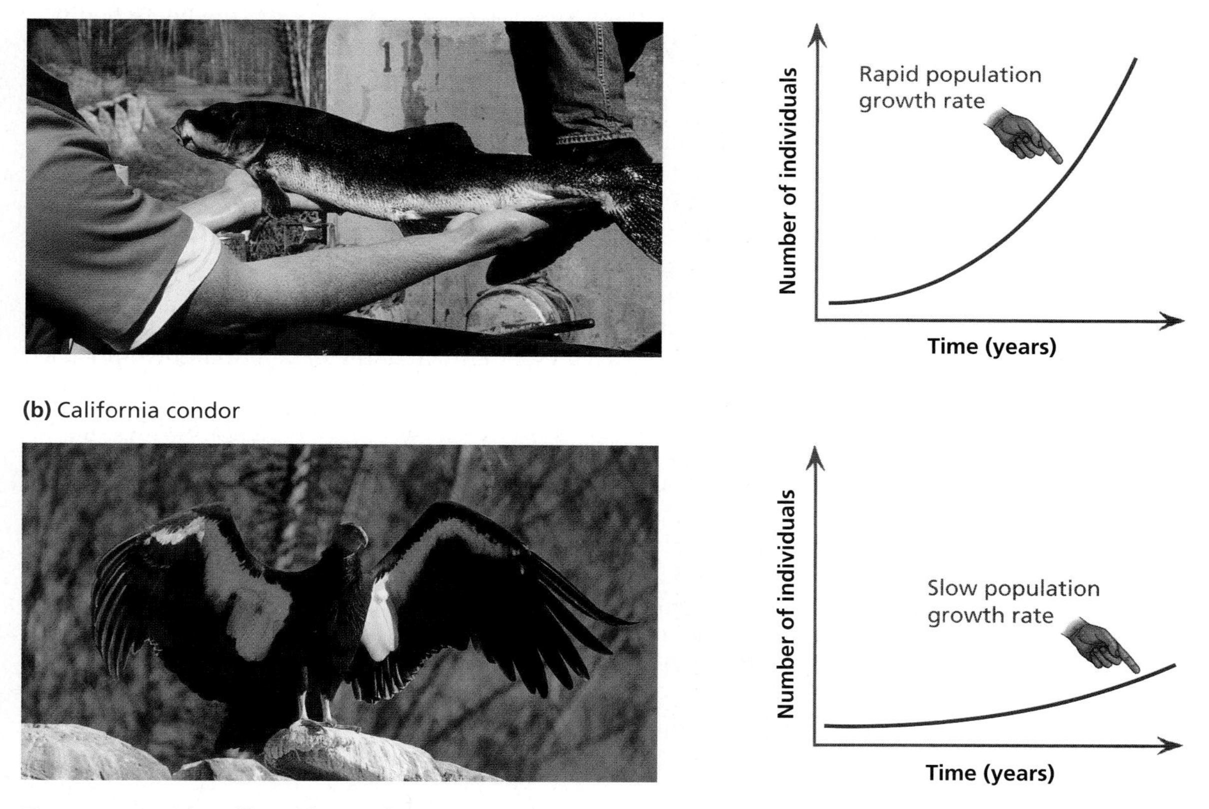

Figure 14.15 The effect of growth rate on species recovery. (a) This graph illustrates the rapid growth of a hypothetical population of quickly reproducing Lost River suckers. (b) The slow growth rate of the California condor has made the recovery of this species a long process. Today, nearly 30 years after recovery efforts began, the population of wild condors is still only in the dozens. Two wild populations of 150 condors each must be established for the bird to be removed from endangered status.

Figure 14.16 A victim of small population size. The heath hen was once abundant throughout the eastern United States. Although it was protected when its population was nearly 50 individuals, a series of unexpected disasters caused its extinction. (© Steven Holt/VIREO)

recover. The rate of a species' recovery is important because the longer a population remains small, the more it is at risk of experiencing a catastrophic environmental event that could eliminate it entirely. The story of the heath hen is a classic example of the dangers facing small populations.

The heath hen was a small relative of the prairie chicken that lived on the East Coast of the United States (Figure 14.16) and was a favorite game bird of early European settlers. Prior to the American Revolution, the heath hen was found from Maine to Virginia. Increased settlement resulted in loss of habitat and increased hunting, noticeably lowering heath hen populations by the time of the Revolutionary War. In the 1870s, the only heath hens that were left occupied a tiny island called Martha's Vineyard off the coast of Cape Cod in Massachusetts. Human development on the island further reduced the suitable habitat for heath hen breeding, and in 1907 there were only 50 heath hens left on Martha's Vineyard. A 1,600-acre sanctuary was established for their protection the following year.

The sanctuary seemed to be successful—the original 50 heath hens reproduced rapidly and there were 2,000 individuals on Martha's Vineyard by 1915. Unexpectedly, a fire in 1916 wiped out much of the habitat that the birds used for breeding. In addition, the next winter was unusually harsh and food was scarce, and an influx of goshawks, predatory birds that preyed on the heath hens, reduced the population further. Finally, many of the remaining heath hens fell victim to a poultry disease brought to the island by domestic turkeys. There were only 14 heath hens left by 1927, and most of them were males. The last living heath hen was seen on March 11, 1932. He died that year.

Why did the heath hen become extinct? The last birds were wiped out by a series of relatively common and entirely natural events: fire, starvation, predation, and disease. The heath hen's continued existence as a species would not have been so vulnerable to these occurrences if the population size had not been severely reduced by habitat loss and overhunting. A small population is very vulnerable to normal fluctuations in its numbers, which are the consequence of disease and disasters. A population of 1,000 individuals can survive a population drop of 100; the same fluctuation dooms a population that starts with only 100 individuals. In the case of the heath hen, even when hunting and habitat destruction were halted, the species' survival was still extremely precarious.

The population goal of 500 individuals for both species of suckers in Upper Klamath Lake is still quite small, but in the short term it will help these fish avoid the same fate as the heath hen.

Genetic Variability and Survival Small populations of endangered species can still be protected from the fate that befell the heath hen. Having additional populations of the species at sites other than Martha's Vineyard would have nearly eliminated the risk that *all* members of the population would be exposed to the same series of environmental disasters. This is the rationale behind placing captive populations of endangered species at several different sites. For instance, the captive whooping crane population is located at the U.S. National Biological Service's Patuxent Wildlife Research Center in Maryland, the International Crane Foundation in Wisconsin, the Calgary Zoo in Canada, and the Audubon Center for Endangered Species Research in New Orleans. However, if endangered species populations remain small in number, they are subject to a subtler but potentially equally damaging disaster—the loss of genetic variability.

A species' **genetic variability** is the sum of all of the *alleles* and their distribution within the species. Differences among alleles produce the variety of traits within a population. For example, the gene that determines your blood type comes in three different forms, and the combination of alleles that you possess determines whether your blood type is O, A, B, or AB. Thus, a population containing all three blood-type alleles contains more genetic variability (for this gene) than a population that contains only two alleles.

The loss of genetic variability in a population is a problem for two reasons: (1) On an individual level, low genetic variability leads to low fitness; and (2) on a population level, rapid loss of genetic variability may lead to extinction.

Individual Genetic Variability As we discussed in Chapter 9, *fitness* refers to an individual's ability to survive and reproduce in a given set of environmental conditions. There are two reasons that high genetic variability on an individual level increases fitness. We will use an analogy to illustrate the costs of low genetic variability in individuals. First, imagine that you could own only two sets of footwear (Figure 14.17a). If both pairs are dressy shoes, you might be prepared to meet a potential employer, but if you had to walk across campus to your job interview

(a) Heterozygote has higher fitness than either homozygote.

Homozygote: Relatively low fitness

Homozygote: Relatively low fitness

Heterozygote: Relatively high fitness

(b) Heterozygote masks the deleterious allele.

Homozygote: Relatively high fitness

Homozygote: Relatively low fitness

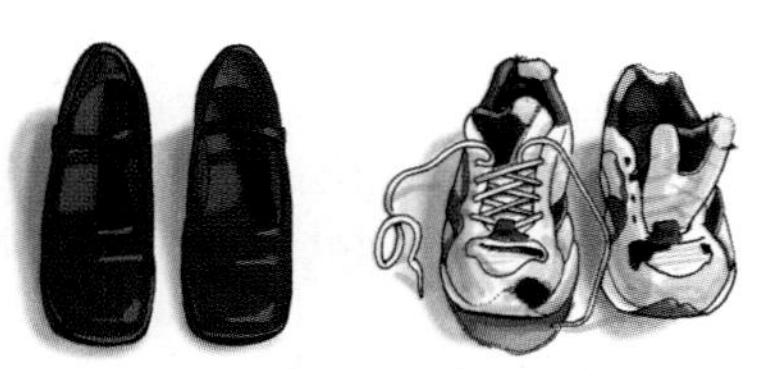

Heterozygote: Relatively high fitness

Figure 14.17 The benefits of heterozygosity. In this analogy, each pair of shoes represents an allele. (a) Heterozygotes may better prepared for a diversity of life experiences than homozygotes. (b) Heterozygotes may have higher fitness than some homozygotes because certain alleles are deleterious and recessive. In this case, homozygotes for the normal allele also have higher fitness than homozygotes for the recessive allele.

in a snowstorm, you would be pretty uncomfortable. If you own two sets of winter boots, your feet will always be protected from the cold, but you would look pretty silly at a dinner party. However, if you own both dress shoes and winter boots, you are ready for slush and snow as well as a nice date. In a way, individuals experience the same advantages when they carry two different alleles for a gene—that is, when they are *heterozygous*. If a protein produced by each allele works best in different environments, heterozygous individuals are able to function efficiently over a wider range of conditions.

The second reason that high individual genetic variability increases fitness is that, in many cases, one allele for a gene is deleterious—that is, it produces a protein that is not very functional. In our shoe analogy, this might be sneakers with blown-out toes. If you have these sneakers and dress shoes, at least you have one pair of shoes that covers your feet (Figure 14.17b). In the case of a deleterious allele, a heterozygous individual still carries one functional copy of the gene. Genetic variability can help mask the effects of deleterious alleles, because the functional allele is *dominant*—that is, it tends to drown out the deleterious allele (see Chapter 4). An individual who is *homozygous* (carries two identical copies of a gene) for the deleterious allele will have low fitness—in our analogy, two pairs of blown-out sneakers and nothing else. When individuals are heterozygous for many genes, the cumulative effect is often greater fitness relative to individuals who are homozygous for many genes.

In a small population, where mates are more likely to be related to each other than in a very large population simply because there are fewer mates to choose from, heterozygosity declines. When related individuals mate—known as **inbreeding**—the chance that their offspring will be homozygous for any allele (one that both parents inherited from a shared ancestor) is relatively high. The negative effect of homozygosity on fitness is known as **inbreeding depression**. This is seen in humans as well as other species—numerous studies consistently show that the children of first cousins have higher mortality rates (thus, lower fitness) than children of unrelated parents. In a population of an endangered species, the low rates of survival and reproduction associated with high rates of inbreeding can seriously hamper its ability to recover from endangerment.

We should note that in some populations, inbreeding does not lead to lower fitness. This appears to be the case in populations in which inbreeding has been historically common. Here, deleterious alleles have been "purged" from the population because inbreeding exposes these alleles to natural selection (that is, allows them to be expressed in homozygotes). Because these homozygotes have low fitness, the alleles they carry are rarer in subsequent generations and are lost over time. However, this appears to be a relatively rare occurrence—in an examination of 25 captive mammal species, only one showed clear evidence that deleterious alleles had been purged. For most species, inbreeding seems to be a significant threat to survival.

Genetic Variability in Populations Small populations also lose genetic variability as a result of **genetic drift**, a change in the frequency of an allele in a population occurring simply by chance. Genetic drift was discussed as a process for causing evolutionary change in Chapter 10. However, genetic drift in a small population can have detrimental consequences.

Imagine a population in which the frequency of blood-type allele *A* is 1%—that is, only one out of every 100 blood-type genes in the population is the *A* form (we use the symbol I^A). In a population of 20,000 individuals, we calculate the total number of I^A alleles as follows:

$$\text{total number of blood-type alleles in population} = \text{total population} \times 2 \text{ alleles/person}$$

$$20{,}000 \times 2 = 40{,}000$$

total number of I^A alleles =
total number of alleles in population $\times$ frequency of allele I^A

$$40{,}000 \times 1\% = 40{,}000 \times 0.01 = 400.$$

If a few of the individuals who carry the I^A allele die accidentally before they reproduce, the number of copies of the allele drops slightly in the next generation of 20,000 people, say to 385 out of 40,000 alleles. The chance occurrences that led to this drop result in a new allele frequency:

frequency of I^A alleles in population =
total number of I^A alleles / total number of blood-type alleles in population

$$\frac{385}{40{,}000} = 0.0096 = 0.96\%.$$

The change in frequency from 1% to 0.96% is the result of genetic drift.

A change in allele frequency of 0.04% is relatively minor. There will still be hundreds of individuals who carry the I^A allele. It is not unlikely that in a subsequent generation, a few individuals carrying allele I^A will have an unusually large number of offspring, thus increasing the allele's frequency in the next generation.

Now imagine the effects of genetic drift on a small population. In a population of only 200 individuals and with an I^A frequency at 1%, only four of the individuals in the population carry the allele. If two of these individuals fail to pass it on, the frequency will drop to 0.5%. Another chance occurrence in the following generation could completely eliminate the two remaining I^A alleles from the population. Thus, genetic drift occurs more rapidly in small populations and is much more likely to result in the complete loss of alleles (Figure 14.18). Typically, the alleles that are lost via genetic drift have little current effect on fitness—after all, if the protein produced by the allele significantly

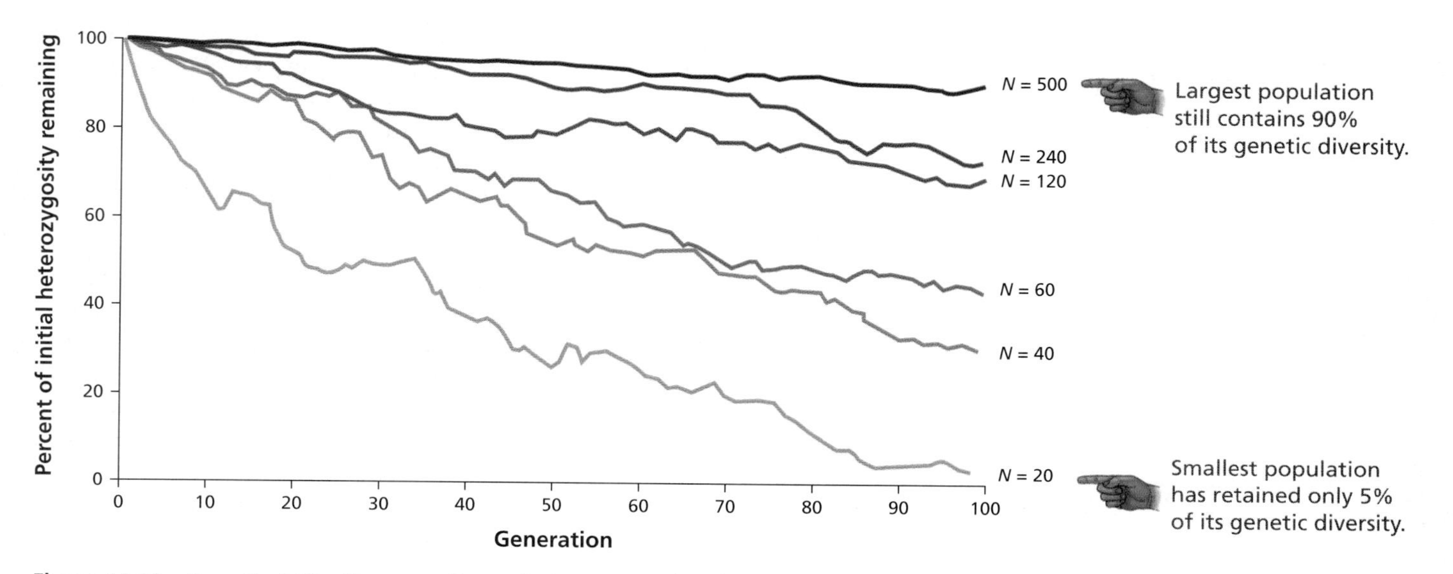

Figure 14.18 Genetic drift affects small populations more than large populations. In this graph, each line represents the average of 25 computer simulations of genetic drift for a given population size. After 100 generations, a population of 500 individuals still contains 90% of its genetic variability. In contrast, a population of 20 individuals has less than 5% of its original variability.

increased fitness, natural selection should result in the allele increasing in the population. However, many alleles that appear to be neutral with respect to fitness in one environment may have positive fitness in another environment. For example, there is some evidence that individuals with type A blood are more resistant to cholera and bubonic plague than people with blood-type O or B. Therefore, possessing the I^A allele may be neutral relative to other blood-type alleles in areas where these diseases are rare, but it could be an advantage where the diseases are common.

Populations with low levels of genetic variability have an insecure future for two reasons. First, when alleles are lost, the level of inbreeding depression in a population increases, which means lower reproduction and higher death rates, leading to smaller populations that are susceptible to all of the other problems of small populations. Second, populations with low genetic variability may be at risk of extinction because they cannot respond to changes in the environment. When few alleles are available for any given gene, it is possible that no individuals in a population possess a trait that allows them to survive an environmental challenge. For example, if blood-type A really does protect against some infectious diseases, a population with no individuals carrying the I^A allele could have no survivors after exposure to bubonic plague.

As always, there are some exceptions to the "rule" described above. For example, widespread hunting of northern elephant seals in the 1890s reduced the population to 20 individuals—this probably wiped out much of the genetic variation in the species. However, elephant seal populations have rebounded to include about 150,000 individuals today. Although genetic variability is quite low in this species today, elephant seals continue to thrive, but there are many more examples of the costs of low genetic variability. The Irish potato is perhaps the most dramatic example of this cost.

Potatoes were a staple crop of rural Irish populations until the 1850s—a healthy adult man consumed about 10 potatoes, or 14 pounds, each day. Although the population of Irish potatoes was high, it had remarkably low genetic variability for two reasons. First, potatoes are not native to Ireland (in fact, they originated in South America), meaning the crop was limited to few varieties that were originally imported—and the majority of potatoes grown on the island were of a single variety, called Lumper for its bumpy shape. Second, new potato plants are grown from potatoes produced by the previous year's plants, and thus are genetically identical to their parents. This agricultural practice ensured that all of the potatoes in a given plot had identical alleles for every gene. All of the available evidence indicates that the genetic variability of potatoes grown in Ireland during the nineteenth century was extremely low.

When the organism that causes potato blight arrived in Ireland in September 1845, nearly all of the planted potatoes became infected and rotted in the fields. The few potatoes that had escaped the initial infection were used to plant the following year's crops. Some varieties of potatoes in South America carry alleles that allow them to resist potato blight and escape an infestation unaffected. However, apparently very few or no Irish potatoes carried these alleles, and in 1846 the entire Irish potato crop failed. As a result of this failure and another in 1848, along with harsh policies instituted by the ruling British government in Ireland, nearly 1.5 million Irish peasants died of starvation and disease and another 1 million left home for North America.

Irish potatoes descended from a small group of plants that were missing the allele for blight resistance, so even an enormous population of these plants could not escape the catastrophe caused by this disease. Similarly, since small populations lose genetic variability rapidly via genetic drift, keeping endangered species from declining to very small population levels may be critical for avoiding a similar genetic disaster. This is why preserving adequate numbers of Lost River and shortnose suckers, even at the expense of crop production in the Klamath Basin, is such a high priority if we wish to save the species from extinction.

Fish versus Humans?

Saving the Lost River and shortnose suckers from extinction requires totally protecting the remaining fish and dramatic action to restore the habitat they need for reproduction. These actions cause economic and emotional suffering for humans who make their living in the Klamath Basin. In fact, many of the actions necessary to save endangered species usually result in some immediate problems for people. As the Oregon State University student we quoted in our introduction inquired, what can we do to balance the costs and benefits of preserving endangered species?

The Endangered Species Act contains a provision that allows members of a committee to weigh the relative costs and benefits of actions taken to protect endangered species. This Endangered Species Committee, a group of Cabinet-level political appointees—including the secretaries of Agriculture and the Interior, and the chairman of the Council of Economic Advisors—has been convened a number of times for just this purpose. This so-called "God Squad" decides if they should overrule a federal action meant to save an endangered species to protect the economic benefits of the people involved. Farmers in the Klamath Basin have advocated for a God-Squad ruling on the diversion of water from Upper Klamath Lake, but history suggests that the decision is not likely to be in their favor. The Endangered Species Committee convened four times from 1973 through 2001; however, only once did concern for human needs outweigh the decision to protect a species—and that ruling was subsequently overturned in court.

If the debate in the Klamath Basin follows the pattern set by other ESA controversies, a political decision that causes some economic hardship while ensuring at least the immediate survival of the fish will prevail. Biologists working on the problem agree that the recovery goal for the shortnose and Lost River suckers is high enough to ensure short-term survival (50 years), but it is not high enough to ensure both species' long-term survival (500 years). Their assessment is based on computer models that take into account population fluctuations and predicted environmental changes. The recovery population size is large enough to withstand these threats in the short term but in the long term, continued loss of genetic variability and uncertainty about future environmental change results in the extinction of most of the "recovered" populations in their models. The risk to the long-term survival of the fish helps balance the cost to the farmers of the Klamath Basin. The cost to farmers will be immediately alleviated when they receive federal disaster assistance from fellow American taxpayers to help them adjust to the loss of lake-derived irrigation water. Some farmers will use the money to drill wells and provide groundwater to their thirsty crops, while others will use it to establish new income-producing occupations. The federal government will also purchase farmland from willing sellers in the Basin at fair market prices to protect and restore the fishes' habitat. The Fish and Wildlife Service hopes that this long-term solution will help provide habitat for many more than 500 individuals of both shortnose and Lost River suckers.

The ESA has been a successful tool for bringing species such as the peregrine falcon, bald eagle, and gray wolf back from the brink of extinction, but all of these successes have come with some cost to citizens and taxpayers. If the solution to these and other endangered species controversies is any guide, most Americans will feel that the price of saving the shortnose and Lost River suckers ultimately enriches the United States by restoring our natural heritage (Figure 14.19).

Figure 14.19 Protecting our natural heritage. Snow geese alight on a pond in the Lower Klamath Wildlife Refuge. A solution to the water crisis on the Klamath must balance the needs of people and wildlife.

CHAPTER REVIEW

Summary

- The loss of biological diversity through species extinction is apparently exceeding historical rates by 50 to 100 times.
- Species–area curves help us predict how many species will become extinct due to human destruction of natural habitat.
- The additional threats of habitat fragmentation, introduction of exotic species, overexploitation, and pollution also contribute to species extinction.
- Species may be important to us as resources, either directly as consumed products or indirectly as organisms used to provide potential medicines or genetic resources.
- Species are members of ecosystems; their loss as mutualists, predators, and competitors may change an ecosystem such that it is less valuable or even harmful to humans.
- Biological diversity may fulfill a human need to experience natural landscapes.
- If habitat protection is focused upon a few well-defined biodiversity hotspots, the rate of extinction can be markedly reduced.
- When species are already endangered, restoring larger populations is critical for preventing extinction.
- Small populations are at higher risk of extinction due to environmental catastrophes.
- Small populations are at risk when individuals have low fitness due to inbreeding and are thus less able to rapidly increase population size.
- Small populations lose genetic variability as a result of genetic drift, the loss of alleles from a population due to chance events, and thus may have a reduced ability to evolve in response to environmental change.
- The political process enables people to craft plans that help endangered species recover from the brink of extinction while minimizing the negative affects of these actions on people.

Key Terms

biological diversity p. 358
competition p. 373
ecosystem p. 371
ecosystem services p. 371
Endangered Species Act (ESA) p. 358
exotic species p. 364
extinction p. 358
genetic drift p. 382
genetic variability p. 381
habitat p. 361
habitat destruction p. 362
habitat fragmentation p. 364
inbreeding p. 382
inbreeding depression p. 382
mass extinction p. 366
mutualism p. 371
overexploitation p. 364
parasite p. 372
pollution p. 366
predator p. 372
species–area curve p. 362

Learning the Basics

1. How is the estimate of historical rates of extinction generated? What are the criticisms of these estimates?

2. Describe how habitat fragmentation endangers certain species. Which types of species do you think are most threatened by habitat fragmentation?

3. Compare and contrast the species interactions of mutualism, predation, and competition.

4. Why is genetic drift a more serious problem for small populations than large populations?

5. What is inbreeding depression and why does it occur in small populations?

6. Current rates of species extinction appear to be approximately _____ historical rates of extinction.
   a. equal to
   b. 10 times lower than
   c. 10 times higher than
   d. 50–100 times higher than
   e. 1,000–10,000 times higher than

7. The relationship between the size of a natural habitat and the number of species the habitat supports is described by a _____.
   a. habitat fragmentation measure
   b. inbreeding depression matrix
   c. species–area curve
   d. overexploitation scale
   e. ecosystem services cost

8. According to the generalized species–area curve in Figure 14.5b, when 50% of a habitat area is lost, approximately _____ of species originally found there will become extinct.
   a. 90%
   b. 75%
   c. 50%
   d. 10%
   e. 0%

9. Loss of habitat is a primary cause of the extinction and endangerment of biodiversity. Other human-induced causes of extinction and endangerment include _____.
   a. habitat fragmentation
   b. introduction of exotic species
   c. overexploitation
   d. pollution
   e. all of the above

10. A mass extinction _____.
    a. is global in scale
    b. affects many different groups of organisms
    c. is only caused by human activity
    d. a and b are correct
    e. a, b, and c are correct

11. The web of life refers to the _____.
    a. evolutionary relationships among living organisms
    b. connections between species in an ecosystem
    c. complicated nature of genetic variability
    d. flow of information from parent to child
    e. predatory effect of humans on the rest of the natural world

12. According to many scientists, the most effective way to reduce the rate of extinction is to _____.
    a. preserve habitat, especially in highly diverse areas
    b. focus on a single species at a time
    c. eliminate the risk of genetic drift
    d. produce less trash by recycling more
    e. encourage people to rely more on agricultural products and less on wild products

13. The risks faced by small populations include _____.
    a. erosion of genetic variability through genetic drift
    b. decreased fitness of individuals as a result of inbreeding
    c. increased risk of experiencing natural disasters
    d. a and b are correct
    e. a, b, and c are correct

14. One advantage of preserving more than one population and more than one location of an endangered species is ______.
   a. lower risk of extinction of the entire species if a catastrophe strikes one location
   b. higher levels of inbreeding in each population
   c. higher rates of genetic drift in each population
   d. lower numbers of heterozygotes in each population
   e. higher rates of habitat fragmentation in the different locations

15. Recovery plans for endangered species crafted under the Endangered Species Act ideally represent a balance between ______.
   a. long-term survival of the species and short-term survival of the species
   b. the costs to people in the species' habitat and the risks for these people's future survival
   c. long-term survival of the species and short-term disruption and hardship of people in the species' habitat
   d. the financial benefits provided by the endangered species and the costs of their protection
   e. saving species and allowing them to go extinct

Analyzing and Applying the Basics

1. Review Figure 14.5a. The graph depicts the relationship between island size and number of amphibian and reptile species found on an island chain in the West Indies. How many species of reptiles and amphibians would you expect to find on an island that is 15,000 square kilometers in area? Imagine that humans colonize this island and dramatically modify 10,000 square kilometers of the natural habitat. What percentage of the species that were originally found on the island would you expect to go extinct?

2. Examine the web of relationships among organisms depicted in Figure 14.9. Which of the following species pairs are likely competitors? In each case, describe what they compete for.
   a. badger, jackrabbit
   b. bison, coyote
   c. rattlesnake, badger
   d. ground squirrel, deer mouse
   e. jackrabbit, prairie dog

   How could you test your hypothesis that these animals are in competition?

3. Widespread use of the pesticide DDT caused a reduction in the populations of many species, including songbirds such as the American robin and raptors such as the bald eagle. Once this effect was recognized, DDT use was banned and these bird populations began to recover. However, only the bald eagle, and not the American robin, was ever considered endangered as a result of DDT use. Use your understanding of the effects of a species' growth rate on its recovery and the problems of small populations to explain why the eagles were more at risk of extinction than the robins.

4. What could have been done with the potato crop in Ireland to reduce the risk of widespread crop failure from potato blight? Today, the number of varieties of crop plants (that is, different genetic strains) in the United States is much lower than it was in the early 1900s. What are the risks of this change to our food supply?

5. The piping plover is a small shorebird that nests on beaches in North America. The plover population in the Great Lakes is endangered and consists of only around 30 breeding pairs. Imagine that you have been charged with developing a recovery plan for the piping plover in the Great Lakes. What sort of information about the bird and the risks to it would help you determine the population goal for this species as well as the method of reaching this goal?

Connecting the Science

1. One of the most controversial of all endangered species recovery programs is that of the gray wolf. Wolves were once common over much of North America and northern Mexico, but they were almost completely eliminated from the continental United States by a systematic extermination program in the nineteenth and early twentieth centuries. State governments supported the extermination of the wolves because they directly compete with humans for prey, including wild deer and domestic animals such as cattle, sheep, and turkeys. The gray wolf came under the protection of the Endangered Species Act in 1974 and has since expanded in numbers and range in several regions of the country. Since wolves can live in close proximity to humans and in highly human-modified landscapes, they increasingly prey on livestock and other domestic animals as their numbers increase. Predation incidents have revived the debate over whether wolves should be protected by the ESA. Should wolves be allowed to recolonize areas of the United States where there is significant human settlement, or is the cost of returning this predator to its former range too great relative to its benefits? Explain your reasoning.

2. From your perspective, which is a more convincing reason for preserving biodiversity: nonhuman species have roles in ecosystems and should be preserved to preserve the ecosystems that support us, or nonhuman species have a fundamental right to existence? Explain your choice.

3. If a child asks you the following question 20 or 30 years from now, what will be your answer and why?

   "When it became clear that humans were causing a mass extinction, what did you do about it?"

Media Activities

Media Activity 14.1 Species–Area Curve
Estimated Time: 15 minutes
Explore the relationship between the destruction of habitat and the loss of biodiversity, and how that relationship can be used to predict the effects of land development on species diversity.

Media Activity 14.2 Conserving Global Biodiversity Hotspots
Estimated Time: 15 minutes
Explore the significance of biodiversity hotspots, and learn why scientists are so interested in preserving these areas.

Media Activity 14.3 Biodiversity Inventories for Conservation and Economic Development
Estimated Time: 15 minutes
Investigate the role business plays in preserving biodiversity and facilitating economic development in third-world nations.

CHAPTER 15

Can Earth Support the Human Population?

Population and Plant Growth

From space, Earth doesn't look too crowded.

15.1 Is the Human Population Too Large?

15.2 Feeding the Human Population

But Earth's human population is 6.1 billion . . . and rising.

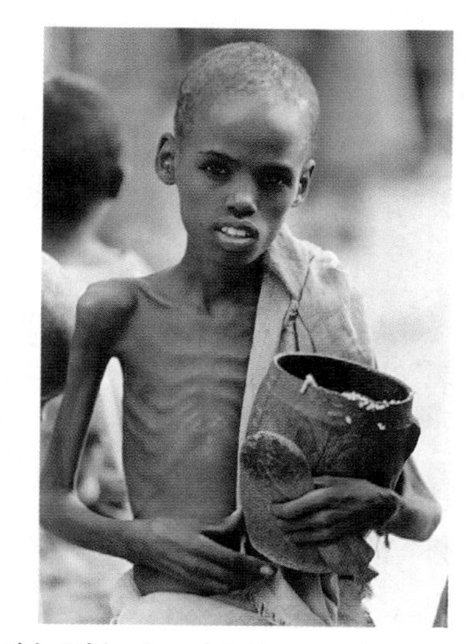

Is this Ethiopian child hungry because the planet is overpopulated?

Or can Earth support everyone more than adequately, as this supermarket suggests?

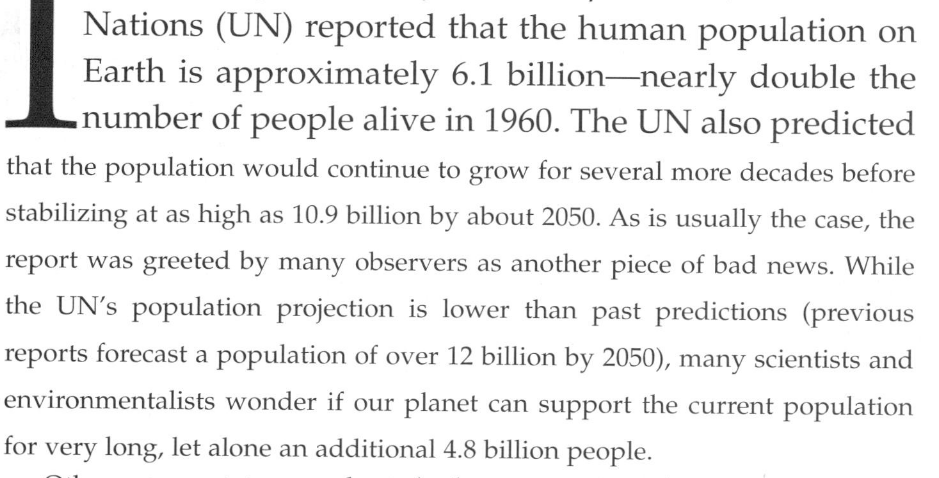

In its *State of World Population Report 2001*, the United Nations (UN) reported that the human population on Earth is approximately 6.1 billion—nearly double the number of people alive in 1960. The UN also predicted that the population would continue to grow for several more decades before stabilizing at as high as 10.9 billion by about 2050. As is usually the case, the report was greeted by many observers as another piece of bad news. While the UN's population projection is lower than past predictions (previous reports forecast a population of over 12 billion by 2050), many scientists and environmentalists wonder if our planet can support the current population for very long, let alone an additional 4.8 billion people.

Other commentators, such as the late economist Julian Simon, a former senior fellow at the influential Cato Institute, are skeptical of environmentalists' statements about population growth. They point to predictions made in Paul Ehrlich's best-selling book *The Population Bomb*, which in 1968 forecast worldwide food and water shortages by the year 2000. In fact, most measures of human health have become more upbeat since 1970, including global declines in infant mortality rates, increases in life expectancy, and a 20%

increase in per-capita income, despite a near doubling in population since the book's publication. By most measures, the average person is better off today than in 1970. Paul Ehrlich was clearly wrong in 1968; why should we believe his doom-and-gloom predictions about the future now?

Ehrlich and his colleagues counter that while they were wrong about how soon it would happen, there are some indications that the large human population is rapidly reaching a real limit to growth. For example, the UN released another report, *The State of Food and Agriculture, 2001*, which described a number of food crises around the world. According the the UN, as of September 2001 there were 34 countries and over 62 million people facing food emergencies, meaning starvation could be imminent. Worldwide, 815 million people—including 150 million children under the age of five—regularly do not get enough food for a healthy existence. A staggering 55% of the nearly 12 million deaths each year among children under five in the developing world are associated with inadequate nutrition. Despite years of international attention and billions of dollars spent to address this problem, the situation has not improved dramatically—there are only 10% fewer children suffering from malnutrition today than there were in 1980.

Economists have countered that the primary cause of famine is political, not biological. Julian Simon said that policies interfering with the ability of individuals to own land and control the means of agricultural production, along with inadequate transportation systems (which prevent the movement of food from areas where production is temporarily high to areas where it is temporarily low), result in food crises. Indeed, these economists argue, Earth currently has a surplus of food. As societies become more democratic and free, the number and impact of food crises will continue to decline.

So what is the truth? Is the human population larger than Earth can support for much longer? Are we headed into a global food crisis and massive famine? Or are we gradually moving toward an era where the entire human population will enjoy the inexpensive and diverse foods available in North American supermarkets?

15.1 Is the Human Population Too Large?

The field of biology that focuses on the interactions between organisms and their environment is known as **ecology**. Many ecologists are interested in the factors that limit the abundance and distribution of particular organisms. In their study of nonhuman species, ecologists see clear limits to growth and can observe the sometimes awful fates of populations that outgrow these limits. For this reason, many professional ecologists join *environmental scientists* who study the effect of humans on the environment in their concern about the consequences of a rapidly growing human population.

You may know of several instances of nonhuman populations outgrowing their food supplies. The population of elk in Yellowstone National Park suffered enormous mortality throughout the 1970s after it grew so large that it degraded its own rangeland; and the massive migrations of Norway lemmings that result in significant numbers of these animals dying occur every five to seven years, after large populations run out of high-quality food in an area. Let us explore what ecology can tell us about the likelihood of human populations suffering the same fate as elk or lemmings.

Human Population Growth

www
Media Activity 15.1 **Human Population Growth and Regulation**

For most of our history, the population of humans has remained at very low levels. The most widely accepted estimates indicate that there were about 5 million humans at the beginning of the agricultural era, about 10,000 years ago; about 100 million during the Egyptian empire 7,000 years later; and about 250 million at the dawn of the Christian religion in 1 C.E. (C.E. refers to Common Era, the year designation used by most Western countries). The population was growing, but at a very slow rate—approximately 0.1% per year. Beginning around 1750, the rate at which the human population was growing jumped to about 2% per year. The human population reached 1 billion in 1800, had doubled to 2 billion by 1930, and had doubled again to 4 billion by 1970. Although the current growth rate is slower, about 1.2% per year, the rapid increase in population looks quite dramatic on a graph of human population over time (Figure 15.1).

The graph of human population growth is a striking illustration of **exponential growth**—growth that occurs in proportion to the current total. The larger a population, the more rapidly it grows, because an increase in numbers depends on reproduction by individuals in the population. So while a growth rate of 1.2% per year may seem rather small (most financial advisors would tell their clients to drop an investment that was growing at a paltry 1.2%), the number of individuals added to the 6 billion-strong human population every year at this rate of growth is a mind-boggling 77 million (approximately the entire population of Germany). Put another way, every second three people are added to the world population, and about a quarter of a million are added every day.

What has fueled this enormous population growth? The *annual growth rate* of a population is the percent change in population size over a single year. Growth rate is a function of the *birth rate* of the population (the number of births averaged over the population as a whole) minus the *death rate* (the number of deaths averaged over the population as a whole). For example, 22 babies are

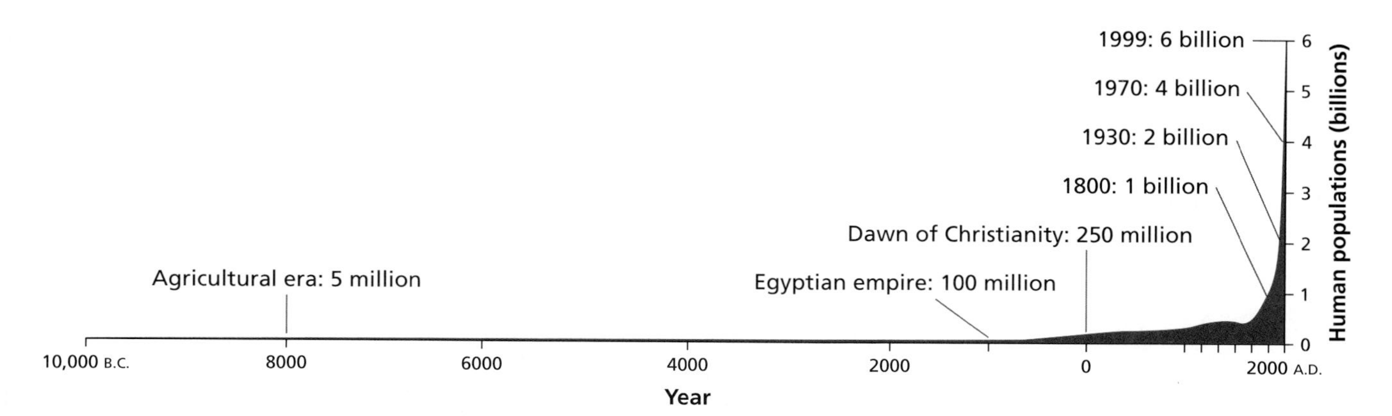

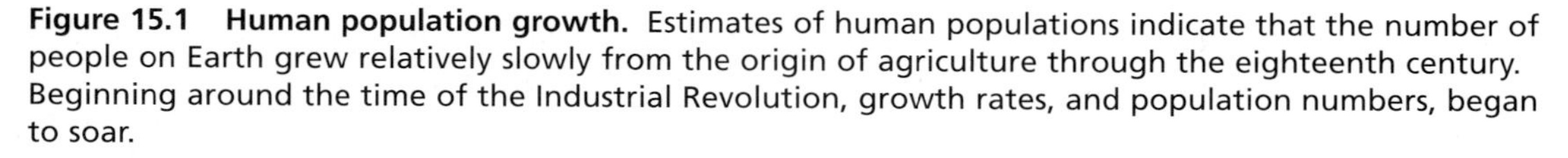

Figure 15.1 Human population growth. Estimates of human populations indicate that the number of people on Earth grew relatively slowly from the origin of agriculture through the eighteenth century. Beginning around the time of the Industrial Revolution, growth rates, and population numbers, began to soar.

born per year, on average, in a group of 1,000 people—that is, the birth rate for the population is 2.2%:

$$\frac{22}{1000} = 0.022 = 2.2\%;$$

and 10 individuals die out of every 1,000 people, resulting in a death rate of 1.0%:

$$\frac{10}{1000} = 0.01 = 1\%$$

This results in the current growth rate of 1.2%:

$$\text{growth rate} = \text{birth rate} - \text{death rate}$$
$$1.2\% = 2.2\% - 1.0\%$$

Today's relatively high growth rate, compared to the historical average of 0.1%, is the result of a large difference between birth and death rates.

Prior to the Industrial Revolution, both birth rates and death rates were high in most human populations. Although women gave birth to many children, relatively few children lived to adulthood. The rapid increase in population growth rate that occurred in the eighteenth century resulted from an equally rapid decrease in *infant mortality* (death rate of infants and children) in industrializing countries. New knowledge of how deadly infectious diseases were transmitted, and thus how they could be prevented, greatly reduced the number of children who suffered from these illnesses. With birth rates high and death rates declining, the population growth rate increased. Not long after death rates declined in these countries, birth rates followed suit, lowering growth rates again. Countries that began the process of industrialization in the eighteenth century, and countries that now have a high per capita income relative to other countries, are called *developed countries*. Nearly all developed countries have low population growth rates.

However, global human population growth rates have remained high as more countries become industrialized and their infant mortality rates decline while birth rates remain high. In addition, several more recent changes have decreased infant mortality even more dramatically in the *developing world* (countries that are early in the process of industrial development and have low per capita incomes). These changes include the use of the pesticide DDT to eliminate mosquito-borne malaria, childhood immunization programs against cholera, diphtheria, and other often-fatal diseases, and the widespread availability of antibiotics to stop bacterial infection. While birth rates are gradually declining in developing countries, they still remain higher than in developed countries. The vast majority of future population growth will occur within populations in the developing world, but the developing world is where the vast majority of food crises are occurring. Are the populations of these countries already too large to support themselves?

Limits to the Growth of Nonhuman Populations

The examples of the elk in Yellowstone and the Norway lemming illustrate a basic biological principle. While populations have the capacity to grow exponentially, their growth is limited by the *resources*—food, water, shelter, and space—individuals in the population need to survive and reproduce. The maximum population that can be supported *indefinitely* in a given environment is known as the **carrying capacity** of that environment.

The growth of a population in an environment where resources are limited appears to be exponential at first, but the effects of declining resources gradually take their toll on its growth rate. A simplified graph of population size over time in these populations is S-shaped (Figure 15.2). This model, called a *logistic growth curve*, shows the growth rate of a population declining to zero as it approaches the carrying capacity. In other words, birth rate and death rate become equal, and the population stabilizes at its maximum size. Not long after ecologists first predicted this pattern of growth, populations of organisms as diverse as flour beetles, water fleas, and single-celled protists were shown to conform with this projected growth curve.

The declining growth rate near a population's carrying capacity is caused by either a decline in birth rate, an increase in death rate, or a little of both. In organisms such as fruit flies growing in laboratory culture bottles, high populations lead to increased mortality of the flies as food supplies dwindle. Water fleas living in crowded aquariums do not have enough food to support egg production, so birth rates drop. Females of white-tailed deer populations living in crowded natural habitats are less likely to be able to carry a pregnancy to term than deer in less-crowded environments. Can we expect similar factors to reduce growth rates in a human population? That is, are humans nearing the carrying capacity of the Earth for our population? If we are, will death rates increase as food resources dwindle and more people starve? Or will birth rates decline because fewer women will have enough food to support themselves *and* a developing baby (Figure 15.3)?

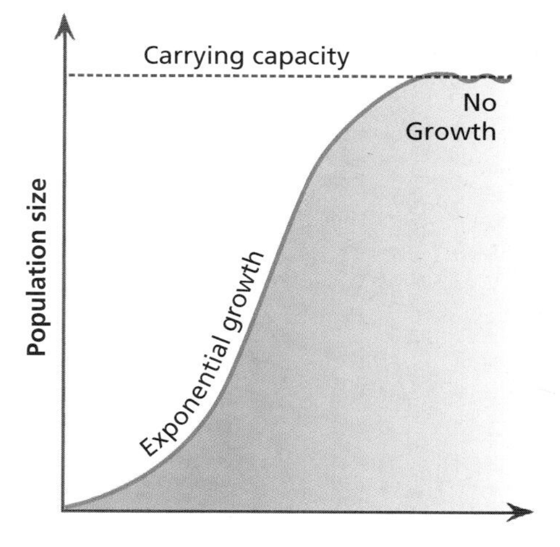

Figure 15.2 The logistic growth curve. This graph illustrates the change in size of an idealized population over time. The S-shaped curve is due to a gradual slowing of the population's growth rate as it approaches the carrying capacity of the environment.

(a) Fruit flies

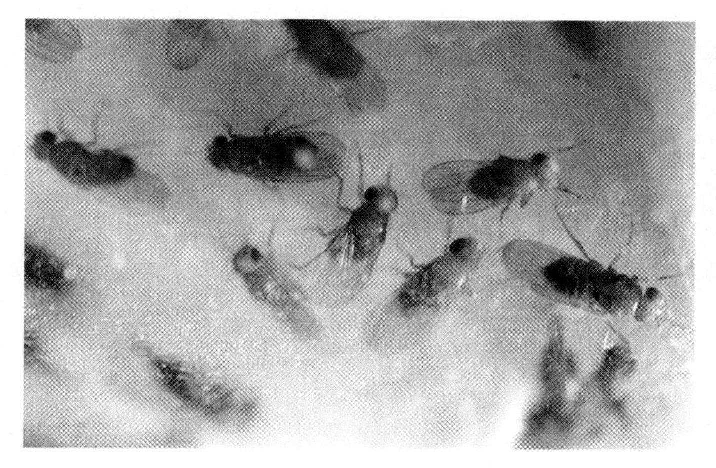

(b) Water fleas

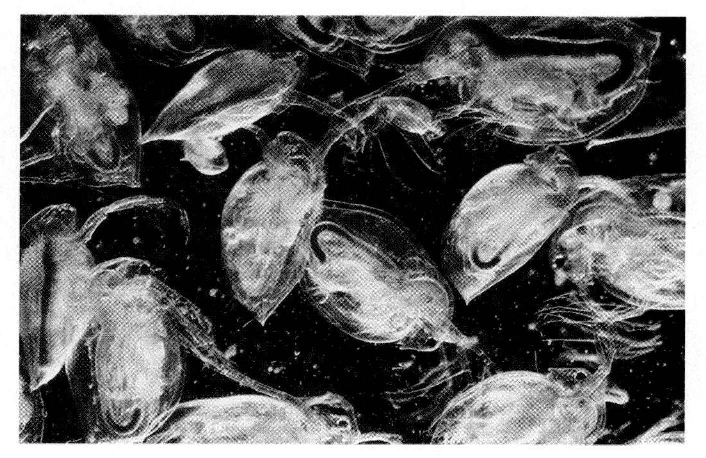

(c) White-tailed deer

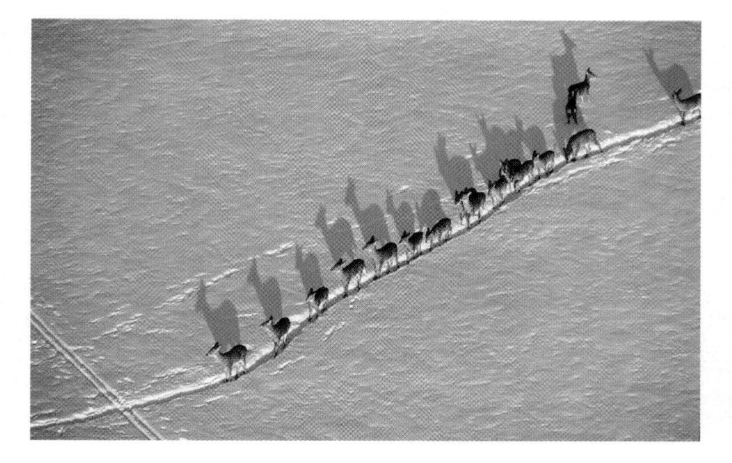

(d) Humans

Figure 15.3 Limits to growth. Populations of (a) fruit flies in a laboratory culture, (b) water fleas in an aquarium, and (c) white-tailed deer in the northeastern United States all experience high death rates and/or low birth rates as their populations approach the carrying capacity of the environment. (d) Do human populations face these same limits?

www
Media Activity 15.2 Human Growth Patterns

Humans and Earth's Carrying Capacity

One way to determine if the human population is reaching Earth's carrying capacity is to examine whether, and how rapidly, the growth rate is declining. As we saw above, the logistic growth curve is a result of a gradually declining growth rate as the population approaches the carrying capacity of the environment.

Human population growth rates were at their highest in the early 1960s, about 2.1% per year, but have since declined to the current rate of 1.2%—this steady decline is one indication that the population is nearing a stable number. Uncertainty about the future rate of decline leads to differing estimates of this number and how soon stability will be reached (Figure 15.4). However, the unique characteristics of humanity make it difficult to determine exactly what number represents Earth's carrying capacity for humans.

The growth rate of fruit flies and water fleas in laboratory populations slowed as these populations neared the limit of their resources, but the growth rates of fruit flies and water fleas are essentially forced down by environmental factors—low levels of resources cause increased death rates or decreased birth rates. However, this is not the case in human populations. Even as the number of people has rapidly increased, death rates continue to decline—an indication that people are not limited by food resources. Growth rates are declining because rates of birth are falling *faster* than death rates. Unlike the water fleas and white-tailed deer, where females are unable to have offspring when their populations are near carrying capacity, birth rates in human populations are falling because women and families are *choosing* to have fewer children, not because they do not have the resources to support a pregnancy. In fact, birth rates are generally lowest in regions where resources are most abundant and where the education of women is encouraged (Figure 15.5). Although the growth rate is slowing, rising living standards around the globe appear to indicate that humans are not nearing Earth's environmental carrying capacity.

Another way to determine if the human population is nearing Earth's carrying capacity is to estimate the amount of resources that are currently being used by humans and to use that estimate to calculate the theoretical limit to

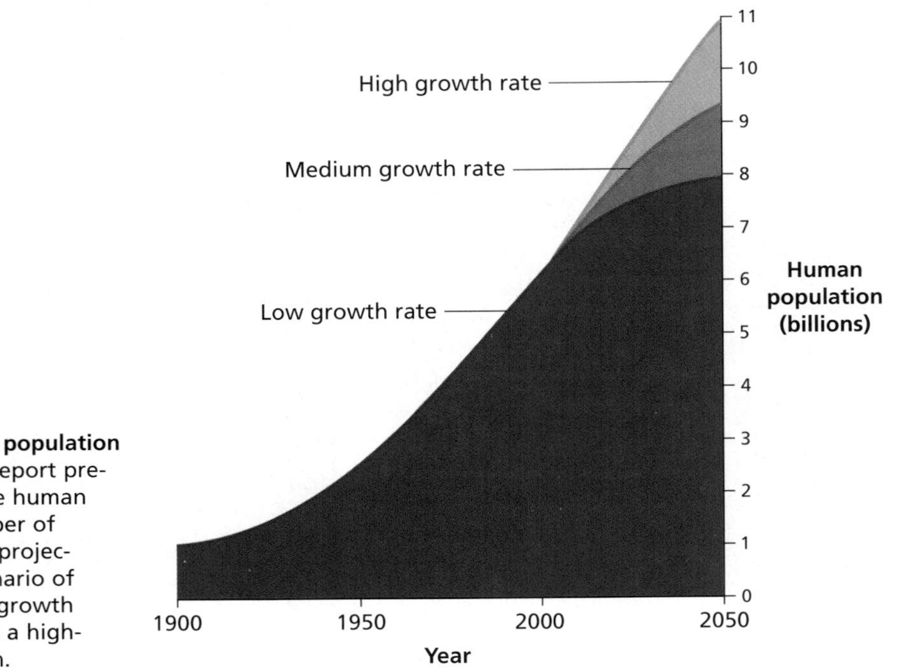

Figure 15.4 Projected human population growth. The United Nations' report predicting the eventual size of the human population is based on a number of uncertainties leading to three projections—from a low-growth scenario of 7.9 billion people, to medium growth resulting in 9.3 billion, or even a high-growth estimate of 10.9 billion.

human population size. The amount of food energy available on the planet is referred to as the **Net Primary Production (NPP)**. NPP is the amount of solar energy converted to chemical energy via plant photosynthesis, minus the amount of this chemical energy plants need to support themselves—in other words, NPP is a measure of plant growth, typically over the course of a single year. Several different analyses of the global extent of agriculture, forestry, and animal grazing have estimated that humans use about 32% of the total land NPP, although this estimate is based on a number of uncertainties. If we accept that the current population is using about 32%, or one-third Earth's NPP, we can estimate that the maximum population is three times the present population, or approximately 18 billion. This theoretical maximum is the total number of humans that could be supported by all the plant life on the planet—leaving no resources for millions of other species that are not utilized by humans. Given the dependence of human agricultural production on natural ecosystems (Chapter 14), it is unlikely that our species could survive on a planet where no natural systems remained. However, even the largest population projection by the UN, 10.9 billion, falls well short of this theoretical maximum.

Ecologists caution that the resources that are required to sustain a population include more than simply food, so the theoretical maximum population supported by the NPP estimates may be much too high. Humans also need a supply of clean water, clean air, and energy for essential tasks such as heating, food production, and food preservation. The relationship between population size and the supply of these resources is not as straightforward as the relationship between population and food. For instance, every new person added to the population requires an equivalent amount of clean water, but every new person also causes a certain amount of pollution to the water supply. We cannot simply divide the annual supply of clean water by 10.9 billion to determine if enough freshwater will be available for everyone, since the total amount of clean water is probably smaller with a larger population.

Furthermore, many of the resources that sustain the current human population are *nonrenewable*, meaning that they are a one-time supply and cannot be easily replaced. The most prominent nonrenewable resource is *fossil fuel*, the

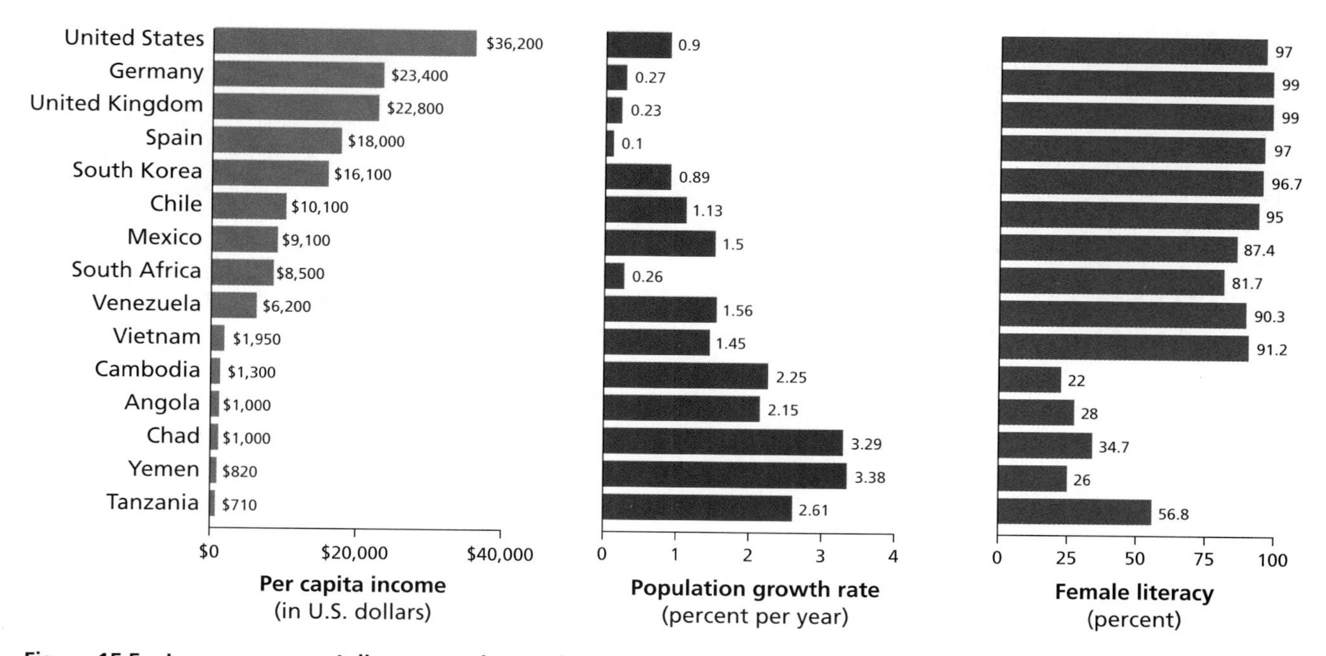

Figure 15.5 Income, women's literacy, and growth rate. These three graphs illustrate the relationships among income, population growth, and female literacy. Note that higher income and literacy are correlated with decreased birth rates and thus population growth in most countries.

Essay 15.1 *I = PAT*

Paul Ehrlich and John Holdren popularized this equation in 1971 as a way to evaluate the environmental impact (I) of a growing human population. According to their model, environmental impact is a function of not only population size (P) but also the affluence (A) of the population as measured by per capita income and the technology (T) employed by the population as measured by the amount of pollution produced per unit of income. The $I = PAT$ model allows comparisons of the environmental damage caused by populations in different countries. According to this model, the population of North America is much more destructive to the planet than some more highly populated areas (Figure E15.1).

A few comparisons between the United States and other countries bear this out. Americans constitute only 5% of the world's population but consume 24% of the world's energy. On average, one American consumes as many resources as two Japanese, six Mexicans, 13 Chinese,

Figure E15.1 The difference between an average American family (above) and an average Bhutanese family (next page) in their worldly possessions. These images dramatically illustrate the greater level of consumption, and thus environmental impact, of American families.

buried remains of ancient plants that have been transformed by heat and pressure into coal, oil, and natural gas. The use of fossil fuel and other nonrenewable resources is not simply a function of the number of people but also of their lifestyles, which we discuss in Essay 15.1. Much of modern food production relies on fossil fuel and other nonrenewable resources, and when these resources begin to run out, we might find that we need far more of Earth's NPP, and that the actual carrying capacity of our planet is much lower than we thought.

31 Indians, 128 Bangladeshis, 307 Tanzanians, or 370 Ethiopians. Americans also collectively eat 815 billion calories of food each day—that is roughly 200 billion more calories than needed, or enough to feed an additional 80 million people. The average daily individual consumption of fresh water in the United States is 159 gallons (not just for washing, cooking, and watering lawns but also for the production of food and consumer goods), while more than half the world's population lives on 25 gallons. Despite the large geographic area of the United States and its relatively small population, 50% of the wetlands, 90% of the northwestern old-growth forests, and 99% of the tall-grass prairie within its borders have been destroyed over the last 200 years.

The populations of more affluent countries tend to have higher per capita environmental impacts than those of low-income countries for two reasons: Wealth gives people the ability to consume more resources, and it provides them with insulation from the effects of that consumption. Because many Americans have the resources to move away from environmentally damaged areas, most Americans are unaware of the environmental stresses caused by their lifestyle. However, wealth also has a potentially positive side—it gives people the opportunity to learn about these impacts and make lifestyle choices that reduce their impact on Earth's ecosystems. The quality of life for future generations depends largely on how those of us in more affluent countries use our resources now.

The use of nonrenewable resources presents a clear risk of the human population **overshooting** a still unknown environmental limit. Ecologists have long known that when populations have high growth rates, they may continue to add new members even after resources have run out. This causes the population to grow larger than the carrying capacity of the environment. The members of this large population are now competing for far too few resources, and the death rate soars while the birth rate plummets. This results in a **population**

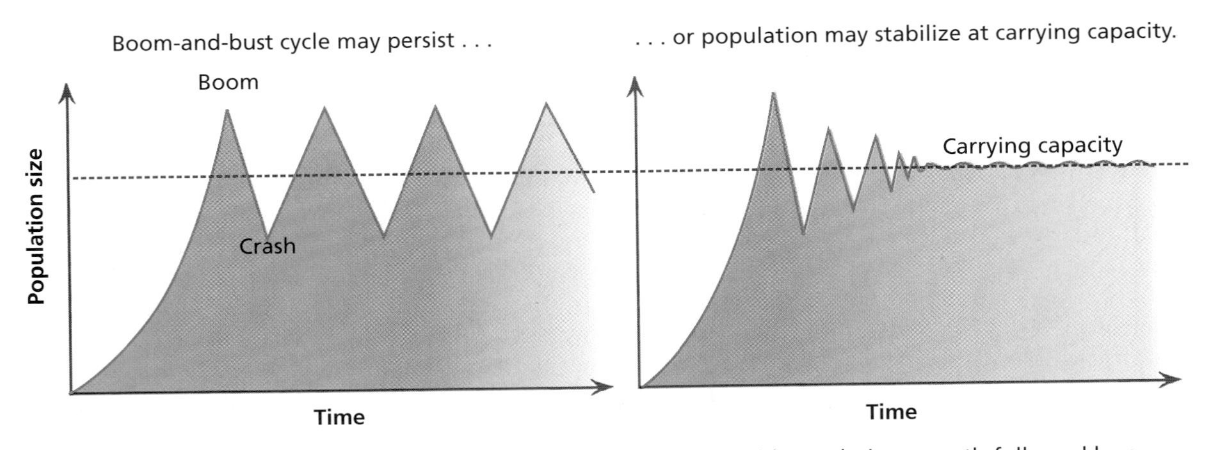

Figure 15.6 Overshooting and crashing. These graphs illustrate rapid population growth followed by a population crash. Over time, the population may stay in a "boom-and-bust" cycle, or it may stabilize at its carrying capacity.

crash, a steep decline in number (Figure 15.6). For instance, in some species of water flea, healthy babies continue to be born for several days after the food supply becomes inadequate because females use the energy stored in their bodies to produce these young. The population continues to rise, but when the young water fleas run out of stored food, there is not enough food available in the environment and most individuals die. A population overshoot and subsequent crash affected the human population on the Pacific island of Rapa Nui during the eighteenth century (Figure 15.7). This 150-square-mile island is separated from the nearest island or mainland by thousands of miles of ocean; therefore, its people were limited to only the resources on their island.

Figure 15.7 The crash of a human population. The population of Rapa Nui, also known as Easter Island, created these large statues. The large population completely deforested this small island and soon after suffered a severe population crash.

Archeological evidence suggests that at one time, the human population on Rapa Nui was at least 7000—apparently far greater than the carrying capacity of the island. The subsequent overuse and loss of Rapa Nui's formerly lush palm forest resulted in an apparently rapid decline to fewer than 700 people by 1775. It is possible that humanity's current use of the stored energy in fossil fuels may be allowing the human population to overshoot the true carrying capacity of Earth.

Biological populations may also overshoot carrying capacity when there is a significant time lag between when the population approaches an environmental limit to growth and when it actually responds to that environmental limit. Scientists who study human populations note a lag between the time humans reduce birth rates and when population numbers respond. They call this lag **demographic momentum**. In this case, parents may be reducing their family size, but because their children will begin having children before the parents die, the population continues to grow. Even when families have an average of two children, just enough to replace the parents, demographic momentum causes the human population to grow for another 60–70 years before reaching a stable level (Figure 15.8). Whether or not demographic momentum in human populations will result in the same sort of overshoot of the planet's actual carrying capacity, followed by a severe crash as on the island of Rapa Nui, remains to be seen.

Finally, determining Earth's carrying capacity for humans as simply a function of whether food will be available also ignores what are known as quality-of-life issues, or what some scientists refer to as *cultural carrying capacity*. An Earth that was wholly given over to the production of food for the human population would lack wild, undisturbed places and the presence of species that bring pleasure and a sense of wonder. With human populations at the limits of growth, much of our creative energy would be used for survival, taking away a large share of the joy of music, art, literature, and discovery.

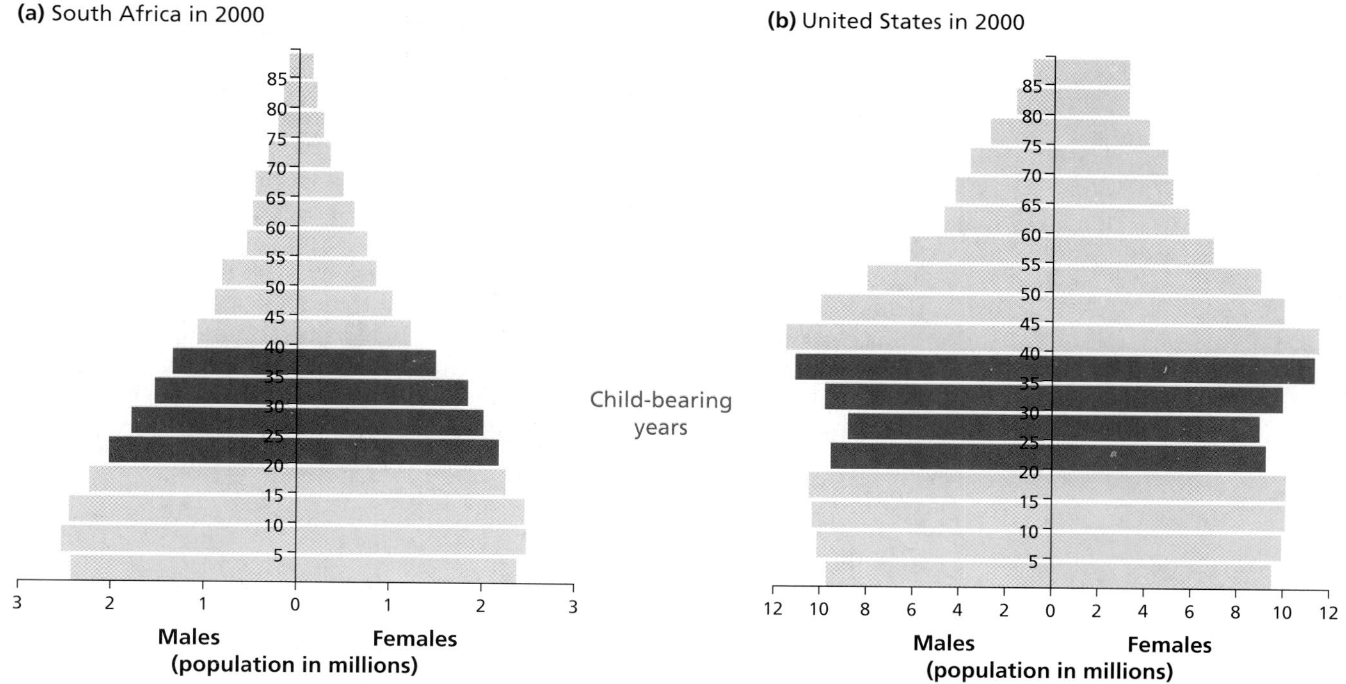

Figure 15.8 Demographic momentum. In a rapidly growing human population, such as the one in (a) South Africa in 2000, the majority of the population is young and the population will continue to grow as these children reach child-bearing age. In a population that is growing more slowly, the distribution of ages is more even, as in the population of (b) the United States in 2000.

What we have learned here is that scientists cannot tell us exactly how many people Earth can support, partly because humans make choices that are not predictable and partly because humans have an incredible capacity to innovate and adjust seemingly fixed biological limits. Ultimately, the question of how many people Earth *should* support—and at what quality—is not solely a question of science but also of values and ethics. For instance, the most effective way to reduce human population growth rates quickly is to provide for the education and improved social status of women—including increasing access to birth control, a strategy that many people oppose on religious grounds.

While we do not know the maximum number of humans Earth can support, we do know that a significant proportion of the current human population receives inadequate food. We also know that our methods of food production rely on the use of nonrenewable resources, a practice that cannot continue for very many more generations. In the next section we investigate what science can tell us about how to feed large human populations now and in the future.

WWW
Media Activity 15.3 Sustainable Agriculture in the 21st Century

15.2 Feeding the Human Population

"And then the dispossessed were drawn west . . . Car-loads, caravans, homeless and hungry; twenty thousand and fifty thousand and a hundred thousand and two hundred thousand. They streamed over the mountains, hungry and restless—restless as ants, scurrying to find work to do—to lift, to push, to pull, to pick, to cut—anything, any burden to bear, for food. The kids are hungry. We got no place to live."

This passage from John Steinbeck's classic novel, *The Grapes of Wrath*, describes the movement of people from the southern Great Plains of the United States to California during the devastating period known as the Dust Bowl. From 1931 to 1936, a combination of intense drought, severe storms, and inappropriate farming practices led to crop failures and widespread **soil erosion**, the loss of soil. Impoverished residents of the panhandles of Oklahoma and Texas faced a desperate choice of trying to ride out the bad times or pulling up stakes and leaving their homes. Nearly 25% of the population in these regions left over a period of just three or four years.

The farming practices that led to the Dust Bowl were initially heralded as great improvements over those of the past. Slow horse-drawn plows and harvesters were replaced with gasoline-powered tractors and combines, allowing more and more of the native prairie ecosystem to be turned into wheat fields—but plowing prairie grasses destroyed the root systems that held the prairie's soil in place. Once these grasses were gone, and between growing seasons for crops, soil could be picked up by fierce midwestern winds and blown thousands of miles in tremendous dust storms (Figure 15.9). Aggressive efforts to reverse soil erosion combined with the return of normal rainfall levels finally ended the Dust Bowl era. Thanks to the advances of modern agriculture, including the use of fertilizers and irrigation, this former disaster area is now part of the most productive corn- and wheat-producing region on Earth.

Fast-forward to March 2001, when a giant dust cloud born on the northern plains of China was blown thousands of miles across the Pacific Ocean and obscured the view of the Rocky Mountains from downtown Denver, Colorado. What caused this dust cloud? Drought, severe storms, and inappropriate agricultural practices—a familiar story. Hundreds of thousands of people are moving away from this Asian Dust Bowl to the cities along China's Pacific coast. Similar migrations have occurred in west-central Africa, parts of the Middle East, southern Russia, and elsewhere around the world as agriculture expands

Figure 15.9 The Dust Bowl. The Dust Bowl of the 1930s was named for the enormous dust storms created by drought, high winds, and soil erosion caused by poorly planned agricultural development. Storms like this one occurred regularly over several years.

to ever-larger areas of Earth's surface. The expansion of agriculture and the environmental problems associated with this expansion are an outgrowth of the rapidly increasing human population.

Is the inevitable cost of agriculture massive environmental damage? Can these newly destroyed regions recover the way areas depleted in the American Dust Bowl did? Or is the Dust Bowl recovery an illusion maintained by huge inputs of nonrenewable resources that will eventually run out?

Agriculture Seeks to Maximize Photosynthesis

Recall from Chapter 2 that plants, along with algae and some bacteria, transform light energy into chemical energy via the process of **photosynthesis**. Photosynthesis converts the energy from sunlight, carbon dioxide from the atmosphere, and water into energy-rich molecules called *carbohydrates*. Plants then use carbohydrates to run the activities of their cells. In other words, plants make their own food. In fact, plants are so effective at converting light energy into chemical energy that they produce enough energy to store and to grow larger. Other organisms, such as humans, harvest this excess energy. Our agricultural practices are intended to maximize the photosynthesis of crop plants so they will produce a surplus of carbohydrates we can harvest.

If we look only at the chemical reaction that occurs during photosynthesis, it is clear that plants require carbon dioxide, water, and light to produce carbohydrate—but the effective production of chemical energy from light requires a few more components. For instance, photosynthesis requires nitrogen-containing proteins in the *choloroplast* and magnesium atoms that are present in the green pigment called *chlorophyll*. Elements such as nitrogen and magnesium, typically called **nutrients**, must be obtained from the plant's environment for

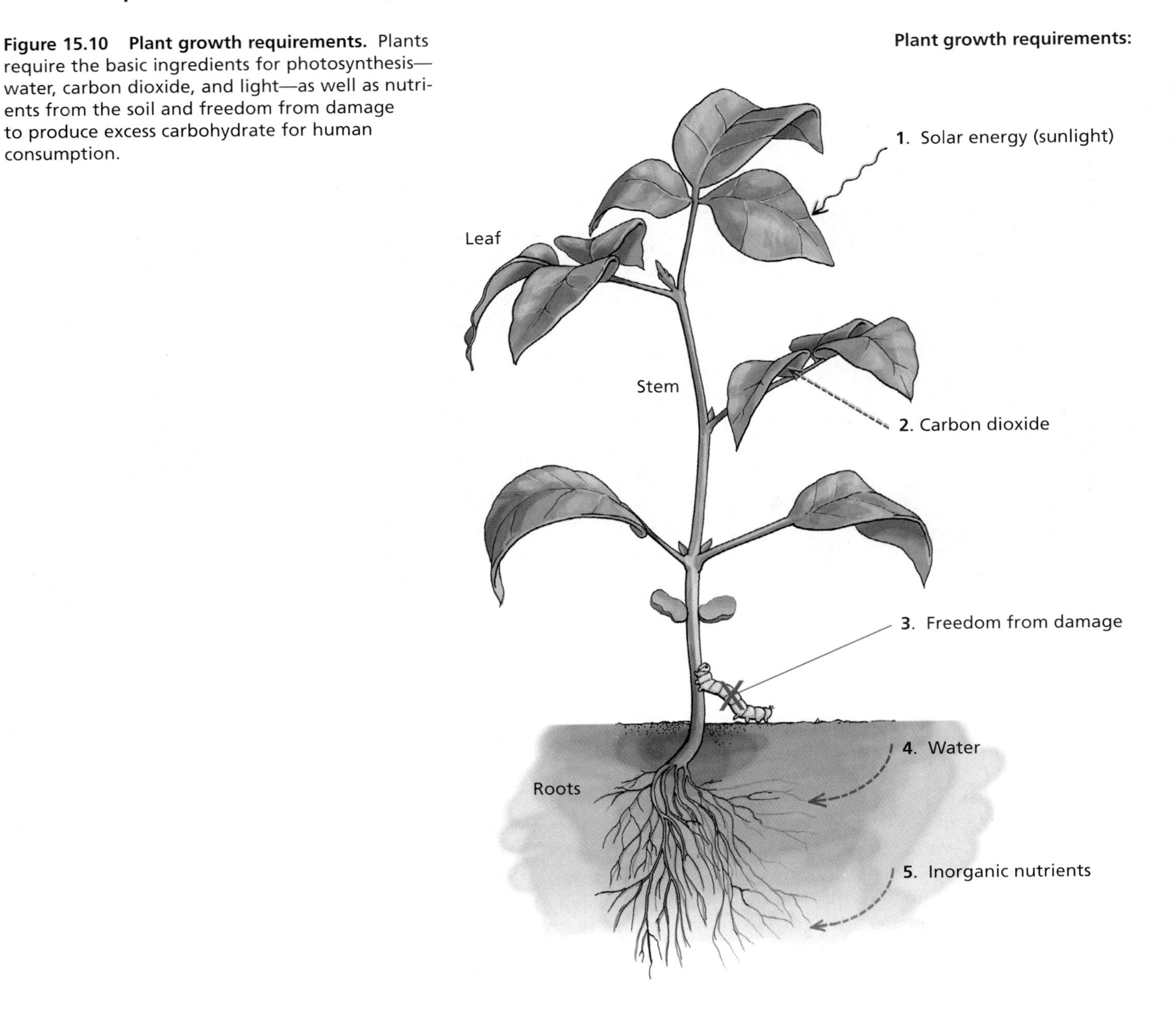

Figure 15.10 Plant growth requirements. Plants require the basic ingredients for photosynthesis—water, carbon dioxide, and light—as well as nutrients from the soil and freedom from damage to produce excess carbohydrate for human consumption.

photosynthesis to occur (Figure 15.10). In addition, plants must be able to support themselves against the pull of gravity; surprisingly, they use water as one of their main methods of support. Thus, plants require much more water than just that needed for the chemical reactions of photosynthesis. Modern farming practices attempt to maximize the amount of light, water, carbon dioxide, and nutrients devoted to crop production.

Maximizing Exposure to Light and Water There are two components to maximizing a plant's exposure to the sun. One is to ensure that the plant can effectively orient its leaves to intercept incoming light. The other is to maximize the amount of light that the plant can potentially intercept.

To hold its leaves perpendicular to the sun's rays, a plant's cells must be full of water. Unlike an animal cell, the membrane of a plant cell is surrounded by a stiff **cell wall**. The cell wall is tough, but it is also elastic, meaning that

Figure 15.11 Water in plant cells. (a) When a plant is fully hydrated, the cell's vacuoles balloon with water and the cells press against each other, so the plant remains "crisp." (b) As the plant dries out, the vacuoles lose water and the cells no longer support each other, so the plant wilts.

it can stretch. As a plant cell's vacuole fills with water, the cell wall balloons slightly, becoming turgid—it is the pressure of adjoining ballooned cell walls that holds the leaves upright (Figure 15.11a). When water levels drop and the cells begin to lose water, the pressure of cells pushing against each other decreases, the cells become flaccid, and the plant wilts (Figure 15.11b). Thus, part of maximizing a plant's exposure to light is ensuring that it has plenty of water.

Plants are continually losing water, especially during hot, sunny weather. To minimize the chance of wilting during these conditions, farmers try to provide an optimal amount of water to their crops. In many regions, adding water through *irrigation* is necessary. The enormous production of corn, wheat, and soybeans in the former Dust Bowl region is a result of an equally enormous amount of irrigation. The water for irrigation in this rather dry region comes from a huge underground reservoir called the *Ogallala aquifer* (Figure 15.12).

Another way to maximize the amount of water reaching a preferred plant—that is, one that a farmer or gardener desires to grow—is to reduce the amount of competition for water. Farmers do this by trying to minimize the number of nonpreferred plants, commonly called *weeds*, in agricultural fields. Any plant can be a weed if it grows in the wrong place—just ask a gardener who maintains a turf lawn and must continually pull turf grass out of the flower bed! Minimizing the number of weeds growing with a crop has the additional benefit of minimizing the chance that a weed will physically block sunlight from the preferred plants. Thus, reducing the number of weeds benefits preferred crops by allowing them access to the maximum amount of light as well as water.

Farmers have several techniques for controlling weed growth. One is to remove competitors before the crop is planted. This process is referred to as *tilling* and involves turning over the soil to kill weeds that have sprouted from seed but are still small and vulnerable to damage. After a crop begins growing, it is possible to reduce the growth of competitors, and in addition keep soil from losing water, by *mulching* the crop. To mulch, farmers spread straw or other dead plant material, or sometimes dark-colored plastic, around the base of preferred plants. A thick layer of mulch keeps sunlight from penetrating

(a) Irrigation

(b) Water source

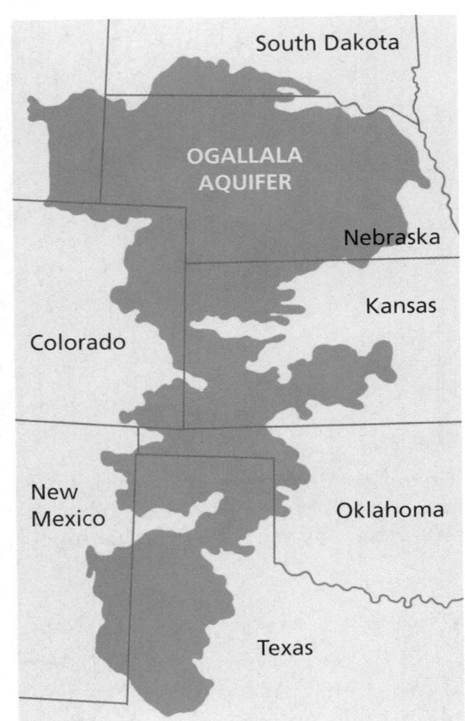

Figure 15.12 Irrigation. (a) The bright green circles in this aerial photo are formed by lush plant growth. The circle pattern derives from center-pivot irrigation sprayers, which spray water in a complete arc from the center point. (b) The Ogallala aquifer is the largest single pool of freshwater in the world; it underlies and supplies the United States' "bread basket" region.

the soil and stimulating weed seeds to germinate. Mulching is impractical for most large crop producers, but it is common in smaller gardens and for certain crops.

Although tilling sets back the growth of weeds, it cannot kill all of the weeds, and some spring back to life quickly. In these cases farmers use an *herbicide*, a chemical that kills plants, to eliminate weeds. Until recently, the use of herbicides was limited to relatively few crops because it is difficult to target just the weeds. Flowering plants can be divided into two large groups—*dicots*, also known as broad-leaved plants, and *monocots*, or narrow-leaved plants. In the 1940s, scientists studying a growth hormone in plants (first discovered by Charles Darwin in the 1880s) found that certain forms of this hormone were deadly to broad-leaved plants but harmless to narrow-leaved plants. They developed herbicides that are used to control weeds in narrow-leaved crops, such as corn, wheat, and rye. You may be familiar with this class of herbicides as well—many homeowners use it to kill dicot dandelions on their monocot grass lawns. However, these herbicides are ineffective against narrow-leaved weeds and cannot be used to control weeds in broad-leaved crops, and other herbicides are deadly to all plants and thus kill the crop along with the weeds.

The development of *genetically modified organisms* (*GMOs*) has increased the applicability of herbicides to many more crop-and-weed combinations in the past decade. As we described in Chapter 7, some crops, such as soybeans and corn, have had a gene inserted in them that makes them resistant to a broad-spectrum herbicide—that is, one that kills all plants. Thus, a field planted with a resistant crop could be sprayed with this herbicide, which would kill everything except the crop.

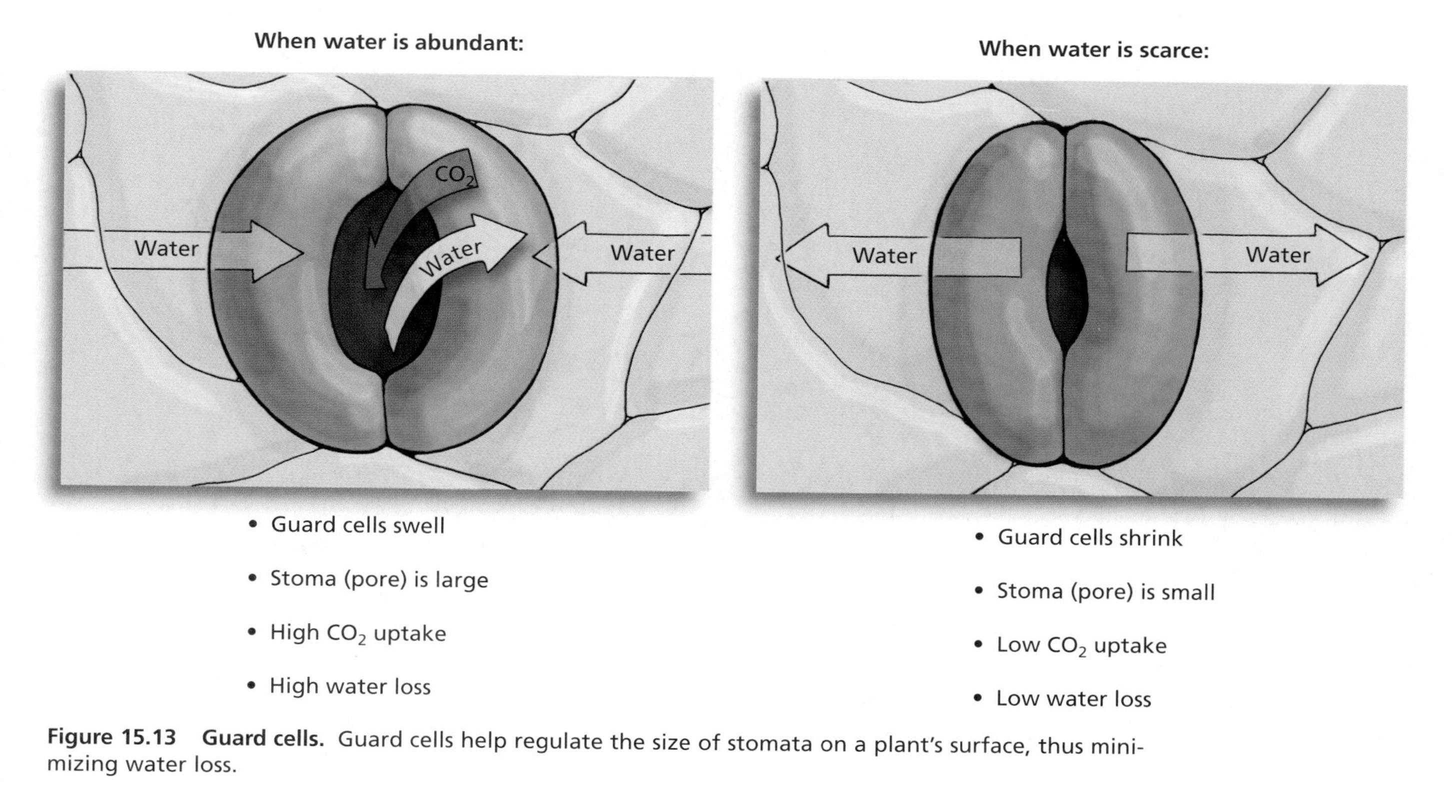

Figure 15.13 Guard cells. Guard cells help regulate the size of stomata on a plant's surface, thus minimizing water loss.

Maximizing Uptake of Carbon Dioxide Carbon dioxide is a gas in the atmosphere. For plant cells to obtain this gas for photosynthesis, the cells must be exposed to the atmosphere. The photosynthetic surfaces of plants are covered with tiny pores called **stomata** that allow air into the internal structure of leaves and green stems. However, the porosity of leaves and stems comes with a price—stomata also provide portals through which water can escape.

In most plants, each stoma is encircled by a pair of **guard cells**, which serve to regulate the size of the pore (Figure 15.13). When the guard cells have abundant water, the pore is a large space. When the guard cells are water-deprived, the pore is tiny. The guard cells thus help minimize water loss under dry conditions and maximize carbon dioxide uptake under wet conditions. Therefore, maximizing the carbon dioxide uptake of crop plants requires the same inputs as maximizing light exposure—abundant water and minimal competition from weeds.

Maximizing the Nutrients Needed for Plant Growth Aside from carbon dioxide, all of the nutrients required for plant growth are available in the soil. A plant's roots take up these nutrients, and the less competition between a crop-plant's roots and a weed-plant's roots, the better the crop will grow. Thus, techniques that reduce competition for light, water, and carbon dioxide between crops and weeds also reduce nutrient competition.

In most environments, nutrients become available to plants through **nutrient cycling**. When materials pass through a *food web*, the pathway that biotic energy follows in an environment (Figure 15.14a), the nutrients are generally not lost from the environment—hence the term nutrient *cycling*.

You should note as you review Figure 15.14a that plants absorb simple molecules from the soil and make them into more complex molecules. These complex molecules move through the food web with relatively minor changes until they reach the soil. Here, complex molecules are broken down into simpler

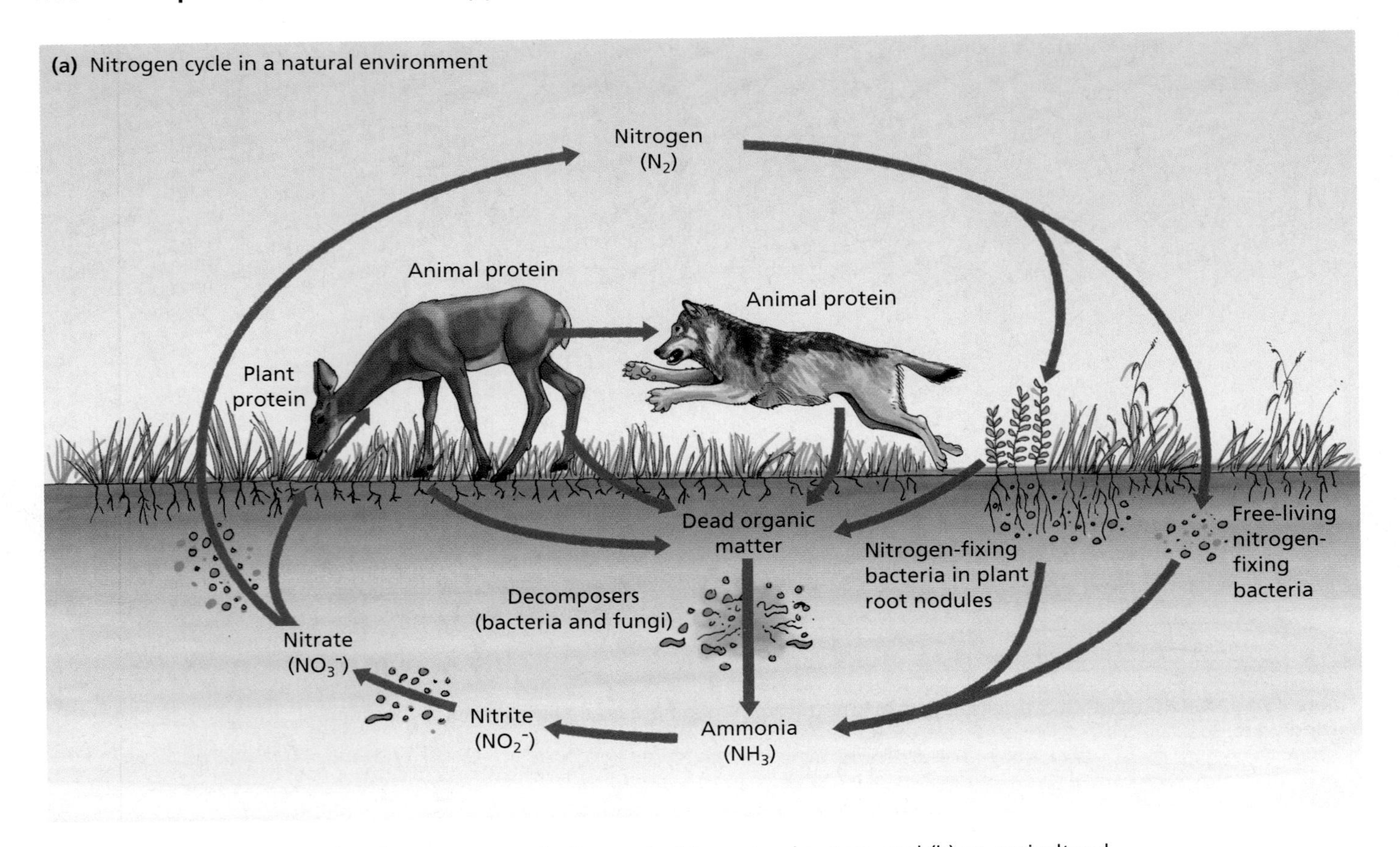

Figure 15.14 A nutrient cycle. The movement of nitrogen in (a) a natural system, and (b) an agricultural system, is illustrated here. As energy in the form of food is transferred from one organism to another in a food web, the nutrients are recycled. Notice that nutrients are not recycled to the soil in an agricultural system; they are lost because plants are removed from the system. Fertilizer is added to soil to make up the loss.

ones by the action of **decomposers**, typically bacteria and fungi. Nutrient cycling in a natural environment relies on a healthy community of these organisms in the soil.

In agricultural systems, the nutrient cycle is often broken, or at least disconnected. Instead of plants being eaten on-site by animals, which are then eaten in the same location by other animals and where their wastes and bodies decompose, the plants from agricultural fields are removed and trucked to other locations where they will be consumed and their nutrients released. In the case of food for human consumption, most of the nutrients that are consumed end up in human waste, which often ends up in a waterway flowing into lakes or oceans. Therefore, in agricultural systems nutrients are not recycled back into the soil, but are essentially mined from it (Figure 15.14b).

Farmers add nutrients to replace what is lost from the soil in the form of *fertilizer*. Fertilizer that is applied to the soil can be *organic*—that is, complex molecules made up of the partially decomposed waste products of plants and animals—or *inorganic*—meaning simple molecules produced by an industrial process. Plants require nutrients in simple, inorganic form—thus organic fertilizer must first be further decomposed before it is available for plant growth. The vast majority of farmers in the United States fertilize primarily with industrially produced inorganic *nitrogen*, but also may apply large amounts of inorganic *phosphorus* (an essential component of DNA), or *potassium* (which is important for maintaining the water balance of plant cells), as well as occasionally other nutrients such as *calcium*, an important component of cell walls.

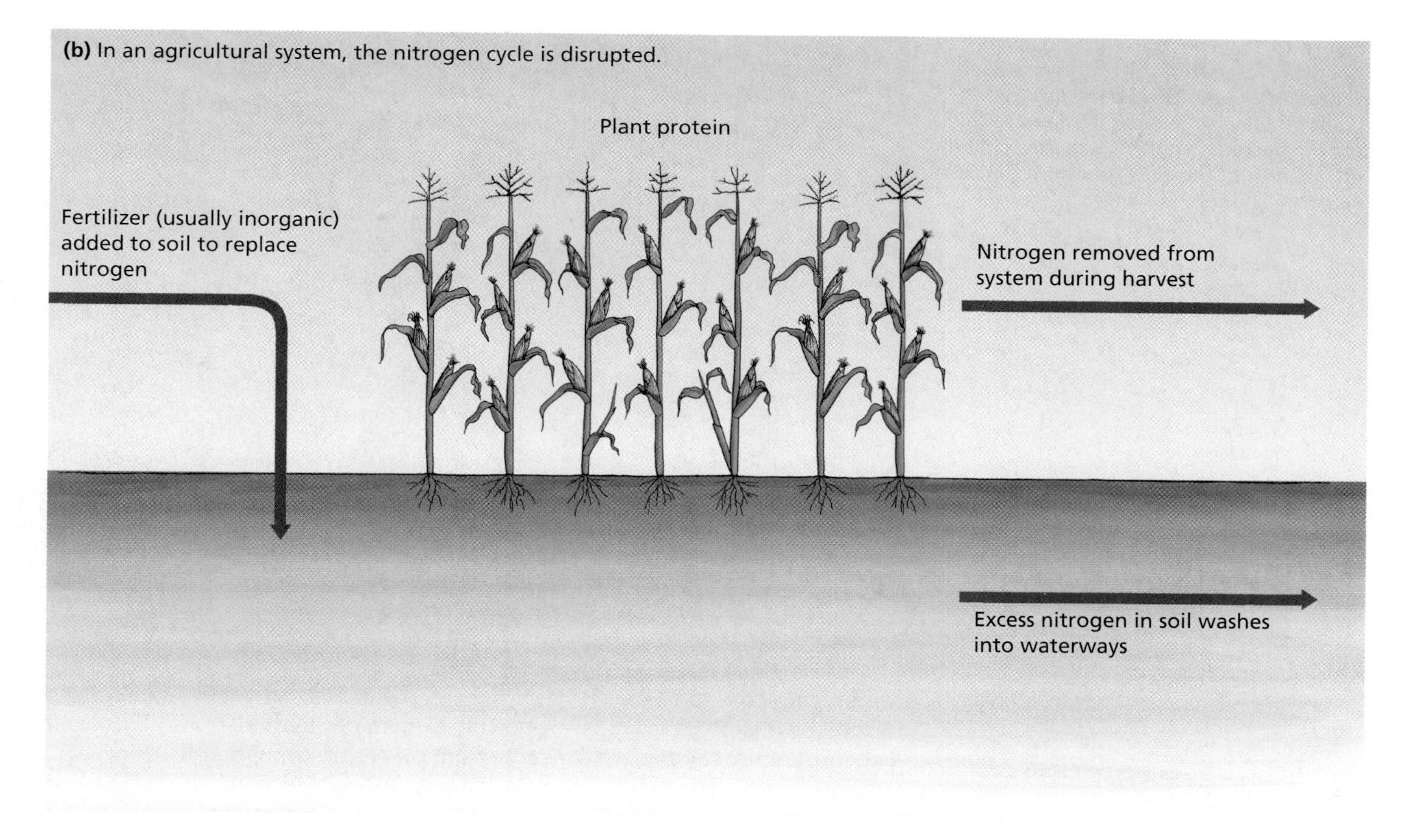
(b) In an agricultural system, the nitrogen cycle is disrupted.
Plant protein
Fertilizer (usually inorganic) added to soil to replace nitrogen
Nitrogen removed from system during harvest
Excess nitrogen in soil washes into waterways

Figure 15.14 ***(continued)***

Farmers choose to use inorganic fertilizer because it is more concentrated and thus much easier to transport, store, and apply than most organic fertilizers and also because plants can utilize it immediately.

Farmers can also replace some nutrients in soil through *crop rotation*; that is, by not putting the same crop in the same field year after year. Surprisingly, some plants actually increase levels of nutrients in the soil. **Nitrogen** is among the most important nutrients for plants in terrestrial systems, because it is the fourth most abundant chemical in plant cells (after carbon, hydrogen, and oxygen, all of which are components of carbon dioxide and water). Plants that can add nitrogen to the soil do this by having a *mutualistic* relationship with **nitrogen-fixing bacteria** (Figure 15.15). These bacteria have the ability to convert nitrogen gas, an abundant component of the atmosphere, into a form that can be taken up by plant roots. *Legumes*, plants that produce seeds in pods, such as soybeans, alfalfa, clover, and peanuts, have a complex symbiosis with nitrogen-fixing bacteria. These plants develop structures called *nodules* on their roots in response to chemical signals from certain species of nitrogen-fixing bacteria. The nodules house the bacteria and supply them with energy in the form of carbohydrate. In return, these plants receive the excess fixed nitrogen the bacteria produce. When a legume is growing well, the bacteria produce more fixed nitrogen than the plant requires, and this excess is released into the surrounding soil. If legumes are planted in alternating years with another crop that uses up soil nitrogen (such as corn or wheat), the need for nitrogen fertilizer can be greatly reduced.

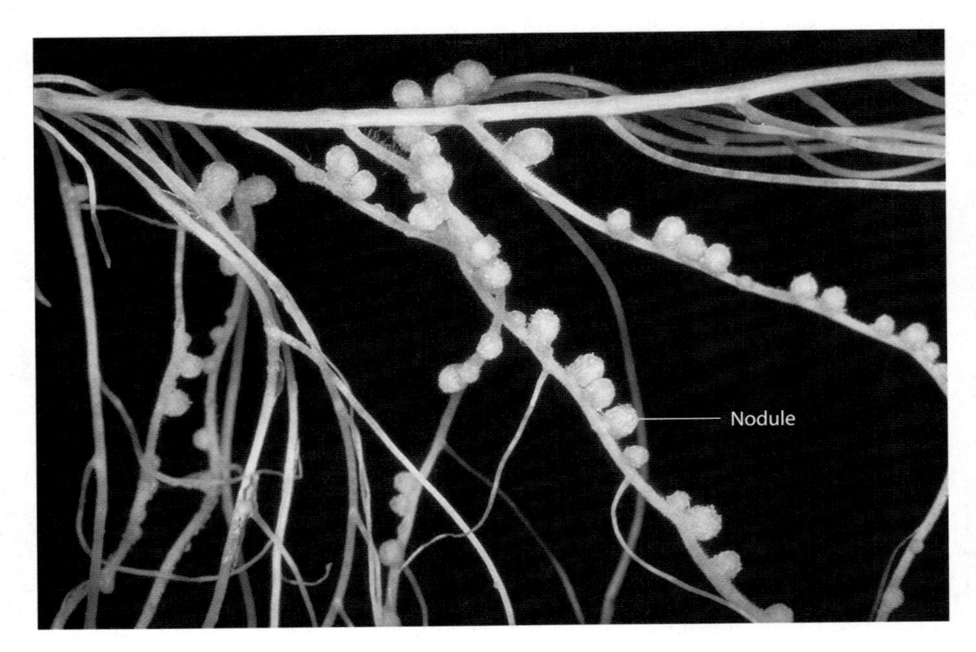

Figure 15.15 Nitrogen-fixing crop. The nodules on the roots of this alfalfa plant contain nitrogen-fixing bacteria. The plant provides a home and a food source for the bacteria, and the bacteria convert nitrogen from the air into a form that is easily taken up by the plant.

Minimizing Loss to Pests Wheat grains, corn kernels, rice grains, and millet—four of the major staple crops supporting human populations—are all *fruits*. In fact, most of the calories we consume from plants are from their fruits or seeds (such as beans). The major exceptions are potatoes, which are tubers (a thickened underground stem); cassava, a thickened root; and sugar, the product of either sugar beets (a root) or sugar cane (a stem). Of course, our diet is made up of other fruits as well, such as apples and cucumbers; and we also consume other plant parts, such as the leaves of lettuce or the flowers of broccoli (Figure 15.16). *Seeds* are small packages containing a plant embryo and an energy-rich food source to support the growth of that embryo. Most fruits function as carriers so the seeds can disperse far from the parent plant. Some fruits help seeds disperse on the wind or water, while others help seeds disperse by using animals to carry the fruit and dropping it, or its seeds, elsewhere (Figure 15.17). Since the seeds and fruits we eat are so packed with energy, they are attractive to many different consumers—including other mammals, birds, insects, fungi, and bacteria. Competing consumers of agricultural products are generally referred to as *pests*.

Some crop pests directly consume the fruits and seeds we seek to harvest. Other pests reduce the growth of plants, limiting the amount of energy the plant can put into seeds, fruits, and tubers; these pests include insects, fungi, and bacteria that damage the leaves, stems, and roots of plants. Some studies estimate that 42% of potential food production in major crops is lost to pests every year. Thus, farmers have a strong incentive to reduce the number and effect of pests to improve their crop yields.

The primary tools farmers use to reduce pest impact are *pesticides*, chemicals designed to kill or reduce the growth of a target pest. (Herbicides are also commonly considered to be a class of pesticides. In this book, we use pesticide to refer only to chemicals that control insect, fungal, and bacterial pests.) Pesticide use became widespread in the years following World War II, when the scientists who were developing chemical weapons for the war effort turned these weapons into potent insect killers. Today, millions of tons of these chemicals are applied to crops all over the world (Figure 15.18).

Surprisingly, pesticide use has not reduced the overall loss of crops to pests much. Instead, these chemicals have had a dramatic impact on farming methods—for instance, allowing farmers to plant a single crop over a

Beans (seeds)
Cucumber (fruit)
Lettuce (leaf)
Potato (tuber)
Broccoli (flower)
Sugar beet (root)
Coconut (endosperm)
Onion (bulb)

Figure 15.16 Plants as food. Our diet includes a diversity of plant parts, such as fruits, seeds, roots, and flowers.

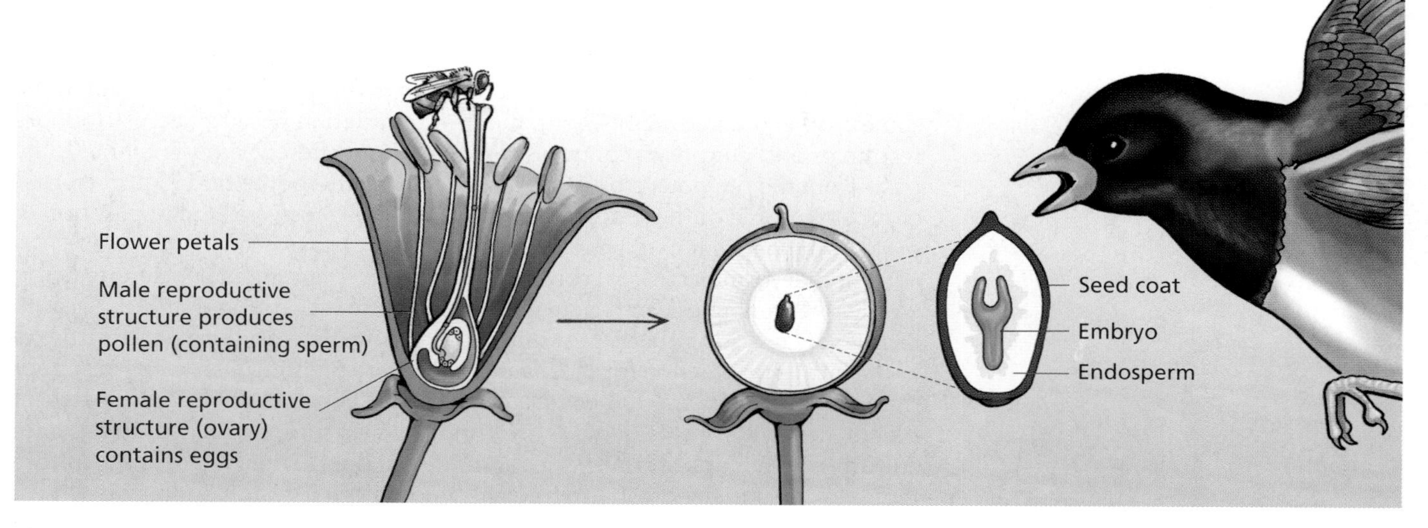
1. **Flowers** attract insects that move pollen from one flower to another, helping fertilization to occur (other flowers are wind pollinated).
2. **Fruits** package seeds in a structure that aids their dispersal, such as a tasty fruit or a parachute.
3. **Seeds** contain an embryo and endosperm (a food source). The seed coat enables some seeds to survive the digestive tracts of some animals.
Flower petals
Male reproductive structure produces pollen (containing sperm)
Female reproductive structure (ovary) contains eggs
Seed coat
Embryo
Endosperm

Figure 15.17 Flowers, fruits, and seeds. Fruits are the ripened ovary of a flower and contain seeds. From the plant's perspective, fruits serve as a delivery system for seeds.

Figure 15.18 Pesticides. Farmers reduce insect damage to crop plants by spraying pesticides, chemicals that may be toxic to both insects and people.

wide acreage. This practice is called *monoculture*. Monocultural production of crops is very efficient. Farmers have one planting, fertilizing, and pest-control schedule; they only require the planting and harvesting equipment specific to their crop; and they can use enormous mechanical harvesters to quickly collect enormous amounts of product (Figure 15.19). Monoculture and its accompanying mechanization has greatly reduced the cost of food to consumers by decreasing the amount of labor that is required to produce it. The percentage of the population working on farms in the United States has dropped from 25% to less than 2% in the past 50 years due to this change in farming practice.

Before the availability of pesticides, farmers used *cultural control* to minimize pest populations. Since most pests have evolved to attack a single crop, the cultural control of crop rotation moves plants away from their pests. A small number of pests in a population of plants can quickly grow into a large population of pests in a single growing season. After the growing season, the pest population spends the winter in the soil or on plant waste left in the field. If the same crop is planted in this field, the pest population has easy access to more resources and continues to grow. However, if a different crop is planted, the pests from the previous crop must disperse in an attempt to find their preferred crop. Only some of these dispersers will find their host, and the population of pests in the new site will be relatively small.

Another form of cultural control is *polyculture*: planting many different crop plants over a single farm's acreage. This minimizes crop loss and the risk to farmers by ensuring that a pest outbreak on a single crop does not destroy all of the farm's production for that year. Polyculture planting also keeps pest populations relatively small, because after the pests have consumed a patch of plants, they must disperse widely to find another source of food—in a monoculture, another appropriate plant is right next door. Thus, while monoculture increases the efficiency of agricultural production, it results in increased pest populations and an increased need for pesticides just to keep pest damage at historical levels.

Figure 15.19 Monoculture. Planting the same crop over many hundreds of acres allows for the efficient use of massive planting and harvesting equipment.

Designing Better Plants In addition to providing resources to crops to maximize production, farmers have turned to scientists to provide crop plants that respond strongly to these resources. Traditionally, agricultural scientists have used plant breeding to produce better crops. Plant breeding generally consists of creating **hybrids**, the offspring of two different varieties, or *races*, of an agricultural crop. For instance, one of the first hybrids produced on a large scale was a wheat plant. One parent variety was a high-producing standard-sized wheat that was difficult to harvest because it tended to fall over when the wheat grains were ripe. The other parent variety was a lower-producing dwarf plant. The hybrid of these two varieties was high producing but short, making it much easier to harvest with the large machines necessary to effectively work on very large farms. Widespread planting of the hybrids sold by seed companies, rather than the numerous varieties that had been maintained on different farms, also ensures that the fruits produced on different farms are relatively uniform in character—this is a benefit especially in the case of wheat, where similar levels of starch makes refining the wheat grains into flour much more efficient. The production of hybrid varieties that produced well and were easy to harvest and process was the foundation of the Green Revolution described in Essay 15.2.

More recently, agricultural scientists have designed plants, called *genetically modified organisms*, or *GMOs*, using DNA technology (inserting genes from one organism into the genetic instructions of another organism, see Chapter 7). DNA technology has resulted in the production of genetically modified corn

Essay 15.2 The Green Revolution

The *Green Revolution* is the name given to the dramatic increase in crop yields that occurred in Mexico, India, Pakistan, and elsewhere in Asia during the 1960s and 1970s. These increases occurred primarily through the application of agricultural techniques developed in the United States, which were almost single-handedly exported by plant breeder Norman Borlaug (Figure E15.2).

While he was pursuing his graduate degree in plant biology during the Dust Bowl years, Borlaug noted that soil erosion and crop failures appeared to be lowest where high-yield approaches to farming were being tried. He decided that his life's work would be to spread the benefits of high-yield farming to the many nations where crop failures as terrible as those in the Dust Bowl were regular facts of life.

Dr. Borlaug was able to pursue his dream when the Rockefeller Foundation established a research institute in 1943 to assist poor farmers in Mexico. The program's initial goal was to teach Mexican farmers new farming ideas, but Borlaug was soon seeking agricultural innovations. His most consequential research achievement was the perfection of dwarf spring wheat, a high-yielding, easily harvested hybrid variety.

The most dramatic achievement of the Green Revolution did not take place in Mexico, however, but in southern Asia. In 1963, the Rockefeller Foundation sent Borlaug to Pakistan and India, which were then experiencing widespread malnutrition. Although Borlaug initially encountered resistance, by 1965 famine in the two countries was so bad that their governments decided to allow widespread planting of dwarf wheat. By 1968, Pakistan was self-sufficient in wheat production. By 1974, India was self-sufficient in the production of all cereals. Food production since the 1960s has increased in both nations faster than the rate of population growth. For his work in these two countries, Borlaug received the 1970 Nobel Peace Prize.

The form of agriculture that Borlaug exported may have prevented a billion deaths in the developing world (although surprisingly, the number of undernourished people in India has not changed since the 1960s). However, dwarf spring wheat does not prosper without fertilizer and irrigation. High-yield crops grow well, but the better they grow, the more moisture they demand, and the faster they deplete soil nutrients. Supplying these crops with enough nutrients from organic fertilizer, such as manure, is nearly impossible. Only inorganic fertilizers based on petroleum and other minerals can renew nutrients in the soil on the scale needed to sustain the Green Revolution varieties. Fertilizer and irrigation is expensive, so these high-yielding varieties are inaccessible to poor farmers. The continuing challenge of the Green Revolution is to provide food for a growing human population while minimizing the environmental costs of its production and sharing its benefits among all people.

Figure E15.2 Norman Borlaug.

containing a bacterial gene that produces its own insecticide, soybeans containing a gene that confers resistance to an herbicide, and a host of other, novel organisms (Figure 15.20)—potentially representing an agricultural revolution on the order of the Green Revolution. However, as we noted in Chapter 7, many questions remain about GMOs, including their safety as human food, the likelihood their widespread use will cause the evolution of resistance in pests, their effects on nontarget organisms, and the consequences of the "escape" of engineered genes into noncrop plants.

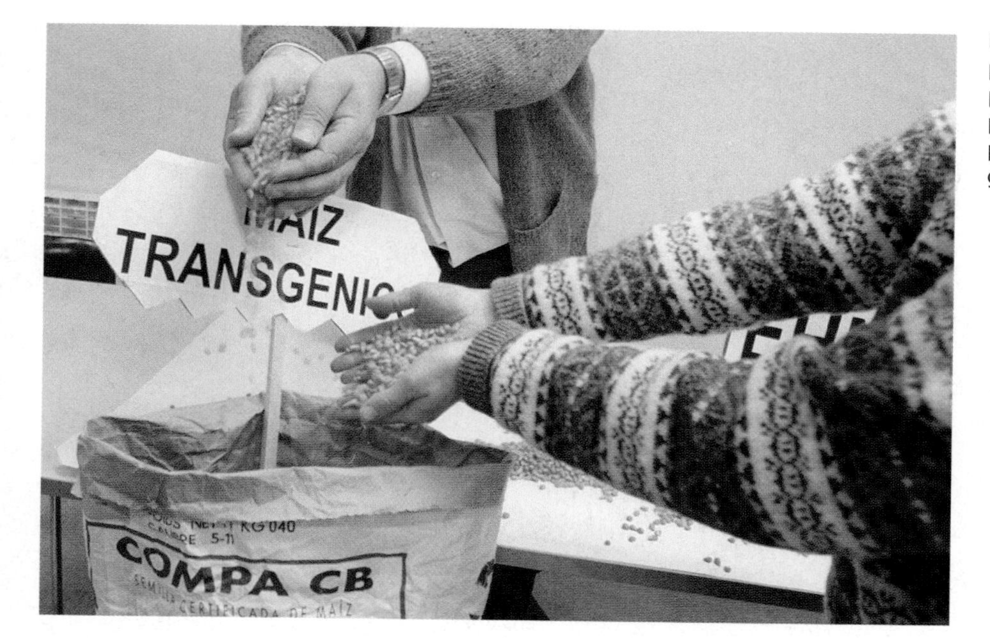

Figure 15.20 Genetically modified plants. These corn seeds contain a gene produced by bacteria. The bacterial gene produces a protein that is toxic to caterpillars, and corn plants containing this gene produce their own pesticide.

Modern Agriculture and Future Food Production

The use of pesticides, herbicides, fertilizer, and irrigation has increased food production so dramatically that even with twice as large a human population and a higher level of food consumption per person, food prices are lower today than they were in the 1940s. Low food prices are a sign that food is abundant. The steady decline in food prices was the basis for some economists' assessment that Earth's potential agricultural production is more than enough to support the growing human population. However, many biologists argue that current high levels of production are *unsustainable*—they compromise the ability of future generations to grow enough to feed a large human population.

What evidence supports the hypothesis that current agricultural practices are unsustainable? First, the production and application of agricultural chemicals such as pesticides and fertilizer requires large amounts of fossil-fuel energy, which is in limited supply; second, the water sources used to provide irrigation water, such as the Ogallala aquifer, are being used faster than they can be replaced by natural precipitation; and third, the environmental costs of modern agricultural practices may be greater than a growing population is able to bear.

The Environmental Costs of Fertilizer One environmental cost of modern agriculture stems from the use of inorganic fertilizer. The nutrients in inorganic fertilizer are in a very simple chemical form; they are available to plants immediately after their application to soil. While this method provides an immediate boost to plant growth, the simple molecules cannot remain in the soil for long. Some of the minerals in inorganic fertilizer that are not taken up by growing plants quickly are carried off by water moving through the soil. This *runoff* accumulates in water reservoirs—either lakes, streams, and oceans or deep underground in reservoirs called *groundwater*. When the nutrients from inorganic fertilizer reach high levels in the water, it becomes unsafe for human consumption and many of the organisms that live in it die. Increasing levels of nitrate fertilizer in water wells in the American Midwest correlate with increasing rates of miscarriage and bladder cancer among women in the surrounding communities, and some forms of nitrogen in drinking water can cause brain damage or death among infants who drink it.

Figure 15.21 The Mississippi River's dead zone. This satellite image shows the delta of the Mississippi River. The cloudy water marks the area of the ocean where fertilizer-rich sediment from the river mixes with seawater, resulting in oxygen depletion and the near absence of living organisms.

Fertilizer runoff from farms in the midwestern United States has also created an enormous *dead zone* in the Gulf of Mexico near the outlet of the Mississippi River (Figure 15.21). This area of very productive ocean, about the size of the state of New Jersey, experiences periods of dangerously low oxygen levels during the growing season. Increased levels of nitrogen and phosphorus (another common fertilizer) increase the growth of algae in the Gulf and other waterways; when the algae die, their decomposers flourish and rapidly use up the available oxygen. This process, called **eutrophication**, results in large fish kills. If the use of fertilizer increases as human populations grow, the loss of drinking water sources and fishing grounds will probably also increase.

The Environmental Costs of Pesticide The use of pesticides also carries an environmental cost. Pesticides directed toward a specific crop pest may lead to an increase in another pest whose populations were held in check by *competition* with the target pest. (This is similar to the role bacteria play in the intestines of chickens, discussed in Chapter 14.) These infestations of secondary pests then result in the use of more pesticides. Similarly, within populations of a target pest, there may exist a few individuals that are genetically resistant to a pesticide. When this pesticide is applied to a crop, the resistant individuals survive to begin the next outbreak. As with the *natural selection* of drug-resistant HIV described in Chapter 9, this results in the evolution of a pest population that is resistant to the pesticide. To control subsequent outbreaks of resistant pests, higher levels of pesticide or new, more toxic pesticides must be applied. The process of continually increasing pesticide use is termed the *pesticide treadmill*—because, like on an exercise treadmill, farmers continually expend energy and resources simply to "stay in place" and keep pest populations under control. In fact, the pesticide treadmill is partially responsible for a 10-fold increase in the amount and toxicity of pesticides applied to crops in the United States over the last 50 years.

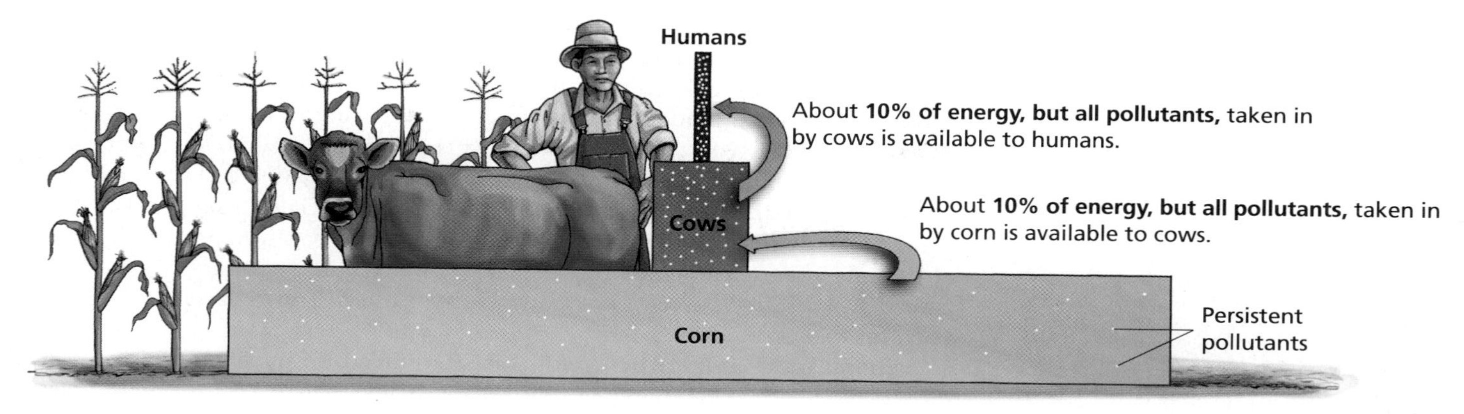

Humans
About **10% of energy, but all pollutants,** taken in by cows is available to humans.
Cows
About **10% of energy, but all pollutants,** taken in by corn is available to cows.
Corn
Persistent pollutants

Figure 15.22 A trophic pyramid. The width of each row in this pyramid is proportional to the amount of energy available for the next step of the pyramid. Pollutants found in the lower levels of the food chain become concentrated as they move up the trophic pyramid.

An additional problem associated with some types of pesticides is **biomagnification**, the concentration of toxic chemicals in higher levels of a food web. Biomagnification of slow-to-degrade chemicals (also called *persistent organic pollutants*, or *POPs*) is the result of a basic rule of biological systems: Most of the energy consumed at one level of a food web is used to support that level and is not put into growth and reproduction. This rule leads to the principle of the **trophic pyramid**, the bottom-heavy relationship between the *biomass*, or total weight, of populations at each level of a food web. Most of the calories used by organisms are for maintenance, not growth, and individuals at one level of a food web must consume many times their biomass of the level below to survive and reproduce.

When a pesticide or other chemical does not break down quickly, it accumulates in photosynthetic organisms as they absorb it from the soil and water in a process known as *bioaccumulation*. **Primary consumers** ingest the chemicals and accumulate them as they use up the accompanying energy from the plants they eat. The pesticide is thus in much higher concentrations in the tissues of consumers than plants. **Secondary consumers** then ingest the primary consumers and accumulate the persistent pesticides stored in each of their prey. This same process continues until there are very high levels of pesticide in the top consumer levels (Figure 15.22).

The problems of bioaccumulation and biomagnification were recognized in the late 1960s—particularly the negative effects of accumulations of high concentrations of the pesticide DDT in fish-eating birds such as bald eagles—leading to the development of much less persistent pesticides. However, many POPs are still used in the United States and elsewhere. Most humans have measurable levels of persistent pesticides in their bodies, including DDT, which is present in measurable quantities in the breast milk of most women in the United States even 30 years after it was banned in this country. The long-term health effects of low-level exposure to these chemicals are still relatively unknown, but some evidence suggests that these effects include increased risks of various forms of cancer, birth defects, and permanent nerve damage. The pesticide treadmill demands ever more powerful and toxic chemicals, and high pesticide levels will negatively affect more and more of the human population.

Wise use of fertilizers and pesticides may help minimize the environmental effects of modern agriculture—the negative consequences fade when their use is reduced. However, one consequence of modern agriculture results in environmental damage that is not easily recovered—the loss of soil via erosion.

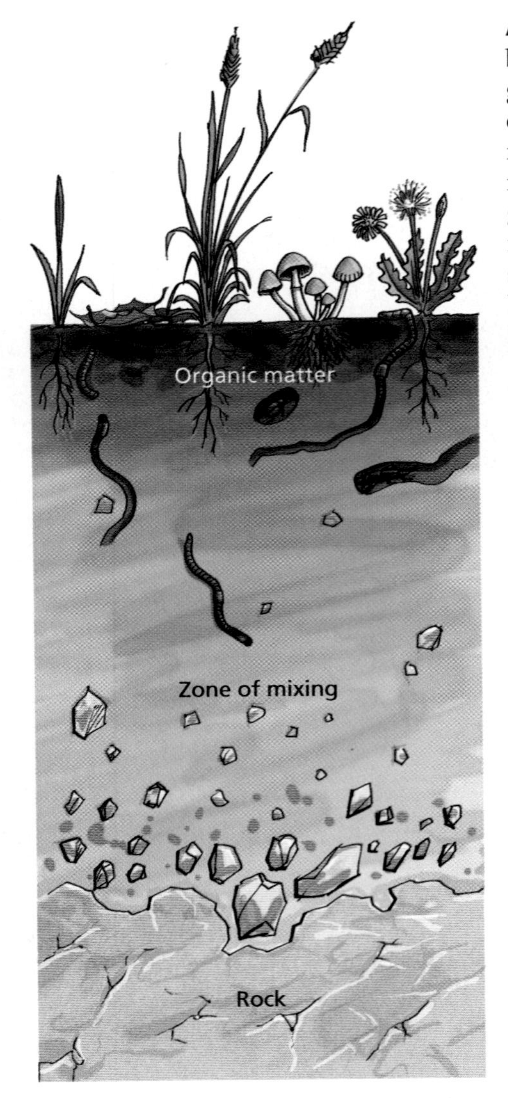

Figure 15.23 Soil structure and development. Soil forms from the breakdown of rock and the accumulation of organic matter from dead and decaying plants. Organisms in the soil decompose the organic matter and "mix" the soil through their movements. These actions make nutrients available for plant growth.

A Longer-Term Cost: The Loss of Soil Soil is a unique material, made up of both living and nonliving components, developing as a result of the action of growing plants on rock. Healthy soil consists of a base layer of recently eroded rock; a top layer made up of the dead and decaying remains of plants, animals, and their wastes; and a middle layer where these two components are mixed by earthworms and other soil organisms (Figure 15.23). Soil is held together both by the organisms that live within it, especially fungi, and by the roots of the plants growing throughout it. Modern agricultural techniques destroy soil by regularly removing the plants (through tilling or harvest) and also by causing the death of soil organisms. Inorganic fertilizer, in particular, is very harmful to soil organisms.

When soil loses its structure, it is easily picked up by strong winds and blown away—the Dust Bowl and the recent dust storms in China are dramatic examples of soil erosion caused by agricultural practices. Soil erosion occurs frequently even when dust clouds do not blot out the Sun. People in the northern plains of the United States still experience localized dust storms that result from topsoil blowing and drifting off of agricultural fields.

Irrigation damages soil by causing *salinization*—literally, the "salting" of soil. Water used in irrigation systems contains tiny amounts of dissolved minerals in the form of salts. As water is applied to the soil, the water is taken up by plants or evaporates and the salts are left behind. Eventually, so much salt accumulates in the soil that plants cannot survive. The development of soil takes many years; soil lost through erosion and salinization is not easily replaced. Every year, millions of acres of agricultural soil are lost to **desertification**: they become too degraded by erosion or salinization to use.

If we also include the uncertainty about how future climate change will affect agriculture (see Essay 14.2), it seems clear that current agricultural practices will not be sufficient to feed a growing human population without incurring very large environmental costs. Fortunately, some alternatives to current agricultural practices *may* help provide enough food for 10 billion people without as much environmental damage.

Can We Feed the World Today and Tomorrow?

There are two strategies for producing enough food to feed a growing population while using fewer unsustainable agricultural practices. The most obvious strategy is to change some of the ways farming is practiced. Another strategy is for consumers to change their eating and food-buying habits.

Many of the techniques farmers can use to reduce the environmental costs of agriculture have already been mentioned. Minimizing monocultural plantings may be the most effective. Not only will planting polycultures reduce the need for pesticides, it will help minimize fertilizer inputs and preserve soil by increasing crop rotation. Moving away from monocultural production will not be an easy task—it would probably have to be encouraged by changes in government policies and agricultural support systems. In the United States, the tendency toward larger and larger corporate-owned farms will also make this transition more difficult.

Even without a switch to polyculture, farmers have an incentive to reduce their use of fertilizer and pesticides—after all, these inputs cost money. Many farmers have begun employing technology to monitor the production of their agricultural fields and now apply fertilizer and pesticides in a more targeted manner. In crops that require high levels of pesticide, farmers have employed a technique called *integrated pest management* (*IPM*), which utilizes releases of predator insects, the introduction of pest competitors, and changes in planting techniques among other strategies to reduce pesticide use. However, IPM is time consuming and labor intensive—since its introduction in conventional agriculture, pesticide and fertilizer use have continued to rise in the United States.

Figure 15.24 Overuse of pesticides. (a) This apple was produced on a conventional farm, with high levels of pesticides to control superficial damage, but the apple in (b) was produced on an organic farm, with no pesticides. If consumers were willing to accept the level of superficial "damage" seen on the organic apple, the amount of pesticides used on fruit and vegetable crops would greatly decline.

Farmers are only one-half of the food-production equation. Consumers also determine what food is produced and how. If consumers consistently select certain types of foods, farmers will try to supply them. In many cases, consumer demand is encouraging some of the more damaging environmental consequences of modern agriculture.

The majority of the pesticides sprayed on vegetable and fruit crops are not applied to reduce the actual consumption of these foods by pests. Instead, pesticides are used to reduce crop "losses" that are caused by consumers who generally refuse to purchase fresh fruits and vegetables with superficial signs of pest damage. Examine the piles of fruit and vegetables displayed at an average grocery store. Most are perfect—no blemishes, large, and shiny (Figure 15.24). The only way to produce such pristine products is to apply hundreds of pounds of pesticides to completely eliminate pest insects. If consumers were willing to accept a visible but small amount of pest damage to fresh vegetables, the amount of pesticides used on these crops could drop tremendously, even without abandoning monocultural production.

Consumer demand also fuels unsustainable farming practices in the United States through the average American's appetite for meat. Most of the beef, pork, and poultry produced in the United States is grain-fed. Unlike the pastoral scene most of us envision, few cows spend their entire lives on farms grazing in a meadow and eating hay—90% spend at least part of their lives in enormous feed lots, where they are fed corn to quickly increase their size before slaughter. The same production process holds for hogs, chickens, and turkeys. In fact, 66% of the grain consumed in the United States, including 80% of the corn and 95% of the oats, is used to feed these animals.

Review the trophic pyramid in Figure 15.22. The grain used as animal feed produces a much smaller amount of calories for human consumption in the form of beef, pork, and poultry. Put another way, a 10-acre farm can support about 10 people by growing corn for human consumption, but it can support only two people by growing corn to feed to cattle raised for human consumption, since most of the corn eaten by cattle provides energy for maintenance, not for building muscle. When consumers consume many of their calories from products that are higher on the agricultural food web, they are forcing the high-intensity farming that causes so many of the environmental problems discussed in this chapter. Reducing meat consumption would also reduce the number of acres farmed, and soils that are highly susceptible to degradation through erosion or salinization could be treated more lightly. Less land needed for agricultural production also leaves more natural habitat for other species to survive (Chapter 14).

There is little doubt that consumer demand can change agricultural practice. In recent years, enough consumers have changed their buying habits that many farms have changed from conventional practices (the use of pesticide and inorganic fertilizer) to **organic farming,** which is supported by practices that do not require industrially produced inputs or the use of GMOs. A report by the U.S. Department of Agriculture in 2002 describes a 20% growth in retail sales of organically grown food every year since 1990. As a result, organic farm acreage in the United States *doubled* between 1992 and 1997. Organic farms are still very much the exception in American agriculture, but the rise in interest in organically grown food is having a noticeable impact on farming practices.

While it is difficult to answer the question of whether or not Earth can support the human population, it appears that it is possible to feed a human population of 10 billion—perhaps not all of them with the diet enjoyed by a typical American today, but enough for basic survival. However, feeding the human population in a way that preserves environmental quality and leaves room for the diversity of other life that shares the planet will require thoughtful decision making about agricultural policy. These decisions include those we make as individuals, such as how much meat to consume and what sort of produce we buy, and those we make collectively, such as federal policies and subsidies that favor one form of agriculture over another (Table 15.1). As with all of the topics we have explored in this book, your knowledge of the science behind the reports and studies and your understanding of the power and limitations of scientific knowledge will help you become an effective participant in this decision making.

CHAPTER REVIEW

Summary

- The human population has grown very rapidly over the last 150 years.
- All populations eventually reach the carrying capacity of their environment when resources begin to run out and the death rate increases.
- The growth of the human population is slowing, but because women are choosing to have fewer children, not due to an increasing death rate.
- It is possible that the human population will overshoot Earth's carrying capacity because resources other than food are also important to human survival and because of demographic momentum.
- Feeding a human population of many billions requires the mass production of food in agricultural systems.
- The addition of water and the removal of competing plants allow plants to maximize light and carbon dioxide uptake.
- Farmers make up for the loss of nutrients from agricultural systems by adding fertilizer.
- Farmers minimize the loss of plants to pests by using cultural control and pesticides.
- The production of hybrid strains and genetic engineering has led to improvements in crop production.
- Fertilizer runs off into waterways, causing human health problems and damage to fish populations.
- Pesticide use continually increases as pests become resistant to them. Persistent pesticides also accumulate in organisms at higher levels of the food web.
- Soil is being lost faster than it is being replaced in many areas through erosion and salinization.
- Planting many crops in an area rather than planting large monocultures can help reduce pest populations by requiring pests to disperse.
- Fertilizer and pesticide use can be minimized through the use of crop rotation and by careful monitoring of the agricultural environment.
- Consumers can help reduce the environmental cost of modern agriculture by accepting a small amount of pest damage on produce, by eating less meat, and by purchasing organically grown products.

Environmental problem	Solution	Actions we can take
Bioaccumulation of persistent pesticides 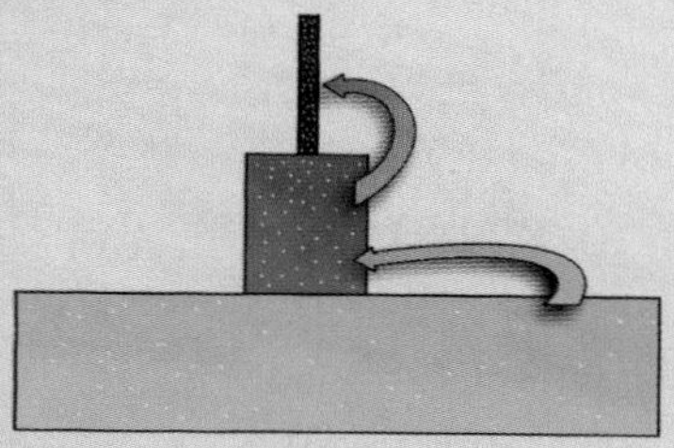	• Minimize monocultural plantings. • Reduce use of pesticides. • Increase use of integrated pest management.	• Accept produce that has superficial pest damage. • Purchase organic foods. • Add diversity of foods to our diets to promote polyculture. • Reduce meat consumption, and thus overall crop production. • Support policies that reduce use of monocultural production (for example, elimination of price supports for crops). • Support integrated pest management research.
Eutrophication of surface water and nitrate pollution of groundwater 	• Reduce use of inorganic fertilizer.	• Purchase organic foods. • Purchase foods grown in polyculture, where crop rotation is employed. • Reduce meat consumption.
Salinization of soil 	• Reduce number of acres irrigated. • Increase efficiency of irrigation.	• Reduce meat consumption. • Support research and national agricultural policies to reduce agricultural water use.
Soil erosion and desertification 	• Reduce use of tilling on marginal lands. • Reduce amount of land used for crop production.	• Reduce meat consumption. • Support research on effective non-chemical means of weed control, and see suggestions below.
Loss of habitat for nonhuman species 	• Minimize the amount of land converted to agricultural production.	• Reduce meat consumption. • Support policies that effectively lead to decreased human population growth rates. • Support policies that provide developing countries with the tools to increase crop yield sustainably.

Table 15.1 Reducing the environmental impact of modern agriculture. This table summarizes the environmental costs of modern agriculture and the individual and collective actions we can employ to reduce these costs.

Key Terms

biomagnification p. 417
carrying capacity p. 394
cell wall p. 404
decomposers p. 408
demographic momentum p. 401
desertification p. 418
ecology p. 392
eutrophication p. 416
exponential growth p. 393
guard cells p. 407
hybrid p. 413
net primary production (NPP) p. 397
nitrogen p. 409
nitrogen-fixing bacteria p. 409
nutrient cycling p. 407
nutrients p. 403
organic farming p. 420
overshoot p. 399
photosynthesis p. 403
population crash p. 399
primary consumer p. 417
secondary consumer p. 417
soil erosion p. 402
stoma p. 407
trophic pyramid p. 417

Learning the Basics

1. What factors have led to the explosive increase in the human population over the past 200 years?

2. Explain why a decrease in population growth rate is expected as a nonhuman population approaches carrying capacity.

3. Describe why demographic momentum may cause the human population to overshoot Earth's carrying capacity.

4. Describe briefly the agricultural practices that maximize the amount of light and water available to crop plants.

5. Why do persistent pesticides biomagnify in a biological system?

6. The growth of human populations over the past 150 years has increased primarily due to _____.
   a. increases in death rate
   b. increases in birth rate
   c. decreases in death rate
   d. decreases in birth rate
   e. increases in net primary production

7. According to Figure 15.25, the carrying capacity of this environment is _____.
   a. 0 flies
   b. 100 flies
   c. 150 flies
   d. between 100 and 150 flies
   e. impossible to determine

8. In contrast to nonhuman populations, human population growth rates have begun to decline due to _____.
   a. voluntarily increasing death rates
   b. voluntarily decreasing birth rates
   c. involuntary increases in death rates
   d. involuntary decreases in birth rates
   e. voluntarily increasing birth rates

9. Which of the following factors is associated with declines in a country's population growth rate?
   a. an increase in per capita income
   b. an increase in female educational attainment
   c. an increase in women's social status
   d. a and c are correct
   e. a, b, and c are correct

10. Demographic momentum refers to the tendency for _____.
   a. low population growth rates to continue to decline
   b. high population growth rates to continue to increase
   c. populations to continue to grow in number even when growth rates reach zero
   d. populations to continue to grow in number, even when women are reducing the number of children they bear
   e. women to continue to have children even though they no longer wish to

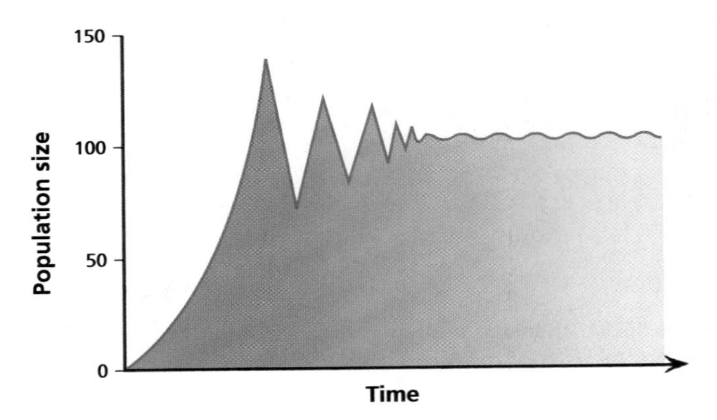

Figure 15.25 Learning the Basics, Question 7; Analyzing and Applying the Basics, Question 1.

11. All of the following agricultural practices are designed to increase the maximum level of photosynthesis in crop plants, EXCEPT _____.
    a. irrigation
    b. tilling
    c. adding fertilizer
    d. mulching
    e. spraying pesticides

12. Soils in agricultural systems require fertilizer because _____.
    a. farming mines nutrients from the soil
    b. weeds compete with crop plants for soil nutrients
    c. most crop plants have nitrogen-fixing bacteria in their roots and need lots of nitrogen
    d. most nutrients run off soils into waterways, causing eutrophication
    e. pests remove most nutrients from crop plants

13. When inorganic fertilizer ends up in waterways _____.
    a. the waterways become eutrophic
    b. algae populations explode in surface waters
    c. fish kills result
    d. drinking water may become contaminated
    e. all of the above

14. There is a lower biomass of organisms at higher levels of the food web than at lower levels because _____.
    a. organisms at lower levels are smaller
    b. organisms at higher levels require more energy to survive
    c. at each level, the majority of the energy is used simply to sustain individuals
    d. only excess energy at each level can be used to support the next trophic level
    e. more than one of the above is correct

15. Consumers can help reduce the environmental effects of modern agriculture by _____.
    a. insisting that produce is free of superficial damage from pests
    b. purchasing foods that require low levels of fertilizer and pesticide
    c. eating less grain-fed meat
    d. b and c are correct
    e. a, b and c are correct

Analyzing and Applying the Basics

1. Review Figure 15.25 on the preceding page. How would you expect the carrying capacity of the population to change if the flies are supplied with a greater amount of food? What other factors might influence carrying capacity in this environment?

2. Imagine two human populations, each of which is made up of 5 million individuals. In one population, over 50% of the members are in the age group 0–20 years and about 2% are over 65. In the other, about 20% are 0–20 years old, and about 20% are over 65. Which of these populations will probably stabilize at a larger number, and why?

3. Draw arrows on Figure 15.26 to indicate the path nutrients take in this environment.

4. Your friend has been planting the same variety of tomatoes in the same spot in his garden for the past eight years. In recent years, he has noticed that his plants are less healthy and that pests damage more of the tomato fruits. Apply what you have learned in this chapter to explain why his tomato crop is declining and suggest several ways he can improve his tomato yield.

5. Rock-and-roller and game hunter Ted Nugent once said, "If everyone became a vegetarian, we would run out of plants." Use your understanding of trophic pyramids to explain why "the Nuge" is incorrect in his reasoning.

Figure 15.26 Analyzing and Applying the Basics, Question 3.

Connecting the Science

1. Review your answer to question 1 in "Analyzing and Applying the Basics." How are the factors that limit fruit fly populations in a culture bottle similar to the factors that limit human populations on Earth? How are they different?

2. Africa is the only continent where increases in food production have not outpaced human population growth. Many of the most severe food crises are in African countries. Should those of us in the developed world assist African populations? How? What factors influence your thoughts on this question?

3. Have you ever considered the environmental impact of your diet? Are you willing to change your diet to reduce its environmental impact? Why or why not?

4. This chapter examined some of the environmental costs of modern agriculture. Are there other effects of modern agriculture, both positive and negative, that deserve discussion in our society? If so, what are they?

Media Activities

Media Activity 15.1 Human Population Growth and Regulation

Estimated Time: 10 minutes

This activity allows students to explore some of the many parameters influencing human population growth and the potential issues associated with unrestrained growth.

Media Activity 15.2 Human Growth Patterns

Estimated Time: 30 minutes

In this activity, you will use Census Bureau databases to analyze human population growth in recent decades.

Media Activity 15.3 Sustainable Agriculture in the 21st Century

Estimated Time: 10 minutes

Here you will explore the utilization of sustainable agricultural practices that are intended to reduce our increasing consumption of nonrenewable resources, encouraging farmers to turn instead to more environmentally friendly sources.

Appendix A
Metric System Conversions

To Convert Metric Units:	Multiply by:	To Get English Equivalent:
	Length	
Centimeters (cm)	0.3937	Inches (in.)
Meters (m)	3.2808	Feet (ft)
Meters (m)	1.0936	Yards (yd)
Kilometers (km)	0.6214	Miles (mi)
	Area	
Square centimeters (cm^2)	0.155	Square inches ($in.^2$)
Square meters (m^2)	10.7639	Square feet (ft^2)
Square meters (m^2)	1.1960	Square yards (yd^2)
Square kilometers (km^2)	0.3831	Square miles (mi^2)
Hectare (ha) (10,000 m^2)	2.4710	Acres (a)
	Volume	
Cubic centimeters (cm^3)	0.06	Cubic inches ($in.^3$)
Cubic meters (m^3)	35.30	Cubic feet (ft^3)
Cubic meters (m^3)	1.3079	Cubic yards (yd^3)
Cubic kilometers (km^3)	0.24	Cubic miles (mi^3)
Liters (L)	1.0567	Quarts (qt), U.S.
Liters (L)	0.26	Gallons (gal), U.S.
	Mass	
Grams (g)	0.03527	Ounces (oz)
Kilograms (kg)	2.2046	Pounds (lb)
Metric ton (tonne) (t)	1.10	Ton (tn), U.S.
	Speed	
Meters/second (mps)	2.24	Miles/hour (mph)
Kilometers/hour (kmph)	0.62	Miles/hour (mph)

To Convert English Units:	Multiply by:	To Get Metric Equivalent:
	Length	
Inches (in.)	2.54	Centimeters (cm)
Feet (ft)	0.3048	Meters (m)
Yards (yd)	0.9144	Meters (m)
Miles (mi)	1.6094	Kilometers (km)
	Area	
Square inches ($in.^2$)	6.45	Square centimeters (cm^2)
Square feet (ft^2)	0.0929	Square meters (m^2)
Square yards (yd^2)	0.8361	Square meters (m^2)
Square miles (mi^2)	2.5900	Square kilometers (km^2)
Acres (a)	0.4047	Hectare (ha) (10,000 m^2)
	Volume	
Cubic inches ($in.^3$)	16.39	Cubic centimeters (cm^3)
Cubic feet (ft^3)	0.028	Cubic meters (m^3)
Cubic yards (yd^3)	0.765	Cubic meters (m^3)
Cubic miles (mi^3)	4.17	Cubic kilometers (km^3)
Quarts (qt), U.S.	0.9463	Liters (L)
Gallons (gal), U.S.	3.8	Liters (L)
	Mass	
Ounces (oz)	28.3495	Grams (g)
Pounds (lb)	0.4536	Kilograms (kg)
Ton (tn), U.S.	0.91	Metric ton (tonne) (t)
	Speed	
Miles/hour (mph)	0.448	Meters/second (mps)
Miles/hour (mph)	1.6094	Kilometers/hour (kmph)

Metric Prefixes

Prefix		Meaning
giga-	G	$10^9 = 1{,}000{,}000{,}000$
mega-	M	$10^6 = 1{,}000{,}000$
kilo-	k	$10^3 = 1{,}000$
hecto-	h	$10^2 = 100$
deka-	da	$10^1 = 10$
		$10^0 = 1$
deci-	d	$10^{-1} = 0.1$
centi-	c	$10^{-2} = 0.01$
milli-	m	$10^{-3} = 0.001$
micro-	m	$10^{-6} = 0.000001$

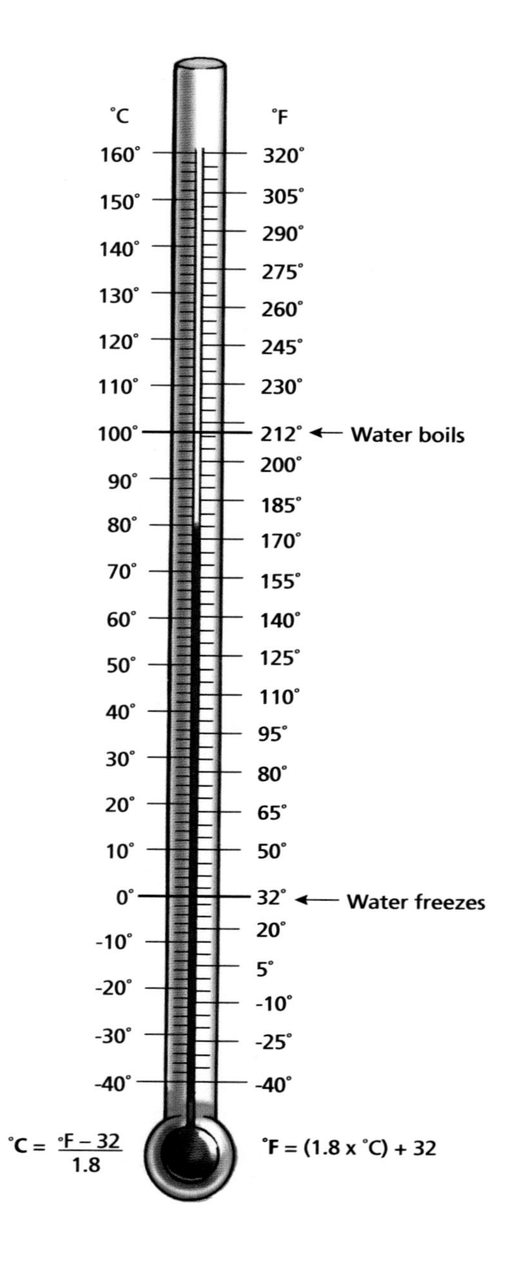

Appendix B

Basic Chemistry for the Biology Student

Elements

All things, living and nonliving, are composed of chemical substances that cannot be broken down, called **elements**. Each element has its own name, symbol, and properties. Of the approximately 100 known elements, 25 are essential for living organisms. Four of these—hydrogen (H), oxygen (O), nitrogen (N), and carbon (C)—are by far the most common in living organisms. The periodic table is a chart found in every chemistry text and classroom. It lists the elements, including their name, symbol, and mass number (discussed below). Figure B.1 shows the first three rows of the periodic table.

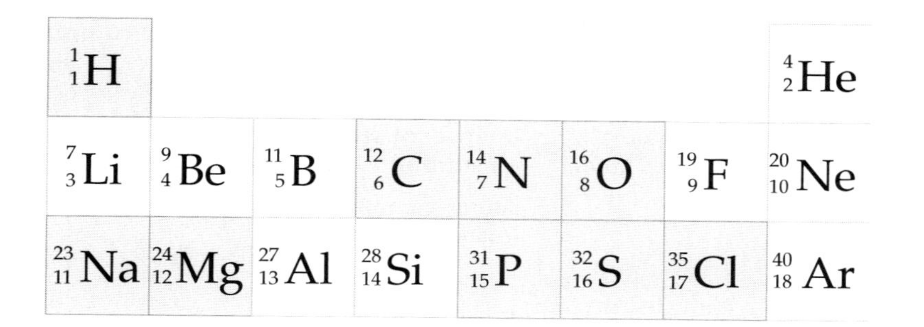

$^{1}_{1}H$							$^{4}_{2}He$
$^{7}_{3}Li$	$^{9}_{4}Be$	$^{11}_{5}B$	$^{12}_{6}C$	$^{14}_{7}N$	$^{16}_{8}O$	$^{19}_{9}F$	$^{20}_{10}Ne$
$^{23}_{11}Na$	$^{24}_{12}Mg$	$^{27}_{13}Al$	$^{28}_{14}Si$	$^{31}_{15}P$	$^{32}_{16}S$	$^{35}_{17}Cl$	$^{40}_{18}Ar$

Figure B.1
The first three rows of the periodic table. Highlighted elements are most commonly found in organisms.

The Structure of Atoms

Atoms are composed of **protons**, **neutrons**, and **electrons**. Together, protons and neutrons make up the nucleus. Protons are positively charged, and neutrons are electrically neutral and thus have no charge. Electrons are found outside the nucleus in the electron cloud. Electrons, symbolized e^-, have a negative charge. Electrons are attracted to the positively charged nucleus. Atoms are electrically neutral and therefore have the same number of electrons as protons (Figure B.2).

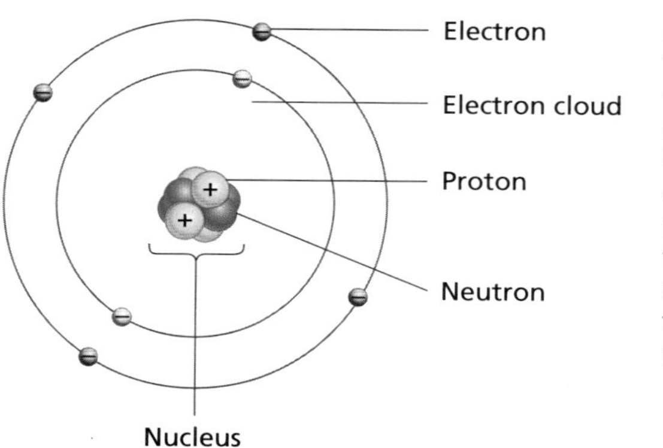

Figure B.2
The carbon atom, showing protons and neutrons in the nucleus and electrons orbiting the nucleus. Because the number of electrons equals the number of protons, there are six protons and six positive charges associated with the nucleus. In this drawing, some of the protons and neutrons are blocked from view.

The hydrogen atom (Figure B.3) has the least complex structure of any atom. Its nucleus has only one proton. Balancing the positive charge of the nucleus is one negatively charged electron.

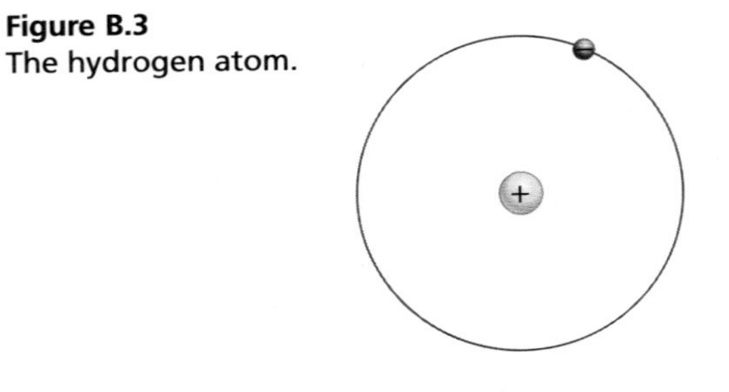

Figure B.3
The hydrogen atom.

The number of protons in the nucleus of an atom is called the **atomic number**. Look back at the abbreviated periodic table. The atomic number is written as a subscript to the left of the abbreviation for the element. Since the number of protons in the nucleus equals the number of electrons in the electron cloud, the atomic number also indicates the number of electrons.

It is possible to deduce the number of neutrons from the **mass number**, which is the sum of the protons and the neutrons in the nucleus of an atom. This number is written as a superscript to the left of the symbol for the element. Hydrogen, which has no neutrons, can be symbolized as in Figure B.4. Note that hydrogen has no neutrons, because its mass number is the same as its atomic number.

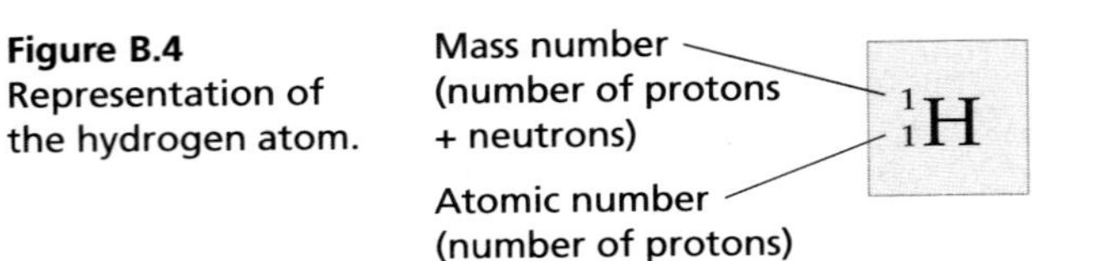

Figure B.4
Representation of the hydrogen atom.

Electrons have almost no mass, so the mass of the atom does not take into account the contributions of the electron cloud. Subtracting the atomic number (the number of protons) from the mass number (number of protons and neutrons) indicates the number of neutrons.

Sometimes two atoms have different mass numbers and the same atomic number. These are called **isotopes**. For example, carbon (C) has 6 protons. ^{12}C has an atomic number of 6 and a mass number of 12. Its nucleus is composed of 6 protons and 6 neutrons. It has 6 electrons in its electron cloud. An isotope of carbon, ^{13}C, has a nucleus composed of 6 protons and 7 neutrons. There are 6 electrons in its electron cloud. Both of these isotopes have 6 protons; otherwise they would not be carbon, because it is the number of protons that determines the identity of an atom. Both of these isotopes are stable, meaning that they do not have a tendency to lose particles. **Radioisotopes** are isotopes that are not stable; they give off particles and energy.

Electrons and Energy

The electrons in the electron cloud that surrounds the nucleus have different energy levels based on their distance from the nucleus. The first energy level, sometimes called a **shell**, is the closest to the nucleus and the electrons located there have the lowest energy. The second energy level is a little farther away and the electrons located in the second shell have a little more energy. The third energy level is even farther away and electrons located in the third shell have even more energy and so on. The periodic table gets its name because the elements are arranged in rows, or periods, that correspond to the number of electron shells each atom of an element contains.

Each energy level can hold a maximum number of electrons. The first shell holds 2 electrons, and the second shell holds a maximum of eight. This is called the *octet rule*. Electrons fill the lowest energy shell before advancing to fill a higher-energy-level shell. Atoms with the same number of electrons in their outermost, **valence shell** exhibit similar chemical behaviors. When the valence shell is full of electrons, the atom is unreactive. Figure B.5 shows some examples, the electron configurations of hydrogen, carbon, nitrogen, and oxygen.

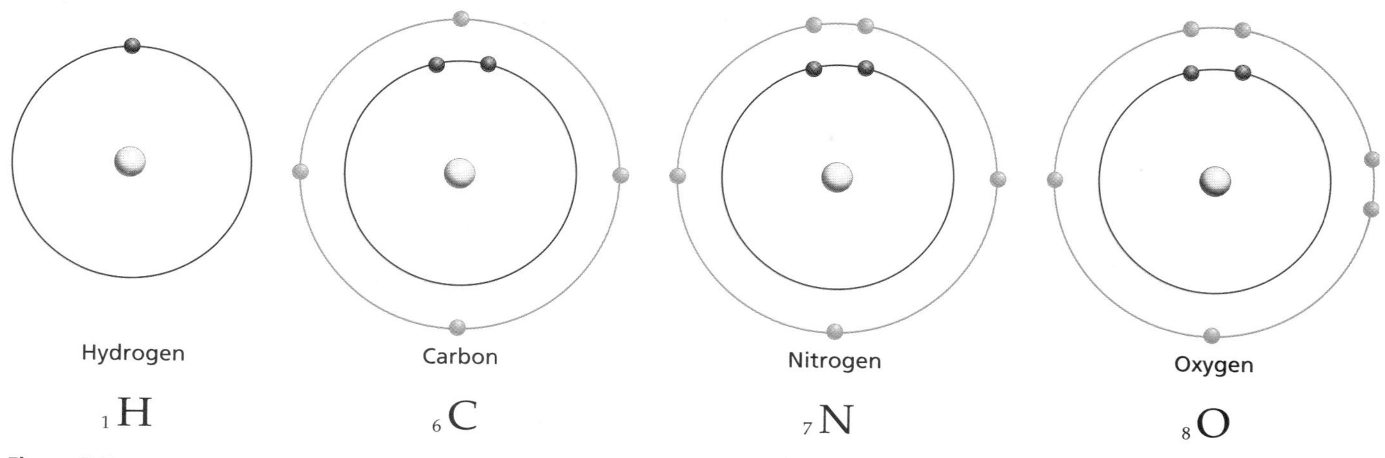

Figure B.5
Sample electron configurations.

Molecules

Atoms combine to form molecules. Molecules consist of 2 or more atoms of the same or different elements. For example, oxygen (O_2) consists of 2 atoms of oxygen while carbon dioxide (CO_2) consists on 1 atom of carbon and 2 of oxygen. One molecule of glucose ($C_6H_{12}O_6$) contains 6 atoms of carbon, 12 of hydrogen and 6 of oxygen. If you have more than 1 molecule, the number of molecules is written to the left of the molecule. For example, 6 water molecules is written 6 H_2O. This means that there are actually 12 hydrogens and 6 oxygens in these 6 molecules combined.

Ions

Ions are charged atoms. A positively charged atom arises when an electron is lost from a neutral atom. If an atom gains an electron, it becomes negatively charged. When a hydrogen atom loses an electron it becomes a hydrogen ion, H^+, due to the loss of its electron from the electron cloud. A hydrogen ion is also called a *proton*.

The sodium atom (Na) has 1 electron in its valence shell (Figure B.6).

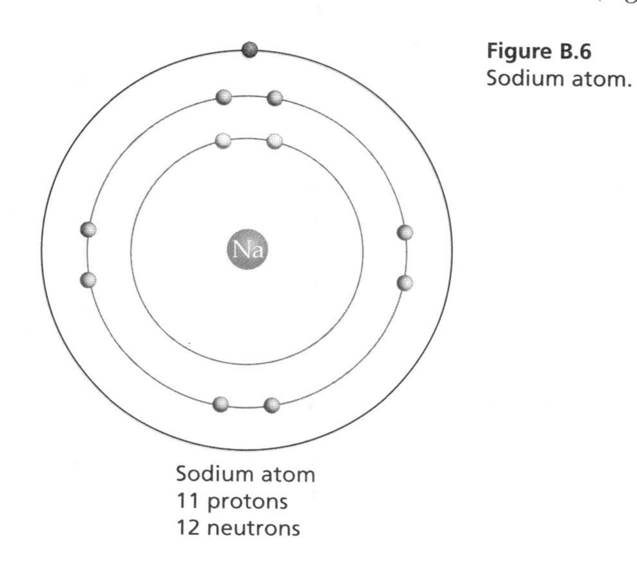

Figure B.6
Sodium atom.

When the sodium atom gives up the electron in the outer shell, it becomes stable, because its valence shell is filled with electrons. In doing so, it becomes a positively charged ion, Na^+. Likewise, a chlorine atom that gains an electron becomes a chloride ion, Cl^-. The chlorine atom (Figure B.7) can gain an electron because its valence shell needs 1 electron for the atom to become stable.

Figure B.7
Chlorine atom.

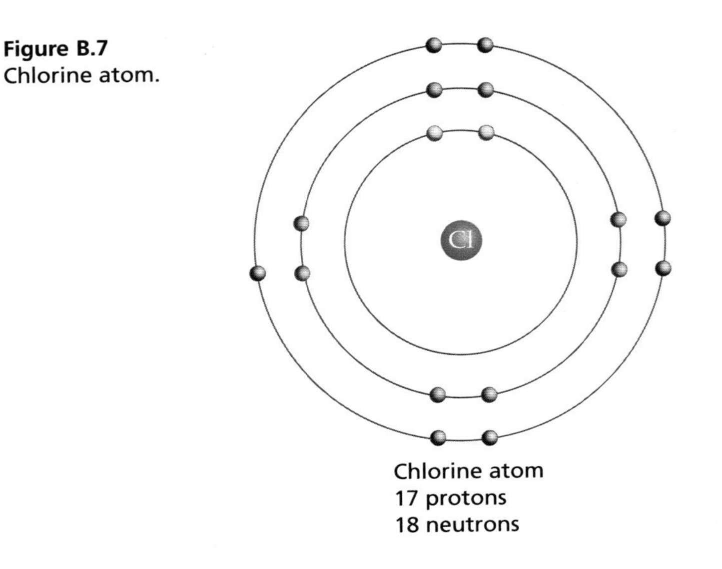

The chlorine atom gains an electron when some other atom loses an electron. Oppositely charged ions form bonds with each other called **ionic bonds**. When a sodium ion loses an electron and a chloride ion gains an electron, the two ions make an ionic bond to form NaCl (sodium chloride, or table salt) (Figure B.8).

Figure B.8
Sodium chloride (NaCl).

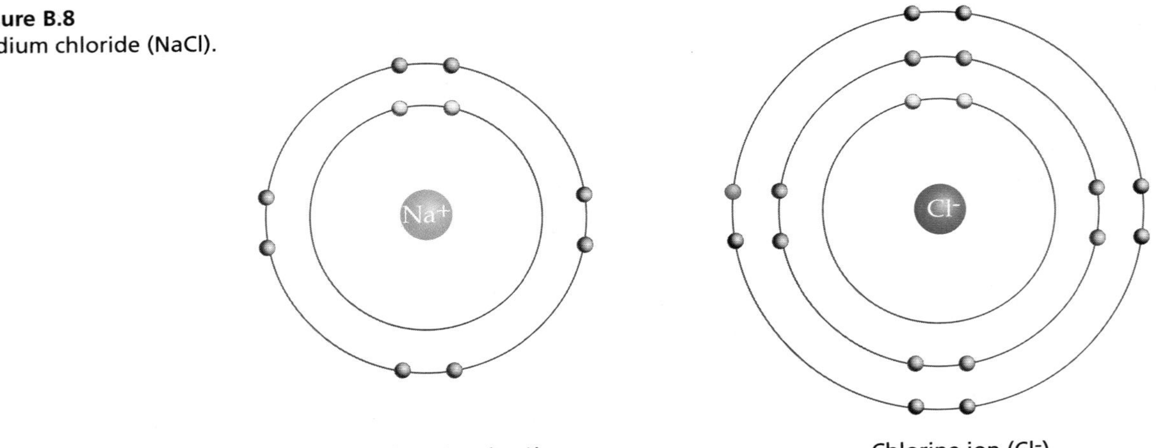

When sodium chloride is placed in water, the ionic bonds are weakened, and the two ions can *dissociate*, or separate, from each other.

Ions dissolve in water easily because water can dissociate also. Water dissociates to produce $H^+ + OH^-$. These ions are very reactive and form chemical bonds with many different ions.

Chemical Bonds

Atoms with one or two electrons in the valence shell tend to lose electrons while atoms with six or seven electrons in the valence shell tend to gain electrons. Atoms with four or five electrons in the valence shell tend to share electrons to complete their valence shells. When atoms share electrons, a type of bond called a **covalent bond** is formed. Carbon atoms are often involved in covalent bonding.

Sharing electrons is different from gaining and losing electrons. Gain or loss of electrons produces ions. Sharing electrons does not produce ions. The carbon atom, atomic number 6, has four electrons in its valence shell. According to the octet rule, it needs four more electrons to complete its valence shell and become stable. Most atoms that are involved in electron-sharing covalent bonds reach a stable configuration by having eight electrons in the valence shell. However, this is not the case for hydrogen, which is stable when it has two electrons in its valence shell; it only has electrons in the first shell and the first shell holds only two electrons. Carbon can form four bonds with hydrogen because carbon needs four electrons to complete its valence shell (producing CH_4, or methane, Figure B.9).

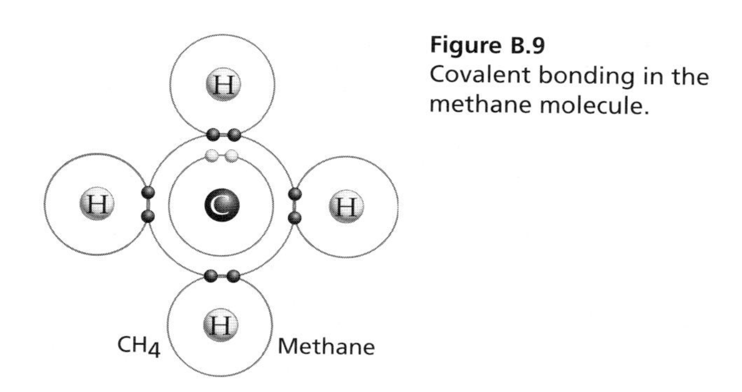

Figure B.9
Covalent bonding in the methane molecule.

Note that in methane there are a total of eight electrons around the valence shell of the carbon atom and two electrons around each hydrogen atom. Thus, both carbon and hydrogen have their valence shells filled by sharing electrons.

Covalent bonds are symbolized by a short line indicating a shared pair of electrons.

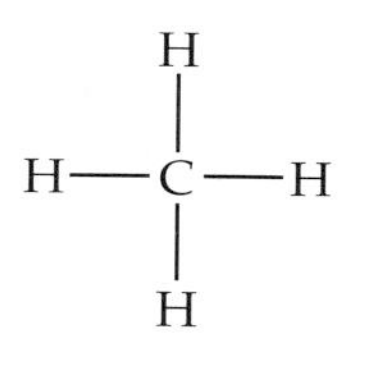

When carbon enters into bonds involving two *pairs* of shared electrons, this is called a *double bond*. A carbon-to-carbon double bond is symbolized as follows:

$$C{=}C$$

Carbon can make four bonds to fill its valence shell. Note that even though the molecule below has double bonds, each carbon atom is still involved in four chemical bonds total, since a double bond counts as two bonds.

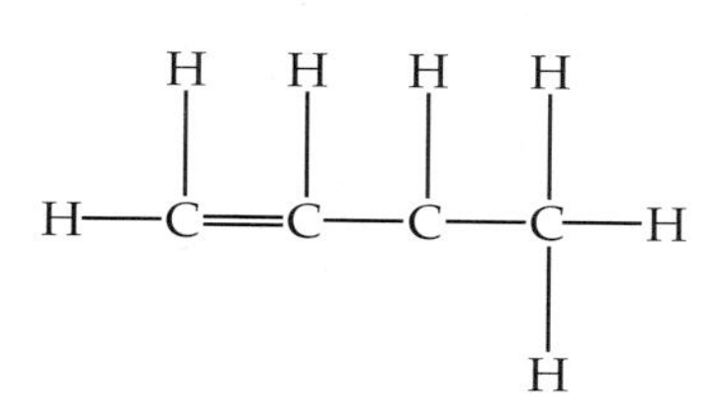

Atoms have characteristic levels of attraction to electrons. Atoms with high levels of attraction to electrons have a high **electronegativity**. When two atoms have the same electronegativity, electrons are shared equally between them, that is, the electrons are not pulled more toward one atom than another. When electrons involved in a covalent bond are shared equally, the bond is considered to be a **nonpolar covalent bond**. When electrons are not shared equally, that is, when they are pulled more toward the more electronegative atom, the bond is said to be a **polar covalent bond**. Water (H_2O) consists of hydrogen and oxygen atoms. The oxygen atom is more electronegative than most other atoms, including hydrogen. This means that the electrons shared in the covalent bonds of water are shifted toward the oxygen atoms. When shared electrons are closer to one atom than another, the atom to which they are the closest will have a partial negative charge, symbolized by the Greek delta, δ^-. The atom from which the electrons are pulled away will have a partial positive charge, symbolized δ^+ (Figure B.10).

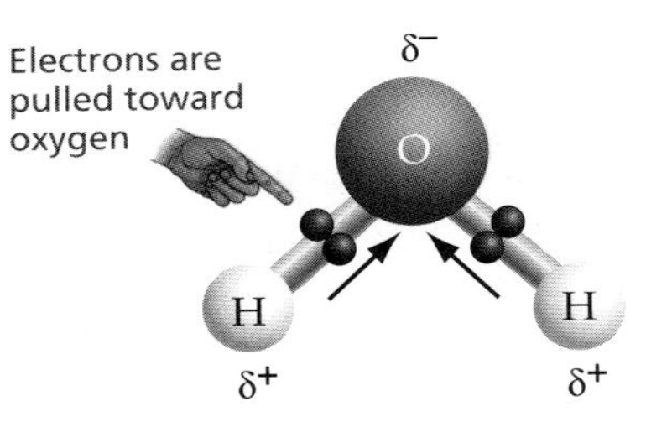

Figure B.10
Polarity in the water molecule.

One other type of chemical bond is the **hydrogen bond**. Hydrogen bonding is a type of weak chemical bond that forms between hydrogen and another atom. This bonding is based on the attraction of partial charges for each other. Figure B.11 shows hydrogen bonding that occurs in water molecules. Hydrogen bonds are represented by dashed lines.

Figure B.11
Hydrogen bonds in the water molecule.
(a) Bond between two water molecules.
(b) Bonds among many water molecules.

(a)

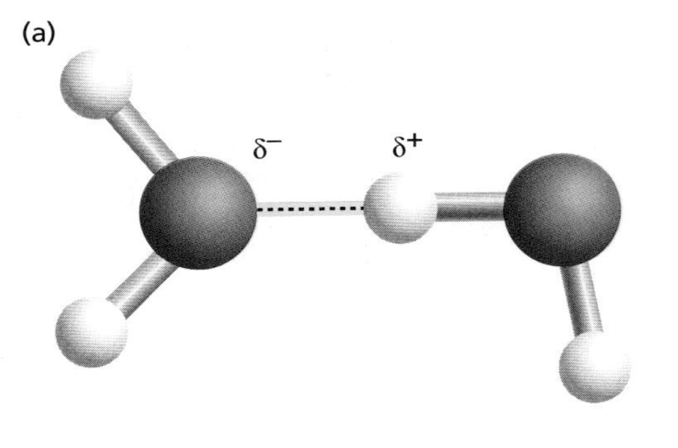

(b)

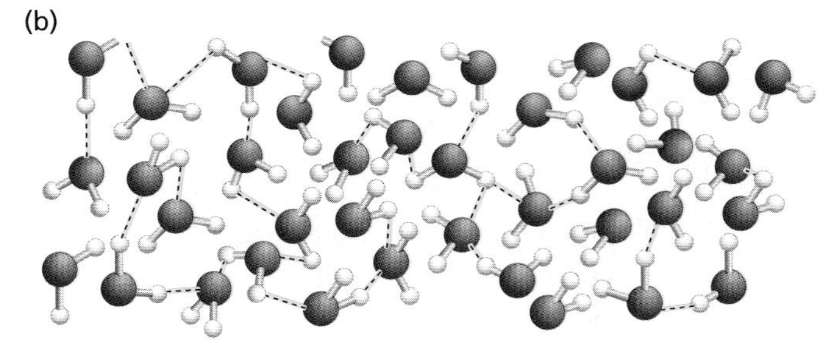

Answers to Learning the Basics

CHAPTER 1

1. A scientific hypothesis must be able to be evaluated via observations of the natural world.
2. In a controlled experiment a second group of subjects is treated exactly the same as experimental subjects except they are not given the experimental treatment.
3. Double-blinding ensures that participants do not influence the results by reporting what they think is expected based on the hypothesis, and it ensures that data collectors do not err in the measurement of results by emphasizing supportive data and minimizing nonsupportive data.
4. A statistically significant result is one that shows a difference between the experimental and control groups larger than the difference expected due to chance variations between the groups.
5. A correlation allows researchers to test hypotheses that are difficult to test via controlled experiments by looking for a relationship between two factors. They have the disadvantage of lacking controls, meaning that not all alternative hypotheses for the relationship can be excluded.
6. b
7. a
8. d
9. e
10. b
11. c
12. a
13. c
14. a

CHAPTER 2

1. Sugars; amino acids; glycerol and fatty acids.
2. An apple has vitamins, minerals, and fiber; not just Calories.
3. An enzyme binds to a substrate that can fit in its active site. Binding changes the shape of the enzyme and stresses the bonds holding the substrate together. This stress speeds the conversion of substrate to product. The enzyme then releases the product and assumes its original shape. Enzymes function in speeding up chemical reactions.
4. The nutrients in an avocado move from the mouth, to the esophagus, and then to the stomach. From the stomach, the digested nutrients move into the bloodstream and are transported to individual cells. From there they must traverse the plasma membrane and enter the mitochondria. Sugars are converted into pyruvate and then into ATP. The amino acids of proteins, and the glycerol and fatty acids of fats enter the mitochondria and are also converted into ATP.
5. See Table 2.5.
6. The cell membrane serves as a permeable barrier to the cell. It is composed of phospholipids arranged in a bilayer, proteins, and cholesterol. Proteins help hydrophilic substances gain access to the cell. Cholesterol helps maintain the fluidity of the membrane.
7. d
8. a
9. e
10. b
11. d
12. c
13. d
14. c
15. c
16. c

CHAPTER 3

1. The tree of life is an illustration of the relationships among living organisms. This is effective because all organisms share a common ancestor and similar organisms share more recent common ancestry, resulting in a branching pattern that very much looks like a tree.
2. Bacteria and archaea are small in size, easy to manipulate, can be grown in large numbers in a small space, and span a diversity of ecological roles, which make them sources for useful chemicals.
3. Double fertilization, which triggers increased food supply to the embryo; the development of fruit, aiding in dispersal; development of the flower, increasing the probability of fertilization; the development of defensive compounds, deterring herbivores.
4. Competition between organisms and the risk of predation favors the evolution of defensive chemicals, some of which may be valuable to humans. Organisms that live on or in other living organisms must have some way to evade their host's immune system, another potentially valuable characteristic.

5. By comparing the DNA sequences of organisms to see if the pattern of similarity and difference matches what is predicted; and by examination of the fossil record, looking for the record of evolutionary change in the group of organisms.
6. d
7. b
8. e
9. a
10. d
11. d
12. c
13. e
14. a
15. b

CHAPTER 4

Answers to Learning the Basics:

1. A gene is a segment of DNA containing information about making a protein. A protein is a chemical that either makes up part of the structure of a cell or helps perform an essential function of the cell. A trait is the physical outcome of the activity of proteins.
2. Each child is produced by the fusion of a single egg and a single sperm. Independent assortment and crossing over during meiosis ensures that virtually every sperm and egg produced by a single individual contains a unique set of alleles.
3. Quantitative variation can occur when many genes, each with more than one allele, influence a single trait and/or when the environment influences how a trait develops.
4. A value for heritability tells us what amount of variation present for a trait in a population is due to genetic differences among individuals in that population.
5. We cannot exactly predict the phenotypes of offspring from looking at the phenotypes of their parents, but in many cases we can determine the probability that a particular offspring will have a particular phenotype.
6. d
7. a
8. b
9. d
10. b
11. b
12. c
13. e
14. a
15. d

Answers to Genetics Problems

1. All have genotype *Tt* and are tall.
2. $\frac{1}{4}$ have genotype *TT* and are tall,

   $\frac{1}{2}$ have genotype *Tt* and are tall,

   $\frac{1}{4}$ have genotype *tt* and are short.
3. Both must be heterozygote (*Aa*).
4. 50% of their offspring are expected to develop Huntington's disease.
5. (a) 0%; (b) 25%.
6. (a) yellow is dominant; (b) *YY* (yellow) and *yy* (green).
7. *Yy* (yellow) and *yy* (green).
8. (a) 50%; (b) 50%; (c) 25%; (d) yes.
9. i is recessive, I^A and I^B are codominant.
10. (a) *BRCA1* is dominant; (b) It is unlike the typical pattern because not all individuals with *BRCA1* have the trait of early breast or ovarian cancer.

CHAPTER 5

1. Eight.
2. Proto-oncogenes and tumor suppressor genes.
3. Cancer cells that have lost their contact inhibition and anchorage dependence making them more likely to metastasize.
4. Chemotherapeutic agents selectively kill dividing cells.
5. Microtubules.
6. d
7. c
8. a
9. a
10. e
11. d
12. b

CHAPTER 6

1. DNA polymerase.
2. DNA is isolated and cut with restriction enzymes. The fragments are loaded on a gel and an electrical current is applied to separate them; they are transferred to a filter, and then probed. The probe binds in complementary regions. X-ray film is placed over filter to generate a fingerprint.
3. A probe is a single-stranded DNA molecule that is complementary to some sequence present in humans. It has radioactive phosphates.
4. (a) diploid; (b) diploid; (c) haploid; (d) haploid.
5. Sugar, phosphate, nitrogen-containing base.
6. c
7. b
8. c
9. b
10. d
11. d
12. e

CHAPTER 7

1. GCUAAUGAAU.
2. gln, arg, ile, leu.
3. Because it is easy to insert foreign genes into plasmids, then allow the plasmids to replicate themselves and produce the desired protein.

4. mRNA, ribosome, amino acids, tRNAs.
5. This gene would not be expressed because the RNA polymerase would not be able to bind the promoter to initiate transcription.
6. b
7. c
8. a
9. b
10. d
11. c
12. b
13. c
14. d
15. d
16. d

CHAPTER 8

1. The theory of common descent describes that the similarities among living species can be explained as a result of their descent from a common ancestor.
2. Darwin observed that organisms on different islands in the archipelago were similar, but not identical to each other, and were clearly related to each other and to similar organisms on the South American mainland.
3. A vestigial structure is one that has little or no function in an organism but appears to be similar to a more useful structure in other organisms. These structures provide support for the theory of common descent because they are best explained as having evolved from a functional structure in a common ancestor.
4. Fossils provide insight into change in a group of species over time. The fossil record supports the theory of common descent because fossil ancestors of modern species have characteristics that are similar to those found in modern relatives of the species.
5. A "missing link" is a common ancestor of two related species. They are difficult to locate and identify because they will not necessarily look like a cross between both modern species—the evolutionary history of both species must be known in order to predict what the common ancestor looked like.
6. d
7. d
8. b
9. e
10. b
11. d
12. a
13. e
14. e
15. d

CHAPTER 9

1. AIDS is caused by infection with HIV. However, individuals can be infected with HIV and not have AIDS, which is the end point on a continuum of HIV infection. Without drug treatment, the vast majority of HIV-infected people will progress to AIDS.
2. Fitness is a term that describes the survival and reproductive output of an individual in a population relative to other members of the population.
3. Individuals in a population vary and some of that variation is passed from parent to offspring. There are more individuals born than survive, and survival and reproduction of individuals is not random—those with particular traits (called adaptations) are more likely to survive and reproduce than offspring with other traits.
4. Artificial selection is a process of selection of plants and animals by humans who control the survival and reproduction of members of a population to increase the frequency of human-preferred traits. Artificial selection is like natural selection in that it causes evolution; however, it differs because humans are directly choosing which organisms reproduce, whereas in natural selection, environmental conditions cause one variant to have higher fitness than other variants.
5. Within a single patient, an HIV population can consist of up to 1 billion different virus variants. These variants differ in a number of traits, including their susceptibility to a particular anti-HIV drug. When the patient takes an anti-HIV drug, the individual viruses that are more resistant to the drug have higher fitness, and natural selection leads to the evolution of a virus population that is resistant to the drug.
6. b
7. c
8. d
9. e
10. a
11. e
12. c
13. d
14. d
15. a

CHAPTER 10

1. The gene pools of populations must become isolated from each other, they must diverge (as a result of natural selection or genetic drift) while they are isolated, and reproductive incompatibility must evolve.
2. Yes, as long as the gene pools of the populations are isolated. An example is when the timing of reproduction in two populations of a species is different.
3. Biological races form when the gene pools of populations of a species become mostly isolated from each other. As long as reproductive isolation has not occurred, these two populations remain the same species. We can find evidence of gene-pool isolation by looking for unique alleles that appear in some races but not others, and by tracing patterns of allele frequency that are common among populations classified in the same race but different from populations classified as a different race.
4. Allele frequency is calculated by dividing the total number of one type of allele found in the gene pool of a population by the total number of copies of the gene in the same population.

5. Genetic drift can occur as a result of founder's effect, the bottleneck effect, or by the chance loss of alleles from small, isolated gene pools.
6. d
7. b
8. a
9. e
10. b
11. e
12. d
13. c
14. d
15. b

CHAPTER 11

1. See Table 11.1.
2. See Table 11.1.
3. See Figures 11.14 and 11.15.
4. Rearrangements of the genes that encode antigen receptors.
5. Refolding of normal prion proteins.
6. c
7. d
8. b
9. c
10. b
11. c
12. b

CHAPTER 12

1. The hypothalamus produces GnRH. GnRH acts on the pituitary gland, which secretes FSH and LH. FSH and LH act on the testes and ovaries. The testes produce testosterone; the ovaries produce estrogen and progesterone. The adrenal glands produce sex hormones in small amounts.
2. Undifferentiated gonads become testes when the *SRY* gene is present, or ovaries when the *SRY* gene is absent and *Gpbox* genes are present.
3. Ducts become able to carry sperm when the Wolffian structure persists, or eggs when the Müllerian structure persists.
4. Differences in muscle mass, fat deposition, and skeletal structure.
5. When estrogen is high, FSH and LH are stimulated. When progesterone is high, FSH and LH are inhibited. LH stimulates ovulation.
6. d
7. b
8. b
9. e
10. a

CHAPTER 13

1. The CNS is composed of the brain and spinal cord and is responsible for integrating, processing, and coordinating information taken in by the senses. It is also the seat of functions such as intelligence, learning, memory, and emotion. The PNS includes the network of nerves outside the brain and spinal cord and functions to link the CNS with senses.
2. Cerebrum (language, memory, sensation, and decision making); cerebellum (balance, movement, coordination); brain stem (reflexes, spontaneous functions).
3. A neurotransmitter is a chemical (stored in vesicles and released from a presynaptic neuron) that diffuses across a synapse to the postsynaptic neuron, and allows the propagation of a nerve impulse to the postsynaptic cell or effector organ or gland.
4. Receptors for neurotransmitters are located on the pre- and postsynaptic neuron of many brain structures.
5. The binding of neurotransmitter to the postsynaptic membrane stimulates a rapid change in the uptake of sodium ions by the postsynaptic cell. Sodium channels open and sodium ions flow inward, causing another depolarization and generating another action potential.
6. d
7. c
8. e
9. a
10. e

CHAPTER 14

1. By examining the fossil record and determining the average life span of various species. The most serious problem with these estimates is that the fossil record over-represents long-lived species and thus overestimates the life span of these species.
2. Habitat fragmentation endangers species because it interferes with their ability to disperse from unsuitable to suitable habitat and because it increases their exposure to humans and human-modified environments. Species that have very specialized habitat requirements, those that cannot move rapidly, and those that are very susceptible to human disturbance are most negatively affected by fragmentation.
3. Mutualism is a relationship among species where all partners benefit. Predation is a relationship among species where one benefits and others are consumed. Competition is a relationship among species where all partners are harmed by the presence of the others.
4. The loss of alleles via genetic drift occurs more rapidly in small populations than in large populations.
5. Inbreeding depression is a reduction in fitness that occurs as a result of increased homozygosity in a population. Increased homozygosity is a result of mating among close relatives, which is more common in small, isolated populations.

6. d
7. c
8. d
9. e
10. d
11. b
12. a
13. d
14. a
15. c

CHAPTER 15

1. A decrease in death rate, especially a decrease in infant mortality, has led to increasing growth rates. In addition, the process of exponential growth lends itself to these population explosions, as the number of people added in any year is a function of the population of the previous year.
2. In most populations, growth rate declines because death rate increases (or birth rate decreases) as resources are "used up" by the population.
3. Demographic momentum occurs because a population continues to grow in number even if birth rates drop, as large numbers of individuals who have not yet reached reproductive maturity add offspring to the population in subsequent years.
4. Tilling, mulching, and herbicides physically block the growth of weeds that would compete with crops for light and water. Irrigation adds additional water to the soil, which also helps the crop plant keep its shape and maximize light interception.
5. Because each trophic level must consume large amounts of its supporting trophic levels to survive. Thus, a pesticide can be in low concentrations in primary consumers, but because secondary consumers must eat large numbers of primary consumers simply to survive, they accumulate all of the toxin contained in each primary consumer within their own bodies.
6. c
7. b
8. b
9. e
10. d
11. e
12. a
13. e
14. e
15. d

Glossary

acetylcholine Impaired transmission of this neurotransmitter is thought to contribute to Alzheimer's disease. (Chapter 13)

Acquired Immune Deficiency Syndrome (AIDS) Syndrome characterized by severely reduced immune system function and numerous opportunistic infections. Results from infection with HIV. (Chapter 9)

action potential Wave of depolarization in a neuron propagated to the end of the axon; also called a *nerve impulse*. (Chapter 13)

active site Substrate-binding region of an enzyme. (Chapter 2)

adaptation Trait that is favored by natural selection and increases an individual's fitness in a particular environment. (Chapters 9, 10)

adaptive radiation Diversification of one or a few species into large and very diverse groups of descendant species. (Chapter 3)

adenine Nitrogenous base in DNA, a purine. (Chapter 6)

adenosine triphosphate (ATP) Form of energy that cells can use. (Chapter 2)

aerobic Requiring oxygen. (Chapter 2)

algae Photosynthetic protists. (Chapter 3)

alleles Alternate versions of the same gene; produced by mutations. (Chapters 4, 6, 9)

alternative hypotheses Factors other than the tested hypothesis that may explain observations. (Chapter 1)

Alzheimer's disease Progressive mental deterioration in which there is memory loss along with the loss of control of bodily functions, ultimately resulting in death. (Chapter 13)

amenorrhea Cessation of menstruation. (Chapter 2)

amino acid Monomer subunit of a protein. (Chapter 2)

anabolic steroids Tissue-building synthetic versions of testosterone. (Chapter 12)

anaerobic Not requiring oxygen. (Chapter 2)

analogy Anatomical similarity due to convergent evolution.

anaphase Stage of mitosis during which microtubules contract and separate sister chromatids. (Chapter 5)

anchorage dependence Phenomenon that holds normal cells in place. Cancer cells can lose anchorage dependence and migrate into other tissues or metastasize. (Chapter 5)

androgens Masculinizing hormones, such as testosterone, secreted by the adrenal glands. (Chapter 12)

anecdote Information or advice based on one person's personal experience. (Chapter 1)

anemia Illness resulting from the loss of red blood cells. (Chapter 10)

angiogenesis Formation of new blood vessels. (Chapter 5)

Animalia Kingdom of Eukarya containing organisms that ingest others and are typically motile for at least part of their life cycle. (Chapter 3)

anorexia Self-starvation. (Chapter 2)

annual growth rate Proportional change in population size over a single year. Growth rate is a function of the birth rate minus the death rate of the population. (Chapter 15)

antibiotic Chemicals that kill or disable bacteria. (Chapter 3)

antibody Protein made by the immune system in response to the presence of foreign substances or antigens. Can serve as a receptor on a B cell or be secreted by plasma cells. (Chapters 5, 9, 11)

anticodon Region of tRNA that binds to an mRNA codon. (Chapter 7)

antigen Short for antibody-generating substances, an antigen is a molecule that is foreign to the host and stimulates the immune system to react. (Chapters 9, 11)

antigen receptors Proteins in B- and T-cell membranes that bind to specific antigens. Called antibodies when found on B cells, and T-cell receptors when found on T cells. (Chapter 11)

anti-Müllerian hormone Hormone produced by the testicles in an embryo that causes regression of the Müllerian ducts and allows development of male ductal structures. (Chapter 11)

antiparallel Feature of DNA double helix in which nucleotides face "up" on one side of the helix and "down" on the other. (Chapter 6)

antisense nucleotide sequence Nucleotide sequence that, when transcribed, produces an mRNA complementary to the normally transcribed mRNA of another gene. Binding of antisense and sense mRNA decreases gene expression. (Chapter 7)

applied research Research that has an immediate and profitable application. (Chapter 7)

Archaea Domain of prokaryotic organisms made up of species known from extreme environments. (Chapter 3)

arrector pili Muscles at the base of hairs that raise them above the level of the skin. In humans, stimulation of these muscles results in "goosebumps." (Chapter 8)

artificial insemination The practice of collecting semen from a male and manually injecting it into a female's reproductive tract. (Chapter 4)

artificial selection Selective breeding of domesticated animals and plants to increase the frequency of desirable traits. (Chapters 4, 9)

assortative mating Tendency for individuals to mate with someone who is like themselves. (Chapter 10)

asymptomatic Stage in an infection that is characterized by relatively unnoticeable, or no symptoms of illness. (Chapter 9)

atherosclerotic Cholesterol-lined arterial walls. (Chapter 2)

attention deficit disorder (ADD) Syndrome characterized by forgetfulness, distractibility, fidgeting, restlessness, impatience, difficulty sustaining attention in work, play, or conversation, or difficulty following instructions and completing tasks in more than one setting. (Chapter 13)

autoimmune Referring to diseases resulting from an attack by the immune system on normal body cells. (Chapter 11)

autosomes Nonsex chromosomes, of which there are 22 pairs in humans. (Chapter 6)

axon Long, wirelike portion of the neuron that ends in a terminal bouton. (Chapter 13)

AZT Drug that inhibits replication of HIV's genetic material while having relatively little effect on the normal replication and function of a patient's cells. (Chapter 9)

B lymphocytes (B cells) White blood cells that develop in bone marrow and recognize and react to small, free-living microorganisms such as bacteria and the toxins they produce. B lymphocytes provide an immune response called *humoral immunity*. (Chapter 11)

Bacteria Domain of prokaryotic organisms. (Chapter 3)

basal metabolic rate Resting energy use of an awake, alert person. (Chapter 2)

base pairing rules In DNA, A pairs with T, C pairs with G. (Chapter 6)

basic research Research for which there is not necessarily a commercial application. (Chapter 7)

benign Tumor that stays in one place and does not affect surrounding tissues. (Chapter 5)

bias Influence of research participants' opinions on experimental results. (Chapter 1)

bilayer Membrane that surrounds cells and organelles, composed of lipids and proteins. (Chapter 2)

bile Mixture of substances produced in the liver that aids digestion. (Chapter 2)

binary fission Asexual form of bacterial reproduction. (Chapter 11)

binomial Name composed of two parts. (Chapter 8)

bioaccumulation Process occurring when a pesticide or other chemical does not break down quickly and accumulates in photosynthetic organisms as they absorb it from the soil and water. (Chapter 15)

biodiversity Variety within and among living organisms. (Chapter 3)

bioelectrical impedance A method for measuring the resistance of body tissues to the flow of an electrical signal to determine body-fat percentage. (Chapter 2)

biological classification Field of science attempting to organize biodiversity into discrete, logical categories. (Chapter 3)

biological diversity Entire variety of living organisms. (Chapter 14)

biological evolution See evolution.

biological race Populations of a single species that have diverged from each other. Biologists do not agree on a definition of "race." See also subspecies, variety. (Chapter 10)

biological species concept Definition of a species as a group of individuals that can interbreed and produce fertile offspring, but typically cannot breed with members of another species. (Chapter 10)

biomagnification Concentration of toxic chemicals in the higher levels of a food web. (Chapter 15)

biomass The mass of all individuals of a species, or of all individuals on a level of a food web, within an ecosystem. (Chapter 15)

biome Major habitat type. (Chapter 14)

biophilia Humans' innate desire to be surrounded by natural landscapes and objects. (Chapter 14)

biopiracy Using the knowledge of the native people in developing countries to discover compounds for use in developed countries. (Chapter 3)

bioprospecting Hunting for new organisms and new uses of old organisms. (Chapter 3)

biopsy Surgical removal of some cells, tissue, or fluid to determine if cells are cancerous. (Chapter 5)

bipedal Walking upright on two limbs. (Chapter 8)

birth rate Number of births averaged over the population as a whole. (Chapter 15)

blind experiments Tests in which subjects are not aware of exactly what they are predicted to experience. (Chapter 1)

body mass index (BMI) Calculation using height and weight to determine a number that correlates an estimate of a person's amount of body fat with health risks. (Chapter 2)

botanist Plant biologist. (Chapter 3)

brain Region of the body where most neural activity takes place, where decisions are reached, and where bodily activities are directed and coordinated. (Chapter 13)

brain stem Region of the brain that lies below the thalamus and hypothalamus that governs reflexes and some involuntary functions such as breathing and swallowing. (Chapter 13)

bulimia Binge eating followed by purging. (Chapter 2)

calcium Nutrient required in plant cells for the production of cell walls, and for bone strength and blood clotting in humans. (Chapters 2, 15)

calorie Amount of energy required to raise the temperature of one gram of water by 1 °C. (Chapter 2)

Calorie A kilocalorie, or 1,000 calories. (Chapter 2)

Cambrian explosion Relatively rapid evolution of the modern forms of multicellular life that occurred approximately 550 million years ago. (Chapter 3)

cancer vaccine Experimental cancer treatment that involves injecting a cancer patient with proteins specific to their tumor to increase immune response. (Chapter 5)

capillary Microscopic blood vessel. (Chapter 2)

capsid Protein coat that surrounds a virus. (Chapter 11)

capsule Gelatinous outer covering of bacterial cells that aids in attachment to host cells during an infection. (Chapter 11)

carbohydrate Energy-rich molecule that is the major source of energy for the cell. (Chapter 2)

carbon dioxide Abundant molecule in the atmosphere (CO_2). (Chapter 2)

carcinogens Substances that damage DNA or chromosomes. (Chapter 5)

carrier Individual who is heterozygous for a recessive disease allele. (Chapter 4)

carrying capacity Maximum population that the environment can support. (Chapter 15)

catalyze Speed up a chemical reaction. (Chapter 2)

caudate nucleus Structure within the cerebral hemispheres that functions as part of the pathway that coordinates movement patterns. (Chapter 13)

CD4+ cell Class of immune-system cells that are susceptible to HIV infection. Most are T4 lymphocytes; they are named for the CD4 receptor on the cell surface. (Chapter 9)

cell Basic unit of life, an organism's fundamental building-block units. (Chapters 2, 9)

cell body Portion of the neuron that houses the nucleus and organelles. (Chapter 13)

cell cycle All events that occur when two cells are created from one cell. (Chapter 5)

cell division Process a cell undergoes when it makes copies of itself. Production of daughter cells from an original parent cell. (Chapter 5)

cell-mediated immunity Immunity conferred by T-cell response. (Chapter 11)

cell wall Tough but elastic structure surrounding plant and bacterial cell membranes. (Chapter 11)

cellular respiration Process requiring oxygen by which cells use food to make ATP. (Chapter 2)

central nervous system (CNS) Includes brain and spinal cord and is responsible for integrating, processing, and coordinating information taken in by the senses. It is the seat of functions such as intelligence, learning, memory, and emotion. (Chapter 13)

centriole Anchor for microtubules during mitosis. (Chapter 5)

centromere Region of a chromosome where sister chromatids are attached and to which microtubules bind. (Chapter 5)

cerebellum Region of the brain that controls balance, muscle movement, and coordination. (Chapter 13)

cerebral cortex Deeply wrinkled outer surface of the cerebrum where conscious activity and higher thought originate. (Chapter 13)

cerebrospinal fluid Protective liquid bath that surrounds the brain within the skull. (Chapter 13)

cerebrum Portion of the brain in which language, memory, sensations, and decision making are controlled. The cerebrum has two hemispheres, each of which has four lobes. (Chapter 13)

cervix Located at the bottom of the uterus and top of the vagina, enlarges during childbirth. (Chapter 12)

checkpoint Stoppage during cell division that occurs to verify that division is proceding correctly. (Chapter 5)

chemical element Basic building block of matter. (Chapter 8)

chemotherapy Using chemicals to try to kill rapidly dividing (cancerous) cells. (Chapter 5)

chlorophyll Green pigment found in plant cells. (Chapter 2)

chloroplast Plant cell structure that gives plants a green color. (Chapter 2)

chromosome Housed inside the nucleus, subcellular structures composed of DNA and proteins. (Chapters 4, 5, 6)

classification systems Methods for organizing biological diversity. (Chapter 8)

clinical trial Controlled scientific experiment to determine the effectiveness of novel treatments. (Chapter 5)

clonal population Population of identical cells copied from the immune cell that first encounters an antigen. The entire clonal population has the same DNA arrangement and all cells in a clonal population carry the same receptor on their membrane. (Chapter 11)

cloning Making copies of a gene or an organism that are genetically identical. (Chapter 7)

codominant Alleles that result in a new protein with a different, but not dominant, activity compared to the normal protein. (Chapter 4)

codons Three-nucleotide sequences of mRNA that tRNA binds with to add an amino acid to the growing protein. The genetic code is read from the mRNA codon. (Chapter 7)

coenzyme Molecule that helps enzymes catalyze chemical reactions. (Chapter 2)

combination drug therapy Treatment with at least three different anti-HIV drugs, from two different classes of drugs. The therapy of choice for HIV patients. (Chapter 9)

competition Interaction that occurs when two species of organisms both require the same resources within a habitat; competition tends to limit the size of populations. (Chapter 14)

competitive exclusion Process of establishing harmless organisms in an ecosystem that serves to reduce levels or harmful organisms. (Chapter 14)

competitors Species that survive on the same food source or otherwise compete for the same resources. (Chapter 3)

complement proteins Proteins in the blood with which an antibody–antigen complex can combine in order to kill bacterial cells. (Chapter 11)

complex carbohydrate Highly branched polysaccharide. (Chapter 2)

contact inhibition Property of cells that prevents them from invading surrounding tissues. Cancer cells may lose this property. (Chapter 5)

contagious Spreading from one organism to another. (Chapter 11)

continuous variation A range of slightly different values for a trait in a population. (Chapter 4)

control Subject for an experiment who is similar to experimental subject, except is not exposed to the experimental treatment, used as baseline values for comparison. (Chapter 1)

convergent evolution Evolution of same trait or set of traits in different populations as a result of shared environmental conditions rather than shared ancestry. (Chapter 10)

corpus callosum Bundle of nerve fibers at the base of the cerebral fissure that provides a communication link between the cerebral hemispheres. (Chapter 11)

corpus luteum Hormone-producing tissue (the ovarian follicle after ovulation) that makes progesterone and estrogen, and degenerates about 12 days after ovulation if fertilization does not occur. (Chapter 12)

correlation Describes a relationship between two factors. (Chapters 1, 4)

corticosteroids Hormones secreted by the adrenal glands that help the body deal with stress and promote synthesis of glucose from noncarbohydrate sources such as proteins. (Chapter 12)

critical habitat Defined by the Endangered Species Act as habitat designated as crucial to the survival of an endangered species. (Chapter 14)

crop rotation Practice of not putting the same crop in the same field in consecutive years. (Chapter 15)

cross Mating of two organisms. (Chapter 4)

crossing over Exchange of genetic information between members of a homologous pair of chromosomes. (Chapter 6)

cultural carrying capacity Maximum human population of Earth that provides not only adequate food for all but an adequate quality of life as well. (Chapter 15)

cultural control Agricultural practices that do not require chemical inputs designed to minimize pest populations such as crop rotation and polyculture. (Chapter 15)

cyst Noncancerous fluid-filled growth. (Chapter 5)

cytokinesis Part of the cell cycle during which two daughter cells are formed by the cytoplasm splitting. (Chapter 5)

cytoplasm Gel-like medium enclosed by the plasma membrane. (Chapter 2)

cytosine Nitrogenous base in DNA, a pyrimidine. (Chapter 6)

cytotoxic T cells Immune-system cells that attack and kill body cells that have become infected with a virus before the virus has had time to replicate. Cytotoxic T cells release a chemical that causes the plasma membranes of the target cells to leak. (Chapter 11)

data Information collected by scientists during hypothesis testing. (Chapter 1)

daughter cells Cells that result from the process of cell division. (Chapter 4)

dead zone Area in the Gulf of Mexico at the mouth of the Mississippi River that experiences very low oxygen levels in the summer, caused by eutrophication due to high levels of fertilizer runoff from farms in the midwestern United States. (Chapter 15)

death rate Number of deaths averaged over the population as a whole. (Chapter 15)

decomposers Organisms, typically bacteria and fungi in the soil, whose action breaks down complex molecules into simpler ones. (Chapter 15)

deductive reasoning Making a prediction about the outcome of a test; "if . . . then" statements. (Chapter 1)

demographic momentum Lag between the time humans reduce birth rates and the time population numbers respond. (Chapter 15)

dendrite Portion of the neuron radiating from the cell body. (Chapter 13)

deoxyribonucleic acid (DNA) Molecule of heredity that stores the information required for making all of the proteins required by the cell. (Chapters 2, 6)

deoxyribose The five-carbon sugar in DNA. (Chapter 6)

depolarization Reduction in the charge difference across the neuronal membrane. (Chapter 13)

depression Disease that involves feelings of helplessness and despair, and thoughts of suicide. (Chapter 13)

descriptive statistics Helps scientists summarize their data. (Chapter 1)

desertification Process by which soils become too degraded to support plant growth, often through salinization and erosion. (Chapter 15)

developed countries Countries that have completed the process of industrial development and have a high per capita income level. (Chapter 15)

developing world Countries beginning the process of industrial development. (Chapter 15)

diabetes Disorder of carbohydrate metabolism characterized by impaired ability to produce or respond to the hormone insulin. (Chapter 2)

dicot One of the two classes of flowering plants, often called *broad-leaved plants*. Seeds of dicots contain two leaves (cotyledons). (Chapter 15)

differentiation Structural and functional divergence of cells as they become specialized. (Chapter 12)

diploid Cell containing homologous pairs of chromosomes (2n). (Chapter 6)

disaccharide Two sugars joined together. (Chapter 2)

diverge In evolution, divergence occurs when gene flow is eliminated between two populations. Over time, traits found in one population begin to differ from traits found in the other population. (Chapter 10)

dizygotic twins Fraternal twins (non-identical) that develop from separate zygotes. (Chapter 4)

DNA See deoxyribonucleic acid.

DNA bases The four building-block chemicals of DNA molecules: A, C, G, and T. (Chapter 9)

DNA fingerprinting Powerful identification technique that takes advantage of differences in DNA sequence. (Chapter 6)

DNA polymerase Enzyme that facilitates base pairing during DNA synthesis. (Chapter 6)

DNA sequence The linear order of nucleotides in a DNA molecule. (Chapter 3)

domain Most inclusive biological category. Life is grouped by many biologists into three major domains. (Chapter 3)

dominant Applies to an allele with an effect that is visible in a heterozygote. (Chapter 4)

dopamine Neurotransmitter that controls emotions and complex movements. (Chapter 13)

double blind Experimental design protocol when both research subjects and scientists performing the measurements are unaware of either the experimental hypothesis or who is in the control or experimental group. (Chapter 1)

drug resistance In pathogens, occurs when the pathogen is no longer susceptible to the effects of a drug, thus infections are no longer controlled by drug treatment. (Chapter 9)

ecology Field of biology that focuses on the interactions between organisms and their environment. (Chapters 3, 15)

ecosystem All of the organisms and natural features in a given area. (Chapter 14)

ecosystem services Proper functioning of the natural world's ecosystems. (Chapter 14)

effector Muscle, gland, or organ stimulated by a nerve. (Chapter 13)

egg Gamete produced by a female organism. (Chapter 4)

element See chemical element.

embryo Developing individual. (Chapter 4)

encephalopathy Pathology, or disease, of the brain. (Chapter 13)

Endangered Species Act (ESA) U.S. law intended to protect and encourage the population growth of threatened and endangered species enacted in 1973. (Chapter 14)

endocrine system Hormones and the organs and glands that produce them. (Chapter 12)

endometrium Lining of the uterus, shed during menstruation. (Chapter 12)

endosymbiotic hypothesis Theory that organelles such as mitochondria and chloroplasts in eukaryotic cells evolved from prokaryotic cells that took up residence inside ancestral eukaryotes. (Chapter 3)

environmental scientist Scientist who studies the affect of human activity on the natural world. (Chapter 15)

enzyme Protein that catalyzes and regulates the rate of metabolic reactions. (Chapter 2)

epidemic Contagious disease that spreads rapidly and extensively among many individuals. (Chapter 11)

epidemiologist Scientist who attempts to determine who is prone to a particular disease, where risk of the disease is highest, and when the disease is most likely to occur. (Chapter 11)

epididymis Sperm-carrying duct. (Chapter 12)

epiglottis Flap that blocks the windpipe so food goes down the pharynx, not into the lungs. (Chapter 2)

esophagus Tube that connects the mouth to the stomach. (Chapter 2)

estrogens Feminizing hormones secreted by the ovary in females and adrenal glands in both sexes. (Chapter 12)

eugenics Science of "improving" the human species through selective breeding. (Chapter 4)

eukaryotes Cells that have a nucleus and membrane-bounded organelles. (Chapter 2)

eutrophication Process resulting in periods of dangerously low oxygen levels in water, sometimes caused by high levels of nitrogen and phosphorus from fertilizer runoff that result in increased growth of algae in waterways. (Chapter 15)

evolution Changes in the features (traits) of individuals in a biological population that occur over the course of generations. (Chapters 3, 8)

evolutionary classification System of organizing biodiversity according to the evolutionary relationships among living organisms. (Chapter 3)

exotic species Organisms introduced by human activity to a region where they had never been found. (Chapter 14)

experiments Contrived situations designed to test specific hypotheses. (Chapter 1)

exponential growth Growth that occurs in proportion to the current total. (Chapter 15)

extinction Complete loss of a species. (Chapter 14)

falsifiable Able to be proved false. (Chapter 1)

fat Energy source that contains more calories than an equal weight of carbohydrates or proteins. (Chapter 2)

femurs Bones extending from the pelvis to the knee, also called *thighbones*. (Chapter 12)

fertility Ability to produce healthy sperm and egg cells. (Chapter 12)

fertilization Fusion of gametes (egg and sperm). (Chapter 4)

fertilizer Organic or inorganic substance used by farmers to replace nutrients lost from the soil. (Chapter 15)

fissure Groove in the brain that divides the cerebrum and its cortex from front to back, into the right and left cerebral hemispheres. (Chapter 13)

fitness Relative survival and reproduction of one variant compared to others in the same population. (Chapter 9)

flagellum Whiplike structure found on some bacterial cells, which aids in motility. (Chapter 12)

flower Reproductive structure of a flowering plant. (Chapter 3)

flowering plants Division of the kingdom Plantae containing members that produce flowers and fruit. (Chapter 3)

follicle Primary oocyte surrounded by layer of nourishing cells. (Chapter 12)

follicle cell Nourishing cells that surround the primary oocyte. (Chapter 12)

follicle-stimulating hormone (FSH) Hormone secreted by the pituitary gland involved in sperm production, regulation of ovulation, and regulation of menstruation. (Chapter 12)

food web The feeding connections between and among organisms in an environment. (Chapter 15)

foramen magnum Hole in the skull that allows passage of the spinal cord. (Chapter 8)

fossil fuel Nonrenewable resource consisting of the buried remains of ancient plants that have been transformed by heat and pressure into coal and oil. (Chapter 15)

fossil record Physical evidence left by organisms that existed in the past. (Chapter 14)

fossils Remains of plants or animals that once existed, left in soil or rock. (Chapter 8)

founder effect Type of sampling error that occurs when a small subset of individuals emigrates from the main population to found a new population. Results in differences in the gene pools of source population and the new population. (Chapter 10)

frontal bone Upper front portion of the cranium, or the forehead. (Chapter 12)

fruit Structure produced by flowering plants that is usually adapted for seed dispersal. (Chapter 15)

functional group Chemical group on an amino acid that has particular chemical properties. (Chapter 2)

Fungi Kingdom of eukaryotes made up of members that are immobile, rely on other organisms as their food source, and are made up of hyphae that secrete digestive enzymes into the environment and that absorb the digested materials. (Chapter 3)

gall bladder Organ that stores bile and empties into small intestine. (Chapter 2)

gamete Specialized sex cells (sperm and egg in humans) that contain half as many chromosomes as other body cells. (Chapters 4, 6)

gene Sequence of DNA that contains information about genetic traits, thus they code for specific proteins. (Chapters 4, 6, 7, 9)

gene expression Turning a gene on or off. A gene is expressed when the protein it encodes is synthesized. (Chapter 7)

gene flow Spread of an allele throughout a species' gene pool. (Chapter 10)

gene gun Device used to shoot DNA-coated pellets into plant cells. (Chapter 7)

gene pool All of the alleles found in the individuals of a species. (Chapter 10)

Generally Recognized As Safe (GRAS) A food that does not need to undergo FDA testing. (Chapter 7)

gene therapy Replacing defective genes (or their protein products) with functional ones. (Chapter 7)

genetic code Table showing which mRNA codons code for which amino acids. (Chapter 7)

genetic drift Change in allele frequency that occurs as a result of chance. (Chapter 10)

genetic engineering Using technology to change one or more genes in an organism. (Chapter 7)

genetic variability All of the forms of genes, and the distribution of these forms, found within a species. (Chapter 14)

genetically modified organisms (GMOs) Transgenic organisms or organisms that have been genetically engineered. (Chapter 7)

genome Entire suite of genes present in an organism. (Chapter 7)

genotype Genetic composition of an individual. (Chapter 4)

genus Broader biological category to which several similar species may belong. (Chapter 10)

glial cells Cells within the brain that do not carry messages but rather support neurons by supplying nutrients, repairing the brain after injury, and attacking invading bacteria. (Chapter 13)

gonadotropin-releasing hormone (GnRH) Hormone produced by the hypothalamus that stimulates the pituitary gland to release FSH and LH, thereby stimulating the activities of the gonads. (Chapter 12)

gonads Testicles in males or ovaries in females. (Chapter 12)

***Gpbox* genes** Group of genes required for ovarian development. (Chapter 12)

gray matter Unmyelinated axons, combined with dendrites and cell bodies of other neurons that appear gray in cross section. (Chapter 13)

Green Revolution Name given to the change in agricultural techniques that started around 1940 and includes the introduction of hybrids and increased use of irrigation and inorganic fertilizer. (Chapter 15)

groundwater Water available in reservoirs beneath the surface of the soil. (Chapter 15)

growth factor Protein that stimulates cell division. (Chapter 5)

growth rate Annual death rate in a population subtracted from the annual birth rate. A species' growth rate is influenced by how long the species takes to reproduce, how often it reproduces, the number of offspring produced each time, and the death rate of individuals under ideal conditions. (Chapter 14)

guanine Nitrogenous base in DNA, a purine. (Chapter 6)

guard cells Paired cells encircling stomata that serve to regulate the size of the stomatal pore, helping to minimize water loss under dry conditions and maximize carbon dioxide uptake under wet conditions. (Chapter 15)

habitat Place where an organism lives. (Chapter 14)

habitat destruction Modification and degradation of natural forests, grasslands, wetlands, and waterways by people; primary cause of species loss. (Chapter 14)

habitat fragmentation Threat to biodiversity caused by humans that occurs when large areas of intact natural habitat are subdivided by human activities. (Chapter 14)

half-life Amount of time required for one-half the amount of a radioactive element that is originally present to decay into the daughter product. (Chapter 8)

haploid Cells containing only one member of each homologous pair of chromosomes (n). (Chapter 6)

Hardy-Weinberg theorem Theorem that holds that allele frequencies remain stable in populations that are large in size, randomly mating, and experiencing no migration or natural selection; used as a baseline to predict how allele frequencies would change if any of its assumptions were violated. (Chapter 10)

helper T cells Immune-system cells that enhance cell-mediated immunity and humoral immunity by secreting a substance that increases the strength of the immune response (see T4 cell). (Chapters 9, 11)

hemoglobin Protein that carries oxygen in red blood cells. (Chapter 4)

hemophilia Rare genetic disorder that prevents normal blood clotting. (Chapter 6)

herbicide Chemical that kills plants. (Chapter 15)

heritability The amount of variation for a trait in a population that can be explained by differences in genes among individuals. (Chapter 4)

heterozygote Individual carrying two different alleles for a particular gene. (Chapter 10)

heterozygous Genotype containing two different alleles for a gene. (Chapter 4)

high-density lipoproteins (HDL) High-protein, low-cholesterol lipoproteins. (Chapter 2)

HIV See Human Immunodeficiency Virus.

hominin Humans and human ancestors. (Chapter 8)

homologous pairs Sets of two chromosomes of the same size and shape with centromeres in the same position. Homologous pairs of chromosomes carry the same genes in the same locations, but may carry different alleles. (Chapter 6)

homology Similarity in characteristics as a result of common ancestry. (Chapter 8)

homozygous Genotype containing identical alleles for a gene. (Chapter 4)

hormones Substances that travel through the circulatory system and act as signals that turn on genes in their target organs. (Chapter 12)

host Organism infected by a pathogen or parasite. (Chapter 9)

Human Genome Project Effort to determine the nucleotide base sequences and chromosomal locations of all human genes. (Chapter 7)

Human Immunodeficiency Virus (HIV) Agent identified as causing the transmission and symptoms of AIDS. (Chapter 9)

humoral immunity B-cell-mediated immunity that occurs when a B-cell receptor binds to an antigen. The B cell divides to produce a clonal population of memory cells; the cell also produces plasma cells. (Chapter 11)

hybrid Offspring of two different strains of an agricultural crop (see also interspecies hybrid). (Chapter 15)

hybridization and assimilation hypothesis Hypothesis about the origin of modern humans stating that *Homo sapiens* evolved in Africa and spread around the world, interbreeding with *H. erectus* populations already present in Asia and Europe. (Chapter 10)

hydrocarbon A molecule consisting of only carbons and hydrogens. (Chapter 2)

hydrogen bond A type of weak chemical bond. In DNA this type of bond forms between nitrogenous bases across the width of the helix. (Chapter 6)

hydrophilic Water-loving molecule. (Chapter 2)

hydrophobic Water-hating molecule. (Chapter 2)

hypertension High blood pressure. (Chapter 2)

hyphae Thin, stringy fungous material that grows over and within a food source. (Chapter 3)

hypothalamus Gland that helps regulate body temperature; influences behaviors such as hunger, thirst, and reproduction; and secretes a hormone (GnRH) that stimulates the activities of the gonads. (Chapter 13)

hypothesis Tentative explanation for an observation that requires testing to validate. (Chapters 1, 8)

IDDM Type of diabetes that requires insulin injections, Type I. (Chapter 2)

immortal Property of cancer cells that allows them to divide more times than normal cells, possibly due to activation of a telomerase gene. (Chapter 6)

immune deficiency Poor immune-system function, usually resulting in increased opportunistic infections. (Chapter 9)

immune response Ability of the immune system to respond to an infection resulting from increased production of B cells and T cells. (Chapters 9, 11)

inbreeding Mating between related individuals. (Chapter 6)

inbreeding depression Negative effect of homozygosity on the fitness of members of a population. (Chapter 14)

independent assortment Physical process of placing one member of each homologous pair into a gamete. Likewise, one allele of each gene is placed into each gamete. (Chapters 4, 6)

infant mortality Death rate of infants and children under the age of 5. (Chapter 15)

infectious When a pathogen finds a tissue inside the body that will support its growth. (Chapter 11)

inferential statistics Allows scientists to extend the results they summarize from their sample to the entire population. (Chapter 1)

inheritance of acquired characteristics Lamarck's discarded hypothesis for how evolution occurred that postulated that traits developed over the lifetime of an individual were passed on to their offspring. (Chapter 8)

inorganic Chemical compounds that do not contain carbon. Inorganic fertilizers consist of simple molecules produced by an industrial process. (Chapters 2, 15)

integrated pest management (IPM) Agricultural technique that uses releases of predator insects and changes in planting techniques, among other strategies, to reduce pesticide use. (Chapter 15)

intermembrane space Space between two membranes. (Chapter 2)

interneurons Neurons located between sensory and motor neurons that function to integrate information. (Chapter 13)

interphase Part of the cell cycle when a cell is preparing for division and the DNA is duplicated. (Chapters 5, 6)

interspecies hybrid Organism with parents from two different species. (Chapter 10)

invertebrates Animals without backbones. (Chapter 3)

irrigation Artificial addition of water to land for crop production. (Chapter 15)

IQ test Tool for measuring intelligence that compares an individual's performance with that of peers. (Chapter 4)

karyotype Picture of chromosomes prepared from blood cells. (Chapter 6)

kingdom In some classifications, the most inclusive group of organisms; usually five or six. In other classification systems, the level below domain on the hierarchy. (Chapter 3)

large intestine Colon, portion of the digestive system located between the small intestine and the anus that absorbs water and forms feces. (Chapter 2)

latent Inactive or dormant phase of a disease. (Chapter 11)

legume Common name for a family of flowering plants that has a mutualistic relationship with nitrogen-fixing bacteria. They produce seeds in pods, such as beans, peas, and alfalfa. (Chapter 15)

Leydig cells Cells scattered between the seminiferous tubules of the testicles that produce testosterone and other androgens. (Chapter 12)

life cycle Description of the growth and reproduction of an individual. (Chapter 4)

linked genes Genes present on the same chromosome. (Chapter 6)

lipoproteins Cholesterol-carrying proteins. (Chapter 2)

liver Organ that produces bile to aid in the absorption of fats. (Chapter 2)

logistic growth Pattern of growth seen in populations that are limited by resources available in the environment. A graph of logistic growth over time typically takes the form of an S-shaped curve. (Chapter 15)

low-density lipoproteins (LDLs) High-cholesterol, low-protein lipoproteins. (Chapter 2)

luteinizing hormone (LH) Hormone involved in sperm production, regulation of ovulation, and regulation of menstruation. (Chapter 12)

lymphocyte White blood cells that make up part of the immune system. (Chapters 9, 11)

macroevolution Evolution of one or more species from an ancestral form; speciation. (Chapter 10)

macrophage Phagocytic white blood cell that swells and releases toxins to kill bacteria. (Chapter 11)

malignant Tumor that invades surrounding tissues. (Chapter 5)

mandible Bone of the lower jaw. (Chapter 12)

mass extinctions Losses of species that are rapid, global in scale, and affect a wide variety of organisms. (Chapter 3)

matrix Semifluid substance inside the inner mitochondrial membrane, which contains some enzymes. (Chapter 2)

mean A quantity intermediate to a set of quanties; average. (Chapters 1, 4)

medulla oblongata Region of the brain stem that is a continuation of the spinal cord and conveys information between the spinal cord and other parts of the brain. (Chapter 13)

meiosis Process of making sperm and eggs during which parents pass half of their DNA to their children. (Chapters 4, 6)

memory cells Cells of a clonal population, programmed to respond to a specific antigen, that help the body respond quickly if the infectious agent is encountered again. (Chapter 11)

menopause Cessation of menstruation. (Chapter 12)

menstrual cycle Changes that occur in the uterus and depend on intricate interrelationships among the brain, ovaries, and lining of the uterus. (Chapter 12)

messenger RNA (mRNA) Complementary RNA copy of a DNA gene produced during transcription. The mRNA undergoes translation to synthesize a protein. (Chapter 7)

metabolic rate Measure of an individual's energy use. (Chapter 2)

metabolism All chemical reactions occurring in the body. (Chapter 2)

metaphase Stage of mitosis during which duplicated chromosomes align across the middle of the cell. (Chapter 5)

metastasis When cells from a tumor break away and start new cancers at distant locations. (Chapter 5)

microbe Microscopic organism, especially Bacteria and Archeae. (Chapter 3)

microevolution Changes that occur in the characteristics of a population. (Chapter 8)

microorganism See microbe.

microtubule Protein structures that move chromosomes around during mitosis and meiosis. (Chapters 5, 6)

midbrain Uppermost region of the brain stem, which adjusts the sensitivity of the eyes to light and the ears to sound. (Chapter 13)

mineral Inorganic nutrient essential to many cell functions. (Chapter 2)

mitochondria Organelles in which products of the digestive system are converted to ATP. (Chapters 2, 3)

mitosis Portion of the cell cycle in which DNA is apportioned into two daughter cells. (Chapter 5)

model organisms Nonhuman organisms used in the Human Genome Project that are easy to manipulate in genetic studies and help scientists understand human genes because they share genes with humans. (Chapter 7)

molecular clock Principle that DNA mutations accumulate in the genome of a species at a constant rate, permitting estimates of when the common ancestor of two species existed. (Chapter 8)

monocot One of the two classes of flowering plants, also called *narrow-leaved plants*. Monocot seeds contain one leaf (cotyledon). (Chapter 15)

monoculture Practice of planting a single crop over a wide acreage. (Chapter 15)

monomer Individual subunit. (Chapter 2)

monosaccharide Simple sugar. (Chapter 2)

monozygotic twins Identical twins that developed from one zygote. (Chapter 4)

morphological species concept Definition of species that relies on differences in physical characteristics among them. (Chapter 10)

morphology Appearance or outward physical characteristics. (Chapter 10)

motor neurons Neurons that carry information away from the brain. (Chapter 13)

mulching Spreading straw or other dead plant material, or sometimes dark-colored plastic, around the base of preferred plants to reduce the growth of competitors and prevent the soil from losing water. (Chapter 15)

Müllerian duct system Embryonic duct system that persists in females due to the action of gene products that prevent testosterone synthesis and secretion. (Chapter 12)

multicellular Made up of many coordinated cells. (Chapter 3)

multiregional hypothesis Hypothesis about the origin of modern humans that states that *Homo sapiens* evolved from *H. erectus* separately in Africa, Asia, and Europe. (Chapter 10)

mutagens Substances that increase the likelihood of mutation occurring; increases the likelihood of cancer. (Chapter 5)

mutations Changes to DNA sequences that occur during DNA synthesis (Chapters 4, 5)

mutualism Interaction between two species that provides benefits to both species. (Chapter 14)

myelin sheath Protective layer that coats many axons, formed by supporting cells, such as Schwann cells. The myelin sheath increases the speed at which the electrochemical impulse travels down the axon. (Chapter 13)

natural experiments Situations where unique circumstances allow a hypothesis test without prior intervention by researchers. (Chapter 4)

natural landscape Landscape that is not strongly modified by humans. (Chapter 14)

natural selection Process by which individuals with certain traits have greater survival and reproduction than individuals who lack these traits, resulting in an increase in the frequency of successful alleles and a decrease in the frequency of unsuccessful ones. (Chapters 3, 9, 10)

nerve Bundle of neurons; nerves branch out from the brain and spinal cord to eyes, ears, internal organs, skin, and bones. (Chapter 13)

nerve impulses Electrochemical signals that control the activities of muscles, glands, organs, and organ systems. (Chapter 13)

nervous system Brain, spinal cord, sense organs, and nerves that connect organs and link this system with other organ systems. (Chapter 13)

net primary production (NPP) Amount of solar energy converted to chemical energy by plants, minus the amount of this chemical energy plants need to support themselves. A measure of plant growth, typically over the course of a single year. (Chapter 15)

neurobiologist Biologist who studies the nervous system. (Chapter 13)

neurons Specialized message-carrying cells of the nervous system. (Chapter 13)

neurotransmitters Chemicals released by the presynaptic neuron into the synapse, which then diffuse across the synapse and bind to receptors on the membrane of the postsynaptic neuron. Each type of neuron secretes one type of neurotransmitter. Each cell responding to the neurotransmitter has receptors specific to that neurotransmitter. (Chapter 13)

NIDDM Type of diabetes that does not require insulin injections, Type II. (Chapter 2)

nitrogen Important nutrient for plants in terrestrial systems (N). The fourth most abundant chemical in plant cells. (Chapter 15)

nitrogen-fixing bacteria Organisms that convert nitrogen gas from the atmosphere into a form that can be taken up by plant roots; some species live in the root nodules of legumes. (Chapter 15)

nitrogenous bases Nitrogen-containing bases found in DNA: A, C, G, and T. (Chapter 6)

nodes of Ranvier Small indentations separating segments of the myelin sheath. Nerve impulses "jump" successively from one node of Ranvier to the next. (Chapter 13)

nonrenewable Resources that are a one-time supply and cannot be easily replaced. (Chapter 15)

normal distribution Bell-shaped curve, as for the distribution of quantitative traits in a population. (Chapter 4)

nuclear transfer Transfer of a nucleus from one cell to another cell that has had its nucleus removed. (Chapter 7)

nucleotides Building blocks of a DNA strand that include a sugar, a phosphate, and a nitrogenous base. (Chapter 6)

nucleus Cell structure that houses DNA; found in eukaryotes. (Chapters 2, 3)

null hypothesis Conclusion that there is zero difference between an experimental group and a control group, and therefore the experimental treatment has no effect. (Chapter 1)

nutrient cycling Process by which nutrients become available to plants. Nutrient cycling in a natural environment relies upon a healthy community of decomposers within the soil. (Chapter 15)

nutrients Atoms other than carbon, hydrogen, and oxygen that must be obtained from a plant's environment for photosynthesis to occur. (Chapter 15)

obesity Condition of having a BMI of 30 or greater. (Chapter 2)

objectivity Without bias, as in data collection. (Chapter 1)

observer bias Systematic errors in measurement and evaluation of results made by researchers. (Chapter 1)

oncogenes Mutant versions of proto-oncogenes. (Chapter 5)

oogenesis Formation and development of female gametes, which occurs in the ovaries and results in the production of egg cells. (Chapter 12)

oogonia Developing egg cells that begin meiosis but pause at prophase I. (Chapter 12)

opportunistic infection Diseases that only occur when a weakened immune system allows access. (Chapter 9)

organelle Subcellular structure that performs a specific job. (Chapter 2)

organic Carbon-containing compound. Alternatively, a fertilizer consisting of complex molecules made up of the partially decomposed waste products of plants and animals. (Chapters 2, 15)

organic farming Agricultural technique that requires no manufactured chemical inputs (such as inorganic fertilizer or chemical pesticides), and relies on developing and maintaining the health of soil. (Chapter 15)

ossa coxae Paired bones that form the bony pelvis. (Chapter 12)

osteoporosis Elevated risk of bone breakage from weakened bones. (Chapter 2)

out-of-Africa hypothesis Hypothesis about the origin of modern humans that states that modern *Homo sapiens* evolved in Africa and replaced *H. erectus* populations. (Chapter 10)

overexploitation Threat to biodiversity caused by humans that encompasses overhunting and overharvesting. Overexploitation occurs when the rate of human destruction or use of a species outpaces the ability of the species to reproduce. (Chapter 14)

overshoot Occurs when a population exceeds the carrying capacity of the environment. Typically followed by a population crash. (Chapter 15)

oviduct Egg-carrying duct that brings egg cells from ovaries to uterus. (Chapter 12)

ovulation Release of an egg cell from the ovary. (Chapter 12)

ovum Larger of the two cells produced during oogenesis, and the cell that receives the majority of the cytoplasmic nutrients and organelles. (Chapter 12)

paleontologist Scientist who searches for, describes, and studies ancient organisms. (Chapter 8)

pancreas Gland that secretes digestive enzymes and insulin. (Chapter 2)

parasites Organisms that feed on other living organisms. (Chapter 14)

pathogens Disease-causing organisms. (Chapters 9, 11)

pedigree Family tree that follows the inheritance of a genetic trait for many generations. (Chapter 6)

peptide bond Chemical bond that joins adjacent amino acids. (Chapter 2)

peripheral nervous system (PNS) Network of nerves outside the brain and spinal cord that links the CNS with sense organs. (Chapter 13)

peristalsis Rhythmic muscle contractions that move food through the digestive system. (Chapter 2)

persistent organic pollutants (POPs) Pollutants that do not break down readily, and thus can become very concentrated in an environment. (Chapter 15)

pesticide treadmill Process of continually increasing pesticide use as a result of the evolution of resistance to pesticides among target pests. (Chapter 15)

pesticides Chemicals designed to kill or reduce the growth of a target pest. Herbicides are often considered a class of pesticides. (Chapter 14)

pests Competing consumers of agricultural products, including mammals, birds, insects, and fungi and bacteria. (Chapter 15)

phagocytosis Ingestion of food or pathogens by cells. (Chapter 2)

pharynx Tube and muscles connecting the mouth to the esophagus; throat. (Chapter 2)

phenotype Physical traits of an individual. (Chapter 4)

phenotypic plasticity Ability of a genotype to produce a range of phenotypes depending on outside influences (such as the environment). (Chapter 4)

phosphodiester bond Chemical bond that joins nucleotides in DNA. (Chapter 6)

phospholipid Molecule that makes up the plasma membrane, with a hydrophilic head and a hydrophobic tail. (Chapter 2)

phosphorus Element required for plant growth that is an important component of proteins and DNA (P). (Chapter 15)

phosphorylating Adding a phosphate group, thereby energizing some other substance. (Chapter 2)

photosynthesis Process by which plants, along with algae and some bacteria, transform light energy to chemical energy. (Chapter 2)

phylogeny Evolutionary history of a group of organisms. (Chapter 3)

pituitary gland Gland located at the base of the skull that, once stimulated by GnRH, synthesizes and releases pituitary gonadotropins. (Chapter 12)

pituitary gonadotropins Follicle-stimulating hormone (FSH) and luteinizing hormone (LH). (Chapter 12)

placebos Sugar pills or fake treatment. (Chapter 1)

placenta Membrane produced by a developing fetus that releases a hormone to extend the life of the corpus luteum. (Chapter 12)

Plantae Multicellular photosynthetic eukaryotes, excluding algae. (Chapter 3)

plasma cells Cells produced by a clonal population that secrete antibodies specific to an antigen. (Chapter 11)

plasma membrane Structure that encloses a cell, defining the cell's outer boundary. (Chapter 2)

plasmid Circular piece of bacterial DNA that normally exists separate from the bacterial chromosome and can make copies of itself. (Chapter 7)

polar body Smaller of the two cells produced during oogenesis, and therefore does not have enough nutrients to undergo further development. (Chapter 12)

pollinators Organisms, such as bees, that transfer sperm (pollen grains) from one flower to the female reproductive structures of another flower. (Chapter 14)

pollution Human-caused threat to biodiversity involving the release of poisons, excess nutrients, and other wastes into the environment. (Chapter 14)

polyculture Practice of planting many different crop plants over a single farm's acreage. (Chapter 15)

polygenic traits Traits influenced by many genes. (Chapter 4)

polymer Combination of monomers. (Chapter 2)

polysaccharide Complex carbohydrate. (Chapter 2)

pons Region of the brain stem that relays messages between the brain and spinal cord. (Chapter 13)

population bottleneck Dramatic but short-lived reduction in population size followed by an increase in population. (Chapter 10)

population crash Steep decline in number that may occur when a population grows larger than the carrying capacity of its environment. (Chapter 15)

population goal Defined by the Endangered Species Act to be the population of an endangered species that would allow it to be removed from the endangered species list. (Chapter 14)

populations Subgroups of a species that are somewhat independent. (Chapters 8, 10)

postzygotic Barrier to reproduction that occurs when fertilization occurs as a result of mating between two members of different species, but the resulting offspring does not survive or is sterile. (Chapter 10)

potassium Element required for plant growth that is essential in regulating water balance within cells (K). (Chapter 15)

predation Act of capturing and consuming an individual of another species. (Chapter 14)

predator Organism that eats other organisms. (Chapter 3)

prediction Result expected from a particular test of a hypothesis in the hypothesis were true. (Chapter 1)

prezygotic Barrier to reproduction that occurs when individuals from different species either do not attempt to mate, or if they do mate, fail to produce a fertilized egg. (Chapter 10)

primary consumers Organism that eats plants. (Chapter 15)

primary follicle Primary oocyte plus surrounding follicle cells. (Chapter 12)

primary oocyte Developing egg cell paused at prophase I of meiosis. (Chapter 12)

primary sources Articles written by researchers and reviewed by the scientific community. (Chapter 1)

primary spermatocyte Diploid cell that begins meiosis in the production of sperm. (Chapter 12)

prion Normally occurring protein produced by brain cells that, when misfolded, causes spongiform encephalopathy; prion is the shortened form of proteinaceous infectious particle. (Chapter 11)

probe Single-stranded nucleic acid that has been radioactively labeled. (Chapter 6)

progesterone Hormone produced by the ovary, high levels of which have a negative feedback effect on the hypothalamus, causing GnRH secretion to decrease. (Chapter 12)

prokaryotes Cells that do not have a nucleus or membrane-bounded organelles. (Chapters 2, 3)

promoter Sequence of nucleotides to which the polymerase binds to start transcription. (Chapter 7)

prophase Stage of mitosis during which duplicated chromosomes condense. (Chapter 5)

protein Cellular constituents made of amino acids coded for by genes. Proteins can have structural, transport, or enzymatic roles. (Chapters 2, 4, 9)

protein synthesis Joining amino acids together, in an order dictated by a gene, to produce a protein. (Chapter 7)

Protista Kingdom in the domain Eukarya containing a diversity of eukaryotic organisms, most of which are unicellular. (Chapter 3)

proto-oncogenes Genes that encode proteins that regulate the cell cycle. Mutated proto-oncogenes (oncogenes) can lead to cancer. (Chapter 5)

Punnett square Table that lists the different kinds of sperm or eggs parents can produce relative to the gene or genes in question and predicts the possible outcomes of a cross between these parents. (Chapter 4)

purine Nitrogenous base (A or G) of the two-ring structure in DNA. (Chapter 6)

pyloric sphincter Contractile tissue that regulates movement of foods from the stomach to the small intestine. (Chapter 2)

pyrimidine Nitrogenous base (C or T) of the single-ring structure in DNA. (Chapter 6)

Q angle Angle of the femur in relation to a horizontal line drawn through the kneecap. (Chapter 12)

qualitative traits Traits that come in distinct categories. (Chapter 4)

quantitative traits Traits that have many possible values. (Chapter 4)

race See biological race.

racism Idea that some groups of people are naturally superior to others. (Chapter 10)

radiation therapy Focussing beams of reactive particles at a tumor to kill the dividing cells. (Chapter 5)

radioactive decay Natural, spontaneous breakdown of radioactive elements into different elements, or "daughter products." (Chapter 8)

radio-immunotherapy Experimental cancer treatment with the goal of delivering radioactive substances directly to tumors without affecting other tissues. (Chapter 5)

radiometric dating Technique that relies on radioactive decay to estimate a fossil's age (Chapter 8)

random alignment Members of a homologous pair line up randomly with respect to maternal or paternal origin during metaphase I of meiosis, thus increasing the genetic diversity of offspring. (Chapter 6)

random assignment Placing individuals into experimental and control groups randomly to eliminate systematic differences between the groups. (Chapter 1)

receptor Protein on the surface of a cell that recognizes and binds to a specific chemical signal. (Chapter 9)

recessive Applies to an allele with an effect that is not visible in a heterozygote. (Chapter 4)

recombinant Produced by manipulating a DNA sequence. (Chapter 7)

recovery plan Defined by the Endangered Species Act as the plan of action put in place designed to remove a species from the endangered species list. (Chapter 14)

reflex Automatic response to a stimulus. (Chapter 13)

reflex arc Nerve pathway followed during a reflex consisting of a sensory receptor, a sensory neuron, an interneuron, a motor neuron, and an effector. (Chapter 13)

replications Repeats of an experiment or hypothesis test. (Chapter 1)

reproductive isolation Prevention of gene flow between different biological species due to failure to produce fertile offspring; can include prezygotic barriers and postzygotic barriers. (Chapter 10)

resources Food, water, shelter, and area required for the survival of a population. (Chapter 15)

restriction enzymes Enzymes that cleave DNA at specific nucleotide sequences. (Chapters 6, 7)

reticular formation Extensive network of neurons that runs through the medulla of the brain and projects toward the cerebral cortex. It functions as a filter for sensory input and activates the cerebral cortex. (Chapter 13)

reuptake Process by which neurotransmitters are reabsorbed by the neuron that secreted them. (Chapter 13)

reverse transcriptase Enzyme in RNA viruses that participates in copying the viral DNA. It performs the reverse of transcription by producing DNA from RNA. (Chapters 9, 11)

ribose The five-carbon sugar in RNA. (Chapter 7)

ribosomes Cellular structures that help translate genetic material into proteins by anchoring and exposing small sequences of mRNA. (Chapters 3, 7)

risk factor Exposures or behaviors that increase the likelihood of disease. (Chapter 5)

Ritalin Stimulant used to treat ADD. (Chapter 13)

RNA Information-carrying molecule composed of nucleotides. (Chapters 3, 7)

RNA polymerase Enzyme that synthesizes mRNA from a DNA template during transcription. (Chapter 7)

runoff Movement of water through the soil. (Chapter 15)

safer sex Practice of minimizing contact with a partner's bodily fluids during sexual activity as prevention against the transmission of HIV and other sexually transmitted diseases. (Chapter 9)

salinization Salt buildup in soil, often caused by irrigation. (Chapter 15)

sample Small subgroup of a population used in an experimental test. (Chapter 1)

sample size Number of individuals in both the experimental and control groups. (Chapter 1)

sampling error Effect of chance on experimental results. (Chapter 1)

saturated fat Fat in which carbons are bound to as many hydrogens as possible. (Chapter 2)

Schwann cell Cells that form the myelin sheath along the axons of nerve cells. (Chapter 13)

scientific theory Body of scientifically accepted general principles that explain natural phenomena. (Chapter 8)

secondary consumers Animals that eat primary consumers; predators. (Chapter 15)

secondary oocyte Egg cell that is ovulated, having entered metaphase II of meiosis. If the secondary oocyte is fertilized, meiosis begins again from metaphase II. (Chapter 12)

secondary sources Books, news media, and advertisements as sources of scientific information. (Chapter 1)

secondary spermatocyte Haploid cell produced by the first meiotic division in the production of sperm. (Chapter 12)

seed Reproductive structure in some plants that contain a plant embryo and its food source in a protective coat. (Chapter 3)

selective breeding Controlling the breeding of individual organisms to influence the phenotype of the next generation. (Chapter 4)

semen Sperm and energy-rich associated fluids. (Chapter 12)

seminal vesicle Gland of the male reproductive system that produces a fructose-rich fluid that helps supply sperm with energy. (Chapter 12)

seminiferous tubules Highly coiled tubes in the testicles where sperm are formed. (Chapter 12)

sense strand DNA strand of the double helix that codes for a protein. (Chapter 7)

sensory neurons Neurons that carry information toward the brain. (Chapter 13)

separate types hypothesis Hypothesis that numerous types of organisms (e.g., birds, cats, ferns) appeared on Earth separately and each type diversified into many species via evolutionary processes. (Chapter 8)

Sertoli cells Cells that secrete substances that remove some of the excess cytoplasm from spermatids, converting them to spermatozoa (sperm). (Chapter 12)

severe combined immunodeficiency disorder (SCID) Illness caused by a genetic mutation that results in the absence of an enzyme, and a severely weakened immune system. (Chapter 7)

sex chromosomes The X and Y chromosomes in humans. (Chapter 6)

sex hormones Estrogen and testosterone. (Chapter 12)

sex-linked genes Genes linked to and inherited along with the X and Y chromosomes. (Chapter 6)

sexual selection Form of natural selection that occurs when a trait influences the likelihood of mating. (Chapter 10)

shaman Indigenous healer. (Chapter 3)

sister chromatids Duplicated identical copies of a chromosome. (Chapter 5)

small intestine Large tubular structure in the digestive system that helps break down food and absorbs nutrients into the blood. (Chapter 2)

social construct Product of history and learned attitudes. (Chapter 10)

soil erosion Loss of topsoil. (Chapter 15)

speciation Evolution of one or more species from an ancestral form; macroevolution. (Chapter 10)

species A group of individuals that regularly breed together and are generally distinct from other species in appearance or behavior. In Linnaeus' classification system, a group in which members have the greatest resemblance. (Chapters 3, 8, 10)

species–area curve Graph describing the relationship between the size of a natural landscape and the relative number of species it contains. (Chapter 14)

specific immune response Cells and antibodies that respond to the antigens on a pathogen. (Chapter 9)

specificity Phenomenon of enzyme shape determining the reaction the enzyme catalyzes. (Chapter 2)

sperm Gametes produced by males. (Chapters 4, 12)

spermatids Cells produced when secondary spermatocytes undergo meiosis II to produce haploid cells that no longer have duplicated chromosomes. (Chapter 12)

spermatogenesis Production of sperm. (Chapter 12)

spermatozoa Mature sperm composed of a small head containing DNA, a midpiece that contains mitochondria, and a tail (flagellum). (Chapter 12)

spinal cord Main nerve pathway between the brain and the rest of the body. (Chapter 13)

spores Resistant cells that are analogous to plant seeds and can germinate into new individuals; present in plants, fungi and bacteria. (Chapters 3, 11)

SRY gene Gene located on the Y chromosome that helps determine maleness. (Chapter 12)

static model Discarded hypothesis about the origin of living organisms that states that they appeared in the recent past and have not changed over time. (Chapter 8)

statistical tests Tests that help scientists evaluate whether the results of a single experiment demonstrate the effect of treatment. (Chapter 1)

statistically significant Low probability that experimental groups differ simply by chance. (Chapter 1)

statistics Specialized branch of mathematics used in the evaluation of experimental data. (Chapter 1)

stem cell Cells that can divide indefinitely and can differentiate into other cell types. (Chapter 7)

steroid hormones The synthesis of these hormones requires cholesterol as a precursor; they are fat soluble and can cross cell membranes readily. (Chapter 12)

stimulant Drugs that increase CNS activity. (Chapter 13)

stomata Pores on the photosynthetic surfaces of plants that allow air into the internal structure of leaves and green stems. Stomata also provide portals through which water can escape. (Chapter 15)

stop codon An mRNA codon that does not code for an amino acid and causes the amino acid chain to be released into the cytoplasm. (Chapter 7)

strong inference A strong statement about the truth of a given hypothesis possible when an experimental protocol greatly minimizes the number of alternative hypotheses that can explain a result. (Chapter 1)

subject expectation Conscious or unconscious modeling of behavior the subject thinks a researcher expects. (Chapter 1)

subspecies Subdivision of a species that is not reproductively isolated but represents a population or set of populations with a unique evolutionary history. See also biological race, variety. (Chapter 10)

substrate The chemicals metabolized by an enzyme-catalyzed reaction. (Chapter 2)

sugar-phosphate backbone Series of alternating sugars and phosphates along the length of the DNA helix. (Chapter 6)

superfamily Taxonomic category between family and order. (Chapter 8)

supernatural Not constrained by the laws of nature. (Chapter 1)

symbiosis A relationship between two species. (Chapter 3)

synapse Gap between neurons consisting of the terminal boutons of the presynaptic neuron, the space between the two adjacent neurons, and the membrane of the postsynaptic neuron. (Chapter 13)

systematist Biologist who specializes in describing and categorizing a particular group of organisms. (Chapter 3)

T4 cell Immune-system cell that helps coordinate the body's specific response to a pathogen; also called a *helper T cell*. (Chapter 9)

T cell receptor Receptors found in T cells, produced in response to an antigen. (Chapter 11)

T lymphocytes (T cells) Immune-system cells that develop in the thymus gland and recognize and respond to body cells that have gone awry, such as cancer cells or cells that have been invaded by viruses, as well as transplanted tissues and organisms such as fungi and parasitic worms. T cells provide an immune response called *cell-mediated immunity*. (Chapter 11)

tectonic plates Slabs of Earth's crust that float on the underlying liquid mantle and have been slowly moving over billions of years. (Chapter 14)

telophase Stage of mitosis during which the nuclear envelope forms around the newly produced daughter nucleus, and chromosomes decondense. (Chapter 5)

temporal bones Paired bones found near the temple. (Chapter 12)

terminal bouton Knoblike structure at the end of an axon. (Chapter 13)

testable Possible to evaluate through observations of the measurable universe. (Chapter 1)

testimonial Statement made by an individual asserting the truth of a particular hypothesis because of personal experience. (Chapter 1)

testosterone Masculinizing hormone secreted by the testes. (Chapter 12)

thalamus Main relay center between the spinal cord and the cerebrum. (Chapter 13)

thymine Nitrogenous base in DNA, a pyrimidine. (Chapter 6)

Ti plasmid Tumor-inducing plasmid used to genetically modify crop plants. (Chapter 7)

tilling Turning over the soil to kill weed seedlings, with the goal of removing competitors before a crop is planted. (Chapter 15)

theory See scientific theory.

toxin Causes disease when produced by bacterial cells. (Chapter 11)

transcription Production of an RNA copy of the protein-encoding DNA gene sequence. (Chapter 7)

transfer RNA (tRNA) Amino-acid-carrying RNA structure with an anticodon that binds to an mRNA codon. (Chapter 7)

transformation hypothesis Hypothesis about the origin of living organisms stating that each arose separately in the past and has changed over time, but not into new species. (Chapter 8)

transgenic organism Organism whose genome incorporates genes from another organism; also called *genetically modified organism* (*GMO*). (Chapter 7)

translation Process by which an mRNA sequence is used to produce a protein. (Chapter 7)

trophic pyramid Relationship among the mass of populations at each level of a food web. (Chapter 15)

tumor Mass of tissue that has no apparent function in the body. (Chapter 5)

tumor suppressors Proteins that stop tumor formation by suppressing cell division, but when mutated lead to increased likelihood of cancer. (Chapter 5)

undifferentiated gonads Two masses in an embryo that can become either the paired ovaries of a female or the paired testicles of a male. (Chapter 12)

unicellular Made up of a single cell. (Chapter 3)

unsaturated fat Fat with many carbon–carbon double bonds that prevent the carbons from binding with hydrogens. (Chapter 2)

unsustainable Relating to practices that may compromise the ability of future generations to support a large human population with an adequate quality of life. (Chapter 15)

urethra Urine-carrying duct that also carries sperm in males. (Chapter 12)

uterus Pear-shaped muscular organ in females that can support pregnancy and that undergoes menstruation when its lining is shed. (Chapter 12)

uracil Nitrogenous base in RNA, a pyrimidine. (Chapter 7)

vagina Tubular structure in females located between the cervix and vulva. (Chapter 12)

variance Mathematical term for the amount of variation in a population. (Chapter 4)

variant An individual in a population that differs genetically from other individuals in the population. (Chapter 9)

variety Subgroup of a species with unique traits relative to other subgroups of the species. Equivalence of this term to biological race or subspecies is disputed by biologists. (Chapter 10)

vas deferens Sperm-carrying duct in males. (Chapter 12)

vascular tissue Cells that transport water and other materials within a plant. (Chapter 3)

vertebrae Bones of the spine. (Chapter 13)

vertebrates Animals with backbones. (Chapter 3)

vesicles Membrane bounded sac-like structures. In neurons, these structures are found in the terminal bouton and store neurotransmitters. (Chapter 13)

vestigial trait Modified with no, or relatively minor function compared to the function in other descendants of the same ancestor. (Chapter 8)

villi Small fingerlike projections on the in side of the small intestine. (Chapter 2)

viral envelope Layer formed around some virus protein coats (capsids) that is derived from the cell membrane of the host cell and may also contain some proteins encoded by the viral genome. (Chapter 11)

virus Infectious intracellular parasite composed of a strand of genetic material and a protein or fatty coating that can only reproduce by forcing its host to make copies of it. (Chapters 9, 11)

vitamin Organic nutrient needed in small amounts. Most vitamins function as coenzymes. (Chapter 2)

weeds Common term for nonpreferred plants. Any plant can be a weed if it grows in the wrong place. (Chapter 15)

white matter Nervous system tissue made of myelinated cells. (Chapter 13)

Wolffian duct system Embryonic duct system that persists in males occurring only when testosterone is present. (Chapter 12)

***x* axis** Horizontal axis of a graph. (Chapter 1)

***y* axis** Vertical axis of a graph. (Chapter 1)

zygote Single cell resulting from the fusion of gametes (egg and sperm). (Chapter 4)

Credits

Front matter

Photographs

Authors	Courtesy of Deborah Shubat
Title page	David McGlynn/Getty Images, Inc.

Chapter 1

Photographs

Opener 1	Carol Guenzi Agents/Index Stock Imagery, Inc.
Opener 2	Antonio Mo/Taxi/Getty Images, Inc.
Opener 3	Taxi/Getty Images, Inc.
Opener 4	Ryan McVay/Getty Images, Inc.
1.3a	Custom Medical Stock Photo, Inc.
1.4	Ed Reschke/Peter Arnold, Inc.

Illustrations

1.5a	Data from Lindenmuth, G.F., and E.B. Lindenmuth, 2000. The Efficacy of Echinacea compound herbal tea preparation on the severity and duration of upper-respiratory and flu symptoms: A randomized, double-blind, placebo-controlled study. *Journal of Alternative and Complementary Medicine* 6(4): 327–334.
1.7	Data from Cohen, S., D.A. Tyrrell, and A.P. Smith, 1991. Psychological stress and susceptibility to the common cold. *New England Journal of Medicine* 325(9): 606–612.
1.9	Data from Mossad, S.B., M.L. Macknin, S.V. Medendorp, and P. Mason, 1996. Zinc Gluconate lozenges for treating the common cold: A randomized, double-blind, placebo-controlled study. *Annals of Internal Medicine* 125(2): 81–88.

Chapter 2

Photographs

Opener 1	Michael Newman/PhotoEdit
Opener 2	Jason Homa/Getty Images, Inc.
Opener 3	Bruce Ayres/Getty Images Inc. - Stone Allstock
Opener 4	Express Newspapers/Getty Images, Inc - Liaison
2.5a	Brian Hagiwara/FoodPix/Getty Images, Inc.
2.5b	Felicia Martinez/PhotoEdit
2.15aL,R	© The New York Times/NYT Graphics
2.15bL	AP/Wide World Photos
2.15bR	Getty Images, Inc - Liaison
2.16	S. Valkeajoki/Health 2000 Survey/National Public Health Institute

Chapter 3

Photographs

Opener 1	Courtesy of Evan S. Whitaker
Opener 2	Robert Plowes/Robert and Beth Plowes Photography
Opener 3	James L. Amos/CORBIS
Opener 4	William Campbell/Homefire Productions, Inc.
3.1	WHOI/Edmund/Visuals Unlimited
3.2	D. Cavagnaro/DRK Photo
3.3a	Michel Viard/Peter Arnold, Inc.
3.3b	Mauro Fermariello/SPL/Science Source/Photo Researchers, Inc.
3.6	J. William Schopf, University of California at Los Angeles
3.8a	Photo Researchers, Inc.
3.8b	Professor J. L. Kerwin
3.8c	David M. Phillips/Visuals Unlimited
3.8d	Walter H Hodge/Peter Arnold, Inc.
3.9	Chip Clark/National Museum of Natural History, © 2003 Smithsonian Institution
3.10	Wayne Lynch/DRK Photo
3.11a	Michael Fogden/DRK Photo
3.11b	David Wrobel/Visuals Unlimited
3.11c	David Barron/Animals Animals/Earth Scenes
3.12	Gregory Ochocki/Photo Researchers, Inc.
3.14	Photo Researchers, Inc.
3.15a	Ed Reschke/Peter Arnold, Inc.
3.15b	Frans Lanting/Minden Pictures
3.15c	Murray, Patti/Animals Animals/Earth Scenes
3.17a	George D Lepp/Photo Researchers, Inc.
3.17b	Joe McDonald/Visuals Unlimited
3.17c	Bill Dyer/Photo Researchers, Inc.
3.19	Alison Wright/CORBIS
3.20	Getty Images Inc. - Stone Allstock

Chapter 4

Photographs

Opener 1	Dr. Yorgos Nikas/SPL/Photo Researchers, Inc.
Opener 2	Virginia Borden
Opener 3	Getty Images Inc. - Image Bank
Opener 4	Gaetano/CORBIS
4.7	Oliver Meckes & Nicole Ottawa/Photo Researchers, Inc.
4.9	A. Ramey/PhotoEdit
4.12T,B	Photos courtesy of New York plastic surgeon, Dr. Darrick E. Antell. www.Antell-MD.com
4.14a	Getty Images, Inc. - Photodisc
4.14b	Getty Images Inc. - Stone Allstock
4.17	David Young/PhotoEdit
E4.1A	Getty Images Inc. - Hulton Archive Photos
E4.2	Lewis Hine/Getty Images Inc. - Hulton Archive Photos

Illustrations

4.10	Data from Galton, F., and E. Hitchcock, 1889. Anthropometric Statistics from Amherst College, Mass., U.S.A. *Journal of the Anthropological Institute of Great Britain and Ireland*, 18: 192–199.
Table 4.2	Data from Cooper, R. M., and Zubek, J. P. (1958). Effects of enriched and restricted early environments on the learning ability of bright and dull rats. *Canadian Journal of Psychology*, 12: 159–164.
4.13	Data from U.S. Central Intelligence Agency, 2001. *CIA World Factbook 2001.*

Chapter 5

Photographs

Opener 1	Photos by Jim and Mary Whitmer
Opener 2	Photos by Jim and Mary Whitmer
Opener 3	Photo Researchers, Inc.
Opener 4	Photos by Jim and Mary Whitmer
5.3	C. Edelmann/La Villette/Photo Researchers, Inc.
5.4a,b	Biophoto Associates/Photo Researchers, Inc.
5.11a	D. M. Phillips/Visuals Unlimited
5.11b	Biodisc/Visuals Unlimited
5.11c	Biodisc/Visuals Unlimited
5.12L,R	Photos by Jim and Mary Whitmer

Chapter 6

Photographs

Opener 1 Corbis/Sygma
Opener 2 AP/Wide World Photos
Opener 3 Agence No.7/Corbis/Sygma
Opener 4 Agence No7/Corbis/Sygma
6.1 CNRI/SPL/Photo Researchers, Inc.
6.12 Getty Images Inc. - Hulton Archive Photos
6.18 AP/Wide World Photos
6.19 AP/Wide World Photos

Chapter 7

Photographs

Opener 1 Picture Desk, Inc./Kobal Collection
Opener 2 Picture Desk, Inc./Kobal Collection
Opener 3 Picture Desk, Inc./Kobal Collection
Opener 4 Picture Desk, Inc./Kobal Collection
7.1i bSt figure published in Biochemistry, 1991, 30: 4389-4398 by L. Carlacci, K.C. Chou, G.M. Maggiora
7.8 ISOPRESS/SIPA Press
7.9 Robert and Beth Plowes Photography
7.11 J. Christensen/New York Times Pictures
7.12 Brad Mogen/Visuals Unlimited
7.13a Brian Prechtel/Agricultural Research Service/USDA/ARS/Agricultural Research Service
7.14 Antoine Serra/Corbis Sygma
7.15 David Young-Wolff/PhotoEdit
7.16 Jack Fields/Photo Researchers, Inc.
7.17a Scott Camazine/Photo Researchers, Inc.
7.17c Frans Lanting/Minden Pictures
7.18 Gabe Palmer/Applied Biosystems
7.22 Photograph courtesy of The Roslin Institute.
E7.1 The Roslin Institute
E7.2 Pascal Goetgheluck/SPL/Photo Researchers, Inc.

Chapter 8

Photographs

Opener 1 Balog, James/Getty Images Inc. - Image Bank
Opener 2 Michelangelo, Creation of Adam (Detail of Sistine Chapel Ceiling). © Scala/Art Resource, NY.
Opener 3 James Cotier/Getty Images Inc. - Stone Allstock
Opener 4 Phil Schermeister/CORBIS
8.2 ARCHIV/Photo Researchers, Inc.
8.3a Getty Images Inc. - Stone Allstock
8.3b Tui De Roy/The Roving Tortoise Nature Photography
8.5 Frans Lanting/Minden Pictures
8.9bL Charles V. Angelo/Photo Researchers, Inc.
8.9bR Custom Medical Stock Photo, Inc.
8.12R Harry Taylor/University Museum of Zoology, Cambridge/Dorling Kindersley
8.13L Skulls Unlimited International
8.13R Geoff Brightling/ ESPL/Denoyer Geppert/Dorling Kindersley
8.14a © The Natural History Museum, London
8.14b University of New Mexico, Albuquerque, Maxwell Museum of Anthropology
8.17 SPL/Photo Researchers, Inc.
8.21a Fred Bruemmer/Peter Arnold, Inc.
8.21b Stewart Cohen/Index Stock Imagery, Inc.
8.21c Jerry Young/Dorling Kindersley
8.21d R. Calentine/Visuals Unlimited
E8.2a NASA/John F. Kennedy Space Center
E8.2b NASA/SPL/Photo Researchers, Inc.

Illustrations

8.16 Adapted from Krogh, D., 2002. *Biology: A Guide to the Natural World,* Second Edition. Prentice Hall. Figure 19.23, page 395.

Chapter 9

Photographs

Opener 1 Manny Millan/Time Inc. Magazines/Sports Illustrated
Opener 2 Bill Aron/PhotoEdit
Opener 3 Rufus F. Folkks/CORBIS
Opener 4 Joe Raedie/Getty Images, Inc - Liaison
9.4a Lynn Stone/Index Stock Imagery, Inc.
9.4b David Newman/Visuals Unlimited
9.5a–f Academy of Natural Sciences of Philadelphia
9.7L © Doug Cheeseman/Peter Arnold, Inc.
9.14 David Young-Wolff/PhotoEdit

Illustrations

9.2 Adapted from Stine, G. J., 2002. *AIDS Update, 2002*. Prentice Hall. Figure 3–4, page 58.
9.3 Adapted from Minkoff, E.C. and P.J. Baker, 2001. *Biology Today: An Issues Approach,* Second Edition. Garland Publishing. Figure 13.7, page 553.
9.7b Adapted from Krogh, D., 2002. *Biology: A Guide to the Natural World,* Second Edition. Prentice Hall. Figure 17.9, page 348.
9.10 Data from Cavener, D.R., and M.T. Clegg, 1981. Multigenic response to ethanol in *Drosophila melanogaster*. *Evolution* 35:1–10.
E9.3 Adapted from Freeman, S., 2002. *Biological Science,* First Edition. Prentice Hall. Figure 26.7, page 509.

Chapter 10

Photographs

Opener 1 Photo by Jim and Mary Whitmer
Opener 2 Lindy Powers/Index Stock Imagery, Inc.
Opener 3 AP/Wide World Photos
Opener 4 Digital Vision Ltd.
10.1a Barbara Gerlack/Visuals Unlimited
10.1b Bstav Verderber/Visuals Unlimited
10.2a Gerard Lacz/Peter Arnold, Inc.
10.7aL,R Courtesy, Dr. Forrest Mitchell, Dept. of Entomology, Texas A & M University (website www.dragonflies.org)
10.7b Getty Images Inc. - Stone Allstock
10.15a Peter Turnley/CORBIS
10.15b Roger De LaHarpe/Gallo Images/CORBIS
10.19 Jeff Greenberg/PhotoEdit
E10.2 AP/Wide World Photos

Illustrations

10.5 Data from Knowlton, N., L.A. Weigt, L.A. Solorzano, D.K. Mills, and E. Bermingham, 1993. Divergence in proteins, mitochondrial DNA, and reproductive incompatibility across the isthmus of Panama. *Science* 260:1629–1632.
10.9 Adapted from Freeman, S., and J.C. Herron, 2001. *Evolutionary Analysis,* Second Edition. Prentice Hall. Figure 16.12, page 565.
10.11 Adapted from Molnar, S., 1998. *Human Variation: Races, Types and Ethnic groups,* Fourth Edition. Prentice Hall. Figure 4–4, page 153.
10.12a Data from Allison, A.C., and B.S. Blumberg, 1959. Ability to taste phenylthiocarbamide among Alaskan Eskimos and other populations. *Human Biology* 31(4):352–359.
10.12b Data from Giblett, E.R., 1969. *Genetic Markers in Human Blood*. F.A. Davis, Co.
10.12c Data from Jorde, L.B., M.J. Bamshad, W.S. Watkins, R. Zenger, A.E. Fraley, P.A. Krakowiak, K.D. Carpenter, H. Soodyall, T. Jenkins, and A.R. Rogers, 1995. Origins and affinities of modern humans: a comparison of mitochondrial and nuclear genetic data. *American Journal of Human Genetics* **57**:523–538.
10.13 Adapted from Molnar, S., 1998. *Human Variation: Races, Types and Ethnic groups,* Fourth Edition. Prentice Hall. Figure 6–4, page 274.
10.14 Adapted from Molnar, S., 1998. *Human Variation: Races, Types and Ethnic groups,* Fourth Edition. Prentice Hall. Figure 4–4, page 153.
10.16 Data from Jablonski, N., and G. Chaplin, 2000. The evolution of human skin coloration. *Journal of Human Evolution* 39(1): 57–106.

Chapter 11

Photographs

Opener 1 TEK/SPL/Photo Researchers, Inc.
Opener 2 Dr. Robert J. Higgins/ UC Davis
Opener 3 AP/Wide World Photos
Opener 4 SIPA Press
11.3 AP/Wide World Photos
11.5 AP/Wide World Photos
11.6 AP/Wide World Photos
11.17 AP/Wide World Photos

E11.1	Michael Newman/PhotoEdit
E11.2A	CORBIS
E11.2B	March of Dimes Birth Defects Foundation
T11.1a	Gary Gaugler/Visuals Unlimited
T11.1b	Oliver Meckes/Photo Researchers, Inc.
T11.1c	Tina Carvalho/Visuals Unlimited
T11.1d	CNRI/Science Photo Library/Photo Researchers, Inc.
T11.1e	Dr. Linda Stannard, UCT/SPL/Photo Researchers, Inc.

Chapter 12

Photographs

Opener 1	Steve Burmeister/Steve Burmeister
Opener 2	© Howie Hanson/Twin Ports People Newspaper
Opener 3	© Howie Hanson/Twin Ports People Newspaper
Opener 4	Josh Meltzer/Duluth News Tribune
12.7b	Dr. Richard Kessel/Dr. C. Shih/Visuals Unlimited
12.8b	Getty Images Inc. - Stone Allstock
12.12	MIke Powell/Getty Images, Inc. - Allsport Photography
12.14	Ariel Skelley/CORBIS
12.15	Reuters NewMedia Inc/Jeff Mitchell/CORBIS

Chapter 13

Photographs

Opener 1	Michael Newman/PhotoEdit
Opener 2	From Newsweek 3/10/96 ©1996 Newsweek, Inc. All rights reserved. Reprinted by permission.
Opener 3	David Young-Wolff/PhotoEdit
Opener 4	Dex Images/CORBIS
13.10a	SPL/Photo Researchers, Inc.

Chapter 14

Photographs

Opener 1	Tupper Ansell Blake/USFWS
Opener 2	AP/Wide World Photos
Opener 3	© Stephan Anderson/The Nature Conservancy
Opener 4	AP/Wide World Photos
14.1	AP/Wide World Photos
14.2	Prentice Hall School Division
14.6a	NASA Space Center/Visuals Unlimited
14.6b	Buddy Mays/CORBIS
14.6c	David Dennis/Animals Animals/Earth Scenes
14.6d	Belinda Wright/DRK Photo
14.6e	Allen Blake Sheldon/Animals Animals/Earth Scenes
14.8a	Professor John Doebley/University of Wisconsin
14.8b	Fred Bruemmer/Peter Arnold, Inc.
14.10	M. Harvey/DRK Photo
14.11a	Ron Austing/Photo Researchers, Inc.
14.11b	Jerry Gildemeister
14.13	Photo Researchers, Inc.
14.15a	Photo by Rollie White
14.15b	Kenneth W. Fink/Photo Researchers, Inc.
14.19	AP/Wide World Photos

Illustrations

E14.2b	Adapted from Petit, J.R., J. Jouzel, D. Raynaud, N.I. Barkov, J.-M. Barnola, I. Basile, M. Bender, J. Chappellaz, M. Davis, G. Delaygue, M. Delmotte, V.M. Kotlyakov, M. Legrande, V.Y. Lipenkov, C. Lorius, L. Pepin, C. Ritz, E. Saltzman, and M. Stievenard, 1999. Climate and atmospheric history of the past 420,000 years from the Vostok ice core, Antarctica. *Nature* 399: 429–436. Figure 2, page 430.
14.4	Adapted from Primack, R.B., 1993. *Essentials of Conservation Biology*, First Edition. Sinauer Associates, Inc. Figure 4.5, page 81. Additional data from the World Conservation Union (IUCN)
14.5a	Adapted from Begon, M., J.L. Harper, and C.R. Townsend. *Ecology: Individuals, Populations and Communities*, Second Edition. Blackwell Scientific Publications. Figure 22.1c, page 769.
14.7	Adapted from Primack, R.B., 1993. *Essentials of Conservation Biology*, First Edition. Sinauer Associates, Inc. Figure 4.2, page 77.
14.9	Adapted from Audesirk, T., G. Audesirk, and B. Byers, 2000. *Life on Earth*, Second Edition. Prentice Hall. Figure 29-5, page 634.
14.14	Adapted from Myers, N., R.A. Mittermeier, C.G. Mittermeier, G.A.B. daFonseca, and J. Kent, 2000. Biodiversity hotspots for conservation priorities. *Nature* 403:853–858. Figure 1, page 853.
14.18	Adapted from Primack, R.B., 1993. *Essentials of Conservation Biology*, First Edition. Sinauer Associates, Inc. Figure 11.2, page 256.

Chapter 15

Photographs

Opener 1	Reto Stockli/NASA Goddard Laboratory for Atmospheres
Opener 2	AP/Wide World Photos
Opener 3	Pat Hamilton/Reuters New Media, Inc./CORBIS
Opener 4	Getty Images Inc. - Stone Allstock
15.3a	Dung Vo Trung/Corbis/Sygma
15.3b	Robert Pickett/CORBIS
15.3c	Edwin Remsberg/Getty Images Inc. - Stone Allstock
15.3d	Steve Wolper/DRK Photo
15.7	Wolfgang Kaehler/CORBIS
15.9	Courtesy of Emilia Dziuk Moczygemba Collection, Flaine Moczygemba. The UT Institute of Texan Cultures at San Antonio
15.11a,b	Nigel Cattlin/Holt Studios International/Photo Researchers, Inc.
15.12a	James L. Amos/CORBIS
15.15	Science Photo Library/Photo Researchers, Inc.
15.16a	Ranald Mckechnie/Dorling Kindersley Media Library
15.16b	Roger Phillips/Dorling Kindersley Media Library
15.16c	Burke/Triolo Photography/Foodpix
15.16d	Steve Gorton/Dorling Kindersley Media Library
15.16e	Unidentified/Dorling Kindersley Media Library
15.16f	Andy Crawford/Dorling Kindersley Media Library
15.16g	Siede Preis/Getty Images, Inc.
15.16h	Dave King/Dorling Kindersley Media Library
15.18	Jose Carillo/PhotoEdit
15.19	Terry W. Eggers/CORBIS
15.20	AP/Wide World Photos
15.21	NASA Photo/Grant Heilman Photography, Inc.
15.24a,b	Michael Newman/PhotoEdit
E15.1a,b	Peter Menzel/Peter Menzel Photography
E15.2	Pelletier Mecheline/Corbis/Sygma

Illustrations

15.1	Adapted from Freeman, S., 2002. *Biological Science*, First Edition. Prentice Hall. Figure 48.5a, page 936.
15.5	Data from U.S. Central Intelligence Agency, 2001. *CIA World Factbook 2001*.
15.8a	Data from Statistics South Africa. Web site: http://www.statssa.gov.za/
15.8b	Data from U.S. Census Bureau. Web site: http://www.censusscope.org/US/chart_age.html

Index

Page references that refer to figures are in **bold.** Page references that refer to tables are in *italics*.